AF535082

PAPERS ON GENETICS

A book of readings

PAPERS ON GENETICS

A book of readings

Louis Levine

B.S., M.S.Ed., M.A., Ph.D.

Professor of Biology, City College of New York

With 186 illustrations

Saint Louis

The C. V. Mosby Company

1971

Printed in the United States of America
Standard Book Number 8016-2983-7
Distributed in Great Britain by Henry Kimpton, London

To

Gabriella

who suggested this book be written

Preface

This book has been prepared to serve in two learning situations—to supplement a textbook when the instructor wishes to expose his students to the source material that forms the basis of our understanding of genetic phenomena or as a textbook if the instructor feels that his students should learn genetics directly from a study of original articles. With these two approaches in mind, the book is divided into fourteen chapters covering the topics considered in a course in genetics. These include gene structure and functions; principles of inheritance; sex linkage and sex determination; chromosome numbers; extrachromosomal inheritance; concepts of the gene as a mutable, recombinational, and functional unit; regulation of gene action and its control of metabolism, development, and behavior; and the genetics of the evolutionary process.

Each chapter contains an introduction outlining the historical development, when appropriate, and the main principles of the particular topic. This is followed by a references section listing the papers cited in the introduction. Italic reference numbers in this list indicate that the articles cited are reprinted in the book. The remainder of each chapter is composed of a collection of articles that includes both classic and modern papers in the field. In selecting the references and articles for this book, I have omitted many earlier presentations of ideas or facts in favor of those reports that served to establish particular concepts and stimulate further research. Whenever possible, the relationship of genetic phenomena to man has been stressed, not only because of the desire to increase the student's interest but also because human genetics has become a very active field in modern research.

This book has benefited from the efforts of my colleagues and friends who have assisted me through their discussions and encouragement. I am especially indebted to the following associates, each of whom kindly consented to review some of the introductory sections of the book and offered valuable suggestions: Professor Theodosius Dobzhansky, Professor L. C. Dunn, Professor Max Hamburgh, and Professor Norman M. Schwartz. Any shortcomings of the book, however, are my responsibility alone. I should also like to thank Mr. Joseph Fevoli for preparing the illustration for the cover of the book, Mrs. Rita Berkowitz for performing the arduous task of typing the manuscript, and Miss Harriet Rubenstein and Miss Helena Stuler for their aid in proofreading the manuscript. Finally I wish to express my deep appreciation to the authors and publishers who graciously gave their permission to have their articles reprinted in this book.

Louis Levine

Contents

PAPERS ON GENETICS

A book of readings

chapter 1

Nature and functions of the hereditary material

IDENTIFICATION OF THE HEREDITARY MATERIAL

Information on the chemical nature of the genetic material was available long before its significance was appreciated. In 1871, Miescher (Ref. 1-1) reported his discovery of an acid material he called *nuclein* from the nuclei of pus cells. He noted an unusually large amount of phosphorus in the substance and expressed the view that in the cell's metabolism it "... perhaps will deserve equal consideration with the proteins." However, the nucleic acids, as they were later called, were considered to be of too simple a structure to be the genetic material. The phenomenon whose explanation resulted in the identification of nucleic acid as the hereditary material was reported by Griffith in 1928 (Ref. 1-2). He demonstrated that heat-killed virulent bacteria, injected into a mouse, could confer their virulence on a living nonvirulent related type which was present in the animal. This acquired virulence was inherited by the descendents of the *transformed* bacteria. Later investigations showed that transformation could occur when the different types of heat-killed and live bacteria were mixed in test tubes. A search for the "transforming principle" of Griffith's experiments was conducted in many laboratories and resulted in the demonstration by Avery and his co-workers in 1944 (Ref. 1-3) that the transforming principle consists entirely of DNA.

Another line of research that identified the nucleic acids as the hereditary material involved the viruses. In one such investigation, Hershey and Chase in 1952 (Ref. 1-4) reported on their experiments with a virus (phage) that consists solely of DNA and protein, and attacks the bacterium, *Escherichia coli.* By the use of radioactive isotopes, they were able to demonstrate that upon infection, nearly all the protein of the bacteriophage was left outside the bacterial cell, whereas the viral DNA entered the cell and resulted in the production of new phage.

In a second type of viral investigation, the virus studied consisted solely of RNA and protein and attacked the cells of the tobacco plant. The virus is called tobacco mosaic virus (TMV). The virus can be separated into its RNA and protein components and reconstituted into its complete form. Fraenkel-Conrat and Williams in 1955 (Ref. 1-5) reported that only the whole virus was capable of attacking tobacco plant cells and producing new viruses. However, it was later found and reported by Fraenkel-Conrat in 1956 (Ref. 1-6) that infectivity and productivity of new viruses were intrinsic properties of the RNA itself. The role of RNA as the genetic material in these viruses was further demonstrated in an experiment by Fraenkel-Conrat and Singer whose paper (Ref. 1-7) was published in 1957 and is reprinted in this collection.

NATURE OF THE HEREDITARY MATERIAL

Two experimental programs provided information on the chemical and physical characteristics of DNA and resulted in a formulation of its structure. The significant chemical characteristic of DNA was discovered in 1950 by Chargaff (Ref. 1-8), who reported that, in addition to sugar and phosphate, DNA contains four bases, adenine (A), guanine (G), thymine (T), and cytosine (C), in amounts such that the ratios of A:T and G:C are each equal to one. The necessary information on the physical characteristics of DNA came from x-ray diffraction studies by Wilkins and his co-workers in 1953 (Ref. 1-9) and Franklin and Gosling in 1953 (Ref. 1-10). These studies indicated that DNA was in the form of a helix. Utilizing the above information, Watson and Crick in 1953 (Ref. 1-11) proposed the now famous Watson-Crick model of the molecular structure of DNA. A later paper by these authors, also published in 1953 (Ref. 1-12), is included in this chapter. This paper not only presents the evidence for their DNA model, but also discusses the implications of their model for genetics, DNA replication, and mutation. (It will be noted that in 1953, guanine and cytosine were thought, as seen in Fig. 6 of Ref. 1-12, to be connected by only two hydrogen bonds).

Although most organisms are found to possess their DNA in a double helix form, some do not. An example of a virus with single-stranded DNA was described by Sinsheimer in 1959 (Ref. 1-13). As was pointed out earlier in this discussion, some viruses have RNA as their hereditary material. The RNA in these organisms is in the form of either a single or a double chain. The characteristics of some animal and plant viruses with double-helical RNA have been reviewed in a paper by Gomatos and Tamm in 1963 (Ref. 1-14).

FUNCTIONS OF THE HEREDITARY MATERIAL

Hereditary material must perform two functions: (1) it must replicate itself and (2) it must provide for protein synthesis. In both instances the model of the DNA molecule must permit the smooth performance of the particular function. With regard to self-replication, Watson and Crick hypothesized that the two chains of the DNA double helix separate and that each chain serves as a template for the formation of a complementary chain. This mode of replication is called *semiconservative,* and its occurrence was demonstrated by Taylor and his co-workers, for DNA in a eucaryotic organism *(Vicia faba),* in 1957 (Ref. 1-15) and by Meselson and Stahl, for DNA in a procaryotic organism *(E. coli),* in 1958 (Ref. 1-16). A paper by Filner in 1965 (Ref. 1-17), which reviews the literature on the nature of DNA replication and discusses its method of occurrence in a higher plant cell, is reprinted in this chapter.

As soon as the semiconservative nature of DNA replication was established by the above experiments, the question of how single-stranded ΦX174 replicates itself was raised. Evidence was presented by Sinsheimer and his co-workers in 1962 (Ref. 1-18) that during the intracellular reproduction of this virus, the viral DNA is converted to a double-stranded form referred to as a *replicative form* or RF. From the RF, more single-stranded ΦX174 DNA is produced, using only the newly formed strand as a template.

The method by which DNA provides for protein synthesis proved to be indirect and was found to require various types of RNA as intermediates. It had been noted by Caspersson in 1941 (Ref. 1-19) that there was a direct correlation between the amount of RNA in a cell and protein production. Later, other research workers found that the RNA in the cytoplasm had its origin in the nucleus. An example of this line of research was reported by Goldstein and Plaut in 1955 (Ref. 1-20) in which they labeled the nuclear RNA of amoebae with ^{32}P and transplanted the radioactive nuclei into unlabeled amoebae. They found that a large portion of the labeled nuclear RNA moved into the cytoplasm. Meanwhile, other research revealed that the cellular RNA was actually composed of three different types. There was a large portion (85% to 90%) of the cell's RNA that was associated with the ribosomes of the cell. Siekevitz in 1952 (Ref. 1-21), tracing the path of a radioactive amino acid into protein, discovered that the ribosomes were the site of protein synthesis. A few years later, in 1958, Hoagland and his co-workers (Ref. 1-22) reported the existence of a small molecular species of RNA, comprising 5% of the cell's RNA, which has variously been called *transfer RNA, soluble RNA,* or *adaptor RNA.* They found that it served as an intermediate in the transfer of amino acids into peptide linkage. A third type of RNA was reported by Volkin and Astrachan in 1956 (Ref. 1-23). They found an unstable RNA fraction that had a high turnover rate. It was later hypothesized by others that this unstable RNA fraction acted as a messenger and carried information from DNA to ribosomes for protein synthesis. The existence of messenger RNA was further demonstrated by Hall and Spiegelman in 1961 (Ref. 1-24) whose paper is included in this chapter.

GENETIC CODE

The final topic that we shall consider in this chapter is the nature of the genetic code. Since both proteins and nucleic acids are polymeric compounds, it was logical to assume that the amino acid sequences of proteins were specified by the nucleotide sequence of corresponding nucleic acids. With only four bases (A, G, T, and C) available to code for twenty amino acids, it became obvious that at least three nucleotides would be required to specify an amino acid. A general discussion of the problem of genetic coding and the experimental evidence in support of the triplet nature of the code was presented by Crick and his co-workers in 1961 (Ref. 1-25).

Although the theoretical requirements and the experimental evidence for a triplet code were well founded, the actual code was unknown, and its "breaking" required the availability of synthetic messenger RNA units of known composition. Interestingly enough, the ability to synthesize messenger RNA had been achieved by Grunberg-Manago and Ochoa in 1955 (Ref. 1-26) who reported the existence of *polynucleotide phosphorylase,* the first isolated enzyme capable of synthesizing a polymeric nucleic acid. It was, however, Nirenberg and Matthaei in 1961 (Ref. 1-27) who demonstrated that a synthetic polyribonucleotide, polyuridilic acid, stimulated the production of a polypeptide chain consisting of the amino acid phenylalanine. However, the assignment of every triplet to a particular amino acid did not occur until Nirenberg and Leder in 1964 (Ref. 1-28) reported on their experiments involving trinucleotides of known nucleotide sequence. A later paper by Nirenberg and his co-workers in 1966 (Ref. 1-29) summarizes the work in this field and is the final paper in this chapter.

REFERENCES

Identification of the hereditary material

1-1. Miescher, F. 1871. On the chemical composition of pus cells. Hoppe-Seyler's Med.-Chem. Untersuch. **4:**441-460. Abridged and translated *in* Gabriel, M. L., and Fogel, S. [eds.] 1955. Great experiments in biology, Prentice-Hall, Inc., Englewood Cliffs, N. J.

1-2. Griffith, F. 1928. The significance of pneumococcal types. J. Hygiene **27:**113-159.

1-3. Avery, D. T., MacLeod, C. M., and McCarty, M. 1944. Studies on the chemical nature of the substance inducing transformation of pneumococcal types. Induction of transformation by a deoxyribonucleic acid fraction isolated from pneumococcus type III. J. Exp. Med. **79:**137-158.

1-4. Hershey, A. D., and Chase, M. 1952. Independent functions of viral protein and nucleic acid in growth of bacteriophage. J. Gen. Physiol. **36:** 39-56.

1-5. Fraenkel-Conrat, H., and Williams, R. C. 1955. Reconstitution of active tobacco mosaic virus from its inactive protein and nucleic acid components. Proc. Natl. Acad. Sci. U. S. A. **41:**690-698.

1-6. Fraenkel-Conrat, H. 1956. The role of the nucleic acid in the reconstitution of active tobacco mosaic virus. J. Am. Chem. Soc. **78:**882-883.

1-7. Fraenkel-Conrat, H., and Singer, B. 1957. Virus reconstitution. II. Combination of protein and nucleic acid from different strains. Biochim. Biophys. Acta **24:**540-548.

Nature of the hereditary material

1-8. Chargaff, E. 1950. Chemical specificity of nucleic acids and the mechanism of their enzymatic degradation. Experientia **6:**201-209.

1-9. Wilkins, M. H. F., Stokes, A. R., and Wilson, H. R. 1953. Molecular structure of deoxypentose nucleic acids. Nature **171:**738-740.

1-10. Franklin, R. E., and Gosling, R. 1953. Molecular configuration in sodium thymonucleate. Nature **171:**740-741.

1-11. Watson, J. D., and Crick, F. H. C. 1953. Molecular structure of nucleic acids. Nature **171:**737-738.

1-12. Watson, J. D., and Crick, F. H. C. 1953. The structure of DNA. Cold Spring Harbor Symp. Quant. Biol. **18:**123-131.

1-13. Sinsheimer, R. L. 1959. A single-stranded deoxyribonucleic acid from bacteriophage ΦX174. J. Mol. Biol. **1:**43-53.

1-14. Gomatos, P. J., and Tamm, I. 1963. Animal and plant viruses with double-helical RNA. Proc. Natl. Acad. Sci. U. S. A. **50:**878-885.

Functions of the hereditary material

1-15. Taylor, J. H., Woods, P. S., and Hughes, W. L. 1957. The organization and duplication of chromosomes as revealed by autoradiographic studies using tritium-labeled thymidine. Proc. Natl. Acad. Sci. U. S. A. **43:**122-128.

1-16. Meselson, M., and Stahl, F. W. 1958. The replication of DNA in *Escherichia coli.* Proc. Natl. Acad. Sci. U.S.A. **44:**671-682.

1-17. Filner, P. 1965. Semi-conservative replication of DNA in a higher plant cell. Exptl. Cell Research **39:**33-39.

1-18. Sinsheimer, R. L., Starman, B., Nagler, C., and Guthrie, S. 1962. The process of infection with bacteriophage ΦX174. I. Evidence for a "replicative form." J. Mol. Biol. **4:**142-160.

1-19. Caspersson, T. 1941. Studien über den Eiweissumsatz der Zelle. Naturwissenschaften **29:**33-43.

1-20. Goldstein, L., and Plaut, W. 1955. Direct evidence for nuclear synthesis of cytoplasmic ribose nucleic acid. Proc. Natl. Acad. Sci. U. S. A. **41:**874-880.

1-21. Siekevitz, P. 1952. Uptake of radioactive alanine *in vitro* into the proteins of rat liver fractions. J. Biol. Chem. **195:**549-565.

1-22. Hoagland, M. B., Stephenson, M. L., Scott, J. F., Hecht, L. I., and Zamecnik, P. C. 1958. A soluble ribonucleic acid intermediate in protein synthesis. J. Biol. Chem. **231:**241-257.

1-23. Volkin, E., and Astrachan, L. 1956. Phosphorus incorporation in *Escherichia coli* ribonucleic acid after infection with bacteriophage T2. Virology **2:**149-161.

1-24. Hall, B. D., and Spiegelman, S. 1961. Sequence complementarity of T2-DNA and T2-specific RNA. Proc. Natl. Acad. Sci. U. S. A. **47:**137-146.

Genetic code

1-25. Crick, F. H. C., Barnett, L., Brenner, S., and Watts-Tobin, R. J. 1961. General nature of the genetic code for proteins. Nature **192:**1227-1232.

1-26. Grunberg-Manago, M., and Ochoa, S. 1955. Enzymatic synthesis and breakdown of polynucleotides: polynucleotide phosphorylase. J. Am. Chem. Soc. **77:**3165-3166.

1-27. Nirenberg, M. W., and Matthaei, J. H. 1961. The dependence of cell-free protein synthesis in *E. coli* upon naturally occurring or synthetic polyribonucleotides. Proc. Natl. Acad. Sci. U. S. A. **47:**1588-1602.

1-28. Nirenberg, M., and Leder, P. 1964. RNA codewords and protein synthesis: The effect of trinucleotides upon the binding of sRNA to ribosomes. Science **145:**1399-1407.

1-29. Nirenberg, M., Caskey, T., Marshall, R., Brimacombe, R., Kellogg, D., Doctor, B., Hatfield, D., Levin, J., Rottman, F., Pestka, S., Wilcox, M., and Anderson, F. 1966. The RNA code and protein synthesis. Cold Spring Harbor Symp. Quant. Biol. **31:**11-24.

1 Virus reconstitution. II. Combination of protein and nucleic acid from different strains

H. Fraenkel-Conrat
B. Singer
The Virus Laboratory
University of California
Berkeley, California

The *in vitro* formation of typical TMV particles from small molecular fractions of virus protein and virus nucleic acid has been described,[1] as well as the finding that some infectivity similar in nature to that of the original virus was restored in this process. These studies have now been extended to various strains of TMV. Of particular significance, both from a theoretical and a potentially practical standpoint, appeared the incorporation into one virus particle of protein and nucleic acid originating from different strains of the virus. This has been achieved with various combinations of nucleic acid from 4 different strains and of protein from 3 strains. The biological and immunological characteristics of such mixed virus preparations have supplied what appears to be incontrovertible evidence that the infectivity of the reconstituted virus is actually a property of the newly formed virus particles. The biological and chemical nature of the progeny of a number of preparations of virus reconstituted from one or two strains has been studied. Some of the conclusions have been described in a preliminary note.[2]

METHODS AND MATERIALS

Virus preparations and fractions

The different strains of TMV were the same as used in earlier studies from this laboratory.[3,4] All virus preparations were isolated by differential centrifugation. Nucleic acid was prepared from these strains by a slight modification[5] of the detergent method previously used.[1] About 90% of the experiments yielded preparations of biologically active nucleic acid, stable for periods up to several months, if stored at −60°.

For the preparation of native protein, the virus was degraded at 3° and at pH 10.0 to 10.5.[1] Recently 2-amino-2-methylpropanol-1 and ethanolamine have been suggested as advantageous buffers for that purpose.* After dialysis for 16 hours of a 1% solution of virus (20-50 ml) against 1000 ml of an 0.1% solution of the amine adjusted with HCl to pH 10.5, degradation was almost complete, as indicated by the small amount of material sedimented upon ultracentrifugation (1 hour at 40,000 with refrigeration). The clear supernate was brought to 0.28 saturation with ammonium sulfate and centrifuged. The precipitated protein was redissolved in water, freed from small amounts of material precipitating at low ammonium sulfate concentrations, and the bulk of the protein reprecipitated between about 0.15 and 0.25 salt saturation. The nature and the amount of material in each fraction was ascertained spectrophotometrically. The final protein precipitate generally showed a sharp maximum at 280 mμ and an R-value (max/min) of 2.2 to 2.4, and of 2.4 to 2.5 after dialysis. A contamination with 0.1% nucleic acid decreased this ratio by 0.1. After thorough dialysis in the cold, the protein solutions were adjusted to pH 8.0, and subjected to ultracentrifugation (2 hours, 40,000 r.p.m., refrigerated). The marked tendency of the protein solutions to spoilage could be counteracted by storing them in the frozen state. Lyophilization caused some denaturation and increased their suitability for reconstitution.

The preparation of protein from the masked strain was possible by the same method. Protein from the ribgrass strain (HR) could be prepared only with considerable difficulty and in poorer yield. A lower pH was required for splitting (9.8-10), because of the great tendency of this protein to become denatured by alkali. From the YA strain no native protein could be isolated, probably for the same reason.

Antisera

The rabbit antisera and γ-globulin fractions were kindly prepared and placed at our disposal by Dr. R. C. Backus and Mrs. G. Perez-Mendez. The sera were prepared in customary manner by biweekly intramuscular injections of about 1 mg of TMV or HR with mineral oil and aquaphor as adjuvants. After 3 weeks the rabbits were bled and then injected intermittently and bled weekly.

From Biochimica et Biophysica Acta **24**:540-548, 1957. Used with permission.

Aided by a grant from the National Foundation for Infantile Paralysis, and by a grant from the Rockefeller Foundation.

The authors wish to thank Professors W. M. Stanley and R. C. Williams for their stimulating and encouraging interest in this work, and Professor C. A. Knight for his help and advice in the assay and differentiation of TMV strains. The untiring assistance of Miss Sue Veldee in the assays of virus infectivity is also gratefully acknowledged.

*Unpublished results of P. E. and M. Newmark.

Table I. Neutralization and crossreaction of virus strains and antisera

Anti-serum		Percentage neutralization of infectivity* of	
Type	ml/mg virus	HR	TMV
Anti-HR-serum**	0.1	90	0
Anti-HR-γ-globulin**	4	95	15
Anti-HR-γ-globulin	0.4	87	0
Anti-TMV-γ-globulin I**	4	62	97
Anti-TMV-γ-globulin II**	8	26	98

*Average of 2-8 experiments, each tested on about 8 half leaves at levels giving about 20 lesions per half leaf.
**Cross-absorbed with heterologous virus.

The γ-globulin fraction was separated as the trailing component in the analytical electrophoresis cell.

The efficacy of the antisera was tested by means of precipitin and neutralization tests. As an example of the latter, 1 ml of an 0.01% solution (0.075*M* sodium chloride) of TMV or HR virus was treated for 16 hours at 3° with varying amounts of the homologous or heterologous antiserum. Of the homologous sera 0.01 to 0.025 ml were required to reduce the infectivity of the virus (100 γ) by a factor of 5 or 10, as indicated by the number of lesions produced, after suitable dilution, by the usual assay procedure. There was, however, considerable cross reaction in the case of both the unfractionated serum and the γ-globulin fraction, particularly between anti-TMV sera and the HR virus. To decrease this hetero-specificity, the antisera were treated with varying amounts of the heterologous virus (0.16-4.0 mg/ml), and ultracentrifuged after several hours. Anti-HR sera and γ-globulins were thus obtained which had very little if any effect on TMV while reducing the infectivity of HR by about 95%.

From anti-TMV sera no similarly selective antibody could be isolated. Repeated pretreatment with great amounts of HR virus removed all antibody activity from the solutions. Serum preparations cross-absorbed with less HR reduced the infectivity of TMV by about 97%, and that of HR to a somewhat variable extent, averaging 44%. Fortunately the latter antibody preparations were quite adequate to permit clear-cut serological identification of the two virus strains (Table I).

Analytical methods

For amino acid analysis, virus preparations (about 6 mg in 0.2-0.4 ml) were mixed with 2 ml of twice-redistilled constant boiling HCl, sealed *in vacuo*, and heated to 108° for 16 hours. After repeated evaporation of the acid in a desiccator, the hydrolysates were taken up in a 50-fold amount of water (50 γ per mg virus). Aliquots were chromatographed one-dimensionally on paper for the detection and analysis of histidine, methionine, tyrosine, and arginine by a recently described technique.[6] Another aliquot (1 mg) was dinitrophenylated for complete amino acid analysis, in principle according to Levy.[7] The correction factors for the recovery of the amino acids as DNP-derivatives have been reinvestigated, and several were found to differ from those obtained two years ago under seemingly similar conditions.[8] Since only comparative data were required, hydrolyses for varying time periods were not carried out, and the analyses were not corrected for destruction of acid-sensitive amino acids during hydrolysis. Di-DNP-Cystine was almost absent, and not accounted for by a corresponding amount of DNP-cysteic acid. For histidine, methionine, tyrosine and arginine the results of the DNP-analyses were not readily reproducible, and the colorimetric analyses after chromatographic separation[6] were regarded as more reliable.

Tryptophan was determined, in conjunction with tyrosine on unhydrolyzed protein preparations by the spectrophotometric method as applied by Beavan and Holiday.[9] The tyrosine values were generally in good accord with those obtained by colorimetry.[6] The results of the amino acid analyses for TMV and HR are listed on Table II together with those available from the literature.

RESULTS

Reconstitution of virus from common TMV protein and nucleic acid

The technique generally used for reconstitution was as follows: To 1-10 mg of protein (0.5-1.0% solution of pH 8) was added one tenth the amount of nucleic acid, and 3*M* pH 6 acetate (10 γ per ml reaction mixture). Phosphate buffer (pH 6.8*M*, 50 γ per ml) has also often been used and has at times given higher yields of active virus. Below pH 5.0 and above pH 8.5 little or no active virus was formed. The solutions were held at room temperature at least for the first few hours. Aliquots of the reaction mixture, which rapidly gets opalescent, were diluted for assay after various time periods, *e.g.*, 15 minutes, 1 hour, 20 hours. Upon assay, maximal activity was sometimes found after the short reaction periods, but more generally maximal activity was obtained after 20 hours. Activity has also occasionally been observed to decrease or disappear from reaction mixtures. Many of these experiments, however, were performed before the intrinsic infectivity of the nucleic acid, and its instability in the assay medium,[5] were recognized, and

Table II. Comparison of amino-acid composition of TMV and HR with values in the literature*

	TMV					HR			
	Present methods	Microbiol.		Average**	Ion-exchange column***	Present methods	Microbiol.		Average**
Glycine	2.3	1.8	2.5	2.1	2.7	1.6	1.3	1.7	1.5
Valine	9.6	9.2	10.9	10.1	9.1	5.9	6.3	7.3	6.8
Alanine	6.5	5.1	7.4	6.3	7.9	8.5	6.4	9.2	7.8
Leucine + Isoleucine	14.2	15.9	13.9	14.9	15.1	12.2	15.2	13.1	14.2
Proline	5.0	5.8	5.5	5.7	6.3	5.0	5.8	5.5	5.7
Serine	9.0	7.3	9.1	8.2	8.7	8.1	5.7	7.2	6.5
Threonine	8.9	9.9	11.9	10.9	10.5	7.2	8.2	9.8	9.0
Lysine	1.9	1.5	1.4	1.5		2.4	1.5	1.4	1.5
Arginine	9.5	9.8	9.7	9.8		8.9	9.8	9.7	9.8
Histidine	0.0	0.0	0.0	0.0	0.0	0.7	0.7	0.7	0.7
Phenylalanine	7.2	8.4	8.2	8.3	7.6	5.3	5.4	5.3	5.4
Tyrosine	4.1	3.8	3.7	3.8	4.2	6.3	6.7	6.6	6.7
Tryptophan	2.8 §	2.1	1.9	2.0		2.2	1.4	1.4	1.4
Methionine	0.0	0.0	0.0	0.0	0.0	2.0	2.2	2.2	2.2
Glutamic acid	12.4	11.3	11.0	11.2	13.5	16.4	15.5	15.1	15.3
Aspartic acid	13.8	13.5	11.9	12.8	14.5	15.0	12.6	11.2	11.9
	107.2	105.4	108.1			106.9	104.7	107.4	

*All values expressed as g of amino acid per 100 g virus, not corrected for destruction during hydrolysis. It must be noted, however, that the values of Fraser and Newmark are percentages of the material recovered from the ion-exchange column. Cysteine (about 0.6%) was not determined, nor listed.

**The two sets of analyses for TMV published by Knight[3] and by Black and Knight[11] are listed, as well as the average. In the case of HR, the second column represents the expected values, had HR been again analyzed in 1953 and shown changes parallel to those observed with TMV.

***Fraser and Newmark.[10]

§ The possibility of different protein fractions differing in their tryptophan content is still under investigation (2.6-3.2%). See footnote on p. 9.

these experiments are being repeated with due regard to the properties of the two types of infectious agents. When ribonuclease is added at various time intervals to inactivate any free nucleic acid, the infectivity is a true measure of the extent of reconstitution. It appears that at room temperature reaction is quite rapid in phosphate and proceeds to the same point in 24 hours in acetate (Fig. 1).

Considerable effort has been put into establishing the extent of contamination, if any, of the protein and nucleic acid fractions with undegraded virus. The protein preparations usually gave no lesions when tested at 1-5 mg/ml, or at 100- to 500-fold levels of those at which reconstituted preparations gave 20-50 lesions per half leaf. That contaminating TMV would be pathogenic under these conditions was shown when TMV (0.1-0.5 γ/ml) added to the protein (1-5 mg/ml) gave 10-50% of the expected lesions. Assays of typical protein preparations by Dr. W. Takahashi with a particularly sensitive assay technique suggested the presence of less than 0.0001% virus.

A similar search for contaminating virus in nucleic acid preparations has led to the discovery that the nucleic acid *per se* is infectious.[2, 5]* It was shown also that contaminating virus can be removed very effectively by ultracentrifugation of nucleic acid preparations. Upon repeated ultracentrifugation rod-like particles most of which are usually shorter than 300 mμ may at times be sedimented from such solutions. Their appearance seems to be favored by the presence of salts, and is probably due to reconstitution involving traces of contaminating protein. These findings have rendered definitive proof for the absence of the last traces of contaminating virus from the nucleic acid both less crucially important and more difficult. In view of the much higher yields in activity obtained in some recent reconstitution experiments, such hypothetical "last traces" become progressively more irrelevant. Furthermore, the experiments with strain mixtures to be described below have definitely shown that the

*The same conclusion was reached independently by A. Gierer and G. Schramm, Nature **177** (1956) 702.

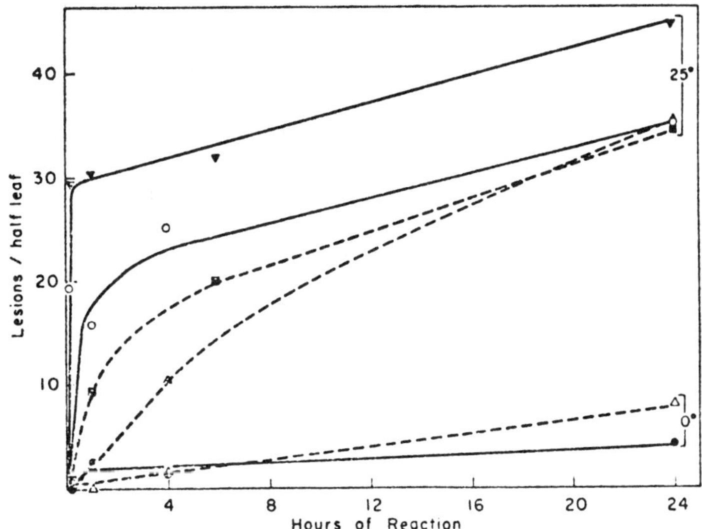

Fig. 1. Rate of reconstitution of active virus in reaction mixtures containing TMV protein (1%), nucleic acid (0.1%) and buffer. All samples are treated with ribonuclease (10 λ 0.1% per ml, 16 hours) prior to assay, to inactivate any uncombined nucleic acid. Solid lines are results obtained with pH 7 phosphate (0.05*M*), broken lines are for pH 6 acetate (0.03*M*). Experiment I (▼, ◘) yielded higher activity and was assayed at a protein concentration of 5 γ/ml. The final activity reached indicated a yield of about 5%. Experiment II (○, ●, ▲, △) was assayed at 25 γ/ml.

activity in reconstitution experiments is inherent in newly formed particles, and cannot be attributed to undegraded virus particles.

Mixed strains

The viruses employed were common TMV, and the masked (M), yellow aucuba (YA) and Holmes ribgrass (HR) strains.[3] Active virus rods were successfully reconstituted from (1) TMV-nucleic acid and M-protein, (2) M-nucleic acid and TMV-protein, (3) YA-nucleic acid and TMV-protein, (4) YA-nucleic acid and M-protein, (5) HR-nucleic acid and TMV-protein, (6) TMV-nucleic acid and HR-protein and (7) HR-nucleic acid and M-protein. The infectivity of these mixed virus preparations was within the same range as that obtained for TMV-nucleic acid + TMV-protein. This was surprising in the case of the HR strain, because the original virus showed only 5% of the activity of TMV on a weight basis (to be referred to as specific infectivity, henceforth). The lesser infectivity of HR seems to be due to the protein component, however, since HR nucleic acid was found to be as infectious as TMV nucleic acid. It has in this case been possible to obtain a reconstituted preparation showing a fourfold higher specific infectivity than that of the original virus supplying the nucleic acid, the HR strain (mixed virus of type 5 above, to be referred to as M. V. HR/TMV).

Of greatest interest in these mixed virus experiments were the biological properties of the reconstituted virus, as compared to those of the two "parent" strains. When each of the reaction products was tested on *N. tabacum* and on *N. sylvestris,* it gave in every case the same symptoms as did the original strain supplying the nucleic acid. Thus in Turkish tobacco, M.V. TMV/M and TMV/HR gave a green mosaic disease, M.V. M/TMV produced virus without visible symptoms, M.V. YA/TMV and YA/M gave a yellow mosaic disease and M.V. HR/TMV and HR/M gave typical ringspot lesions.* These findings strongly suggest that the nucleic acid is the genetic determinant in TMV, and related strains, playing the same decisive role which DNA seems to play in the bacteriophages.

In striking contrast to the nature of the infection which is determined by the nucleic acid component, the serological characteristics of mixed virus preparations resemble those of the protein component. When anti-TMV serum or its γ-globulin (see methods and materials) was added to the two preparations, it neutral-

*The difference between HR- and TMV-like virus is also quite evident from the size of the local lesions on *N. glutinosa.* The differentiation of TMV and M on the one side, and HR and YA on the other is based on the type of response obtained in *N. sylvestris.*

Table III. Neutralization of infectivity of TMV, HR, and the two mixed virus preparations derived from these*

Assay level (γ/ml)	TMV	HR	HR/TMV**	TMV/HR**	TMV and		HR and		HR/TMV	TMV/HR
					TMV Prot.	HR Prot.	TMV Prot.	HR Prot.		
	0.17	1.7	0.44	22	0.11	0.11	1.7	1.7	0.45	9.7
Untreated	22	21	19	29	30	28	37	56	22	14
+ Anti-TMV γ-globulin	*0.9*	8.0	*3.6*	11	*0.2*	*0.8*	17	33	*1*	9
+ Anti-HR γ-globulin	23	*1.1*	16	*2.9*	6	3	*2*	*4*	11	*3*

*All figures in the table represent average number of lesions per half leaf on *N. glutinosa* plants. Most assays were performed 2 or 3 times on 8 or 10 half leaves per sample.

Two experiments selected out of about 30 are listed, all of which showed the same phenomenon though some were less complete, others showed more cross reaction. Of TMV and HR 5γ were used, of the mixed virus preparations 20-30 γ; of the γ-globulin fractions, pretreated with the heterologous virus (see EXPERIMENTAL) 10 or 20 λ were used, and the final volume of the reaction mixtures was 0.3 ml. When protein was added to the reaction mixture, 25 γ was used. Combinations showing specific neutralization have been printed in italics.

**HR/TMV represents mixed virus prepared from HR-nucleic acid and TMV-protein; TMV/HR represents the virus prepared from TMV-nucleic acid and HR-protein.

ized M.V. HR/TMV to a similar extent as it did TMV, but had little effect on M.V. TMV/HR. Anti-HR serum, on the other hand, neutralized M.V. TMV/HR much more effectively than it did M.V. HR/TMV (Table III). At the same time, and in the same assays, the nature of the lesions clearly showed that the latter was of HR character, while M.V. TMV/HR was of TMV character. Control experiments in which the neutralisability of the two viruses by anti-sera was tested in the presence of excess homologous or heterologous protein confirmed the validity of the experimental procedure. Thus there appears to be no doubt that the activity appearing in reaction mixtures containing HR-nucleic acid and TMV-protein is due to a particle containing a genetically determining HR-nucleic acid core, and an immunologically determining TMV-protein coat. No alternate explanations seem able to explain the observed facts. Thus reconstitution of infectious virus particles from two chemical components appears to be definitely established.

Nature of the progeny of mixed virus

The finding that TMV-nucleic acid will combine with HR-protein, and vice versa, is particularly surprising in view of the great differences between the proteins of these two strains. It appears from Knight's analyses,[3] as confirmed by us, that only 2 to 4 amino acids occur in the same amounts in HR and TMV. Histidine and methionine are completely absent from all strains that have been investigated by Knight, with the exception of HR. An important functional property of the virus proteins probably resides in their specific tendency to aggregation in a superhelical array around, if they are present, nucleic acid strands. In view of the exchangeability of different virus proteins, as observed in the present experiments, one must conclude that this activity is dependent only upon a few suitably situated key sites surrounded by nonspecific areas.

In view of the marked difference in amino acid composition of TMV and HR, the mixed virus preparations obtained from these two were of particular interest and value in establishing the nature of the progeny of reconstituted virus. Paper chromatographic comparison of the hydrolysates of TMV, HR, and the progeny of the two mixed virus preparations has clearly demonstrated the presence of about 0.7% histidine and about 2% methionine in HR and the progeny of M.V. HR/TMV, and the absence of these amino acids from TMV and the progeny of M.V. TMV/HR. Complete amino acid analyses were then carried out for these 4 types of preparations (see Table IV). The first impression, based on the presence or absence of histidine and methionine was generally confirmed by these analyses. The protein of each progeny closely resembled that of the virus supplying the nucleic acid to the mixed virus from which it was derived. Only very minor differences were noted (less than 10% of the content in any one amino acid). However, in the case of glycine, that small difference was observed in 8 separate hydrolysates of 5 different progeny preparations from M.V. HR/TMV as compared to HR, and a small

Table IV. Comparison of amino acid composition of HR and TMV with progeny of experimental preparations*

	HR	Progeny of HR/TMV	TMV	Progeny of TMV/HR	Mutant strain from TMV-nucleic acid
Glycine	1.6	*1.8*	2.3	2.3	2.5
Alanine	8.5	8.5	6.5	6.9	5.5
Valine	5.9	6.3	9.6	9.0	9.6
Leucine + Isoleucine	12.2	12.2	14.2	14.3	13.0
Proline	5.0	5.1	5.0	5.1	4.3
Serine	8.1	8.1	9.0	8.8	7.8
Threonine	7.2	7.5	8.9	8.9	8.8
Lysine	2.4	2.3	1.9	1.8	2.1
Arginine	8.9	8.5	9.5	9.7	*7.6*
Histidine	0.70	0.70	0.0	0.0	0.0
Phenylalanine	5.3	5.4	7.2	7.1	6.8
Tyrosine	6.3	6.2	4.1	4.3	*5.4*
Tryptophan	2.2	2.2	2.8	2.6	2.7
Methionine	2.0	2.2	0.0	0.0	0.6
Glutamic acid	16.4	17.3	12.4	12.1	12.4
Aspartic acid	15.0	14.8	13.8	14.2	15.5

*All values expressed as g of amino acid per 100 g virus.
Values are not corrected for destruction during hydrolysis.
The first 2 columns represent averages of 8 hydrolysates each; the next three columns averages of 3 hydrolysates each.
Seemingly significant differences have been printed in italics.

difference in the lysine contents between these has been observed almost as consistently. Yet these differences are too small to be regarded as more than a suggestion that the protein component may slightly influence the genetic message transferred by the nucleic acid.

It must further be noted that the observed differences are definitely less than one amino acid residue per subunit, and their acceptance requires the assumption that not all subunits can be identical. This assumption appears also required to explain the tryptophan content of TMV which is close to 2.5 residues per 18,000 molecular weight subunit. That value was obtained both by a chemical method,[8] and by spectrophotometry both in 0.1*N* sodium hydroxide and at neutral pH,[8,9]* and appears definitely more probable than the lower value obtained by microbiological methods.

Another difference between HR and the progeny of M.V. HR/TMV is shown by the specific infectivity, which has been higher for many of the latter preparations than the highest obtained for any HR sample (20-45% *vs* 12%).

The significance of these differences between the original virus strains and the M.V. preparations will be dependent on whether they recur in successive progeny preparations. This seems to be the case for the specific infectivities, but as yet no sufficient number of amino acid analyses have been completed with successive single-lesion progeny preparations to warrant any definite conclusions in this regard. At the present time, one can only conclude from all this work that the nucleic acid of each strain has the ability to provoke the synthesis, within the host cell, of new virus protein very similar to, if not identical with, its own homologous protein; and that it retains this ability even when packaged, *in vitro,* in the protein of another strain.

Heritable modifications

In observing the plant-pathogenic nature, and the protein composition of the progeny from single lesions of many preparations of mixed virus or of isolated nucleic acids, a marked variation was noted in at least one instance. This apparent mutant was characterized by differences in both disease symptomatology and amino acid composition of progeny virus (Table IV). Its appearance may be regarded as indicative of the labilisation of the genetic

*The spectrum of TMV protein (1 mg/ml) differs markedly in shape from that of an *ad hoc* mixture of the amino acids contributing to its absorption (30 γ tryptophan, 40 γ tyrosine, 65 γ phenylalanine, 7 γ cysteine, and 1.0 mg triglycine per ml). However, the maximum, though at 282 and 278 mμ respectively, is the same (O.D. = 1.25). When solutions of protein and amino acids were prepared in 67% acetic acid, the O.D.'s were the same and the spectra very much more similar in shape (maxima at 279 and 278 mμ, respectively).

material through chemical exposure and manipulation, since no similar variations have been observed from single lesion propagation of the original virus strains. In contrast to other instances where genetically more labile variants were observed, this mutant strain has continued to produce a striking necrotic disease in Turkish tobacco through several passages and isolations, including the separation of its nucleic acid and protein.

SUMMARY

1. A method of preparation of native protein from TMV and other strains has been described.

2. The reconstitution of virus particles from protein and nucleic acid of different strains has yielded very active preparations, one of which showed higher infectivity than one of its parent strains.

3. The nature of the disease provoked by mixed virus preparations resembled in each case that characteristic of the virus supplying the nucleic acid.

4. The chemical nature of the progeny of mixed virus preparations also closely resembled that of the virus supplying the nucleic acid, although the significance of minor differences in amino acid composition has not yet been established.

5. In contrast to these properties the serological characteristics of mixed virus preparations were those of the virus supplying the protein.

6. The dual nature of the activity of reconstituted virus particles has thus been clearly demonstrated.

7. Variants or mutants of different biological and chemical properties have occurred randomly in the course of this work, and are regarded as indications of a labilisation of the genetic material through chemical manipulation.

REFERENCES

1. H. Fraenkel-Conrat and R. C. Williams, Proc. Natl. Acad. Sci. U.S. **41** (1955) 690.
2. H. Fraenkel-Conrat, J. Am. Chem. Soc. **78** (1956) 882.
3. C. A. Knight, J. Biol. Chem. **171** (1947) 297.
4. C. I. Niu and H. Fraenkel-Conrat, Arch. Biochem. Biophys. **59** (1955) 538.
5. H. Fraenkel-Conrat, B. Singer and R. C. Williams, Biochim. Biophys. Acta **25** (1957) (in the press); see also: Symposium on the Chemical Basis of Heredity, McCollum Pratt Institute, Johns Hopkins University, Baltimore, June 1956.
6. H. Fraenkel-Conrat and B. Singer, Arch. Biochem. Biophys., (in the press).
7. A. L. Levy, Nature **174** (1954) 126.
8. H. Fraenkel-Conrat and B. Singer, Arch. Biochem. Biophys. **60** (1956) 64.
9. G. H. Beavan and E. R. Holiday, Advances in Prot. Chem. **7** (1952) 320.
10. D. Fraser and P. Newmark, J. Am. Chem. Soc. **78** (1956) 1588.
11. F. L. Black and C. A. Knight, J. Biol. Chem. **202** (1953) 51.

2 The structure of DNA

J. D. Watson*
F. H. C. Crick**
Cavendish Laboratory
Cambridge, England

It would be superfluous at a Symposium on Viruses to introduce a paper on the structure of DNA with a discussion on its importance to the problem of virus reproduction. Instead we shall not only assume that DNA is important, but in addition that it is the carrier of the genetic specificity of the virus and thus must possess in some sense the capacity for exact self-duplication. In this paper we shall describe a structure for DNA which suggests a mechanism for its self-duplication and allows us to propose, for the first time, a detailed hypothesis on the atomic level for the self-reproduction of genetic material.

We first discuss the chemical and physical-chemical data which show that DNA is a long fibrous molecule. Next we explain why crystallographic evidence suggests that the structural unit of DNA consists not of one but of two polynucleotide chains. We then discuss a stereochemical model which we believe satisfactorily accounts for both the chemical and crystallographic data. In conclusion we suggest some obvious genetical implications of the proposed structure. A preliminary account of some of these data has already appeared in Nature (Watson and Crick, 1953a, 1953b).

I. EVIDENCE FOR THE FIBROUS NATURE OF DNA

The basic chemical formula of DNA is now well established. As shown in Figure 1 it consists of a very long chain, the backbone of which is made up of alternate sugar and phosphate groups, joined together in regular 3′ 5′ phosphate di-ester linkages. To each sugar is attached a nitrogenous base, only four different kinds of which are commonly found in DNA. Two of these—adenine and guanine—are purines, and the other two—thymine and cytosine—are pyrimidines. A fifth base, 5-methyl cytosine, occurs in smaller amounts in certain organisms, and a sixth, 5-hydroxy-methyl-cytosine, is found instead of cytosine in the T even phages (Wyatt and Cohen, 1952).

It should be noted that the chain is unbranched, a consequence of the regular internucleotide linkage. On the other hand the sequence of the different nucleotides is, as far as can be ascertained, completely irregular. Thus, DNA has some features which are regular, and some which are irregular.

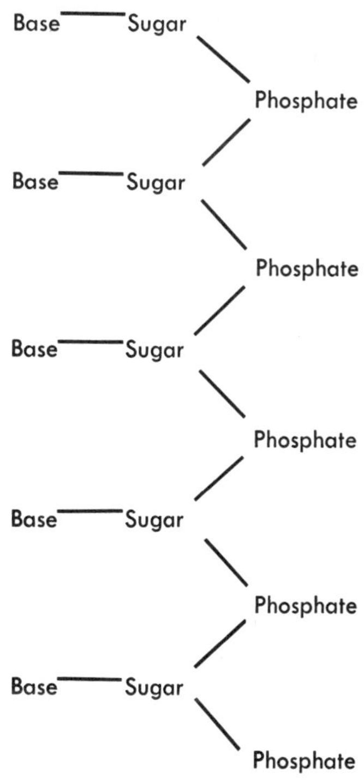

Fig. 1. Chemical formula (diagrammatic) of a single chain of desoxyribonucleic acid.

Reproduced from Vol. XVIII; Cold Spring Harbor Symposia, 1953. Used with permission.

*Aided by a Fellowship from the National Foundation for Infantile Paralysis.

Present address: Director of the Cold Spring Harbor Laboratory for Quantitative Biology, Cold Spring Harbor, N. Y.

**Present address: Medical Research Council, Laboratory of Molecular Biology, Cambridge, England.

A similar conception of the DNA molecule as a long thin fiber is obtained from physicochemical analysis involving sedimentation, diffusion, light scattering, and viscosity measurements. These techniques indicate that DNA is a very asymmetrical structure approximately 20 A wide and many thousands of angstroms long. Estimates of its molecular weight currently center between 5 X 10^6 and 10^7 (approximately 3 X 10^4 nucleotides). Surprisingly each of these measurements tends to suggest that the DNA is relatively rigid, a puzzling finding in view of the large number of single bonds (5 per nucleotide) in the phosphate-sugar backbone. Recently these indirect inferences have been confirmed by electron microscopy. Employing

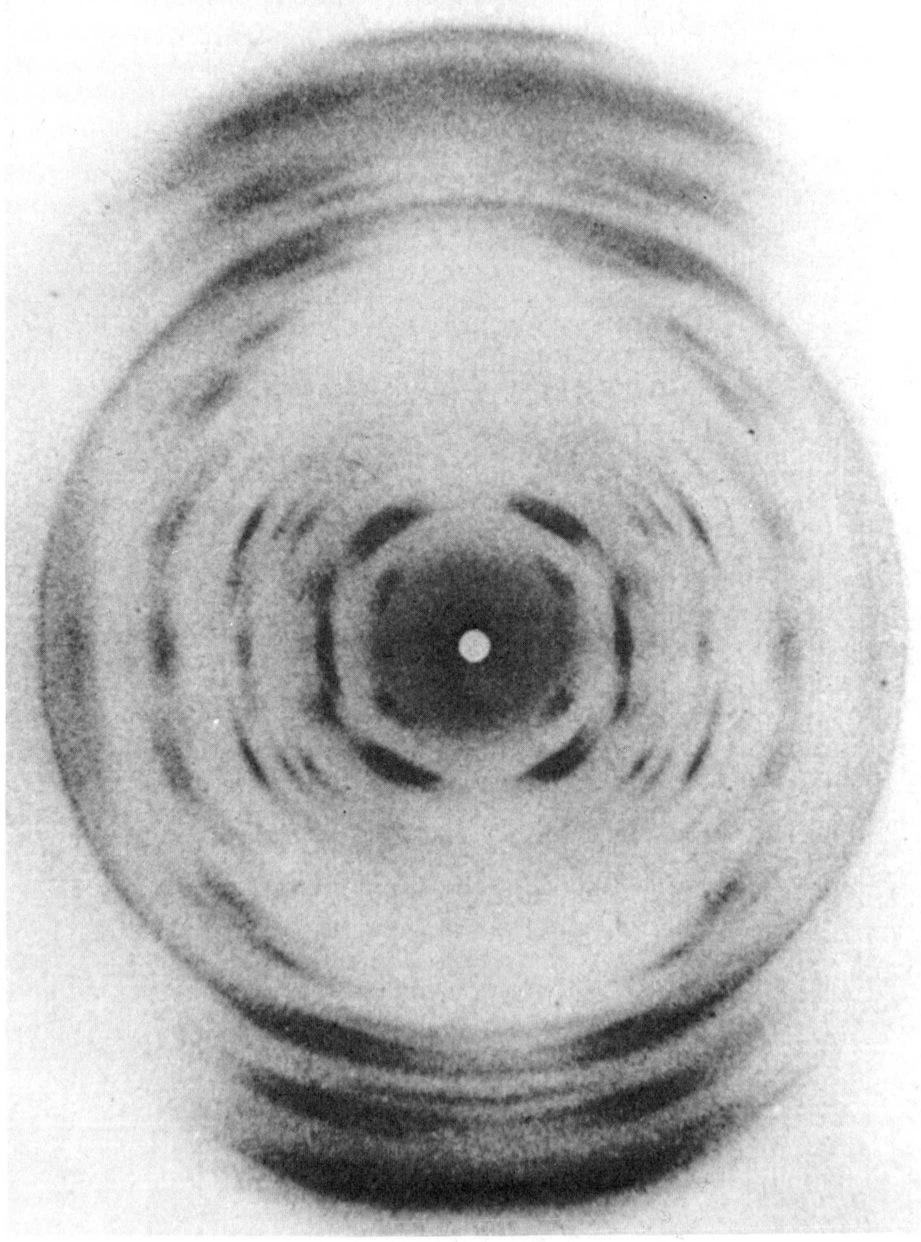

Fig. 2. X-ray fiber diagram of Structure A of desoxyribonucleic acid. (H. M. F. Wilkins and H. R. Wilson, unpub.)

high resolution techniques both Williams (1952) and Kahler *et al.* (1953) have observed, in preparations of DNA, very long thin fibers with a uniform width of approximately 15-20 A.

II. EVIDENCE FOR THE EXISTENCE OF TWO CHEMICAL CHAINS IN THE FIBER

This evidence comes mainly from X-ray studies. The material used is the sodium salt of DNA (usually from calf thymus) which has been extracted, purified, and drawn into fibers. These fibers are highly birefringent, show marked ultraviolet and infrared dichroism (Wilkins *et al.*, 1951; Fraser and Fraser, 1951), and give good X-ray fiber diagrams. From a preliminary study of these, Wilkins, Franklin and their co-workers at King's College, London (Wilkins *et al.*, 1953; Franklin and Gosling 1953a, b and c) have been able to draw certain general conclusions about the structure of DNA. Two important facts emerge from their work. They are:

(1) *Two distinct forms of DNA exist.* Firstly a crystalline form, Structure A, (Figure 2) which occurs at about 75 per cent relative humidity and contains approximately 30 per cent water. At higher humidities the fibers take up more water, increase in length by about 30

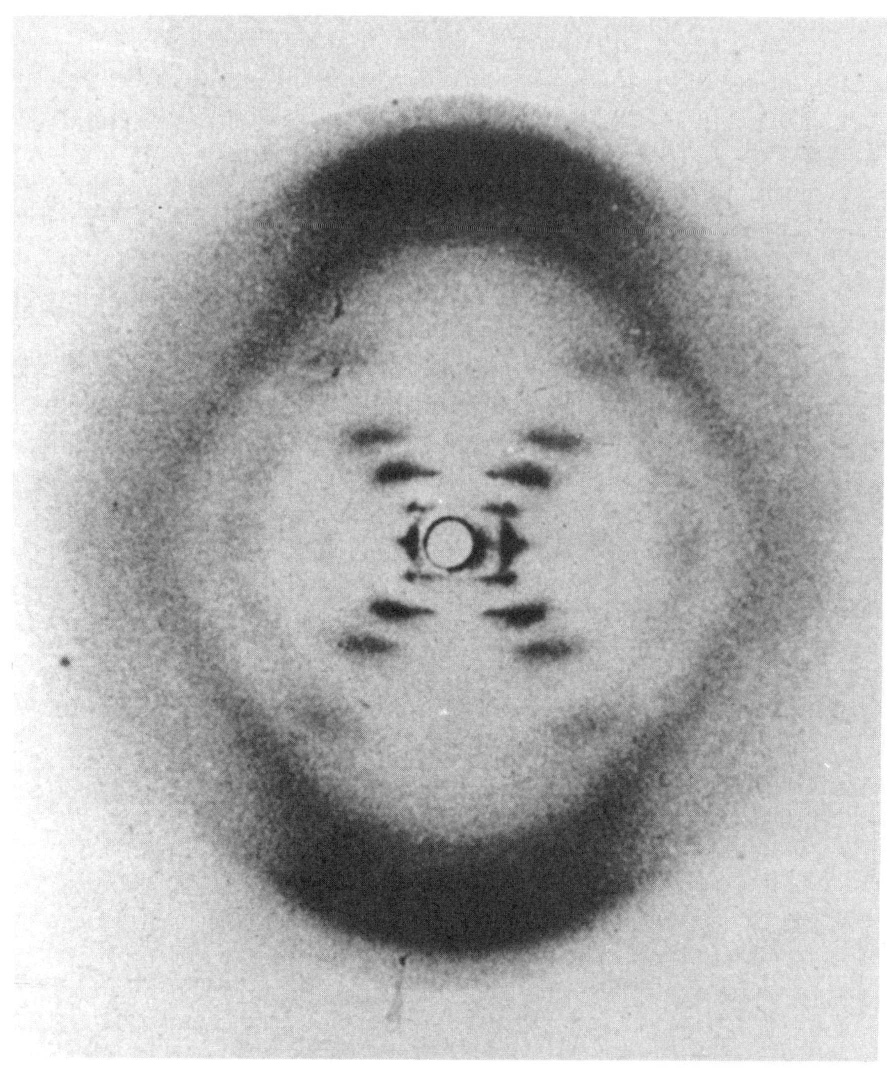

Fig. 3. X-ray fiber diagram of Structure B of desoxyribonucleic acid. (R. E. Franklin and R. Gosling, 1953a.)

Table 1
(From Franklin and Gosling, 1953a, b and c)

	Degree of orientation	Repeat distance along fiber axis	Location of first equatorial spacing	Water content	Number of nucleotides within unit cell
Structure A	Crystalline	28 A	18 A	30%	22-24
Structure B	Paracrystalline	34 A	22-24 A	> 30%	20 (?)

per cent and assume Structure B (Figure 3). This is a less ordered form than Structure A, and appears to be paracrystalline; that is, the individual molecules are all packed parallel to one another, but are not otherwise regularly arranged in space. In Table 1, we have tabulated some of the characteristic features which distinguish the two forms. The transition from A to B is reversible and therefore the two structures are likely to be related in a simple manner.

(2) *The crystallographic unit contains two polynucleotide chains.* The argument is crystallographic and so will only be given in outline. Structure B has a very strong 3.4 A reflexion on the meridian. As first pointed out by Astbury (1947), this can only mean that the nucleotides in it occur in groups spaced 3.4 A apart in the fiber direction. On going from Structure B to Structure A the fiber shortens by about 30 per cent. Thus in Structure A the groups must be about 2.5 per cent A apart axially. The measured density of Structure A, (Franklin and Gosling, 1953c) together with the cell dimensions, shows that there must be *two* nucleotides in each such group. Thus it is very probable that the crystallographic unit consists of two distinct polynucleotide chains. Final proof of this can only come from a complete solution of the structure.

Structure A has a pseudo-hexagonal lattice, in which the lattice points are 22 A apart. This distance roughly corresponds with the diameter of fibers seen in the electron microscope, bearing in mind that the latter are quite dry. Thus it is probable that the crystallographic unit and the fiber are the one and the same.

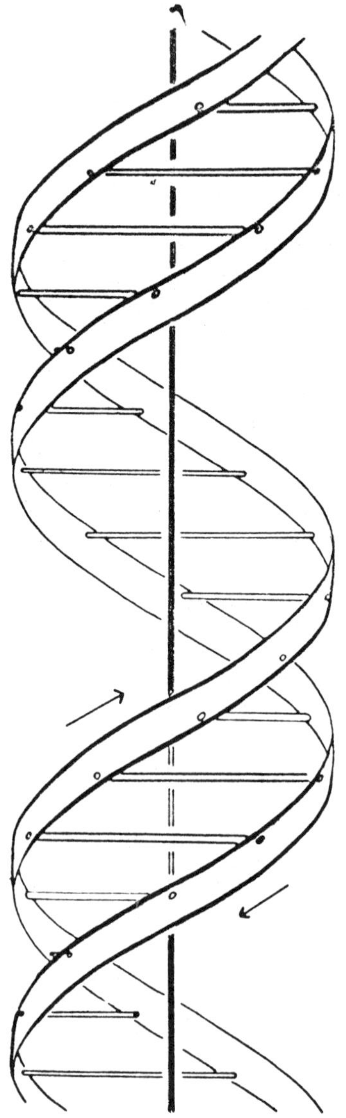

Fig. 4. This figure is diagrammatic. The two ribbons symbolize the two phosphate-sugar chains, and the horizontal rods the paths of bases holding the chain together. The vertical line marks the fiber axis.

III. DESCRIPTION OF THE PROPOSED STRUCTURE

Two conclusions might profitably be drawn from the above data. Firstly, the structure of DNA is regular enough to form a three dimensional crystal. This is in spite of the fact that its component chains may have an irregular sequence of purine and pyrimidine nucleotides. Secondly, as the structure contains two chains, these chains must be regularly arranged in relation to each other.

To account for these findings, we have proposed (Watson and Crick, 1953a) a structure in which the two chains are coiled round a common axis and joined together by hydrogen bonds between the nucleotide bases (see Figure 4). Both chains follow right handed helices, but the sequences of the atoms in the phosphate-sugar backbones run in opposite directions and so are related by a dyad perpendicular to the helix axis. The phosphates and sugar groups are on the outside of the helix whilst the bases are on the inside. The distance of a phosphorus atom from the fiber axis is 10 A. We have built our model to correspond to Structure B, which the X-ray data show to have a repeat distance of 34 A in the fiber direction and a very strong reflexion of spacing 3.4 A on the meridian of the X-ray pattern. To fit these observations our structure has a nucleotide on each chain every 3.4 A in the fiber direction, and makes one complete turn after 10 such intervals, i.e., after 34 A. Our structure is a well-defined one and all bond distances and angles, including van der Waal distances, are stereochemically acceptable.

The essential element of the structure is the manner in which the two chains are held together by hydrogen bonds between the bases. The bases are perpendicular to the fiber axis and joined together in pairs. The pairing arrangement is very specific, and only certain pairs of bases will fit into the structure. The basic reason for this is that we have assumed that the backbone of each polynucleotide chain is in the form of a regular helix. Thus, irrespective of which bases are present, the glucosidic bonds (which join sugar and base) are arranged in a regular manner in space. In particular, any two glucosidic bonds (one from each chain) which are attached to a bonded pair of bases, must always occur at a fixed distance apart due to the regularity of the two backbones to which they are joined. The result is that one member of a pair of bases must always be a purine, and the other a pyrimidine, in order to bridge between the two chains. If a pair consisted of two purines, for example, there would not be room for it; if of two pyrimidines they would be too far apart to form hydrogen bonds.

In theory a base can exist in a number of tautomeric forms, differing in the exact positions at which its hydrogen atoms are attached. However, under physiological conditions one particular form of each base is much more probable than any of the others. If we make the assumption that the favored forms always occur, then the pairing requirements are even more restrictive. Adenine can only pair with thymine, and guanine only with cytosine (or 5-methyl cytosine, or 5-hydroxy-methyl-cytosine). This pairing is shown in detail in Figures 5 and 6. If adenine tried to pair with cytosine it could not form hydrogen bonds, since there would be two hydrogens near one of

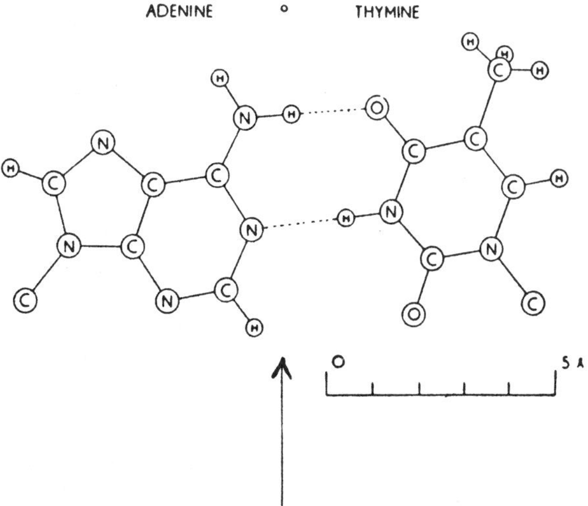

Fig. 5. Pairing of adenine and thymine. Hydrogen bonds are shown dotted. One carbon atom of each sugar is shown.

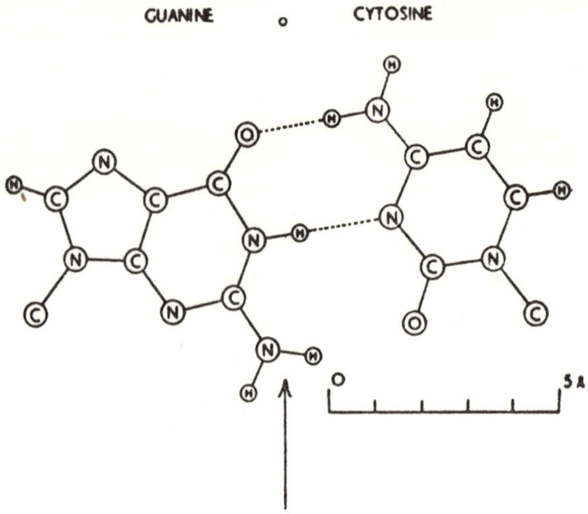

Fig. 6. Pairing of guanine and cytosine. Hydrogen bonds are shown dotted. One carbon atom of each sugar is shown.

the bonding positions, and none at the other, instead of one in each.

A given pair can be either way round. Adenine, for example, can occur on either chain, but when it does its partner on the other chain must always be thymine. This is possible because the two glucoside bonds of a pair (see Figures 5 and 6) are symmetrically related to each other, and thus occur in the same positions if the pair is turned over.

It should be emphasized that since each base can form hydrogen bonds at a number of points one can pair up *isolated* nucleotides in a large variety of ways. *Specific* pairing of bases can only be obtained by imposing some restriction, and in our case it is in a direct consequence of the postulated regularity of the phosphate-sugar backbone.

It should further be emphasized that whatever pair of bases occurs at one particular point in the DNA structure, no restriction is imposed on the neighboring pairs, and any *sequence* of pairs can occur. This is because all the bases are flat, and since they are stacked roughly one above another like a pile of pennies, it makes no difference which pair is neighbor to which.

Though any sequence of bases can fit into our structure, the necessity for specific pairing demands a definite relationship between the sequences on the two chains. That is, if we knew the actual order of the bases on one chain, we could automatically write down the order on the other. *Our structure therefore consists of two chains, each of which is the complement of the other.*

IV. EVIDENCE IN FAVOR OF THE COMPLEMENTARY MODEL

The experimental evidence available to us now offers strong support to our model though we should emphasize that, as yet, it has not been proved correct. The evidence in its favor is of three types:

(1) The general appearance of the X-ray picture strongly suggests that the basic structure is helical (Wilkins *et al.*, 1953; Franklin and Gosling, 1953a). If we postulate that a helix is present, we immediately are able to deduce from the X-ray pattern of Structure B (Figure 3), that its pitch is 34 A and its diameter approximately 20 A. Moreover, the pattern suggests a high concentration of atoms on the circumference of the helix, in accord with our model which places the phosphate-sugar backbone on the outside. The photograph also indicates that the two polynucleotide chains are not spaced equally along the fiber axis, but are probably displaced from each other by about three-eighths of the fiber axis period, an inference again in qualitative agreement with our model.

The interpretation of the X-ray pattern of Structure A (the crystalline form) is less obvious. This form does not give a meridional reflexion at 3.4 A, but instead (Figure 2) gives a series of reflexions around 25° off the meridian at spacings between 3 A and 4 A. This suggests to us that in this form the bases are no longer perpendicular to the fiber axis, but are tilted about 25° from the perpendicular position in a way that allows the fiber to contract 30 per cent and reduces the longitudinal translation of each nucleotide to about 2.5 A. It should be noted that the X-ray pattern of Structure A is much more detailed than that of Structure B and so if correctly interpreted, can yield more precise information about DNA. Any proposed model for DNA must be capable of forming either Structure A or Structure B and so it remains imperative for our very tentative interpretation of Structure A to be confirmed.

(2) The anomolous titration curves of undegraded DNA with acids and bases strongly suggests that hydrogen bond formation is a characteristic aspect of DNA structure. When a solution of DNA is initially treated with acids or bases, no groups are titratable at first between pH 5 and pH 11.0, but outside these limits a rapid ionization occurs (Gulland and Jordan, 1947; Jordan, 1951). On back titration, however, either with acid from pH 12 or with alkali from pH 2½, a different titration curve is obtained indicating that the titratable groups are more accessible to acids and bases than is the untreated solution. Accompanying the initial release of groups at pH 11.5 and in the range pH 3.5 to pH 4.5 is a marked fall in the viscosity and the disappearance of strong flow birefringence. While this decrease was originally thought to be caused by a reversible depolymerization (Vilbrandt and Tennent, 1943), it has been shown by Gulland, Jordan and Taylor (1947) that this is unlikely as no increase was observed in the amount of secondary phosphoryl groups. Instead these authors suggested that some of the groups of the bases formed hydrogen bonds between different bases. They were unable to decide whether the hydrogen bonds linked bases in the same or in adjacent structural units. The fact that most of the ionizable groups are originally inaccessible to acids and bases is more easily explained if the hydrogen bonds are between bases within the same structural unit. This point would definitely be established if it were shown that the shape of the initial titration curve was the same at very low DNA concentrations, when the interaction between neighboring structural units is small.

(3) The analytical data on the relative proportion of the various bases show that the amount of adenine is close to that of thymine, and the amount of guanine close to the amount of cytosine + 5-methyl cytosine, although the ratio of adenine to guanine can vary from one source to another (Chargaff, 1951; Wyatt, 1952). In fact as the techniques for estimation of the bases improve, the ratios of adenine to thymine, and guanine to cytosine + 5-methyl cytosine appear to grow very close to unity. This is a most striking result, especially as the sequence of bases on a given chain is likely to be irregular, and suggests a structure involving paired bases. In fact, we believe the analytical data offer the most important evidence so far available in support of our model, since they specifically support the biologically interesting feature, the presence of complementary chains.

We thus believe that the present experimental evidence justifies the working hypothesis that the essential features of our model are correct and allows us to consider its genetic possibilities.

V. GENETICAL IMPLICATIONS OF THE COMPLEMENTARY MODEL

As a preliminary we should state that the DNA fibers from which the X-ray diffraction patterns were obtained are not artifacts arising in the method of preparation. In the first place, Wilkins and his co-workers (see Wilkins *et al.*, 1953) have shown that X-ray patterns similar to those from the isolated fibers can be obtained from certain intact biological materials such as sperm head and bacteriophage particles. Secondly, our postulated model is so extremely specific that we find it impossible to believe that it could be formed during the isolation from living cells.

A genetic material must in some way fulfil two functions. It must duplicate itself, and it must exert a highly specific influence on the cell. Our model for DNA suggests a simple mechanism for the first process, but at the moment we cannot see how it carries out the second one. We believe, however, that its specificity is expressed by the precise sequence of the pairs of bases. The backbone of our model is highly regular, and the sequence is the only feature which can carry the genetical information. It should not be thought that because in our structure the bases are on the "inside," they would be unable to come into contact with other molecules. Owing to the open nature of our structure they are in fact fairly accessible.

A mechanism for DNA replication

The complementary nature of our structure suggests how it duplicates itself. It is difficult to imagine how like attracts like, and it has been suggested (see Pauling and Delbrück, 1940; Friedrich-Freksa, 1940; and Muller, 1947) that self duplication may involve the union of each part with an opposite or complementary part. In these discussions it has generally been suggested that protein and nucleic acid are complementary to each other and that self replication involves the alternate syntheses of these two components. We should like to propose instead that the specificity of DNA self replication is accomplished without recourse to specific protein synthesis and that each of our complementary DNA chains serves as a template or mould for the formation onto itself of a new companion chain.

For this to occur the hydrogen bonds linking the complementary chains must break and the two chains unwind and separate. It seems likely that the single chain (or the relevant part of it) might itself assume the helical form and serve as a mould onto which free nucleotides (strictly polynucleotide precursors) can attach themselves by forming hydrogen bonds. We propose that polymerization of the precursors to form a new chain only occurs if the resulting chain forms the proposed structure. This is plausible because steric reasons would not allow monomers "crystallized" onto the first chain to approach one another in such a way that they could be joined together in a new chain, unless they were those monomers which could fit into our structure. It is not obvious to us whether a special enzyme would be required to carry out the polymerization or whether the existing single helical chain could act effectively as an enzyme.

Difficulties in the replication scheme

While this scheme appears intriguing, it nevertheless raises a number of difficulties, none of which, however, do we regard as insuperable. The first difficulty is that our structure does not differentiate between cytosine and 5-methyl cytosine, and therefore during replication the specificity in sequence involving these bases would not be perpetuated. The amount of 5-methyl cytosine varies considerably from one species to another, though it is usually rather small or absent. The present experimental results (Wyatt, 1952) suggest that each species has a characteristic amount. They also show that the sum of the two cytosines is more nearly equal to the amount of guanine than is the amount of cytosine by itself. It may well be that the difference between the two cytosines is not functionally significant. This interpretation would be considerably strengthened if it proved possible to change the amount of 5-methyl cytosine in the DNA of an organism without altering its genetical make-up.

The occurrence of 5-hydroxy-methyl-cytosine in the T even phages (Wyatt and Cohen, 1952) presents no such difficulty, since it completely replaces cytosine, and its amount in the DNA is close to that of guanine.

The second main objection to our scheme is that it completely ignores the role of the basic protamines and histones, proteins known to be combined with DNA in most living organisms. This was done for two reasons. Firstly, we can formulate a scheme of DNA reproduction involving it alone and so from the viewpoint of simplicity it seems better to believe (at least at present) that the genetic specificity is never passed through a protein intermediary. Secondly, we know almost nothing about the structural features of protamines and histones. Our only clue is the finding of Astbury (1947) and of Wilkins and Randall (1953) that the X-ray pattern of nucleoprotamine is very similar to that of DNA alone. This suggests that the protein component, or at least some of it, also assumes a helical form and in view of the very open nature of our model, we suspect that protein forms a third helical chain between the pair of polynucleotide chains (see Figure 4). As yet nothing is known about the function of the protein; perhaps it controls the coiling and uncoiling and perhaps it assists in holding the single polynucleotide chains in a helical configuration.

The third difficulty involves the necessity for the two complementary chains to unwind in order to serve as a template for a new chain. This is a very fundamental difficulty when the two chains are interlaced as in our model. The two main ways in which a pair of helices can be coiled together have been called plectonemic coiling and paranemic coiling. These terms have been used by cytologists to describe the coiling of chromosomes (Huskins, 1941; for a review see Manton, 1950). The type of coiling found in our model (see Figure 4) is called plectonemic. Paranemic coiling is found when two separate helices are brought to lie side by side and then pushed together so that their axes roughly coincide. Though one may start with two regular helices the process of pushing them together necessarily distorts them. It is impossible to have paranemic coiling with two regular simple helices going round the same axis. This point can only be clearly grasped by studying models.

There is of course no difficulty in "unwinding" a *single* chain of DNA coiled into a helix, since a polynucleotide chain has so many single bonds about which rotation is possible. The difficulty occurs when one has a pair of simple helices with a common axis. The difficulty is a topological one and cannot be surmounted by simple manipulation. Apart from breaking the chains there are only two sorts of ways to separate two chains coiled plectonemically. In the first, one takes hold of one end of one chain, and the other end of the other, and simply pulls in the axial direction. The two chains slip over each other, and finish up separate and end to end. It seems to us highly unlikely that this occurs in this case, and we shall not consider it further. In the second way the two chains must be directly untwisted. When this has been done they are separate and side by side. The number of turns necessary to untwist them completely is equal to the number of turns of one of the chains round the common axis. For our structure this comes to one turn every 34 A, and thus about 150 turns per million molecular weight of DNA, that is per 5000 A of our structure. The problem of uncoiling falls into two parts:

(1) How many turns must be made, and how is tangling avoided?

(2) What are the physical or chemical forces which produce it?

For the moment we shall be mainly discussing the first of these. It is not easy to decide what is the uninterrupted length of functionally active DNA. As a lower limit we may take the molecular weight of the DNA after isolation, say fifty thousand A in length and having about 1000 turns. This is only a lower limit as there is evidence suggesting a breakage of the DNA fiber during the process of extraction. The upper limit might be the total amount of DNA in a virus or in the case of a higher organism, the total amount of DNA in a chromosome. For T2 this upper limit is approximately 800,000 A which corresponds to 20,000 turns, while in the higher organisms this upper limit may sometimes be 1000 fold higher.

The difficulty might be more simple to resolve if successive parts of a chromosome coiled in opposite directions. The most obvious way would be to have both right and left handed DNA helices in sequence but this seems unlikely as we have only been able to build our model in the right handed sense. Another possibility might be that the long strands of right handed DNA are joined together by compensating strands of left handed polypeptide helices. The merits of this proposition are difficult to assess, but the fact that the phage DNA does not seem to be linked to protein makes it rather unattractive.

The untwisting process would be less complicated if replication started at the ends as soon as the chains began to separate. This mechanism would produce a new two-strand structure without requiring at any time a free single-strand stage. In this way the danger of tangling would be considerably decreased as the two-strand structure is much more rigid than a single strand and would resist attempts to coil around its neighbors. Once the replicating process is started the presence, at the growing end of the pair, of double-stranded structures might facilitate the breaking of hydrogen bonds in the original unduplicated section and allow replication to proceed in a zipper-like fashion.

It is also possible that one chain of a pair occasionally breaks under the strain of twisting. The polynucleotide chain remaining intact could then release the accumulated twist by rotation about single bonds and following this, the broken ends, being still in close proximity, might rejoin.

It is clear that, in spite of the tentative suggestions we have just made, the difficulty of untwisting is a formidable one, and it is therefore worthwhile re-examining why we postulate plectonemic coiling, and not paranemic coiling in which the two helical threads are not intertwined, but merely in close apposition to each other. Our answer is that with paranemic coiling, the specific pairing of bases would not allow the successive residues of each helix to be in equivalent orientation with regard to the helical axis. This is a possibility we strongly oppose as it implies that a large number of stereochemical alternatives for the sugar-phosphate backbone are possible, an inference at variance to our finding, with stereochemical models (Crick and Watson, 1953) that the position of the sugar-phosphate group is rather restrictive and cannot be subject to the large variability necessary for paranemic coiling. Moreover, such a model would not lead to specific pairing of the bases, since this only follows if the glucosidic links are arranged regularly in space. We therefore believe that if a helical structure is present, the relationship between the helices will be plectonemic.

We should ask, however, whether there might not be another complementary structure which maintains the necessary regularity but which is not helical. One such structure can, in fact, be imagined. It would consist of a ribbon-like arrangement in which again the two chains are joined together by specific pairs of bases, located 3.4 A above each other, but in which the sugar-phosphate backbone instead of forming a helix, runs in a straight line at an angle approximately 30° off the line formed by the pair of bases. While this ribbon-like structure would give many of the features of the X-ray diagram of Structure B, we are unable to define precisely how it should pack in a macroscopic fiber, and why in particular it should give a strong equatorial reflexion at 20-24 A. We are thus not enthusiastic about this model though we should emphasize that it has not yet been disproved.

Independent of the details of our model, there are two geometrical problems which *any* model for DNA must face. Both involve the necessity for some form of super folding process and can be illustrated with bacteriophage. Firstly, the total length of the DNA within T2 is about 8×10^5 A. As its DNA is thought (Siegal and Singer, 1953) to have the same very large M.W. as that from other sources, it must bend back and forth many times in order to fit into the phage head of diameter 800 A. Secondly, the DNA must replicate itself without getting tangled. Approximately 500 phage particles can be synthesized within a single bacterium of average dimensions $10^4 \times 10^4 \times 2 \times 10^4$ A. The total length of the newly produced DNA is some 4×10^8 A, all of which we believe was at some interval in contact with its parental template. Whatever the precise mechanism of replication we suspect the most reasonable way to avoid tangling is to have the DNA fold up into a compact bundle as it is formed.

A possible mechanism for natural mutation

In our duplication scheme, the specificity of replication is achieved by means of specific pairing between purine and pyrimidine bases; adenine with thymine, and guanine with one of the cytosines. This specificity results from our

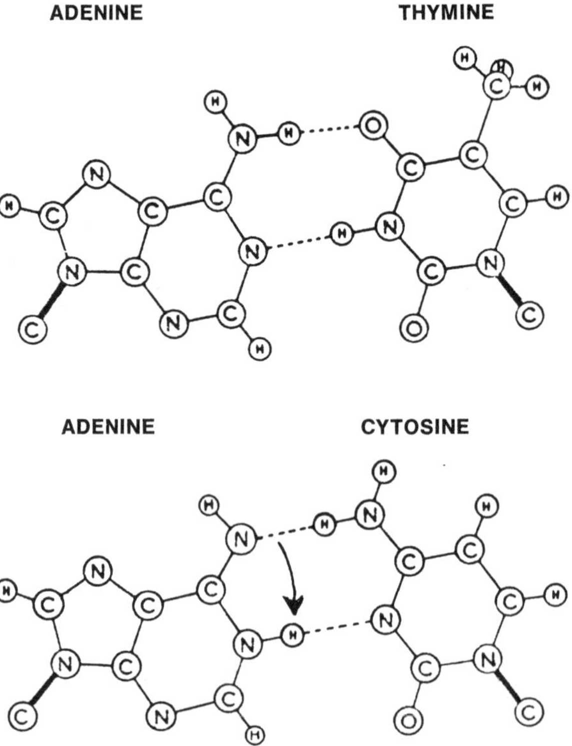

Fig. 7. Pairing arrangements of adenine before (above) and after (below) it has undergone a tautomeric shift.

assumption that each of the bases possesses one tautomeric form which is very much more stable than any of the other possibilities. The fact that a compound is tautomeric, however, means that the hydrogen atoms can occasionally change their locations. It seems plausible to us that a spontaneous mutation, which as implied earlier we imagine to be a change in the sequence of bases, is due to a base occurring very occasionally in one of the less likely tautomeric forms, at the moment when the complementary chain is being formed. For example, while adenine will normally pair with thymine, if there is a tautomeric shift of one of its hydrogen atoms it can pair with cytosine (Figure 7). The next time pairing occurs, the adenine (having resumed its more usual tautomeric form) will pair with thymine, but the cytosine will pair with guanine, and so a change in the sequence of bases will have occurred. It would be of interest to know the precise difference in free energy between the various tautomeric forms under physiological conditions.

GENERAL CONCLUSION

The proof or disproof of our structure will have to come from further crystallographic analysis, a task we hope will be accomplished soon. It would be surprising to us, however, if the idea of complementary chains turns out to be wrong. This feature was initially postulated by us to account for the crystallographic regularity and it seems to us unlikely that its obvious connection with self replication is a matter of chance. On the other hand the plectonemic coiling is, superficially at least, biologically unattractive and so demands precise crystallographic proof. In any case the evidence for both the model and the suggested replication scheme will be strengthened if it can be shown unambiguously that the genetic specificity is carried by DNA alone, and, on the molecular side, how the structure could exert a specific influence on the cell.

REFERENCES

Astbury, W. T., 1947, X-Ray studies of nucleic acids in tissues. Sym. Soc. Exp. Biol. **1**:66-76.

Chargaff, E., 1951, Structure and function of nucleic acids as cell constituents. Fed. Proc. **10**:654-659.

Crick, F. H. C., and Watson, J. D., 1953, Manuscript in preparation.

Franklin, R. E., and Gosling, R., 1953a, Molecular configuration in sodium thymonucleate. Nature, Lond. **171**:740-741.

1953b, Fiber diagrams of sodium thymonucleate. I. The influence of water content. Acta Cryst., Camb. (in press).

1953c, The structure of sodium thymonucleate fibers. II. The cylindrically symmetrical Patterson Function. Acta Cryst., Camb. (in press).

Fraser, M. S., and Fraser, R. D. B., 1951, Evidence on the structure of desoxyribonucleic acid from measurements with polarized infra-red radiation. Nature, Lond. **167**:760-761.

Friedrich-Freksa, H., 1940, Bei der Chromosomen Konjugation wirksame Krafte und ihre Bedeutung für die identische Verdoppling von Nucleoproteinen. Naturwissenshaften **28**:376-379.

Gulland, J. M., and Jordan, D. O., 1946, The macromolecular behavior of nucleic acids. Sym. Soc. Exp. Biol. **1**:56-65.

Gulland, J. M., Jordan, D. O., and Taylor, H. F. W., 1947, Electrometric titration of the acidic and basic groups of the desoxypentose nucleic acid of calf thymus. J. Chem. Soc. 1131-1141.

Huskins, C. L., 1941, The coiling of chromonemata. Cold Spr. Harb. Symp. Quant. Biol. **9**:13-18.

Jordan, D. O., 1951, Physiochemical properties of the nucleic acids. Prog. Biophys. **2**:51-89.

Kahler, H., and Lloyd, B. J., 1953, The electron microscopy of sodium desoxyribonucleate. Biochim. Biophys. Acta **10**:355-359.

Manton, I., 1950, The spiral structure of chromosomes. Biol. Rev. **25**:486-508.

Muller, H. J., 1947, The Gene. Proc. Roy. Soc. Lond. Ser. B. **134**:1-37.

Pauling, L., and Delbrück, M., 1940, The nature of the intermolecular forces operative in biological processes. Science **92**:77-79.

Siegal, A., and Singer, S. J., 1953, The preparation and properties of desoxypentosenucleic acid. Biochim. Biophys. Acta **10**:311-319.

Vilbrandt, C. F., and Tennent, H. G., 1943, The effect of pH changes upon some properties of sodium thymonucleate solutions. J. Amer. Chem. Soc. **63**:1806-1809.

Watson, J. D., and Crick, F. H. C., 1953a, A structure for desoxyribose nucleic acids. Nature, Lond. **171**:737-738.

1953b, Genetical implications of the structure of desoxyribose nucleic acid. Nature, Lond. (in press).

Wilkins, M. H. F., Gosling, R. G., and Seeds, W. E., 1951, Physical studies of nucleic acids–nucleic acid: an extensible molecule. Nature, Lond. **167**: 759-760.

Wilkins, M. H. F., and Randall, J. T., 1953, Crystallinity in sperm-heads: molecular structure of nucleoprotein in vivo. Biochim. Biophys. Acta **10**:192 (1953).

Wilkins, M. H. F., Stokes, A. R., and Wilson, H. R., 1953, Molecular structure of desoxypentose nucleic acids. Nature, Lond. **171**:738-740.

Williams, R. C., 1952, Electron microscopy of sodium desoxyribonucleate by use of a new freeze-drying method. Biochim. Biophys. Acta **9**:237-239.

Wyatt, G. R., 1952, Specificity in the composition of nucleic acids. In The Chemistry and Physiology of the Nucleus, pp. 201-213, N. Y. Academic Press.

Wyatt, G. R., and Cohen, S. S., 1952, A new pyrimidine base from bacteriophage nucleic acid. Nature, Lond. **170**:1072.

3 Semi-conservative replication of DNA in a higher plant cell

P. Filner*
Division of Biology
California Institute of Technology
Pasadena, California

The experiments of Meselson and Stahl [6] have demonstrated that in *Escherichia coli* DNA is replicated by a semi-conservative mechanism [2]. By means of ^{15}N density labeling of the DNA and cesium chloride density gradient equilibrium ultracentrifugation [7], it was found that after one replication in ^{14}N medium, the [^{15}N] DNA had disappeared and that a hybrid DNA with density intermediate between those of heavy [^{15}N] and light [^{14}N] DNA had appeared. A second replication in ^{14}N medium resulted in the appearance of light DNA equal in quantity to the persisting hybrid DNA. During subsequent replications, the light DNA fraction increased, while the hybrid DNA persisted, although it became a progressively smaller fraction of the total DNA. Furthermore, denaturation of the hybrid DNA produced two denatured species, one with the density of denatured light DNA, and one with the density of denatured heavy DNA. These results are consistent with the Watson-Crick hypothesis [14] that DNA is a double-stranded molecule, and that in the course of replication the strands separate, each serving as template for the synthesis of a new complementary strand.

The experiment has also been performed with the alga *Chlamydomonas reinhardii* by Sueoka [10]. The results were similar, with the reservation that some heavy DNA persisted beyond one doubling of the cells, indicating that not all of the DNA in the culture replicated before the onset of the second replication.

Mammalian cell DNA has been shown to replicate semi-conservatively by experiments which used 5-bromodeoxyuridine rather than ^{15}N as the density label [1, 3, 9].

Chromosome replication in higher plant cells has been shown to be semi-conservative by Taylor, Woods and Hughes [13]. Two DNA-containing chromosomal subunits which extend the full length of the chromosome separate and are preserved in the daughter chromosomes.

It would seem likely that the conserved DNA-containing chromosomal subunits of Taylor, Woods and Hughes are subunits of the DNA molecules themselves, as with bacterial, algal, and mammalian cells. That this is indeed true will be shown in the present paper. The method of Meselson and Stahl has been applied to exponentially dividing tobacco cells in suspension culture [8].

MATERIAL AND METHODS

Cells. In November 1961 the cells of sterile stem sections of *Nicotiana tabacum* Var. Xanthi were induced to proliferate on a complex agar medium, M-31 (Table I). The cell mass produced consisted of loosely connected, highly vacuolated spherical cells about 400 μ in diameter. After several serial subcultures on agar, the cells were dispersed in liquid M-31 medium and shaken. Proliferation occurred, and the resulting suspension of cell clusters and single cells could be readily subcultured in liquid medium. In suspension culture the cells are smaller, about 80 μ in diameter, and more tightly bound to each other. This cell line has been maintained through more than fifty serial subcultures, and is referred to as the "X" strain.

A derived cell line, "XD", was established by subculturing X cells in a completely defined medium, M-2D (Table I). The XD line is distinct both in its morphology and in its superior ability to grow on medium M-2D. The XD cells grow as linear chains of cells rather than clusters. The cells are cylindrical rather than spherical. On agar, they become large and spherical like X cells. XD cells can also proliferate in the minimal medium, M-1D (Table I) which contains nitrate as the nitrogen source. The XD cell line, grown in medium M-1D, was used for all of the experiments described below.

Media. The compositions of the media are given in Table I. The basal medium is a modification of White's medium [15]. In all cases pH was adjusted to 6.2 with NaOH before sterilization.

Culture conditions. The cells were grown in one liter flasks containing 500 ml of medium. The flasks were cotton stoppered and sealed with aluminum foil. They were shaken on a reciprocal shaker operating at 100 cycles per minute, with a horizontal displacement

From Experimental Cell Research **39**:33-39, 1965. Used with permission.

This work was supported by United States Public Health Service Training Grant T1GM86.

*Present address: MSU/AEC Plant Research Laboratory, Michigan State University, East Lansing, Michigan.

Table I

Basal medium		17 Amino acids[a]	
Sucrose	20,000 mg/l	L-Glu	51 mg/l
$MgSO_4 \cdot 7\ H_2O$	360	L-Phe	32
Na_2SO_4	200	L-Asp	18
$Ca(NO_3)_2 \cdot 4\ H_2O$	200	L-Pro	16
KNO_3	80	L-Lys	13
KCl	65	L-His	11
$NaH_2PO_4 \cdot H_2O$	16.5	L-Ser	10
$MnSO_4 \cdot 4\ H_2O$	4.5	L-Ala	8
Ferric citrate	2.0	L-Thr	8
$ZnSO_4 \cdot 7\ H_2O$	1.5	L-Val	8
H_3BO_3	1.5	L-Cys	7
KI	0.75	L-Met	7
2,4-Dichlorophenoxyacetic acid	0.5	L-Arg	5
Nicotinic acid	0.5	Gly	5
Pyridoxin	0.1	L-Ileu	5
Thiamin	0.1	L-Leu	5
		L-Try	4

Medium	Contents
M-1D	Basal medium
M-2D	Basal medium 17 amino acids
M-31	Basal medium Casein acid hydrolysate,[b] 200 mg/l Malt extract,[c] 500 mg/l

[a]CalBiochem Corp.
[b]Nutritional Biochemical Co.
[c]Difco.

of 4.5 cm. The culture room was maintained at 27°C. Subcultures were routinely made every ten days by diluting an aliquot of the parent culture 25-fold into fresh medium.

For the DNA replication experiments, cells were grown in ^{15}N-M-1D medium for eight generations. The ^{15}N-M-1D was prepared with 98.4 atom per cent ^{15}N-nitric acid purchased from Volk Chemical Co. The ^{15}N-labeled cells were harvested by filtration on Miracloth (Chicopee Manufacturing Co.), then resuspended and diluted in ^{14}N-M-1D to a concentration of 5 g fresh wt/l. The time of resuspension in ^{14}N medium constitutes time 0 in the replication experiments. At least five grams fresh weight of cells was harvested from the ^{14}N cultures daily for nucleic acid extractions.

Nucleic acid extractions. Cells were harvested by filtration on Miracloth. The cells were resuspended in ice cold 0.15 *M* NaCl–0.015 *M* Na citrate pH 7.0 (saline-citrate) 2 ml per gram fresh weight of cells. The suspension was homogenized by 50 strokes of a motor driven Thomas teflon-glass homogenizer. The cell walls were removed by filtering the homogenate through two layers of Miracloth. The filtrate was centrifuged at 1500 × *g* for 30 min. The supernatant was decanted and the pellet resuspended in 3.0 ml saline-citrate. The pellet was then extracted by a procedure based on that of Marmur [4]. To the suspension was added 0.3 ml 25 per cent sodium dodecyl sulfate, and solid NaCl to bring the aqueous phase NaCl concentration to 1 *M*. An equal volume of 24:1 chloroform:isoamyl alcohol was added and the emulsion shaken for 30 min at 27°C. The emulsion was broken by centrifuging at 12,000 × *g* for 10 min. The aqueous supernatant was pipetted into two volumes of ice cold 95 per cent EtOH. The fibrous, white nucleic acid precipitate was wound up on a glass rod and redissolved in 1.0 ml of saline-citrate. It was approximately 50 per cent DNA and 50 per cent RNA. The nucleic acid solution was stored over chloroform at 2°C.

The DNA was further purified for physical characterization by RNase treatment according to Marmur [4], followed by chloroform:isoamyl alcohol (24:1 V:V) extraction, precipitation with two volumes of ice cold ethanol, and then redissolved in saline-citrate.

Analytical ultracentrifugation. Approximately 2 μg of nucleic acid was centrifuged in ρ = 1.711 CsCl solution buffered with 0.05 *M* tris at pH 8.0 in a Spinco Model E. Analytical Ultracentrifuge for 24 hr at 44,700 rev./min, 25°C. *Micrococcus lysodeikticus* DNA was used as a density reference.

Growth curves. For dry weight growth curves a set of cultures was started from one parent culture. Each day two were harvested. The cells were dried at 60°C for 48 hr, and their dry weight determined.

For cell titer growth curves, one culture was sampled periodically. Cell counts were made on 0.2 ml aliquots containing 1000 to 2000 cells. An aliquot was placed on a microscope slide and covered with a 22 × 50 mm coverglass. The slide was systematically scanned at 100 × magnification, and the cells in each aggregate were counted. The number of cells in large aggregates was estimated since they could not be counted accurately. The fraction of total cells in large aggregates was small so that the error introduced is negligible.

RESULTS

The growth curves by dry weight and cell titer are similar (Figs. 1, 2). Both indicate a generation time of two days, with little lag period when the initial cell concentration is 5000 cells/ml. Exponential growth proceeds for between four and five generations.

The purified DNA from XD cells has a T_m = 85.5°C in 0.15 *M* NaCl, 0.015 *M* Na citrate, with a hyperchromicity of 35 per cent. The sedimentation constant is $S_{20,w}$ = 21.9. Its buoyant density in CsCl was computed to be ρ = 1.696, using the formula of Sueoka [11], and ρ = 1.731 for the reference DNA, *Micrococcus lysodeikticus.* The T_m and buoyant density both correspond to a G–C content of about 38 per cent [5, 12].

Since the generation time is two days, nucleic acid was prepared from cells harvested each 24 hr following the transfer from ^{15}N–M-1D to ^{14}N–M-1D. The results of density gradient equilibrium centrifugation of such DNA preparations are shown in Fig. 3. The fully ^{15}N-labeled DNA band disappears between days 2 and 3. Accompanying the disappearance of this band is the appearance of a band of intermediate density due to hybrid DNA. Totally light [^{14}N] DNA appears during the second replication, the hybrid persisting. During subsequent replications, as more [^{14}N] DNA is produced, the hybrid continues to persist but becomes a progressively smaller fraction of the total.

Attempts to demonstrate the presence of separable ^{15}N and ^{14}N strands in the hybrid DNA were unsuccessful, since denatured ^{14}N and denatured ^{15}N DNA yielded very broad unresolvable bands in the CsCl density gradient. This was probably due to DNase damage during isolation of DNA, as well as density heterogeneity in the DNA.

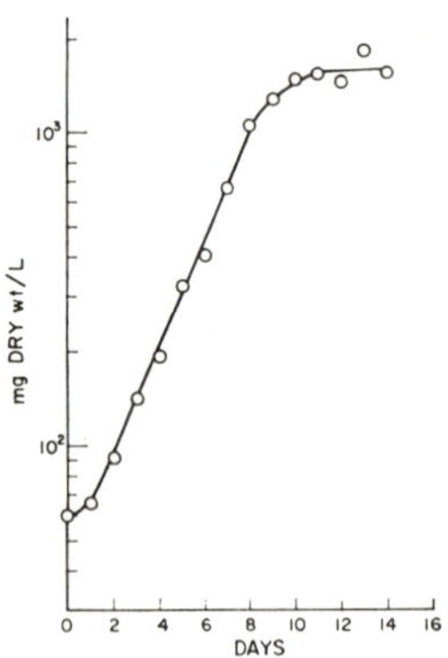

Fig. 1. Growth curve of XD cells in M-1D, dry weight assay.

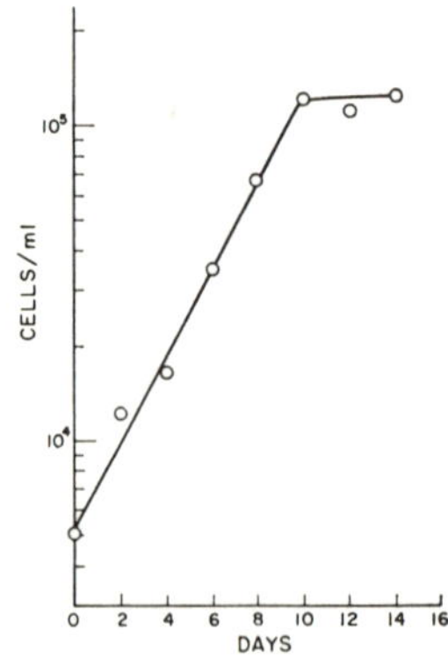

Fig. 2. Growth curve of XD cells in M-1D, cell titer assay.

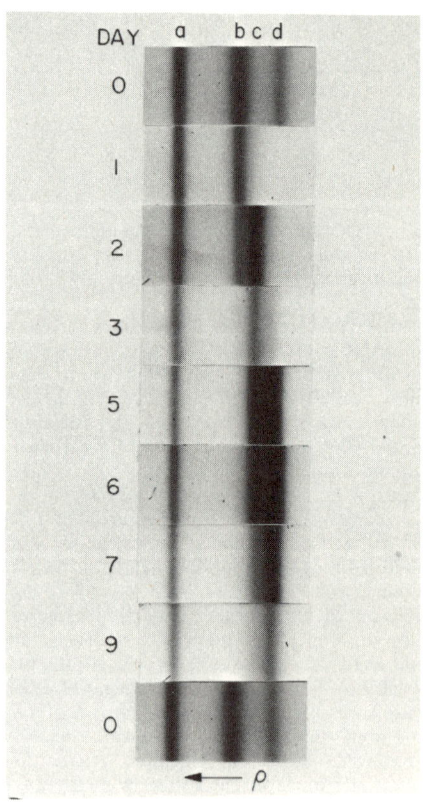

Fig. 3. Cesium chloride density gradient pictures of native XD-DNA bands. Band A, *Micrococcus lysodeikticus* DNA for reference; Band B, ^{15}N XD-DNA; Band C, hybrid XD-DNA; Band D, ^{14}N XD-DNA. The frames at top and bottom are synthetic mixtures of reference DNA, ^{15}N XD-DNA and ^{14}N XD-DNA. DNA was isolated at 24-hr intervals following transfer from ^{15}N to ^{14}N medium.

DISCUSSION

These experiments have shown that the cultured tobacco cell replicates all of its DNA, that the replication takes slightly more than a generation, and that the replication is semi-conservative.

The completion of DNA replication in about one generation indicates that the rate of DNA synthesis is uniform both intercellularly and intracellularly. If there were a fast DNA synthesizing subpopulation, the ^{14}N band would have appeared sooner, while a slow DNA synthesizing subpopulation would have caused the ^{15}N band to linger. In the intracellular case this may be interpreted to mean that a cycle of DNA replication must be complete before the next replicate can begin to be synthesized. In intercellular terms, it means the cell population is homogeneous by this criterion. This agrees with the fact that the generation time has consistently been two days for the more than 200 generations the cells have been carried in suspension culture. These observations are of interest because the line has not been cloned.

The persistence of the hybrid density DNA band indicates a semi-conservative mode of replication. Thus higher plant cell DNA replication exhibits the same semi-conservative characteristics which are found in DNA replication of bacteria, algae and animal cells. This result brings into focus an important aspect of chromosome structure as well as the mechanism of synthesis. There must be subunit continuity in the DNA of individual plant chromosomes since, according to the experiment presented above, the conserved chromosomal subunit is itself a subunit of the DNA molecule, and according to Taylor *et al.* [13], the chromosomal subunit extends the full length of the chromosome. It is not known whether the DNA of the plant chromosome consists of many small molecules or one very large one. If it is one long molecule, DNA subunit continuity is implicit. If it is many small ones, then they must be linked so that the DNA subunits produced during a given round of replication are associated.

SUMMARY

Tobacco cells in suspension culture multiplied exponentially with a generation time of two days. The replication of DNA in these cells was studied. Cells were grown on ^{15}N-nitrate, then transferred to ^{14}N-nitrate medium, and DNA was isolated at daily intervals thereafter. The CsCl density gradient analysis indicated that all the DNA replicated in slightly more than two days, and the mode of replication was semi-conservative.

REFERENCES

1. Chun, G. H. L. and Littlefield, J. W., J. Mol. Biol. **3**:668 (1961).
2. Delbrück, M. and Stent, G. S., *in* W. D. McElroy and B. Glass (eds.), The chemical basis of heredity, p. 699. Johns Hopkins Press, Baltimore, 1957.
3. Djordjevic, B. and Szybalski, W., J. Exptl. Med. **112**:509 (1960).
4. Marmur, J., J. Mol. Biol. **3**:208 (1961).
5. Marmur, J. and Doty, P., Nature **183**:1427 (1959).
6. Meselson, M. and Stahl, F. W., Proc. Natl. Acad. Sci. U.S. **44**:671 (1958).
7. Meselson, M., Stahl, F. W. and Vinograd, J., *ibid.* **43**:581 (1957).
8. Nickell, L. G., Proc. Natl. Acad. Sci. U.S. **42**:848 (1956).
9. Simon, E. H., J. Mol. Biol. **3**:101 (1961).
10. Sueoka, N., Proc. Natl. Acad. Sci. U.S. **46**:83 (1960).
11. ——— J. Mol. Biol. **3**:31 (1961).
12. Sueoka, N., Marmur, J. and Doty, P., Nature **183**:1429 (1959).
13. Taylor, J. H., Woods, P. S. and Hughes, W. L., Proc. Natl. Acad. Sci. U.S. **43**:122 (1957).
14. Watson, J. D. and Crick, F. H. C., Cold Spring Harbor Symp. Quant. Biol. **18**:123 (1953).
15. White, P. R., *in* A handbook of plant tissue culture, p. 103. The Ronald Press, New York, 1943.

4 Sequence complementarity of T2-DNA and T2-specific RNA

Benjamin D. Hall*
S. Spiegelman**
Departments of Chemistry and Microbiology
University of Illinois
Urbana, Illinois

Investigations of the functional interrelations among DNA, RNA, and protein are most conveniently performed under conditions which limit the synthesis of each macromolecular class to a few chemical species. A situation of this type obtains in *E. coli* cells infected with bacteriophage T2. Volkin and Astrachan[1] examined the nature of the RNA synthesized in the T2-coli complex by means of P^{32}-labeling. Estimation of the relative P^{32} content of the 2′, 3′-nucleotides isolated from an alkaline hydrolysate led Volkin and Astrachan to deduce that the RNA formed in the infected cell possessed an apparent base ratio analogous to that of T2-DNA. Subsequently, Volkin[2] obtained data suggesting that the synthesis of a specific RNA is a prerequisite for the intracellular production of bacteriophage.

Nomura, Hall, and Spiegelman[3] confirmed the observations on the apparent base ratios. In addition, they offered independent evidence for the existence of a "T2-Specific RNA" by demonstrating that RNA molecules synthesized after infection differed from the bulk of the *E. coli* RNA in electrophoretic mobility and average sedimentation coefficient. Because the procedures employed (zone electrophoresis and sedimentation) led to a selective separation of T2-specific RNA from the normal RNA of *E. coli,* they open up possibilities of further experiments relevant to an understanding of the nature of T2-RNA.

The fact that "T2-RNA" possesses a base ratio analogous to that of T2-DNA is of interest because it suggests that the similarity may go further and extend to a detailed correspondence of base sequence. The central issue of the significance and meaning of "T2-RNA" is whether or not this is in fact the case. A direct attack on this problem by complete sequence determination is technically not feasible at the moment. However, some recent findings of Marmur, Doty, *et al.*[4,5] suggest the possibility for an illuminating experiment. These authors demonstrated the specific reformation of active double-stranded DNA when heat-denatured DNA is subjected to a slow-cooling process. Such reconstitution of the double-stranded structure occurs only between DNA strands which originate from the same or closely related organisms. Presumably, the specificity requirement for a successful union of two strands reflects the need for a perfect, or near-perfect, complementarity of their nucleotide sequences. We have here then a possible method for detecting complementary nucleotide sequences. The formation of a double-stranded hybrid during a slow cooling of a mixture of two types of polynucleotide strands can be accepted as evidence for complementarity of the input strands.

We have used this procedure to examine for complementarity of sequence between "T2-RNA" and T2-DNA. Purified T2-RNA was used in order to provide an optimal opportunity for the T2-RNA to combine with its DNA complement, unhindered by non-specific interactions involving irrelevant RNA. Since the hybrid would have a lower density than uncombined RNA, a separation of the two might be attainable by equilibrium centrifugation in CsCl gradients.[6] To insure a sensitive and unambiguous detection of the hybrid, should it occur, double labeling was used. The T2-RNA was marked with P^{32} and the T2-DNA with H^3. Two isotopes emitting β-particles differing in their energies are conveniently assayed in each other's presence in a scintillation spectrometer.[7] This device, coupled with

---

From Proceedings of the National Academy of Sciences **47**:137-146, 1961. Used with permission.

This investigation was aided by grants in aid from the U.S. Public Health Service, National Science Foundation, and the Office of Naval Research.

We are indebted to Drs. Noboru Sueoka and John Drake for providing the T5 DNA and *Pseudomonas* DNA used in this work and to Miss Cherith Watson for her assistance in performing the experiments.

*Present address: Department of Genetics, University of Washington, Seattle, Wash.

**Present address: Institute of Cancer Research, Columbia University, New York, N. Y.

the use of the swinging-bucket rotor for the equilibrium centrifugation, permits the actual isolation of the pertinent fractions along with a ready and certain identification of any hybrids formed.

The primary purpose of the present paper is to present the results of such experiments. The data obtained demonstrate that specific complexes are indeed formed between "T2-RNA" and its homologous DNA. Their occurrence offers strong presumptive evidence for a detailed complementarity of the nucleotide sequences in these two macromolecules.

1. *Preparation and denaturation of DNA:* Tritiated phage were prepared by the addition of H^3-thymidine to a T2-infected culture of *E. coli* B. The cells were treated with 5-fluorouracil deoxyriboside (0.5 μg/ml) prior to infection. The phage were purified by treatment of the lysate with DNAase and RNAase followed by three cycles of high- and low-speed centrifugation. DNA was extracted from the purified phage by treatment with sodium dodecyl sulfate followed by chloroform-iso-amyl alcohol deproteinization and ethanol precipitation of the DNA.[8] This preparation will be designated by H^3-DNA(T2). DNA from other sources was similarly purified.

Tritiated *E. coli* DNA was prepared from cells of a thymineless mutant ($15T^-$) grown in a synthetic medium supplemented with tritiated thymidine. This preparation will be designated by H^3-DNA *(E. coli).*

DNA used for complex formation with RNA was first denatured by heating for 15 minutes at 95°C in 0.15 *M* NaCl + 0.01 *M* sodium citrate (pH 7.8), after which the tube containing the DNA was quickly placed in an ice bath. In all cases, the denaturation was carried out at a DNA concentration of 130 μg/cc.

2. *Preparation of T2-specific RNA labeled with P^{32}:* P^{32}-labeled ribosome RNA was obtained from *E. coli* B grown in synthetic medium, infected with T2 at a multiplicity of 3.8, and labeled with 10 millicuries of P^{32} between three and eight minutes after infection. The infection and radioisotope incorporation were done at 37°C in medium C (Roberts *et al.*)[9] modified to include 5 gm NaCl, 0.37 gm KCl, and 1 gm casamino acids per liter. The phosphate concentration was lowered to 10^{-3} *M* and 0.1 *M* tris (hydroxymethyl) aminomethane (tris), pH 7.3, was used for buffering. The number of infective centers and uninfected survivors (2.5%) agreed with the multiplicity of infection. The procedures used for stopping incorporation, washing and disrupting cells, and preparing ribosomal RNA were those described previously (Nomura, Hall, and Spiegelman[3]).

3. *Purification of T2-specific RNA:* Enrichment of the ribosomal RNA preparation in its content of T2-specific RNA (as judged by an eightfold increase in specific activity of P^{32}) was obtained by zone centrifugation through a sucrose gradient. One ml of a 1.5% sucrose solution (w/w) + one ml of P^{32} ribosome RNA solution (1 mg/ml, 106,000 cpm/ml) were layered, with an inverted gradient of RNA on 20 ml of a 2 to 15 per cent sucrose gradient. All solutions were 0.05 *M* in KCl and 10^{-2} *M* in tris buffer at a pH of 7.3. Following centrifugation for eight hours at 25,000 rpm in the SW-25 rotor of the Spinco preparative ultracentrifuge, the contents of the tube were removed by dripping through a hole punctured in the bottom of the tube. Fractions of 1.2 ml were collected by drop counting. The ultraviolet absorption at 260 mμ and P^{32} content of the fractions are shown in Figure 1. The two fractions at the peak of P^{32} activity (corresponding to 18 and 19.8 ml) were used for hybrid formation. These will be referred to in the text as P^{32}-RNA(T2). In the experiments described below, the two fractions exhibited identical properties.

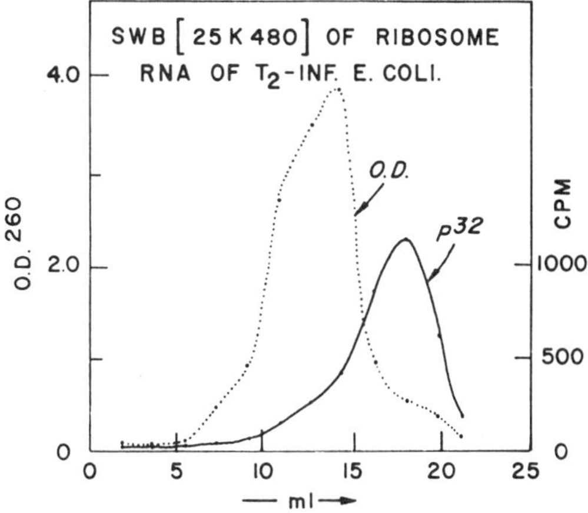

Fig. 1. Separation of P^{32}-RNA(T2) from *E. coli* RNA by sucrose-gradient sedimentation. One ml of ribosome RNA solution containing 1 mg RNA and 106,000 cpm P^{32} + 1 ml 1.5% sucrose solution were layered on 20 ml of a 2-15% sucrose gradient. Centrifugation: 8 hours at 25,000 rpm. Cpm shown refer to 0.05 ml fractions of the swinging-bucket fractions.

4. *Slow cooling of DNA and RNA:* Slow cooling of RNA and DNA was done in solutions 0.03 *M* in sodium citrate and 0.3 *M* in NaCl at a pH of 7.8. An insulated water bath having a capacity of 40 liters was used to provide slow cooling as follows:

Time (hr.)	Temperature (°C)
0	65
3.5	52
7.5	44
13.5	36
24.5	28.5
30.5	26

When the bath temperature reached 26°, the tubes containing RNA and DNA were removed and brought to a volume of 5.1 cc and a density of 1.74 gm/cc by addition of suitable amounts of water and saturated CsCl solution. Twenty-five μg of unlabeled, undenatured DNA were added to the solution as a reference density marker.

5. *Separation of RNA from DNA by density-gradient centrifugation:* The solutions of RNA and DNA containing CsCl were centrifuged at 33,000 rpm in the SW-39 rotor at a temperature of 25°C. At the end of each run, fractions corresponding to various

density levels in the tube were obtained by piercing the bottom of the tube and collecting drops, 30 for each fraction. These were diluted to a volume of 1.2 cc for measurement of ultraviolet absorption and radioisotope concentration.

6. *Counting of H^3-DNA and P^{32}-DNA:* To an aliquot from each swinging-bucket fraction 250 μg herring sperm DNA was added as carrier. The nucleic acid was then precipitated with trichloracetic acid (final concentration 10%) in the cold, collected, and washed on a millipore filter (course, 50 mm dia.). The filter was air-dried for one hour and placed in a cylindrical glass vial filled with 15 ml redistilled toluene containing 1.5 mg of 1,4-bis-2-(5-phenyloxazolyl)benzene (POPOP) and 60 mg of 2,5-diphenyloxazole (PPO). P^{32} and H^3 were counted in a Packard Tri-Carb liquid scintillation counter.

SEPARATION OF T2-SPECIFIC RNA IN A CsCl GRADIENT

It was first necessary to establish the conditions required for an adequate separation of T2-specific RNA from T2-DNA. Whereas *E. coli* ribosome RNA formed a narrow band within two days, T-2 RNA, because of its smaller size, required five days to form a band near the bottom of the tube.

Figure 2 shows the result of a five-day run carried out under the conditions specified above. The mixture being separated consisted of 6.5 μg of heat-denatured H^3-DNA(T2), 25 μg of unlabeled and undenatured T2-DNA, and 14 μg of the purified P^{32}-RNA(T2). Here, the three nucleic acids were not exposed to a slow-cooling operation but were mixed at 25°C, immediately put in the CsCl solution, and centrifuged. It will be noted that there is no appreciable interaction between the RNA and DNA as evidenced by the absence of any appreciable overlapping of the P^{32}- and H^3-containing regions. The small "tail" of P^{32} which extends to the top of the tube is presumably a consequence of the low molecular weight of the T2-specific RNA.

HYBRID FORMATION BETWEEN DENATURED T2-DNA AND T2-SPECIFIC RNA

The results described in Figure 2 show that CsCl density gradient centrifugation permits a clear separation of H^3-DNA(T2) from P^{32}-RNA(T2) and provides, therefore, a test for interactions leading to the formation of RNA-DNA hybrids. Any distortion of the distribution of H^3-DNA or P^{32}-RNA from that observed in Figure 2 which leads to regions of overlap between H^3 and P^{32} would be indicative of such interactions.

1. *The effect of temperature during slow cooling on hybrid formation:* The influence of the starting temperature of the slow-cooling process was examined in a number of runs. In all cases, the nucleic acid mixture incubated consisted of 6.5 μg of heat-denatured H^3-DNA(T2) and 14 μg of P^{32}-RNA(T2). The rate and conditions of the cooling were as described earlier.

Three tubes containing this RNA-DNA mixture were placed in the slow-cooling bath at starting temperatures of 65°, 52°, and 40°C respectively. Slow cooling was followed by CsCl gradient centrifugation.

Figure 3 shows the optical density profiles

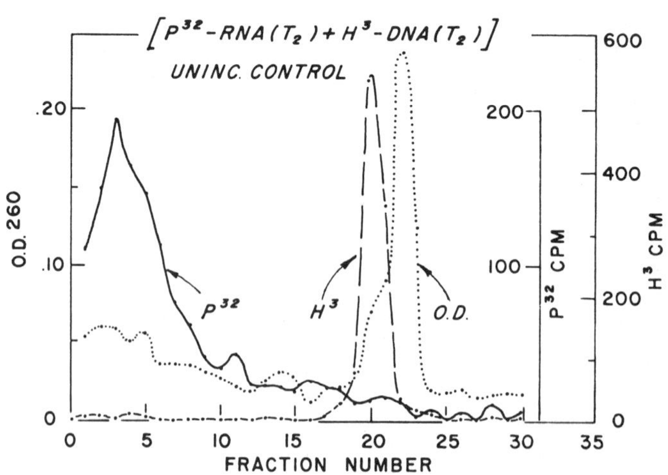

Fig. 2. Separation of P^{32}-RNA(T2) from H^3-DNA(T2) by CsCl-gradient centrifugation. A mixture of 6.5 μg heat-denatured H^3-DNA, 14 μg P^{32}-RNA and 25 μg undenatured, unlabeled T2 DNA was made at 25°C, immediately diluted with CsCl, and then centrifuged for five days at 33,000 rpm.

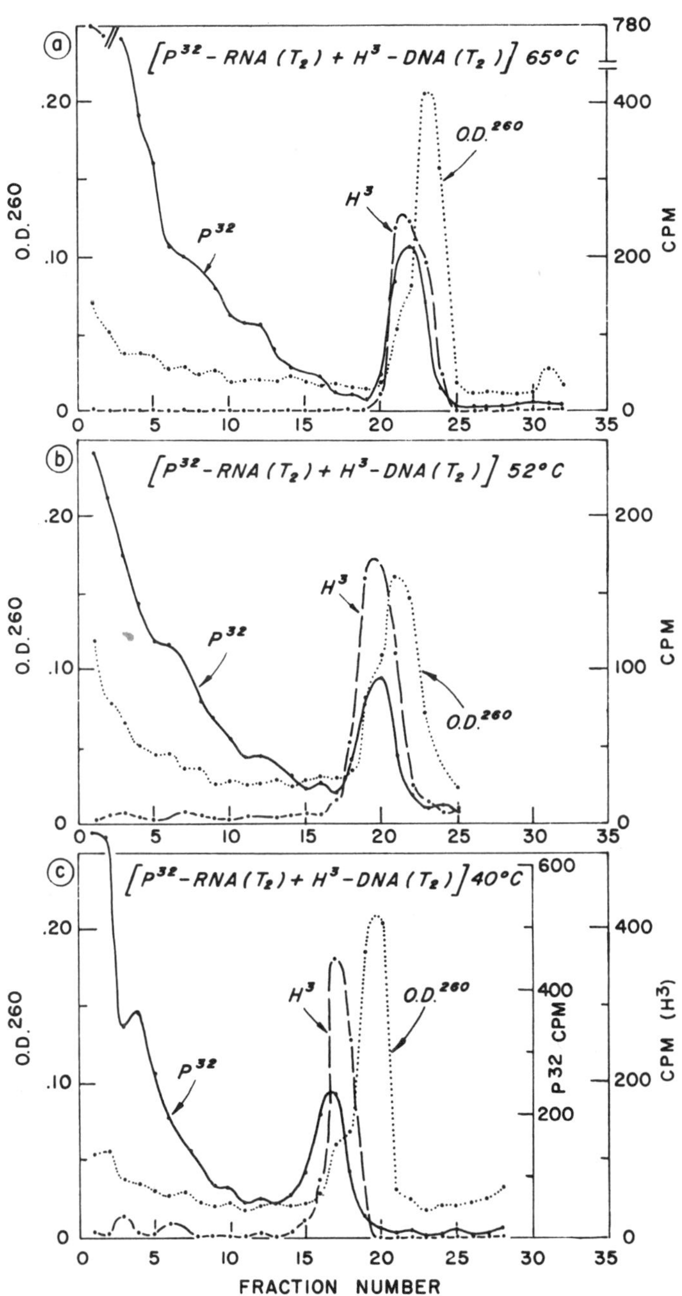

Fig. 3. Formation of DNA-RNA hybrid at various temperatures. CsCl-gradient centrifugation analysis. P^{32}-RNA(T2) (14 μg) and H^3-DNA(T2) (6.5 μg) were mixed in 0.6 ml 0.3 *M* NaCl and 0.03 *M* Na citrate (pH 7.8); then the solution was immediately placed in the slow-cooling bath. Three identical solutions were made; **a** was placed in the bath at 65°, **b** at 52°, and **c** at 40°C. When the bath temperature reached 26°, CsCl and 25 μg T2 DNA were added to each solution; then they were centrifuged for five days at 33,000 rpm.

and distributions of H^3 and P^{32} obtained at the three temperatures. Comparison of the profiles of H^3 and P^{32} with those of the control (Fig. 2) shows clearly that in all three cases, slow cooling of the DNA and RNA has produced a new peak of P^{32} approximately centered on the band of H^3 (denatured DNA). This new P^{32}-containing band must contain an RNA-DNA hybrid having approximately the same density as denatured T2-DNA. The amount of complex formed on cooling from the three temperatures was the same within experimental error. The three differ slightly in the density of the complex relative to DNA, the complex formed at low temperature being apparently more dense. This may be explained by the occurrence of partial renaturation of the H^3-DNA at the higher temperatures.

2. *Requirement for presence of single-stranded DNA during cooling:* In order to successfully complex with T2-RNA, the molecules of T2-DNA must be present in the single-stranded state. This was shown by an experiment in which a mixture of native H^3-DNA(T2) (13 μg) and P^{32}-RNA(T2) (15 μg) was subjected to slow cooling, starting from 40°C. No evidence of hybrid formation is observed (Fig. 4). In a companion run (a repetition of the experiment of Fig. 3c) with denatured H^3-DNA(T2), approximately 10 per cent of the P^{32}-RNA was included in the hybrid region.

3. *Stoichiometry of hybrid formation:* Assuming the specific activity of P^{32} to be equal in all RNA molecules, one can estimate the amount of RNA which formed hybrid. From the data of Figure 3, this would be 1.4 μg RNA. (This figure is a maximum value, for some of the RNA which failed to form hybrid may be preexisting *E. coli* RNA and, therefore, devoid of P^{32}.) The amount of DNA in the hybrid cannot exceed 6.5 μg, the total amount of denatured T2-DNA present. Because the hybrid and denatured T2-DNA have the same density, no more precise estimate can be made. From these considerations, it appears probable that the ratio of DNA to RNA in the hybrid does not exceed 5. That the complex does in fact contain considerably more DNA than RNA is suggested by its density, which is very nearly that of T2-DNA. A further indication that the entire 6.5 μg of T2-DNA has participated in complex formation is the proportionality observed between the amount of hybrid formed and the amount of DNA present when the ratio of DNA to RNA is varied. Experiments completely comparable to those described by Figure 3 were carried out with the same concentration of P^{32}-RNA(T2) but with 1/5 the quantity of denatured T2-DNA. In these cases, the amount of P^{32}-RNA found associated with the denatured DNA band was approximately two per cent of the input RNA which compares with the average of ten per cent observed when five times as much T2-DNA is included in the cooling mixture.

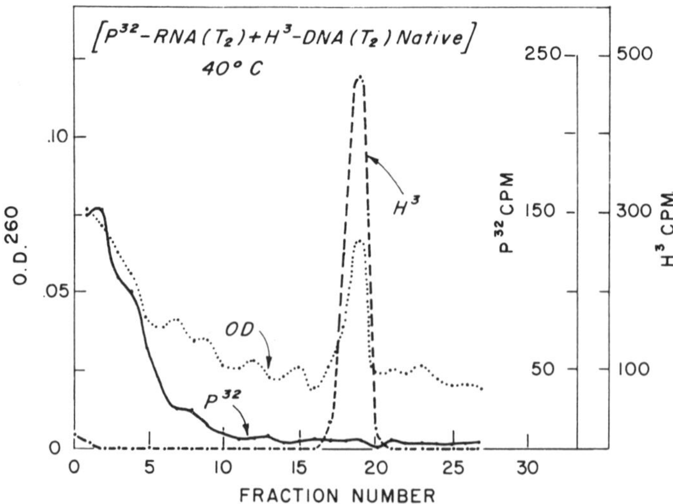

Fig. 4. CsCl-gradient centrifugation of a slowly cooled mixture of native H^3-DNA(T2) with P^{32}-RNA(T2). 13 μg DNA and 15 μg RNA were mixed in 1.2 ml 0.3 *M* NaCl, 0.03 *M* Na citrate, slowly cooled from 40° to 26°, diluted with saturated CsCl solution, and centrifuged.

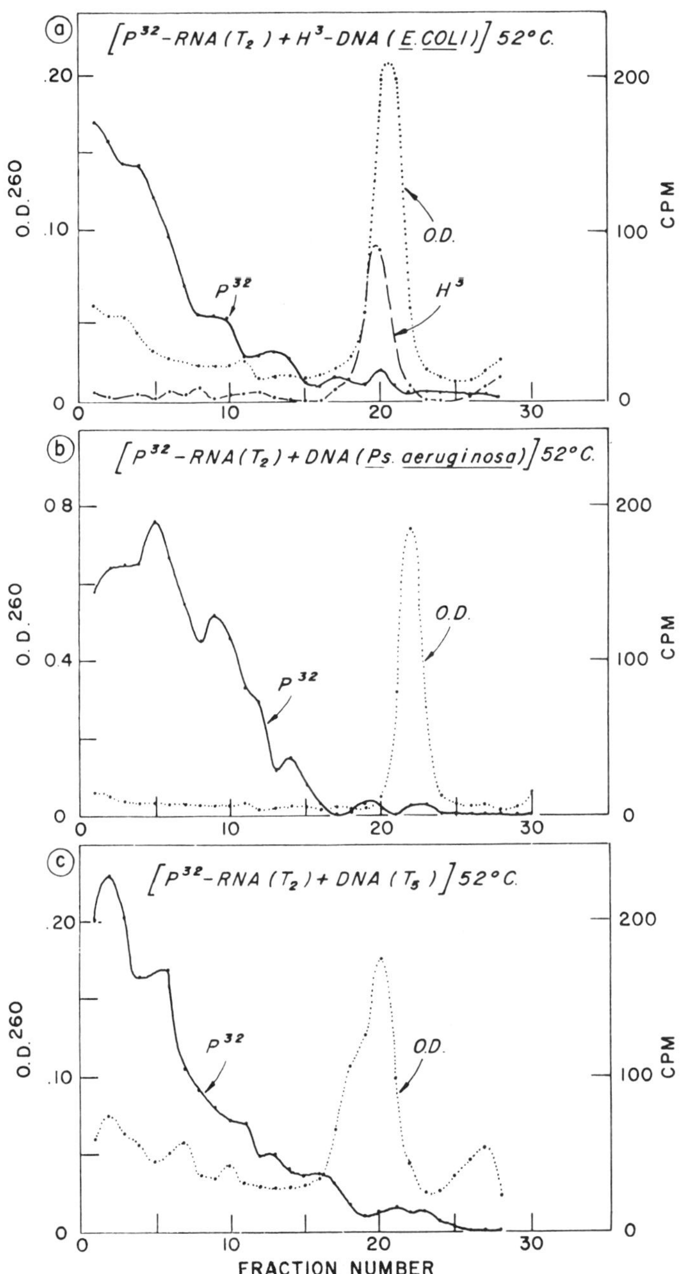

Fig. 5. CsCl-gradient centrifugation of slowly cooled heterologous DNA-RNA mixtures. **a,** P^{32}-RNA(T2) + H^3-DNA *(E. coli);* **b,** P^{32}-RNA(T2) + DNA *(Ps. aeruginosa);* **c,** P^{32}-RNA(T2) + DNA(T5).

In each case, 14 μg P^{32}-RNA and 6.5 μg heat-denatured heterologous DNA were mixed in 0.6 ml 0.3 *M* NaCl, 0.03 *M* sodium citrate and slowly cooled from 52° to 26° C. Before CsCl-gradient centrifugation, 25 μg T2 DNA was added to **a** and **c** and 25 μg *Pseudomonas* DNA to **b.**

4. *On the specificity of the interaction leading to hybrid formation:* Having established the existence of the phenomenon and the conditions required for its occurrence, it became of obvious interest to examine the specificity requirements of hybrid formation. This was tested by carrying out the cooling process with mixtures of P^{32}-RNA(T2) and denatured DNA from heterologous sources. These included DNA of *Ps. aeruginosa, E. coli,* and bacteriophage T5. The DNA of the *E. coli* was labeled with H^3-thymidine whereas the others were unlabeled. The mixtures of P^{32}-RNA(T2) and denatured DNA preparations were subjected to a 52°C slow-cooling incubation under conditions identical to those described for the experiment of Figure 3b. Upon completion of the incubation, unlabeled native DNA was added to each tube as a density marker.

Figure 5 gives the optical density profiles and distributions of radioactivities. There is a suggestion of a very slight peak of P^{32} in the DNA region of the mixture containing *E. coli* DNA. It corresponds to 1/30 the amount of hybrid produced with T2-DNA in a similar experiment. This may reflect the presence of the small amount of non-infected cells present during the P^{32}-labeling of the material from which the T2-RNA was obtained. However, it is too small to be considered seriously without further investigation. None of the other heterologous mixtures tested yielded detectable amounts of hybrid. It is of interest to note that although T5 has the same over-all base ratio as T2, no evidence of interaction between T2-RNA and T5-DNA was observed.

INTERPRETATION OF THE RESULTS

The data presented here show that RNA molecules synthesized in bacteriophage-infected cells have the ability to form a well-defined complex with denatured DNA of the virus. That this interaction is unique to the homologous pair is shown by the virtual absence of such complexes when T2-specific RNA is slowly cooled with heterologous DNA. The fact that T2-RNA and DNA do satisfy the specificity requirement must reflect a correspondence in structure between the two. Structural specificity of this order in single polynucleotide strands can only reside in definite sequences of nucleotides. We conclude that the most likely interrelationship of the nucleotide sequences of T2-DNA and RNA is one which is complementary in terms of the scheme of hydrogen bonding proposed by Watson and Crick.[10]

EXTENSION TO OTHER SYSTEMS

The bulk of the RNA in *E. coli* corresponds to the 18S and 25S[11] components of the ribosomes. These are metabolically stable,[12] remain firmly attached to ribosome protein at 10^{-4} *M* Mg^{++},[13] and have a base composition[14] not related in any obvious way to the DNA of the cells. In addition to the lack of correspondence in base ratio, two other reasons can be advanced for doubting the suitability of the large ribosomal RNA molecules for directing the synthesis of proteins as specified by the genetic material. First, the experiments of Riley *et al.*[15] suggest that the intermediary between the genome and the protein-synthesizing mechanism is metabolically unstable. Second, the formation of the larger RNA components is virtually absent in T2-infected cells (cf. Fig. 1) despite the fact that they are actively synthesizing a variety of new protein species.

It seems more likely that the RNA molecules directly concerned with specifying protein synthesis in normal cells would have a base ratio corresponding to DNA and would possess other properties analogous to those found for T2-specific RNA. Its principal characteristics may be summarized as follows:[1,3] (1) a weak linkage with the ribosome fractions since it can be broken by dialysis against 10^{-4} *M* Mg^{++}, (2) an active metabolic turnover, (3) an average sedimentation coefficient of about 8S, (4) a base composition which is closely analogous to its homologous DNA (considering thymidine equivalent to uridine and similarly for cytidine and hydroxymethyl cytidine), and (5) a sequence complementary to its homologous DNA.

The detection of the complementary RNA in T2-infected cells was greatly facilitated by the fact that the larger ribosomal components are not synthesized. Indeed, it would appear as if RNA synthesis in the T2-coli complex is largely confined to the class which is complementary to DNA. This advantage is not present in uninfected cells. Consequently the search for normal complementary RNA will be technically more difficult. That it is nevertheless feasible is suggested by the experiments of Yčas and Vincent[16] with yeast. These authors used P^{32} in a manner comparable to the procedures of Volkin and Astrachan and, despite surprisingly long pulses, they were able to detect the formation of a fraction with a high metabolic turnover and possessing a base composition analogous to yeast DNA.

Ultimately, attempts at establishing the presence in normal cells of RNA complementary to the genetic material will require that it be separated from the other RNA components. If all complementary RNA molecules possess physical chemical characteristics analogous to those of T2-specific RNA, the same methods which effected a successful isolation in this case may well serve in others. Once isolated, sequential complementarity to relevant DNA can be examined by the methods described above.

SOME IMPLICATIONS OF COMPLEMENTARY RNA

An increasing amount of attention is currently being focused on the possibility of forming hybrid helical complexes composed of DNA and RNA strands. Interest in this originates quite naturally from its obvious implication for translating the genetic information coded in DNA to a functional RNA complement. Previous experiments[17,18] had already demonstrated that paired helices were generated in mixtures of the synthetic polyribonucleotides of uridylate and adenylate. More recently,[19,20] this has been extended to combinations involving synthetic polydeoxyribonucleotides and polyribonucleotides. It is of some interest that the experiments reported in the present paper lend support to the concepts underlying such model experiments by exhibiting hybrid formation between natural polynucleotides which are complementary and biologically related.

The demonstration of sequence complementarity between homologous DNA and RNA is happily consistent with an attractively simple mechanism of informational RNA synthesis in which a single strand of DNA acts as a template for the polymerization of a complementary RNA strand.

SUMMARY

Experiments are described showing specific complex formation between single-stranded T2-DNA and the RNA synthesized subsequent to infection of *E. coli* with bacteriophage T2. No such hybrid formation is observed with heterologous DNA even if it has the same over-all base composition as T2-DNA. It is concluded that T2-DNA and T2-specific RNA form hybrids because they possess complementary nucleotide sequences. The generality of the existence of complementary RNA and its possible role as a carrier of information from the genetic material to the site of protein synthesis is briefly discussed.

REFERENCES

1. Volkin, E., and L. Astrachan, Virology **2:**149 (1956).
2. Volkin, E., these Proceedings **46:**1336 (1960).
3. Nomura, M., B. D. Hall, and S. Spiegelman, J. Mol. Biol. **2:**306 (1960).
4. Marmur, J., and D. Lane, these Proceedings **46:**453 (1960).
5. Doty, P., J. Marmur, J. Eigner, and C. Schildkraut, *ibid.* **46:**461 (1960).
6. Meselson, M., F. W. Stahl, and J. Vinograd, *ibid.* **43:**581 (1957).
7. Okita, G. T., J. J. Kabara, F. Richardson, and G. V. Le Roy, Nucleonics **15:**111 (1957).
8. Marmur, J., personal communication.
9. Roberts, R. B., P. H. Abelson, D. B. Cowie, E. T. Bolton, and R. J. Britten, "Studies of Biosynthesis in *Escherichia coli,*" Carnegie Institution of Washington, Publication #607 (1957), p. 5.
10. Watson, J. D., and F. H. C. Crick, Nature **171:**964 (1953).
11. Kurland, C. G., J. Mol. Biol. **2:**83 (1960).
12. Davern, C. I., and M. Meselson, *ibid.* **2:**153 (1960).
13. Tissières, A., J. D. Watson, D. Schlessinger, and B. R. Hollingworth, *ibid.* **1:**221 (1959).
14. Spahr, P. F., and A. Tissières, *ibid.* **1:**237 (1959).
15. Riley, M., A. B. Pardee, F. Jacob, and J. Monod. *ibid.* **2:**216 (1960).
16. Yčas M., and W. S. Vincent, these Proceedings **46:**804 (1960).
17. Warner, R. C., Fed. Proc. **15:**379 (1956).
18. Rich, A., and D. R. Davies, J. Am. Chem. Soc. **78:**3548 (1956).
19. Rich, A., these Proceedings **46:**1044 (1960).
20. Schildkraut, C. L., J. Marmur, J. R. Fresco, and P. Doty, J. Biol. Chem. (in press).

5 The RNA code and protein synthesis

M. Nirenberg
T. Caskey
R. Marshall
R. Brimacombe
D. Kellogg
B. Doctor*
D. Hatfield
J. Levin
F. Rottman
S. Pestka
M. Wilcox
F. Anderson

Laboratory of Biochemical Genetics
National Heart Institute
National Institutes of Health
Bethesda, Maryland
and
*Division of Biochemistry
Walter Reed Army Institute of Research
Walter Reed Army Medical Center
Washington, D. C.

Many properties of the RNA code which were discussed at the 1963 Cold Spring Harbor meeting were based on information obtained with randomly ordered synthetic polynucleotides. Most questions concerning the code which were raised at that time related to its fine structure, that is, the order of the bases within RNA codons. After the 1963 meetings a relatively simple means of determining nucleotide sequences of RNA codons was devised which depends upon the ability of trinucleotides of known sequence to stimulate AA-sRNA binding to ribosomes (Nirenberg and Leder, 1964). In this paper, information obtained since 1963 relating to the following topics will be discussed:

(1) The fine structure of the RNA code

(2) Factors affecting the formation of codon-ribosome-AA-sRNA complexes

(3) Patterns of synonym codons for amino acids and purified sRNA fractions

(4) Mechanism of codon recognition

(5) Universality

(6) Unusual aspects of codon recognition as potential indicators of special codon functions

(7) Modification of codon recognition due to phage infection.

FINE STRUCTURE OF THE RNA CODE

Formation of codon-ribosome-AA-sRNA complexes

The assay for base sequences of RNA codons depends, first upon the ability of trinucleotides to serve as templates for AA-sRNA binding to ribosomes prior to peptide bond formation, and second, upon the observation that codon-ribosome-AA-sRNA complexes are retained by cellulose nitrate filters (Nirenberg and Leder, 1964). Results shown in Table 1 illustrate characteristics of codon-ribosome-sRNA complex formation. Ribosomes, Mg^{++}, and poly U are required for the binding of C^{14}-Phe-sRNA to ribosomes. The addition of deacylated sRNA to reactions at zero time greatly reduces the binding of C^{14}-Phe-sRNA (Table 1), since poly U specifically stimulates the binding of both deacylated $sRNA^{Phe}$ and C^{14}-Phe-sRNA to ribosomes. Ribosomal bound C^{14}-Phe-sRNA is

---

Reproduced from Vol. XXXI; Cold Spring Harbor Symposia, 1966. Used with permission.

It is a pleasure to thank Miss Norma Zabriskie, Mrs. Theresa Caryk, Mr. Taysir M. Jaouni, and Mr. Wayne Kemper for their invaluable assistance. D. Kellogg is a Postdoctoral fellow of the Helen Hay Whitney Foundation. J. Levin is supported by USPHS grant 1-F2-GM-6369-01. F. Rottman is supported by grant PF-244 from the American Cancer Society.

The following abbreviations are used: Ala-, alanine-; Arg-, arginine-; Asn-, asparagine-; Asp-, aspartic acid-; Cys-, cysteine-; Glu-, glutamic acid-, Gln-, glutamine-, Gly-, glycine, His-, histidine-, Ile-, isoleucine-, Leu-, leucine-, Lys-, lysine-, Met-, methionine-, Phe-, phenylalanine-, Pro-, proline-, Ser-, serine-, Thr-, threonine-, Trp-, tryptophan-, Tyr-, tyrosine-, and Val-, valine-sRNA; sRNA, transfer RNA; AA-sRNA, aminoacyl-sRNA; $sRNA^{Phe}$, deacylated phenylalanine-acceptor sRNA; Ala-$sRNA^{Yeast}$, acylated alanine-acceptor sRNA from yeast. U, uridine; C, cytidine; A, adenosine; G, guanosine; I, inosine; rT, ribothymidine; ψ, psuedo-uridine; DiHU, dihydro-uridine; MAK, methylated albumin kieselguhr; F-Met, N-formylmethionine. For brevity, trinucleoside diphosphates are referred to as trinucleotides. Internal phosphates of trinucleotides are (3′,5′)-phosphodiester linkages.

Table 1. Characteristics of AA-sRNA binding to ribosomes

Modifications	C^{14}-Phe-sRNA bound to ribosomes (μμmole)
Complete	5.99
– Poly U	0.12
– Ribosomes	0.00
– Mg^{++}	0.09
+ deacylated sRNA at 50 min	
0.50 A^{260} units	5.69
2.50 A^{260} units	5.39
+ deacylated sRNA at zero time	
0.50 A^{260} units	4.49
2.50 A^{260} units	2.08

Complete reactions in a volume of 0.05 ml contained the following: 0.1 M Tris acetate (pH 7.2) (in other experiments described in this paper 0.05 M Tris acetate, pH 7.2 was used), 0.02 M magnesium acetate, 0.05 M potassium chloride (standard buffer); 2.0 A^{260} units of *E. coli* W3100 70 S ribosomes (washed by centrifugation 3 times); 15 mμmoles of uridylic acid residues of poly U; and 20.6 μμmoles C^{14}-Phe-sRNA (0.71 A^{260} units). All components were added to tubes at 0°C. C^{14}-Phe-sRNA was added last to initiate binding reactions.

Incubation was at 0°C for 60 min (in all other experiments described in this paper, reactions were incubated at 24° for 15 min). Deacylated sRNA was added either at zero time or after 50 min of incubation, as indicated. After incubation, tubes were placed in ice and each reaction was immediately diluted with 3 ml of standard buffer at 0° to 3°C. A cellulose nitrate filter (HA type, Millipore Filter Corp., 25 mm diameter, 0.45 μ pore size) in a stainless steel holder was washed with gentle suction with 5 ml of the cold standard buffer. The diluted reaction mixture was immediately poured on the filter under suction and washed to remove unbound C^{14}-Phe-sRNA with three 3-ml and one 15-ml portions of standard buffer at 3°. Ribosomes and bound sRNA remained on the filter (Nirenberg and Leder, 1964). The filters were then dried, placed in vials containing 10 ml of a scintillation fluid (containing 4 gm 2,5-diphenyloxazole and 0.05 gm 1,4-bis-2-(5-phenyloxazolyl)-benzene per liter of toluene) and counted in a scintillation spectrometer.

not readily exchangeable with unbound Phe-sRNA or deacylated $sRNA^{Phe}$ except at low Mg^{++} concentrations (Levin and Nirenberg, in prep.). Later in this volume Dr. Dolph Hatfield discusses the characteristics of exchange of ribosomal bound with unbound AA-sRNA when trinucleotides are present.

Two enzymatic methods were devised for oligonucleotide synthesis, since most trinucleotide sequences had not been isolated or synthesized earlier. One procedure employed polynucleotide phosphorylase to catalyze the synthesis of oligonucleotides from dinucleoside monophosphate primers and nucleoside diphosphates (Leder, Singer, and Brimacombe, 1965; Thach and Doty, 1965); the other approach (Bernfield, 1966) was based upon the demonstration (Heppel, Whitfeld, and Markham, 1955) that pancreatic RNase catalyzes the synthesis of oligonucleotides from uridine- or cytidine-2′,3′-cyclic phosphate and acceptor moieties. Elegant chemical procedures for oligonucleotide synthesis devised by Khorana and his associates also are available.

Template activity of oligonucleotides with terminal and internal substitutions

The trinucleotides, UpUpU and ApApA, but not the corresponding dinucleotides, stimulate markedly the binding of C^{14}-Phe- and C^{14}-Lys-sRNA, respectively. Such data directly demonstrate a triplet code and also show that codons contain three *sequential* bases. The template activity of triplets with 5′-terminal phosphate, pUpUpU equals that of the corresponding tetra- and pentanucleotides; whereas, oligo U preparations with 2′,3′-terminal phosphate are much less active. Hexa-A preparations, with and without 3′-terminal phosphate, are considerably more active as templates than the corresponding pentamers; thus, one molecule of hexa-A may be recognized by two Lys-sRNA molecules bound to adjacent ribosomal sites (Rottman and Nirenberg, 1966).

An extensively purified doublet with 5′-terminal phosphate, pUpC, serves as a template for Ser-sRNA (but not for Leu- or Ile-sRNA), whereas a doublet without terminal phosphate, UpC, is inactive (see Figs. 1a and b). However, the template activity of pUpC is considerably lower than that of the triplet, UpCpU. The relation between Mg^{++} concentration and template activity is shown in Fig. 1b. pUpC and UpCpU stimulate Ser-sRNA binding in reactions containing 0.02–0.08 M Mg^{++}. These results demonstrate that a doublet with 5′-terminal phosphate can serve as a specific, although relatively weak, template for AA-sRNA. It is particularly intriguing to relate recognition of a doublet to the possibility that only two out of three bases in a triplet may be recognized occasionally during protein synthesis, and also to the possibility that a triplet code evolved from a more primitive doublet code.

Further studies on template activities of oligonucleotides with terminal and internal modifications are summarized in Table 2. At limiting oligonucleotide concentrations, the

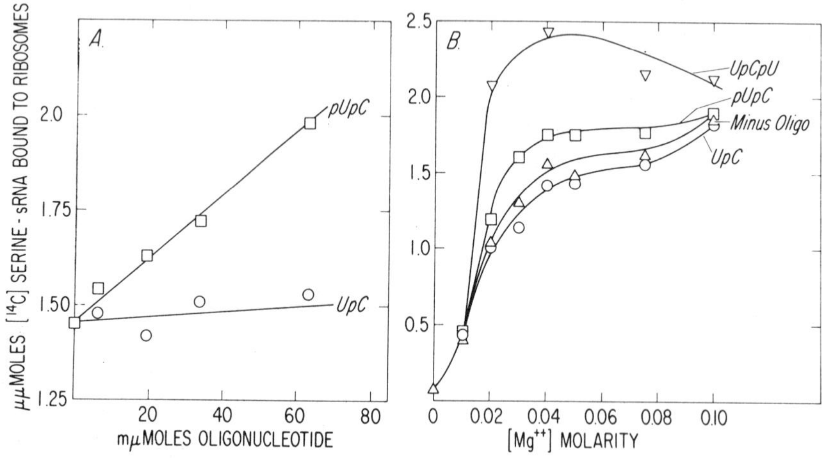

Fig. 1a, b. The effects of UpC and pUpC on the binding of C^{14}-Ser-sRNA to ribosomes. The relation between oligonucleotide concentration and C^{14}-Ser-sRNA binding to ribosomes at 0.03 M Mg^{++} is shown in Fig. 1a. It should be noted that the ordinate begins at 1.25 μμmoles of C^{14}-Ser-sRNA. The relation between Mg^{++} concentration and C^{14}-Ser-sRNA binding to ribosomes is shown in Fig. 1b. As indicated, 50 mμmoles of UpC or pUpC, or 15 mμmoles of UpCpU, were added to each reaction. Each point in parts **a** and **b** represents a 50 μl reaction containing the components described in the legend to Table 1 except for the following: 14.3 μμmoles C^{14}-Ser-sRNA (0.42 A^{260} unit); 1.1 A^{260} units of ribosomes. Incubations were for 15 min at 24°C. (Data from Rottman and Nirenberg, 1966.)

relative template activities of oligo U preparations are as follows: p-5′-UpUpU > UpUpU > CH_3O-p-5′-UpUpU > UpUpU-3′-p > UpUpU-3′-p-OCH_3 > UpUpU-2′,3′-cyclic phosphate. Trimers with (2′-5′) phosphodiester linkages, (2′-5′)-UpUpU and (2′-5′)-ApApA, do not serve as templates for Phe- or Lys-, sRNA respectively. The relative template efficiencies of oligo A preparations are as follows: p-5′-ApApA > ApApA > ApApA-3′-p > ApApA-2′-p.

These studies led to the proposal that RNA and DNA contain three classes of codons, differing in structure; 5′-terminal, 3′-terminal, and internal codons (Nirenberg and Leder, 1964). Certainly the first base of a 5′-terminal codon and the third base of a 3′-terminal codon may be recognized with less fidelity than an internal codon, for in the absence of a nucleotide neighbor a terminal base may have a greater freedom of movement on the ribosome. Substitution of 5′- or 3′-terminal hydroxyl groups may impose restrictions upon the orientation of terminal bases during codon recognition. 5′-Terminal and perhaps also 3′-terminal codons possibly serve, together with neighboring codons, as operator regions.

Since many enzymes have been described which catalyze the transfer of nucleotides, amino acids, phosphate, and other molecules to or from terminal ribose or deoxyribose of nucleic acids, modification of sugar hydroxyl groups was proposed as a possible mechanism for regulating the reading of RNA or DNA (Nirenberg and Leder, 1964).

Nucleotide sequences of RNA codons

A summary of nucleotide sequences of RNA codons by *E. coli* AA-sRNA is shown in Table 3 and patterns of degeneracy in Table 4. Almost every trinucleotide was assayed for template specificity with 20 AA-sRNA preparations (unfractionated sRNA acylated with one labeled and 19 unlabeled amino acids). It is important to test trinucleotide template specificity with 20 AA-sRNA preparations, since relative responses of AA-sRNA are then quite apparent. In surveying trinucleotide specificity, unfractionated AA-sRNA should be used initially because altering ratios of sRNA species often influences the fidelity of codon recognition.

Almost all triplets correspond to amino acids; furthermore, patterns of codon degeneracy are logical. Six degenerate codons correspond to serine, five or six to arginine and also to leucine, and from one to four to each of the remaining amino acids. Alternate bases often occupy the third positions of triplets comprising degenerate codon sets. In all cases triplet pairs with 3′-terminal pyrimidines (XYU and XYC, where X and Y represent the first and

Table 2. Relative template activity of substituted oliogonucleotides

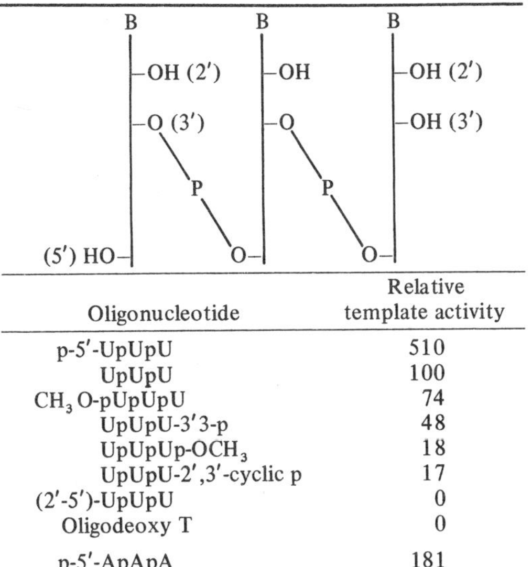

Oligonucleotide	Relative template activity
p-5′-UpUpU	510
UpUpU	100
CH_3 O-pUpUpU	74
UpUpU-3′3-p	48
UpUpUp-OCH_3	18
UpUpU-2′,3′-cyclic p	17
(2′-5′)-UpUpU	0
Oligodeoxy T	0
p-5′-ApApA	181
ApApA	100
ApApA-3′-p	57
ApApA-2′-p	15
(2′-5′)-ApApA	0
Oligodeoxy A	0

Relative template activities are approximations obtained by comparing the amount of AA-sRNA bound to ribosomes in the presence of limiting concentrations of oligonucleotides (0.50 or 0.12 mμmoles of oligonucleotides containing U or A, respectively) compared to either UpUpU, for C^{14}-Phe-sRNA; or ApApA, for C^{14}-Lys-sRNA (each assumed to be 100%). Data are from Rottman and Nirenberg (1966) except results with oligodeoxynucleotides which are from Nirenberg and Leder (1964).

Table 3. Nucleotide sequences of RNA codons

1st base	2nd base U	C	A	G	3rd base
U	PHE*	SER*	TYR*	CYS*	U
	PHE*	SER*	TYR*	CYS	C
	leu*?	SER	TERM?	cys?	A
	leu*, f-met	SER*	TERM?	TRP*	G
C	leu*	pro*	HIS*	ARG*	U
	leu*	pro*	HIS*	ARG*	C
	leu	PRO*	GLN*	ARG*	A
	LEU	PRO	gln*	arg	G
A	ILE*	THR*	ASN*	SER	U
	ILE*	THR*	ASN*	SER*	C
	ile*	THR*	LYS*	arg*	A
	MET*, F-MET	THR	lys	arg	G
G	VAL*	ALA*	ASP*	GLY*	U
	VAL	ALA*	ASP*	GLY*	C
	VAL*	ALA*	GLU*	GLY*	A
	VAL	ALA	glu	GLY	G

Nucleotide sequences of RNA codons were determined by stimulating binding of *E. coli* AA-sRNA to *E. coli* ribosomes with trinucleotide templates. Amino acids shown in capitals represent trinucleotides with relatively high template activities compared to other trinucleotide codons corresponding to the same amino acid. Asterisks (*) represent base compositions of codons which were determined previously by directing protein synthesis in *E. coli* extracts with synthetic randomly-ordered polynucleotides (Speyer et al., 1963; Nirenberg et al., 1963). F-Met, represents N-formyl-Met-sRNA which may recognize initiator codons. TERM represents possible terminator codons. Question marks (?) indicate uncertain codon function. Data are from Nirenberg et al., 1965; Brimacombe et al., 1965.

Table 4. Patterns of degenerate codons for amino acids

••U/C/A/G; ••U/C	••U/C/A/G; ••G/(A?)	••U/C/A/G	••U/C/(A)	••U/C	••A/G	••G	U/C/A/(G)••
SER	ARG LEU	GLY ALA VAL THR PRO	CYS ILE	ASP ASN HIS TYR PHE	GLU GLN LYS TERM?	MET TRP	F-MET

Solid circles represent the first and second bases of trinucleotides; U, C, A, and G indicate bases which may occupy the remaining position of degenerate codons. In the case of F-Met (N-formyl-methionine), circles represent the second and third bases. Parentheses indicate codons with relatively low template activities.

second bases, respectively, in the triplet) correspond to the same amino acid; often XYA and XYG correspond to the same amino acid; sometimes XYG alone corresponds to an amino acid. For eight amino acids, U, C, A, or G may occupy the third position of synonym codons. Alternate bases also may occupy the first position of synonyms, as for N-formyl-methionine.

One consequence of logical degeneracy is that many single base replacements in DNA may be silent and thus not result in amino acid replacement in protein (cf. Sonneborn, 1965). Also, the code is arranged so that the effects of some errors may be minimized, since amino acids which are structurally or metabolically related often correspond to similar RNA codons (for example, Asp-codons, GAU, and GAC, are similar to Glu-codons, GAA, and GAG). When various amino acids are grouped according to common biosynthetic precursors, close relationships among their synonym codons sometimes are observed. For example, codons for amino acids derived from aspartic acid begin with A: Asp, GAU, GAC; Asn, AAU, AAC; Lys, AAA, AAG; Thr, ACU, ACC, ACA, ACG; Ile, AUU, AUC, AUA; Met, AUG. Likewise, aromatic amino acids have codons beginning with U; Phe, UUU, UUC; Tyr, UAU, UAC; Trp, UGG. Such relationships may reflect either the evolution of the code or direct interactions between amino acids and bases in codons.

At the time of the 1963 meeting at Cold Spring Harbor, 53 base compositions of RNA codons had been estimated (14 tentatively) in studies with randomly-ordered synthetic polynucleotides and a cell-free protein synthesizing system derived from *E. coli* (Speyer et al., 1963; Nirenberg et al., 1963). Forty-six base composition assignments now are confirmed by base sequence studies with trinucleotides (shown in Table 3). Thus, codon base compositions and base sequence assignments, obtained by assaying protein synthesis and AA-sRNA binding, respectively, agree well with one another. In addition, codon base sequences are confirmed by most amino acid replacement data obtained in vivo.

Patterns of synonym codons recognized by purified sRNA fractions

Table 5 contains a summary of synonym codons recognized by purified sRNA fractions

Table 5. Codon patterns recognized by purified sRNA fractions

Alternate acceptable bases in 3rd or 1st positions of triplet											
	C U		A G		G		U C A		A G (U)		Possibly only 2 bases recognized
TYR$_{1,2}$	UA$^{C}_{U}$	LYS	AA$^{A}_{G}$	LEU$_2$	CUG	ALA^{yeast}	U GCC A	ALA$_1$	A GCG (U)	LEU$_3$	CU$^{(U)}_{(C)}$
VAL$_3$	GU$^{C}_{U}$			LEU$_5$	UUG	SER$^{yeast}_{2,3}$	U UCC A	VAL$_{1,2}$	A GUG (U)	LEU$_{4a,b}$	UU$^{(U)}_{(C)}$
				MET$_2$	AUG	F-MET$_1$	U C UG A			LEU$_1$	(U)UG
						TRP$_2$	U CGG (A)				

Patterns of degenerate codons recognized by purified AA-sRNA fractions. sRNA fractions are from *E. coli* B, unless otherwise specified. At the top of the table are shown the alternate bases which may occupy the third or first positions of degenerate codon sets. Purified sRNA fractions and corresponding codons are shown below. Parentheses indicate codons with relatively low template activity. sRNA fractions were obtained by counter-current distribution (Kellogg et al., 1966), unless otherwise specified. Yeast Ser-sRNA fractions 2 and 3 (Connelly and Doctor, 1966) are thought to be equivalent to yeast Ser-sRNA fractions 1 and 2, respectively. Yeast Ala-sRNA was the gift of R. W. Holley; results are from Leder and Nirenberg (unpubl.). Results obtained with Val-, Met-, and Ala-sRNA$^{E.\ coli}$ fractions are from Kellogg et al. (1966). Leu-sRNA fractions (see Fig. 6) and Lys-sRNA (Kellogg, Doctor, and Nirenberg, unpubl.) were obtained by MAK column chromatography. Three Leu-sRNA fractions also were obtained by counter-current distribution (Nirenberg and Leder, 1964). Reactions contained the usual components (see legend to Table 1) and 0.01 or 0.02 M Mg^{++}. Incubation was at 24° for 15 min.

obtained either by countercurrent distribution or by MAK column chromatography. The following patterns of codon recognition involving alternate bases in the third positions of synonym codons were found; C = U; A = G; G; U = C = A; A = G = (U). For example, Val-sRNA$_3$ recognized GUU and GUC, whereas the major peak of Val-sRNA (fractions 1 and 2) recognizes GUA, GUG and, to a lesser extent, GUU. The possibility that the latter Val-sRNA fraction contains two or more Val-sRNA components has not been excluded. Met-sRNA$_1$ responds to UUG, CUG, AUG and, to a lesser extent, GUG, and can be converted enzymatically to N-formyl-Met-sRNA, whereas, Met-sRNA$_2$ responds primarily to AUG and does not accept formyl moieties (see later discussion). Unfractionated Trp-sRNA responds only to UGG; however one fraction of Trp-sRNA, after extensive purification, responds to UGG, CGG and AGG. Possibly the latter responses depend upon the removal of sRNA for other amino acids (e.g., Arg-sRNA) which also may recognize CGG or AGG. Yeast Ala- and Ser-sRNA$_{2,3}$ fractions recognize synonyms containing U, C, or A in the third position. Leu-sRNA$_{1,3,4}$ bind to ribosomes in response to polynucleotide templates but not to trinucleotides. Possibly, only two of the three bases are recognized by these Leu-sRNA fractions.

Mechanism of codon recognition

Crick (1966) has suggested that certain bases in anticodons may form alternate hydrogen bonds, via a wobble mechanism, with corresponding bases in mRNA codons. This hypothesis and further experimental findings are discussed below.

Yeast Ala-sRNA of known base sequence and of high purity (>95%) was the generous gift of Dr. Robert Holley. In Figs. 2 and 3 are shown the responses of purified yeast and unfractionated *E. coli* C^{14}-Ala-sRNA, respectively, to synonym Ala-codons as a function of Mg^{++}

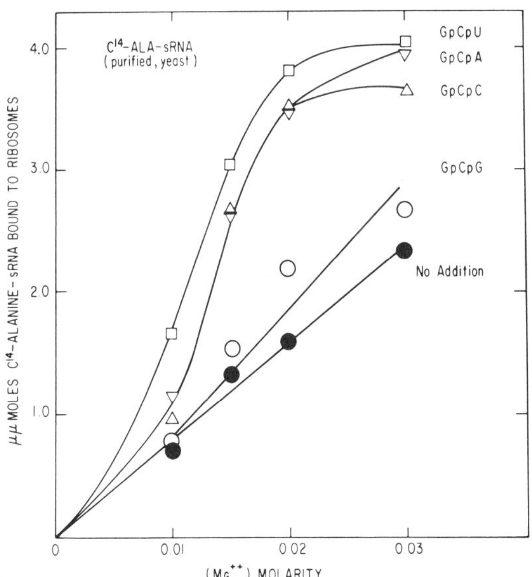

Fig. 2. The relation between Mg^{++} concentration and binding to ribosomes of purified yeast C^{14}-Ala-sRNA of known base sequence (Holley et al., 1965) in response to trinucleotides. Each point represents a 50 μl reaction containing the components described in the legend to Table 1 except for the following: 1.5 A^{260} units of *E. coli* ribosomes, 11.2 $\mu\mu$moles of purified yeast C^{14}-Ala-sRNA (0.038 A^{260} units); and 0.1 A^{260} unit of trinucleotide as specified. Reactions were incubated at 24° for 15 min (Leder and Nirenberg, unpubl.).

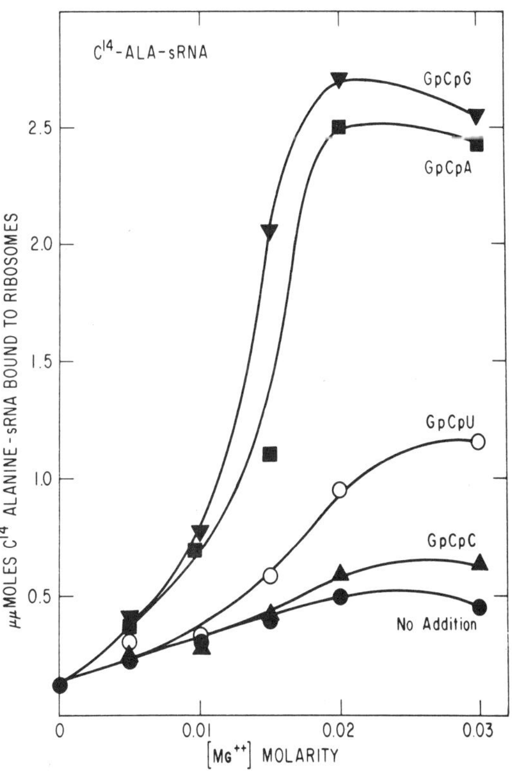

Fig. 3. Relation between Mg^{++} concentration and binding of unfractionated *E. coli* C^{14}-Ala-sRNA to ribosomes in response to trinucleotides. Each point represents a 50 μl reaction containing the components described in the legend to Table 1, 2.0 A^{260} units of ribosomes; 18.8 $\mu\mu$moles of unfractionated *E. coli* C^{14}-Ala-sRNA (0.54 A^{260} unit); and 0.1 A^{260} unit of trinucleotide, as specified (Leder and Nirenberg, unpubl.).

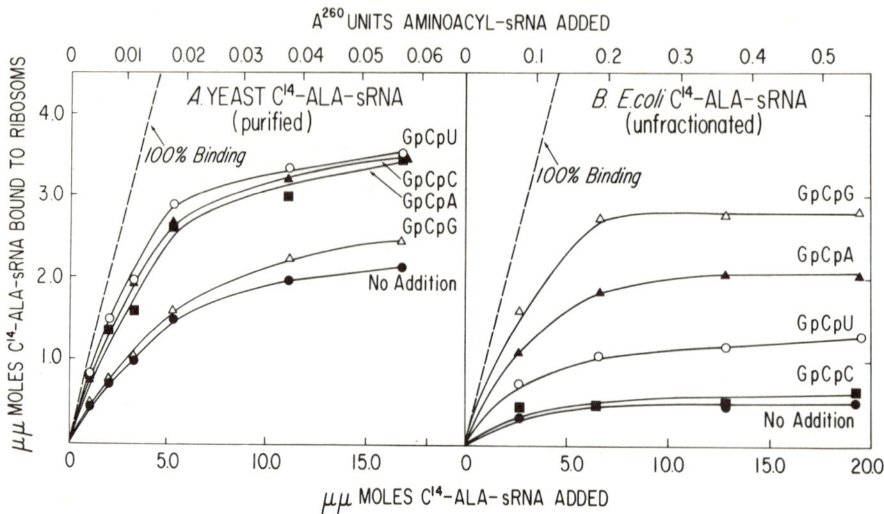

Fig. 4a, b. Relation between the template activities of trinucleotides and the concentrations of purified yeast C^{14}-Ala-sRNA (part **a**) and unfractionated *E. coli* C^{14}-Ala-sRNA (part **b**). Each point represents a 50 μl reaction containing the components described in the legend of Table 1, and the following components: 0.02 M magnesium acetate; 0.1 A^{260} unit of trinucleotide as specified; 1.1 A^{260} units of *E. coli* ribosomes (part **a**) and 2.0 A^{260} units of *E. coli* ribosomes (part **b**); and C^{14}-Ala-sRNA as indicated on the abscissa (Leder and Nirenberg, unpubl.).

concentration. Purified yeast C^{14}-Ala-sRNA responds well to GCU, GCC, and GCA, but only slightly to GCG. Similar results were obtained with unfractionated Ala-sRNA^{Yeast}. In contrast, unfractionated *E. coli* C^{14}-Ala-sRNA responds best to GCG and GCA, less well to GCU, and only slightly to GCC.

In Fig. 4a and b, the relation between concentration of yeast or *E. coli* C^{14}-Ala-sRNA and response to synonym Ala-codons is shown. At *limiting* concentrations of purified yeast C^{14}-Ala-sRNA, at least 59, 45, 45, and 3% of the available C^{14}-Ala-sRNA molecules bind to ribosomes in response to GCU, GCC, GCA, and GCG, respectively. The response of unfractionated *E. coli* C^{14}-Ala-sRNA to each codon was 18, 2, 38, and 64%, respectively. Similar results have been obtained by Keller and Ferger (1966). Since the purity of the yeast Ala-sRNA was greater than 95%, the extent of binding at limiting Ala-sRNA concentrations indicates that one molecule of Ala-sRNA recognizes 3, possibly 4, synonym codons. In addition, the data demonstrate marked differences between the relative responses of yeast and *E. coli* Ala-sRNA to synonym codons.

Correlating the base sequences of yeast Ala-sRNA with corresponding mRNA codons also provides insight into the structure of the Ala-sRNA anticodon and the mechanism of codon recognition. Possible anticodon or enzyme recognition sequences in Ala-sRNA^{Yeast} are –IGC MeI– and DiHU–CGG–DiHU (Fig. 5; Holley et al., 1965). Each site potentially comprises a single-stranded loop region at the end of a hairpin-like double-stranded segment. If CGG were the anticodon, *parallel* hydrogen bonding with GCU, GCC, GCA codons would be expected. If IGC were the anticodon, *antiparallel* Watson-Crick hydrogen bonding between GC in the anticodon and GC in the first and second positions of codons, and

RECOGNITION OF ALA-CODONS BY YEAST ALA-sRNA

	ME	DiH DiH
sRNA	CUUIGCIψGG	UAGUCGGUAGC
mRNA	GCU	GCU
	GCC	GCC
	GCA	GCA
	(GCG)	(GCG)

Fig. 5. Base sequences from yeast Ala-sRNA shown in the upper portion of the figure represent possible anticodons. Base sequences of synonym RNA Ala-codons are shown in the lower portion of the figure. The first and second bases of Ala-codons on the left would form antiparallel Watson-Crick hydrogen bonds with the anticodon, while those on the right would form parallel hydrogen bonds. See text for further details.

Table 6. Alternate base pairing

sRNA Anticodon	mRNA Codon
U	A G
C	G
A	U
G	C U
I	U C A
rT	A G
ψ	A G (U)
DiHU	No base pairing

The base in an sRNA anticodon shown in the left-hand column forms antiparallel hydrogen bonds with the base(s) shown in the right-hand column, which usually occupy the third position of degenerate mRNA codons. Relationships for U, C, A, G, and I of anticodons are "wobble" hydrogen bonds suggested by Crick (1966). See text for further details.

alternate pairing of inosine in the anticodon with U, C, or A, but not G, in the third position of Ala-codons, would be expected. All of the available evidence is consistent with an IGC Ala-anticodon. Zachau has shown that Ser-sRNA$_{1\ \mathrm{and}\ 2}^{\mathrm{Yeast}}$ contain, in appropriate positions, IGA sequences (Zachau, Dütting, and Feldmann, 1966), and we find that Ser-sRNA^{Yeast} fractions 2 and 3 (believed to correspond to fractions 1 and 2 of Zachau) recognize UCU, UCC, and UCA, but not UCG (see Table 5). A purified Val-sRNA^{Yeast} fraction contains the sequence IAC which corresponds to three Val-codons, GUU, GUC, and GUA (Ingram and Sjöquist, 1963). In addition, the sequence, GψA, is found at the postulated anticodon site of Tyr-sRNA^{Yeast} which corresponds to the Tyr-codons, UAU and UAC (Madison, Everett, and Kung, 1966).

Crick's wobble hypothesis and patterns of synonym codons found experimentally are in full agreement. In Table 6 are shown bases in anticodons which form alternate hydrogen bonds, via the wobble mechanism, with bases usually occupying the third positions of mRNA codons. U in the sRNA anticodon may pair alternately with A or G in mRNA codons; C may pair with G; A with U; G with C or U; and I with U, C, or A. In addition, we suggest that ribo T in the anticodon may hydrogen bond more strongly with A, and perhaps with G also, than U; and ψ in the anticodon may hydrogen bond alternately with A, G or, less well, U.

Dihydro U in an anticodon may be unable to hydrogen bond with a base in mRNA but may be repelled less by pyrimidines than by purines. Possibly, hydrogen bonds then form between the two remaining bases of the codon (bases 1 and 2, or 2 and 3) and the corresponding bases in the anticodon. Only two out of three bases in a codon would then be recognized. This possibility is supported by the studies of Rottman and Cerutti (1966) and Cerutti, Miles, and Frazier, (1966). Possibly, some synonym codon patterns may be due to the formation of two rather than three base pairs per triplet, particularly if both are (C) · (G) pairs (also see earlier discussion concerning template activity of pUpC).

In summary, patterns for amino acids often represent the sum of two or more codon patterns recognized by different sRNA species. Specific sRNA patterns, in turn, often result from alternate pairing between bases in the codon and anticodon or, possibly, from the formation of only two base pairs if the remaining bases do not greatly repel one another.

UNIVERSALITY

The results of many studies indicate that the RNA code is largely universal. However, translation of the RNA code can be altered in vivo by extragenic suppressors and in vitro by altering components of reactions or conditions of incubation. Thus, cells sometimes differ in specificity of codon translation.

To investigate the fine structure of the code recognized by AA-sRNA from different organisms, nucleotide sequences and relative template activities of RNA codons recognized by bacterial, amphibian, and mammalian AA-sRNA (*E. coli, Xenopus laevis* and guinea pig liver, respectively) were determined (Marshall, Caskey, and Nirenberg, submitted for publication). Acylation of sRNA was catalyzed in all cases by aminoacyl-sRNA synthetases from corresponding organisms and tissues. *E. coli* ribosomes were used for binding studies. Therefore, the specificities of sRNA and AA-sRNA synthetases were investigated.

The results are shown in Table 7. Almost identical translations of nucleotide sequences to

Table 7. Nucleotide sequences of RNA codons recognized by AA-sRNA from bacteria and amphibian and mammalian liver

	U	C	A	G	
U	PHE	SER	TYR	cys	U
	PHE	SER	TYR	cys	C
	leu?	SER	TERM?	[cys]	A
	leu, F-MET	[SER]	TERM?	trp	G
C	leu	PRO	HIS	ARG	U
	leu	PRO	HIS	ARG	C
	leu	PRO	gln	ARG	A
	leu	PRO	gln	[ARG]	G
A	ILE	THR	asn	[SER]	U
	ILE	THR	asn	[SER]	C
	[ILE]	THR	LYS	[ARG†]	A
	MET, F-MET?	THR	[LYS]	[ARG]	G
G	VAL	ALA	ASP	GLY	U
	VAL	ALA	ASP	GLY	C
	VAL	ALA	GLU	gly	A
	VAL	[ALA]	GLU	gly	G

Universality of the RNA code. Nucleotide sequences and relative template activities of RNA codons determined with trinucleotides and AA-sRNA from *E. coli, Xenopus laevis* and guinea pig liver. Rectangles represent trinucleotides which are active templates for AA-sRNA from one organism, but not from another. Assignments in capitals indicate that the trinucleotide was assayed with AA-sRNAs from *E. coli, Xenopus laevis* liver, and guinea pig liver. Assignments in lower case indicate that the trinucleotide was assayed only with *E. coli* AA-sRNA (with the exception of cys-codons which were assayed with both *E. coli* and guinea pig liver Cys-sRNA).

†Söll et al. (1965) reported that both AGA and AGG stimulate yeast Arg-sRNA binding to ribosomes. The trinucleotide, AGA, however, has little or no effect upon the binding of *E. coli, Xenopus laevis* or guinea pig Arg-sRNA to ribosomes.

Reactions contained components described in the legend to Table 1, 0.01 or 0.02 M Mg^{++}, *E. coli* ribosomes, and 0.150 A^{260} units of trinucleotides (data from Marshall, Caskey, and Nirenberg, in prep.).

Table 8. Species dependent differences in response of AA-sRNA to trinucleotide codons

Codon		sRNA: Bacterial (*E. coli*)	Amphibian (*Xenopus laevis*)	Mammalian (Guinea pig liver)
ARG	AGG	±	+ + + +	+ + +
	CGG	±	+ + + +	+ + + +
MET	UUG	+ +	±	±
ALA	GCG	+ + + +	±	+ +
ILE	AUA	±	+ +	+ +
LYS	AAG	±	+ + + +	+ + + +
SER	UCG	+ + + +	±	+ +
	AGU	±	+ + +	+ + +
	AGC	±	+ + +	+ + +
CYS	UGA	±		+ + +

Possible differences: ACG, THR; AUC, ILE; CAC, HIS; GUC, VAL; and GCC, ALA

No differences found: ASP, GLY, GLU, PHE, PRO, and TYR.

The following scale indicates the approximate response of AA-sRNA to a trinucleotide relative to the responses of the same AA-sRNA preparation to all other trinucleotides for that amino acid (except Gly-sRNA which was assayed only with GGU and GGC).

+ + + +	70-100%
+ + +	50-70%
+ +	20-50%
±	0-20%

amino acids were found with bacterial, amphibian, and mammalian AA-sRNA. In addition, similar sets of synonym codons usually were recognized by AA-sRNA from each organism. However, *E. coli* AA-sRNA sometimes differed strikingly from *Xenopus* and guinea pig liver AA-sRNA in relative response to synonym codons. Differences in codon recognition are shown in Table 8. The following trinucleotides had little or no detectable template activity for unfractionated *E. coli* AA-sRNA but served as active templates with *Xenopus* and guinea pig AA-sRNA: AGG, CGG, arginine; AUA, isoleucine; AAG, lysine; AGU, AGC, serine; and UGA, cysteine. Those trinucleotides with high template activity for *E. coli* AA-sRNA but low activity for *Xenopus* or guinea pig liver AA-sRNA were: UUG, N-formyl-methionine; GCG, alanine; and UCG, serine. Possible differences also were observed with ACG, threonine; AUC, isoleucine; CAC, histidine; GCC, alanine; and GUC, valine. No species dependent differences were found with Asp-, Gly-, Glu-, Phe-, Pro-, and Tyr-codons.

Thus, some degenerate trinucleotides were active templates with sRNA from each species studied, whereas others were active with sRNA from one species but not from another.

UAA and UAG do not appreciably stimulate binding of unfractionated *E. coli* AA-sRNA (AA-sRNA for each amino acid tested); *Xenopus* Arg-, Phe-, Ser-, or Tyr-sRNA; or guinea pig Ala-, Arg-, Asp-, His-, Ile-, Met-, Pro-, Ser-, or Thr-sRNA.

Nucleotide sequences recognized by *Xenopus* skeletal muscle Arg-, Lys-, Met-, and Ser-sRNA were determined and compared with sequences recognized by corresponding *Xenopus* liver AA-sRNA preparations. No differences between liver and muscle AA-sRNA were detected, either in nucleotide sequences recognized or in relative responses to synonym codons.

Fossil records of bacteria 3.1 billion years old have been reported (Barghoorn and Schopf, 1966). The first vertebrates appeared approximately 510 million years ago, and amphibians and mammals, 355 and 181 million years ago, respectively. The presence of bacteria 3 billion years ago may indicate the presence of a functional genetic code at that time. Almost surely the code has functioned for more than 500 million years. The remarkable similarity in codon base sequences recognized by bacterial, amphibian, and mammalian AA-sRNA suggest that most, if not all, forms of life on this planet use almost the same genetic language, and that the language has been used, possibly with few major changes, for at least 500 million years.

UNUSUAL ASPECTS OF CODON RECOGNITION AS POTENTIAL INDICATORS OF SPECIAL CODON FUNCTIONS

Most codons correspond to amino acids; however, some codons serve in other capacities, such as initiation, termination or regulation of protein synthesis. Although only a few codons have been assigned special functions thus far, we think it likely that many additional codons eventually may be found to serve special functions. Unusual properties of codon recognition sometimes may indicate special codon functions. For example, the properties of initiator and terminator codons, during codon recognition, are quite distinctive (see below). We find that approximately 20 codons have unusual properties related either to codon position, template activity, specificity, patterns of degeneracy, or stability of codon-ribosome-sRNA complexes. Until more information is available these observations will be considered as *possible* indicators of special codon functions.

Conclusions will be stated first to provide a frame of reference for discussion:

(1) A codon may have alternate meanings. (For example, UUG at or near the 5′-terminus of mRNA may correspond to N-formyl-methionine; whereas, an internal UUG codon may correspond to leucine.)

(2) A codon may serve multiple functions simultaneously. (For example, a codon may specify both initiation and an amino acid, perhaps via AA-sRNA with high affinity for peptidyl-sRNA sites on ribosomes.)

(3) Codon function sometimes is subject to modification.

(4) Degenerate codons for the same amino acid often differ markedly in template properties.

Codon frequency and distribution

Often, multiple species of sRNA corresponding to the same amino acid recognize different synonym codons. Degenerate codon usage in mRNA sometimes is nonrandom (Garen, pers. comm.). The possibility that different sets of sRNA may be required for the synthesis of two proteins with the same amino acid composition suggests that protein synthesis sometimes may be regulated by codon frequency and distribution coupled with differential recognition of degenerate codons. Possibly, the rates of synthesis of certain proteins may be regulated simultaneously by alterations which affect the apparatus recognizing one degeneracy but not another (see reviews by Ames and Hartman, 1963; and Stent, 1964).

Codon position

As discussed in an earlier section, the template properties of 5′-terminal-, 3′-terminal-, and internal- codons may differ. Regulatory mechanisms based on such differences have been suggested. Reading of mRNA probably is initiated at or near the 5′-terminal codon and then proceeds toward the 3′-terminus of the RNA chain (Salas, Smith, Stanley, Jr., Wahba, and Ochoa, 1965). It is not known whether mechanisms of 5′-terminal and internal initiation in polycistronic messages are similar. Also, internal- and 3′-terminal mechanisms of termination remain to be defined.

N-formyl-Met-sRNA may serve as an initiator of protein synthesis in *E. coli* (Clark and Marcker, 1966; Adams and Capecchi, 1966; Webster, Englehardt, and Zinder, 1966; Thach, Dewey, Brown, and Doty, 1966). Met-sRNA$_1$ can be converted enzymatically to N-formyl-Met-sRNA$_1$ and responds to UUG, CUG, AUG and, to a lesser extent, GUG. Met-sRNA$_2$ does not accept formyl-moieties and responds primarily to AUG (Clark and Marcker, 1966; also Kellogg, Doctor, Loebel, and Nirenberg, 1966). In *E. coli* extracts protein synthesis is initiated in at least two ways: by initiator codons specifying N-formyl-Met-sRNA or, at somewhat higher Mg^{++} concentrations, by another means, probably not dependent upon N-formyl-Met-sRNA since many synthetic polynucleotides without known initiator codons direct cell-free protein synthesis (Nakamoto and Kolakofsky, 1966). Poly U, for example, directs di- as well as polyphenylalanine synthesis (Arlinghaus, Schaeffer, and Schweet, 1964). Probably codons for N-formyl-Met-sRNA initiate protein synthesis with greater accuracy than codons which serve as initiators only at relatively high Mg^{++}concentrations.

UAA and UAG may function as terminator codons (Brenner, Stretton, and Kaplan, 1965; Weigert and Garen, 1965). The trinucleotides UAA and UAG do not stimulate binding appreciably of *unfractionated E. coli* AA-sRNA to ribosomes. However, sRNA fraction(s) corresponding to UAA and/or UAG are not ruled out.

Extragenic suppressors may affect the specificity of UAA and/or UAG recognition (see review by Beckwith and Gorini, 1966). The efficiencies of ochre suppressors (UAA) are relatively low compared to that of amber suppressors (UAG). Since amber suppressors do not markedly affect the rate of cell growth, and ochre suppressors with high efficiency have not been found, UAA may specify chain termination in vivo more frequently than UAG. In a study of great interest, Newton, Beckwith, Zipser and Brenner (1965) have shown that the synthesis of protein (probably mRNA also) is regulated by the relative position in the RNA message of codons sensitive to amber suppressors. Therefore, a codon may perform a regulatory function at one position but not at another.

Template activity

Trinucleotides with little activity for AA-sRNA (in studies thus far) are: UAA, UAG, and UUA, (perhaps CUA also). In addition, the following trinucleotides are active templates with AA-sRNA from one organism, but not from another: AGG, AGA, CGG, arginine; UUG, (N-formyl-)-methionine; GCG, alanine; AUA, isoleucine; AAG, lysine, UCG, AGU, AGC, serine; and UGA, cysteine (see Universality Section and Table 9). However, some inactive trinucleotides possibly function as active codons at internal positions. For example, the following codon base compositions were estimated with synthetic polynucleotides and a cell-free protein synthesizing system from *E. coli;* AUA, isoleucine; AGA, arginine; and

Table 9. Template activity of trinucleotides in 0.01 or 0.03 M Mg^{++}

	U	C	A	G	
	PHE	[SER]	TYR	[CYS]	U
	PHE	SER	TYR	[CYS]	C
U					
		(SER)			A
	[F-MET]	[SER]		(TRP)	G
		PRO	HIS	[ARG]	U
		PRO	HIS	ARG	C
C					
		(PRO)	GLN	[ARG]	A
	LEU	(PRO)	[GLN]	ARG	G
	ILE	[THR]	ASN	SER,CYS	U
	ILE	THR	ASN	SER,CYS	C
A					
		THR	[LYS]		A
	[MET]	[THR]	LYS		G
	[VAL]	ALA	[ASP]	[GLY]	U
	[VAL]	ALA	[ASP]	[GLY]	C
G					
	[VAL]	[ALA]	[GLU]	(GLY)	A
	[VAL]	[ALA]	GLU	(GLY)	G

Legend:	0.01 M Mg	0.03 M Mg
[ ] =	+	+
No Box =	–	+
( ) =	not tested	

Relative template activities of trinucleotides in reactions containing 0.01 or 0.03 M Mg^{++}. A plus (+) sign in the legend means that the trinucleotide stimulates AA-sRNA binding to ribosomes at that magnesium concentration; a minus (–) sign means it is relatively inactive as a template. The results refer to AA-sRNA from *E. coli* strains B and/or W3100. The data are from Anderson, Nirenberg, Marshall, and Caskey (1966).

AGC, serine (Nirenberg et al., 1963; Speyer et al., 1963; also see Jones, Nishimura, and Khorana, 1966, for results with AGA). Among the many possible explanations for low template activities of trinucleotides in binding assays are: special codon function; codon position; appropriate species of sRNA absent or in low concentration; competition for codons or for ribosomal sites by additional species of sRNA; high ratio of deacylated to AA-sRNA; cryptic (non-acylatable) sRNA; reaction conditions, e.g., low concentration of Mg^{++} or other components, time or temperature of incubation.

Codon specificity

Often synonym trinucleotides differ strikingly in template specificity. Such observations may indicate that template specificities of terminal- and internal-codons differ, or that special function codons or suppressors are present. At 0.010–0.015 M Mg^{++}, trinucleotide template specificity is high, in many cases higher than that of a polynucleotide; for example, poly U, but not UUU, stimulates binding of Ile-sRNA to ribosomes. However, at 0.03 M Mg^{++} ambiguous recognitions of tri- and polynucleotides are observed more frequently.

Relative template activities of synonym trinucleotides in reactions containing 0.01 or 0.03 M Mg^{++} are shown in Table 9. In some cases, only one or two trinucleotides in a synonym set are active templates at 0.01 M Mg^{++}; whereas all degeneracies are active at 0.03 M Mg^{++} (e.g., Glu, Lys, Ala, Thr). In other cases either all synonym trinucleotides are active at 0.01 M Mg^{++} as well as at 0.03 M Mg^{++} (e.g., Val), or none are active at the lower Mg^{++} concentration (e.g., Tyr, His, Asn). Such data suggest that codon-ribosome-AA-sRNA complexes formed with degenerate trinucleotides often differ in stability.

MODIFICATION OF CODON RECOGNITION DUE TO PHAGE INFECTION

N. and T. Sueoka (1964) have shown that infection of *E. coli* by T2 bacteriophage results, within one to three minutes, in the modification of one or more species of Leu-sRNA present in the *E. coli* host. Concomitantly, *E. coli,* but not viral protein synthesis is inhibited. Protein synthesis is required, however, for modification of Leu-sRNA.

In collaboration with N. and T. Sueoka, modification of Leu-sRNA has been correlated with codon recognition specificity. sRNA preparations were isolated from *E. coli* before phage infection and at 1 and 8 minutes after

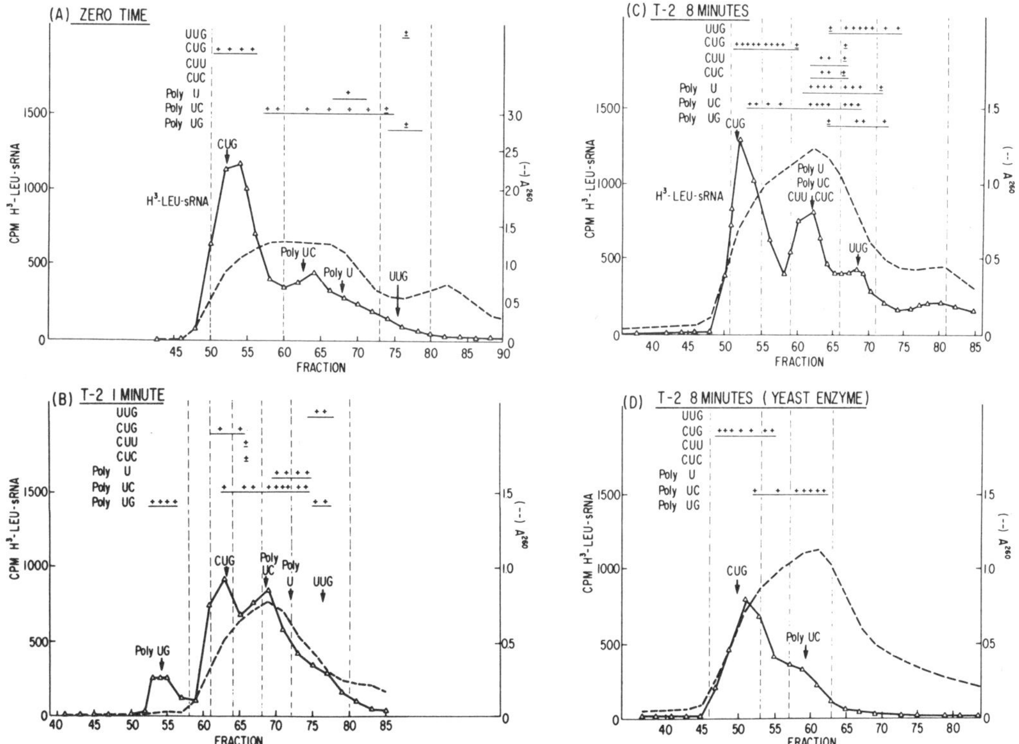

Fig. 6. The graphs represent MAK column fractions of H^3-Leu-sRNA from *E. coli* B before infection, **a,** and at 1 min, **b,** and 8 min, **c** and **d,** after infection with T2 phage. sRNA was acylated prior to chromatography with H^3-leucine using *E. coli* **(a, b, c)** or yeast **(d)** synthetase preparations. Column eluates were pooled as indicated by the vertical broken lines; dialyzed against 5×10^{-4} M potassium cacodylate, pH 5.5, and lyophilized. Then binding of each fraction to ribosomes in response to tri- or polynucleotide templates was determined. At the top of each graph relative responses of Leu-sRNA fractions to templates are shown. Approximate relative responses are indicated as follows: No symbol, no detectable response of Leu-sRNA; ±, possible response; +, slight response; and ++ to +++++, moderate to strong responses. Profiles represented by broken lines indicate A^{260} units; △-△, represent H^3-Leu-sRNA. Data are from Kano-Sueoka, Nirenberg, and Sueoka (unpubl.). Also see Sueoka et al., this volume.

infection. After acylation, Leu-sRNA preparations were purified by MAK column chromatography and the binding of each pooled Leu-sRNA fraction to ribosomes in response to templates was determined (Fig. 6). The profile of Leu-sRNA (8 minutes after infection) acylated with yeast, rather than *E. coli.* Leu-sRNA synthetase is shown also (Fig. 6D); thus, both anticodon and enzyme recognition sites were monitored. In Fig. 7 the approximate chromatographic mobility on MAK columns of each Leu-sRNA fraction is shown diagrammatically, together with the relative response of each fraction to tri- and polynucleotide templates and acylation specificity of *E. coli* and yeast Leu-sRNA synthetase preparations.

Within 1 minute after infection, a marked decrease was observed in Leu-sRNA$_2$, responding to CUG, and a corresponding increase was seen in Leu-sRNA$_1$, responding to poly UG, but not to the trinucleotides, UUU, UUG, UGU, GUU, UGG, GUG, GGU, CUU, CUC, CUG, UAA, UAG, UGA, or to poly U or poly UC. However, Leu-sRNA$_1$ was not detected 8 minutes after infection.

A marked increase in the response of Leu-sRNA$_5$ to UUG was observed 1 minute after infection, and an even greater increase was seen 8 minutes after infection.

Greater responses of Leu-sRNA$_3$ and Leu-sRNA$_{4a,b}$ to poly UC also were observed 8 minutes after phage infection. Leu-sRNA fractions 3 and 4 differ in chromatographic mobility and in acylation specificity by yeast and *E. coli* Leu-sRNA synthetase preparations. Thus, Leu-sRNA$_3$ and a component in fraction 4 differ, although both fractions 3 and 4 respond to poly UC. The multiple responses of Leu-sRNA$_{4a,b}$ to poly U, poly UC, and the trinucleotides, CUU and CUC, suggest that fraction 4 may contain two or more Leu-sRNA species. Striking increases in response of fraction 4 to poly U were observed 1 and 8 minutes after infection.

Leu-sRNA fractions 1, 2, and 3 are related, for each is recognized by yeast as well as by *E. coli* Leu-sRNA synthetase preparations. In contrast, Leu-sRNA$_{4a,b}$ and Leu-sRNA$_5$ are recognized by *E. coli*, but not yeast Leu-sRNA synthetase; thus, fraction 4 is related to

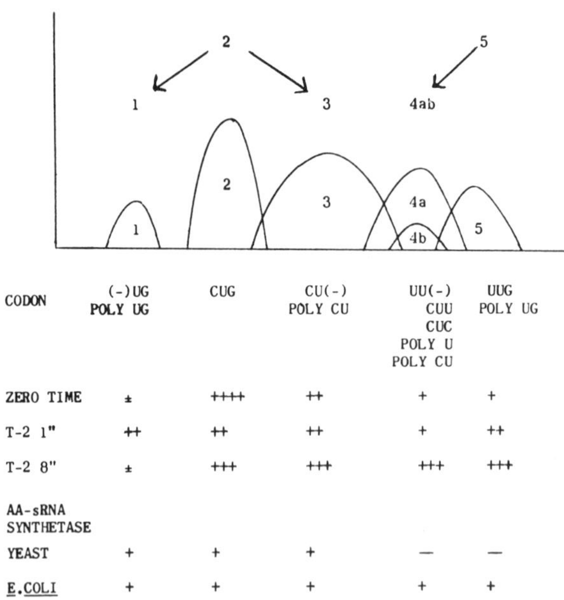

Fig. 7. Diagrammatic representation of the data shown in Fig. 6. The relative mobilities of multiple species of Leu-sRNA, before and after phage infection, fractionated by MAK column chromatography, are shown at the top. Leu-sRNA peaks are numbered. Arrows represent predicted Leu-sRNA precursor-product relationships (Fractions 2 and 5 possibly are products of different cistrons).

Tri- and polynucleotide codons recognized by each Leu-sRNA peak are shown below. Approximate relative responses of Leu-sRNA$_{1\text{-}5}$ to codons are indicated as follows: ±, possible response; + to ++++, slight to strong responses.

On the bottom are shown the specificities of *E. coli* (zero time, 1 and 8 min after infection) and yeast (8 min after infection only) Leu-sRNA synthetase preparations for sRNA^{Leu} fractions.

fraction 5. Two *different* cistrons of Leu-sRNA are predicted: Leu-sRNA fractions 1, 2, and 3 may be products of one cistron; whereas, fractions 4 and 5 may be products of a different cistron. In this regard, Berg, Lagerkvist, and Dieckman (1962) have shown that *E. coli* Leu-sRNA contains two base sequences at the 4th, 5th, and 6th base positions from the 3′-terminus of the sRNA.

The data suggest the following sRNA precursor-product relationships. Leu-sRNA$_2$ is a product of "cistron A"; the decrease in Leu-sRNA$_2$ and the simultaneous increase in Leu-sRNA$_1$ (within one minute after infection) suggests that Leu-sRNA$_2$ is the precursor of Leu-sRNA$_1$. The data also suggest that Leu-sRNA$_2$ is a precursor of Leu-sRNA$_3$. The following anticodons and mRNA codons are suggested for Leu-sRNA fractions 2, 3, and 1, respectively (note: asterisks represent modifications of a nucleotide base; codon and anticodon sequences are written with 3′,5′-phosphodiester linkages; antiparallel hydrogen bonding between codon and anticodons is assumed): Leu-sRNA$_2$-product of "cistron A", CAG anticodon, [CUG codon]; Leu-sRNA$_3$- derived from fraction 2, C*AG anticodon, [CU(—) codon]; Leu-sRNA$_1$- derived from fraction 2, CAG** anticodon, [(—)UG codon].

Leu-sRNA$_5$ is a product of "cistron B", and differs from Leu-sRNA$_2$ in anticodon and Leu-sRNA synthetase recognition sites. The sequence, CAA, is suggested for the Leu-sRNA$_5$ anticodon, corresponding to a UUG mRNA codon. Leu-sRNA$_{4a,b}$ are derived from fraction 5. Possible anticodons and codons are: C*AA anticodon, [UU(—) codon]; C*IA anticodon, [UU(—), UC(—), UA(—) codons]; C*AI anticodon, [UU(—), CU(—), AU(—) codons].

Since modification of Leu-sRNA after phage infection is dependent upon protein synthesis, enzyme(s) may be needed to modify bases in Leu-sRNA fractions.

The inhibition of host *E. coli,* but not viral protein synthesis following viral infection may result from modification of Leu-sRNA fractions. N-formyl-Met-sRNA$_1$ serves as an initiator of protein synthesis in *E. coli* and responds to two trinucleotides, UUG and CUG, which are also recognized by Leu-sRNA fractions (see previous discussion on special function codons). Possibly, initiation or termination of *E. coli,* but not viral protein synthesis is affected. Further studies are needed, however, to elucidate the mechanism of viral induced inhibition of host protein synthesis.

REFERENCES

Adams, J. M., and M. R. Capecchi. 1966. N-formylmethionyl-sRNA as the initiator of protein synthesis. Proc. Natl. Acad. Sci. **55**:147-155.

Ames, B. N., and P. E. Hartman. 1963. The histidine operon. Cold Spring Harbor Symp. Quant. Biol. **28**:349-356.

Anderson, W. F., M. W. Nirenberg, R. E. Marshall, and C. T. Caskey. 1966. RNA codons and protein synthesis: Relative activity of synonym codons. Fed. Proc. **25**:404.

Arlinghaus, R., J. Shaeffer, and R. Schweet. 1964. Mechanism of peptide bond formation in polypeptide synthesis. Proc. Natl. Acad. Sci. **51**:1291-1299.

Barghoorn, E. S., and J. W. Schopf. 1966. Microorganisms three billion years old from the precambrian of South Africa. Science **152**:758-763.

Beckwith, J. R., and L. Gorini. 1966. Suppression. Ann. Rev. Microbiol., in press.

Berg, P., U. Lagerkvist, and M. Dieckmann. 1962. The enzymic synthesis of amino acyl derivatives of ribonucleic acid. VI. Nucleotide sequences adjacent to the . . .pCpCpA end groups of isoleucine- and leucine-specific chains. J. Mol. Biol. **5**:159-171.

Bernfield, M. 1966. Ribonuclease and oligoribonucleotide synthesis. II. Synthesis of oligonucleotides of specific sequence. J. Biol. Chem. **241**:2014-2023.

Brenner, S., A. O. W. Stretton, and S. Kaplan. 1965. Genetic Code: the 'nonsense' triplets for chain termination and their suppression. Nature **206**: 994-998.

Brimacombe, R., J. Trupin, M. Nirenberg, P. Leder, M. Bernfield, and T. Jaouni. 1965. RNA codewords and protein synthesis. VIII. Nucleotide sequences of synonym codons for arginine, valine, cysteine and alanine. Proc. Natl. Acad. Sci. **54**:954-960.

Cerutti, P., H. T. Miles, and J. Frazier. 1966. Interaction of partially reduced polyuridylic acid with a polyadenylic acid. Biochem. Biophys. Res. Commun. **22**:466-472.

Clark, B., and K. Marcker. 1966. The role of N-formyl-methionyl-sRNA in protein biosynthesis. J. Mol. Biol. **17**:394-406.

Connelly, C. M., and B. P. Doctor. 1966. Purification of two yeast serine transfer ribonucleic acids by counter-current distribution. J. Biol. Chem. **241**: 715-719.

Crick, F. H. C. 1966. Codon-anticodon pairing: the wobble hypothesis. J. Mol. Biol. **19**:548-555.

Heppel, L. A., P. R. Whitfield, and R. Markham. 1955. Nucleotide exchange reactions catalyzed by ribonuclease and spleen phosphodiesterase. 2. Synthesis of polynucleotides. Biochem. J. **60**:8-15.

Holley, R. W., J. Apgar, G. A. Everett, J. T. Madison, M. Marquisee, S. H. Merrill, J. R. Penswick, and A. Zamir. 1965. Structure of a ribonucleic acid. Science **147**:1462-1465.

Ingram, V. M., and J. A. Sjöquist. 1963. Studies on the structure of purified alanine and valine transfer RNA from yeast. Cold Spring Harbor Symp. Quant. Biol. **28**:133-138.

Jones, D. S., S. Nishimura, and H. G. Khorana. 1966. Studies on polynucleotides LVI. Further syntheses, *in vitro,* of copolypeptides containing two amino acids in alternating sequence dependent upon DNA-like polymers containing two nucleotides in alternating sequence. J. Mol. Biol. **16**:454-472.

Keller, E. B., and M. F. Ferger. 1966. Alanyl-sRNA in the aminoacyl polymerase system of protein synthesis. Fed. Proc. **25**:215.

Kellogg, D. A., B. P. Doctor, J. E. Loebel, and M. W. Nirenberg. 1966. RNA codons and protein synthesis. IX. Synonym codon recognition by multiple species of valine-, alanine-, and methionine-sRNA. Proc. Natl. Acad. Sci. **55**:912-919.

Leder, P., M. F. Singer, and R. L. C. Brimacombe. 1965. Synthesis of trinucleoside diphosphates with polynucleotide phosphorylase. Biochem. **4**:1561-1567.

Madison, J. T., G. A. Everett, and H. Kung. 1966. Nucleotide sequence of a yeast tyrosine transfer RNA. Science **153**:531-534.

Nakamoto, T., and D. Kolakofsky. 1966. A possible mechanism for initiation of protein synthesis. Proc. Natl. Acad. Sci. **55**:606-613.

Newton, W. A., J. R. Beckwith, D. Zipser, and S. Brenner. 1965. Nonsense mutants and polarity in the *Lac operon of Escherichia coli.* J. Mol. Biol. **14**:290-296.

Nirenberg, M. W., O. W. Jones, P. Leder, B. F. C. Clark, W. S. Sly, and S. Pestka. 1963. On the coding of genetic information. Cold Spring Harbor Symp. Quant. Biol. **28**:549-557.

Nirenberg, M., and P. Leder. 1964. RNA codewords and protein synthesis. I. The effect of trinucleotides upon the binding of sRNA to ribosomes. Science **145**:1399-1407.

Nirenberg, M., P. Leder, M. Bernfield, R. Brimacombe, J. Trupin, F. Rottman, and C. O'Neal. 1965. RNA codewords and protein synthesis. VII. On the general nature of the RNA code. Proc. Natl. Acad. Sci. **53**:1161-1168.

Rottman, F., and P. Cerutti. 1966. Template activity of uridylic acid-dihydrouridylic acid copolymers. Proc. Natl. Acad. Sci. **55**:960-966.

Rottman, F., and M. Nirenberg. 1966. Regulatory mechanisms and protein synthesis. XI. Template activity of modified RNA codons. J. Mol. Biol., in press.

Salas, M., M. A. Smith, W. M. Stanley, Jr., A. J. Wahba, and S. Ochoa. 1965. Direction of reading of the genetic message. J. Biol. Chem. **240**:3988-3995.

Söll, D., E. Ohtsuka, D. S. Jones, R. Lohrmann, H. Hayatsu, S. Nishimura, and H. G. Khorana. 1965. Studies on polynucleotides, XLIX. Stimulation of the binding of aminoacyl-sRNA's to ribosomes by ribotrinucleotides and a survey of codon assignments for 20 amino acids. Proc. Natl. Acad. Sci. **54**:1378-1385.

Sonneborn, T. M. 1965. Degeneracy of the genetic code: Extent, nature and genetic implications. pp. 377-397. *In:* V. Bryson and H. J. Vogel (ed.) Evolving Genes and Proteins. Academic Press, New York.

Speyer, J., P. Lengyel, C. Basilio, A. Wahba, R. Gardner, and S. Ochoa. 1963. Synthetic polynucleotides and the amino acid code. Cold Spring Harbor Symp. Quant. Biol. **28**:559-567.

Stent, G. S. 1964. The operon: On its third anniversary. Science **144**:816-820.

Sueoka, N., and T. Kano-Sueoka. 1964. A specific modification of Leucyl-sRNA of *Escherichia coli* after phage T2 infection. Proc. Natl. Acad. Sci. **52**:1535-1540.

Thach, R. E., K. F. Dewey, J. C. Brown, and P. Doty. 1966. Formylmethionine codon AUG as an initiator of polypeptide synthesis. Science **153**:416-418.

Thach, R. E., and P. Doty. 1965. Enzymatic synthesis of tri- and tetranucleotides of defined sequence. Science **148**:632-634.

Webster, R. E., D. L. Engelhardt, and N. D. Zinder. 1966. *In vitro* protein synthesis: Chain initiation. Proc. Natl. Acad. Sci. **55**:155-161.

Weigert, M., and A. Garen. 1965. Base composition of nonsense codons in *E. coli;* evidence from amino-acid substitutions at a tryptophan site in alkaline phosphatase. Nature **206**:992-994.

Zachau, H., D. Dütting, and H. Feldmann. 1966. Nucleotide sequences of two serine-specific transfer ribonucleic acids (1). Angew. Chem. **5**:422, English Edition.

chapter 2

Physical basis of inheritance

THE VIRAL CHROMOSOME

The hereditary material in viruses consists of a *single* molecule of DNA, or in some cases RNA, which in most viruses is in the form of a *circle.* The single nature of the virus ΦX174 chromosome was indicated by the findings of Sinsheimer in 1959 (Ref. 1-13). By a light-scattering technique, he determined the molecular weight of the DNA in a ΦX174 particle to be 1.7×10^6. This was in good agreement with the estimated weight of DNA in a ΦX174 particle as calculated from the total virus particle weight, 6.2×10^6, with a known DNA content of 25.5%. The correspondence of DNA molecular weight by these two methods indicated to him that there was one molecule of DNA per ΦX174 particle. Substantiating evidence for the single nature of viral DNA was presented in 1960 by Streisinger and Bruce (Ref. 2-1), who demonstrated, by a method similar to the one to be discussed in Chapter 5, that the known genetic characteristics of bacteriophage T2 are linked to one another in a single "genetic map." Further support came from the experiments of Rubinstein and his co-workers in 1961 (Ref. 2-2). Using autoradiography, they also concluded that each T2 virus contains a single DNA molecule.

Circularity of the ΦX174 viral chromosome was proposed by Fiers and Sinsheimer in 1962 (Ref. 2-3); this proposal was based on enzymatic experiments. It was found that a bacterial enzyme capable of attacking a broken, single strand of *Escherichia coli* DNA was not effective against intact single-stranded ΦX174 DNA. This implied that the viral chromosome has no free end for the enzyme to attack and suggested that the viral DNA is in circular form. Visual proof of the circular arrangement of the viral chromosome came from electron micrographs of intact single-stranded ΦX174 chromosomes. These were obtained by Freifelder and his co-workers in 1964 (Ref. 2-4). Supporting genetic evidence for the circularity of another viral chromosome was presented in 1964 by Streisinger and his co-workers (Ref. 2-5), who reported that the single genetic map of phage T4 is in the form of a circle.

While in the host's cell, viruses can transfer some of their genetic material to one another. This was first observed by Delbrück and Bailey in 1946 (Ref. 2-6) whose paper is included in this chapter. The exchange of genetic material among viruses constitutes a form of sexual recombination but is generally recognized to be quite different from the sexual reproduction found in multicellular organisms.

THE BACTERIAL CHROMOSOME

The hereditary material of all bacteria thus far studied, like that of most viruses, is in the form of a single and circular DNA chromosome. Evidence presented by Clowes and Rowley in 1954 (Ref. 2-7) showed that the known genetic characteristics of *E. coli* are linked to one another in a single unit. Substantiating evidence for the single nature of the bacterial chromosome came from the experiments of Forro and Wertheimer in 1960 (Ref. 2-8); they exposed growing cells of *E. coli* to tritium-labeled thymidine and observed the distribution of the subsequently labeled DNA in daughter cells. The pattern of segregation of the labeled DNA indicated that two large DNA-containing units separate regularly at cell division. These observations are consistent with the view that the two units represent the two polynucleotide chains of a single DNA chromosome.

Circularity of the bacterial chromosome was reported in 1957 by Jacob and Wollman (Ref. 2-9), who found that all of the known genetic characteristics of *E. coli* are linked together in a ring form. Visual proof of the single and circular nature of the bacterial chromosome was provided in 1963 by Cairns (Ref. 2-10), who used autoradiography. A later paper by Cairns, also published in 1963 (Ref. 2-11), is included in this collection.

Bacteria are capable of transferring part of, or all, their genetic material to other bacteria. There are three known mechanisms by which bacteria can do this: *transformation, conjuga-*

tion, or *transduction.* These processes differ from the sexual process of "higher" organisms in that a true fusion cell (zygote) is not formed.

Transformation was discussed in Chapter 1 in connection with the identification of DNA as the genetic material in bacteria. Further research on transformation has yielded additional information about the process. A study by Ottolenghi and Hotchkiss in 1960 (Ref. 2-12) revealed that normal, growing populations of pneumococcus release into the culture medium deoxyribonucleate-containing material with transforming activity. This implies that transformation may provide a natural mechanism for the exchange of genetic material in this organism. A report by Catlin in 1964 (Ref. 2-13) announced the occurrence of reciprocal genetic transformation of streptomycin-susceptible cells to streptomycin-resistant cells between two bacterial species, *Neisseria catarrhalis* (coccus) and *Moraxella nonliquefaciens* (bacillus).

The mechanism by which transforming DNA becomes incorporated in the recipient cell's chromosome was reported in 1964 by Fox and Allen (Ref. 2-14), whose paper is reprinted in this book. A general review of bacterial transformation was published by Schaeffer in 1964 (Ref. 2-15).

Conjugation is the transfer of chromosomal material from one bacterial cell to another by direct cell-to-cell contact. Its occurrence was discovered by Lederberg and Tatum in 1946 (Ref. 2-16), whose paper is included in this chapter. Later research demonstrated that there are two mating types in *E. coli.* One of these acts as a donor of chromosomal material. It is designated as *male* and is either in an F^+ or Hfr condition. The other mating type acts as a recipient of chromosomal material and is designated as *female* or F^-. Only if the male cell is in the Hfr condition will there be a transfer of chromosomal material. In the chromosomal transfer the known genetic characteristics of the male cell penetrate the female cell in a linear order. The transfer of the entire donor chromosome takes about 110 minutes, but the process is usually interrupted by spontaneous breakage of the donor chromosome. The studies that elucidated these facts were reviewed in a paper by Wollman and his co-workers in 1956 (Ref. 2-17).

It is of interest to know whether conjugation is a restricted or widespread mechanism of genetic exchange among bacteria. Conjugation has been reported within the following genera: *Escherichia, Pseudomonas, Salmonella,* and *Vibrio.* In addition, intergeneric conjugation has been reported for the following pairs of genera: *Escherichia-Salmonella, Escherichia-Shigella, Salmonella-Shigella,* and *Salmonella-Vibrio.* Another question raised about conjugation related to the time of formation of the transfer chromosome. The answer came from the experiment of Jacob and his co-workers in 1963 (Ref. 2-18). By use of radioactive isotopes, they were able to show that the transferred chromosome is formed after the bacteria have come in contact with one another and that chromosome transfer is directly geared to chromosome replication. A general review of bacterial conjugation was published by Gross in 1964 (Ref. 2-19).

Transduction is the transfer of a portion of a bacterial chromosome from a donor cell to a recipient by a phage which acts as a vector. The phenomenon was discovered by Zinder and Lederberg in 1952 (Ref. 2-20). A paper by Zinder in 1953 (Ref. 2-21), describing the transduction phenomenon, is included in this collection. A direct demonstration of the phage-vector role was provided in 1960 by Kaiser and Hogness (Ref. 2-22), who were able to accomplish transformation with DNA isolated from a transducing phage preparation. It was soon found that in some host-bacteriophage systems, all genetic markers have a roughly equal chance of being transduced. This is called *generalized transduction,* and an example of this was reported by Lennox in 1955 (Ref. 2-23). In contrast to the generalized transduction system, there are other host-bacteriophage systems in which only one particular region of the donor chromosome can be transduced. This is called *restricted transduction,* and an example of this was reported by Morse in 1954 (Ref. 2-24). As with transformation and conjugation, transduction also appears to be a widespread phenomenon. It has been shown to occur in the genera *Bacillus, Escherichia, Micrococcus, Pseudomonas, Salmonella,* and *Shigella.* A review of the subject was published by Campbell in 1964 (Ref. 2-25).

THE CHROMOSOME OF HIGHER ORGANISMS

In higher organisms (protozoa, algae except blue-green, and all multicellular forms) the DNA is associated with protein and is organized into more than a single chromosome. It is not known whether the DNA in each chromosome is in the form of a circle. The chromosomes, which vary in number from 2 to 768 (depending on the species), are located in a special cellular structure, the nucleus. Higher organisms exhibit two types of cell division: *mitosis,* which leads to an increase either in number of individuals or in the size of an organism, and *meiosis,* which provides either the nuclei or the cells needed for sexual reproduction. In both of these types of cell cycles, it is of interest to know when the DNA duplicates itself. This became known through the work of Swift in 1950 (Ref. 2-26), who made microspectrophotometric measurements of the DNA content of individual Feulgen-stained cells at various times during mitotic and meiotic cell cycles.

The following phases were found for cells undergoing mitosis: (1) at the beginning of interphase, after mitosis, a period during which DNA synthesis does not take place (G_1); (2) a subsequent period of DNA synthesis during which the DNA content of the interphase nucleus is doubled (S); (3) another period during which DNA is not synthesized (G_2); and finally (4) a period of mitosis (M) that occurs without DNA synthesis. For cells undergoing meiosis, it was found, in the case of spermatogenesis, that mature spermatogonial cells contained all the DNA necessary for the production of four spermatids. The above findings were confirmed by other workers using autoradiography. An example of this type of study involving the mitotic cell cycle of a protozoan was reported by Prescott in 1966 (Ref. 2-27); it is the final paper reprinted in this chapter.

The significance of mitosis for genetics is that it involves an exact division and distribution of the cell's chromosome material to its daughter cells. Mitosis therefore leads to the production of cells with identical hereditary material. Meiosis, on the other hand, involves the reduction of the chromosome number in the daughter cells to one half the amount in the original cell. Meiosis results in a random distribution of the cell's parental chromosomes to the daughter cells and represents the physical basis for the new combinations of hereditary material that occur in every generation.

REFERENCES

The viral chromosome

2-1. Streisinger, G., and Bruce, V. 1960. Linkage of genetic markers in phages T2 and T4. Genetics **45:**1289-1296.

2-2. Rubinstein, I., Thomas, A. C., and Hershey, A. D. 1961. The molecular weights of T2 bacteriophage DNA and its first and second breakage products. Proc. Natl. Acad. Sci. U. S. A. **47:** 1113-1122.

2-3. Fiers, W., and Sinsheimer, R. L. 1962. The structure of the DNA of bacteriophage ΦX174. I. The action of exopolynucleotidases. J. Mol. Biol. **5:**408-419.

2-4. Freifelder, D., Kleinschmidt, A. K., and Sinsheimer, R. L. 1964. Electron microscopy of single-stranded DNA: Circularity of DNA of bacteriophage ΦX174. Science **146:**254-255.

2-5. Streisinger, G., Edgar, R. S., and Denhardt, G. H. 1964. Chromosome structure in phage T4. I. Circularity of the linkage map. Proc. Natl. Acad. Sci. U. S. A. **51:**775-779.

2-6. Delbrück, M., and Bailey, W. T., Jr. 1946. Induced mutations in bacterial viruses. Cold Spring Harbor Symp. Quant. Biol. **11:**33-37.

The bacterial chromosome

2-7. Clowes, R. C., and Rowley, D. 1954. Some observations on linkage effects in genetic recombination in *Escherichia coli* K-12. J. Gen. Microbiol. **11:**250-260.

2-8. Forro, F., Jr., and Wertheimer, S. A. 1960. The organization and replication of deoxyribonucleic acid in thymine-deficient strains of *Escherichia coli.* Biochim. Biophys. Acta **40:**9-21.

2-9. Jacob, F., and Wollman, E. L. 1957. Analyse des groupes de liaison génétique de différentes souches donatrices d'*Escherichia coli* K-12. Compt. Rend. Acad. Sci., Paris **245:**1840-1843.

2-10. Cairns, J. 1963. The bacterial chromosome and its manner of replication as seen by radioautography. J. Mol. Biol. **6:**208-213.

2-11. Cairns, J. 1963. The chromosome of *Escherichia coli.* Cold Spring Harbor Symp. Quant. Biol. **28:**43-46.

2-12. Ottolenghi, E., and Hotchkiss, R. D. 1960. Appearance of genetic transforming activity in pneumococcal cultures. Science **132:**1257-1258.

2-13. Catlin, B. W. 1964. Reciprocal genetic transformation between *Neisseria catarrhalis* and *Moraxella nonliquefaciens.* J. Gen. Microbiol. **37:**369-379.

2-14. Fox, M. S., and Allen, M. K. 1964. On the mechanism of deoxyribonucleate integration in pneumococcal transformation. Proc. Natl. Acad. Sci. U. S. A. **52:**412-419.

2-15. Schaeffer, P. 1964. Transformation, vol. 5, p. 87-153. *In* I. C. Gunsalus and R. Y. Stanier [eds.] The bacteria. Academic Press, Inc., New York.

2-16. Lederberg, J., and Tatum, E. L. 1946. Gene recombination in *Escherichia coli.* Nature **158:** 558.

2-17. Wollman, E. L., Jacob, F., and Hayes, W. 1956. Conjugation and genetic recombination in *Escherichia coli* K-12. Cold Spring Harbor Symp. Quant. Biol. **21:**141-162.

2-18. Jacob, F., Brenner, S., and Cuzin, F. 1963. On the regulation of DNA replication in bacteria. Cold Spring Harbor Symp. Quant. Biol. **28:**329-348.

2-19. Gross, J. D. 1964. Conjugation in bacteria, vol. 5, p. 1-48. *In* I. C. Gunsalus and R. Y. Stanier [eds.] The bacteria. Academic Press, Inc., New York.

2-20. Zinder, N. D., and Lederberg, J. 1952. Genetic exchange in *Salmonella.* J. Bact. **64**:679-699.

2-21. Zinder, N. D. 1953. Infective heredity in bacteria. Cold Spring Harbor Symp. Quant. Biol. **18**:261-269.

2-22. Kaiser, A. D., and Hogness, D. J. 1960. The transformation of *Escherichia coli* with deoxyribonucleic acid isolated from bacteriophage λ*dg.* J. Mol. Biol. **2**:392-415.

2-23. Lennox, E. S. 1955. Transduction of linked genetic characters of the host by bacteriophage P1. Virology **1**:190-206.

2-24. Morse, M. L. 1954. Transduction of certain loci in *Escherichia coli* K-12. Genetics **39**:984-985.

2-25. Campbell, A. 1964. Transduction, vol. 5, p. 49-85. *In* I. C. Gunsalus and R. Y. Stanier [eds.] The bacteria. Academic Press, Inc., New York.

The chromosome of higher organisms

2-26. Swift, H. H. 1950. The deoxyribose nucleic acid content of animal nuclei. Physiol. Zool. **23**:169-198.

2-27. Prescott, D. M. 1966. The syntheses of total macronuclear protein, histone, and DNA during the cell cycle in *Euplotes eurystomus.* J. Cell Biol. **31**:1-9.

6 Induced mutations in bacterial viruses

M. Delbrück
W. T. Bailey, Jr.
California Institute of Technology
Pasadena, California

In another paper of this Symposium, to which ours is closely related and to which we refer the reader for a description of material and for terminology, Dr. Hershey *(4)* has described a variety of spontaneously occurring mutations of bacterial viruses. One class of mutations affects the type of plaque. These mutations occur in only one group of serologically related viruses, the group to which belong T2, T4, and T6. The most conspicuous of these mutations is the *r* mutation. Our observations are concerned exclusively with this *r* mutation. We have seen also some of the other mutations which affect the type of plaque and which Hershey has described, but we have not made systematic experiments concerning them.

We have infected bacteria simultaneously with mixtures of wild type and *r* mutant of the viruses T2, T4, and T6, and have investigated the yields of virus from such mixedly infected bacteria. These experiments are a sequel to previous studies of mixed infections with pairs of different viruses. The chief result of the studies to be reported here is the fact that the yield of virus from mixedly infected bacteria may contain a high proportion of one or more new types of virus—i.e., of a type that was not used for the infection. In all cases the new types exhibit combinations of the genetic markers of the infecting types.

MUTUAL EXCLUSION

We will begin with a recapitulation of earlier work on mixed infections *(3, 6, 2)*. The chief finding of these earlier studies was the mutual-exclusion effect. It was found that any mixedly infected bacterium yields upon lysis only one of the infecting types of virus. The other virus does not multiply; even the adsorbed particles of the excluded type are not recovered upon lysis. Which one of the two types of virus used for infection is excluded and which one multiplies depends on the pair used and on the conditions of the experiment, such as timing and multiplicity of infection. For the pair T1, T2 the virus T1 is always excluded except when it is given a head start of at least four minutes *(3)*. If T1 is added more than four minutes earlier than T2, then T2 will be excluded in an appreciable proportion of the infected bacteria. Lysis of any one bacterium always occurs after a time interval corresponding to the latent period of intracellular virus multiplication of the virus type which does multiply in that particular bacterium.

A similar situation was encountered *(2)* in the study of the pair T1, T7. Here, too, the mutual-exclusion mechanism operates perfectly; in practically every bacterium either one or the other of the two viruses is excluded from multiplication. The exclusion powers of these two viruses are nearly balanced. In a bacterial culture simultaneously infected with T1 and T7 there is a clean split into T1 yielders and T7 yielders, the two types occurring with comparable frequency.

During a closer study of this pair it became evident that the excluded virus is not without effect on the course of events. The excluded virus may reduce the number of virus particles liberated upon lysis of the bacterium. This has been called the depressor effect *(2)*.

A cursory survey of other pairs of virus particles seemed only to confirm these findings and, in particular, seemed to point to the mutual-exclusion effect as a very general phenomenon.

In these earlier investigations an attempt was made also to test whether mutual exclusion occurs when a bacterium is infected with two particles of the *same* kind. Such an assumption ("self-interference") seemed to suggest itself from the observation that bacteria infected with several particles of one kind are lysed after exactly the same latent period as are bacteria infected with only one particle. The test of

---

Reproduced from Vol. XI; Cold Spring Harbor Symposia, 1946. Used with permission.

This work was supported by grants-in-aid from The Rockefeller Foundation and from the John and Mary R. Markle Foundation.

mutual exclusion requires that one find out whether the yield of virus from any one bacterium is the offspring of only one or of several of the infecting particles. To make such a test one must be able to differentiate the offspring of the various infecting particles; in other words, one has to mark the infecting particles with hereditary markers. One is thus naturally led to the study of mutual exclusion between a virus and one of its mutants.

The first attempt in this direction was made by Luria *(5)*, who studied interference between T2 and T2*h*. The difficulty with this pair lies in the fact that no indicator strain resistant to T2*h* and sensitive to T2 is available. Luria succeeded nevertheless in showing that a large proportion of the bacteria infected with T2 *and* T2*h* did not liberate any T2*h*. At least a partial functioning of the mutual-exclusion mechanism seemed to be indicated by these results.

THE BREAKDOWN OF MUTUAL EXCLUSION FOR THE PAIRS (T2r^+, T2r) AND (T4r^+, T4r)

The first definite indication of a new phenomenon came in March, 1945, when Hershey tried mixed infections with a wild-type strain (T2) and its *r* mutant. Hershey observed that the mixedly infected bacteria give rise to mottled plaques, and he verified that the mottled plaques contain a mixture of the two types used for infection. Dr. Hershey communicated his discovery to us and we have since been following this promising lead.

Mixed infections with T2r^+, T2r

Hershey's finding that the majority of the mixedly infected bacteria give mixed yields was confirmed by two methods; *viz.,* (1) by plating mixedly infected bacteria before lysis, (2) by plating single bursts after lysis *(1)*.

In these experiments the infecting doses of wild type were slightly higher than those of the *r* mutant. Each kind of virus was in at least threefold excess over the bacteria. In the case of simultaneous infection, about one-third of the bacteria gave pure wild-type bursts. Most of the remaining bursts were mixed. These mixed bursts contained wild type and mutant in all proportions. On the average, however, wild type was predominant. The predominance may be due to an inherent advantage of wild type, or it may be due to the fact that in these experiments the infecting doses of wild type were slightly greater than those of the mutant. We have found that the wild type of this strain of virus is somewhat more rapidly adsorbed than its *r* mutant, and the predominance of wild type in the bursts may be due in part to the more rapid adsorption of the wild type.

If the two viruses are not given simultaneously, the ratios are shifted in favor of the virus which precedes. Thus, if wild type precedes by six minutes, almost all bursts are pure wild type. If the *r* mutant precedes by six minutes, there is a majority of pure *r* bursts and mixed bursts, but there is still a fair proportion of pure wild type.

Mixed infections with T4r^+, T4r

The results for this pair were similar to those for mixed infections with wild type and *r* mutant of T2, with the following minor differences:

(1) A greater proportion of the single bursts showed mixed yields (23 out of 25).

(2) The *r* mutants predominated in the mixed bursts, although all proportions were encountered. Fig. 1 shows the correlations between wild type and *r* mutant in the individual bursts of one large experiment, in which sixty samples were plated for bursts.

For this pair, too, it was found that wild type is adsorbed slightly more rapidly than is its *r* mutant.

These experiments substantiated Hershey's findings. They showed, moreover, that in the bacteria giving mixed yields all proportions of wild type to *r* mutant could be found. The results seemed to prove an almost complete breakdown of the mutual-exclusion mechanism, nearly every bacterium yielding virus particles of both the infecting types. However, modifications of the experimental set-up to be reported presently revealed unexpected new features, which throw the interpretation of Hershey's experiment into doubt.

It may be recalled that the first experiments on mixed infection had been undertaken in the hope of obtaining lysis of bacteria at an intermediate stage of intracellular virus multiplication. Our expectation had been that in mixed infection with (T1, T2) the bacteria would be lysed after 13 minutes, the latent period of T1, and that thus an intermediate stage in the multiplication of T2 would be revealed. T2 by itself does not lyse the bacteria until 21 minutes after infection. This hope had been frustrated when mutual exclusion was discovered. Hershey's discovery of an apparent

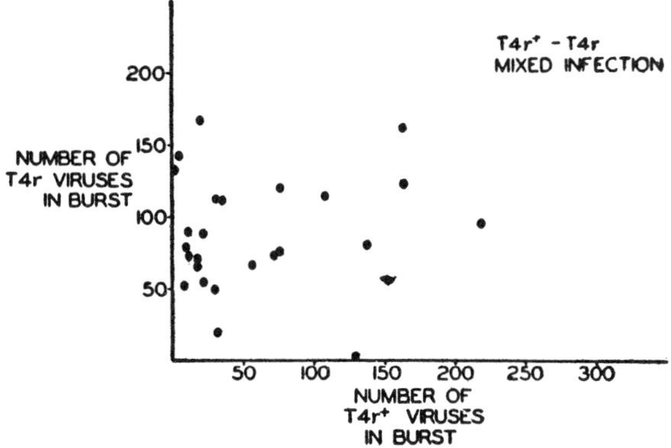

Fig. 1. T4r^+ and T4r content of individual bursts of bacteria infected with these viruses. Each dot represents one burst. The abscissa is the T4r^+ content, the ordinate the T4r content.

ly complete breakdown of the mutual-exclusion mechanism for infections with the pair (T2r^+, T2r) revived the hope of studying intermediate stages of intracellular multiplication of viruses. However, the study of pairs like (T2r^+, T2r) suffers from the handicap that indicator strains are not available for obtaining separate counts of the two types. Instead, one has to rely on plaque appearance. The majority of the plaques are clearly differentiated, but there are always a few plaques whose classification is a little uncertain. These would require subculture for verification, a very laborious procedure when large numbers are involved. For this reason we eventually (October, 1945) decided to try the pairs (T2, T4), (T6, T4), and (T2, T6). These viruses are closely related to each other serologically and morphologically, but they are otherwise independent wild types, and indicator strains which sharply distinguish between them are available. The r mutation was used as an additional genetic marker.

THE BREAKDOWN OF MUTUAL EXCLUSION FOR THE PAIRS (T4, T2), (T4, T6), AND (T2, T6)

In Hershey's case of mixed infection with wild type and r mutant, a breakdown of mutual exclusion is indicated by the appearance of mottled plaques when mixedly infected bacteria are plated. For pairs with host-range differences another criterion can be employed, the appearance of clear plaques in platings on mixed indicator strains *(2)*. Any bacterium that liberates at least one particle of each of the infecting types can lyse both indicator strains, while any bacterium that liberates only particles of one of the infecting types will form a plaque which is overgrown by the indicator strain for the other type and which will therefore be turbid. A comparison of plaque counts on mixed indicators and on strain B gives the fraction of the bacteria with mixed yields—i.e., the fraction of the bacteria in which the mutual-exclusion mechanism failed to operate.

Table 1. The breakdown of the mutual-exclusion principle

Infecting pair		% Mixed yielders
T6	T4	
r^+	r^+	Few
r	r	10
r	r^+	20
r^+	r	12
T2	T4	
r^+	r^+	60-90
r^+	r	60-90
r	r^+	80
T2	T6	
r^+	r	80

Table 1 lists the fraction of mixed yielders for a variety of combinations between the wild types and r mutants of T2, T4, and T6. The breakdown of the mutual-exclusion mechanism is particularly marked for the combinations involving T2. These findings would seem to fit well with Hershey's findings, and to suggest the generalization that mutual exclusion operates the more perfectly the more dissimilar the two infecting viruses. Wild type and r mutant give no mutual exclusion, serologically closely related viruses give partial mutual exclusion, and unrelated viruses give complete mutual exclusion.

The nature of the T2 particles liberated by mixedly infected bacteria involving T2

A paradoxical feature occurred in all the combinations involving T2. When mixedly infected bacteria were plated, the number of *clear* plaques on mixed indicators was in all cases much greater than the number of plaques on the indicator for T2. The number of clear plaques *on mixed indicators* gives the number of bacteria which liberate a mixture of the infecting types. The number of plaques *on the T2 indicator* gives the number of bacteria which liberate T2, irrespective of whether or not such a bacterium also liberates the other type. The latter plaque count comprises two classes, the mixed yielders plus the pure T2 yielders. The mixed indicators show only one of these classes, the mixed yielders. It follows that the count on mixed indicators should always be smaller than, or at most equal to, the count on the T2 indicator, contrary to the actual finding.

The possibility was considered that the clear plaques found on mixed indicators might be due, not to a mixed yield of T2 and T4 particles, but to host-range mutants. Conceivably, host-range mutants might arise during the growth of the plaques and these mutants might be responsible for the lysis of both indicator strains. It must be realized that mixed indicators constitute an ideal enrichment medium for host-range mutants.

This possibility was ruled out by two tests. First, the viruses in question were plated separately on the same pair of mixed indicators. No clear plaques were found. This shows that host-range mutants in sufficient number to cause clear plaques on mixed indicators do not occur when the viruses are plated separately. Second, the contents of ten of the clear plaques from platings of mixedly infected bacteria on mixed indicators were analyzed for their virus content. Each of these plaques contained a mixture of particles with host ranges characteristic of the infecting types, and no particles with extended host range.

It follows that the clear plaques on mixed indicators are due to genuine mixed yields of T2 and T4 particles. The low count of T2 yielders on the indicator for T2 must mean that a considerable proportion of the putative T2 particles liberated from mixedly infected bacteria fail to form plaques on the T2 indicator. The T2 particles used for infection in these experiments do not have this property; they register with the same efficiency on the T2 indicator as on mixed indicators. The low efficiency of plating of the T2 particles in the yields from mixedly infected bacteria is not a hereditary property. Their progeny registers with the same efficiency on the T2 indicator as on mixed indicators.

It might be mentioned in passing that there are many T2 indicator strains of the type B/4 on which the normal T2 exhibits a low efficiency of plating. In fact the strain of B/4 employed in these experiments is unique in registering T2 with full efficiency of plating. This strain was isolated by Dr. Hershey. The question arose whether the low efficiency of plating of the normal T2 on the other strains of B/4 is also increased by the addition of B/2. Experiments showed that this is not the case. Therefore, low efficiency of plating of the T2 particles liberated from mixedly infected bacteria must be caused by a specific property of these particles.

The findings may be summarized as follows. A large proportion of bacteria mixedly infected with T4 and T2, or T6 and T2, give mixed yields. The liberated T2 particles have a low efficiency of plating on the T2 indicator but not on mixed indicators. The low efficiency of plating of these particles on the T2 indicator is not a hereditary property.

More experiments will be needed to clarify this situation. Specifically, the role of the second indicator strain in raising the efficiency of plating of the liberated T2 particles has to be clarified. This indicator strain by itself is totally resistant to T2. Preliminary experiments have not given any clue to the factor contributed by the second indicator strain.

INDUCED MUTATIONS

We now turn to the principal point of this paper, the occurrence in mixed infections of induced mutations at the *r* locus. In the experiments described in the two preceding sections, the mixedly infected bacteria were regularly plated on strain B, on each of the indicator strains separately, and on mixed indicators. In some of these experiments one of the infecting viruses was genetically marked by using the *r* mutant instead of the wild type. It was noted that the platings on one or the other of the indicator strains, or on both, gave a high proportion of mottled plaques. For instance, in a mixed infection with the viruses (T2r^+, T4r) the plating on B/2 gave plaques the majority of which had mottled halos, indicating that the majority of the bacteria had liberated a mixture of wild-type and *r*-type particles which could attack B/2. Wild-type particles able to attack B/2 had not been used in this experiment. The wild-type particles, therefore, represented a new type, created during the mixed infection.

The creation of wild-type particles with the host range characteristic of T4 was verified in three ways. First, by analyzing the contents of the mottled plaques. These analyses confirmed the assumption of the presence of a mixture of T4r^+ and T4r particles. Second, by plating on B/2 after lysis. These plates, too, showed the presence of T4r^+, though in smaller proportion, indicating that induced mutants occur in small numbers, though they occur in the yields of the majority of the mixedly infected bacteria. Third, by plating single bursts on B/2. Small numbers of induced mutants were found in the majority of the bursts, confirming the inference of the previous test.

A priori, this new type could have arisen as a modification of either one of the infecting types: either as a modification of T4r by a mutation

T4$r \longrightarrow$ T4r^+ (under the influence of T2r^+),

or as a modification of T2r^+ by a mutation

T2$r^+ \longrightarrow$ T4r^+ (under the influence of T4r).

The first assumption implies a mutation at the *r* locus, the second assumption a mutation at the genetic site (or sites) determining host range. We prefer the first hypothesis for three reasons. First, the mutation at the *r* locus is known to occur spontaneously and is known to require only one step. Second, a change from the host range characteristic for T2 to that characteristic for T4, as implied in the second assumption, may be expected to require several mutational steps, since T2 and T4 are independent wild types, and since at least two phenotypic changes are involved—namely, loss of activity on B/4 and gain of activity on B/2. Third, when the new type was tested serologically, it was found to be indistinguishable from the infecting type, with the same host range. We therefore believe that the new types arise by a mutation at the *r* locus, without change of host range.

Table 2 is a summary of the combinations of infecting types which have been tested for the occurrence of induced mutations at the *r* locus. The data point to three generalizations:

(1) Mutations occur from wild type to *r* type or from *r* type to wild type.

Table 2. Mixed infections yielding induced mutations

Infecting pair		Induced types
T6	T4	
r^+	r^+	none
r	r	none
r	r^+	T6r^+, T4r
r^+	r	T4r^+
T2	T4	
r^+	r^+	none
r^+	r	T2r, T4r^+
T2	T6	
r^+	r	T2r

(2) Mutations occur only if one of the infecting types is wild type, the other *r* type. When both the infecting types are wild type, or both *r* type, no mutations are found.

(3) In the same mixed infection both infecting types may be changed, wild type to *r* type, and *r* type to wild type.

A discussion of possible theoretical interpretations of these findings does not seem warranted at this point, since our studies are far from complete. Perhaps one might dispute the propriety of calling the observed changes "induced mutations." In some respects they look more like transfers, or even exchanges, of genetic materials. We do not pretend to be able to put forward convincing arguments for either point of view.

A comment might be added with respect to Hershey's original discovery of mixed yields of wild type and *r* type. We now know that mutual exclusion may break down in infections with closely related viruses. Hershey's finding may therefore be interpreted as a lack of mutual exclusion. On the other hand, we know that in mixed infections with wild type and *r* type induced mutations do occur. To explain Hershey's findings one might assume, therefore, that only one of the infecting types multiplies while the other type induces mutations in it. A much closer study of the interrelations between the breakdown of mutual exclusion and the occurrence of induced mutations will be necessary to settle this ambiguity. For this purpose detailed studies of the contents of single bursts, and of the numerical relations between the different types of viruses in such bursts, should prove of great value.

SUMMARY

We will briefly retrace in historical order the steps that have led to our present state of knowledge regarding mixed infections of bacteria with bacterial viruses.

Mixed infections with pairs of *unrelated* viruses, like (T1, T2) or (T1, T7), result in mutual exclusion. Only one of the infecting types multiplies, the other is lost *(3, 6, 2)*.

The excluded virus may greatly reduce the yield of successful virus (depressor effect) *(2)*.

Mixed infections with wild-type and *r*-type particles of the *same* strain do not, apparently, give rise to mutual exclusion (Hershey, *4*).

Mixed infections with pairs of *related* viruses of the group T2, T4, T6 give partial mutual exclusion. The T2 particles liberated in mixed infections of this type exhibit certain nonheritable peculiarities, the nature of which has not yet been ascertained.

Mixed infections with pairs of the group T2, T4, T6, in which one of the pair is used in the wild-type form, the other in the *r*-type form, give rise to the liberation of new types, which can be characterized as mutants of the infecting types. The mutations occur at the *r* locus.

REFERENCES

1. Delbrück, M. The burst size distribution in the growth of bacterial viruses (bacteriophages). J. Bact. **50**:131-135. 1945.
2. Delbrück, M. Interference between bacterial viruses. III. The mutual exclusion effect and the depressor effect. J. Bact. **50**:151-170. 1945.
3. Delbrück, M., and Luria, S. E. Interference between bacterial viruses. I. Interference between two bacterial viruses acting upon the same host, and the mechanism of virus growth. Arch. Biochem. **1**:111-141. 1942.
4. Hershey, A. D. Spontaneous mutations in bacterial viruses. Cold Spring Harbor Symp. Quant. Biol. **11**:67-77. 1946.
5. Luria, S. E. Mutations of bacterial viruses affecting their host range. Genetics **30**:84-99. 1945.
6. Luria, S. E., and Delbrück, M. Interference between bacterial viruses. II. Interference between inactivated bacterial virus and active virus of the same strain and of a different strain. Arch. Biochem. **1**:207-218. 1942.

7 The chromosome of Escherichia coli

John Cairns
Cold Spring Harbor Laboratory of Quantitative Biology
Cold Spring Harbor, New York

INTRODUCTION

Autoradiography of *Escherichia coli*, labeled with tritiated thymidine and lysed with duponol, has shown that the bacterial chromosome comprises a single piece of DNA which is probably duplicated at a single growing point (Cairns, 1963). Further, it seemed likely from the variety of structures seen that this DNA is in the form of a circle while it is being replicated, even though no intact replicating circles had, at that time, been found.

There is no immediate prospect of proving that the bacterial chromosome is simply a continuous DNA double helix; the existence, for example, of protein linkers scattered along the chromosome (Freese, 1958) could be disproved only by a degree of purification of the intact chromosome that, at the moment, is technically impossible. So, rather than attempt any further purification, an effort was made to extract the chromosome as an intact but replicating circle by lysing labeled bacteria with lysozyme instead of duponol; extraction with lysozyme, it was thought, might leave the DNA complexed with basic proteins and polyamines and therefore less liable to breakage by turbulence (Kaiser, Tabor and Tabor, 1963) or by tritium decay (Hershey, unpublished). Whether or not this reasoning is correct, intact and replicating circles have now been found.

RESULTS

E. coli K12#3000 Hfr thy$^-$ was labeled by growth for about two generations in medium containing H^3-thymidine (10 C/mM), as described previously (Cairns, 1963), and was lysed at a concentration of 10^4/ml by incubation at 37°C, in the usual dialysis chamber, in a medium containing 1.5 M sucrose, 0.01 M KCN, 0.01 M EDTA (pH 8), 5 μg/ml calf thymus DNA, and 200 μg/ml lysozyme. Following incubation for 6 hr, the lysed bacteria were dialyzed against repeated changes of 0.005 M EDTA for 18 hr at room temperature. Finally the dialysis chambers were drained and the membranes (VM Millipore Filters) were subjected to autoradiography.

This procedure displays up to 1% of the chromosomes as more or less tangled circles which, when fully extended, have a circumference of 1100-1400 μ. Usually these circles are seen to be engaged in duplication. Here, one such example will be described in considerable detail. First, however, it is simplest to describe the model of chromosome duplication to which it gives rise.

Figure 1 shows diagrammatically two rounds of duplication of a circular chromosome, following the introduction of a labeled precursor at some arbitrary point in the cycle. The diagram is based on the assumption that duplication always starts at the same point (in this case, at 12 o'clock) and always advances in the same direction (in this case, counter-clockwise). To make the diagram in a sense complete, some provision must be made for free rotation of the unduplicated part of the circle with respect to the rest, so that the parental double helix can unwind as it is duplicated; this provision, which we may noncommittally refer to as a swivel, has been placed at the junction of starting and finishing point and is itself marked as being duplicated just before the completed daughter chromosomes separate and begin the next round of duplication. Aside from any question of the swivel, we see that each daughter chromosome (and two of the four granddaughters) shows the stage in the duplication cycle at which label was originally introduced.

Figure 2 shows the autoradiograph of an unbroken replicating circular chromosome that is almost entirely untangled. For the purpose of grain counts and length measurements it was divided, according to the inset diagram, into three sections (A, B, and C) which all meet at

---

Reproduced from Vol. XXVIII; Cold Spring Harbor Symposia, 1963. Used with permission.

I am greatly indebted to V. J. Paral and R. Westen, of the Australian National University, who brought their technical skill to bear on the problem of photographing the autoradiographs of DNA; without their help such pictures could not have been taken.

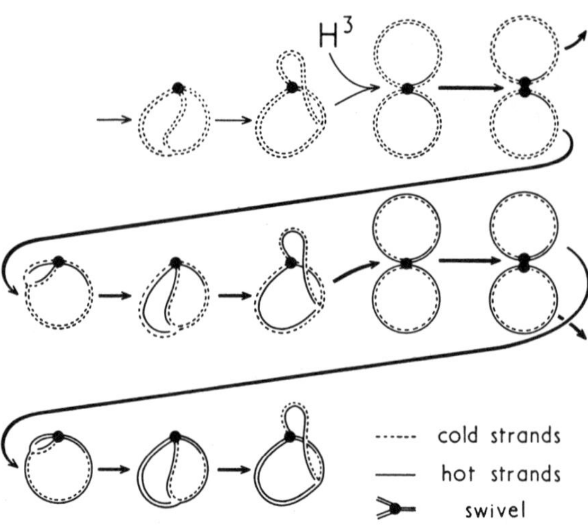

Fig. 1. A diagrammatic representation of the replication of a circle, based on the assumption that each round of replication begins at the same place and proceeds in the same direction.

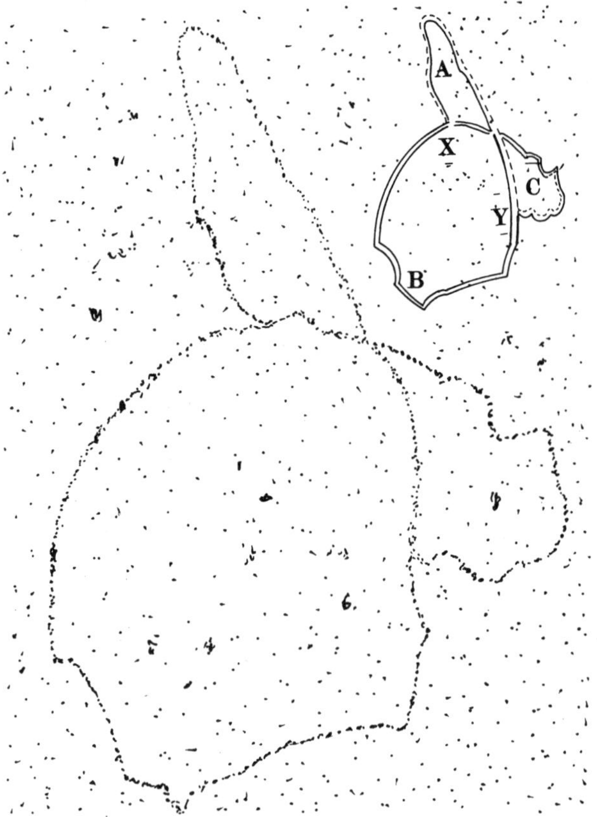

Fig. 2. Autoradiograph of the chromosome of *E. coli* K12 Hfr, labeled with tritiated thymidine for two generations and extracted with lysozyme. Exposure time two months. The scale shows 100 μ. Inset, the same structure is shown diagrammatically and divided into three sections (A, B, and C) that arise at the two forks (X and Y).

Table 1. Grain counts and length measurements for the three sections that unite the forks, X and Y, of Fig. 2

Section	Grains	Length (μ)	Grains/μ
A	714	670	1.1
B	1298	680	1.9
Y to C	213	215	1.0
C			
C to X	359	205	1.8

two forks (X and Y). The grain counts and length measurements are given in Table 1.

We see here a predominately half-hot chromosome that has completed about two-thirds of the process of duplication. Part of the still-unduplicated section is half-hot (from Y to C) and part is hot-hot (from C to X); as shown in Fig. 1, this situation arises when the moment of introduction of the label does not coincide with the start of a round of replication. And it is for this reason that one can, in this instance, identify one of the forks (X) as the starting and finishing point of duplication and the other (Y) as the growing point. Thus the history of this chromosome is taken to be as pictured in Fig. 1.

Discounting the excess due to replication, the total length of the chromosome seen here is 1100 μ (420 μ plus 670-680 μ) or about 22 times the length of T2 DNA–i.e., equivalent to about 2.8×10^9 daltons of DNA. This value is slightly higher than that reported earlier (Cairns, 1963) and agrees well with the maximum value of 23 T2-equivalents, obtained when the reported total DNA content of 32 T2-equivalents (Hershey and Melechen, 1957) is multiplied by *ln* 2 to correct for continuous duplication.

The process of duplication portrayed here is, in most respects, merely the physical embodiment of the conclusions of others. For it has become clear from quite unrelated experiments that the bacterial chromosome is duplicated at a single growing point (Bonhoeffer, 1963) which, at least in *E. coli* Hfr and in certain strains of *B. subtilis,* always starts at the same place and moves in the same direction (Nagata, 1963; Yoshikawa and Sueoka, 1963).

More problematical than the process of DNA synthesis itself, about which there is such satisfactory agreement, are the processes occurring between one round of duplication and the next. These have been represented diagrammatically in Fig. 1 as a separate stage in the duplication of the chromosome during which the swivel is supposed to be duplicated. Presumably, it is at this time that RNA and protein synthesis become obligatory (Maaløe and Hanawalt, 1961), that the Hfr chromosome becomes available for transfer during mating (Bouck and Adelberg, 1963), and that thymineless death may be consummated.

At first sight it seemed surprising to find that the chromosome is physically in the form of a circle even while it is being replicated, for this arrangement demands that somewhere in the circle there must be something that acts as a swivel. However, in view of the apparent importance of the structure that unites the ends of the chromosome and so completes the circle, one must now consider the possibility that the structure actively drives DNA replication by rotating one end of the chromosome relative to the other; in this way, single-stranded DNA might be continually produced at the replicating fork to act as primer for the polymerase. In short, it now seems conceivable that rapid DNA synthesis is possible only for circles.

REFERENCES

Bonhoeffer, F. 1963. Personal communication.

Bouck, N., and E. A. Adelberg. 1963. The relationship between DNA synthesis and conjugation in *Escherichia coli.* Biochem. Biophys. Res. Commun. **11**:24-27.

Cairns, J. 1963. The bacterial chromosome and its manner of replication as seen by autoradiography. J. Mol. Biol. **6**:208-213.

Freese, E. 1958. The arrangement of DNA in the chromosome. Cold Spring Harbor Symp. Quant. Biol. **23**:13-18.

Hershey, A. D., and N. E. Melechen. 1957. Synthesis of phage-precursor nucleic acid in the presence of chloramphenicol. Virology **3**:207-236.

Kaiser, D., H. Tabor, and C. W. Tabor. 1963. Spermine protection of coliphage λ DNA against breakage by hydrodynamic shear. J. Mol. Biol. **6**:141-147.

Maaløe, O., and P. C. Hanawalt. 1961. Thymine deficiency and the normal DNA replication cycle, I. J. Mol. Biol. **3**:144-155.

Nagata, T. 1963. The molecular synchrony and sequential replication of DNA in *Escherichia coli.* Proc. Natl. Acad. Sci. **49**:551-559.

Yoshikawa, H., and N. Sueoka. 1963. Sequential replication of *Bacillus subtilis* chromosome. 1. Comparison of marker frequencies in exponential and stationary growth phases. Proc. Natl. Acad. Sci. **49**:559-566.

DISCUSSION

Butler: I should like to mention an idea put forward at a meeting of the British Biophysical Society December 1962 in a paper by Godson, Barr, and myself (see following Fig. 1). The DNA polymerase is pictured as a disc with two

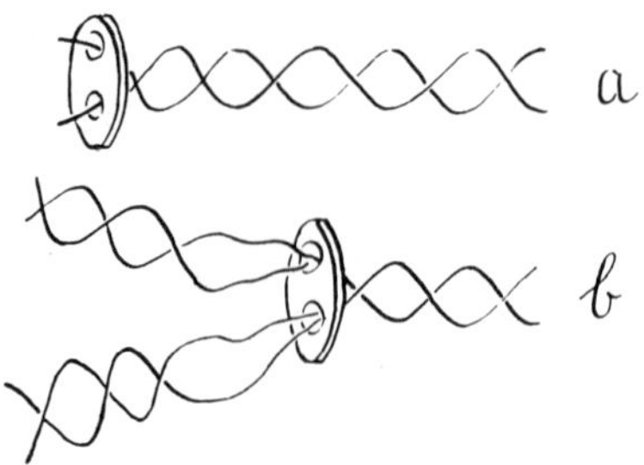

Fig. 1

holes or slots, through each of which one of the strands of the DNA passes. As it passes up the primer, this disc is pictured as rotating relative to it and so separating and unwinding the strands. The energy required for this could probably be provided by the energy of condensation of the triphosphates in the condensation process (about 8 kcals per nucleotide pair). The two new double fibers of DNA would initially be loosely wound round each other, as is often seen in chromosomes, but thermal agitation in the cell would tend to unwind them. The idea suggested by Dr. Cairns that the primer strand itself may rotate would get over the need of the two new fibers of DNA to be wound round each other.

8 # On the mechanism of deoxyribonucleate integration in pneumococcal transformation

Maurice S. Fox
Marcia K. Allen*
Department of Biology
Massachusetts Institute of Technology
Cambridge, Massachusetts

In the transformation of *Diplococcus pneumoniae,* fragments of the transforming deoxyribonucleate (DNA) are inserted into the genome of recipient transformable bacteria. This entire process requires little or no net DNA synthesis.[1,2]

Immediately following its fixation, the transforming DNA can be reisolated from the transformed bacterial population and can be demonstrated to be without biological activity. The newly introduced DNA recovers its activity with a half time of about 3 min and thereafter replicates in synchrony with the bulk DNA of the recipient bacteria.[1,3] Lacks[4] has examined the inactive transforming DNA extracted from transformed bacteria. Some of the material was degraded and the remainder was denatured. He concluded that the denatured DNA was single-stranded and proposed an integration mechanism on this basis.

We will present evidence demonstrating that the transforming DNA extracted, prior to its replication, from transformed bacteria is a hybrid, which is apparently formed with the DNA of the recipient bacteria and which extends over a region of about one or two million daltons. Furthermore, the newly introduced DNA appears to be covalently linked to the DNA of the recipient bacteria.

EXPERIMENTAL

Transforming DNA carrying the density labels deuterium and nitrogen-15 was isolated from a streptomycin-resistant, p-nitrobenzoic acid-sensitive strain of pneumococcus,[5] RF_6S. The bacteria were allowed many generations of growth in a heavy-isotope-labeled medium containing P^{32} in the form of phosphate. The nutrients in the medium were in the form of sugar and amino acid extracts from algae that had been grown in a deuterium, nitrogen-15-substituted medium.[6] The algal extracts were generously provided by H. Crespi.

The bacteria were centrifuged, washed two times in 0.15 *M* NaCl, 0.01 *M* versene, pH 7.5, and then lysed with a mixture of sodium dodecyl sulphate and deoxycholate at a final concentration of 0.05% and 0.025%, respectively. The lysate was shaken with chloroform and isoamyl alcohol[2] and either *(a)* treated with boiled ribonuclease, alcohol precipitated, redissolved in *M*/40 phosphate buffered saline, pH 7.5, and filtered through an HA Millipore filter before using, or *(b)* DNA reisolated from a preparative equilibrium density gradient centrifugation in CsCl.

Equilibrium density gradient centrifugation was performed by adding DNA samples to a concentrated solution of CsCl (Trona) to give a final density of 1.70, a final volume of 1.5 ml, and a DNA concentration of less than 10 μg/ml. Samples were centrifuged at 36,000 rpm and 21°C for 42 hr. The centrifuge tubes were punctured with a needle, and 26-32 fractions of two drops each were collected. The drops were diluted with 5 vol of *M*/40 phosphate buffered saline, pH 7.5, and aliquots were scanned for radioactivity and assayed for biological activity. Both the donor marker, streptomycin resistance, and the recipient marker, p-nitrobenzoic acid resistance, were assayed on the test strain RF6 which is sensitive to both drugs. Under these conditions, native "heavy" DNA may be separated from native light DNA by 10 fractions as is shown in Figure 1. Denaturation in boiling water for 10 min, followed by rapid chilling, shifts the position of a given DNA preparation by 5 fractions. This shift is consistent with a density increase of 0.015 gm/cc. The heavy DNA, therefore, has a buoyant density in CsCl of about 0.03 gm/cc greater than that of native light DNA.

In order to examine the fate of the heavy transforming DNA, transformation was carried out in the following manner. About 300 ml of a frozen transformable culture[7] of strain R6 (resistant to p-nitrobenzoic acid and streptomycin-sensitive) were incubated at 30° for 15 min, centrifuged in the cold, and resuspended in 10 ml of casein hydrolyzate medium[8] supplemented with 10^{-3} *M* $CaCl_2$. The labeled DNA was added to the culture at 30°C. After 8 min, DNase (Worthington, IX crystallized) was added to give a final concentration of 10 μg/ml and the culture was transferred to 37°C. After 1, 4, and 15 min of incubation, samples were chilled to 0°C,

From Proceedings of the National Academy of Sciences **52**:412-419, 1964. Used with permission.

This work was supported by grants from the National Science Foundation (GB-644), and the National Institutes of Health (AI-05388).

The authors wish to thank Dr. P. F. Davison for his advice and suggestions, and Mr. J. Shamoun for his technical assistance.

*National Institutes of Health postdoctoral fellow.

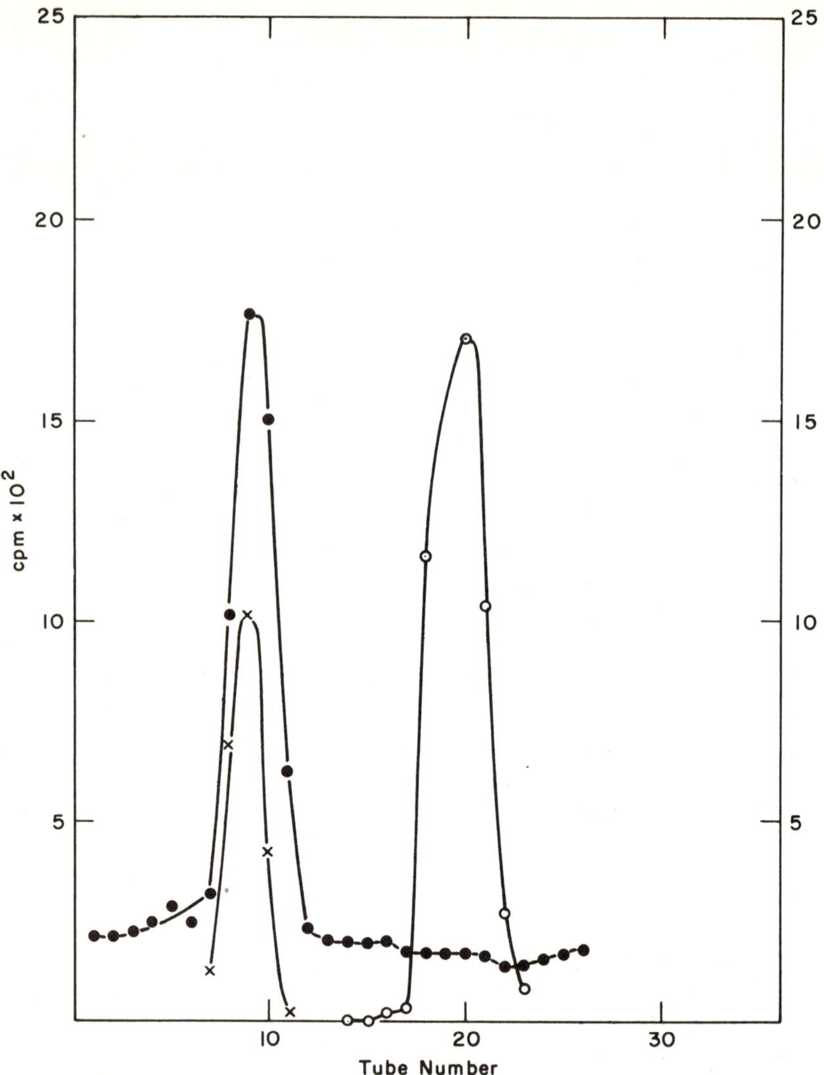

Fig. 1. Distribution of pneumococcal transforming activities of light (O) and heavy (X) DNA after CsCl density gradient centrifugation. The heavy DNA is also labeled with P^{32} (solid dots). Transformants: X, heavy marker (in 10^{-3} ml); O, light marker (in 10^{-3} ml).

centrifuged in the cold, and resuspended in an equal volume of casein hydrolyzate medium containing 10 μg/ml DNase. The samples were further washed three times by centrifuging and resuspending in 2 ml of 0.15 *M* NaCl, with 0.01 *M* versene, and 0.02% albumin. This washing was sufficient to remove residual traces of transforming DNA that had not been fixed by the treated bacteria. After final resuspension, the culture was lysed with 0.05% sodium dodecyl sulfate and 0.025% deoxycholate. The suspensions cleared in about 1 min at 37°C and were shaken for 15 min at room temperature with an equal volume of chloroform and 1/20 vol of isoamyl alcohol. The aqueous layers were recovered and examined by density gradient centrifugations. Small variations in the above procedure are described in the text.

The procedures that have been described permit the recovery in the density gradient of between 80 and 90% of the radioactivity added to the cesium chloride solution, and this in turn accounts for at least 90% of the radioactivity that has been irreversibly fixed by the transformable bacterial population.

RESULTS

The density gradient patterns of DNA extracted from transformed bacteria that had been treated in the manner described above are illustrated in Figure 2. A small amount of cold heavy transforming DNA was added to the gradients as a position marker. At 1 min, only

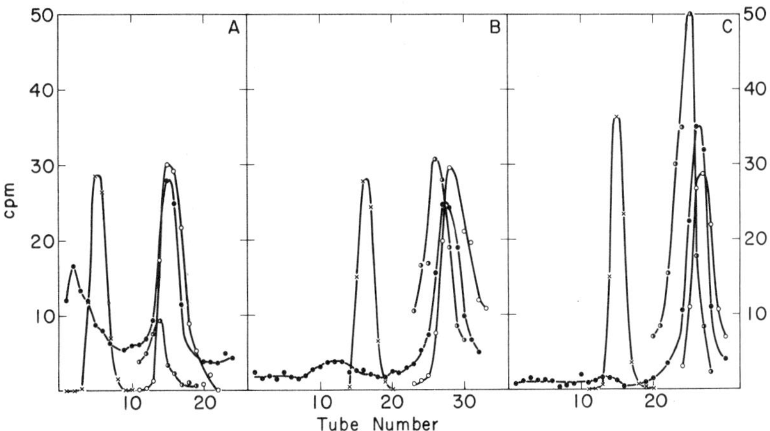

Fig. 2. Density gradient distributions of extracts from transformed bacteria that had been allowed 8-min exposure to labeled DNA followed by, **A**, 1-min; **B**, 4-min; and, **C**, 15-min incubation with 10 μg/ml DNase at 37°C. The distributions show the radioactivity (●) and biological activity (◐) of the donor DNA, the biological activity of the recipient DNA (○), and the biological activity of nonradioactive heavy DNA added as a position marker (X). Gradient **A** contains 30% more extract than the others. Transformants: ○, resident marker (in 4×10^{-4} ml); ◐, input marker (in 0.25 ml); X, heavy marker (in 4×10^{-2} ml).

Table 1. Distribution of radioactivity in extracts of transformed bacteria

Incubation time following DNase addition (min)	% TCA-soluble	% Heavy denatured	% Resembling light native	% Recovery from eclipse
1	43	21	36	21
4	37	5	54	60
15	19	0	81	90

21 per cent of the biological activity of newly introduced DNA has recovered from eclipse; by 4 min, about 60 per cent has recovered; and by 15 min, the newly introduced DNA has almost reached equilibrium. The distribution of radioactivity among the DNA fractions, heavy denatured, light nativelike, and soluble in 5 per cent trichloroacetic acid (TCA), is given in Table 1. It can be seen that there exists little or no material in the gradients that physically resembles the original heavy transforming DNA. At a time when the input marker has recovered only 21 per cent of its biological activity, all of this activity and more than a third of the DNA radioactivity has been so altered as to band in a position very near that of the light recipient DNA. The remaining DNA radioactivity is distributed so that 20 per cent appears in a position to be expected of heavy denatured DNA and 40 per cent is in degraded TCA-soluble fragments. These latter components are lost on incubation, and their P^{32} becomes associated with light native DNA. Lacks[4] has similarly reported the denaturation and degradation of newly fixed transforming DNA. The quantitative differences that exist between Lacks' results and those reported here are probably the consequence of a difference in methods of DNA isolation.

In order to increase the amount of material available for study, the remaining experiments were carried out on extracts from bacteria that had been allowed 15 min exposure at 30°C to the transforming DNA, followed by 3 min at 37° with 10 μg/ml of DNase. Under these conditions, most of the newly introduced DNA has recovered from eclipse, and there has been no detectable multiplication of the DNA element responsible for the transformation.[2] The density gradient pattern of such an extract is shown in Fig. 3*A*, and the distribution of radioactivities is given in Table 2. All of the biological activity and most of the radioactivity of the newly introduced DNA is physically associated with sufficiently large elements of light native DNA so as to mask about 90 per cent of the density label. A similar association between the transforming DNA and the DNA of the recipient bacteria has been observed in *B. subtilis* by Szybalski[9] and Bodmer and Gane-

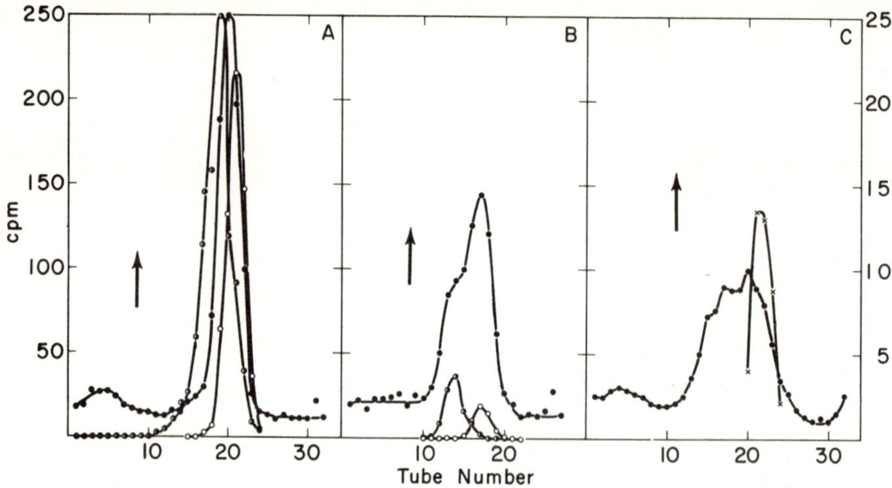

Fig. 3. Density gradient distributions of extracts from transformed bacteria subjected to, **A**, no sonication; **B**, 1-min sonication; and, **C**, 6-min sonication. The arrows indicate the gradient position to be expected for heavy DNA, and the biological activity (X) of added light DNA constitutes a position marker in gradient **C**. The distributions show the radioactivity (●) and biological activity (◑) of the donor DNA, and the biological activity of the recipient DNA (○). Transformants: ○, resident marker (in 10^{-5} ml); ◑, input marker (in 5×10^{-3} ml); X, light marker (in 10^{-4}).

Table 2. Distribution of radioactivity in gradient fractions

Treatment of extract	% TCA-soluble	% Heavy denatured	% Light denatured	% Resembling native
None	36	17	...	46
Heat-denatured	39	19	42	...
Alkali-denatured	30	25	45	...
Sonicated	38	11	...	51
Sonicated and heat-denatured	37	39	23	...
Sonicated and alkali-denatured	34	39	27	...

san.[10] In addition, it can be seen that there is no biological or physical material resembling the native heavy DNA with which these bacteria had been transformed. The irreversible fixation of transforming DNA must, therefore, require a reaction more complex than the mere passage across the bacterial membrane.

Zone sedimentation in a sucrose gradient demonstrates that the macromolecular P^{32} in the transformed bacterial extracts sediments at the same rate and with the same distribution as does freshly isolated H^3-labeled T7 DNA. Since T7 DNA has a molecular weight of 20 million,[11] the P^{32} must be associated with elements having approximately the same molecular weight. Sonication of the extract for 1 min and 6 min, in the manner described by Freifelder and Davison,[12] reduces the sedimentation coefficient from 30*S* to 13*S* and 11*S*, respectively (on the basis of linear distance moved in the gradient). These sedimentation coefficients correspond, according to Doty *et al.*,[13] to molecular weights of about two and one million, respectively.

The distribution of material in density gradients of extracts that had been sonicated for 1 and 6 min are shown in Figure 3*B* and *C*. As a consequence of the sonication, the biological activity of the input DNA shifts away from that of the resident DNA toward a position of higher density and approaches a density that is halfway between that of heavy and light native DNA. Moreover, the P^{32}-containing material, with increasing sonication, becomes spread over a broader and broader density region which ultimately appears to extend from the position of light DNA to the position where one might expect to find "hybrid" DNA. Fractions taken from various positions in the broad band of the sonicated

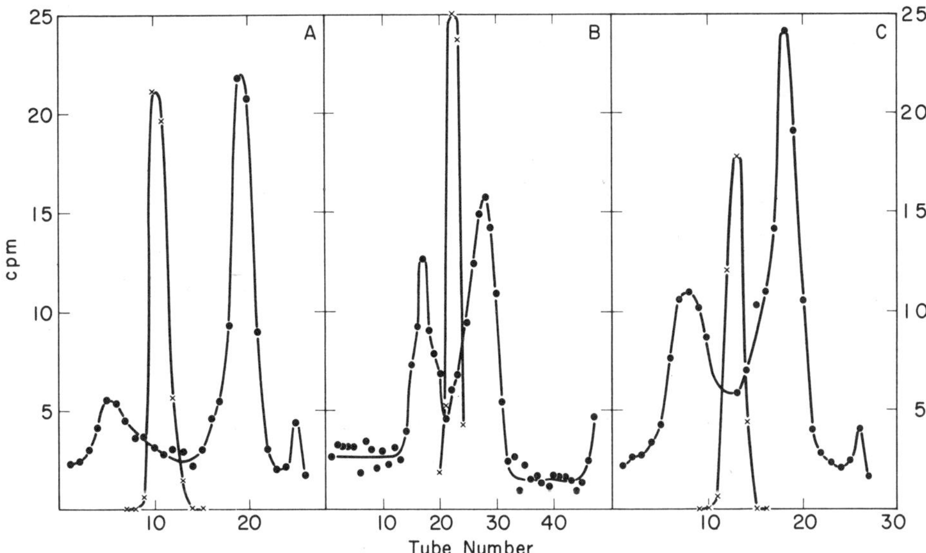

Fig. 4. The extract from transformed bacteria showing the distribution of radioactivity (●) present in the donor DNA. The extract was, **A**, untreated; **B**, heat-denatured; and, **C**, alkali-denatured. The biological activity of added heavy DNA (X) constitutes a position marker. Transformants: X, heavy marker (in 4×10^{-2} ml).

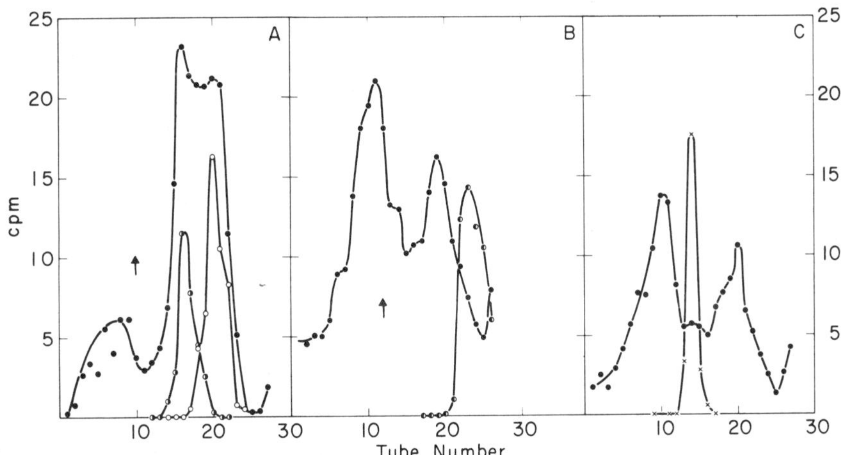

Fig. 5. Density gradient distributions of an extract from transformed bacteria that had been sonicated for 4 min, **A**, then, **B**, heat-denatured or, **C**, alkali-denatured. The arrows indicate the position that native heavy DNA would be expected to assume. The distributions show the radioactivity (●) and biological activity (◑) of the donor DNA and the biological activity of the recipient DNA (○), as well as the biological activity of a light position marker (◐) in **B** and a heavy position marker (X) in **C**. Transformants: ○, resident marker (in 10^{-3} ml); ●, input marker (in 1.25 ml); X, heavy marker (in 5×10^{-2} ml); ◐, light marker (in 5×10^{-5} ml).

sample retain their positions with respect to light DNA when centrifuged again in a density gradient. It is thus demonstrated that the DNA molecules produced by sonication are heterogeneous with respect to density. In spite of the very substantial reduction in molecular weight to one or two million daltons, little, if any, radioactivity appears in the density position of native fully heavy DNA.

Sonication of an artificial mixture of P^{32}-labeled heavy DNA and a large excess of light DNA produces no P^{32} in the position of light native DNA, nor does the P^{32}-band pattern manifest any heterogeneity.

The structure of the complex between the heavy transforming DNA and the light DNA of the recipient bacteria is further elucidated by examining its fate upon heat and alkali denaturation. The heat denaturation was accomplished by placing the DNA extract in boiling water for 7 min and chilling rapidly. Alkali denaturation was brought about by adding enough alkaline phosphate to raise the pH to 12.3 (Beckman combination electrode) incubating at 45°C for 5 min, chilling, and neutralizing with HCl. The density gradient pattern of the DNA extracted from transformed bacteria is shown in Figure 4, which compares the untreated extract with the heat-denatured and the alkali-denatured samples. On denaturation a small additional amount of P^{32}-containing material moves to the position of heavy denatured DNA, but most of the P^{32} is still associated with large elements of light, now denatured, DNA.

The same DNA extract described in Figure 4 was sonicated for 4 min, and aliquots were heat- and alkali-denatured. The density gradients are shown in Figure 5. Following sonication and denaturation, a large portion of the macromolecular P^{32} behaves like heavy denatured DNA. Table 2 summarizes the distribution of radioactivity among the various fractions.

Sonication creates no fully heavy native DNA, only hybrid material which can, by denaturation, be demonstrated to be largely if not entirely an association of fully heavy and light subunits.

DISCUSSION

In extracts from transformed bacteria, the heavy transforming DNA that has been fixed is associated with elements of recipient DNA sufficiently large to obscure most of its characteristic buoyant density. Since most of these elements remain associated with the light DNA after either thermal or alkali denaturation, they would appear to be covalently linked to the light DNA of the recipient bacteria. After sonication, to reduce the molecular weight of the DNA in the extracts from twenty million to about one million, the elements with which the biological activity of the heavy DNA are associated increase in density, approaching the density to be expected of hybrid DNA.

Extracts also contain components of the heavy transforming DNA that have been denatured and that have been degraded. These components are rapidly converted into material that resembles light native DNA.

When the DNA of a sonicated extract is denatured by heat or exposure to high pH, the bulk of the P^{32} is then found associated with heavy denatured DNA. A substantial amount is also found in association with light denatured DNA. This light denatured band is probably best explained as the product of new synthesis of DNA using predominantly light precursors but reutilizing the acid soluble P^{32}-containing products of that portion of the transforming DNA that has been degraded. The heavy denatured band represents material which could not have been more than hybrid in the native configuration and which becomes fully heavy on denaturation. The fraction of the native material that was hybrid must therefore have been half heavy because it was heavy in one strand and light in the other. It is with this hybrid that all of the donor-type biological activity of the reisolated DNA is associated. These findings demonstrate that the transforming DNA that has recovered from eclipse in transformed bacteria is in the form of a duplex that contains only a single strand of the physical material of the native donor transforming DNA. The single strand is present in elements whose molecular weight is of the order of one to two million daltons and is both chemically and genetically[1] coupled with the DNA of the recipient bacteria.

A number of hypotheses will be considered regarding the origin of the hybrid that has been described. One might assume that only a unique strand of the transforming DNA is used. If this unique strand were to form a hybrid with the complementary DNA of the recipient bacterium, then after one bacterial doubling there would have been no increase in the transforming activity of the newly introduced marker—a strand of DNA would have been synthesized that would not be active in transformation. This model can be excluded, since the transforming activity of the newly introduced DNA multiplies in synchrony with that of the recipient DNA.[2-4]

A second possibility is that either strand can be used in transformation and that the strand used forms a hybrid with a rapidly synthesized new complementary strand of itself. Evaluation of this hypothesis can be made by re-examining the results of experiments in which the sensitivity of transforming activity to the disintegration of incorporated P^{32} was measured in preparations of DNA heavily labeled with P^{32} [2] The rate of loss of the biological activity of DNA that has been reisolated very soon after fixation by transformable bacteria is not distinguishable from that of the original DNA.

This model can, therefore, also be excluded, since the newly synthesized strand would carry biological activity and would not contain lethal P^{32} atoms.

A third possibility is that a unique strand of the transforming DNA forms a hybrid with a complementary, rapidly synthesized new strand. This model can only be assessed by examining the sensitivity to P^{32} disintegrations of transforming DNA that has been reisolated after one replication in the transformed bacteria. At this time, the transforming activity should be equally distributed between stable and fully sensitive elements. After infinite decay of the P^{32}, DNA isolated from such a population would be expected to retain at least half of its original transforming activity. Since it has been observed[2] that substantially more than half of the replicated products are in fact subject to inactivation by P^{32} disintegration, this possibility seems to be excluded.

Since it is clear that, as Lacks suggested,[4] only a single strand of the transforming DNA is being used intact, the P^{32} target must be a single strand. The survival, after infinite P^{32} decay, of less than half of the transforming activity, reisolated after one doubling, must be the consequence of a disintegration in one strand inactivating the complementary strand. The data that have been presented[2] are consistent with a cross-strand inactivation at a level of one in every five or six inactivating disintegrations. The absence of such a cross-strand effect to be expected for DNA reisolated soon after recovery from eclipse would reduce its sensitivity to P^{32} disintegrations by 17-20 per cent. Such a reduction was observed but was assumed to be the consequence of a small amount of DNA replication.

The remaining and most appealing model is that in which the hybrid is formed by either strand of the transforming DNA with a complementary region in the DNA of the recipient bacterial genome. This hybrid element would be half as active in transformation as an element which carries the marker on both strands. Observations on annealed mixtures of DNA are consistent with such an assumption.[14]

Further support is lent such a model by the observations of Guild and Robison[15] which suggest that for at least one transformable marker, there exist, in alkaline denatured DNA, two density species, apparently complementary strands of about equal activity. The authors concluded that either strand could carry transforming activity and suggested a similar single-strand displacement mechanism. Furthermore, observations reported by Hotchkiss[16] demonstrate the existence of phenotypically transformed bacteria capable of yielding on replication both transformed and untransformed progeny, with respect to a single selective marker. This would indicate that the transformants must have been heterozygous with respect to the selective property. It would therefore appear that the hybrid that has been described is a hybrid between the newly introduced strand of DNA and its bacterial complement. This hybrid has been formed without a substantial amount of DNA synthesis and at a time when there has been little or no detectable replication of the DNA entity responsible for the transformation.[2]

It may be that fixation of elements of transforming DNA is initiated by one of its strands pairing with an open unpaired region of its bacterial complement. As this pairing proceeds, bacterial exonucleases would hydrolyze the unused strand of the transforming DNA as it is released. This would explain the absence of native heavy DNA in extracts of transformed bacteria as well as the presence of TCA-soluble fragments and what appears to be heavy denatured DNA. Upon completion of this pairing, the enzyme or enzymes responsible for repairing various DNA damages might excise that region of the recipient DNA whose complementary region is now occupied by a strand of the transforming DNA. This step might possibly be the step responsible for recovery from eclipse. Following this recovery, the establishment of chemical continuity between the recipient and the donor DNA could occur. This step might reasonably be considered identical with that in which genetic linkage is established.[1] Reactions such as those that have been described are not substantially different from those that have been observed as being responsible for the dark reactivation of bacteria that have been irradiated with ultraviolet light.[17,18]

In order to explain recombination within the region of the incorporated element of transforming DNA, it would be necessary to assume that the pairing process might be interrupted, perhaps by a break in the strand of donor DNA that is being incorporated. This might result in a switch, in which the other strand of donor DNA begins to pair with its complement in the host genome (later to be excised), and then a switch back to incorporate the remainder of the element of transforming DNA.

Whatever the final description of the mechanism of recombination, it is clear that, at least in the case of DNA-mediated transformation of *D. pneumoniae,* the genetic exchange occurs by the physical insertion of a single strand of transforming DNA into the genetic material of the recipient bacterium.

REFERENCES

1. Fox, M. S., Nature **187:**1004 (1960).
2. Fox, M. S., these Proceedings **48:**1043 (1962).
3. Fox, M. S., and R. D. Hotchkiss, Nature **187:**1002 (1960).
4. Lacks, S., J. Mol. Biol. **5:**119 (1962).
5. Hotchkiss, R. D., and A. H. Evans, in Exchange of Genetic Material: Mechanisms and Consequences, Cold Spring Harbor Symposia on Quantitative Biology, vol. 23 (1958), p. 85.
6. Crespi, H. L., J. Marmur, and J. J. Katz, J. Am. Chem. Soc. **84:**3489 (1962).
7. Fox, M. S., and R. D. Hotchkiss, Nature **179:**1322 (1957).
8. Fox, M. S., J. Gen. Physiol. **42:**737 (1959).
9. Szybalski, W., Journal de Chimie Physique et de Physico-Chimie Biologique **58:**1098 (1961).
10. Bodmer, W. F., and A. T. Ganesan, in Proceedings of the Eleventh International Congress of Genetics, ed. S. J. Geerts (Oxford: Pergamon Press, 1963), p. 30.
11. Davison, P. F., and D. Freifelder, J. Mol. Biol. **5:**643 (1962).
12. Freifelder, D., and P. F. Davison, Biophys. J. **2:**235 (1962).
13. Doty, P., B. McGill, and S. A. Rice, these Proceedings **44:**432 (1958).
14. Suzuki, K., H. Yamagami, and N. Rebeyrotte, J. Mol. Biol. **5:**577 (1962).
15. Guild, W. R., and M. Robison, these Proceedings **50:**106 (1963).
16. Hotchkiss, R. D., in Enzymes: Units of Biological Structure and Function, ed. O. H. Gaebler (New York: Academic Press, 1956), p. 119.
17. Setlow, R., and W. Carrier, these Proceedings **51:**226 (1964).
18. Boyce, R. P., and P. Howard-Flanders, these Proceedings **51:**293 (1964).

9 Gene recombination in Escherichia coli

Joshua Lederberg*,**
E. L. Tatum
Osborn Botanical Laboratory
Yale University
New Haven, Connecticut

Analysis of mixed cultures of nutritional mutants has revealed the presence of new types which strongly suggest the occurrence of a sexual process in the bacterium, *Escherichia coli.*

The mutants consist of strains which differ from their parent wild type, strain *K*-12, in lacking the ability to synthesize growth-factors. As a result of these deficiencies they will only grow in media supplemented with their specific nutritional requirements. In these mutants single nutritional requirements are established at single mutational steps under the influence of *X*-ray or ultra-violet.[1,2] By successive treatments, strains with several requirements have been obtained.

In the recombination studies here reported, two triple mutants have been used: *Y*-10, requiring threonine, leucine and thiamin, and *Y*-24, requiring biotin, phenylalanine and cystine. These strains were grown in mixed culture in 'Bacto' yeast-beef broth. When fully grown, the cells were washed with sterile water and inoculated heavily into synthetic agar medium, to which various supplements have been added to allow the growth of colonies of various nutritional types. This procedure readily allows the detection of very small numbers of cell types different from the parental forms.

The only new types found in 'pure' cultures of the individual mutants were occasional forms which had reverted for a single factor, giving strains which required only two of the original three substances. In mixed cultures, however, a variety of types has been found. These include wild-type strains with no growth-factor deficiencies and single mutant types requiring only thiamin or phenylalanine. In addition, double requirement types have been obtained, including strains deficient in the syntheses of biotin and leucine, biotin and threonine, and biotin and thiamin respectively. The wild-type strains have been studied most intensively, and several independent lines of evidence have indicated their stability and homogeneity.

In other experiments, using the triple mutants mentioned, except that one was resistant to the coli phage *Tl* (obtained by the procedure of Luria and Delbrück[3]), nutritionally wild-type strains were found both in sensitive and in resistant categories. Similarly, recombinations between biochemical requirements and phage resistance have frequently been found.

These types can most reasonably be interpreted as instances of the assortment of genes in new combinations. In order that various genes may have the opportunity to recombine, a cell fusion would be required. The only apparent alternative to this interpretation would be the occurrence in the medium of transforming factors capable of inducing the mutation of genes, bilaterally, both to and from the wild condition. Attempts at the induction of transformations in single cultures by the use of sterile filtrates have been unsuccessful.

The fusion presumably occurs only rarely, since in the cultures investigated only one cell in a million can be classified as a recombination type. The hypothetical zygote has not been detected cytologically.

These experiments imply the occurrence of a sexual process in the bacterium *Escherichia coli;* they will be reported in more detail elsewhere.

From Nature **158**:558, 1946. Used with permission.

This work was supported in part by a grant from the Jane Coffin Childs Memorial Fund for Medical Research.

*Fellow of the Jane Coffin Childs Memorial Fund for Medical Research.

**Present address: Genetics Department, Stanford University Medical School, Palo Alto, California.

REFERENCES

1. Tatum, E. L., Proc. Nat. Acad. Sci. **31**:215 (1945).
2. Tatum, E. L., Cold Spring Harbor, Symposia Quant. Biol., vol. 11 (in the press, 1946).
3. Luria, S. E., and Delbrück, M., Genetics **28**:491 (1943).

10 Infective heredity in bacteria

Norton D. Zinder
The Rockefeller Institute for Medical Research
New York, New York

The study of mechanisms of heredity in bacteria was long hindered by the lack of methods to perform the standard mating tests used so successfully to analyze the genetics of higher forms. The discovery of sexual recombination in the bacterium *Escherichia coli* by Tatum and Lederberg (1947) provided a method for genetic analysis with this organism. Recombinational analysis of hereditary mechanisms in *E. coli* (Lederberg *et al.,* 1951) has provided evidence that many bacterial characteristics are controlled by segregating factors not unlike the genes of higher forms; that these factors are organized into a linear or a number of linear structures analogous to chromosomes; and that each cell is haploid having only a single representative of any particular kind of genetic factor. These are all inferences from the indirect genetic studies. In the history of classical genetics it was the merger of genetic results with cytological events of meiosis and mitosis that provided the firm foundation upon which genetics stands. Because of the small size of bacteria and the technical limitations of instruments now available, this correlation of genetic events with morphological events is not yet possible. Some modern bacterial cytologists, however, have seen what appears to be a bacterial nucleus (Robinow, 1945) and perhaps even chromosomes (DeLamater, 1951).

In addition to genetic recombination in *E. coli,* bacteria have provided other unique mechanisms of intercellular character transfer. Some of these instances of apparent character transfer—the transfer of Vi type (sensitivity or resistance to a series of typing bacteriophages) in *Salmonella typhi* (Anderson, 1952), and the transfer of virulence in *Corynebacterium diphtheriae* (Freeman and Morse, 1951)—seem to result from the lysogenization of appropriate hosts by particular temperate bacteriophages with, as yet, no implication of the previous host's genome. Another instance of character transfer, the ambulatory fertility factor in *Escherichia coli* (Lederberg *et al.,* 1952; Hayes, 1953) and its relation to recombination, was discussed by other workers at this Symposium. We shall restrict our discussion to those instances of character transfer which appear to involve the transfer of fragments of bacterial genetic material from strain to strain in an infective process, namely, transformation in the pneumococcus and *Hemophilus influenzae,* and transduction in *Salmonella.*

The history of transformation reactions is well known. Discovered in 1928 by Griffith, it awaited the now classical work of Avery and his co-workers (1944) for its entrance into the realm of biologically fundamental phenomena. Two years ago at this Symposium, at the time when Ephrussi-Taylor (1951) and Hotchkiss (1951) developed genetic analysis by transformation reactions, transduction in *Salmonella* was announced (Lederberg *et al.,* 1951). Since that time the analysis of both of these phenomena has progressed sufficiently to warrant a direct comparison. It is to that end that this discussion is directed.

TRANSFORMATION AND TRANSDUCTION

The hypothesis underlying our analysis is that transformation and transduction are the result of a similar process, the transfer of restricted amounts of genetic material from donor bacteria to recipient bacteria. There are two aspects to be considered: the alteration of the recipient strain, and the agent which brought it about. It is in the area of the alterations produced that we find the most similarity, while there are, at least, superficial differences between the agents used, accounting in part for the difference in terminology.

It is also our view that the results obtained with transformation and transduction are, in some measure, comparable to genetic systems in more completely hybridizing forms. However, reasoning from them to these incomplete sexual processes has certain dangers as long as one lacks any independent means of verifica-

Reproduced from Vol. XVIII; Cold Spring Harbor Symposia, 1953. Used with permission.

tion. The value of such speculation lies in its furnishing working hypotheses for subsequent tests and in providing an organizational frame for the analysis.

The agents involved may be different, specific desoxyribonucleates (TA) in transformation and phage borne material (FA) of unknown composition in transduction, but the phenomena share the following aspects. Sterile preparations from donor bacteria contain material which infects recipient bacteria and causes the heritable alteration of discrete bacterial characters. Those alterations which have been observed reflect genetic differences between the donor and recipient strains. Individual bacteria are altered, for the most part, in a single character, although among the treated populations different bacteria have different characters affected. The new characters are stable and the agent which brought about the character is incorporated into the bacterial genome by a process involving the replacement of homologous material.

Although *Salmonella* offer a wider variety of kinds of characters for study than the other genera, there seems little doubt that any genetic trait can be transferred. To put it another way, active preparations contain material representative of all or most of the bacterial genome. However, in order to study any particular character efficient selective procedures must be available.

Single factor analysis

In general, individual bacterial characters are altered independently. For example, when rough, penicillin-sensitive pneumococci are treated with transforming agent (TA) from smooth, penicillin-resistant cells there is transformation to both smooth, penicillin-sensitive and rough, penicillin-resistant types. The frequency of the double transformation, to smooth, penicillin-resistant, is comparable to the product of the frequencies of the singles (10^{-2}). Streptomycin resistance is transformed independently of the penicillin and capsule factors (Hotchkiss, 1952). Similarly in *Salmonella* individual characters are transduced one at a time (Zinder and Lederberg, 1952). Strains have been obtained in the laboratory that have as many as eight independent marker mutations serving to differentiate donor and recipient bacteria, including nutritional requirements, fermentative abilities and drug responses. Transduction of any one leaves the other markers unaltered. The frequency of the single transductions, about one per million treated bacteria, is such that independent double transductions could not be found. Also, different *Salmonella* serotypes, e.g., *S. typhimurium* and *S. typhi,* have many natural differentiating characters, and although each in turn may be transducible, they are transduced one at a time.

Both transformation and transduction are limited in their effects upon individual bacteria and it may be asked whether this is due to a restricted material transfer and, if so, is this material a fragment of the gene complement. When a cell mutates, the potentialities of the agents that it can produce reflect the mutation. The fact that drug resistance transfers are limited to those instances wherein mutant cells have been used as the donor illustrates this point. Drug resistance markers also afford us the opportunity for direct comparison of genetic analyses by mutational, recombinational and infective processes. At the first level of analysis, a unit genetic factor is recognized only when it has an alternative or mutant factor. Thus, one definition of a gene is that it is a unit of mutation. Mutational resistance to different drugs falls into two classes: those in which maximum resistance can be obtained in a single selective step and those in which maximum resistance is obtained gradually or step-wise. Streptomycin resistance is illustrative of the former and penicillin and chloromycetin of the latter. Demerec (1948) postulated that streptomycin resistance appeared in this wise because there were one or more genes which could mutate to give the highest level of resistance, while with penicillin resistance there were many additive equipotent genes, the mutation of any one of which would confer only a low level of resistance. The gene as a unit of mutation may become a unit of segregation when mating procedures are available. Newcombe and Nyholm (1951) using *E. coli* recombination found that crosses of streptomycin-resistant strains with sensitive strains segregated only parental types: complete resistants, and sensitives. In fact many independent isolates of either streptomycin-resistant or streptomycin-dependent mutations were referable to a single genetic site on an *E. coli* chromosome. On the other hand, Cavalli and Maccacaro (1952) found that when chloromycetin-resistant strains, obtained by gradual exposure to increasing concentrations of the drug, were crossed with sensitive strains, many intermediate resistant strains appeared among the progeny. Different resistance factors were referable to different sites on the *E. coli* chromosome. Transformation of streptomycin resistance has been studied in both pneumococci (Hotchkiss, 1952) and *Hemophilus influenzae* (Alexander and Leidy, 1953). Maximum resistance is obtained in a single selective step. When sensitives are treated with agent from resistant strains only complete resistants appear among the progeny. Penicillin resistance, in pneumococci is obtained step-wise and transformation recapitulates the steps (Hotchkiss, 1952). For example, a sensitive strain treated with agent from a fourth step resistant mutant will be transformed to the first step. It can subsequently be brought up to higher levels of resistance. As yet only streptomycin resistance has been studied in *Salmonella* transduction and the results obtained were similar to those described above for the transformations (Zinder and Lederberg, 1952). These results support the view that similar genetic units are involved whether studied by mutation, recombination or infective processes.

One might suppose that only overt bacterial characters could be transferred. However, there are instances wherein characters unexpressed in either donor or recipient bacteria appear among the treated cells. In *Salmonella* independent isolates of galactose non-fermenting mutants can often transduce each other to galactose fermentation (Zinder and Lederberg, 1952) and different isolates of non-motile *Salmonella* can often transduce each other to motility (Stocker *et al.*, in press). In pneumococci two partially encapsulated strains can transform each other to more complete encapsulation (Ephrussi-Taylor, 1951). Similarly in *Hemophilus* the treatment of rough cells, a mutant of type *b*, with agent from smooth type *a*, results in some smooth *ab* cells. These results will be elaborated upon later but it may be here noted that it is possible by recombination of non-homologous genetic factors to do more than reassort parental characteristics. It thus appears that the agents of transformation and transduction reflect directly the genetic constitution of the donor cells.

Is the transforming or transducing material simply added to the gene complement of the cell or does it also replace homologous material? Lederberg (1951) has shown that diploid *E. coli* strains having both streptomycin resistant and sensitive genetic factors (heterozygotes) are streptomycin-sensitive, that is to say, sensitivity is dominant. If the same dominance relationship exists in these other genera, the occurrence of drug resistance transfers would indicate that a dominant sensitivity factor had been lost following the acquisition of the recessive resistance factor. However, more direct evidence is at hand. When either rough or extreme-rough pneumococci, derived from each other by mutation, are treated with an agent from the other they are transformed to the donor type. Extracts prepared from these newly transformed cells can only transform to their new type; thus the factors for rough and extreme-rough are not maintained in the same cell (Taylor, 1949). In *Salmonella* the factors controlling a number of flagellar antigens are similarly interchangeable with the transduced cells retaining none of the original agent, e.g., mutual exclusion between the *i* flagellar antigen factor of *S. typhimurium* and the *b* factor of *S. paratyphi* B (Stocker *et al.*, in press). It would seem that, "These transformations are, then, not only transformations of the bacterium, but transformation of a particular element of the bacterium. . ." (Ephrussi-Taylor, 1951).

An example of what appears to be simple addition of genetic factors without substitutive replacement has been found in *Salmonella* (Stocker *et al.*, in press). When non-motile cells are inoculated on the surface of semi-solid agar, they remain fixed at the site of inoculation. When treated with appropriate FA, transduction to motility results and the cells migrate across the plate (a swarm). Besides swarms there appear linear groups (trails) of microcolonies well removed from the inoculum. These seem to be abortive transductions. Thus if a genetic factor for motility is added to a cell and if it is dominant the cell would become motile. However, one may suppose that the exogenous factor is not stabilized or reproduced and is segregated at each cell division given a motile descendent which swims ahead and a non-motile descendent which forms a microcolony in the agar. This process continues until there is a loss of the factor, for the trails rarely, if ever, produce swarms. This inability to stabilize, in these instances, implies a further requirement for transduction, possibly incorporation into a bacterial chromosome. Since cells that obtain an FA or TA genetic factor seem to lose the homologues of the genetic factor, incorporation into a chromosome would provide one means of accomplishing a replacement and a stabilization simultaneously.

Genetic control of bacterial surface properties

Although restricted in general to single factor transfer, transformation and transduction can provide some information on the activity and organization of genetic material in bacteria. Intensive analysis of the genetic control of particular characteristics is possible. The analysis of drug resistance has already been mentioned. Other characteristics that have been studied are motility and its attendant flagella and flagellar antigen in *Salmonella* (Stocker *et al.*, in press), encapsulation and type specificity in *Hemophilus influenzae* (Leidy *et al.*, 1953), and the degree of encapsulation in Type III pneumococci (Ephrussi-Taylor, 1951). Although capsules and flagella are surface properties of bacteria there is little other similarity between them; the former a polysaccharide coat conferring type specificity, the latter a tail-like structure composed of protein and related to motility. The genetic control of these two properties also seems quite different.

Motile *Salmonella* possess flagella which have associated with them a characteristic antigen (H) which has been used as a diagnostic criterion in differentiating the many *Salmonella* serotypes. By mutation flagella may be lost with the consequent loss of motility and H antigen. Many strains are isolated in nature as non-motile ("O" forms in contrast to the motile flagellated "H" forms); others have been observed to occur in the laboratory. Mutation from non-motile to motile has also been observed to occur.

Transduction of motility is readily effected. When non-motile cells are inoculated on the surface of soft agar plates, they multiply but remain fixed at the inoculum site. Motile cells, formed either by spontaneous mutation or transduction, leave the inoculum and swarm over the plate. The test of such swarms shows them to be composed of cells that are motile, flagellated and reactive with H antiserum. Some strains produce too many spontaneous swarms or produced them later and with lower frequency than the induced swarms. In order to determine if the same factor controlled motility and antigenicity a careful examination was made of those swarms produced when the donor and recipient differed originally in their H antigen. The H antigen of the non-motile recipient was inferred either from its spontaneous swarms, or from its origin (e.g., a mutant in the laboratory, or a strain associated with a motile strain identical in all respects except for motility). In all instances the predominant component of the induced swarms exhibited the H antigen characteristic of the recipient, not the donor. In other words, the same antigen appeared as with the spontaneous swarms. For example, the treatment of a non-motile mutant of *S. typhi* 0901 (the parent strain had flagellar antigen d) with FA from *S. typhimurium* (antigen i) resulted in swarms with antigen d. Thus antigenicity and motility are genetically separable. Non-motile *Salmonella* have a latent antigen factor which is revealed when single factor transduction provides the necessary flagellar factor.

As was mentioned previously, different isolates of non-motile cells can transduce each other to motility. This would indicate that there are a number of non-homologous loci, each of which when mutated leads to the loss of motility. The treatment of seven different non-motile strains with FA from the others, whenever possible, indicated that there are at least five different loci affecting motility.

There are also strains that are non-motile although possessing flagella. These flagella may be considered to be paralyzed (Friewer and Leifson, 1952) and can be transduced to active motility. The two known paralysis loci are non-homologous with any of the five motility loci.

One non-motile strain when transduced to motility swarmed at a much slower rate than the others. In turn it could be transduced to rapid swarming. The factor controlling slow swarming is non-homologous to any of the other loci affecting motility. Thus mutation at many different loci can bring about both qualitative and quantitative alterations in flagellar function. By transduction it has been possible to demonstrate the complex genetic control of this bacterial organelle.

Transduction of motility factors has also provided the first evidence for two-factor or linked transduction. When non-motile cells are transduced to motility the usual finding is the revealing of the latent flagellar antigen of the recipient strain. However, in a number of instances not only were these found, but also some transduced cells with the H antigen of the donor. For example, a non-motile strain of *S. paratyphi* B occasionally mutates to a motile form with the flagellar antigen b, characteristic of this serotype. When this strain is transduced to motility with FA from *S. typhimurium* (H antigen i) about 75 per cent of the swarms are composed of b reactive cells and 25 per cent of i reactive cells. Treatment with FA from *S. heidelberg* (H antigen r) gives similar results with the r antigen substituted for the i. Thus some cells are transduced to motility and retain their latent antigen factor and others become motile and have their antigen factor replaced as well. The linkage between motility and antigen factors was verified by the "progeny testing" of the transduced cells. FA from the doubly transduced *S. paratyphi* B, with flagellar antigen i, was applied to the non-motile *S. paratyphi* B and both b and i reactive motile transduced cells were found. Thus the factors for antigenicity and motility again exhibit incomplete linkage. It seems reasonable to assume that the *S. paratyphi* B strain was non-motile because of a mutation at a locus linked to its H antigen factor and that these have homologues in the two donor strains. Transduction sometimes provides only the homologue of the mutated flagella gene giving rise to motile cells with flagellar antigen b and sometimes provides both the flagellar factor and its linked antigen factor to yield motile cells with antigen i. The recovery of the other predicted type, the non-motile with the suppressed antigen of the donor (i or r) is not feasible with current techniques.

So far the genetic analysis of the flagellar factors in *Salmonella* has given rather readily explicable results. However, the capsule factors in pneumococci and *Hemophilus* seem to have a more obscure kind of genetic control which perhaps prompted Ephrussi-Taylor's (1951) remark, "It *(transformation)* is a system in which, as yet, no bridge can be seen leading over into classical genetics . . ." The difficulties in both cases arise from somewhat similar sources and will be discussed together.

Serological type specificity is conferred on pneumococci and *Hemophilus* by special polysaccharide capsular material. Mutation can result in the total loss of the capsule, giving rise to rough (*R* as contrasted to *S* for smooth encapsulated cells) cells or to partial losses with intermediate amounts of capsule production. Transformation is accomplished by the seeding of recipient cells into broths containing antiserum to the recipient and TA from the donor. The further reaction is stopped after a short interval by the addition of desoxyribonuclease. The untransformed recipient cells grow in clumps due to the presence of the antiserum. The tubes are incubated and observed for diffuse growth, indicative of a transformation. Single clones for test are obtained by dilution and plating. Such selection procedures may be inefficient especially in indicating the relative extent of two or more kinds of transformation.

Tables 1 and 2 summarize the pertinent published data of *Pneumococcus* and *Hemophilus* transformation. In general when TA from any smooth type is applied to rough cells the transformed cells are similar to the donor cells; i.e., in amount of capsule production in pneumococci and in type specificity in *Hemophilus.* Thus there is no evidence for latent specificities as there had been in non-motile *Salmonella* strains. This might be taken to indicate that all capsule factors are referable to a single locus. However, in *Hemophilus* the treatment of *R* cells, derived from type *b,* with TA from smooth type *a,* leads to the formation of smooth type *ab* cells, apparently unveiling a latent *b* factor which is not a homologue of the *a* factor. Again, in pneumococci, different smooth types with only a trace encapsulation

Table 1. Composite* of some of the results of intertype transformation in *Hemophilus influenzae*

Type treated	TA from	Transformation to
**R_n	†S_x	S_x
R_b	S_a	S_{ab}
S_b	S_a	Sab (Sa)
††Iab	Sd	Sad
Rd	Sab	Sa, Sb, Sab
Rd	Sa + Sb	Sa, Sb
Rd	Rb	—

*Alexander and Leidy, 1951; Leidy *et al.,* 1953.
**Rn refers to a rough mutant of a smooth type.
†Sx refers to any smooth type.
††Iab partially encapsulated mutant of Sab.

Table 2. Abstract* of some of the results of transformation between partially encapsulated mutants of Type III pneumococci

Strain treated	TA from				
	***S-1	S-1a	†S-2	††S-N	S-2(a)
**R	S-1	S-1a	S-2	S-N	S-2
S-1	—	S-2(a)	S-2, S-N	S-N	S-2
S-1a	S-2	—	S-2, S-N	S-N	S-2
S-1b	S-2	—	S-2, S-N	S-N	S-2
S-1c	S-N	S-N	S-2, S-N	S-N	S-2, S-N

*Ephrussi-Taylor, 1951.
**R rough unencapsulated mutant.
***S-1 trace capsule.
†S-2 nearly normal capsule.
††S-N normal capsule.

can transform each other to normal or near normal capsule production. *Hemophilus* transformation also provides an instance of what appears to be linked transformation. The treatment of *Rd* with agent from *Sab* leads to the formation of *Sa, Sb* and *Sab* cells. Thus there are homologies and non-homologies among the capsule factors and they therefore cannot be attributed to a single set of alleles; however, they are also not independent of each other. Ephrussi-Taylor (1951) presented one possible spacial model for the pneumococcus factors which requires a kind of recombination of overlapping factors; others are possible. In the absence of further progeny testing of some of the synthesized types it is difficult to develop a consistent genetic picture. It does seem that the capsule factors are controlled by a single complex locus with the rough condition involving a considerable chromosomal region, the intermediate types in *Pneumococcus* and the different serotypes in *Hemophilus* being related to one or more sub-units within these regions, and which travel both together and separately in transformation. Similar complex loci have been found in corn (Stadler, 1951) and *Drosophila* (Lewis, 1951) and, because of technical problems, are difficult to analyze. Transformation and transduction which deal directly with fragments of genetic material acting upon large populations might then provide the tools for genetic analyses at precisely the level wherein the analysis of higher forms becomes difficult.

TRANSFORMING AGENT (TA) AND TRANSDUCING AGENT (FA)

The preparation and the purification of TA and its properties has been documented (Avery *et al.*, 1944; McCarty, 1946; Zamenhof, 1952). Lysis of bacterial cultures by bile salts or penicillin results in active preparations. The activity resides in the nucleic acid fraction of these lysates. Specific and irreversible inactivation of TA is produced by the enzyme desoxyribonuclease. It is resistant to the application of proteolytic enzymes and ribonuclease. Highly purified preparations have been shown to have less than 0.02 per cent contaminating protein, a maximum value fixed by the sensitivity of the chemical procedures (Hotchkiss, 1952a). There can be but little doubt that polymerized desoxyribonucleates are sufficient for transformation; however, ascertaining the possible role of contaminant materials must await the determination of the minimum amount of TA required for transformation. The average molecular weight of nucleic acid molecules in an active preparation was reported to be about half a million (Avery *et al.*, 1944). However, by studying the radiation sensitive volume, Fluke *et al.* (1952) estimated a weight of six million.

The agent of transduction is far from chemical resolution. Zinder and Lederberg (1952) tentatively related FA to the bacteriophage particles present in active preparations. Much evidence has been accumulated to support this view and here we shall elaborate upon this and its consequences on transductive analyses.

FA and bacteriophage

FA is produced when bacteria are lysed by bacteriophage. The phages which are involved are in general temperate, and biologically similar to the temperate phages discussed at this Symposium. Boyd (1950) classified the temperate phages of *S. typhimurium* and the phage PLT-22 (Zinder and Lederberg, 1952) with which we are most concerned belongs to his A1 class. It produces a haloed turbid plaque three to four mm in diameter.

In order to demonstrate the identity of the particles producing plaques on appropriate indicator strains and those producing transductions, evidence was sought from many sources. Plaque producing titers were assayed by the agar layer method with *S. gallinarum* as the indicator. Transduction was assayed, except when otherwise noted, by the simultaneous plating of lysate with strain LA-22 (carrying PLT-22), which requires for growth phenylalanine, tyrosine and tryptophane (obtained as two successive mutations but transduced to prototrophy in a single step; Zinder and Lederberg, 1952). This is done on a minimal medium and the number of prototrophic transductions is counted.

It was noted consistently that there was about a 10^7 ratio of phage to FA in different lysates, indicating a correlation in their production. The amount of lysis occurring when sensitive bacteria are infected with a temperate phage is related to the multiplicity of infection (Boyd, 1951). By following, in time, the appearance of phage and FA after infection at different multiplicities, it was found that they were produced simultaneously and in constant ratio. The phage lyses only a fraction of the infected cells the rest becoming lysogenic, but because the appearance of FA is a function of the amount of lysis it must be concluded that it is contributed by the lysing fraction. Such lysates are heavily contaminated with inactive materials. FA can be purified by low speed centrifugation and filtration to remove bacterial debris. The lysate may then be subjected to deproteinization procedures by shaking with chloroform and amyl alcohol. All of the activity remains in the aqueous phase. Further purification and concentration can be achieved by centrifugation in the Spinco ultracentrifuge and resuspension in buffered saline. Throughout such procedures the phage to FA ratio remains constant, although some percentage of each may be lost at each step. The final concentrated solution is opalescent, with a titer of 10^{11} to 10^{12} plaque producing particles per ml. The phage and the FA are resistant to the application of ribonuclease and desoxyribonuclease.

Heat inactivated phage does not adsorb to bacterial surfaces (Adams and Lark, 1950). Active filtrates, in broth, were subjected to heat treatment. Detectible inactivation occurs at 70° C with a rate for phage and FA of about 2.5 per cent per minute. Thus FA may not be more sensitive to heat than the phage. However, if FA is an agent carried by a phage, it may of course be intrinsically more heat-resistant than its carrier.

Neither the phage nor the FA is affected by antibacterial serum. Antiserum prepared by the injection of purified lysates into rabbits inactivates the phage and the FA at the same rate. Assuming phage and FA were not related to the same particle, it would be most remarkable for two discrete antigens, when injected in unknown proportion, to stimulate the formation of a serum of identical titer for each. The antiserum was prepared by the injection of a lysate having no streptomycin resistance transducing ability (phage grown on streptomycin-sensitive cells). However, the serum neutralized, at the same rate, transduction to streptomycin resistance of the same phage grown on streptomycin-resistant cells. Thus many genetic effects referable to the same kind of phage are neutralized with the phage regardless of whether they were present in the original injected antigen.

Ultra-filtration through gradocol membranes provides one means of determining particle size. Ninety-nine per cent of phage and FA are retained by a membrane of A.P.D. 120 mμ indicating that they have a common particle size (Zinder and Lederberg, 1952).

FA and phage have common sites of adsorption. By the testing of many different *Salmonella* serotypes it had been shown that FA was adsorbed by cells with somatic antigen XII (Zinder and Lederberg, 1952). The phage, PLT-22, is adsorbed by these same serotypes although many of them are not phage sensitive. In following the adsorption on *S. typhimurium* strain LT-22 (PLT-22 carrying) both phage and FA reach saturation at the same level of 8 to 10 particles per bacterium; a factor that limits the maximum frequency of transduction. Some phage sensitive strains will adsorb many times more phage-FA and their maximum transduction frequencies are correspondingly raised. The ability to adsorb the phage in an active lysate provides a necessary condition for transduction.

It is evident that those operations which remove FA remove phage, those which do not affect FA do not affect phage and those which prevent FA adsorption similarly prevent the adsorption of phage. The hypothesis that FA is phage bound seems quite valid and as a consequence we are not determining FA properties *per se* but only as a reflection of phage properties.

It has been implicit in the foregoing analysis that the occurrence of phage infection need not result in a particular transduction. The phage and the FA may be considered different biological entities sharing a common adsorptive mechanism.

The growth of PLT-22 on any of a number of different genotypes of *Salmonella* results, in a single phage generation, in a lysate with activity comparable to the genotype of the secondary host and so on. It is not yet known whether this loss of material from the previous hosts is by simple dilution, destruction, or re-working. However, it is clear that an essentially homogeneous phage becomes endowed in each generation with a heterogeneity of genetic properties.

The phage, as reflected by plaque production, can only be assayed on sensitive bacteria which includes only a part of the phage-FA adsorbing group of strains. Transduction, on the other hand, can be assayed on all receptive bacteria (terminology of Jacob *et al.*, 1953) which includes sensitives, immunes (phage carrying) and those in which an abortive infection results.

When phage sensitive bacteria are transduced, cells are almost always lysogenic. However, occasionally transduced cells are found that are non-lysogenic and phage-sensitive (Stocker *et al.*, in press) indicating the possibility of independent stabilization of FA and phage.

One agent that does not interfere with phage adsorption but does prevent phage reproduction is ultraviolet irradiation. The phage is inactivated exponentially as the dose of ultraviolet increases while the transductive titer rises at first and then falls with a slope some three to four times slower than that of the phage. This activation of transduction at low ultraviolet doses can be demonstrated with many different characters in many different strains. The peak of the activation is comparable to one per cent phage survival. When phage sensitive strains are transduced with irradiated phage there is usually still enough viable phage to secondarily infect the transduced cells, however, Lederberg (personal communication) has found occasional non-lysogenic transduced cells. Again it is difficult to say whether we are determining an FA property or only the ability of the phage to deliver the FA. If the slow but definite inactivation by ultraviolet is an FA property then it might be taken to indicate a nucleic acid component.

When immune bacteria (carrying pro-phage) are transduced with an homologous phage there is no possible change in the overall phage content and one does not know whether the transducing phage has become established. Bertani (1953) has shown that related pro-phages will exclude each other so that serially infected strains will eventually carry one or the other, not both. He has also shown that virulent phage mutants are excluded by cells carrying related pro-phages. The only mutant that has so far been available to us is a virulent one. It can, however, be used to transduce when the assay organisms are immune. The transductions are stable and the cells carry the original pro-phage. Similarly by previously preparing lysogenic derivatives of sensitive strains they can be transduced with virulent phage. Thus it appears that FA can dissociate itself from its phage carrier and, although the phage is excluded, the transduction proceeds.

Some strains of *Salmonella* adsorb the phage and are transduced but neither lyse nor become lysogenic. The phage infected cells do not act as infective centers so that it must be concluded that the infection is abortive and the phage is destroyed. The evidence presented above indicates that although phage and FA share a common particulate existence they are not so interdependent that they cannot be separated, at least within the bacterial cell.

DISCUSSION

The systems of bacterial gene transference discussed herein lead fairly directly to the conclusion that bacteria have hereditary determinants differentiable from the bacteria as a whole. These determinants are so analogous to what are called genes in higher forms that it seems unnecessary to distinguish between the two at this time. The mode of organization of the determinants may not be so readily assumed. In higher forms the genes are organized in linear structures called chromosomes. The evidence from transformation and transduction tends to show that there are both homologies and non-homologies among the various genetic factors. That is, added genetic factors always replace their homologues and thus factors are not added *ad infinitum.* Hence it would seem that at any one time a cell has only a single representative of any one kind of factor. *A priori,* this singleness of kinds of factors and the necessity of distribution of each kind at cell division speaks for some kind of more tightly knit organization than free distribution in the cell. Again there is evidence that the transferable units have a spatial differentiation, which in turn could be part of a larger structure, a chromosome. It then would become necessary to explain the unit nature of the transfers. Fragmentation of chromosomes could occur when there is a dissolution of pneumococci or *Hemophilus* and phage lysis of *Salmonella.* Also larger fragments may not be able to penetrate the recipient cells in transformation reactions or be incorporated into the phage carrier in transductive reactions. Limitation upon the effective agents could also be determined by the inability of larger fragments to be stabilized once they have entered the recipient cell. The examples of linked transfers have thus far all involved factors related to the same character. In other forms numerous instances have been found where genes, on known chromosomes, affecting the same character are adjacent to each other (Stadler, 1951; Lewis, 1951). Thus in systems wherein there may be physical limitations determining which particles can participate, linkage could only be found for closely linked genes and, by analogy to the other forms, be more often related to the control of the same character. The number of independent characters that have been studied is still small and there may yet be found unrelated ones whose factors exhibit linkage. The evidence from these systems of gene transfer may neither be taken to support the notion that there are bacterial chromosomes nor can it be construed to disprove this point of view.

When the bacteriophage T2 infects its host it sheds its protein coat and mainly its nucleic acid enters the host (Hershey and Chase, 1952). The genetic continuity of the phage is presumed to depend upon the injected nucleic acid and in this respect phage infection resembles transformation reactions. *Salmonella* transduction might then be thought to combine the features of phage infection with those of transformation; the simultaneous injection of both viral nucleic acid and unaltered bacterial nucleic acid acquired from the previous bacterial host. However, the verification of this must await the demonstration that the transducing phages also inject a part of themselves into the host cell and, if so, the determination of the chemical nature of the material injected.

The transfer of the genetic material in transformation is then more direct than in transduction. Artificial lysis of the cell releases the active TA in the former case which in turn may penetrate recipient cells and bring about the transformation. In *Salmonella* lysis of the cells is not readily brought about and those extracts artificially obtained have been inactive. Only when bacteriophage is present in sufficient amount is any activity found. The evidence indicates that the active material is incorporated into the virus particle itself.

Three developmental stages of temperate bacteriophage have been postulated; infective, vegetative and pro-phage. It is the pro-phage stage that is peculiar to the temperate bacteriophages. Studies on the inheritance of the temperate phage *lambda,* in crosses of *E. coli,* of carrying by non-carrying lines has shown that *lambda* segregates as a unit and exhibits linkage to certain genetic markers (Lederberg and Lederberg, 1953; Wollman, 1953). Taken with Bertani's (1953) showing of mutual exclusion between related pro-phages, these results indicate an intimate relationship between the bacterial hereditary determinants and the pro-phage state such that there might be only a single site for pro-phage maintenance and this upon the bacterial chromosome. However, *Salmonella* transduction seems to be only indirectly, if at all, dependent upon this relationship.

The phage does not have the genetic activity conferred upon it at the pro-phage stage. Were this to be so it would restrict the transduction to those characters that were related to the pro-phage and transduction seems to encompass the entire bacterial genome. The amount of FA produced after infection of sensitive cells with temperate phage is a function of the amount of lysis with the phage presumably by-passing the pro-phage stage. Similarly virulent phage has transductive activity associated with it. Although PLT-22 production, from lysogenic strains, is not readily induced with ultraviolet irradiation those induced lysates that have been obtained have as wide a range of genetic effect as those obtained following direct phage lysis. There is also no evidence, as yet, that phage obtained by ultraviolet induction of cells that had become lysogenic, simultaneously with transduction of a particular character has any increased propensity for transduction of that particular character. It may therefore be concluded that the bacterial genetic material, even when obtained following induction, becomes associated with the phage between the vegetative and infective stages.

Transduction also does not seem to be contingent upon the reduction of infective phage to pro-phage. The fact that virulent phages, lacking this stage, and that phage and FA can be separated within the transduced cell, support this view. It therefore seems that although transduction could only be demonstrated with a lysogenic system it is more of a consequence of the nature of phage reproduction than of any relationship between pro-phage and bacterial hereditary determinants. Evans (1953) in a discussion of the origin of the components of coliphage has pointed out that a proportion of the components of host nucleic acid is incorporated directly into the phage progeny. It is perhaps by this mechanism that the genetic activity is conferred upon the phage.

The ability to effect transduction with virulent phages promises to make available the many refined techniques of phage study such that a more precise determination of the inter-relationships of phage and FA should be possible.

REFERENCES

Adams, M. E., and Lark, G., 1950, Mutation to heat resistance in coliphage T5. J. Immunol. **64**:335-347.

Alexander, H. E., and Leidy, G., 1951, Determination of inherited traits of *Hemophilus influenzae* by desoxyribonucleic acid fractions from type specific cells. J. Exp. Med. **93**:345-359.

1953, Induction of streptomycin resistance in sensitive *Hemophilus influenzae* by extracts containing desoxyribonucleic acid from resistant *Hemophilus influenzae.* J. Exp. Med. **97**:17-31.

Anderson, E. S., 1951, The significance of Vi-phage types F1 and F2 of *Salmonella typhi.* J. Hyg., Camb. **49**:458-470.

Avery, O. T., MacLeod, C. M., and McCarty, M., 1944, Studies on the chemical nature of the substance inducing transformation of pneumococcal types. J. Exp. Med. **79**:137-158.

Bertani, G., 1953, Infections bactériophagiques secondaire des bactéries lysogènes. Ann. Inst. Pasteur **84**:273-280.

Boyd, J. S. K., 1950, The symbiotic bacteriophages of *Salmonella typhimurium.* J. Path. Bact. **62**:501-517.

1951, Observations on the relationship of symbiotic and lytic bacteriophage. J. Path. Bact. **63**:445-447.

Cavalli, L. L., and Maccacaro, G. A., 1952, Polygenic inheritance of drug-resistance in the bacterium *Escherichia coli.* Heredity **6**:311-331.

DeLamater, E. D., 1951, A new cytological basis for bacterial genetics. Cold Spr. Harb. Symp. Quant. Biol. **16**:381-409.

Demerec, M., 1948, Origin of bacterial resistance to antibiotics. J. Bact. **56**:63-74.

Ephrussi-Taylor, H., 1951, Genetic aspects of transformations of pneumococci. Cold Spr. Harb. Symp. Quant. Biol. **16**:445-455.

Evans, E. A., Jr., 1953, Origin of the components of the bacteriophage particle. Ann. Inst. Pasteur **84**:129-142.

Fluke, D., Drew, R., and Pollard, E., 1952, Ionizing particle evidence for the molecular weight of the pneumococcus transforming principle. Proc. Nat. Acad. Sci., Wash. **38**:180-187.

Freeman, V. J., and Morse, I. U., 1951, Further observations on the change to virulence of bacteriophage-infected avirulent strains of *Corynebacterium diphtheriae.* J. Bact. **63**:407-414.

Friewer, F., and Leifson, E., 1952, Non-motile flagellated variants of *Salmonella typhimurium.* J. Path. Bact. **64**:223-224.

Griffith, F., 1928, The significance of pneumococcal types. J. Hyg., Camb. **27**:113-159.

Hayes, W., 1953, Observations on a transmissible agent determining sexual differentiation in *Bacterium coli.* J. Gen. Microbiol. **8**:72-88.

Hershey, A. D., and Chase, M., 1952, Independent functions of viral protein and nucleic acid in growth of bacteriophage. J. Gen. Physiol. **36**:39-56.

Hotchkiss, R. D., 1951, Transfer of penicillin resistance in pneumococci by the desoxyribonucleate derived from resistant culture. Cold Spr. Harb. Symp. Quant. Biol. **16**:457-460.

1952, The biological nature of the bacterial transforming factors. Exp. Cell Res. Supplement, **2**:383-389.

1952a, The role of desoxyribonucleates in bacterial transformations. Symp. on Phosphorus Metabolism **2**:426-436.

Jacob, F., Lwoff, A., Siminovitch, A., and Wollman, E., 1953, Définition de quelques termes relatifs à la lysogénie. Ann. Inst. Pasteur **84**:222-224.

Lederberg, E., and Lederberg, J., 1953, Genetic studies of lysogenicity in *Escherichia coli.* Genetics **38**:51-64.

Lederberg, J., 1951, Streptomycin resistance: a genetically recessive mutation. J. Bact. **61**:549-550.

Lederberg, J., Cavalli, L. L., and Lederberg, E., 1952, Sex compatibility in *Escherichia coli.* Genetics **37**:720-730.

Lederberg, J., Lederberg, E., Zinder, N. D., and Lively, E., 1951, Recombination analysis of bacterial heredity. Cold Spr. Harb. Symp. Quant. Biol. **16**:413-441.

Leidy, G., Hahn, E., and Alexander, H., 1953, *In vitro* production of new types of *Hemophilus influenzae.* J. Exp. Med. **97**:467-482.

Lewis, E. B., 1951, Pseudoallelism and gene evolution. Cold Spr. Harb. Symp. Quant. Biol. **16**:159-172.

McCarty, M., 1946, Chemical nature and biological specificity of the substance inducing transformations of pneumococcal types. Bact. Rev. **10**:63-71.

Newcombe, H. B., and Nyholm, M. H., 1950, The inheritance of streptomycin resistance and dependence in crosses of *Escherichia coli.* Genetics **35**:603-611.

Robinow, C. F., 1945, Addendum: Nuclear apparatus and cell structure of rod-shaped bacteria. In: "The Bacterial Cell," by R. J. Dubos, Cambridge, Mass., Harvard Univ. Press, pp. 355-377.

Stadler, L. J., 1951, Spontaneous mutation in maize. Cold Spr. Harb. Symp. Quant. Biol. **16**:49-63.

Stocker, B. D., Zinder, N. D., and Lederberg, J., Transduction of flagellar characters in *Salmonella.* In press.

Tatum, E. L., and Lederberg, J., 1947, Gene recombination in the bacterium *Escherichia coli.* J. Bact. **53**:673-684.

Taylor, H. E., 1949, Transformations reciproques des formes R et ER chez le pneumocoque. C. R. Acad. Sci. **228**:1258-1259.

Wollman, E., 1953, Sur le déterminisime génétique de la lysogénie. Ann. Inst. Pasteur **84**:282-293.

Zamenhof, S., 1952, Newer aspects of the chemistry of nucleic acids. Symp. on Phosphorus Metabolism **2**:301-328.

Zinder, N. D., and Lederberg, J., 1952, Genetic exchange in *Salmonella.* J. Bact. **64**:679-699.

DISCUSSION

Alexander: There is now proof that desoxyribonucleic acid (DNA) can change certain inherited traits in three different bacterial families, pneumococci, *H. influenzae* and meningococci, and the indirect evidence suggests that genetic units of viruses are influenced by a comparable force.

The possibility is therefore raised that a comparable control of inheritance may apply to cells in general. The nature of the reaction which brings about a change in inheritance in bacteria therefore assumes great importance.

Our investigations on highly purified DNA's of *H. influenzae* in collaboration with Dr. Zamenhof suggest that while the DNA itself plays the dynamic role of the gene or hereditary determinant, the expression of the genetic change induced requires a substrate which appears to be present in only a small proportion of the total population exposed. The data now available suggest that the cells susceptible to change by DNA represent specialized cells with functions different from other members.

The *H. influenzae* family is a useful system for studying this subject. Seven hereditary determinants are available for use as genetic markers since each has been shown to induce heritable changes with predictable uniformity. Six of them control the respective specific types of *H. influenzae* a through f. The seventh, the most valuable for purposes of this study, became available when it was demonstrated in our laboratory that resistance to streptomycin could be induced in sensitive *H. influenzae* cells by a single exposure for less than ten minutes to DNA extract of *H. influenzae* cells, highly resistant to streptomycin.

Even though the type specific DNA's are not suitable tools for study of quantitative aspects needed to explore the role of the susceptible cell the following principles have been established by their use:

I. When susceptible cells are brought into contact with active DNA, the reaction which changes heritable traits is an immediate one.

II. Only a very small proportion of the total cells exposed are susceptible to any one of these DNA's.

III. This proportion appears to be characteristic of a species, i.e., the specific type from which a cell originates; the frequency pattern of susceptible cells is a heritable trait and persists after a number of transformations.

Rd (cells derived from type d) shows the highest proportion of susceptible cells, Ra the lowest, and the other types range between the two.

IV. In a given population type, the cells susceptible to the action of type-specific DNA's and the streptomycin resistance DNA show comparable frequency patterns. Therefore, the possibility was raised that the same cell is susceptible to each.

Two groups of experiments were designed to answer this question: Is the same cell susceptible to change by each one of the seven DNA's?

(1) The first examined mixtures of type-specific DNA's (in different proportions of the total DNA) for signs of competition for the same cell.

(2) The second group of experiments examined the capacities of type-specific DNA's to exclude the action of the DNA controlling streptomycin resistance.

The results show that in *H. influenzae* the same cell appears to be susceptible to each of the seven genetic markers and suggest that these cells are equipped for a special function not possessed by the other members.

In an attempt to explore the relationship of the susceptible cell to the other members of the population, the effect on the frequency of the susceptible cell, of growth of the population as a whole, was studied. The data fail to demonstrate evidence of growth of the susceptible cell under circumstances providing optimal growth for other members. When the total population reaches a certain size the cells in which new heritable traits can be induced, emerge and reach peak frequency within a short period; they remain unchanged in incidence for several hours and then slowly disappear.

The results suggest that in *H. influenzae* a number of DNA's induce their heritable changes through the same cells which appear to be specially equipped for this purpose. They exhibit certain traits which are also characteristic of gametes.

11 The syntheses of total macronuclear protein, histone, and DNA during the cell cycle in Euplotes eurystomus

David M. Prescott*
Department of Anatomy
University of Colorado
Denver, Colorado

Abstract. *The syntheses of histone, total protein, and DNA during the cell cycle were measured in the macronucleus of* Euplotes eurystomus *by assaying the incorporation of tritiated amino acids and tritiated thymidine in groups of 800 to 1000 synchronized cells. The synthesis of DNA begins at 30% completion of the cell cycle, proceeds at a constant rate, and ends very shortly before the beginning of macronuclear division. Histone labeling is absent during G_1, begins in phase with DNA synthesis, continues at an unchanging rate during the S phase, and ends with the completion of DNA synthesis. The results support the view that the syntheses of histone and DNA are closely coupled events. Label in total protein accumulates at a constant rate during G_1 and appears to shift to a slightly higher rate when histone synthesis begins. At division, radioactive DNA, histone, and total protein are distributed equally between the daughter macronuclei without loss of radioactivity. Radioautographic analysis showed that protein labeling occurs throughout the macronucleus during the entire life cycle. There was no clear difference in the degree of protein labeling between replicated and unreplicated regions of the macronucleus. The distribution of label suggests that most of macronuclear protein labeling during the cell cycle is concerned with the events of transcription rather than replication.*

From The Journal of Cell Biology **31:**1-9, 1966. Reprinted by permission of the Rockefeller University Press.

This work was supported by the National Science Foundation.

It is a pleasure to acknowledge the expert and conscientious assistance of Mrs. Geraldine Stevens.

*Present address: Institute of Developmental Biology, University of Colorado, Boulder.

INTRODUCTION

Progress through the cell life cycle is probably centered on a continuum of nuclear events, some of which (particularly in G_1) must be modulated in various ways by the state of the cytoplasm. The only two currently identifiable events which normally form part of nuclear cycle continuity are DNA synthesis and nuclear division. There are several fairly obvious gaps in knowledge about sections of the nuclear cycle, for example, the steps leading to initiation and continuation of DNA synthesis, the events which form the causal relation between the end of DNA synthesis, and the segregation of DNA into separate nuclei. The experiments required to provide direct and specific information on these points are difficult to devise, and it is still necessary to proceed with somewhat more oblique approaches. In this vein, the work in this paper concerns the synthesis of total macronuclear protein, histone, and DNA during the cell cycle in *Euplotes eurystomus,* the turnover of total nuclear protein and histone during several cell cycles.

Euplotes has proven especially suited for this type of study because the cells and macronuclei are large, synchronization of a thousand cells is easily accomplished by selection of dividers, the progress and site of DNA synthesis are marked cytologically by the replication bands, and the macronuclei can be isolated and assayed individually or in small groups.

METHODS AND MATERIALS

Euplotes was cultured at 29°C on *Aerobacter aerogenes* grown in Cerophyll (300 mg/liter) (Cerophyll Laboratories, Inc., Kansas City, Missouri) plus *Tetrahymena.* The generation time of *Euplotes* under such conditions is about 12 hr.

To study nuclear synthesis over the complete cycle, 400 to 500 cells in a late stage of division were picked out of a single, log phase culture with a braking pipette. Such selection requires about 20 min and the synchrony of this population is usually extremely good. Four hours after division, 95% of the cells enter DNA synthesis over a 15 min period, and 100% of the cells begin synthesis over a 30 min period. At the next division all the cells divide between approximately 11 and 13 hr with a major peak of division usually close to 12 hr.

Macronuclear protein synthesis was followed by measuring the incorporation of tritiated amino acids with time in synchronized groups of cells. The radioactivity was introduced by feeding *Tetrahymena* which had been cultured on a mixture of tritiated leucine, arginine, lysine, phenylalanine, tryptophan, alanine, histidine, and isoleucine. All tritiated amino acids were the L form with specific activities ranging between 0.5 and 20 c/mmole. All were obtained from various commercial sources.

DNA synthesis in the macronucleus was followed over a complete cell cycle by feeding the *Euplotes* on *Tetrahymena* previously labeled very heavily with tritiated thymidine (15 c/mmole).

Incorporated radioactivity was measured in groups of 30 to 60 isolated macronuclei using a windowless gas flow counter. The macronuclei were isolated free of cytoplasm individually in a solution of a nonionic detergent, triton X-100 plus spermidine, washed in 70% alcohol, and air-dried on a cover glass. The isolation technique has been described elsewhere (1). The air dried nuclei were washed with an acetone-ether mixture (3:1) at 0°C. Acid-soluble proteins were extracted by a 3 hr treatment with 100 μliter of 0.1 N H_2SO_4 at 0°C. The extract was neutralized with NaOH, air-dried on a planchet, spread with formic acid, and assayed for radioactivity. The extraction procedure was tested on isolated macronuclei in which the proteins were labeled with tritiated tryptophan only. Consistently, 1 to 2% of the incorporated tryptophan was soluble in acid. This is accepted as reflecting the level of nonhistone protein removed by acid. After a full cell cycle of feeding with mixed tritiated amino acids, about 15% of the incorporated radioactivity is acid soluble, but with the tritiated tryptophan the amount remains at 1 to 2%. This suggests that about 10% of the protein removed by the sulfuric acid is nonhistone. The amount of tryptophan-containing protein removed by 0.1 N HCl was much higher, and the H_2SO_4 procedure was therefore adopted.

The amount of nonhistone protein in acid extracts can, in some situations, be reduced by prior extraction of the material with 0.14 M NaCl (see reference 2). A 12 hr treatment of isolated macronuclei of *Euplotes* with 0.14 M NaCl at 0°C did not reduce the amount of acid-soluble, tryptophan-containing protein, so this step was omitted.

To reduce self absorption, the macronuclei were finally dispersed into a more-or-less even film on the cover glass with a few drops of formic acid. The cover glasses were placed in planchets and assayed in a windowless gas flow counter.

Additional groups of macronuclei were radioautographed in order to determine the relationship between the sites of protein and DNA syntheses over the cycle. For radioautography of tritiated amino acid labeling, the isolated macronuclei were rinsed in 70 and 100% alcohol and air-dried. The nuclei labeled with tritiated thymidine were washed in 1 N HCl for 5 min at 25°C to remove unincorporated radioactivity, rinsed in 100% alcohol several times and air-dried. The radioautography was done with NTB3 emulsion by a conventional liquid emulsion technique (3).

RESULTS

Synthesis of DNA

In the macronucleus of *Euplotes,* DNA synthesis occurs in two waves, one wave originating at each end of the elongated nucleus. The positions of the two waves of synthesis are marked cytologically by the presence of the two replication bands in which DNA synthesis is localized (4, 5). The series of radioautographs in Fig. 1 shows, with continuous tritiated thymidine labeling, the progress of DNA synthesis through the macronucleus. No labeling occurs in the region between the two replication bands, but a trail of intense labeling occurs in the nuclear region behind each band. The bands ultimately meet and fuse at the center of the nucleus. At this time the macronucleus has already begun to shorten in preparation for amitotic splitting. A measurable G_2 period is absent unless it be considered as the period of amitotic splitting.

The incorporation of tritiated thymidine was measured in groups of 30 to 60 macronuclei isolated from synchronized cells at many points of the cycle. The results are recorded in Figs. 2 and 3. DNA synthesis begins at about 30% completion of the cycle and continues until a few minutes before cytokinesis. These results agree precisely with the cytological and radioautographic detection of the synthesis period. The rate of synthesis shows no deviation from a linear course. Such constancy in the rate of DNA synthesis has previously been described for *Paramecium* (6). Cytological observations of the replication bands in *Euplotes* gave the impression that the rate of DNA synthesis increased toward the end of S, but this proves not to be so. The observations of Ringertz and Hoskins (7) suggest that, under conditions of slower growth, DNA synthesis accelerates toward the end of the S phase. At cell division the DNA is distributed, by amitosis, to the two daughter nuclei with approximate equality. Two sets of data are given (Figs. 2 and 3) in order to give a measure of the consistency of the results.

Synthesis of total macronuclear protein

The incorporation of tritiated amino acids into total macronuclear protein is shown in Figs. 2 and 3. The incorporation during G_1 is

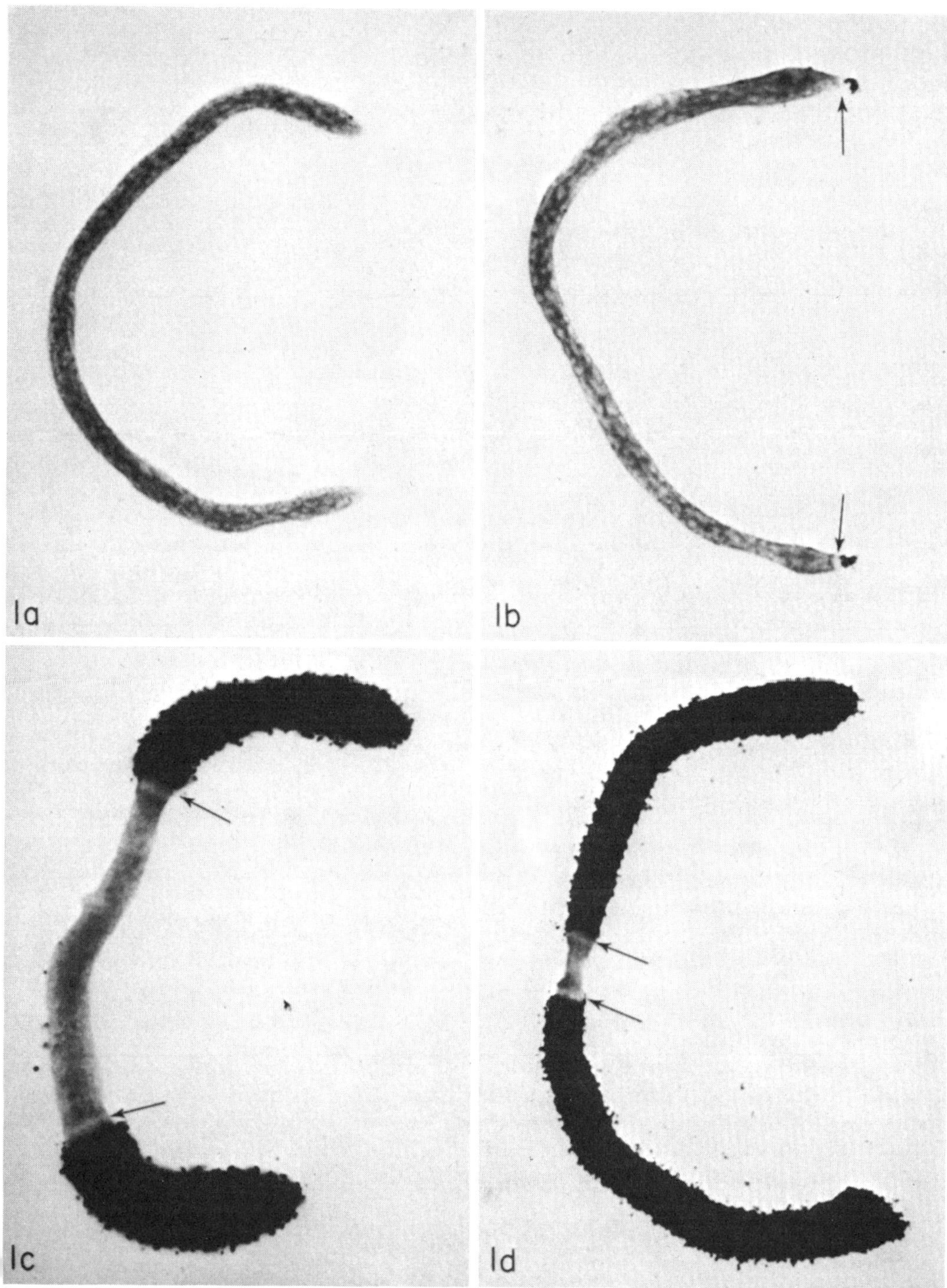

Fig. 1. A series of radioautographs of isolated macronucleus of *Euplotes* labeled with tritiated thymidine. Arrows ndicate DNA replication bands. **1a,** A macronucleus from a cell 25% through the cell cycle and still in the G_1 stage. The replication bands have not yet appeared at the tips of the nucleus, and no thymidine incorporation has occurred. **1b,** A macronucleus 30% through the cell cycle; the radioautograph shows the first few minutes of DNA synthesis (arrows). **1c,** A macronucleus about 50% through DNA synthesis: about 65% cycle completion. **1d,** A macronucleus in which the two bands have almost met in the center. DNA synthesis is almost finished: about 90% cycle completion. The exposure time for these radioautographs was 12 hr. Actual length of macronuclei is about 140 μ. X 1000.

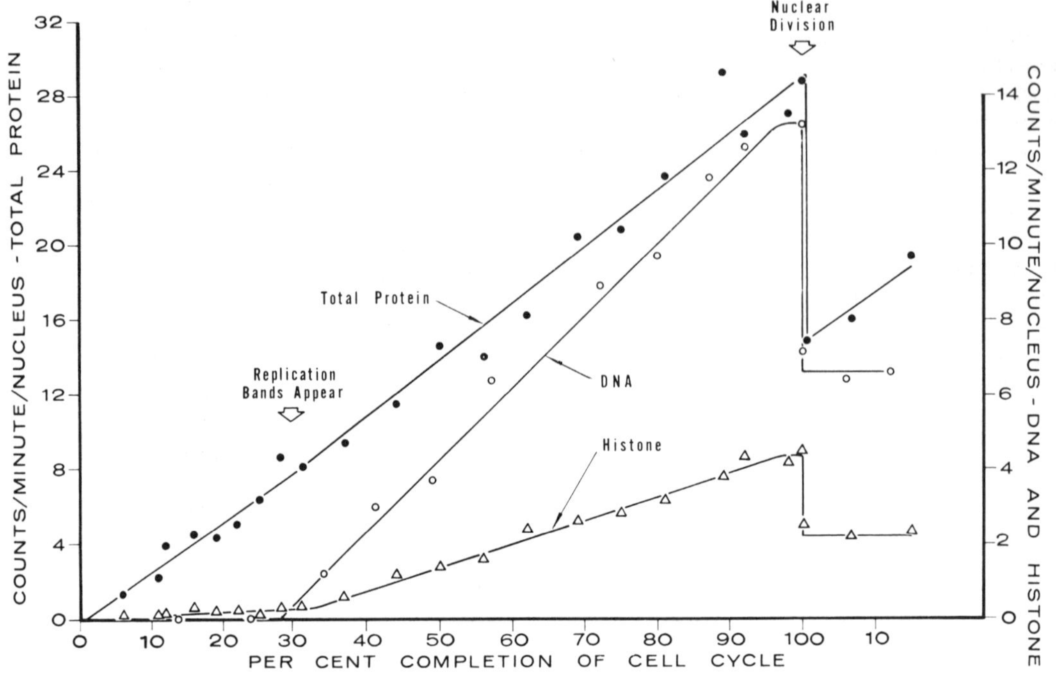

Fig. 2. Incorporation of tritiated thymidine into DNA and of tritiated amino acids into total protein and histone in the macronucleus of *Euplotes* during the full cell life cycle. Each point is the mean count for 30 to 60 nuclei.

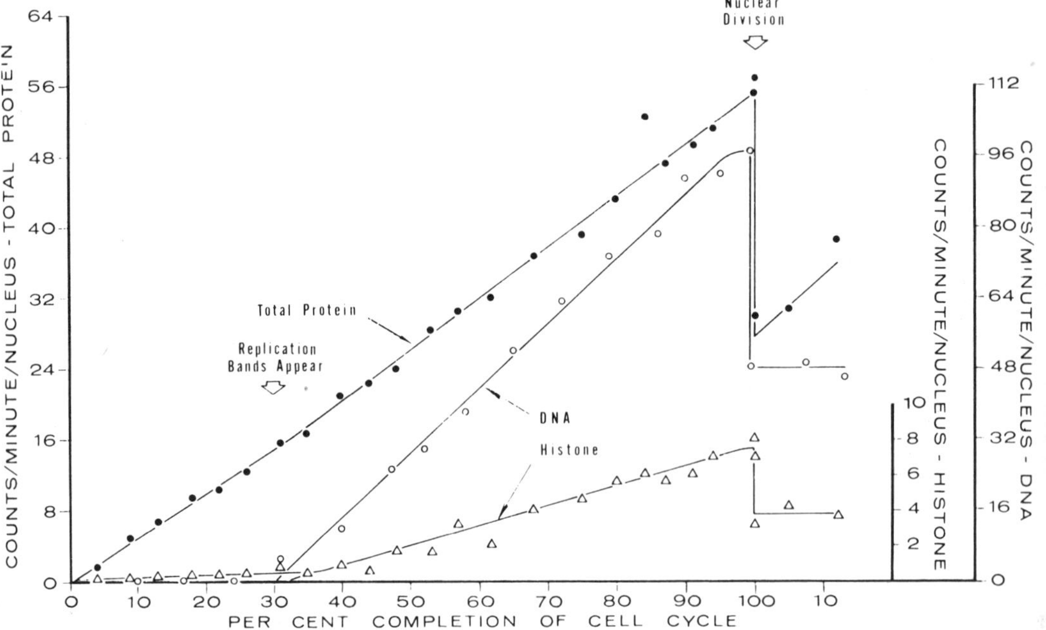

Fig. 3. This is a repeat of the experiment in Fig. 2. The average incorporation per nucleus of tritiated thymidine into DNA and of tritiated amino acids into total protein and histone are increased over the values in Fig. 2 because a longer labeling period was used for the food organisms *(Tetrahymena)*.

linear, indicating no major changes in the rate of synthesis. At the beginning of S there may be a small increase in rate, and the curves in Figs. 2 and 3 have been drawn with a slight upward shift in slope. Such a small increase might be expected at the G_1 to S transition, either because histone begins to accumulate at this time or because the nucleus may require addition of new protein to support DNA synthesis, or both. The addition of radioactive protein to the macronucleus during S follows a linear course, as it does in G_1. At division, total nuclear protein is divided equally between the two daughter macronuclei. The pattern of increase in labeled protein is somewhat similar to the course of increase of macronuclear proteins measured microspectrophotometrically in *Paramecium* (6). The primary difference is a rapid increase in rate of protein increase just before division in *Paramecium*.

The radioautographs in Fig. 4 show the distribution of accumulated radioactive protein at two points in the cycle. Labeled protein occurs generally throughout the nucleus during G_1 and S. There is no obvious difference in the concentration of radioactivity between the replicated and the nonreplicated regions of the macronucleus. It would be reasonable to expect a higher concentration of radioactive protein in the region behind the replication band because of addition of new (radioactive) histone at the band. Since histone contributes only 15% of the total nuclear protein in this cell, however, its contribution would not be easy to detect radioautographically against the background of accumulated G_1 labeling.

Synthesis of histone

A relatively small amount of tritiated amino acids is accumulated into acid-extractable material during G_1 (Figs. 2 and 3). This is believed to represent nonhistone protein because the percentage (1 to 2%) of the total radioactivity that is soluble in 0.1 N H_2SO_4 during G_1 is the same whether labeling is achieved with a mixture of tritiated amino acids or with tritiated tryptophan alone. Tryptophan has been found to be absent from histones of all

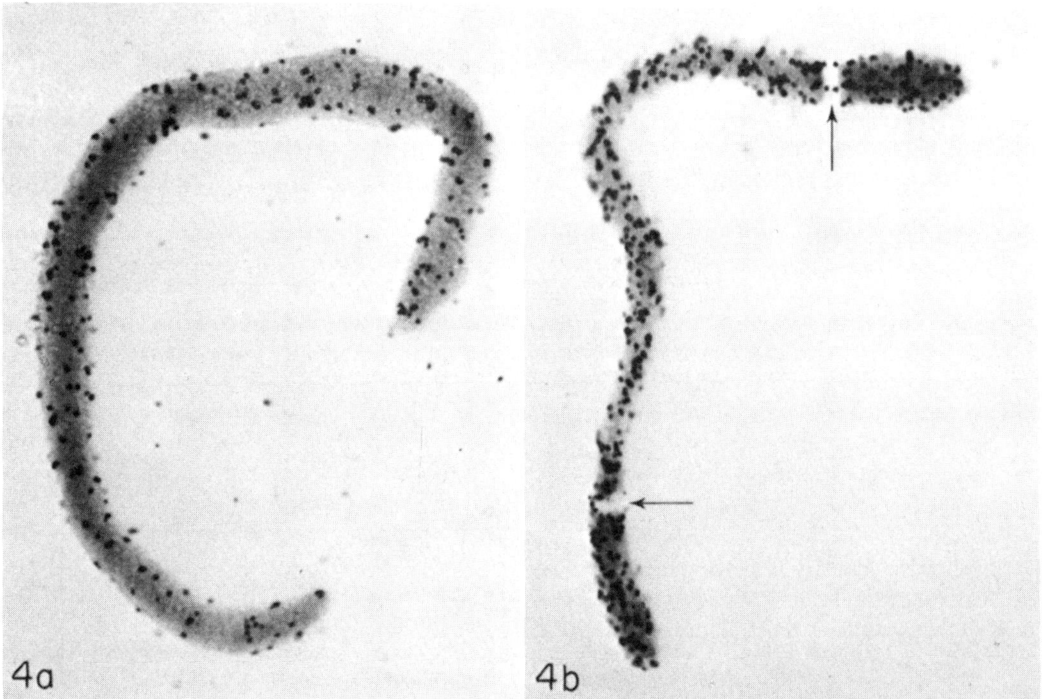

Fig. 4. Radioautographs of macronuclei of *Euplotes* labeled continuously from the previous division with tritiated amino acids. **4a,** A macronucleus 25% through the cell cycle and still in the G_1 stage. **4b,** A macronucleus about 30% through DNA synthesis: about 45% cycle completion. Arrows indicate DNA replication bands. The nucleus is generally labeled with no clear difference in concentration of radioactivity between the replicated and nonreplicated regions. The exposure time for these radioautographs was 6 hr. Actual length of macronuclei is about 140μ. × 1000.

cell types so far examined (see reference 2). The acid-soluble radioactivity of G_1 is not extracted by 20% trichloroacetic acid, and therefore does not represent amino acid pool material.

At the G_1 to S transition the rate of incorporation of tritiated amino acids into acid-soluble protein increases sharply. No increase occurs with tritiated tryptophan labeling. The acid-soluble radioactivity is considered to be primarily histone, and the course of this histone labeling is shown in Figs. 2 and 3. As already discussed, the small incorporation into acid-soluble protein in G_1 is interpreted as a contaminant protein, and nuclear histone labeling is concluded to be absent during G_1. At the beginning of DNA synthesis, labeled histone begins to appear in the nucleus. The accumulation of radioactive histone continues at a constant rate, concurrently with DNA synthesis, and continues up to the time of division. At division, histone is segregated equally between the two daughter nuclei. The curves in Figs. 2 and 3 show that the amount of radioactivity in histone remains constant as the cells start through the next G_1 period. Histone and DNA accumulation, therefore, each proceed with linear courses over the same interval of the cell cycle.

The fate of macronuclear protein during growth and division

Euplotes were labeled with tritiated amino acids for one full cycle and at division were washed and cultured on nonradioactive *Tetrahymena.* At intervals over the next two cell cycles, macronuclei were isolated and total labeled protein and labeled histone were determined. The results are plotted in Fig. 5. During roughly the first 25% of the next cycle (first 3 hr of G_1) the radioactivity in total nuclear protein continues to increase. This represents the time required to use up the tritiated amino acids previously ingested. Very shortly before DNA synthesis begins, the amount of radioactivity in total protein ceases to rise. During the

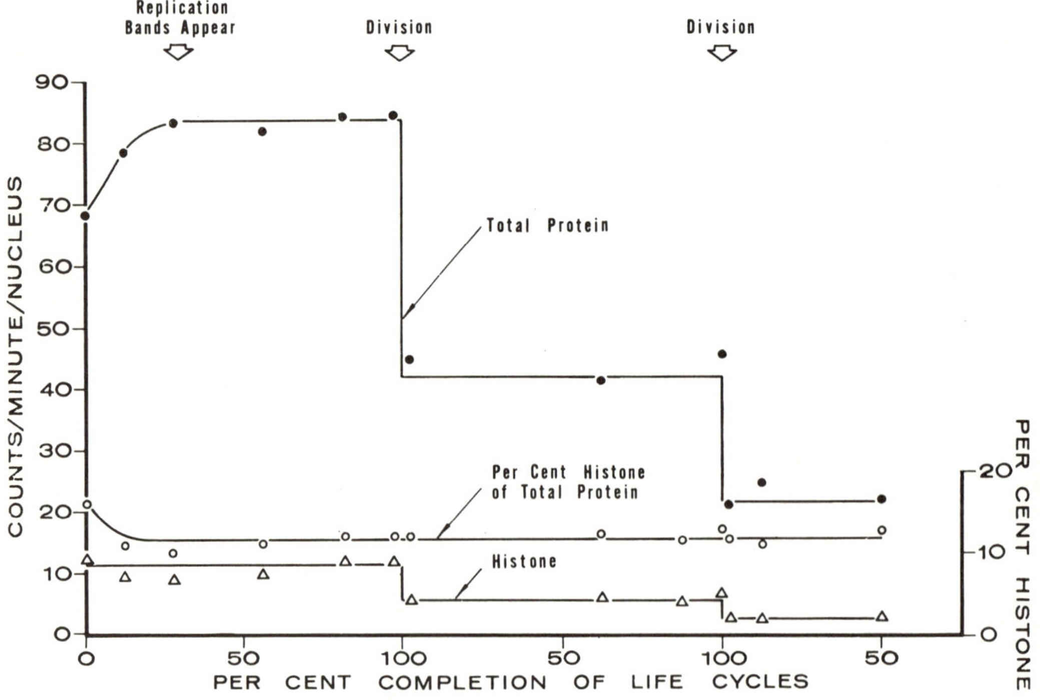

Fig. 5. Fate of radioactivity in macronuclear protein of *Euplotes* during growth and division on cold nutrient. Cells were labeled for one cell cycle by feeding *Tetrahymena* cultured on tritiated amino acids. At time zero, the dividing *Euplotes* were washed and the daughter cells cultured on nonradioactive food. Radioactivity in total protein continues to rise until preingested radioactive food is exhausted. No rise occurs in radioactivity in histone, and the amount of total labeled protein and labeled histone per nucleus is halved at each of the next two nuclear divisions. Each point is the mean for 30 to 60 nuclei.

remainder of the cycle no change occurs in total labeled protein, and the radioactivity is divided equally between the two daughter nuclei at division. At this division, synchrony was reimposed by discarding those cells which did not divide over the 30 min period occupied by the majority of divisions. In the next cycle the lack of any decrease in total protein radioactivity suggests conservation of nuclear protein. The labeled protein per nucleus was reduced to one-half with the next division.

Histone radioactivity remained unchanged during the initial G_1 period when the previously ingested radioactive *Tetrahymena* were being used up. When DNA synthesis began, there was still no significant change in the absolute amount of radioactive histone. An increase would be expected if histones had been synthesized in the cytoplasm during G_1 and then transferred, in any substantial amount, to the nucleus during subsequent DNA synthesis. During the remainder of the experiment the amount of histone labeling did not decrease, except for halving at each of the two subsequent divisions.

DISCUSSION AND CONCLUSIONS

The specific functions of most nuclear proteins are still unknown, limiting the possibilities for relating in any precisely detailed manner, the macronuclear protein labeling over the cell cycle to the two known macronuclear functions of genetic replication and transcription. In addition, interpretations of the data must take into account the fact that labeling of nuclear proteins with tritiated amino acids will detect both turnover and net increase of protein. Turnover may be composed partially of protein breakdown and resynthesis and may also involve an exchange of unlabeled proteins of the nucleus for newly synthesized cytoplasmic proteins. Continuous shifting of protein between nucleus and cytoplasm during interphase has, for example, been demonstrated decisively in ameba by Goldstein (8), and a large fraction of total nuclear protein is involved in this shuttling movement. At least some of this protein is synthesized in the cytoplasm (9) and migrates into the nucleus. In these experiments on *Euplotes* there is no way of assessing how much of the macronuclear protein is synthesized in the cytoplasm.

It does seem likely that a substantial part of the increase in radioactivity in proteins of the *Euplotes* macronucleus over the cycle represents a net increase since the total protein content of the macronucleus doubles with each cell cycle. It could be reasonably assumed that the labeling of proteins during the S phase represents addition of new enzymes associated with DNA synthesis or new proteins necessary to cope with the increased potential for RNA synthesis stemming from DNA increase. Preliminary measurements on *Euplotes,* however, have failed to show any change in the rate of RNA labeling in the nucleus (D. Evenson, unpublished), and this interpretation cannot be invoked. Nevertheless, the synthesis of messenger RNA, even at an unchanging rate, might be accompanied by nuclear protein turnover, although the reasons why this should be so cannot be specified at this point. In addition, the synthesis of ribosomal RNA may require addition of new ribosomal protein to the nucleus. Such activities could conceivably account for a major part of labeling of nuclear protein in both the G_1 and the S phase. Part of the S phase protein labeling is clearly due to histone addition, but histone accounts for only 15% of the total macronuclear protein in *Euplotes.*

Presumably, the progress of the macronucleus through G_1 involves qualitative and/or quantitative changes in the macronucleus; some of the protein labeling may be a part of these changes. Advancement out of G_1 into DNA synthesis apparently requires protein synthesis (10, 11), and certainly this protein synthesis must in some degree be associated with the nucleus.

The radioautographs in Fig. 4 show that protein labeling over the cycle occurs generally throughout the macronucleus. Although Gall (4) has shown that histone increase occurs strictly at the site of DNA synthesis (at the replication band), and short pulse labeling with tritiated histidine results in greater radioactivity at the band (5), there is no indication in Fig. 4 of a pronounced increase in protein labeling at the replication site. Since histone makes up only 15% of the total nuclear protein, and since any increase, in the form of histone labeling, takes place in these experiments against a background of extensive G_1 labeling of proteins, histone would not be expected to produce much intensification of the radioautograph at the replication band. The general distribution of labeling throughout the macronucleus in the experiment described here (Fig. 4) might be interpreted as evidence that the bulk of the new protein is associated in various ways with genetic transcription rather than genetic replication.

The histone labeling experiments can be interpreted in a more straightforward manner. Since the paper of Bloch and Godman (12), there have been a number of confirming reports (for example, references 4, 13, 14) that net increase of nuclear histone during the cell cycle occurs only in parallel with DNA synthesis. None of this work, including the original paper of Bloch and Godman, allows a decision about the site of histone synthesis (cytoplasm versus nucleus) or the time of histone synthesis. These reports do not eliminate the formal hypothesis that some (or even all) histone is synthesized in the cytoplasm during G_1, to enter the nucleus only during S. Bloch and Brack (15) have, more recently, published evidence that special histone is synthesized in the cytoplasm (in the absence of DNA synthesis) and moves into the nucleus during differentiation of sperm cells in the grasshopper. This is one of several reports (16-18) that histone synthesis can be uncoupled from DNA synthesis, but the synthesis of DNA without concomitant histone increase has not been convincingly demonstrated, except of course in systems (bacteria and viruses) which normally lack DNA-associated histones.

The curves in Figs. 2 and 3 confirm that histone is added to the nucleus only during S and show, in addition, that there is at most only a trace of nuclear histone labeling during G_1. Robbins, Borun, and Maizel (19) have obtained very similar results on synchronized HeLa cells. Histone labeling is absent in nuclei during G_1 and begins with the initiation of DNA synthesis. In the *Euplotes* experiments, not only do the periods of macronuclear DNA and histone labeling coincide but both macromolecules are added to the macronucleus at constant rates, suggesting close temporal coordination of the processes.

The experiment in Fig. 5 provides evidence that the histone that is added to the macronucleus during the S phase is synthesized during S itself. Total cell proteins were labeled through one full cell cycle and up to the end of the G_1 phase of the next cycle (first cycle shown in Fig. 5). If any significant amount of histone, destined for the nucleus, had been synthesized and retained in the cytoplasm during the previous cycle or during the G_1 period in question, an increase in macronuclear histone labeling would have been expected during the first S period on nonradioactive food. No increase was detectable, and it is concluded that the addition of histone to the nucleus during S is closely linked to the event of histone synthesis. This conclusion is also supported by previously reported work (5) on histidine labeling at the replication band. These findings indicate that the regulation of histone *synthesis* during the cell cycle must be closely coupled to the synthesis of DNA. There is no evidence yet to indicate whether this regulation of synthesis is at the level of transcription of DNA or translation of RNA into protein.

These experiments on *Euplotes* have no direct bearing on the question of a nuclear versus a cytoplasmic site of histone synthesis. The most pertinent data on the point are the description of cytoplasmic synthesis during spermiogenesis (15) and the report of histone synthesis in isolated nuclei (20) and in the nucleolus (21), but the problem needs more study.

The data in Fig. 5 for labeled macronuclear proteins beyond the first G_1 period show that total protein and histone are divided equally between daughter nuclei. There is no indication of nuclear protein turnover, but this probably would not be detectable in this type of experiment. Chalkley and Maurer (18) have reported that some histone undergoes turnover in the absence of DNA synthesis but lysine-rich histone becomes labeled only in concert with DNA synthesis.

Finally, as a general conclusion, although the data on *Euplotes* are a step toward describing nuclear progress during the cell cycle, an understanding of the cause and effect continuum of the nuclear cycle will obviously require much more refined and qualitative data on the progressive changes in those proteins involved in DNA replication and transcription and in their regulation.

REFERENCES

1. Prescott, D. M., Rao, M. V. N., Evenson, D. P., Stone, G. E., and Thrasher, J. D., *in* Methods in Cell Physiology, (D. M. Prescott, editor), New York, Academic Press Inc., **2**, 1966.
2. Murray, K., Ann. Rev. Biochem., 1965, **34**:209.
3. Prescott, D. M., *in* Methods in Cell Physiology, (D. M. Prescott, editor) New York, Academic Press Inc., **1**, 1964.
4. Gall, J. G., J. Biophysic, and Biochem. Cytol., 1959, **5**:295.
5. Prescott, D. M., and Kimball, R. F., Proc. Nat. Acad. Sc., 1961, **47**:686.
6. Woodard, J., Gelber, B., and Swift, H., Exp. Cell Research, 1961, **23**:258.

7. Ringertz, N. R., and Hoskins, G. C., Exp. Cell Research, 1965, **38:**160.
8. Goldstein, L., Exp. Cell Research, 1958, **15:**635.
9. Byers, T. J., Platt, D. B., and Goldstein, L., J. Cell Biol., 1963, **19:**467.
10. Lieberman, I., Abrams, R., Hunt, N., and Ove, P., J. Biol. Chem., 1963, **238:**3955.
11. Maaløe, O., J. Cell Comp. Physiol., 1963, **62,** suppl. 1:31.
12. Bloch, D. P., and Godman, G. C., J. Biophysic. and Biochem. Cytol., 1955, **1:**17.
13. Alfert, M., Exp. Cell Research, 1958, suppl. 6, 227.
14. Niehaus, W. G., and Barnum, C. P., Exp. Cell Research, 1965, **39:**435.
15. Bloch, D. P., and Brack, S. D., J. Cell Biol., 1964, **22:**327.
16. Flamm, W. G., and Birnstiel, M. L., Exp. Cell Research, 1964, **33:**616.
17. Ontko, J. A., and Moorehead, W. R., Biochim. et Biophysica Acta, 1964, **91:**658.
18. Chalkley, G. R., and Maurer, H. R., Proc. Nat. Acad. Sc., 1965, **54:**498.
19. Robbins, E., Barun, T. W., and Maizel, J. V., Fed. Proc., 1966, **25:**439.
20. Reid, B. R., and Cole, R. D., Proc. Nat. Acad. Sc., 1964, **51:**1044.
21. Birnstiel, M. L., and Flamm, W. G., Science, 1964, **145:**1435.

chapter 3

Fundamentals of genetics

MENDELIAN PRINCIPLES

In relatively few instances is the origin of a branch of science so firmly linked to an individual and a date as is that of genetics. The individual is Gregor Mendel and the date is 1900. In that year Correns, de Vries, and Tschermak reported the results of their independent experiments, each of which served to confirm Mendel's earlier (1865) discoveries. A collection of the pertinent publications relating to Mendel's experiments was edited by Stern and Sherwood in 1966 (Ref. 3-1). A translation of Correns's paper is reprinted in this chapter (Ref. 3-2). The conclusions that Mendel drew from his experiments can be summarized as follows:

1. An individual receives from each parent a "unit factor" (gene) that can determine a trait of the individual.
2. When genes from the two parents cause different effects, the individual will resemble one parent but not the other *(principle of dominance).*
3. In the formation of gametes, the paternal gene for each characteristic segregates from the maternal gene for that same characteristic *(principle of segregation).*
4. The distribution of segregated genes in the gametes is random with respect to one another *(principle of independent assortment).*

As we shall see, later research demonstrated that the inheritance of many characteristics follows a more complicated model than that proposed by Mendel.

It is interesting to note that Mendel's original report in 1865 went virtually unnoticed and completely unappreciated. Yet 35 years later, its confirmation brought an immediate and warm recognition of the importance of these principles in explaining the mode of inheritance of certain traits and the implications of these rules of inheritance for further research. The proof that the Mendelian principles do in fact have general application was produced independently by many research workers. An example of this type of study, in man, was reported by Garrod in 1902 (Ref. 3-3) whose paper on alkaptonuria is included in this collection. There is a continuing study of the genetic basis of observed variations in organisms. A more recent discovery of an inherited human defect was reported by Rimoin and his co-workers in 1966 (Ref. 3-4); this report is reprinted in this chapter.

The rediscovery of Mendel's findings in 1900 stimulated a search to identify the cell structure that contained the hereditary material. This was accomplished in 1903 by Sutton (Ref. 3-5), who presented evidence that the chromosomes were the carriers of the hereditary material. His studies on meiosis demonstrated the parallelism between the behavior of the postulated Mendelian "unit factors," or genes, and the directly observable nuclear components, the chromosomes.

The name *Genetics* was given to this branch of biology by Bateson in 1906 in an address to the Third Conference on Hybridization and Plant Breeding whose proceedings were published in 1907 (Ref. 3-6). In his address Bateson said:

> I suggest for the consideration of this congress the term Genetics, which sufficiently indicates that our labours are devoted to the elucidation of the phenomena of heredity and variation: in other words to the physiology of descent, with implied bearing on the theoretical problems of the evolutionist and the systematist, and application to the practical problems of the breeder, whether of animals or of plants.*

EXCEPTIONS TO MENDEL'S RULES OF INHERITANCE

Although the rules of inheritance proposed by Mendel apply in many cases, exceptions to them appeared immediately. Correns himself, in announcing his verification of Mendel's findings (Ref. 3-2), was quick to point out that *dominance* did not occur in every case of contrasting characteristics. He showed that a

*From Bateson, W. 1907. Report of the Third International Conference, 1906, on Hybridization and Plant Breeding. Royal Horticultural Society, London. Inaugural address, pp. 90-97.

blending type of inheritance occurred in many crosses and one could distinguish the hybrid from each of the parental types. Another exception to dominance was found to occur in the inheritance of human blood group genes, *M* and *N*. In this case, Landsteiner and Levine reported in 1927 (Ref. 3-7) that the red blood cells of a hybrid exhibited both M and N antigenic properties. The phenomenon in which both alleles are expressed in the hybrid is called *codominance* and has been found to occur in many instances.

Mendel's discovery of contrasting unit factors (alleles) led to the formulation of a number of postulates, each of which, in turn, was found to have many exceptions. One such concept was that a single gene could have only two alternate forms. Breeding experiments with mice reported by Cuénot in 1905 (Ref. 3-8) demonstrated that a mutant gene for yellow coat color was allelic with both agouti and nonagouti genes. A general discussion of multiple alleles was presented by Sturtevant in 1913 (Ref. 3-9) and is reprinted in this collection. Another implication of Mendel's report was that each trait of an individual was decided by a single pair of genes. This concept was also found to have many exceptions. In a series of papers by Bateson and his co-workers, it was demonstrated that such characteristics as comb shape in poultry and flower color in sweet peas were the result of nonallelic gene interaction. These experiments have been summarized in a collection of Bateson's papers edited by Punnett in 1928 (Ref. 3-10). An important instance of nonallelic gene interaction was reported by Cuénot in 1903 (Ref. 3-11), who showed that albino mice could transmit nonallelic coat color genes to their offspring. This demonstrated that the expression of one gene could be suppressed by the action of a different and independent gene. The phenomenon was called *epistasis* by Bateson and has been found to operate in many genetic traits.

The application of the principle of *independent assortment* to all genes came into question very soon after the rediscovery of Mendel's findings. Evidence for linkage between two coat-color genes in the house mouse was presented by Darbishire in 1904 (Ref. 3-12). This discovery was followed by reports of other examples of linkage, and it became generally accepted that the principle of independent assortment could only apply to genes in different chromosomes. We shall return to a more detailed consideration of this topic in Chapter 5.

Of the three rules of inheritance postulated by Mendel, only the principle of *segregation* of paternal from maternal genes has been shown to be almost universal. However, exceptions are known to have occurred here, too. In these exceptional cases one finds a nonseparation of one or more pairs of homologous chromosomes in meiosis. The phenomenon is called *nondisjunction* and results in the formation of gametes either with an excess or a deficient number of chromosomes. The occurrence of nondisjunction in *Drosophila* was first reported by Bridges in 1916 (Ref. 3-13). The first report of nondisjunction in man was made by Lejeune and his co-workers in 1959 (Ref. 3-14) and involved children suffering from Down's syndrome (mongolism). We shall consider this subject in greater detail in Chapter 7.

In concluding this discussion of Mendel and his work, it is worth noting that despite the many exceptions found to his rules of inheritance, both the man and his discoveries remain the cornerstones of the science of genetics. The reasons for this are to be found in the impact that his findings had on the thinking and research efforts of the biologists at the beginning of the twentieth century. His report, though belatedly recognized, stimulated some of the best minds of the day to turn their energies to the elucidation of the factors involved in heredity. It is also a tribute to the man and his work to realize that the type of research he did continues to the present day, aided by our sophisticated technology.

REFERENCES

Mendelian principles

3-1. Stern, C., and Sherwood, E. R. (Eds.). 1966. The origin of genetics. A Mendel source book. W. H. Freeman & Co., San Francisco.

3-2. Correns, C. 1900. G. Mendel's Regel über das Verhalten der Nachkommenschaft der Rassenbastarde. Berichte der deutschen botanischen Gesellschaft **18:**158-168. (Translated by Leonie Kellen Piternick. 1950. Genetics **35** (5, pt. 2):33-41.)

3-3. Garrod, A. E. 1902. The incidence of alkaptonuria: A study in chemical individuality. Lancet **ii:**1616-1620.

3-4. Rimoin, D. L., Merimee, T. J., and McKusick, V. A. 1966. Growth-hormone deficiency in man: An isolated, recessively inherited defect. Science **152:**1635-1637.

3-5. Sutton, W. S. 1903. The chromosomes in heredity. Biol. Bull. **4:**231-251.

3-6. Bateson, W. 1907. Report of the Third International Conference, 1906, on Hybridization and

Plant Breeding. Royal Horticultural Society, London. Inaugural address. pp. 90-97.

Exceptions to Mendel's rules of inheritance

3-7. Landsteiner, K., and Levine, P. 1927. Further observations on individual differences of human blood. Proc. Soc. Exptl. Biol. Med. **24**:941-942.

3-8. Cuénot, L. 1905. Les races pures et les combinaisons chez les souris. Arch. Zool. Exp. et Génet. **3**:123-132.

3-9. Sturtevant, A. H. 1913. The Himalayan rabbit case, with some considerations on multiple allelomorphs. Am. Nat. **47**:234-239.

3-10. Punnett, R. C. (Ed.). 1928. The scientific papers of William Bateson. 2 vols. Cambridge University Press, London.

3-11. Cuénot, L. 1903. L'hérédité de la pigmentation chez les souris. Arch. Zool. Exp. et Génet. **2**:33-38.

3-12. Darbishire, A. D. 1904. On the result of crossing Japanese waltzing with albino mice. Biometrika **3**:1-51.

3-13. Bridges, C. B. 1916. Nondisjunction as proof of the chromosome theory of heredity. Genetics **1**:1-51, 107-163.

3-14. Lejeune, J., Gautier, M., and Turpin, R. 1959. Étude des chromosomes somatiques de neuf enfants mongoliens. Compt. Rend. Acad. Sci. Paris **248**:1721-1722. (Translated *in* S. H. Boyer, IV [ed.] Papers on human genetics. 1963. Prentice-Hall, Inc., Englewood Cliffs, N. J.)

12 G. Mendel's law concerning the behavior of progeny of varietal hybrids

Carl Correns

Translated by
Leonie K. Piternick*
Department of Zoology
University of California
Berkeley, California

The latest publication of Hugo de Vries: Sur la loi de disjonction des hybrides[1] which through the courtesy of the author reached me yesterday, prompts me to make the following statement:

In my hybridization experiments with varieties of maize and peas, I have come to the same results as de Vries, who experimented with varieties of many different kinds of plants, among them two varieties of maize. When I discovered the regularity of the phenomena, and the explanation thereof—to which I shall return presently—the same thing happened to me which now seems to be happening to de Vries: I thought that I had found *something new.*[2] *But then I convinced myself that the Abbot Gregor Mendel in Brünn, had, during the sixties, not only obtained the same result through extensive experiments with peas, which lasted for many years, as did de Vries and I, but had also given exactly the same explanation, as far as that was possible in 1866.*[3] Today one has only to substitute "egg cell" or "egg nucleus" for "germinal cell" or "germinal vesicle" and perhaps "generative nucleus" for "pollen cell." An identical result was obtained by Mendel in several experiments with Phaseolus, and thus he suspected that the rules found might be applicable in many cases.

Mendel's paper, which although mentioned, is not properly appreciated in Focke's "Pflanzenmischlingen," and which otherwise had hardly been noticed, is among the best that has ever been written about hybrids, in spite of some objections which one might raise with respect to matters of secondary importance, e.g. terminology.

At the time I did not consider it necessary to establish my priority for this "re-discovery" by a preliminary note, but rather decided to continue the experiments further.

In the following I shall limit myself to an account of the experiments with varieties of peas.[4] Inter-varietal hybrids of maize show identical behavior in all essential points, but are more difficult to experiment with, and I have not yet elucidated to my satisfaction several points of secondary importance. They will be discussed somewhere else.

Varieties of peas are invaluable for the problem which interests us here, as Mendel had emphasized correctly, since the flowers are not only autogamous, but are very rarely fertilized by insects. On the basis of experiments on the formation of xenia—which, in the case of peas, yielded only negative results—I came across this material. When I realized, that the rules here are much clearer than they are in maize, where I had first discovered them, I continued the observations.

The *characters* which differentiate the varieties of peas, can, as in all other cases, be grouped into *pairs,* each member having an effect on the same trait, one in one and the other in the other one of the varieties, e.g. the color of the cotyledons, the color of the flower, the color of the seed coat, the hilum of the

From Piternick, L. K.: Genetics **35** (pt. 2):33-41, 1950. Used with permission.

Berichte der Deutschen Botanischen Gesellschaft **18**:158-168. Reprinted in: Carl Correns Gesammelte Abhandlungen zur Vererbungswissenschaft aus periodischen Schriften 1899-1924. (Fritz v. Wettstein, ed.) Berlin, Julius Springer, 1924, pp. 9-16.

*Present address: Department of Zoology, University of Washington, Seattle, Washington.

[1] Compt. rend. de l'Acad. des Sciences, (Paris) **130**:1900, 26. mars.

[2] See the postscript. (footnote added later).

[3] Gregor Mendel, Versuche über Pflanzen-Hybriden. Verh. des Naturf. Vereines in Brünn **4.** 1866.

[4] The names of the varieties given in this paper are those which I received from Haage and Schmidt in Erfurt.

seed, etc. In many pairs one character, or rather the anlage thereof is so much stronger than the other character, or its anlage, that the former alone appears in the hybrid plant, while the latter does not show up at all. This one may be called the *dominant,* the other one the *recessive* anlage. Mendel named them in this way, and, by a strange coincidence, de Vries now does likewise. For example the yellow color of the cotyledons is dominant over the green color, and red flower color over white flower color.

I can not understand why de Vries assumes that in all pairs of characters which differentiate two strains, one member must always be dominant.[5] Even in peas, where some characters completely conform to this rule, other character pairs are also known, in which neither character is dominant, as for instance the color of the seed coat, being either reddish-orange or greenish-hyaline.[6] In this case the hybrid may show all transitions (this is true especially for the seed coat of peas), or it shows either more of one or of the other character (for example in hybrids of stocks; here a certain hybrid may be *just barely* distinguished from one parental form by its hardly noticeable slighter covering of hairs, although with some care, separation is always possible, while it is *highly distinct* from the other, i.e., the glabrous parental type).

The following holds only for pairs of characters which have a dominant and a recessive member; there is no reason to believe that it may not hold for other types of pairs of characters as well, but at present we know of no example.[7] Let us first consider a single pair of characters. It is immaterial whether the varieties to be crossed are only differentiated by this one pair, or by others as well. The specific pair of characters we may select is the *color of the embryo, either yellow or green.* It is very easy to obtain large numbers for this trait.

The facts, which Mendel found, I can fully confirm. They also agree with the findings of de Vries for his experimental objects. They are as follows:

1. In the *first* generation, all hybrid individuals are uniform and only the *dominant* character appears. In our special case the cotyledons are yellow.

2. When these seeds with yellow embryos are sown, plants are obtained, whose pods, which were produced by self-fertilization, contain seeds with *yellow* embryos and seeds with *green* embryos (the second generation) and on the average, there are *three* yellow ones for each green *one.* If there are four or more seeds in each pod, one containing a green embryo will usually be among them.

3. When the seeds with a *green* embryo are sown, plants are obtained, whose pods, which were produced by self-fertilization, contain *only* seeds with *green* embryos, (the *third* generation). These, in turn, produce only seeds with *green* embryos, (the *fourth* generation), etc. With respect to this character, the *recessive* one, they behave like the *pure* variety, which carries it.

4. If the seeds with *yellow* embryo are sown, plants are produced which may be grouped into *two classes,*

Class A, those plants, whose pods, which were obtained by self-fertilization, contain *only* seeds with *yellow* embryos (the *third* generation) and

Class B, those plants, whose pods, which were produced by self-fertilization, contain seeds with *yellow* as well as seeds with *green* embryos (the *third* generation). Numerically, there are again on the average *three* seeds with *yellow* embryos for each *one* with a *green* embryo, just as in the second generation (see paragraph 2).

The *number of individuals* in classes A and B is approximately *one* to *two.*

Let me emphasize again, that *embryos* of Class A do not differ in their *appearance* in any way from those in Class B, only after the pods which were produced by self-fertilization, have been harvested, can it be decided to which one of the classes the seed belonged.

5. Seeds with *yellow* embryos, which descended from plants of *Class A* (paragraph 4), produce plants, whose pods, which originated by self-fertilization, again contain *only* seeds with *yellow* embryos (the *fourth* generation). Plants which develop from them in turn produce *only* seeds with *yellow* embryos, etc.

---

[5] For instance "D'autre part, l'étude des caractères simples des hybrides peut fournir la preuve la plus directe du principe énonce. *L'hybride montre toujours le caractère d'un des deux parents, et cela dans toute sa force;* jamais le caractère d'un parent, manquant a l'autre, ne se trouve réduit de moitié." (l.c. paragraph 3, italics mine).

[6] The color of the hilum on the other hand (whether black, brownish, etc.) represents a dominant character.

[7] In the meantime I have found an example (Footnote added later).

As regards this character, the *dominant* one, they behave like the *pure* variety which carries it.

6. The seeds with *green* embryos, which are obtained from plants of *Class B* (paragraph 4, B) produce plants, whose pods, which originated by self-fertilization again contain only seeds with *green* embryos (the *fourth* generation). Plants which develop from them in turn produce *only* seeds with *green* embryos (the *fifth* generation), etc.;–just as did the green embryos of the *second* generation (paragraph 3).

7. The seeds with *yellow* embryos, which are obtained from plants of *Class B* (paragraph 4, B) again produce, just as it was described in paragraph 4, *two types of plants,* in the ratio *one* to *two,* whose seeds behave in the same way as described in paragraphs 5 and 6 and so forth.

Table 1 explains and summarizes the results discussed above, it also gives the numerical ratios. The sign ∞ indicates that all of the seeds of the progeny in this group contained like embryos.

Tables 2 and 3 show the results obtained in two of my experimental series. The generations are given in vertical sequence. The upper *figure in bold face* denotes in each generation the *number of embryos obtained,* the *figure in light face* the *number of individuals,* which were raised from these embryos and produced fruits; ye- yellow, gr- green. The rest is self-explanatory.

Table 1

Parents	Hybrid					
	I. Gen.	II. Gen.	III. Gen.	IV. Gen.	V. Gen.	VI. Gen.
green		1 green . . .	∞ green . . .	∞ green . . .	∞ green . .	∞ green
			1 green . . .	∞ green . . .	∞ green . .	∞ green
				1 green . . .	∞ green . .	∞ green
					1 green . .	∞ green
						1 green
					3 { 2 yellow	
	∞ yellow	3 { 2 yellow	3 { 2 yellow	3 { 2 yellow		3 yellow
					1 yellow . .	
						∞ yellow
				1 yellow . . .	∞ yellow . .	∞ yellow
			1 yellow . . .	∞ yellow . . .	∞ yellow . .	∞ yellow
yellow		1 yellow . . .	∞ yellow . . .	∞ yellow . . .	∞ yellow . .	∞ yellow

Table 2

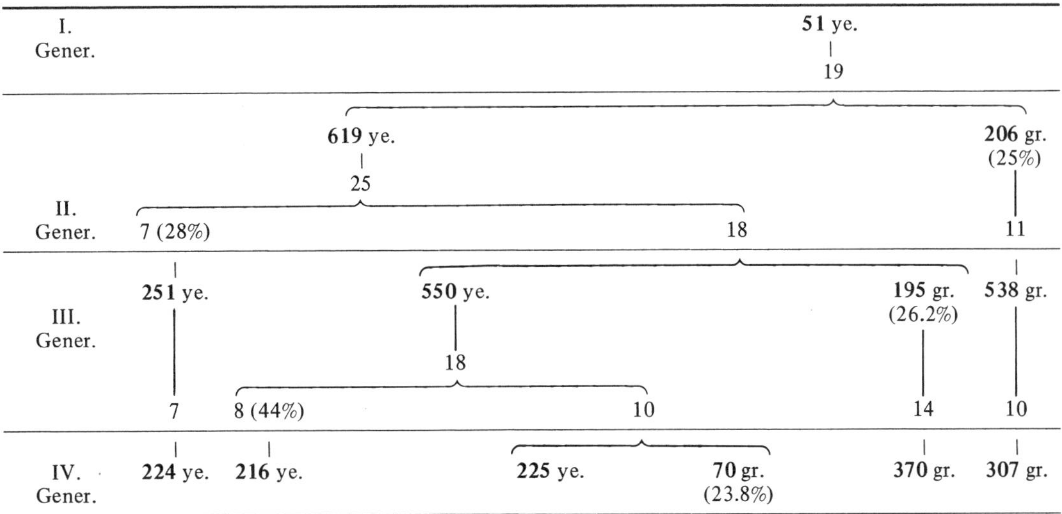

I. Gener.: **51** ye. — 19

II. Gener.: **619** ye. — 25; **206** gr. (25%) — 11
7 (28%) 18

III. Gener.: **251** ye. — 7; **550** ye. — 18; **195** gr. (26.2%) — 14; **538** gr. — 10
8 (44%) 10

IV. Gener.: **224** ye. **216** ye. **225** ye. **70** gr. (23.8%) **370** gr. **307** gr.

Table 3

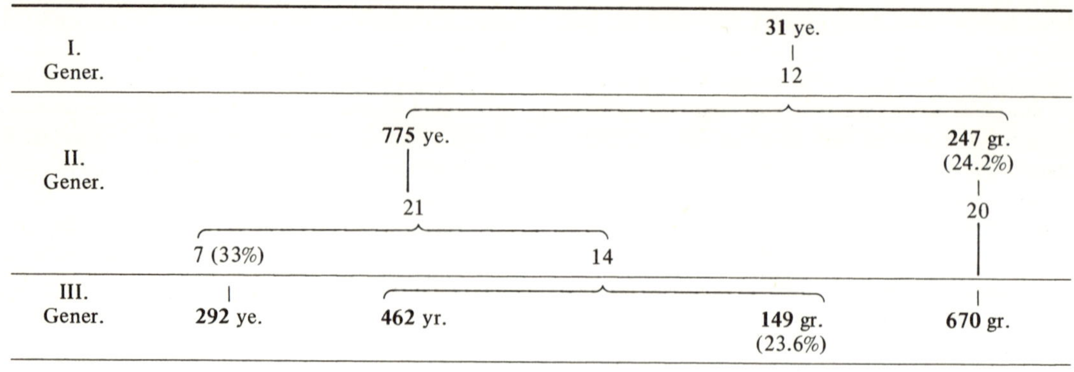

Experiment I

Hybrid between the *"green late [variety] Erfurter Folgererbse"* with *green* embryos and the *"[variety] Kneifelerbse with purple-violet pods,"* having *yellow* embryos.[8]

Experiment II

Hybrid between the *"green, late [variety] Erfurter Folgererbse,"* with *green* embryos and the *[variety] "Bohnenerbse,"* with *yellow* embryos.

The numerical ratios of yellow embryos to green embryos are quite variable in individual plants. In experiment I the smallest percentages for green being 14.9 and 7.7 and the largest ones 44.2 and 40.0.–It is of no importance whether the dominant character was introduced by the paternal or by the maternal plant; in *all* varieties, which possess a specific pair of characteristics, the latter behaves in the same manner.

Experiment II shows by chance the exact numerical ratios between the two classes of individuals which are produced by seeds with yellow embryos (7:14=1:2), while this ratio can be determined only from the mean of generations III and IV in experiment I:15 [=7(III)+8(IV)] individuals in one class as opposed to 28 [=18(III)+10(IV)] of the other (34.9:65.1 instead of 33.3:66.6).

In order to *explain* the facts, one must assume (as did Mendel) that following fusion of the reproductive nuclei,[9] the anlage for one character, the recessive one, (*green* in our case) is suppressed by the other character, the dominant one, therefore all embryos are *yellow.* The anlage, however, although "latent" is preserved, and prior to *the definitive formation of the reproductive nuclei a complete separation of the two anlagen occurs, so that one half of the reproductive nuclei receive the anlage for the recessive character, i.e. green, the other half the anlage for the dominant character, i.e. yellow.* The earliest time at which this separation might occur is the time of formation of the primordial anlage of both the seed and the anthers.[10] The numerical ratio 1:1 strongly suggests that the separation occurs during a *nuclear division,* the reduction division of Weisman,[11] but, because of the numerous problems involved, a more detailed discussion would lead too far.

Thus among ovules, 500 contain the anlage for the dominant character (yellow), 500 the anlage for the recessive character (green), and among 1000 generative pollen tube nuclei there are also 500 each with the dominant (yellow) and with the recessive character (green). If the reproductive nuclei are brought together by chance, then the probability that among 1000

[8] Under identical conditions the plants produced an average of 43.3, 47.7, and 28.8 seeds in successive generations; this is a good example to show the consequences of self-fertilization, and also furnishes an explanation of the "giant growth" of some hybrids (footnote added later).

[9] Mendel, of course, does not mention nuclei, but only "germinal cells" and "pollen cells."

[10] and at the latest at the time of the first division of the pollen grain nucleus from which the primary nucleus of the embryo sac is formed. For, in maize, it is shown by the similarity between the hybrid endosperm and the hybrid embryo that the two generative pollen tube nuclei and all of the eight nuclei of the embryo sac contain only one of the two anlagen. (Footnote added later.)

[11] See also "Keimplasma" p. 392 ff.

nuclear fusions two anlagen *of the same kind* will meet (either two dominant or two recessive ones) is equal to the probability of *two different* anlagen meeting (one dominant, one recessive). Thus each type of combination will occur 500 times, or in 50 percent of all of the combinations.

In the first case, i.e. when *like* anlagen meet, the probability that they will be two *recessive* ones is as great as that they will be two *dominant* ones, again one half, or each one will occur 250 times or in 25 per cent of all combinations. For the pair of characters under investigation, the result is here the same, as if two reproductive nuclei of either one of the pure varieties would unite.

In the second case—when *different* anlagen are combined the result of self-fertilization must be the same as that found in hybrids of the first generation, which were produced by experiment. The dominant anlage suppresses the recessive one, but later, preceding definitive formation of the reproductive nuclei, the two anlagen separate again, as was described for the artificially produced hybrid. "There occurs, accordingly a repeated hybridization" (Mendel).

The progeny of the first generation must consequently be separable into three classes, 25 percent having *only* the recessive, 25 percent having *only* the dominant and 50 percent having *both* characters, although [in the latter] only the dominant character may be recognized.—It follows from this assumption that in the first two cases all future generations will breed true for one of the two characters, while in the third case segregation will occur again.

If the hybrid (in the first generation) is pollinated with pollen of that parental variety which has the *dominant* character, instead of with its *own* pollen, only plants which show the *dominant* character are obtained, but among their progeny *one half* will in turn produce *only* individuals with the *dominant* character, while *the other half* produces some plants with the *dominant* and others with the *recessive* character, in a ratio of 3:1. If, on the other hand, the hybrid (first generation) is fertilized with the pollen of the parental variety with the *recessive* character, then *one half* of the plants obtained will show the *recessive* character, while the *other half* shows the *dominant* character, and the progeny of the *latter* again show the *dominant and* the *recessive* characters in a ratio of 3:1.

This theoretically derived rule also holds in the hybrids of maize.

Since two classes of individuals, i.e. those with the dominant anlage only and those with both the dominant and the recessive anlagen, cannot be distinguished from one another externally, the correct numerical ratios can only be determined by *self-fertilization.* Since self-fertilization normally occurs in peas, they are such excellent experimental objects.

A further consequence of the above is the following: as long as, because of chance selection, the number of individuals of a plot remains constant in successive generations, the number of individuals in the modal class, i.e., those containing both anlagen, decreases steadily, until they finally disappear completely. In the second generation they make up 50 percent of the total, 25 percent in the third generation, 12.5 in the fourth, 6.25 in the fifth and $100/2^{n-1}$ percent in the nth generation. This numerical decrease of the modal class had already been derived by Mendel.[12]

Thus far we have considered only the behavior of those pairs of characters in which one member is dominant. The case of two or more differentiating characters also was discussed *theoretically* and tested *experimentally* by Mendel. It was shown that all possible combinations occur as frequently as they are expected on the basis of the laws of probability, assuming that their production is due to chance. "It is demonstrated at the same time, *that the relation of each pair of different characters in hybrid union is independent of the other differences in the two original parental stocks*" (Mendel).[13]

In the case of *two* pairs of characters, *nine* different classes of individuals may occur. However, only *four groups* may be distinguished *externally,* the numbers of individuals in the classes must occur in a ratio of 9:3:3:1. Among 1000 individuals, 562.5, 187.5, 187.5 and 62.5 respectively will be grouped together. In a suitable experiment Mendel did obtain the numbers 315, 101, 108, and 32 respectively, which on the basis of 1000 are as 566.6, 181.6, 194.4 and 57.6. This is a good approximation to the ratio. With hybrids of maize I have

[12]One hardly needs to point out, how important this behavior is in regard to the question of species formation from hybrids. (Footnote added later.)

[13]There are again exceptions to this rule; strains with linked characters exist. (Footnote added later.)

obtained the same result, in one case, for instance, the numbers 308, 104, 96, and 37 or, calculated on the basis of 1000, 565, 191, 176, and 68.

Mendel concludes *"that the pea hybrids form egg and pollen cells which, in their constitution, represent in equal numbers all constant*[14] *forms which result from the combination of the characters when united in fertilization."* We may say in the terms used in this paper: *In the hybrid reproductive cells are produced in which the anlagen for the individual parental characteristics are contained in all possible combinations, but both anlagen for the same pair of characters are never combined. Each combination occurs with approximately the same frequency.*—If the parental strains differ only in *one* pair of characters (2 characters: *A, a*) the hybrid will form only *two types* of reproductive nuclei (*A, a*) which are like those of the parents. Each type is 50 percent of the total. If the parents differ in *two* pairs of characters (4 characters: *A, a; B, b*) *four types* of reproductive nuclei will be formed, (*AB, Ab, aB, ab*) and 25 percent of the total will be of each type. If the parents differ in *three* pairs of characters (6 characters:*A, a; B, b; C, c*) *eight types* of reproductive nuclei will be formed (*ABC, ABc, AbC, Abc, aBC, aBc, abC, abc*), and 12.5 percent of the total are of each type.[15]

This I call Mendel's law. It includes the "loi de disjonction" of de Vries, also. Everything else may be derived from this law.

At present, however, this law is applicable only to a certain number of cases, i.e. those where one member of a pair of characters is dominant,[16] and probably only to hybrids between varieties. It seems impossible that all pairs of characters of all hybrids should behave according to this law. Some hybrids of peas bear this out.

In the *first* generation of the combination of the "green, late [variety] Erfurter Folgererbse" with an almost colorless seed coat, and the "[variety] Kneifelerbse with purple-violet pods," or the "[variety] Pahlerbse with purple pods" both having a solid-color, orange-red seed coat, which turns brown on aging, the seed coats within the same pod were sometimes colorless, sometimes intensely red, *but usually more or less tinted with orange,* and also speckled to a variable degree with purplish-black spots. Thus, in addition to a *dilution* of one of the characters, an (apparently) *new* character had originated. In the second generation, however, the seeds which show the two extremes of coloration, i.e. those with orange red, and those with almost colorless seed coats will again produce the extreme types and all of the transitions between them. The speckling was sometimes unchanged, sometimes not present at all or very slight and sometimes somewhat increased. Size and shape of the seed and texture of the seed surface behaved in a similar way.

I will discuss these points at a later time.

Postscript (added in proof)

In the meantime de Vries has published in these proceedings (No. 3 of this year) some more details concerning his experiments. There he refers to Mendel's investigations, which *were not even mentioned in the "Comptes rendus."* I must emphasize again:

1. *that in many pairs of characters there is no dominant member* (p. 96),
2. *that Mendel's law of segregation cannot be applied universally* (p. 99).

[14] Mendel calls a type constant, if it no longer contains the two different anlagen of a pair of characters.

[15] If the pollen grains of the two parental strains differ externally, one may, if Mendel's law holds, expect the hybrid to form two externally different types of pollen grains. That this is true was first observed by Focke.

[16] See footnote 6.

13 The incidence of alkaptonuria: a study in chemical individuality

Archibald E. Garrod†
Physician to the Hospital for Sick Children
Demonstrator of Chemical Pathology at St. Bartholomew's Hospital
London, England

All the more recent work on alkaptonuria has tended to show that the constant feature of that condition is the excretion of homogentisic acid, to the presence of which substance the special properties of alkapton urine, the darkening with alkalies and on exposure to air, the power of staining fabrics deeply, and that of reducing metallic salts, are alike due. In every case which has been fully investigated since Wolkow and Baumann[1] first isolated and described this acid its presence has been demonstrated and re-examination of the material from some of the earlier cases also has led to its detection. The second allied alkapton acid, uroleucic, has hitherto only been found in the cases investigated by Kirk and in them in association with larger amounts of homogentisic acid.[2] By the kindness of Dr. R. Kirk I have recently been enabled to examine fresh specimens of the urines of his patients who have now reached manhood and was able to satisfy myself that at the present time even they are no longer excreting uroleucic acid. After as much of the homogentisic acid as possible had been allowed to separate out as the lead salt the small residue of alkapton acid was converted into the ethyl ester by a method recently described by Erich Meyer[3] and the crystalline product obtained had the melting point of ethyl homogentisate (120° C.). Further observations, and especially those of Mittelbach,[4] have also strengthened the belief that the homogentisic acid excreted is derived from tyrosin, but why alkaptonuric individuals pass the benzene ring of their tyrosin unbroken and how and where the peculiar chemical change from tyrosin to homogentisic acid is brought about, remain unsolved problems.

There are good reasons for thinking that alkaptonuria is not the manifestation of a disease but is rather of the nature of an alternative course of metabolism, harmless and usually congenital and lifelong. Witness is borne to its harmlessness by those who have manifested the peculiarity without any apparent detriment to health from infancy on into adult and even advanced life, as also by the observations of Erich Meyer who has shown that in the quantities ordinarily excreted by such persons homogentisic acid neither acts as an aromatic poison nor causes acid intoxication, for it is not excreted as an aromatic sulphate as aromatic poisons are, nor is its presence in the urine attended by any excessive output of ammonia. However, regarded as an alternative course of metabolism the alkaptonuric must be looked upon as somewhat inferior to the ordinary plan, inasmuch as the excretion of homogentisic acid in place of the ordinary end products involves a certain slight waste of potential energy. In this connexion it is also interesting to note that, as far as our knowledge goes, an individual is either frankly alkaptonuric or conforms to the normal type, that is to say, excretes several grammes of homogentisic acid per diem or none at all. Its appearance in traces, or in gradually increasing or diminishing quantities, has never yet been observed, even in the few recorded temporary or intermittent cases. In cases in which estimations have been carried out the daily output has been found to lie within limits which, considering the great influence of proteid food upon the excretion of homogentisic acid and allowing for differences of sex and age, may be described as narrow. This is well illustrated by Table I, in which the cases are arranged in order of age.

The information available as to the incidence of alkaptonuria is of great interest in connexion with the above view of its nature. That the

From The Lancet **ii:**1616-1620, 1902. Used with permission.

†Deceased.

[1] Wolkow and Baumann. 1891. Z. physiol. Chemie **XV:**228.

[2] R. Kirk. 1889. Journal of Anatomy and Physiology **XXIII:**69; Huppert. 1897. Zeitschrift für physiologische Chemie **XXIII:**412.

[3] E. Meyer. 1901. Deutsches Archiv für klinische Medicin **LXX:**443.

[4] Mittelbach. 1901. *Ibid.* **LXXI:**50.

Table I. Showing the average excretion of homogentisic acid

No.	Sex	Age	Average excretion of homogentisic acid per 24 hours on ordinary mixed diet	Name of observers
1	M	2½ years	3.2 grams	Erich Meyer
2	M	3½ years	2.6 grams	A. E. Garrod
3	M	8 years	2.7 grams	Ewald Stier
4	M	18 years	5.9 grams	P. Stange
5	M	44 years	4.6 grams	Mittelbach
6	M	45 years	4.7 grams	H. Ogden
7	M	60 years	5.3 grams	Hammarsten
8	F	60 years	3.2 grams	H. Emlslen
9	M	68 years	4.8 grams	Wolkow and Baumann

Table II. Showing the proportion of alkaptonuric members to normal members in 9 families

No.	Total number of family (brothers and sisters)	Number of alkaptonuric members	Number of normal members	Observers
1	14	4	10	Pavy
2	4	3	1	Kirk
3	7	3	4	Winternitz
4	2	1	1	Ewald Stier
5	2	2	0	Baumann, Embden
6	1	1	0	Erich Meyer
7	10	1	9	Noccioli and Domenici
8	5	2	3	A. E. Garrod
9	3	2	1	W. Smith, Garrod
Totals	48	19	29	–

peculiarity is in the great majority of instances congenital cannot be doubted. The staining property of the urine allows of its being readily traced back to early infancy. This has been repeatedly done and in one of my cases[5] the staining of the napkins was conspicuous 57 hours after the birth of the child. The abnormality is apt to make its appearance in two or more brothers and sisters whose parents are normal and among whose forefathers there is no record of its having occurred, a peculiar mode of incidence which is well known in connexion with some other conditions. Thus of 32 known examples, which were presumably congenital, no less than 19 have occurred in seven families. One family contained four alkaptonurics, three others contained three, and the remaining three two each. The proportion of alkaptonuric to normal members is of some interest and Table II embodies such definite knowledge upon this point as is at present available regarding congenital cases.

The preponderance of males is very conspicuous. Thus, of the 40 subjects whose cases have hitherto been recorded 29 have been males and only 11 females.

In a paper read before the Royal Medical and Chirurgical Society in 1901 the present writer pointed out that of four British families in which 11 were congenitally alkaptonuric members no less than three were the offspring of marriages of first cousins who did not themselves exhibit this anomaly. This fact has such interesting bearings upon the etiology of alkaptonuria that it seemed desirable to obtain further information about as many as possible of the other recorded cases and especially of those which were presumably congenital. My inquiries of a number of investigators who have recorded such cases met with a most kindly response, and although the number of examples about which information could still be obtained

[5] A. E. Garrod. November 30, 1901. Lancet: 1481; 1902. Transactions of the Royal Medical and Chirurgical Society **LXXXV**:69.

proved to be very limited, some valuable facts previously unknown have been brought to light and indications are afforded of points which may be inquired into with advantage regarding cases which may come under observation in the future. In a number of instances the patients have been lost sight of, or for various reasons information can no longer be obtained concerning them. To those who have tried to help me with regard to such cases, and have in some instances been at great trouble in vain, my hearty thanks are no less due than to those who have been able to furnish fresh information.[6]

The following is a brief summary of the fresh information collected. Dr. Erich Meyer,[7] who mentioned in his paper that the parents of his patient were related, informs me that as a matter of fact they are first cousins. Dr. H. Ogden[8] states that his patient is the seventh of a family of eight members and that his parents were first cousins. The three eldest children died in infancy; the fifth, a female, has three children, but neither is she nor are they alkaptonuric. There is no record of any other examples in the family. The patient, whose wife is not a blood relation, has three children none of whom are alkaptonuric. Professor Hammarsten[9] states that the parents of an alkaptonuric man, whose case he recently described, were first cousins. The patient, aged 61 years, has three brothers and the only brother whose urine has been seen is not alkaptonuric. I have learned from Professor Noccioli[10] that the parents of the woman whose case he investigated with Dr. Domenici were not blood relations. The patient, a twin, who is one of two survivors of a family of ten, states that none of her relations have exhibited the condition. Dr. Ewald Stier[11] informs me that the parents of his patient were not related and it is mentioned in his paper that they were not alkaptonuric. Professor Ebstein[12] states that the parents of the child with "pyrocatechinuria" whose case was investigated by him in conjunction with Dr. Müller in 1875 were not related, but I gather that he would not regard this as an ordinary case of alkaptonuria, the abnormal substance in the urine having been identified as pyrocatechin. Lastly, Professor Osler supplies the very interesting information that of two sons of the alkaptonuric man previously described by Dr. Futcher[13] one is alkaptonuric. This is the first known instance of direct transmission of the peculiarity. The parents of the father, who has an alkaptonuric brother whose case was recorded by Marshall,[14] were not blood relations. The above particulars are embodied with those of the congenital British cases previously recorded in the following tabular epitome (Table III).

It will be seen that the results of further inquiries on the continent of Europe and in America confirm the impression derived from the British cases that of alkaptonuric individuals a very large proportion are children of first cousins. The above table includes 19 cases in all out of a total of 40 recorded examples of the condition, and there is little chance of obtaining any further information on the point until fresh cases shall come under observation. It will be noticed that among the families of parents who do not themselves exhibit the anomaly a proportion corresponding to 60 per cent are the offspring of marriages of first cousins. In order to appreciate how high this proportion is it is necessary to form some idea of the total proportion of the children of such unions to the community at large. Professor G. Darwin,[15] as the outcome of an elaborate statistical investigation, arrived at the conclusion that in England some 4 per cent of all marriages among the aristocracy and gentry are between first cousins; that in the country and smaller towns

[6] To Hofrath Professor Huppert and to Professor Osler my very special thanks are due for invaluable aid in collecting information, and I would also express my most sincere gratitude to Professor Hammarsten, Geheimrath Professor Ebstein, Geheimrath Professor Fürbringer, Geheimrath Professor Erb, Professor Noccioli, and Professor Denigès, as also to Dr. F. W. Pavy, Dr. Kirk, Dr. Maguire, Dr. Futcher, Dr. Erich Meyer, Dr. H. Ogden, Dr. H. Embden, Dr. Mittelbach, Dr. Ewald Stier, Dr. Grassi, Dr. Carl Hirsch, and Dr. Winternitz, all of whom have been kind enough to help the inquiry in various ways.

[7] E. Meyer. *Loc. cit.*

[8] H. Ogden. 1895. Z. physiol. Chemie **XX**:289.

[9] Hammarsten. 1901. Upsala Läkareförenings Förhandlingar **VII**:26.

[10] Noccioli e Domenici. 1898. Gazetta degli Ospedali **XIX**:303.

[11] Ewald Stier. 1898. Berliner klinische Wochenschrift **XXXV**:185.

[12] Ebstein and Müller. 1875. Virchow's Archiv **LXII**:554.

[13] Futcher. 1898. New York Medical Journal **LXVII**:69.

[14] Marshall. 1887. Medical News, Philadelphia **L**:35.

[15] G. Darwin. 1875. Journal of the Statistical Society **XXXVIII**:153.

Table III. Showing the large proportion of alkaptonurics who are the offspring of marriage of first cousins

A

Families the offspring of marriages of first cousins.

No.	Total number of family	Number of known alkaptonuric members	Observers
1	14	4	Pavy
2	4	3	R. Kirk
3	5	2	A. E. Garrod
4	1	1	Erich Meyer
5	8	1	H. Ogden
6	4	1	Hammarsten
Total	36	12	–

B

Families whose parents were not related and not alkaptonuric.

No.	Total number of family	Number of known alkaptonuric members	Observers
1	3	2	Armstrong, Walter Smith, and Garrod
2	2	1	Ewald Stier
3	10	1	Noccioli and Domenici
4*	?	2	Marshall and Futcher
Total		6	–

C

Family in which alkaptonuria was directly inherited from a parent.

No.	Total number of family	Number of known alkaptonuric members	Observers
1*	?	1	Osler and Futcher
Total	–	1	–

*B 4 and C 1 refer to two generations of one family. No information is forthcoming as to the absence of alkaptonuria in previous generations. Ebstein and Müller's case, which is not included in the table for reasons given above, would raise the number of families in list B to 5.

the proportion is between 2 and 3 per cent, whereas in London it is perhaps as low as 1.5 per cent. He suggests 3 per cent as a probable superior limit for the whole population. Assuming, although this is, perhaps, not the case, that the same proportion of these as of all marriages are fruitful, similar percentages will hold good for families, and assuming further that the average number of children results from such marriages they will hold good for individuals also. A very limited number of observations which I have made among hospital patients in London gave results which are quite compatible with the above figures. Thus, among 50 patients simultaneously inmates of St. Bartholomew's Hospital there was one whose parents were first cousins. On another occasion one such was found among 100 patients, and there was one child of first cousins among 100 children admitted to my ward at the Hospital for Sick Children. It is evident, on the one hand, that the proportion of alkaptonuric families and individuals who are the offspring of first cousins is remarkably high, and, on the other hand, it is equally clear that only a minute proportion of the children of such unions are alkaptonuric. Even if such persons form only 1 per cent of the community their numbers in London alone should exceed 50,000, and of this multitude only six are known to be alkaptonuric. Doubtless there are others, but that the peculiarity is extremely rare is hardly open to question. A careful look-out maintained for several years at two large hospitals has convinced me of this, and although the subject has recently attracted much more attention than formerly the roll of recorded examples increases but slowly.

The question of the liability of children of consanguineous marriages to exhibit certain abnormalities or to develop certain diseases has been much discussed, but seldom in a strictly scientific spirit. Those who have written on the subject have too often aimed at demonstrating the deleterious results of such unions on the one hand, or their harmlessness on the other, questions which do not here concern us at all. There is no reason to suppose that mere consanguinity of parents can originate such a condition as alkaptonuria in their offspring, and we must rather seek an explanation in some peculiarity of the parents, which may remain latent for generations, but which has the best chance of asserting itself in the offspring of the union of two members of a family in which it is transmitted. This applies equally to other examples of that peculiar form of heredity which has long been a puzzle to investigators of such subjects, which results in the appearance in several collateral members of a family of a peculiarity which has not been manifested at least in recent preceding generations.

It has recently been pointed out by Bateson[16] that the law of heredity discovered by Mendel offers a reasonable account of such phenomena. It asserts that as regards two mutually exclusive characters, one of which tends to be dominant and the other recessive, cross-bred organisms will produce germinal cells (gametes) each of which, as regards the characters in question, conforms to one or other of the pure ancestral types and is therefore incapable of transmitting the opposite character. When a recessive gamete meets one of the dominant type the resulting organism (the zygote) will usually exhibit the dominant character, whereas when two recessive gametes meet the recessive character will necessarily be manifested in the zygote. In the case of a rare recessive characteristic we may easily imagine that many generations may pass before the union of two recessive gametes takes place. The application of this to the case in question is further pointed out by Bateson, who, commenting upon the above observations on the incidence of alkaptonuria, writes as follows:[17] "Now there may be other accounts possible, but we note that the mating of first cousins gives exactly the conditions most likely to enable a rare, and usually recessive, character to show itself. If the bearers of such a gamete mate with individuals not bearing it the character will hardly ever be seen; but first cousins will frequently be the bearers of similar gametes, which may in such unions meet each other and thus lead to the manifestation of the peculiar recessive characters in the zygote." Such an explanation removes the question altogether out of the range of prejudice, for, if it be the true account of the matter, it is not the mating of first cousins in general but of those who come of particular stocks that tends to induce the development of alkaptonuria in the offspring. For example, if a man inherits the tendency on his father's side his union with one of his maternal first cousins will be no more liable to result in alkaptonuric offspring than his marriage with one who is in no way related to him by blood. On the other hand, if members of two families who both inherit the strain should intermarry the liability to alkaptonuria in the offspring will be as great as from the union of two members of either family, and it is only to be expected that the peculiarity will also manifest itself in the children of parents who are not related. Whether the Mendelian explanation be the true one or not, there seems to be little room for doubt that the peculiarities of the incidence of alkaptonuria and of conditions which appear in a similar way are best explained by supposing that, leaving aside exceptional cases in which the character, usually recessive, assumes dominance, a peculiarity of the gametes of *both* parents is necessary for its production.

Hitherto nothing has been recorded about the children of alkaptonuric parents, and the information supplied by Professor Osler and Dr. Ogden on this point has therefore a very special interest. Whereas Professor Osler's case shows that the condition may be directly inherited from a parent Dr. Ogden's case demonstrates

[16] W. Bateson. 1902. Mendel's Principles of Heredity, Cambridge.

[17] W. Bateson and Miss E. R. Saunders. 1902. Report to the Evolution Committee of the Royal Society (1):133*n*.

that none of the children of such a parent need share his peculiarity. As the matter now stands, of five children of two alkaptonuric fathers whose condition is known only one is himself alkaptonuric. It will be interesting to learn whether this low proportion is maintained when larger numbers of cases shall be available. That it will be so is rendered highly probable by the undoubted fact that a very small proportion of alkaptonurics are the offspring of parents either of whom exhibits the anomaly. It would also be extremely interesting to have further examples of second marriages of the parents of alkaptonurics. In the case of the family observed by Dr. Kirk the only child of the second marriage of the father, not consanguineous, is a girl who does not exhibit the abnormality. The only other available example is recorded by Embden. The two alkaptonurics studied by Professor Baumann and himself were a brother and sister born out of wedlock, and as far as could be ascertained the condition was not present in the children of the subsequent marriages which both parents contracted. The patient of Noccioli and Domenici was a twin, and I gather from Professor Noccioli's kind letter that the other twin was also a female, did not survive, and was not alkaptonuric. Further particulars are wanting, and the information was derived from the patient herself, who is described as a woman of limited intelligence but who was aware that in her own case the condition had existed from infancy. It is difficult to imagine that of twins developed from a single ovum one should be alkaptonuric and the other normal, but this does not necessarily apply to twins developed from separate ova.

It may be objected to the view that alkaptonuria is merely an alternative mode of metabolism and not a morbid condition, that in a few instances, not included in the above tables, it appears not to have been congenital and continuous but temporary or intermittent. In some of the cases referred to the evidence available is not altogether conclusive, and it is obvious that for the proof of a point of so much importance to the theory of alkaptonuria nothing can be regarded as wholly satisfactory which falls short of a complete demonstration of the presence of homogentisic acid in the urine at one time and its absence at another. The degree and rate of darkening of the urine vary at different periods apart from any conspicuous fluctuations in the quantity of homogentisic acid which it contains. The staining of linen in infancy is a much more reliable indication, especially if the mother of the child has had previous experience of alkaptonuric staining. In Geyger's case[18] of a diabetic man the intermittent appearance in the urine of an acid which he identified with the glycosuric acid of Marshall was established beyond all doubt, and the melting point and proportion of lead in the lead salt render it almost certain that he was dealing with homogentisic acid. In Carl Hirsch's case[19] a girl, aged 17 years, with febrile gastro-intestinal catarrh, passed dark urine which gave the indican reaction for three days. Professor Siegfried extracted by shaking with ether an acid which gave the reactions of homogentisic acid and formed a sparingly soluble lead salt. Neither the melting point of the acid nor any analytical figures are given. After three days the urine resumed its natural colour and reactions.

Von Moraczewski[20] also records a case of a woman, aged 43 years, who shortly before her death passed increasingly dark urine, rich in indican, from which he extracted an acid which had the melting point and reactions of homogentisic acid. Such increasing darkening of the urine as was here observed not infrequently occurs with urines rich in indoxyl-sulphate, as Baumann and Brieger first pointed out, and this was probably a contributory factor in the production of the colour which first called attention to the condition. Stange[21] has described a case in which the presence of homogentisic acid was very fully established, but he clearly does not regard the mother's evidence as to the intermittent character of the condition as conclusive. Zimnicki's[22] case of intermittent excretion of homogentisic acid by a man with hypertrophic biliary cirrhosis is published in a Russian journal which is inaccessible to me, and having only seen abstracts of his paper I am unacquainted with the details. Of hearsay evidence the most convincing is afforded in Winternitz's cases.[23] The mother of

[18] A. Geyger. 1892. Pharmazeutische Zeitung:488.

[19] C. Hirsch. 1897. Berliner klinische Wochenschrift **XXXIV**:866.

[20] W. von Moraczewski. 1896. Centralblatt für die Innere Medizin **XVII**:177.

[21] P. Stange. 1896. Virchow's Archiv **CXLVI**:86.

[22] Zimnicki. 1900. Jeschenedelnik. Abstract, Centralblatt für Stoffwechsel und Verdauungs-Krankheiten **I**(4):348.

[23] Winternitz. 1899. Münchener Medizinische Wochenschrift **XLVI**:749.

seven children, three of whom are alkaptonuric, was convinced that whereas two of her children had been alkaptonuric from the earliest days of life this had not been so with the youngest child in whom she had only noticed the peculiarity from the age of five years. This is specially interesting as supplying a link between the temporary and congenital cases. In a somewhat similar case described by Maguire[24] the evidence of a late onset is not so conclusive. Slosse's case[25] in which, as in von Moraczewski's, the condition apparently developed in the last stages of a fatal illness, completes the list of those falling into the temporary class. Evidently we have still much to learn about temporary or intermittent alkaptonuria, but it appears reasonable to suppose that those who exhibit the phenomenon are in a state of unstable equilibrium in this respect, and that they excrete homogentisic acid under the influence of causes which do not bring about this result in normal individuals. There is reason to believe that a similar instability plays a not unimportant part in determining the incidence of certain forms of *disease* in which derangements of metabolism are the most conspicuous features. Thus von Noorden,[26] after mentioning that diabetes occasionally develops at an early age in brothers and sisters and comparatively seldom occurs in the children of diabetic parents, adds that in three instances he has met with this disease in the offspring of marriages of first cousins. In one such family two out of six children, in another two out of three, and in the third the only two children became diabetic at ages between one and four years.

The view that alkaptonuria is a "sport" or an alternative mode of metabolism will obviously gain considerably in weight if it can be shown that it is not an isolated example of such a chemical abnormality, but that there are other conditions which may reasonably be placed in the same category. In the phenomenon of albinism we have an abnormality which may be looked upon as chemical in its basis, being due rather to a failure to produce the pigments of the melanin group which play so conspicuous a part in animal colouration than to any defect of development of the parts in which in normal individuals such pigments are laid down. When we study the incidence of albinism in man we find that it shows a striking resemblance to that of alkaptonuria. It, too, is commoner in males than in females, and tends to occur in brothers and sisters of families in which it has not previously appeared, at least in recent generations. Moreover, there is reason to believe that an undue proportion of albinos are the offspring of marriages of first cousins. Albinism is mentioned by most authors who have discussed the effects of such marriages and Arcoleo,[27] who gives some statistics of albinism in Sicily, states that of 24 families in which there were 62 albino members five were the offspring of parents related to each other in the second canonical degree. On the other hand, Bemiss[28] found that of 191 children of 34 marriages of first or second cousins five were albinos. In a remarkable instance recorded by Devay[29] two brothers married two sisters, their first cousins. There were no known instances of albinism in their families, but the two children of the one marriage and the five children of the other were all albinos. After the death of his wife the father of the second family married again and none of the four children of his second marriage were albinos. Again, albinism is occasionally directly inherited from a parent, as in one instance quoted by Arcoleo, but this appears to be an exceptional occurrence. The resemblance between the modes of incidence of the two conditions is so striking that it is hardly possible to doubt that whatever laws control the incidence of the one control that of the other also.

A third condition which suggests itself as being probably another chemical "sport" is cystinuria. Our knowledge of its incidence is far more incomplete and at first sight direct inheritance appears to play here a more prominent part. However, when more information is forthcoming it may turn out that it is controlled by similar laws. In this connexion a most interesting family described by Pfeiffer[30] is very suggestive. Both parents were normal, but all their four children, two daughters and

[24] R. Maguire. 1884. Brit. Med. J. **II**:808.

[25] A. Slosse. 1895. Annales de la Société Royale des Sciences Médicales et Naturelles, Bruxelles **IV**:89.

[26] Von Noorden. 1901. Die Zuckerkrankheit. 3, Aufgabe:47.

[27] G. Arcoleo. 1871. Sull' Albinismo in Sicilia. See notice in Archivio per l'Anthropologia **I**:367.

[28] Bemiss. 1857. J. of Psychol. Med. **X**:368.

[29] Devay. 1857. Du Danger des Mariages Consanguins, & c., Paris.

[30] E. Pfeiffer. 1894. Centralblatt für Krankheiten der Harn-und Sexual-Organe **V**:187.

two sons, were cystinuric. The elder daughter had two children neither of whom was cystinuric. A number of other examples of cystinuria in brothers and sisters are recorded, but information about the parents is wanting, except in the cases of direct transmission. In some of the earlier cases such transmission through three generations was thought to be probable, but the presence of cystin in the urine of parent and child has only been actually demonstrated in two instances. In Joel's[31] often-quoted case it was only shown that the mother's urine contained excess of neutral sulphur. E. Pfeiffer[32] found cystin in the urine of a father and son and in a family observed by Cohn[33] the mother and six of her children shared the peculiarity. As more than 100 cases are on record the proportion of cases of direct inheritance has not hitherto been shown to be at all high and Pfeiffer's first case shows that, as with alkaptonuria, the children of a cystinuric parent may escape. A large majority of the recorded cystinurics have been males. There is as yet no evidence of any influence of consanguinity of parents and in the only two cases about which I have information the parents were not related. Neither has it yet been shown that cystinuria is a congenital anomaly, although in one case, at any rate, it has been traced back to the first year of life. Observations upon children of cystinuric parents from their earliest infancy or upon newly-born brothers or sisters of cystinurics would be of great interest and should in time settle this question. Lastly, it seems certain that, like alkaptonuria, this peculiarity of metabolism is occasionally temporary or intermittent. The so frequent association with cystinuria of the excretion of cadaverine and putrescine adds to the difficulty of the problem of its nature and upon it is based the infective theory of its causation. However, it is possible that, as C. E. Simon[34] has suggested, these diamines may themselves be products of abnormal metabolism. Unlike alkaptonuria and albinism cystinuria is a distinctly harmful condition, but its ill effects are secondary to its deposition in crystalline form and the readiness with which it forms concretions. Its appearance in the urine is not associated with any primary morbid symptoms. All three conditions referred to above are extremely rare and all tend to advertise their presence in conspicuous manners. An albino cannot escape observation; the staining of clothing and the colour of the urine of alkaptonurics seldom fail to attract attention, and the calculous troubles and the cystitis to which cystinurics are so liable usually bring them under observation sooner or later. May it not well be that there are other such chemical abnormalities which are attended by no obvious peculiarities and which could only be revealed by chemical analysis? If such exist and are equally rare with the above they may well have wholly eluded notice up till now. A deliberate search for such, without some guiding indications, appears as hopeless an undertaking as the proverbial search for a needle in a haystack.

If it be, indeed, the case that in alkaptonuria and the other conditions mentioned we are dealing with individualities of metabolism and not with the results of morbid processes the thought naturally presents itself that these are merely extreme examples of variations of chemical behavior which are probably everywhere present in minor degrees and that just as no two individuals of a species are absolutely identical in bodily structure neither are their chemical processes carried out on exactly the same lines. Such minor chemical differences will obviously be far more subtle than those of form, for whereas the latter are evident to any careful observer the former will only be revealed by elaborate chemical methods, including painstaking comparisons of the intake and output of the organism. This view that there is no rigid uniformity of chemical processes in the individual members of a species, probable as it is *a priori,* may also be arrived at by a wholly different line of argument. There can be no question that between the families, genera and species both of the animal and vegetable kingdoms, differences exist both of chemical composition and of metabolic processes. The evidences for this are admirably set forth in a most suggestive address delivered by Professor Huppert[35] in 1895. In it he points out that we find evidence of chemical specificity of important constituents of the body, such as the

[31] Joel. 1855. Annalen der Chemie und Pharmacologie:247.

[32] E. Pfeiffer. 1897. Centralblatt für Krankheiten der Harn-und Sexual-Organe **VIII:**173.

[33] J. Cohn. 1899. Berliner klinische Wochenschrift **XXXVI:**503.

[34] C. E. Simon. 1890. Am. J. Med. Sci. **CXIX:**39.

[35] Huppert. 1896. Ueber die Erhaltung der Arteigenschaften, Prague.

haemoglobins of different animals, as well as in their secretory and excretory products such as the bile acids and the cynuric acid of the urine of dogs. Again, in their behaviour to different drugs and infecting organisms the members of the various genera and species manifest peculiarities which presumably have a chemical basis, as the more recent researches of Ehrlich tend still further to show. To the above examples may be added the results of F. G. Hopkins's[36] well-known researches on the pigments of the pieridae and the recent observations of the precipitation of the blood proteids of one kind of animal by the serum of another. From the vegetable kingdom examples of such generic and specific chemical differences might be multiplied to an almost indefinite extent. Nor are instances wanting of the influence of natural selection upon chemical processes, as for example, in the production of such protective materials as the sepia of the cuttlefish and the odorous secretion of the skunk, not to mention the innumerable modifications of surface pigmentation. If, then, the several genera and species thus differ in their chemistry we can hardly imagine that within the species, when once it is established, a rigid chemical uniformity exists. Such a conception is at variance with all that is known of the origin of species. Nor are direct evidences wanting of such minor chemical diversities as we have supposed to exist within the species. Such slight peculiarities of metabolism will necessarily be hard to trace by methods of direct analysis and will readily be masked by the influences of diet and of disease, but the results of observations on metabolism reveal differences which are apparently independent of such causes, as for example, in the excretion of uric acid by different human individuals. The phenomena of obesity and the various tints of hair, skin, and eyes point in the same direction, and if we pass to differences presumably chemical in their basis idiosyncrasies as regards drugs and the various degrees of natural immunity against infections are only less marked in individual human beings and in the several races of mankind than in distinct genera and species of animals.

If it be a correct inference from the available facts that the individuals of a species do not conform to an absolutely rigid standard of metabolism, but differ slightly in their chemistry as they do in their structure, it is no more surprising that they should occasionally exhibit conspicuous deviations from the specific type of metabolism than that we should meet with such wide departures from the structural uniformity of the species as the presence of supernumerary digits or transposition of the viscera.

[36] F. G. Hopkins. 1895. Philosophical Transactions of the Royal Society **CLXXXVI**(B):661.

14 Growth-hormone deficiency in man: an isolated, recessively inherited defect

David L. Rimoin*
Thomas J. Merimee
Victor A. McKusick
Divisions of Medical Genetics and Endocrinology
Department of Medicine
Johns Hopkins University
School of Medicine
Baltimore, Maryland

Abstract. *A deficiency of human growth hormone not associated with other pituitary deficiencies was observed in midgets with sexual ateliosis, a form of dwarfism inherited as an autosomal recessive trait. Body proportions, sexual development, birth weight, and postpartum lactation are normal in this syndrome.*

Gilford *(1)* termed normally proportioned dwarfs, or midgets, ateliotic, and he distinguished sexual and asexual types, depending on the state of sexual development and function. Autosomal recessive inheritance of sexual ateliotic dwarfism was supported by observations of affected sibs, both male and female, with unaffected parents who frequently were related *(2)*.

At the same time as Gilford's nosologic contributions, the role of the pituitary in growth became apparent from clinical and experimental observations, and deficient production of a growth-promoting factor by the pituitary was suspected in ateliosis. In the absence of specific methods for assay of growth hormone, as well as the lack of potent, nonantigenic material with which a convincing therapeutic test could be made, pituitary insufficiency could be established only in asexual ateliotics, that is, cases of panhypopituitarism in which the additional deficiency of thyrotropic, adrenocorticotropic, and gonadotropic hormones indicated the pituitary basis of the defect in growth. Probably partly because the cases of panhypopituitarism were almost always nonfamilial *(3)*, cases of sexual ateliosis were relegated to an idiopathic group called "primordial dwarfs," the tacit implication being that the defect is not pituitary but is, vaguely conceived, a genetically determined one "at the cellular level."

An isolated growth hormone deficiency has been suspected in some instances *(4)*. The experience reported here documents an isolated deficiency of growth hormone in midgets with autosomal recessive sexual ateliosis.

Six sexually mature individuals with proportionate dwarfism were studied. Three (family A) were members of an inbred West Virginia kindred and three (family B) were husband, wife, and daughter (Fig. 1). Their ages ranged from 39 to 77 years, their heights from 123 to 139.5 cm, and their weights from 30.6 to 42 kg. None had a congenital malformation. Bone age was adult in all six. Birth weights, from 2727 to 4545 g, had been normal in all; the midget daughter of two midget parents had a birth weight of 2727 g, cesarean delivery having been performed about 2 weeks before term. Growth retardation was first noted in the first 2 years of life. Five of the six grew gradually, with halt in growth soon after puberty. One of the males had a 20.5-cm growth spurt during puberty, which did not occur until age 25. Secondary sexual characteristics were normal, although puberty was delayed by 2 to 10 years in each. The skin was smooth and wrinkled and appeared abnormally thick. The voices were

---

From Science **152:**1635-1637, June 17, 1966.

We thank Mrs. B. Smith for technical assistance, Mrs. M. H. Abbott for aid in collecting the genealogies, and Dr. J. O. Humphries for referring case 3. Supported in part by USPHS grant 5-R01-GM 10189. The patients were studied in the Johns Hopkins Clinical Research Center which was supported by USPHS grant 5-M01-FR 35. D.L.R. was supported by a USPHS graduate training grant (5-T1-GM-624) and T.J.M. by a USPHS diabetes training grant (TI-AM-5136).

*Present address: Washington University School of Medicine, Division of Medical Genetics, The Edward Mallinckrodt Department of Pediatrics, St. Louis Children's Hospital, St. Louis, Missouri.

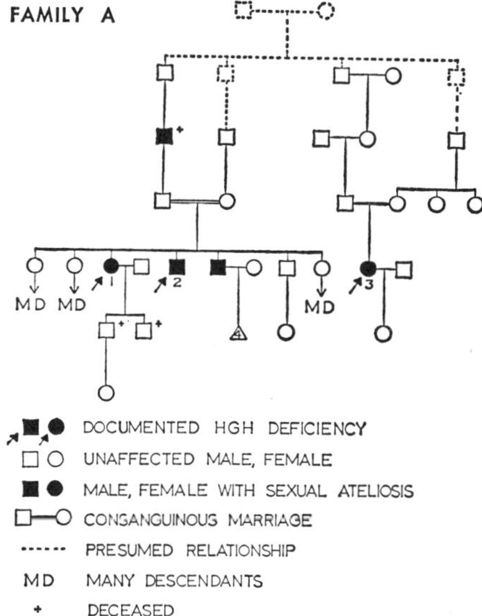

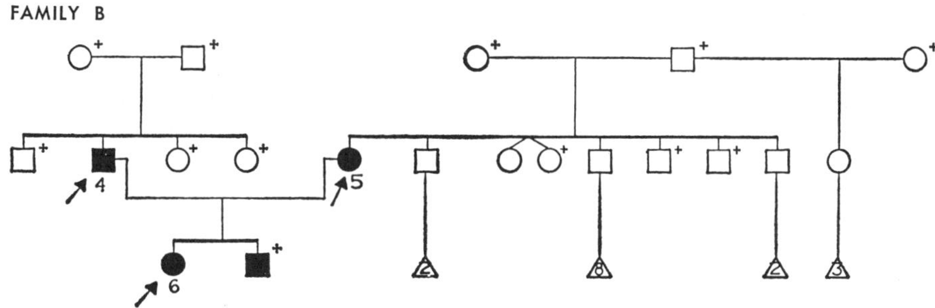

Fig. 1. Pedigrees of families A and B. The evidence for the presumed family lines (---) in family A is excellent, but not yet completely established. One offspring of case 1 died of Hodgkin's disease, the other of pneumonia during infancy. Case 6 of family B married a midget and has three children; two are midgets, and one is apparently normal.

high-pitched, with a "small" timbre. None had experienced hypoglycemic attacks and all were in general good health.

Extensive clinical, radiologic, and endocrinologic investigations were performed in each. Pituitary trophic hormone production was assessed by the following methods. (i) Thyrotropin (TSH) was assessed by the serum protein bound iodine and I^{131} uptake; (ii) gonadotropin (FSH) by biological activity of urinary extracts with the mouse uterine-weight technique *(5)*; (iii) adrenocorticotropin (ACTH) by the presence in the urine of 17-hydroxycorticosteroids and 17-ketosteroids under basal conditions, and on the day of, and the day after the administration of metapirone, 300 mg per 100 pounds of body weight being given every 4 hours for six doses *(5)*; (iv) growth hormone (HGH) by a standard human growth hormone immunoassay *(6)* after insulin-induced hypoglycemia and arginine infusions, both potent means of stimulating growth hormone secretion *(6, 7)*. In all six subjects, insulin, given intravenously in a dose of 0.1 unit per kilogram of body weight, produced marked symptoms of hypoglycemia approximately 30 minutes after the insulin administration. Arginine (20 g in 500 ml of water) was infused intravenously over a 30-minute period. The males were given 2.5 mg of stilbestrol twice a day for 2 days prior to the arginine infusion *(7)*. Control values for growth hormone were obtained from 15 healthy postpubertal individuals of normal stature, who underwent similar insulin and arginine infusions.

All patients studied had normal secretion of TSH, FSH, and ACTH. All the sexual ateliotics, after fasting, showed a low concentration (1.5

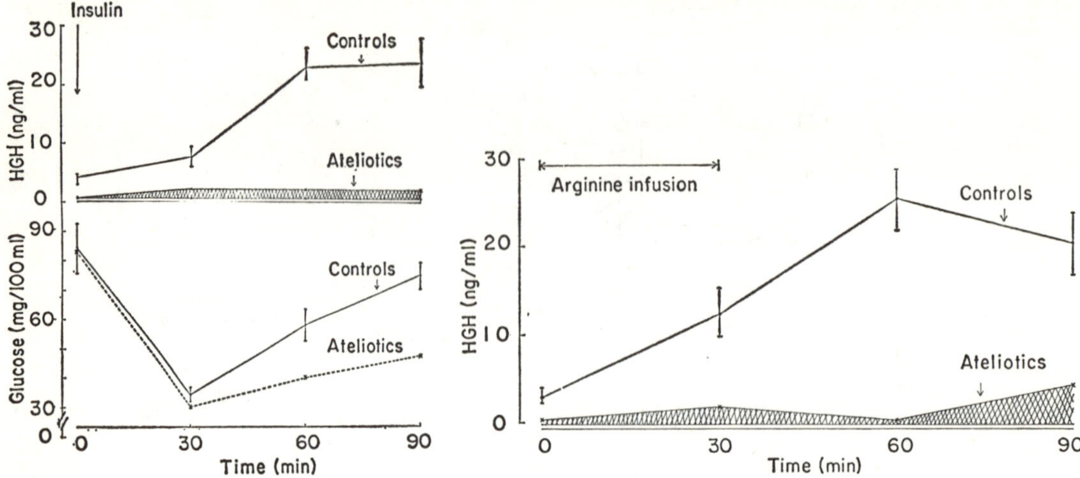

Fig. 2. The crosshatched area represents the total range of response of the six ateliotics. The unbroken line and bars represent the mean (± one standard error of the mean) levels of the controls. (Left) Human growth-hormone and blood-sugar responses after the intravenous administration of insulin (0.1 unit/kg body weight) at zero time. (Right) Analysis of human growth hormone after the intravenous infusion of 20 g arginine from time 0 to 30 minutes. The HGH concentration did not rise above 2.5 ng/ml in five of the six ateliotics.

ng/ml or less) of growth hormone in the serum, and the concentration did not rise significantly after insulin-induced hypoglycemia or arginine infusion (Fig. 2). The highest value was 2.0 ng/ml after insulin and 2.5 ng/ml after arginine, except for a single measurement of 5.0 ng/ml in case No. 6, 90 minutes after arginine. In controls there was a mean rise to 23.8 ± 4.0 ng/ml after insulin and to 26.1 ± 3.50 ng/ml after arginine. Glucose changes were comparable in the two groups. These normally proportioned, sexually mature midgets have thus been shown to have an isolated deficiency of human growth hormone, a deficiency which may be homologous to that associated with recessively inherited pituitary dwarfism in mice. Since puberty is often delayed, gonadotropin deficiency cannot be confirmed in a dwarf with growth-hormone deficiency until he is about age 25.

The findings in the two families (Fig. 1) support the view that sexual ateliotic dwarfism is inherited as an autosomal recessive trait. In addition to multiple, affected sibs from consanguineous marriages, the mating of two affected persons resulted in the birth of two similarly affected offspring (Fig. 1, family B). Monotropic growth hormone deficiency was demonstrated in the two parents and one of the offspring; the other affected offspring died accidentally in childhood.

Since all six subjects deficient in growth hormone had normal birth weights, and since all children born to the three females with growth hormone deficiency, including at least one who was herself growth-hormone deficient, weighed over 2700 g at birth, neither maternal nor fetal pituitary growth hormone is probably the major factor responsible for intrauterine growth. Normal intrauterine growth is also a feature of panhypopituitarism *(3)* and of anencephaly, in which the anterior pituitary appears to be at least functionally defective *(8)*.

All four growth-hormone–deficient females in this series had normal pregnancies, terminated by cesarean section, and normal lactation. Hence, the human growth-hormone molecule (or at least those parts of it which are immunologically active and are involved in growth promotion) is not essential to postpartum lactation. Since lactation was also normal in the midget mother of two midget children, it appears that placental lactogen *(9)* is under genetic control separate from pituitary growth hormone or is not necessary for lactation.

Mutations resulting in isolated deficiency of pituitary gonadotropin *(10)*, adrenocorticotropin *(11)* or thyrotropin *(12)* probably also occur in man.

REFERENCES AND NOTES

1. H. Gilford, Med.-Chirurg. Trans. **85**:305 (1902).
2. H. Rischbieth and A. Barrington, Treasury of Human Inheritance, part VIII, Section XV A, 1912.

3. W. H. Daugherty, in Textbook of Endocrinology, R. H. Williams, Ed. (Saunders, Philadelphia, 1962), p. 53.
4. J. F. Wilber and W. D. Odell, Metabolism **14:**590 (1965); J. A. Brasel, J. C. Wright, L. Wilkins, R. M. Blizzard, Amer. J. Med. **38:**484 (1965); T. R. Bierich, Acta Endocrinol. Suppl. **89:**27 (1964); H. L. Nadler, L. L. Neumann, H. Gershberg, J. Pediat. **63:**977 (1963); O. Trygstad and M. Seip, Acta Paediat. **53:**527 (1964); T. F. Hewer, J. Endocrinol. **3:**397 (1944); J. M. F. Antonin, Helv. Paediat. Acta **16:**267 (1961).
5. Dr. Claude J. Migeon, Baltimore, assayed the hormones.
6. J. Roth, S. M. Glick, R. S. Yalow, J. A. Berson, Science **140:**987 (1963).
7. T. J. Merimee, D. A. Lillicrap, D. Rabinowitz, Lancet **1965-II:**668 (1965).
8. D. M. Angevine, Arch. Pathol. **26:**507 (1938); K. Y. Ch'in, Chinese Med. J. suppl. **2:**63 (1938); C. Kind, Helv. Paediat. Acta **17:**244 (1962).
9. S. L. Kaplan and M. M. Grumbach, J. Clin. Endocrin. **25:**1370 (1965).
10. R. L. Biben and G. S. Gordon, *ibid.* **15:**931 (1955).
11. W. W. Cleveland, O. C. Green, C. J. Migeon, J. Pediat. **57:**376 (1960).
12. A. Querido and J. B. Stanbury, J. Clin. Endocrinol. **10:**1192 (1950).

15 The Himalayan rabbit case, with some considerations on multiple allelomorphs

A. H. Sturtevant*
Columbia University
New York, New York

It has been shown by Castle ('06, '09), Hurst ('06) and Punnett ('12) that the Himalayan pattern in rabbits behaves as a simple recessive to self color, and as a simple dominant to albino. Thus, as Punnett points out, we might suppose self to be the double dominant, Himalayan a recessive in one factor, and albino a double recessive. But, to use Punnett's words:

> The F_2 from self × albino should consequently contain Himalayans as well as true albinos. But among the large number of animals reared from such matings no Himalayans have hitherto been recorded, and for the present the relations between these various forms remain obscure.†

If we suppose that albino may be either the second single recessive or the double recessive we avoid this difficulty, but are then unable to explain why albino × Himalayan should not, at least occasionally, produce selfs by recombination.

Now it seems to me that the facts of the case are fitted equally well by either of two hypotheses. In the first place, we may consider, as above, that Himalayan is a single recessive and albino a double recessive—if we suppose the two factors concerned to be completely linked. The gametic (not zygotic) constitution of the three types would then be represented thus, *C* being the color producer and *S* the factor changing Himalayan to self.

Self	– *CS*
Himalayan	– *Cs*
Albino	– *cs*

If *C* and *S* be completely linked no *cS* individual can be obtained, and *CS* × *cs* would give no *Cs* in F_2.

On the other hand, we may consider the factor for self as allelomorphic to that for Himalayan pattern, and also to that for albinism. Then the three pure types might be represented thus (zygotic formulae):

Self	– *SS*
Himalayan	– *HH*
Albino	– *AA*

S, H, and *A* being allelomorphic each to itself or to either of the others, the crosses would result thus:

Self	– *SS*
Himalayan	– *HH*
F_1	*SH* – self
F_2	*SS*, *SH*, *SH* – 3 self; *HH* – 1 Himalayan

Self	– *SS*
Albino	– *AA*
F_1	*SA* – self
F_2	*SS*, *SA*, *SA* – 3 self; *AA* – 1 albino

Himalayan	– *HH*
Albino	– *AA*
F_1	*HA* – Himalayan
F_2	*HH*, *HA*, *HA* – 3 Himalayan; *AA* – 1 albino

An explanation similar to the second one above has been given by de Meijere ('10) for Jacobson's results with *Papilio Memnon.* The evidence on this case is, however, very incomplete, and there are complications due to sex. Either triple allelomorphs or complete coupling would seem to cover the facts as we have them at present. Shull ('11) has also used a system of three allelomorphs for a case in *Lychnis dioica.* I shall refer to this case again.

It will be seen that triple allelomorphs may be substituted for complete coupling as an explanation of any case where only three of the four combinations possible on the complete coupling scheme are known. But if we have the double dominant, both single recessives, and

From The American Naturalist **47**:234-239, 1913. Used with permission.

*Present address: Division of Biology, California Institute of Technology, Pasadena, California.

†From Punnett, R. C. 1912. Inheritance of coat-color in rabbits. Jour. Genet. **2**, p. 221.

the double recessive, then triple allelomorphism will no longer work. Thus, if a race of albino rabbits is discovered which produces self when mated to Himalayan, complete linkage will be the most likely explanation of the case.

There are certain other cases which fulfil the above requirements. Emerson ('11) has reported a case in beans (green leaves–green pods, green leaves–yellow pods, and yellow leaves–yellow pods are the three races concerned). The similar cases of complete linkage reported for corn by East and by Emerson are probably more easily explainable by linkage than by multiple allelomorphs, as, at least in some cases, all four possible races are found. Baur ('12) has a case in *Aquilegia,* where three types of leaves are found–green, variegated (green and yellowish green), and yellowish green. These behave toward each other in a manner exactly similar to that of the self, Himalayan and albino rabbits. Finally, Morgan ('12) has reported a case in *Drosophila ampelophila.* Red eye is a dominant to eosin and to white, and eosin is also a dominant to white. No two types ever give the third when crossed, either in F_1 or in F_2. The explanation which has been given in beans, columbines and flies has been that of two allelomorphic pairs, completely linked to each other.

The question as to which of these views is the more probable is closely bound up with the presence and absence hypothesis. On a strict application of this idea there is of course no possibility of more than two members of any given allelomorphic group. The presence and absence hypothesis as a universal principle has been criticized by Morgan ('13) in a recent paper, on what seem to me very strong grounds. It seems very unlikely that protoplasm (chromatin?) is such a simple substance that the only possible change in a given unit (molecule?) involves the loss of that unit. On the other hand, if a slight change takes place in a chemically complex gene, is it necessary to suppose that its allelomorphic relations must be upset? That very slight changes in the constitution of a gene might easily affect its behavior in ontogeny will, I think, be readily granted.

It is to be noted that in all the cases cited above the supposed three allelomorphs have similar ontogenetic effects. Thus the three in rabbits, in *Aquilegia,* and in *Papilio* all affect the distribution of pigment (and, in *Papilio,* also the shape of the wings), those in *Lychnis* the sex, those in beans the production of the same color in different organs, and those in *Drosophila* the production of different colors in the same organ. This may perhaps seem to be in favor of the view that we have here different modifications of the same gene, rather than two distinct genes and their absences.

The history of the red-white-eosin group of eye colors in *Drosophila* is interesting when considered from the viewpoint of the presence and absence hypothesis. The first white-eyed fly arose as a mutant in red stock. On presence and absence it must have been caused by the simultaneous loss of two factors, which were called *C* and *O* by Morgan. Then, in white-eyed stock there appeared an eosin-eyed fly. Here the factor called *O,* just lost, must have been put back again. Finally, in one of my own cultures, eosin has given rise to white by mutation.[1] In both these latter cases the flies had miniature wings, and in the white-to-eosin case they also had black body color. These characters give a check on the results, and make it extremely unlikely that any contamination had occurred. Further evidence to this effect is given, in the eosin-to-white case, by the fact that the mutant fly was one of 127 obtained from a single pair, all of her brothers and sisters being of the expected classes (half of the females heterozygous for white), as were likewise the flies from six sister pairs.[2]

The presence and absence hypothesis involving a dropping out of whole genes or addition of entirely new ones does not offer as simple an explanation of this case as does the conception that we have here a relatively unstable gene, which does not drop out entirely, but undergoes various changes, that from white to eosin being reversible.

It should be noted that Shull ('11) has reported a case of what he calls reversible mutation in the sex-determining factor in *Lychnis,* which is very similar to the above red-eosin-white case. He has adopted a system of triple allelomorphs to explain it, though

[1] After this paper went to press it was pointed out to me by Mr. H. J. Muller that there is another possible explanation of this case, which does not involve mutation from eosin to white. This interpretation can not be entered into until certain phenomena observed by Mr. C. B. Bridges have been more fully investigated.

[2] There has been at least one case where eosin has seemed to arise as a mutant in red stock, but as there was no other character to serve as a check against contamination, the case does not carry very great weight.

admitting that complete linkage will also cover the facts. He has also considered the bearing of the case upon the presence and absence hypothesis and upon the nature of mutation, reaching conclusions somewhat similar to those given above.

LITERATURE CITED

Baur, E. 1912. Vererbungs- und Bastardierungsversuche mit *Antirrhinum.* II. Faktorenkoppelung. Zts. ind. Abst.-Vererb.-Lehre **6,** p. 201.

Castle, W. E. 1906. Heredity of Coat Characters in Guinea-pigs and Rabbits. Carnegie Inst. Wash. publ. 23.

______ (and others). 1909. Studies of Inheritance in Rabbits. Carnegie Inst. Wash. publ. 114.

Emerson, R. A. 1911. Genetic Correlation and Spurious Allelomorphism in Maize. Ann. Rept. Nebr. Agr. Sta. **24,** p. 58.

Hurst, C. C. 1906. Mendelian Characters in Plants and Animals. Rep. Conf. Gen., Roy Hort. Soc. London, p. 114.

de Meijere, J. C. H. 1910. Über Jacobsons Züchtungsversuche bezüglich des Polymorphismus von *Papilio Memnon* L. ♀, und über die Vererbung sekundärer Geschlechtsmerkmale. Zts. ind. Abst.-Vererb.-Lehre **3,** p. 161.

Morgan, T. H. 1912. Further Experiments with Mutations in Eye Color in *Drosophila.* Jour. Acad. Nat. Sci. Phila. **15,** p. 323.

______ 1913. Factors and Unit Characters in Mendelian Heredity. Am. Nat. **47,** p. 5.

Punnett, R. C. 1912. Inheritance of Coat-color in Rabbits. Jour. Genet. **2,** p. 221.

Shull, G. H. 1911. Reversible Sex-mutants in *Lychnis dioica.* Bot. Gaz. **42,** p. 329.

chapter 4

Multiple-factor inheritance

It was pointed out in Chapter 3 that some traits like comb shape in poultry and flower color in sweet peas were the results of nonallelic gene interaction (Ref. 3-10). When the number of loci affecting a trait is small, the contribution of each gene to the individual's phenotype can be ascertained. However, as the number of genes affecting a characteristic increases, it becomes difficult to determine how much of the trait is attributable to the action of a given gene. When the number of loci becomes too great, it is no longer possible to specify the genotype of the individual. Under such conditions, one can only characterize the population as to the average size or color of the trait in question and the variability of the trait in the population. The study of multiple-factor inheritance is steadily increasing because many important traits (stature, weight, intelligence; egg, milk, and meat production; yields of fruits and seeds; etc.) are determined by multiple genes.

COLOR IN WHEAT KERNELS

An early example of multiple-factor inheritance that was analyzed in terms of Mendelian principles was reported by Nilsson-Ehle in 1908 (Ref. 4-1). He found that crosses involving a white-kerneled and a red-kerneled variety of wheat yielded an F_1 whose seeds were intermediate in color between those of the two parents. The F_2 was found to be quite variable with regard to kernel color and could be divided into seven different color classes in the ratio of 1:6:15:20:15:6:1. Only 1/64 of the kernels were completely white, and only 1/64 were as red as the red-kerneled variety of wheat used in the first cross. It was postulated that three pairs of genes were involved in color production in wheat kernels. The red parent was assumed to have six genes for pigment production and the white parent to contain the alleles that prevented pigment production. In addition, it appeared that all alleles are equally potent in the production or lack of production of pigment. In essence, this experiment represents the results of a trihybrid cross in which there is an additive effect of the various allelic and nonallelic pigment-producing genes.

Other similar cases of quantitative characters were also found to be analyzable in Mendelian terms. A basic discussion of the multiple gene hypothesis was presented by East in 1910 (Ref. 4-2) and is reprinted in this chapter.

SKIN COLOR IN MAN

Shortly after the Mendelian basis for multiple-factor inheritance had been postulated, attempts were made to analyze the human traits that appeared to follow this pattern. One of the earlier investigations dealt with human skin color and was reported by Davenport in 1913 (Ref. 4-3). Working with Negroes and Caucasians and the hybrids between them, he postulated that two pairs of genes are involved in the production of the skin color pigment, melanin. This would imply that the offspring of racial crosses between Negroes and Caucasians represent dihybrids with additive gene effects. The F_2 of such a cross would yield five different classes of color intensities in the ratio of 1 black: 4 dark: 6 intermediate or mulatto: 4 light: 1 white. The above hypothesis appears to postulate too few segregating alleles. A more recent analysis of human skin color inheritance was prepared by Harrison and Owen in 1964 (Ref. 4-4) and is included in this collection.

TRANSGRESSIVE VARIATION

The study of multiple-factor inheritance revealed some interesting information about strains and varieties of organisms that had been inbred for a particular trait. One case involved breeds of different sized chickens and was summarized by Punnett in 1923 (Ref. 4-5). When a Golden Hamburg chicken (large-sized breed) was crossed with a Sebright Bantam (small-sized breed), the F_1 was found to be intermediate in size between the parents. However, the F_2 was found to contain some birds that were larger and some that were smaller than the parental strains, while most of the F_2 were intermediate between the parental types. On the basis of the frequencies of the extreme types, it was postulated that four pairs of genes were involved in determining size in these chickens. The Golden Hamburg was considered to have the genotype *AABBCCdd* and the Sebright Bantam *aabbccDD.* The F_1 would then be heterozygous for all four genes, *AaBbCcDd,* and of intermediate size. According to Mendelian rules of inheritance, the F_2 would contain some individuals who were homozygous for all the dominant alleles. These birds would be larger than the parental Golden Hamburg chicken. The F_2 would also contain some individuals who were homozygous for all the recessive alleles, and these birds would be smaller than the parental Sebright Bantam type. This situation, in which the extremes of the F_2 exceed those of the parents, is called *transgressive variation.* It is usually found in crossing inbred lines, where the inbred lines do not represent the extremes possible for the species.

POLYGENIC INHERITANCE

The examples of multiple-factor inheritance considered thus far have been relatively simple and easy to analyze. However, as pointed out earlier, the larger the number of loci, the more difficult the analysis. The situation can be even further complicated if the alleles at the various loci have a relatively small effect on the trait being studied. The term *polygene* is used to describe those genes whose alleles produce small phenotypic differences of about the same order of magnitude as that caused by usual environmental fluctuations. It is often impossible to estimate how much of a phenotype is determined by polygenes and how much is determined by environment. Polygenic inheritance can be studied through statistical analysis of the trait in a population. The importance of polygenic inheritance is easily recognized when one realizes that most traits of the organism are controlled by polygenes. A review of the fundamental concepts of quantitative inheritance was prepared by Mather in 1943 (Ref. 4-6). A more recent analysis of a polygenic trait was made in 1960 by Milkman (Ref. 4-7), who studied the genetic control of crossvein formation in the *Drosophila* wing. He found that there were a large number of genes, located in the three major chromosomes of *D. melanogaster,* each of which had a relatively small effect on crossvein formation.

In the next chapter, we shall consider the spatial arrangement of genes in chromosomes, and the various ways in which this arrangement can be changed.

REFERENCES

Color in wheat kernels

4-1. Nilsson-Ehle, H. 1908. Einige Ergebnisse von Kreuzungen bei Hafer und Weisen. Bot. Notiser, Lund **1908-1909:**257-294.

4-2. East, E. M. 1910. A Mendelian interpretation of variation that is apparently continuous. Am. Nat. **44:**65-82.

Skin color in man

4-3. Davenport, C. B. 1913. Heredity of skin color in Negro-white crosses. Carnegie Institution, Washington, pub. no. 188, Washington, D. C.

4-4. Harrison, G. A., and Owen, J. J. T. 1964. Studies on the inheritance of human skin colour. Ann. Hum. Genet., Lond. **28:**27-37.

Transgressive variation

4-5. Punnett, R. C. 1923. Heredity in poultry. The Macmillan Co., New York.

Polygenic inheritance

4-6. Mather, K. 1943. Polygenic inheritance and natural selection. Biol. Rev. **18:**32-64.

4-7. Milkman, R. D. 1960. The genetic basis of natural variation. II. Analysis of a polygenic system in *Drosophila melanogaster.* Genetics **45:**377-391.

16 A Mendelian interpretation of variation that is apparently continuous

Edward M. East†
Harvard University
Cambridge, Massachusetts

There are two objects in writing this paper. One is to present some new facts of inheritance obtained from pedigree cultures of maize; the other is to discuss the hypotheses to which an extension of this class of facts naturally leads. This discussion is to be regarded simply as a suggestion toward a working hypothesis, for the facts are not sufficient to support a theory. They do, however, impose certain limitations upon speculation which should receive careful consideration.

The facts which are submitted have to do with independent allelomorphic pairs which cause the formation of like or similar characters in the zygote. Nilsson-Ehle[1] has just published facts of the same character obtained from cultures of oats and of wheat. My own work is largely supplementary to his, but it had been given these interpretations previous to the publication of his paper.

In brief, Nilsson-Ehle's results are as follows: He found that while in most varieties of oats with black glumes blackness behaved as a simple Mendelian mono-hybrid, yet in one case there were two definite independent Mendelian unit characters, each of which was allelomorphic to its absence. Furthermore, in most varieties of oats having a ligule, the character behaved as a mono-hybrid dominant to absence of ligule, but in one case no less than four independent characters for presence of ligule, each being dominant to its absence, were found. In wheat a similar phenomenon occurred. Many crosses were made between varieties having red seeds and those having white seeds. In every case but one the F_2 generation gave the ordinary ratio of three red to one white. In the one exception—a very old red variety from the north of Sweden—the ratio in the F_2 generation was 63 red to 1 white. The reds of the F_2 generation gave in the F_3 generation a very close approximation to the theoretical expectation, which is 37 constant red, 8 red and white separating in the ratio of 63:1, 12 red and white separating in the ratio of 15:1, 6 red and white separating in the ratio of 3:1, and one constant white. He did not happen to obtain the expected constant white, but in the total progeny of 78 F_2 plants his other results are so close to the theoretical calculation that they quite convince one that he was really dealing with three indistinguishable but independent red characters, each allelomorphic to its absence. Nor can the experimental proof of the two colors of the oat glumes be doubted. The evidence of four characters for presence of ligule in the oat is not so conclusive.

In my own work there is sufficient proof to show that in certain cases the endosperm of maize contains two indistinguishable, independent yellow colors, although in most yellow races only one color is present. There is also some evidence that there are three and possibly four independent red colors in the pericarp, and two colors in the aleurone cells. The colors in the aleurone cells when pure are easily distinguished, but when they are together they grade into each other very gradually.

Fully fifteen different yellow varieties of maize have been crossed with various white varieties, in which the crosses have all given a simple mono-hybrid ratio. In the other cases that follow it is seen that there is a di-hybrid ratio.

No. 5–20, a pure white eight-rowed flint, was pollinated by No. 6, a dent pure for yellow endosperm. An eight-rowed ear was obtained containing 159 medium yellow kernels and 145 light yellow kernels. The pollen parent was evidently a hybrid homozygous for one yellow which we will call Y_1 and heterozygous for another yellow Y_2. The gametes $Y_1 Y_2$ and Y_1

From The American Naturalist **44**:65-82, 1910. Used with permission.

Contributions from the Laboratory of Genetics, Bussey Institution, Harvard University, No. 4. Read before the annual meeting of the American Society of Naturalists, Boston, December 29, 1909.

†Deceased.

[1] Nilsson-Ehle, H. Kreuzungsuntersuchungen an Hafer und Weizen. Lunds Universitets Årsskrift, N. F. Afd. 2., Bd. 5, No. 2, 1909.

Table I.[2] F_2 seeds from cross of no. 5–20, white flint X no. 6 yellow dent, homozygous for Y_1 and heterozygous for Y_2

Dark seeds heterozygous for both yellows planted

Ear No.	Dark Y	Light Y	Total Y	No Y
1	270	56	326	29
2	101	215	316	27
3	261	52	313	28
5	273	284	557	35
10	358	117	475	25
12	296	72	368	19
13	207	156	363	35
14	387	102	489	29
Total	2153	1054	3207	227
Ratio			14.1	1

[2] In these tables only hand pollinated ears are given.

Table II. F_2 seeds from same cross as shown in Table I

Light yellow seeds heterozygous for Y_1 *planted*

Ear No.	Dark Y	Light Y	No Y
1		359	117
2		144	54
3		173	63
4		433	136
6		316	120
8	331		109
8a		229	86
9		325	115
10		227	87
11[3]		4	434
12		318	118
13		256	93
Total		3111	1098
Ratio		2.8	1

[3] Discarded from average. This ear evidently grew from one kernel of the original white mother that was accidentally self-pollinated. The four yellow kernels all show zenia from accidental pollination in the next generation.

Table III. No. 11 yellow X no. 8 white

F_3 generation from yellow seeds of F_2 generation

Ear No.	Dark Y	Light Y	Total Y	No Y	Ratio they approximate
1	116	95	211	19	15Y:1 no Y
14			88	5	15Y:1 no Y
5	181	122			$3Y_1Y_2$:1 $Y_{1 \text{ or } 2}$
4		253		68	3Y:1 no Y
6		193		73	"
8		163		79	"
11		108		35	"
9		456			Constant $Y_{1 \text{ or } 2}$

Table IV. Progeny of ear no. 8 of the same cross as shown in Table III

F_3 generation from yellow seeds of F_2 generation

Ear No.	Dark Y	Light Y	Total Y	No Y	Ratio they approximate
10	101	188	289	25	15Y:1 no Y
11	89	219	308	23	15Y:1 no Y
3		233			constant light Y
9	dark and light		331		3 dark:1 light Y
13	dark and light		350		3 dark:1 light Y
8		294		108	3 light:1 no Y
15		221		87	3 light:1 no Y
1[4]		197		203	

[4] Kernel from which this ear grew was evidently pollinated by no Y.

fertilized the white in equal quantities, giving a ratio of approximately one medium yellow to one light yellow. The F_2 kernels from the dark yellow were as shown in Table I. The ratios of light yellows to dark yellows is very arbitrary, for there was a fine gradation of shades. The ratio of total yellows to white, however, is unmistakably 15:1.

In the next table (Table II) are given the results of F_2 kernels from the light yellows of F_1. Only ear No. 8, which was really planted with the dark yellows, showed yellows dark enough to be mistaken for kernels containing both Y_1 and Y_2. The remaining ears are clearly mono-hybrids with reference to yellow endosperm.

In a second case the female parent possessed the yellow endosperm. No. 11, a twelve-rowed yellow flint, was crossed with No. 8, a white dent. The F_2 kernels in part showed clearly a mono-hybrid ratio, and in part blended gradually into white. Two of these indefinite ears proved in the F_3 generation to have had the 15:1 ratio in the F_2 generation. Ear 7 of the F_2 generation calculated from the results of the entire F_3 crop must have had about 547 yellow to 52 white kernels, the theoretical number being 561 to 31. The hand-pollinated ears of the F_3 generation (yellow seeds) gave the results shown in Table III.

The F_3 generation grown from the other ear, Ear No. 8, showed that the ratio of yellows to whites in the F_2 generation was about 227 to 47. As the theoretical ratio is 257 to 17, the ratio obtained is somewhat inconclusive. A classification of the open field crop could not be made accurately on account of the light color of the yellows and the presence of many kernels showing zenia. Table IV, however, showing the hand-pollinated kernels of the interbred yellows of the F_2 generation, settles beyond a doubt the fact that the two yellows were present.

In a third case an eight-rowed yellow flint, No. 22, was crossed with a white dent, No. 8. Only four selfed ears were obtained in the F_2 generation. Ear 1 had 72 yellow to 37 white kernels. This ear was poorly developed and undoubtedly had some yellow kernels which were classed as whites. Ear 4 had 158 yellow and 42 white kernels. It is very likely that both of these ears were mono-hybrids, but the F_3 generation was not grown. Ear 5 had 148 yellow and 15 white kernels. Ear 7 had 78 yellow and 5 white kernels. It seems probable that both of these ears were di-hybrids, but only Ear 5 was grown another generation. The kernels classed as white proved to be pure; the open field crop from the yellow kernels gave 14 pure yellow ears and 14 hybrid yellow. Theoretically the ratio should be 7 pure yellows (that is, pure for either one or both yellows) and 8 hybrid yellows (4 giving 15 yellows to 1 white and 4 giving 3 yellows to 1 white). Five hand-pollinated selfed ears were obtained. Three of these gave mono-hybrid ratios, with a total of 607 yellows to 185 white kernels. One ear was a pure dark yellow (probably $Y_1 Y_1 Y_2 Y_2$). The other ear was poorly filled, but had 27 dark yellows (probably $Y_1 Y_2$) and 7 light yellow kernels (Y_1 or Y_2). Unfortunately no 15:1 ratio was obtained in this generation, but this is quite likely to happen when only five selfed ears are counted. The gradation of colors and the general appearance of the open field crop, however, led me to believe that we were again dealing with a di-hybrid.

Two yellows appeared in still another case, that of white sweet No. 40♀ X yellow dent No. 3♂. Only one selfed ear was obtained in the F_2 generation giving 599 yellow to 43 white kernels. Of these kernels 486 were starchy and 156 sweet, which complicated matters in the F_3 generation because it was very difficult to separate the light yellow sweet from the white sweet kernels. Among the selfed ears were three pure to the starchy character, and in these ears the dark yellows, the light yellows and whites stood out very distinctly. Ear 12 had 156 dark yellow; 47 light yellow; 14 white kernels. Ear 13 had 347 dark yellow; 93 light yellow; 25 white kernels. The third starchy ear, No. 6, had 320 light yellow; 97 white kernels. Two ears, therefore, were di-hybrids, and one ear a mono-hybrid.

The ears which were heterozygous for starch and no starch and those homozygous for no starch, could not all be classified accurately, but it is certain that some pure dark yellows, some pure light yellows, some showing segregation of yellows and whites at the ratio 15:1, and some showing segregation of yellows and whites at the ratio of 3:1, were obtained.

One other case should be mentioned. One ear of a dent variety of unknown parentage obtained for another purpose was found to have some apparently heterozygous yellow kernels. Seven selfed ears were obtained from them, of which two were pure yellow. The other five ears each gave the di-hybrid ratio. There was a total of 1906 yellow seeds to 181 white seeds, which is reasonably close to the expected ratio, 1956 yellow to 131 white.

It is to be regretted that I can present no other case of this class that has been fully worked out, although several other characters which I have under observation in both maize and tobacco seem likely to be included ultimately. Nevertheless, the fact that we have to deal with conditions of this kind in studying inheritance is established; granting only that they will be somewhat numerous, it opens up an entirely new outlook in the field of genetics.

In certain cases it would appear that we may have several allelomorphic pairs each of which is inherited independently of the others, and each of which is separately capable of forming the same character. When present in different numbers in different individuals, these units simply form quantitative differences. It may be objected that we do not know that two colors that appear the same physically are exactly the same chemically. That is true; but Nilsson-Ehle's case of several unit characters for presence of ligule in oats is certainly one where each of several Mendelian units forms exactly the same character. It may be that there is a kind of biological isomerism, in which, instead of molecules of the same formula having different physical properties, there are isomers capable of forming the same character, although, through difference in construction, they are not allelomorphic to each other. At least it is quite a probable supposition that through imperfections in the mechanism of heredity an individual possessing a certain character should give rise to different lines of descent so that in the F_n generation when individuals of these different lines are crossed, the character behaves as a di-hybrid instead of as a mono-hybrid. In other words, it is more probable that these units arise through variation in different individuals and are combined by hybridization, than that actually different structures for forming the same character arise in the same individual.

On the other hand, there is a possibility of an action just the opposite of this. Several of these quantitative units which produce the same character may become attached like a chemical radical and again behave as a single pair. Nilsson-Ehle gives one case which he does not attempt to explain, where the same cross gave a 4:1 ratio in one instance and 8.4:1 ratio in another instance. In his other work characters always behaved the same way; that is, either as one pair, two pairs, three pairs, etc. In my work, the yellow endosperm of maize has behaved differently in the same strain, but it is probably because the yellow parent is homozygous for one yellow and heterozygous for the other. They were known to be pure for one yellow, but it would take a long series of crosses to prove purity in two yellows.

Let us now consider what is the concrete result of the inter-action of several cumulative units affecting the same character. Where there is simple presence dominant to absence of a number n of such factors, in a cross where all are present in one parent and all absent in the other parent, there must be 4^n individuals to run an even chance of obtaining a single F_2 individual in which the character is absent. When four such units, $A_1A_2A_3A_4$ are crossed with $a_1a_2a_3a_4$, their absence, only one pure recessive is expected in 256 individuals. And 256 individuals is a larger number than is usually reported in genetic publications. When a smaller population is considered, it will appear to be a blend of the two parents with a fluctuating variability on each side of its mode. Of course if there is absolute dominance and each unit appears to affect the zygote in the same manner that they do when combined, the F_2 generation will appear like the dominant parent unless a very large number of progeny are under observation and pure recessives are obtained. This may be an explanation of the results obtained by Millardet; it is certainly as probable as the hypothesis of the non-formation of homozygotes. Ordinarily, however, there is not perfect dominance, and variation due to heterozygosis combined with fluctuating variation makes it almost impossible to classify the individuals except by breeding. The two yellows in the endosperm of maize is an example of how few characters are necessary to make classification difficult. First, there is a small amount of fluctuation in different ears due to varying light conditions owing to differences in thickness of the husk; second, all the classes having different gametic formulae differ in the intensity of their yellow in the following order, $Y_1Y_1Y_2Y_2$, $Y_1y_1Y_2Y_2$ or $Y_1Y_1Y_2y_2$, Y_1Y_1, Y_2Y_2, Y_1y_1, Y_2y_2, y_1y_2. As dominance becomes less and less evident, the Mendelian classes vary more and more from the formula $(3 + 1)^n$, and approach the normal curve, with a regular gradation of individuals on each side of the mode. When there is no dominance and open fertilization, a state is reached in which the curve of variation simulates the fluctuation curve, with the difference that the gradations are heritable.

One other important feature of this class of genetic facts must be considered. If units $A_1A_2A_3a_4$ meet units $a_1a_2a_3A_4$, in the F_2 generation there will be one pure recessive, $a_1a_2a_3a_4$, in every 256 individuals. This explains an apparent paradox. Two individuals are crossed, both seemingly pure for presence of the same character, yet one individual out of 256 is a pure recessive. When we consider the rarity with which pure dominants or pure recessives (for all characters) are obtained when there are more than three factors, we can hardly avoid the suspicion that here is a perfectly logical way of accounting for many cases of so-called atavism. Furthermore, many apparently new characters may be formed by the gradual dropping of these cumulative factors without any additional hypothesis. For example, in *Nicotiana tabacum* varieties there is every gradation[5] of loss of leaf surface near the base of the sessile leaf, until in *N. tabacum fruticosa* the leaf is only one step removed from a petioled condition. If this step should occur the new plant would almost certainly be called a new species; yet it is only one degree further in a definite series of loss gradations that have already taken place. If it should be assumed that in other instances slight qualitative as well as quantitative changes take place as units are added, then it becomes very easy, theoretically,

[5] It is not known at present how this character behaves in inheritance.

to account for quite different characters in the individual homozygous for presence of all dominant units, and in the individual in which they are all absent.

Unfortunately for these conceptions, although I feel it extremely probable that variations in *some* characters that seem to be continuous will prove to be combinations of segregating characters, it is exceedingly difficult to demonstrate the matter beyond a reasonable doubt. As an illustration of the difficulties involved in the analysis of pedigree cultures embracing such characters, I wish to discuss some data regarding the inheritance of the number of rows of kernels on the maize cob.

The maize ear may be regarded as a fusion of four or more spikes, each joint of the rachis bearing two spikelets. The rows are, therefore, distinctly paired, and no case is known where one of the pair has been aborted. This is a peculiar fact when we consider the great number of odd kinds of variations that occur in nature. The number of rows per cob has been considered to belong to continuous variations by DeVries, and a glance at the progeny from the seeds of a single selfed ear as shown in Table V seems to confirm this view.

Table V. Progeny of a selfed ear of leaming maize having 20 rows

Classes of rows..	12	14	16	18	20	22	24	26	28	30
No. of ears.........	1	0	5	4	53	35	19	5	2	1

There is considerable evidence, however, that this character is made up of a series of cumulative units, independent in their inheritance. There is no reason why it should not be considered to be of the same nature as various other size characters in which variation seems to be continuous, but in which relatively constant gradations may be isolated, each fluctuating around a particular mode. But this particular case possesses an advantage not held by most phenomena of its class, in that there is a definite discontinuous series of numbers by which each individual may be classified.

Previous to analyzing the data from pedigree cultures, however, it is necessary to take into consideration several facts. In the first place, what limits are to be placed on fluctuations?[6] From the variability of the progeny of single ears of dent varieties that have been inbred for several generations, it might be concluded that the deviations are very large. But this is not necessarily the case; these deviations may be due largely to gametic structure in spite of the inbreeding, since no conscious selection of homozygotes has been made. There is no such variation in eight-rowed varieties, which may be considered as the last subtraction form in which maize appears and therefore an extreme homozygous recessive. In a count of the population of an isolated maize field where Longfellow, an eight-rowed flint, had been grown for many years, 4 four-rowed, 993 eight-rowed, 2 ten-rowed and 1 twelve-rowed ears were found. Only seven aberrant ears out of a thousand had been produced, and some of these may have been due to vicinism.

On the other hand, a large number of counts of the number of rows of both ears on stalks that bore two ears has shown that it is very rare that there is a change greater than ± 2 rows. If conditions are more favorable at the time when the upper ear is laid down it will have two more rows than the second ear; if conditions are favorable all through the season, the ears generally have the same number of rows; while if conditions are unfavorable when the upper ear is laid down, the lower ear may have two more rows than the upper ear. Furthermore, seeds from the same ear have several times been grown on different soils and in different seasons, and in each case the frequency distribution has been the same. Hence it may be concluded that in the great majority of cases fluctuation is not greater than in ± 2 rows, although fluctuations of ± 4 rows have been found.

A second question worthy of consideration is: Do somatic variations due to varying conditions during development take place with equal frequency in individuals with a large number of rows and in individuals with a small number of rows? From the fact that several of my inbred strains that have been selected for three generations for a constant number of rows, increase directly in variability as the number of rows increases, the question should probably be answered in the negative. This answer is reasonable upon other grounds. The eight-rowed ear may vary in any one of four spikes, the sixteen-rowed ear may vary in any one of eight spikes; therefore the sixteen-rowed ear may vary twice as often as the eight-rowed

[6] The word fluctuation is used to designate the somatic changes due to immediate environment, and which *are not inherited.*

ear. By the same reasoning, the sixteen-rowed ear may sometimes throw fluctuations twice as wide as the eight-rowed ear.

A third consideration is the possibility of increased fluctuation due to hybridization. Shull[7] and East[8] have shown that there is an increased stimulus to cell division when maize biotypes are crossed—a phenomenon apart from inheritance. There is no evidence, however, that increased gametic variability results. Johannsen[9] has shown that there is no such increase in fluctuation when close-pollinated plants are crossed. I have crossed several distinct varieties of maize where the modal number of rows of each parent was twelve, and in every instance the F_1 progeny had the same mode and about the same variability.

Finally, a possibility of gametic coupling should be considered. Our common races of flint maize all have a low number of rows, usually eight but sometimes twelve; dent races have various modes running from twelve to twenty-four rows. When crosses between the two subspecies are made, the tendency is to separate in the same manner.

[7] Shull, G. H., "A Pure-line Method in Corn Breeding," Rept. Amer. Breeders' Assn. **5**:51-59, 1909.

[8] East, E. M., "The Distinction between Development and Heredity in Inbreeding," Amer. Nat. **43**:173-181, 1909.

[9] Johannsen, W., "Does Hybridization Increase Fluctuating Variability?" Rept. Third Inter. Con. on Genetics, 98-113, London, Spottiswoode, 1907.

Attention is not called to these obscuring factors with the idea that they are universally applicable in the study of supposed continuous variation. But there are similar conditions always present that make analysis of these variations difficult, and the facts given here should serve to prevent premature decision that they do not show segregation in their inheritance.

Table VI shows the results from several crosses between maize races with different modal values for number of rows. Several interesting points are noticeable. The modal number is always divisible by four. This is also the case with some twenty-five other races that I have examined but which are not shown in the table. I suspect that through the presence of pure units zygotes having a multiple of four rows are formed, while heterozygous units cause the dropping of two rows. The eight-rowed races are pure for that character, the twelve-rowed races vary but little, but the races having a higher number of rows are exceedingly variable.

When twelve-rowed races are crossed with those having eight rows, the resulting F_1 generation always—or nearly always—has the mode at twelve rows. In one case cited in Table VI, No. 24 × No. 53, nearly all the F_1 progeny were eight-rowed. It might appear from this, either that the low number of rows was in this case dominant, or that the female parent has more influence on the resulting progeny than

Table VI. Crosses between maize strains with different numbers of rows

Parents (female given first)	Gen.	Row classes						
		8	10	12	14	16	18	20
Flint No. 5		100						
Flint No. 11		1	4	387	7	1		
Flint No. 24		100						
Flint No. 15		100						
Dent No. 6				6	31	51	18	4
Dent No. 8			3	54	36	12	2	
Sweet No. 53[10]		1	5	25	4			
Sweet No. 54[10]		25	2	1				
No. 5 × No. 53	F_1	1	7	13				
No. 5 × No. 6	F_1	11	18	27	3			
No. 11 × No. 5	F_1	2	4	18				
No. 11 × No. 53	F_1	2	5	17				
No. 24 × No. 53	F_1	57	8	3				
No. 15 × No. 8	F_1	1	14	26	3	1		
No. 15 × No. 8 (from 10-row ear)	F_2	14	15	28	9	1		
No. 15 × No. 8 (from 12-row ear)	F_2	4	13	25	6	3		
No. 8 × No. 54	F_1	1	6	14				
No. 8 × No. 54 (from 12-row ear)	F_2	11	25	38	2	1		

[10] Approximately.

the male parent. I prefer to believe, however, that the individual of No. 53 which furnished the pollen was due to produce eight-rowed progeny. Unfortunately no record was kept of the ear borne by this plant, but No. 53 sometimes does produce eight-rowed ears.

When a race with a mode higher than twelve is crossed with an eight-rowed race, the F_1 generation is always intermediate, although it tends to be nearer the high-rowed parent. Only one example is given in the table, but it is indicative of the class. These results are rather confusing, for there seems to be a tendency to dominance in the twelve-rowed form that is not found in the forms with a higher number of rows. I have seen cultures of other investigators where 12-row X 8-row resulted in a ten-rowed F_1 generation, so the complication need not worry us at present.

The results of the F_2 generation show a definite tendency toward segregation and reproduction of the parent types. I might add that in at least two cases I have planted extracted eight-rowed ears and have immediately obtained an eight-rowed race which showed only slight departures from the type. Selection from those ears having a high number of rows has also given races like the high-rowed parent without recrossing with it. It is regretted that commercial problems were on hand at the time and no exact data were recorded. It can be stated with confidence, however, that ears like each parent are obtained in the F_2 generation, from which with care *races* like each parent may be produced. *Segregation seems to be the best interpretation of the matter.*

These various items may seem disconnected and uninteresting, but they have been given to show the tangible basis for the following theoretical interpretation. No hard and fast conclusion is attempted, but I feel that this interpretation with possibly slight modifications will be found to aid the explanation of many cases where variation is apparently continuous.

Suppose a basal unit to be present in the gametes of all maize races, this unit to account for the production of eight rows. Let additional independent interchangeable units, each allelomorphic to its own absence, account for each additional four rows; and let the heterozygous condition of any unit represent only half of the homozygous condition, or two rows. Then the gametic condition of a homozygous twenty-rowed race would be 8 + *AABBCC*, each letter actually representing two rows. When crossed with an eight-rowed race, the F_2 generation will show ears of from eight to twenty rows, each class being represented by the number of units in the coefficients in the binomial expansion where the exponent is twice the number of characters, or in this case $(a + b)^6$.

The result appears to be a blend between the characters of the two parents with a normal frequency distribution of the deviants. Only one twenty-rowed individual occurs in 64 instead of the 27 expected by the interaction of three dominant factors in the usual Mendelian ratios. The remainder of the 27 will have different numbers of rows, and, by their gametic formulae, different expectations in future breeding as follows:

1 *AABBCC* = 20 rows.
2 *AaBBCC* = 18 rows.
2 *AABbCC* = 18 rows.
2 *AABBCc* = 18 rows.
4 *AaBbCC* = 16 rows.
4 *AaBBCc* = 16 rows.
4 *AABbCc* = 16 rows.
8 *AaBbCc* = 14 rows.

There are four visibly different classes and eight gametically different classes. It must also be remembered that the probability that the original twenty-rowed ear in actual practise may have had more than three units in its gametes has not been considered. This point is illustrated clearly if we work out the complete ratio for the three characters, and note the number of gametically different classes which compose the modal class of fourteen rows in Table VII. It actually contains seven gametically different classes and not a single homozygote. If this conception of independent allelomorphic pairs affecting the same character proves true, it will sadly upset the biometric belief that the modal class is *the type* around which the variants converge, for there is actually less chance of these individuals breeding true than those from *any other* class.

The conception is simple and is capable theoretically of bringing in order many complicated facts, although the presence of fluctu-

Table VII. Theoretical expectation in F_2 when a homozygous twenty-rowed maize ear is crossed with an eight-rowed ear

Classes	8	10	12	14	16	18	20
No. ears	1	6	15	20	15	6	1

ating variation will be a great factor in preventing analysis of data. I have thought of only one fact that is difficult to bring into line. If $8AA$, $8BB$ and $8CC$ all represent homozygous twelve-rowed ears—to continue the maize illustration—and none of these factors are allelomorphic to each other, sixteen-rowed ears should sometimes be obtained when crossing two twelve-rowed ears. I am not sure but that this would happen if we were to extract all the homozygous twelve-rowed strains after a cross between sixteen-row and eight-row, and after proving their purity cross them. In some cases the additional four-row units would probably be allelomorphic to each other and in other cases independent of each other. On the other hand, this is only an hypothesis, and while I have faith in its foundation facts, the details may need change.

Castle has raised the point that greater variation should be expected in the F_1 generation than in the P_1 generations when crossing widely deviating individuals showing variation apparently continuous. If the parents are strictly pure for a definite number of units, say for size, a greater variation should certainly be expected in the F_1 generation after crossing. But considering the difficulties that arise when even five independent units are considered, can it be said that anything has heretofore been known concerning the actual gametic status of parents which it is known do vary in the character in question and in which the variations are inherited, for the race can be changed by selection within it. It may be, too, that the correct criterion has not been used in size measurements, for, as others have suggested, solids vary as the cube root of their mass, whereas the sum of the weights of the body cells has usually been measured and compared directly with similar sums.

Attention should be called to one further point. Many characters in all probability are truly blending in their inheritance, but there is another interpretation which may apply in certain cases. I have repeatedly tried to cross Giant Missouri Cob Pipe maize (14 feet high) and Tom Thumb pop maize (2 feet high), but have always failed. They both cross readily with varieties intermediate in size, but are sterile between themselves. We may imagine that the gametes of each race, though varying in structure, are all so dissimilar that none of them can unite to form zygotes. Other races may be found where only part of the gametes of varying structure are so unlike that they will not develop after fusion. The zygotes that do develop will be from those more alike in construction. An apparent blend results, and although segregation may take place, no progeny as extreme as either of the parents will ever occur.

I may say in conclusion that the effect of the truth of this hypothesis would be to add another link to the increasing chain of evidence that the word mutation may properly be applied to any inherited variation, however small; and the word fluctuation should be restricted to those variations due to immediate environment which do not affect the germ cells, and which—it has been shown—are not inherited. In addition it gives a rational basis for the origin of *new* characters, which has hitherto been somewhat of a Mendelian stumbling-block; and also gives the term unit-character less of an irrevocably-fixed-entity conception, which is more in accord with other biological beliefs.

17 Studies on the inheritance of human skin colour

G. Ainsworth Harrison
J. J. T. Owen
Anthropology Laboratory
Department of Human Anatomy
University of Oxford
Oxford, England

INTRODUCTION

Variation in human skin colour is remarkable in that the variation between different populations is often very great by comparison with that within populations. The nature of the between population differences has therefore seemed to be particularly open to analysis, although the character shows quantitative variation in hybrid groups.

A number of studies have been made, particularly on the differences between Africans and Europeans (Davenport & Danielson, 1913; Gates, 1949; Stern, 1953), but these have tended to suffer from the inadequacy of the methods previously available for measuring skin colour and the diversity of relationships in the populations studied. Recently, reflectance spectrophotometry has provided an objective method of measuring pigmentation (Weiner, 1951; Lasker, 1954; Harrison & Owen, 1956; Barnicot, 1958; Walsh, 1963) and in various parts of Britain, hybrid populations between Africans and Europeans have been forming in which it is often possible to trace the relationship of individuals back to the original miscegenation. The present analysis is based upon such a population studied in Liverpool during the past 10 years.

MATERIALS AND METHODS

A portable E.E.L.* reflectance spectrophotometer was used in the study. This instrument is fitted with nine different filters which 'sample' the whole of the visual spectrum. Typically, at least three of these filters were used in measuring subjects, viz. filter 601 with a dominant wavelength (d.w.l.) of 425 mμ; filter 605, d.w.l. 545 mμ; and filter 609, d.w.l. 685 mμ; but, in a few instances, and especially in the case of one large group of West African seamen, it was only possible to take readings with one filter. In these cases the 609 filter was chosen. On the other hand, readings on about a half of the subjects were taken using all or most of the filters. In every case measurements were made on the medial aspect of the right upper arm, and except in the case of young children, the position was standardized as precisely as possible by placing the rim of the applicator head against the medial epicondyle of the humerus.

Most of the subjects were residents of Liverpool and were contacted by house to house surveys in areas where there is a large West African population. The European, 105, with a large Irish component, came from the same areas as the hybrids and were of comparable social class; in many cases they were the mothers of hybrid families, or relatives of these mothers. The sample of West Africans (106) is made up of three groups: (1) a group of settled Liverpool residents, (2) a group of seamen who are usually on the West African 'run' but often have homes in Liverpool, and (3) a small group of university students. Analysis of the results has shown that these groups do not differ in skin colour, but there are three important points about this sample of West Africans. First, it covers a wide range of West African populations (mostly from coastal regions of Ghana and Nigeria) and is of diverse tribal origin; Barnicot (1958) has shown that there are considerable differences in colour between such neighbouring tribes as the Yoruba and Ibo. Secondly, it is possible that a few individuals in the group have had some distant white ancestry, since some miscegenation has occurred in West Africa—particularly along the coast. Finally, it is a sample composed almost exclusively

From Annals of Human Genetics, London **28:**27-37, 1964. Used with permission.

Acknowledgement is made to the Eugenics Society for an initial grant which enabled the work to start, and to an anonymous donor in the U. S. A. who provided funds for the continuance of the work through the G.C.L. Bertram Research Account.

*Evans Electro-Selenium Co. Ltd., Halstead.

of males, and it has been shown in a number of studies (Lasker, 1954; Barnicot, 1958) that males tend to have slightly darker skins than females, partly because of a richer blood supply to their skin (Edwards & Duntley, 1939) and partly because they tend to expose themselves to more sunlight. In the present data on Europeans and F_1 hybrids the same tendency is evident but in neither case is the sex difference significant and it has been ignored in the subsequent analysis. Since measurements were made on the medial aspect of the upper arm, rather than on the forearm or forehead as in other investigations there are no *a priori* grounds for expecting a sex difference.

Four types of hybrid have been found: first generation (94) F_1, hybrids between African males and European females; and backcross European (30), backcross African (26) and (14), F_2, hybrids, but the second generation hybrids, particularly F_2s, are still rare. Further, the likelihood of the numbers increasing is remote, since the recent large-scale immigration of West Indians of unknown ancestry has markedly widened the variety of possible mates. All the hybrid samples, and especially the second generation hybrids, are of lower mean age than the parental samples, but there is evidence that after the first few years of life there is no change in pigmentation (Lasker, 1954). No children under 2 years of age were included. The pedigrees of the hybrids were

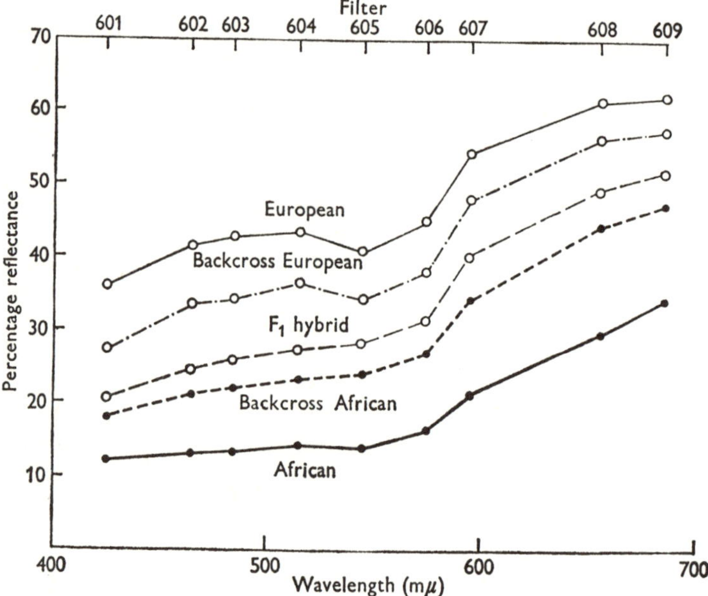

Fig. 1. Mean reflectance curves of European, African and various hybrid groups.

Table 1. Means and standard errors of reflectance values at nine wavelengths of Europeans, West Africans and various hybrid groups

BE, backcross European *BA*, backcross African

Wave-length	European		West African		F_1 hybrid		*BE* hybrid		*BA* hybrid		F_2 hybrid	
(mμ)	No.	Mean±S.E.	No.	Mean±S.E.	No.	Mean±S.E.	No.	Mean±S.E.	No.	Mean±S.E.	No.	Mean±S.E.
425	104	36.1±0.453	40	12.3±0.484	94	20.8±0.508	30	27.2±1.320	21	18.0±0.915	13	22.2±1.566
465	51	41.6±0.710	40	13.2±0.522	86	24.5±0.572	29	33.5±1.319	18	21.1±1.275	6	28.5±1.962
485	49	43.0±0.668	39	13.4±0.567	85	26.0±0.620	26	34.1±1.371	18	22.1±1.553	6	27.7±2.304
515	46	43.7±0.640	37	14.6±0.680	77	27.4±0.590	14	36.7±2.056	16	23.3±1.685	3	25.7±1.517
545	103	41.0±0.453	40	14.4±0.611	94	28.4±0.581	30	34.7±1.122	21	24.2±1.334	12	30.3±1.483
575	51	45.2±0.526	40	16.6±0.704	86	31.7±0.585	28	38.2±1.206	18	27.2±1.515	6	33.5±1.944
595	51	54.8±0.529	40	21.7±0.834	86	40.5±0.633	29	48.2±1.145	18	34.6±1.699	6	43.7±1.943
655	51	61.7±0.436	40	29.9±1.062	87	49.7±0.586	29	56.7±0.947	18	44.7±1.615	6	53.0±1.789
685	105	62.3±0.342	106	34.7±0.591	94	52.0±0.546	30	57.9±0.926	26	47.8±1.205	14	53.4±1.455

ascertained as carefully as possible and in many cases the legitimacy of children was checked by blood-grouping. No cases of undisclosed illegitimacy were found.

RESULTS

Reflectance curves

The mean reflectance values of the parental and various hybrid groups, together with their standard errors and the number of individuals on which they are based are presented in Table 1. The means are also represented graphically in Fig. 1. Although the various reflectance curves are based only on a series of sample wavelengths, their general form is very similar to that obtained with a continuous recording instrument, e.g. the Hardy reflectance spectrophotometer (Weiner, 1951). In particular, they show the characteristic absorption of melanin at the shorter wavelengths, and the absorption band of haemoglobin at 545 mμ which is increasingly obvious with decreasing amounts of melanin. The F_1 hybrid curve is approximately intermediate between the two parental groups, but it is evident that the precise relationship changes with wavelength. In particular the curve is nearer that of the African parent than the European parent at 425 mμ, whilst at 685 mμ the relationship is reversed. In both instances, the curves for the backcrosses fall roughly intermediate between the curves for the F_1 hybrid and the respective parent, but obviously in the case of the backcross African the curve lies nearer the F_1 hybrid than the African parent.

Melanin concentration

It has been shown (Harrison & Owen, 1956) that *in vitro,* melanin concentration is linearly proportional to the reciprocal of the reflectance values. At short wavelengths and high concentrations of melanin the relationship tends to be disturbed but at long wavelengths linearity is evident over a considerable range of concentrations. It has further been found that the effects of scattering of light by the skin do not profoundly affect this relationship.

The reflectance values of the two parental groups and the F_1 hybrids have therefore been transformed to reciprocals and the means of these are plotted in Fig. 2. This clearly shows that the F_1 hybrid is more similar to the European than the African parent in terms of melanin concentration. It is also evident that on the reciprocal of reflectance scale, the relationship of the F_1 hybrid to the two parents is more comparable at the different wavelengths than it is on the reflectance scale. Indeed at

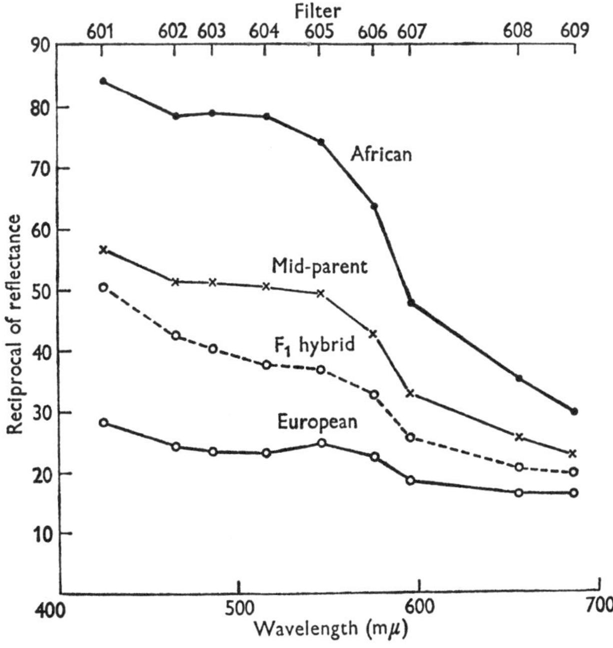

Fig. 2. Mean reciprocal of reflectance curves of Europeans, Africans and F_1 hybrids. The mid-parent values are also shown.

wavelengths longer than 545 mμ the relationship is exactly comparable for the different filters. At shorter wavelengths the comparability is less exact but, as already mentioned, at these wavelengths the linear relationship between concentration and the reciprocal of reflectance exists over a much smaller range of melanin concentration.

The criteria of scaling

For purposes of a genetical analysis it is necessary that the measurements should be made on a scale in which (1) non-heritable variation is independent of that due to gene segregation, and (2) the responsible genes are additive on average in their effects. In analysing the genetical differences between true breeding lines the first criterion can be tested by comparing the variances of the true breeding parents and their F_1 hybrid. Under most circumstances, one would expect to find equality of variances on that scale which removes environmental interaction. In the present situation, of course, the two parental populations cannot strictly be regarded as true breeding. Not only must there be some genetical component to the intraparental variation, but one would also expect the African variance to be larger than the European variance because of the more heterogeneous origin of the African sample. However, the genetical variability within either parental group must be small in comparison with that between the parental groups. A scale has, therefore, been sought in which the parental and F_1 hybrid variances are of essentially the same magnitude and are each small by comparison with the inter-parental variation. Because of the very small number of F_2s available, no attention has been given to this type of hybrid, but it has been assumed that the estimated backcross variances are not too awry.

In Table 2, the variances of the different groups, together with the corresponding means at three wavelengths and on various scales are presented. The comparative magnitudes of these variances on the reflectance scale are obviously related to mean measurement. For instance, the European variance tends to decrease with the rise in mean measurement which occurs with increasing wavelength. Conversely the African variance tends to rise from its value at 425 mμ to that at 685 mμ. It would in fact appear that there is a falling off of variance with high and low reflectance values, as sampled respectively by the European and African groups. This trend is fully confirmed by the results obtained at intermediate wavelengths. It is of interest to note that the estimated F_1 and F_2 variances are fairly constant at the different wavelengths, no doubt because their mean measurements fall intermediate in the reflectance scale, but both backcross hybrids show the same trend as the parental groups. This effect has been noted in other work on the reflectometry of hair and skin colour (Sunderland, 1956; Barnicot, 1958) and seems to be intrinsic in the method of

Table 2. Means and variances of reflectances at different wavelengths and with various transformations
B_E, European backcross B_A, African backcross

	European		African		F_1 hybrid	
Scale (mμ) R	Mean $\bar{E}$	Variance $V\bar{E}$	Mean $\bar{A}$	Variance $V\bar{A}$	Mean $\bar{F}_1$	Variance $V\bar{F}_1$
425	36.1	21.283	12.3	9.354	20.8	20.428
545	41.0	21.098	14.4	14.918	28.4	31.748
685	62.3	12.248	34.7	37.039	52.0	28.011
$1/R \times 1000$						
425	28.2	16.194	85.4	632.423	50.6	125.760
545	24.7	8.829	74.0	356.608	36.5	50.260
685	16.1	0.956	29.5	38.064	19.5	4.494
$\text{Log}_{10} R$						
425	1.556	0.00317	1.082	0.01236	1.308	0.00885
Antilog R						
685	0.421	0.00109	0.225	0.00105	0.334	0.00159

measuring pigmentation by reflectance spectrophotometry.

This factor would seem to account for the anomalous comparative variances that one finds at short and long wavelengths on the reflectance scale. At 425 mμ, not only does one find that the estimated African variance is lower than that of the European—an unlikely situation—but also that the backcross African variance is less than both the European and F_1 variances! Similarly, on reflectance at 685 mμ the backcross European variance is less than the African and F_1 variances. On these grounds alone, these two scales are clearly unsuitable for a genetical analysis, since there is obvious interaction between genotype and the 'environment'. In this case, of course, the environment relates to the method of measurement. On the other hand, at 545 mμ, although the estimated African variance is lower than one would expect in comparison with the European, and is also significantly smaller than the F_1 variance, both backcross variances exceed those of the non-segregating populations. It would seem, therefore, that of the three different wavelengths environmental interaction is least at 545 mμ and it is possible that it is sufficiently small to permit a genetical analysis.

It is evident from Table 2 that although the reciprocal of reflectance is proportional to melanin concentration, and would afford a scale which is more or less independent of wavelength, it is completely unsuitable for a genetical analysis. The transformation is so strong that the African variance becomes not only about 40 times that of the European variance but is also greater than the variances of all the other genotypes. It would be possible to weaken this type of transformation by employing a scale such as $1/(1 - R)$ but in terms of simple transformations, which are likely to more or less equalize the parental variances at 425 and 685 mμ, logarithmic and antilogarithmic scales respectively would seem to be the most obvious choices. Table 2 shows that these transformations meet the criteria of scaling better than the corresponding reflectance scales. Admittedly at 425 mμ the backcross African variance has become smaller than the African variance, indicating that the transformation is too strong and that there is still considerable environmental interaction. However, compared with the European the magnitude of the African variance is in the expected direction, and the backcross African variance, instead of being less than the variances of two of the non-segregating groups, is now only less than one. The antilog of reflectance at 685 mμ is even better. Not only is there approximate equality of the parental and F_1 variances, but both backcross variances are in excess of these. From the point of view of removing environmental interaction this scale would seem to be the best of those tested.

The second criterion for scaling is that the effect of a particular gene substitution should be independent of the rest of the genotype. This can be tested by comparing the means of

B_E hybrid		B_A hybrid		F_2 hybrid	
Mean $\bar{B}_E$	Variance $V\bar{B}_E$	Mean $\bar{B}_A$	Variance $V\bar{E}$	Mean $\bar{F}_2$	Variance $V\bar{F}_2$
27.2	52.305	18.0	17.600	22.2	31.858
34.7	37.766	24.2	37.366	30.3	26.382
57.9	25.697	47.8	37.760	53.4	29.646
39.4	109.822	58.5	189.440	47.4	112.818
29.9	30.099	44.1	150.098	34.0	35.282
17.5	2.499	21.3	10.288	18.9	3.928
1.420	0.01310	1.245	0.01024	1.336	0.01047
0.382	0.00200	0.304	0.00171	0.346	0.00199

the second generation hybrids with those of the parents and F_1 hybrid. Additiveness of genic effect is evidenced, on the one hand, by strict intermediacy of the backcross mean between the F_1 mean and the corresponding parental mean, and on the other, by the relationship of the F_2 mean to the means of the three non-segregating groups. The results of such tests on various scales are shown in Table 3.

It is apparent that the backcross European mean does not deviate significantly from strict intermediacy on any of the scales tested. However, some significant departure of the backcross African mean does occur on all the scales, with the exception of reflectance at 425 mμ, though in all but one case the significance of this departure is only just over the 5% level. Tests using the F_2 mean do not discriminate between the various scales apart from confirming that reflectance measured at 685 mμ gives the poorest fit. Whilst reflectance at 425 mμ would seem on this test of means to afford the best scale, it has already been shown that it is totally inadequate on the variance criteria, and although the logarithmic transformation of reflectance at this wavelength gives a poorer fit for additiveness, on an overall judgement of both criteria it is to be preferred. Although there is some small departure of the backcross African mean from strict intermediacy on reflectance at 545 mμ, this scale seems to fit fairly well the criterion of additiveness, as well as being the reflectance scale on which there is least environmental interaction. The antilog transformation at 685 mμ, not only provides more acceptable variances than reflectance at this wavelength, but also considerably improves the scaling on means.

On the basis of the scaling criteria it seemed justifiable to proceed to a partition of variation on the following three scales: log *R* 425 mμ, *R* 545 mμ and antilog *R* 685 mμ.

Partition of variation

Quantitative variation is typically made up of three components. Following the nomenclature of Mather (1949) these may be represented as (1) *E*, a component which arises from non-heritable causes, (2) *D*, a heritable component arising from the differences in phenotypic expression associated with the two homozygotes for each gene pair, i.e. fixable variation due to the parental difference, and (3) *H*, a heritable component due to the differences in expression between heterozygotes for each gene pair and the average of the two corresponding homozygotes, i.e. unfixable variation due to the average dominance of the gene sets involved.

In the present data an estimate of *E* is provided by the mean of the European, African and F_1 hybrid variances. Genetical differences within the parental groups, but not contributing to the differences between these groups, will necessarily be included within this component and thus will lead to an overestimation of the true effect of the external environment. However, this should not seriously affect the estimations of the heritable inter-population difference. On the assumption that the parental populations are essentially homozygous for the genes responsible for this difference the principles of the method developed by Fisher, Immer & Tedin (1932) and Mather (1949) for partitioning *D* and *H* can be applied. These authors have shown that the sum of the two backcross variances equals $\frac{1}{2}D+\frac{1}{2}H+2E$ and that

Table 3. Scaling tests using the mean values and the corresponding variances as given in Table 2

Scale (mμ) *R*	A	S.E.	B	S.E.	C	S.E.
425	−2.46	2.7196	+2.88	1.9504	− 1.08	6.3560
545	−0.08	2.3618	+5.55*	2.7978	+ 8.75	6.0910
685	+1.53	1.9599	+8.97***	2.5409	+12.80*	5.9616
Log_{10} *R* 425	−0.023	0.04326	+0.100*	0.04817	+ 0.90	0.11602
Antilog *R* 685	+0.008	0.01711	+0.049*	0.01703	+ 0.070	0.04861

$A = 2\bar{B}_E - \bar{F}_1 - \bar{E}$, $V_A = 4V\bar{B}_E + V\bar{F}_1 + V\bar{E}$,
$B = 2\bar{B}_A - \bar{F}_1 - \bar{A}$, $V_B = 4V\bar{B}_A + V\bar{F}_1 + V\bar{A}$,
$C = 4\bar{F}_2 - 2\bar{F}_1 - \bar{E} - \bar{A}$, $V_C = 16V\bar{F}_2 + 4V\bar{F}_1 + V\bar{E} + V\bar{A}$

*Represents significance at the 5% probability level. ***Represents significance at the 0.1% probability level.

the F_2 variance equals ½D+¼H+E. Since in the present data the estimate of F_2 variance is too unreliable to be of use, some other equation relating H and D is required. The deviation of the F_1 mean from the mid-parent expressed as a ratio of half the parental difference provides a measure of the relative potence of the gene sets. Although this potence ratio is not representative of average dominance, unless the dominance is isodirectional, it does provide a minimal estimate of the magnitude of H in relation to D, since with isodirectional dominance $\sqrt{(H/D)}$ equals the potence ratio. It would seem that the estimate of H which can be obtained from the equations relating backcross variance with the potence ratio, however inaccurate, is better than none at all. It must be noted that there are no *a priori* grounds for assuming that dominance is likely to be isodirectional, though one might have expected from the results of other interracial crosses that dominance would be slight.

The various components of variation as measured on the most suitable scales are presented in Table 4. It can be seen that in each instance the potence ratio is negligible in amount, but it is always directed towards the European. It follows from the method of analysis used that the estimate of H is also small in all cases. There is further agreement on the different scales in the relative proportions of the D and E components, with the D component contributing between 63 and 72% of the total variation and conversely E contributing between 36 and 27%. An estimate of the number of effective factors responsible for the inter-parental difference (k_1) can be shown to be equal to the ratio of the square of half the parental difference to D (Mather, 1949). Using the various scales at different wavelengths, the number of effective factors has been found to be between 3 and 4.

DISCUSSION

The deficiencies in the data in this study are keenly appreciated by the writers, but since there appear at present to be no opportunities for improving the data, it seems justifiable to take the analysis as far as possible. In considering where the deficiencies mainly lie the principal factor is the small number of segregating hybrids available. This is of particular importance in view of the reliance which must be placed on the estimated variance of at least the backcross generations. As is well known variances are less robust statistics than means, and as Gilbert (1961) has suggested it would be preferable for an analysis to be based on means alone. However, it may be noted that on the scales used the backcross variances are approximately equal and this is consistent with the small potence ratio found in Table 4. The latter is based upon well substantiated mean measurements.

A second source of error may be a failure to find the best scale. Scales, however, can, in

Table 4. Components of variation, where *D, H, E* are described in the text (p. 132)

Scale	D	H	E	Potence ratio = $\frac{\text{Deviation of } F_1 \text{ from midparent}}{\text{½ Parental difference}}$	$K = \frac{(\text{difference between parents})^2}{4D}$	
R 545mμ	59.746	0.165	22.588	0.053	2.96	E / D, H
$\text{Log}_{10} R$ 425mμ	0.01414	0.00003	0.00813	0.046	3.97	E / D, H
Antilog R 685mμ	0.002417	0.000030	0.001243	0.112	3.97	E / D, H

most cases, only be sought on an empirical basis, and on the basis of the present data there seemed little justification for testing complex transformations. Even the functional representation of reflectance curves is proving an extremely complex problem. The fact that there is general agreement on three scales which more or less meet the required criteria suggests that any error due to scale effect is small.

In the genetical analysis the use of the potence ratio as a measure of average dominance has probably led to an underestimation of H, but it can be said that if, in fact, dominance does exist it must be evenly balanced. Although there are no theoretical limits to the possible error in the estimated magnitude of H, even if the D component were half that calculated, the number of effective factors would remain relatively small, i.e. 6–8.

Effective factors, of course, do not necessarily represent individual genes; they could equally well be segments of chromosomes, or indeed whole chromosomes, but in the absence of third-generation hybrid data, it is impossible to detect any linkage that may exist. The analysis does indicate, however, that probably three or four chromosomes are involved in determining the pigmentary differences between Africans and Europeans. Further, the estimate of effective factors agrees closely with determinations made in other studies of quantitative variation (Mather, 1949).

It is impossible at the moment to say whether these conclusions relate solely to the site of measurement—the medial aspect of the upper arm—or whether they are of general applicability to the body as a whole. Certainly one would expect the environmental component to vary with position, since not only are different parts of the body differentially exposed to ultraviolet radiation, but also it has been shown that different areas have varying capacities to tan (Edwards & Duntley, 1939). The reasons for choosing the medial aspect of the upper arm in this study were that, while reasonably accessible, the region is normally not exposed to much sunlight and anyway has a poor tanning capacity. The differences between the present data and the measurements made by Barnicot (1958) on essentially similar populations may well be largely attributable to the differences in the site of measurement, since Barnicot measured reflectance on the forearm. It is interesting to note, however, that in Barnicot's data the comparative relationships of the mean reflectance values of the F_1 hybrids to the two parental types are different from those found in this investigation: the F_1 hybrid being relatively nearer the African. Since Barnicot measured his Europeans in Britain and his Africans and hybrids in West Africa, comparative tanning might account for this difference.

The nature of this analysis raises very clearly the problems involved in using reflectance spectrophotometry (or for that matter any other measure) in comparing the genetical basis for population differences in skin colour. It is apparent that on the reflectance scale comparative similarity is a function of the wavelength at which measurements are made. For instance, the F_1 mean reflectance is relatively nearer the African than the European at 425 mμ and contrariwise at 625 mμ (see Fig. 1). If the F_1 population had been of unknown origin, its genetical affinities would have appeared different at the two wavelengths and, as the present analysis has shown, neither of these would have been correct. The question of scale in the use of anthropometric characters is of crucial importance, yet rarely does one have the opportunity of directly measuring the same character on different scales. It is very easy to forget that the scale of measurement may bear no relation to the underlying genetical differences. The scales of additiveness which this analysis has revealed may be far from perfect, but it seems reasonable to suppose that they will be generally applicable to comparative surveys of skin colour with at least the E.E.L. spectrophotometer.

SUMMARY

Measures of skin colour have been made using an E.E.L. spectrophotometer of 105 Europeans, 106 West Africans, 94 F_1 hybrids, 30 backcross European and 26 backcross African hybrids and 14 F_2 hybrids in Liverpool. Of the scaling tests applied, additiveness of genic effect and independence of non-heritable variations was greatest with a log transformation of reflectance of incident light of 425 mμ, without transformation at 545 mμ and with an antilog transformation at 685 mμ. Using these scales it has been shown that the relative potence of gene sets is negligible. By making the assumption that such potence as does exist provides a measure of average dominance, the number of effective factors responsible for the difference in skin colour between Europeans and Africans has been estimated from the different scales to be between 3 and 4.

REFERENCES

Barnicot, N. A. (1958). Reflectometry of the skin in southern Nigerians and in some Mulattoes. Hum. Biol. **30:**150.

Davenport, C. B. & Danielson, F. H. (1913). Skin colour in Negro-White crosses. Publ. 188, Carnegie Institution of Washington.

Edwards, E. A. & Duntley, S. Q. (1939). Pigments and colour of living human skin. Amer. J. Anat. **65:**1.

Fisher, R. A., Immer, F. R. & Tedin, O. (1932). The genetical interpretation of statistics of the third degree in the study of quantitative inheritance. Genetics **17:**107.

Gates, R. R. (1949). Pedigrees of Negro Families. Philadelphia: Blakiston.

Gilbert, N. (1961). Polygene analysis. Genet. Res. **2:**96.

Harrison, G. A. & Owen, J. J. T. (1956). The application of spectrophotometry to the study of skin colour inheritance. Acta Genet. **6:**481.

Lasker, G. W. (1954). Photoelectric measurement of skin colour in a Mexican mestizo population. Amer. J. Phys. Anthrop., n.s. **12:**115.

Mather, K. (1949). Biometrical Genetics. London: Methuen.

Stern, C. (1953). Model estimates of the frequency of white and near-white segregants in the American Negro. Acta Genet. **4:**281.

Sunderland, E. (1956). Hair-colour variation in the United Kingdom. Ann. Hum. Genet., Lond., **20:** 312.

Walsh, R. J. (1963). Variations of melanin pigmentation of the skin in some Asian and Pacific peoples. J. Roy. Anthrop. Inst. **93:**126.

Weiner, J. S. (1951). A spectrophotometer for measurement of skin colour. Man **51:**253.

chapter 5

Genes and chromosomes

LINKAGE, CROSSING-OVER, AND GENETIC MAPS

It will be recalled from Chapter 3 that soon after the rediscovery of Mendel's findings, the relationship of genes to chromosomes was recognized (Ref. 3-5). This was followed by evidence that some genes were linked to one another (Ref. 3-12). However, it was also found that linked genes did not always remain together. Complicating the linkage picture was the fact that the frequency of separation of linked genes varied with the pair studied. The problem was resolved by Morgan in 1911 (Ref. 5-1), who hypothesized that genes are arranged in a linear order in a chromosome and that the degree, or strength, of linkage depends on the distance between the genes. His paper is reprinted in this chapter. Subsequently, a tremendous number of experiments involving many diverse species have confirmed the correctness of Morgan's interpretation of linkage. An early example of this type of experiment in *Drosophila* was reported by Sturtevant in 1913 (Ref. 5-2).

The genetic demonstration that linked genes can separate and form new combinations clearly required that, during meiosis, homologous chromosomes must be capable of breakage and reunion in such a way as to result in an exchange of corresponding regions of their chromosome material. The occurrence of such a phenomenon had been reported in 1909 by Janssens (Ref. 5-3), who was studying meiosis in certain amphibians. He observed the presence at synapsis of cross figures, to which he gave the name *chiasmata.* He interpreted each chiasma as involving two of the four strands of the particular tetrad. Morgan refers to Janssens's observations as supporting evidence for his own theory on gene linkage. Unmistakable proof that recombination of linked genes involved a reciprocal exchange of chromosome material was demonstrated independently in 1931 by Stern, working with *Drosophila* (Ref. 5-4), and by Creighton and McClintock, working with maize (Ref. 5-5). The latter paper is included in this collection.

Linkage studies permitted the construction of genetic maps of the known mutations in a species. In these maps the genes are placed in a relative order to one another, and the distance between them is determined by the observed frequency of recombinants formed in specifically designed matings. It was soon found that in every well-studied species the number of linkage groups equaled the haploid number of chromosomes. In viruses the correspondence of linkage groups to chromosomes means that all the known genes are to be found in a single chromosome, as discussed in Chapter 2 (Ref. 2-1). The same situation exists for the bacteria (Ref. 2-7). In *D. melanogaster,* the number of linkage groups is four, whereas in man it is twenty-three.

The first report of linkage in humans was made by Bell and Haldane in 1937 (Ref. 5-6) for the genes causing color-blindness and hemophilia. A paper discussing the general problems involved in mapping human genes was prepared by Lawler and Renwick in 1959 (Ref. 5-7) and is included in this chapter. A more recent review of our knowledge of linkage groups involving human autosomes was presented by Renwick in 1969 (Ref. 5-8).

Recombination is not restricted to chromosomes involved in meiosis. Crossing-over can also occur in somatic cells. Stern in 1936 (Ref. 5-9) demonstrated the occurrence of recombination in somatic cells of *D. melanogaster.* Cytological evidence that crossing-over can occur in human somatic cells was reported by German in 1964 (Ref. 5-10).

FACTORS AFFECTING RECOMBINATION FREQUENCY

The frequency of recombination between genes was found to vary, depending on experimental factors. It was soon realized that variation in crossover frequency could be brought about by changing the environmental conditions under which an experiment was conducted or by choosing organisms of certain genotypes.

Environmental factors that have been found

to affect recombination frequency include, among others, temperature, nutrition, and exposure to ionizing radiations. It was reported by Plough in 1921 (Ref. 5-11) that temperatures both above and below 22° C increase the crossover rate in *D. melanogaster.* Levine in 1955 (Ref. 5-12) found that *Drosophila* fed a high-calcium diet exhibited a decreased rate of crossing-over, whereas flies fed a medium that removes metallic ions showed an increased frequency of recombination. Whittinghill in 1937 (Ref. 5-13) discovered that *Drosophila* exposed to x-ray irradiation exhibited an increase in crossover frequency.

Genetic factors that have been found to affect recombination frequency include, among others, the sex of the individual, the presence of chromosomal rearrangements, and the possession of a particular gene. The influence of the individual's sex on crossover rates is most strikingly seen in all *Drosophila* species, where crossing-over is suppressed in males. This phenomenon was shown to exist by Morgan in 1912 (Ref. 5-14) for genes located in the second chromosome of *D. melanogaster* and was subsequently found to include the other chromosomes of this species and the chromosomes of all other *Drosophila* species.

An example of the effect of a chromosomal rearrangement on crossing-over was first observed by Bridges in 1917 (Ref. 5-15). He found that in *D. melanogaster* a deficiency of a section of a chromosome led to an elimination of crossing-over throughout the length of the deficient section in flies containing both a complete and a partially deficient chromosome. The influence of a single gene on recombination frequency in *D. melanogaster* was reported by Gowen and Gowen in 1922 (Ref. 5-16). They found a third-chromosome gene which, when homozygous, completely suppresses crossing-over in females. A comparable situation was discovered in *Escherichia coli* by Clark and Margulies in 1965 (Ref. 5-17), whose paper is reprinted in this chapter.

CHROMOSOMAL REARRANGEMENTS

The above discussion of crossing-over has been limited to the situation of a mutual exchange of equal sized segments of homologous chromosomes. However, this need not always be the case. Should chromosome breaks occur at different positions in the homologous chromosomes, there would result an "unequal" crossing-over by which genes originally present in the two parental chromosomes would emerge in the same chromosome. This rearrangement of chromosomal material is called a *duplication* and was first reported in *D. melanogaster* by Sturtevant and Morgan in 1923 (Ref. 5-18). A discussion and analysis of this chromosomal rearrangement was prepared by Bridges in 1936 (Ref. 5-19).

In addition to exchanges of genetic material between homologous chromosomes, other types of chromosomal breaks and reunions are possible. These may involve the loss or rearrangement of genetic material of a single chromosome or an exchange of genetic material between nonhomologous chromosomes.

When two breaks occur in the same chromosome, two end fragments and a middle segment result. If the two end fragments join one another, they will form a chromosome that lacks the genetic material of the middle segment. This is called a *deficiency* and was first reported in *D. melanogaster* by Bridges in 1917 (Ref. 5-15). Whenever a deficient chromosome is synapsed with its normal homologue, a "buckling" will occur in the normal chromosome at the point of the deficiency. Should any recessive genes be present in the section of the normal chromosome that corresponds to the deficiency, they will be expressed in the organism's phenotype. This latter phenomenon is called *pseudodominance.*

Another type of chromosomal rearrangement can result from the reunion of fragments produced by two breaks in the same chromosome. All three chromosomal segments may join together with the middle piece remaining in the middle but turned end-for-end. This is called an *inversion.* In an inversion, there is no addition or loss of chromosome material. Only the order of the genes has been changed. The first inversions were detected in *D. melanogaster* as a result of the *suppression of crossing-over* between homologues consisting of a normal and an inversion-bearing chromosome. An analysis of this genetic evidence for inversions was presented by Sturtevant in 1926 (Ref. 5-20). Whenever an inversion-bearing chromosome is synapsed with its normal homologue, a "loop" will be formed. This cytological evidence of the occurrence of an inversion was analyzed by Painter in 1934 (Ref. 5-21).

The final type of chromosomal rearrangement that we shall discuss involves nonhomologous chromosomes. Should a segment from one chromosome become attached to a nonhomologous chromosome, the result is called *transloca-tion.* A "reciprocal translocation" occurs when segments from nonhomologous chromosomes become involved in a mutual interchange. The occurrence of a translocation in *D. melanogaster* was first reported by Bridges in 1923 (Ref. 5-22). A recent discussion of reciprocal translocations, especially as related to man, was prepared by Ford and Clegg in 1969 (Ref. 5-23) and is the last article reprinted in this chapter.

REFERENCES

Linkage, crossing-over, and genetic maps

5-1. Morgan, T. H. 1911. Random segregation versus coupling in Mendelian inheritance. Science **34:** 384.

5-2. Sturtevant, A. H. 1913. The linear arrangement of six sex-linked factors in *Drosophila,* as shown by their mode of association. J. Exptl. Zool. **14:**43-59.

5-3. Janssens, F. A. 1909. La théorie de la chiasmatypie, nouvelle interprétation des cinésis de maturation. La Cellule **25:**387-406.

5-4. Stern, C. 1931. Zytologisch-genetische Untersuchungen als Beweise für die Morgansche Theorie des Faktorenaustausches. Biol. Zentralbl. **51:**547-587.

5-5. Creighton, H. B., and McClintock, B. 1931. A correlation of cytological and genetical crossing-over in *Zea mays.* Proc. Natl. Acad. Sci. U.S.A. **17:**492-497.

5-6. Bell, J., and Haldane, J. B. S. 1937. The linkage between the genes for colour-blindness and haemophilia in man. Proc. Roy. Soc., Lond. B. **123:**119-150.

5-7. Lawler, S. D., and Renwick, J. H. 1959. Blood groups and genetic linkage. Brit. Med. Bull. **15:**145-149.

5-8. Renwick, J. H. 1969. Progress in mapping human autosomes. Brit. Med. Bull. **25:**65-73.

5-9. Stern, C. 1936. Somatic crossing over and segregation in *Drosophila melanogaster.* Genetics **21:**625-630.

5-10. German, J. 1964. Cytological evidence for crossing-over in vitro in human lymphoid cells. Science **144:**298-301.

Factors affecting recombination frequency

5-11. Plough, H. H. 1921. Further studies on the effect of temperature on crossing-over. J. Exptl. Zool. **32:**187-202.

5-12. Levine, R. P. 1955. Chromosome structure and the mechanism of crossing-over. Proc. Natl. Acad. Sci. U.S.A. **41:**727-730.

5-13. Whittinghill, M. 1937. Induced crossing-over in *Drosophila* males and its probable nature. Genetics **22:**114-129.

5-14. Morgan, T. H. 1912. Complete linkage in the second chromosome of the male. Science **36:** 719-720.

5-15. Bridges, C. B. 1917. Deficiency. Genetics **2:**445-465.

5-16. Gowen, M. S., and Gowen, J. W. 1922. Complete linkage in *Drosophila melanogaster.* Am. Nat. **56:**286-288.

5-17. Clark, A. J., and Margulies, A. D. 1965. Isolation and characterization of recombination deficient mutants of *E. coli* K12. Proc. Natl. Acad. Sci. U.S.A. **53:**451-459.

Chromosomal rearrangements

5-18. Sturtevant, A H., and Morgan, T. H. 1923. Reverse mutation of the bar gene correlated with crossing over. Science **57:**746-747.

5-19. Bridges, C. B. 1936. The bar "gene" a duplication. Science **83:**210-211.

5-20. Sturtevant, A. H. 1926. A crossover reducer in *Drosophila melanogaster* due to inversion of a section of the third chromosome. Biol. Zentralbl. **46:**697-702.

5-21. Painter, T. S. 1934. A new method for the study of chromosome aberrations and the plotting of chromosome maps in *Drosophila melanogaster.* Genetics **19:**175-188.

5-22. Bridges, C. B. 1923. The translocation of a section of chromosome II upon chromosome III in *Drosophila.* Anat. Rec. **24:**426-427.

5-23. Ford, C. E., and Clegg, H. M. 1969. Reciprocal translocations. Brit. Med. Bull. **25:**110-114.

18 Random segregation versus coupling in Mendelian inheritance

T. H. Morgan†

Mendel's law of inheritance rests on the assumption of random segregation of the factors for unit characters. The typical proportions for two or more characters, such as 9:3:3:1, etc., that characterize Mendelian inheritance, depend on an assumption of this kind. In recent years a number of cases have come to light in which when two or more characters are involved the proportions do not accord with Mendel's assumption of random segregation. The most notable cases of this sort are found in sex-limited inheritance in *Abraxas* and *Drosophila,* and in several breeds of poultry, in which a coupling between the factors for femaleness and one other factor must be assumed to take place, and in the case of peas where color and shape of pollen are involved. In addition to these cases Bateson and his collaborators (Punnett, DeVilmorin and Gregory) have recently published[1] a number of new ones.

In order to account for the results Bateson assumes not only coupling, but also repulsions in the germ cells. The facts appear to be exactly comparable to those that I have discovered in *Drosophila,* and since these results have led me to a very simple interpretation, I venture to contrast Bateson's hypothesis with the one that I have to offer.

The facts on which Bateson bases his interpretation may be briefly stated in his own words, namely: "that if *A, a* and *B, b* are two allelomorphic pairs subject to coupling and repulsion, the factors *A* and *B* will repel each other in the gametogenesis of the double heterozygote resulting from the union *Ab* × *aB,* but will be coupled in the gametogenesis of the double heterozygote resulting from the union *AB* × *ab,*" and further, "We have as yet no probable surmise to offer as to the essential nature of this distinction, and all that can yet be said is that in these special cases the distribution of the characters in the heterozygote is affected by the distribution in the original pure parents." Bateson further points out that since "sex in the fowls acts as a repeller of at least three other factors, . . . some of them may be found able to take precedence of the others in such a way as to annul the present repulsion with subsequent coupling as a consequence."

In place of attractions, repulsions and orders of precedence, and the elaborate systems of coupling, I venture to suggest a comparatively simple explanation based on results of inheritance of eye color, body color, wing mutations and the sex factor for femaleness in *Drosophila.* If the materials that represent these factors are contained in the chromosomes, and if those factors that "couple" be near together in a linear series, then when the parental pairs (in the heterozygote) conjugate like regions will stand opposed. There is good evidence to support the view that during the strepsinema stage homologous chromosomes twist around each other, but when the chromosomes separate (split) the split is in a single plane, as maintained by Janssens. In consequence, the original materials will, for short distances, be more likely to fall on the same side of the split, while remoter regions will be as likely to fall on the same side as the last, as on the opposite side. In consequence, we find coupling in certain characters, and little or no evidence at all of coupling in other characters; the difference depending on the linear distance apart of the chromosomal materials that represent the factors. Such an explanation will account for all of the many phenomena that I have observed and will explain equally, I think, the other cases so far described. The results are a simple mechanical result of the location of the materials in the chromosomes, and of the method of union of homologous chromosomes, and the proportions that result are not so much the expression of a numerical system as of the relative location of the factors in the chromosomes. *Instead of random segregation in Mendel's sense we find "associations of factors" that are located near together in the chromosomes. Cytology furnishes the mechanism that the experimental evidence demands.*

From Science **34**:384, September 10, 1911. Used with permission.

†Deceased.

[1] Proc. Royal Soc., Vol. **84**, 1911.

19 A correlation of cytological and genetical crossing-over in Zea mays

Harriet B. Creighton*
Barbara McClintock
Botany Department
Cornell University
Ithaca, New York

A requirement for the genetical study of crossing-over is the heterozygous condition of two allelomorphic factors in the same linkage group. The analysis of the behavior of homologous or partially homologous chromosomes, which are morphologically distinguishable at two points, should show evidence of cytological crossing-over. It is the aim of the present paper to show that cytological crossing-over occurs and that it is accompanied by genetical crossing-over.

In a certain strain of maize the second-smallest chromosome (chromosome 9) possesses a conspicuous knob at the end of the short arm. Its distribution through successive generations is similar to that of a gene. If a plant possessing knobs at the ends of both of its 2nd-smallest chromosomes is crossed to a plant with no knobs, cytological observations show that in the resulting F_1 individuals only one member of the homologous pair possesses a knob. When such an individual is back-crossed to one having no knob on either chromosome, half of the offspring are heterozygous for the knob and half possess no knob at all. The knob, therefore, is a constant feature of the chromosome possessing it. When present on one chromosome and not on its homologue, the knob renders the chromosome pair visibly heteromorphic.

In a previous report[1] it was shown that in a certain strain of maize an interchange had taken place between chromosomes 8 and 9. The interchanged pieces were unequal in size; the long arm of chromosome 9 was increased in relative length, whereas the long arm of chromosome 8 was correspondingly shortened. When a gamete possessing these two interchanged chromosomes meets a gamete containing a normal chromosome set, meiosis in the resulting individual is characterized by a side-by-side synapsis of homologous parts. Therefore, it should be possible to have crossing-over between the knob and the interchange point.

In the previous report it was also shown that in such an individual the only functioning gametes are those which possess either the two normal chromosomes *(N,n)* or the two interchanged chromosome *(I,i)*, i.e., the full genom in one or the other arrangement. The functional gametes therefore possess either the shorter, normal, knobbed chromosome *(n)* or the longer, interchanged, knobbed chromosome *(I)*. Hence, when such a plant is crossed to a plant possessing the normal chromosome complement, the presence of the normal chromosome in functioning gametes of the former will be indicated by the appearance of ten bivalents in the prophase of meiosis of the resulting individuals. The presence of the interchanged chromosome in other gametes will be indicated in other F_1 individuals by the appearance of eight bivalents plus a ring of four chromosomes in the late prophase of meiosis.

If a gamete possessing a normal chromosome number 9 with no knob, meets a gamete possessing an interchanged chromosome with a knob, it is clear that these two chromosomes which synapse along their homologous parts during prophase of meiosis in the resulting individual are visibly different at each of their two ends. If no crossing-over occurs, the gametes formed by such an individual will contain either the knobbed, interchanged chromosome (*a*, Fig. 1) or the normal chromosome without a knob (*d*, Fig. 1). Gametes containing either a knobbed normal chromosome (*c*, Fig. 1) or a knobless, interchanged chromosome (*b*,

From Proceedings of the National Academy of Sciences **17**:492-497, 1931. Used with permission.

The authors wish to express appreciation to Dr. L. W. Sharp for aid in the revision of the manuscripts of this and the preceding paper. They are indebted to Dr. C. R. Burnham for furnishing unpublished data and for some of the material studied.

*Present address: Department of Biological Sciences, Wellesley College, Wellesley, Massachusetts.

[1] McClintock, B., Proc. Nat. Acad. Sci. **16**:791-796 (1930).

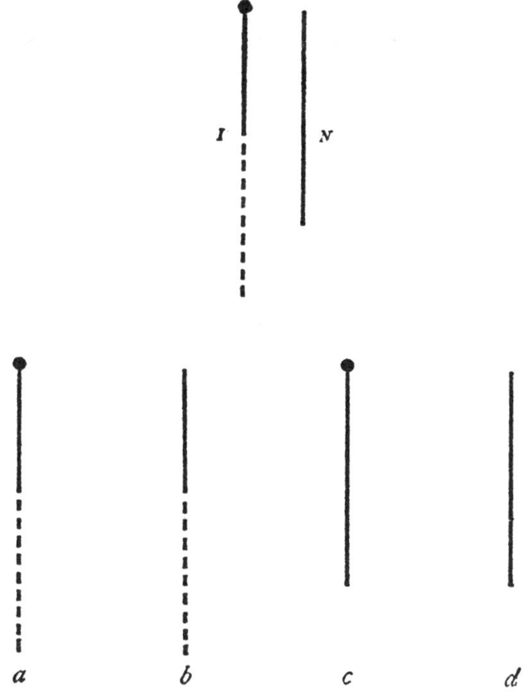

Fig. 1. *Above,* Diagram of the chromosomes in which crossing-over was studied. *I,* Interchange chromosome; *N,* normal chromosome. *Below,* Diagram of chromosome types found in gametes of a plant with the constitution shown above.

a—Knobbed, interchanged chromosome.
b—Knobless, interchanged chromosome.
c—Knobbed, normal chromosome.
d—Knobless, normal chromosome.
a and *d* are non-crossover types.
b and *c* are crossover types.

Fig. 1) will be formed as a result of crossing-over. If such an individual is crossed to a plant possessing two normal knobless chromosomes, the resulting individuals will be of four kinds. The non-crossover gametes would give rise to individuals which show either (1) ten bivalents at prophase of meiosis and no knob on chromosome 9, indicating that a gamete with a chromosome of type *d* has functioned or (2) a ring of four chromosomes with a single conspicuous knob, indicating that a gamete of type *a* has functioned. The crossover types will be recognizable as individuals which possess either (1) ten bivalents and a single knob associated with bivalent chromosome 9 or (2) a ring of four chromosomes with no knob, indicating that crossover gametes of types *c* and *b*, respectively, have functioned. The results of such a cross are given in culture 337, table 1. Similarly, if such a plant is crossed to a normal plant possessing knobs at the ends of both number 9 chromosomes and if crossing-over occurs, the resulting individuals should be of four kinds. The non-crossover types would be represented by (1) plants homozygous for the knob and possessing the interchanged chromosome and (2) plants heterozygous for the knob and possessing two normal chromosomes. The functioning of gametes which had been produced as the result of crossing-over between the knob and the interchange would give rise to (1) individuals heterozygous for the knob and possessing the interchanged chromosome and

Table 1

Knob-interchanged / Knobless-normal × Knobless-normal, culture 337 and knobbed-normal cultures A125 and 340

Culture	Plants possessing 2 normal chromosomes: Non-crossovers	Plants possessing 2 normal chromosomes: Crossovers	Plants possessing an interchanged chromosomes: Non-crossovers	Plants possessing an interchanged chromosomes: Crossovers
337	8	3	6	2
A125	39	31	36	23
340	5	3	5	3
Totals	52	37	47	28

Table 2

Knob-*C-wx* / Knobless-*c-Wx* × Knobless-*c-wx*

C-wx		*c-Wx*		*C-Wx*		*c-wx*	
Knob	Knobless	Knob	Knobless	Knob	Knobless	Knob	Knobless
12	5	5	34	4	0	0	3

(2) those homozygous for the knob and possessing two normal chromosomes. The results of such crosses are given in cultures A125 and 340, table 1. Although the data are few, they are consistent. The amount of crossing-over between the knob and the interchange, as measured from these data, is approximately 39%.

In the preceding paper it was shown that the knobbed chromosome carries the genes for colored aleurone *(C)*, shrunken endosperm *(sh)* and waxy endosperm *(wx)*. Furthermore, it was shown that the order of these genes, beginning at the interchange point is *wx-sh-c*. It is possible, also, that these genes all lie in the short arm of the knobbed chromosome. Therefore, a linkage between the knob and these genes is to be expected.

One chromosome number 9 in a plant possessing the normal complement had a knob and carried the genes *C* and *wx*. Its homologue was knobless and carried the genes *c* and *Wx*. The non-crossover gametes should contain a knobbed-*C*-*wx* or a knobless-*c*-*Wx* chromosome. Crossing-over in region 1 (between the knob and *C*) would give rise to knobless-*C*-*wx* and knobbed-*c*-*Wx* chromosomes. Crossing-over in region 2 (between *C* and *wx*) would give rise to knobbed-*C*-*Wx* and knobless-*c*-*wx* chromosomes. The results of crossing such a plant to a knobless-*c*-*wx* type are given in table 2. It would be expected on the basis of interference that the knob and *C* would remain together when a crossover occurred between *C* and *wx*; hence, the individuals arising from colored starchy *(C-Wx)* kernels should possess a knob, whereas those coming from colorless, waxy *(c-wx)* kernels should be knobless. Although the data are few they are convincing. It is obvious that there is a fairly close association between the knob and *C*.

To obtain a correlation between cytological

Table 3

	Knob-*C*-*wx*-interchanged / Knobless-*c*-*Wx*-normal × Knobless-*c*-*Wx*-normal / Knobless-*c*-*wx*-normal		
Plant number	Knobbed or knobless	Interchanged or normal	
Class I, *C*-*wx* kernels			
1	Knob	Interchanged	
2	Knob	Interchanged	
3	Knob	Interchanged	
Class II, *c*-*wx* kernels			
1	Knobless	Interchanged	
2	Knobless	Interchanged	
Class III, *C*-*Wx* kernels			*Pollen*
1	Knob	Normal	*WxWx*
2	Knob	Normal	....
3	....	Normal	*WxWx*
5	Knob	Normal	....
6	Knob	......	....
7	Knob	Normal	....
8	Knob	Normal	....
Class IV, *c*-*Wx* kernels			
1	Knobless	Normal	*Wxwx*
2	Knobless	Normal	*Wxwx*
3	Knobless	Interchanged	*Wxwx*
4	Knobless	Normal	*Wxwx*
5	Knobless	Interchanged	*WxWx*
6	Knobless	Normal	*WxWx*
7	Knobless	Interchanged	*Wxwx*
8	Knobless	Interchanged	*WxWx*
9	Knobless	Normal	*WxWx*
10	Knobless	Normal	*WxWx*
11	Knobless	Normal	*Wxwx*
12	Knobless	Normal	*Wxwx*
13	Knobless	Normal	*WxWx*
14	Knobless	Normal	*WxWx*
15	Knobless	Normal	*Wx*–

and genetic crossing-over it is necessary to have a plant heteromorphic for the knob, the genes *c* and *wx* and the interchange. Plant 338 (17) possessed in one chromosome the knob, the genes *C* and *wx* and the interchanged piece of chromosome 8. The other chromosome was normal, knobless and contained the genes *c* and *Wx*. This plant was crossed to an individual possessing two normal, knobless chromosomes with the genes *c-Wx* and *c-wx*, respectively. This cross is diagrammed as follows:

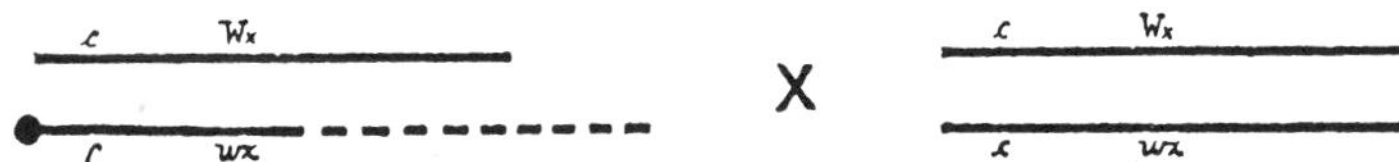

The results of the cross are given in table 3. In this case all the colored kernels gave rise to individuals possessing a knob, whereas all the colorless kernels gave rise to individuals showing no knob.

The amount of crossing-over between the knob and the interchange point is approximately 39% (table 1), between *c* and the interchange approximately 33%, between *wx* and the interchange, 13% (preceding paper). With this information in mind it is possible to analyze the data given in table 3. The data are necessarily few since the ear contained but few kernels. The three individuals in class I are clearly non-crossover types. In class II the individuals have resulted from a crossover in region 2, i.e., between *c* and *wx*. In this case a crossover in region 2 has not been accompanied by a crossover in region 1 (between the knob and *C*) or region 3 (between *wx* and the interchange). All the individuals in class III had normal chromosomes. Unfortunately, pollen was obtained from only 1 of the 6 individuals examined for the presence of the knob. This one individual was clearly of the type expected to come from a gamete produced through crossing-over in region 2. Class IV is more difficult to analyze. Plants 6, 9, 10, 13, and 14 are normal and *WxWx;* they therefore represent non-crossover types. An equal number of non-crossover types are expected among the normal *Wxwx* class. Plants 1, 2, 4, 11 and 12 may be of this type. It is possible but improbable that they have arisen through the union of a *c-Wx* gamete with a gamete resulting from a double crossover in region 2 and 3. Plants 5 and 8 are single crossovers in region 3, whereas plants 3 and 7 probably represent single crossovers in region 2 or 3.

The foregoing evidence points to the fact that cytological crossing-over occurs and is accompanied by the expected types of genetic crossing-over.

CONCLUSIONS

Pairing chromosomes, heteromorphic in two regions, have been shown to exchange parts at the same time they exchange genes assigned to these regions.

20 Blood groups and genetic linkage

Sylvia D. Lawler*
J. H. Renwick
Galton Laboratory
University College
London, England

1. MAPPING OF GENE LOCI IN MAN

One branch of genetic study in man is concerned with determining to which chromosome a given gene belongs and with defining its position on that chromosome. A series of steps is involved. One of the stages envisaged is the discovery, in man, of some chromosomal aberration, such as a small deletion which could be recognized both under the microscope and also functionally through its phenotypic effect. The only mammal in which this has so far been feasible is the house mouse. In that animal, by correlating some observable phenomena in the intact animals with a microscopically visible translocation of one piece of a chromosome to another, it has been possible to assign a group of genes to a particular chromosome (e.g., Carter, Lyon & Phillips, 1954; Slizynski, 1954).

There is little immediate likelihood that this achievement can be directly repeated for man, but there has been recent interest in an immunological approach to the same problem. This is based on the theoretical possibility of locating a gene directly by its products, using the technique of fluorescent antibodies, on the assumption that gene products carrying immunological specificity, remain, at least temporarily, in the vicinity of the gene.

So far, only for the X chromosome can any of the genes on a human chromosome be named. However, some progress can be made in the meantime by collecting genes into groups by means of linkage studies, in which one makes use of the tendency for genes to be transmitted together when they lie on the same chromosome.

2. DETECTION OF LINKAGE

Each chromosome, consisting of a sequence of gene loci, has, in general, a partner with exactly the same loci, but rarely, if ever, the same series of genes (alleles) at these loci. To show that two loci are on the same chromosome a particular gene at one locus must be shown to have a tendency to travel in inheritance with a particular gene at the other. For example, in the section of pedigree illustrated in Fig. 1, the *r* gene complex of the mother has a tendency to be handed on only to those children who have not received the gene for elliptocytosis from her. The word "tendency" rather than "certainty" has been used here, because of the recombination process in consequence of which each single chromosome of the egg or sperm is a composite of the two partner chromosomes present in the primary reproductive cell. Recombination has occurred between the Rhesus and elliptocytosis loci to give rise to the individual III-12 (Fig. 1) who, although normal, has received the R_1 gene complex from his mother rather than the *r* complex. In this sibship of the family there are thus eight non-recombinants and one recombinant (III-12). The frequency of recombinations is roughly dependent on the physical distance between the loci, and can be used as a measure of this distance.

It is clear that, for detection of these recombinations, at least one parent (in this case the mother) must carry, at the two loci, four genes or gene complexes all different (in this case R_1, *r; El, el*); in other words, she must be heterozygous at both loci under consideration.

From British Medical Bulletin **15**:145-149, 1959. Used with permission.

*Present address: Department of Clinical Research, Royal Marsden Hospital, Fulham Road, London SW 3, England.

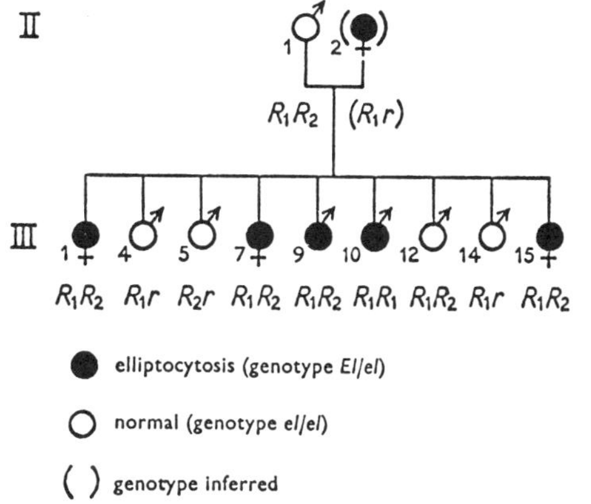

The representation of certain individuals in this family not relevant to the subject of this paper has been omitted. The full pedigree is that of family no. 4, reported by Goodall, Hendry, Lawler & Stephen (1954).

Fig. 1. Illustration of linkage between rhesus and elliptocytosis.

At the same time, it is preferable for the presence of these genes in the children not to be obscured by dominant genes contributed by the other parent.

3. THE NEED FOR MARKER LOCI

Theoretically, it should be possible to test for linkage between any two loci directly, but an individual affected, for example, with two rare abnormalities, say elliptocytosis and osteogenesis imperfecta, may hardly ever occur in the history of man. It is therefore necessary that one, at least, of the loci under test should have a minimum of two distinguishable alleles occurring commonly in the population. Such a locus—for example, one concerned with a blood-group system—can be used as a marker, and any two main loci found linked to the same marker can be allocated to the one chromosome. It is already known that one locus for elliptocytosis is closely linked to the locus of the Rhesus complex (Lawler, 1954; Morton, 1956). If it could be shown that the locus for a particular type of osteogenesis imperfecta was also linked to the Rhesus complex (see some suggestive data of Mohr, 1954), then that elliptocytosis locus and that osteogenesis locus would, by the use of the marker, be shown to be on the same chromosome.

a. Relative efficiencies of two-allele markers

In a standard linkage investigation, it is possible to select, at the main locus, the suitable back-cross mating[1] (e.g., eliptocytosis X normal) from a large population, but for the genes at the marker locus also to constitute a back-cross (in such a way that one parent has four different genes) one must depend on chance. For a two-allele marker locus this chance is maximal when the frequencies p and q of the "dominant" and "recessive" alleles are in the ratio 1:3.

Most of the two-allele marker loci listed in Table I may be regarded as showing dominance, since only two phenotypes are usually distinguished. Column 3 gives, for each marker locus, the proportion of double back-cross matings among those matings which have been chosen for study because of a back-cross at the main locus. The maximum possible figure in this column for a marker with two phenotypes would be about 20%, but because the gene frequencies for most of the markers listed are not optimal (at least in Great Britain) this figure is approached only by the Gm locus (Grubb, 1956), which controls a particular property of γ-globulin.

Another marker character (not actually listed in Table I) which can be regarded for linkage purposes as controlled by a two-allele locus is the haptoglobin type of human sera (Smithies & Walker, 1955) in which three phenotypes can be distinguished. Two informative back-crosses (one to each homozygote) being therefore possible, the frequency of very useful families is increased to 25%. For this type of locus, in which both genes are expressed in the heterozygote, the two genes should ideally have equal frequencies, a situation approached in this example of the haptoglobin alleles.

Of course, in other populations the relative merits of the different markers will depend on the local gene frequencies. For instance, in the Chinese population, the Diego blood-group system could be used as a marker (the frequency of the gene Di^b being 0.97 according to Layrisse & Arends, 1956), whereas the Kell system would be of little value, since Miller, Rosenfield & Vogel (1951) found no K-positive individuals in 103 Chinese tested.

b. Multi-allelic and complex loci as markers

The well-established marker characters with multi-allelic or with complex loci—the Rhesus, MNSs, A_1A_2BO blood groups—are easily the most efficient and they have the added advantage of availability and reliability of the

[1] A back-cross mating is a mating between a heterozygote and a homozygote.

Table I. The frequencies of genes and of back-cross mating types for the two-allele marker loci already in use

Marker locus	Frequency of specified gene in Great Britain		Frequency of double back-cross (%)	Notes
P	(P_2 + p) 0.52		14	There are at least three alleles, P_1, P_2 and p (Sanger, 1958) but p is so rare that it can be ignored for linkage purposes
ABH secretion	(se)	0.48	12	–
PTC tasting	(t)	0.55	15	Harris & Kalmus (1950). Phenotypic discrimination not sharp
Duffy	*(Fy^b)**	0.59	17	–
Kell	*(k)**	0.95	9	–
Lutheran	*(Lu^b)**	0.96	7	–
Kidd	*(Jk^b)**	0.48	12	–
Lewis secretion	*(l)*	0.20	1.4†	Marshall (1958, personal communication)
Gm	*(gm)*	0.63	19	Lawler (1958)

*If anti-Fy^b, -k, -Lu^b, -Jk^b are used and three phenotypes distinguished, the efficiencies of the respective loci are greater than indicated in column 3.

†This figure indicates that the Lewis secretion locus is of very little use as a marker.

Gene frequencies are taken, except where otherwise stated, from Mourant (1954).

standard antisera. Since there are 23 chromosome pairs (Tjio & Levan, 1956; Ford & Hammerton, 1956; Ford, Jacobs & Lajtha, 1958), it is clear that the present battery of about a dozen usable marker loci is still inadequate to label more than a fraction of the genetic material. There is a great need for more marker loci, or for the improvement of known ones, by the discovery of more alleles or by the development of techniques to differentiate heterozygotes from homozygotes.

4. ASSOCIATION DISTINGUISHED FROM LINKAGE

The standard English word "linkage" is used in genetics to indicate the specific type of relationship already discussed. Unfortunately, in the literature on human genetics, the word has sometimes been used in its more general sense, to indicate a tendency to the concurrence of two conditions in an individual or in a family, when the word "association" is really the appropriate term. In an association between a particular blood group and a particular disease (Fraser Roberts, 1959) there is a tendency, over the whole population, for the two associated features to occur together more often than would be expected if the features were occurring independently of one another. This is not so in linkage. Because of the recombining of genes, at the two linked loci, a population survey would show that any particular gene at the one locus is not associated with any particular gene at the other locus. Such a survey cannot therefore be expected to reveal linkage. Table II, for example, shows, in some English data, that the Lutheran phenotypes (hence genotypes) in the general population are independent of the ABH secretor phenotypes (and genotypes), even though, by family studies, Mohr (1951) showed quite clearly that the Lutheran locus was closely linked to the secretor locus (as scored, with almost negligible

Table II. The independence of the Lutheran and ABH secretor phenotypes in a sample of 400 unrelated people tested at the Galton Laboratory

Phenotype	Lu(a+)	Lu(a–)	Totals
ABH secretors	27	270	297
ABH non-secretors	8	95	103
Totals	35	365	400
$\chi^2 = 0.17$		$P \fallingdotseq 0.7$	

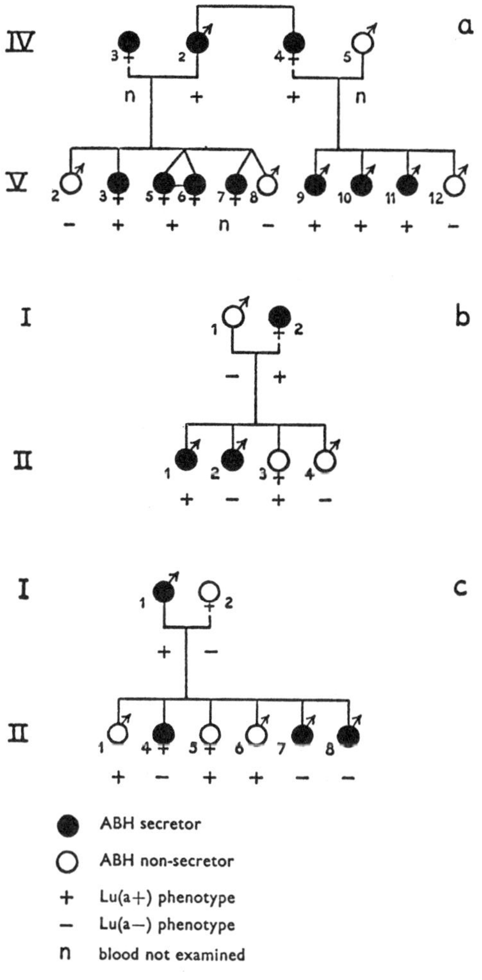

a: part of tylosis family B2 (Lawler & Renwick, unpublished observations)

b: mongol family no. 243 (Lang-Brown, Lawler & Penrose, 1953)

c: family C 38 Be (collected by Silvestroni, Bianco & Siniscalco, in August 1957, unpublished observations)

Fig. 2. Families contributing data on Lutheran: secretor linkage.

error, by the erythrocytic Le^a phenotype). For this linkage, the maximum likelihood value of the recombination frequency is now estimated as 7.5%, with 95% confidence limits of 3% and 15%. In other words, there is a probability, 0.95, that the recombination frequency lies between these limits. The "lod" scores of Smith (1953), tabulated by Morton (1955), were used and the data include those of Mohr (1954, 1956) and of Sanger & Race (1958). The remaining informative families were grouped at the Galton Laboratory, and the relevant data are depicted in Fig. 2.

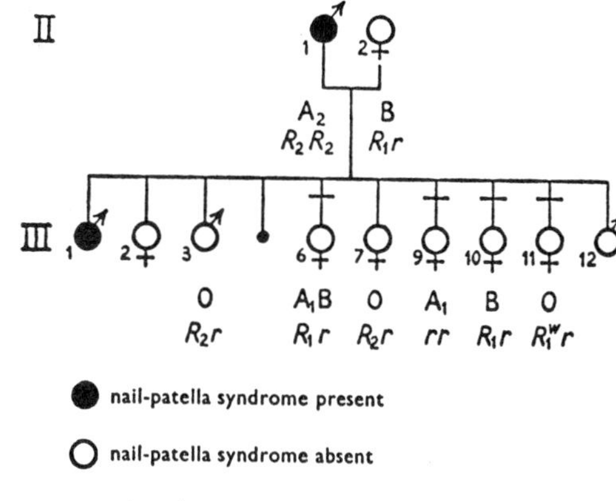

Part of family P (Lawler, Renwick, Hauge, Mosbech & Wildervanck, 1958). The representation of certain individuals not relevant to the subject of this paper has been omitted.

The ABO and Rh blood groups show that III-6, -9, -10 and -11, having II-2 as mother, are not the children of II-1.

Fig. 3. Illustration of the use of marker characters in clarifying parentage.

5. PRACTICAL ASPECTS OF LINKAGE INVESTIGATIONS

The type of linkage search which gives the greatest chance of success is a full pedigree investigation centering round the affected members of a family in which a chosen dominant trait is known to be running through several generations. This provides only the skeleton of the study and, for each affected member, his parents, sibs, spouse and children must also be examined for the marker characters selected. In certain circumstances, to clarify the genotype of a normal relative (say an unaffected sib), it may be helpful to test his children too. These marker characters serve an additional purpose in many genetic investigations in clarifying the parentage of a certain number of individuals, as in the family illustrated in Fig. 3. If the four exclusions of parentage had not been made in this family, the unlikely segregation ratio with respect to the main character would have been misleading from a general genetic point of view as well as for linkage assessment.

The main character can be a recessive instead of a dominant one but, in that case, the back-cross at the main locus will not give useful information. To establish a linkage will therefore require a larger body of data and, hence, a larger collection of primary cases of the condition. It will also entail greater statistical difficulty in testing satisfactorily for homogeneity unless some technique is available for distinguishing the heterozygotes. If heterozygotes could be identified, large pedigrees could be studied and none of these disadvantages would apply.

6. TESTING AND ESTIMATING LINKAGE

Careful inspection of the pedigree will often be sufficient to detect or exclude close linkage involving any of the more efficient markers. This impression can be objectively quantified by the rapid sib-pair test of Penrose (1953) or by use of Fisher's *u* scores (Finney, 1940) or by the method of lod scores (Smith, 1953), usefully tabulated as z scores by Morton (1955, 1956, 1957). Of these, the lod scores have considerable advantages but, in general, none of these relatively quick methods is immediately applicable to pedigrees of more than a single family unit. The information in a large pedigree is considerably greater than the sum of that information gained by treating each family unit separately. To obtain the lod scores on a large pedigree, it is necessary to calculate, for any chosen recombination frequency, the probability of that pedigree's having occurred in all its detail, and to compare this with the probability calculated on the basis of a recombination frequency of 0.5 (i.e., no linkage). The log of this ratio is the lod score, and the curve obtained by plotting lod scores against various chosen recombination frequencies is meaningful in itself; but a significance level and a maximum likelihood estimate can be calculated if required, as by Haldane & Smith (1947). These same lod scores can be used for the sequential test described by Morton (1955), although the appropriateness of a sequential test is a controversial question. The calculation of these scores is simple in principle but very time consuming for a large pedigree, when every possible genotype for each individual has to be considered in the light of population gene frequencies. There is hope that electronic computers will eventually be used fairly widely for this work.

One alternative technique, also applicable to large pedigrees as well as to small, is the counting method of Smith (1957), but this usually turns out to be more laborious. Most statisticians interested in this field appear to be in fairly general agreement that the information from multi-generation pedigrees is best condensed into a series of lod scores calculated for a range of recombination values, as already indicated. Therefore, if one or more such pedigrees occur in the data, the lod scores should also be used for the smaller families, thus allowing all the information to be combined by simple addition.

7. PUBLICATION OF DATA

Even when full calculations are published on a particular investigation, with or without positive findings, it is still necessary to make the detailed findings generally available (preferably published in tabular form to allow others to combine them with further data). Publication of data is, of course, more important than the calculations or the interpretations offered. The names, ages and addresses of the families studied should ideally be filed for each country in some accessible office (such as the archives of the Galton Laboratory in Great Britain) to allow for a later generation to be traced in the future. This becomes particularly important when the test condition is very rare or when there is doubt about whether it is always controlled by the same locus. For example, the close linkage described by Renwick & Lawler (1955) between the ABO and nail-patella loci, although not showing a significant heterogeneity between families (Lawler, Renwick, Hauge, Mosbech & Wildervanck, 1958), does seem to be somewhat loose in at least one family, C. This family was originally described in 1911 and only its relative immobility in a Scottish village allowed its rediscovery. If, in some years' time, other families with high recombination frequencies make a reassessment necessary, it may be advisable to return once more to this family to test the children born since the recent investigation, and then the necessity of having records of family names, ages and addresses filed in a central office will be felt.

8. PRESENT ASSESSMENT OF LINKAGE WORK

The first proved linkage in man—that between colour blindness and haemophilia (Bell & Haldane, 1937)—concerned the sex chromosomes, but the other chromosomes are now being studied with success. Morton (1957), in a review of some of the linkages for which claims have been made, has shown that it is not too surprising that three quite definite linkages ($P<10^{-6}$) have so far been discovered with the blood groups as markers. These three all have recombination values of 10% or less and it will be noted that two of the linkages concern the highly efficient marker loci, ABO and Rhesus.

Lutheran:secretor	7.5%	(Mohr, 1951)
One elliptocytosis locus:Rhesus	3.3%	(Lawler, 1954; Morton, 1956)
Nail-patella:ABO	9.7%	(Renwick & Lawler, 1955; Lawler *et al.* 1958)

Some deviations in favour of linkage have been noted in work on numerous other main loci and, as further data on these accumulate, it is fairly certain that some of the deviations will be proved to indicate true linkages. Gates (1954) gives a useful summary of the literature, though without attempting an independent assessment.

9. SUMMARY

Three certain autosomal linkages are now known in man: a Lutheran:secretor linkage, a linkage between one elliptocytosis locus and the Rhesus complex locus, and the nail-patella: ABO linkage. All of them concern blood-group systems. Some others cannot be regarded as convincing but it is very probable that, with added data, a proportion of them could be established beyond doubt. For the majority of pairs of loci tested, all that can be said is that linkage closer than a certain value has been shown to be unlikely.

REFERENCES

Bell, J. & Haldane, J. B. S. (1937) Proc. roy. Soc. B **123**:119.
Carter, T. C., Lyon, M. F. M. & Phillips, R. J. S. (1954) Nature, Lond. **174**:309.
Finney, D. J. (1940) Ann. Eugen., Lond. **10**:171.
Ford, C. E. & Hammerton, J. L. (1956) Nature, Lond. **178**:1020.
Ford, C. E., Jacobs, P. A. & Lajtha, L. G. (1958) Nature, Lond. **181**:1565.
Gates, R. R. (1954) Genetic linkage in man. Junk, The Hague.
Goodall, H. B., Hendry, D. W. W., Lawler, S. D. & Stephen, S. A. (1954) Ann. Eugen., Lond. **18**:325.
Grubb, R. (1956) Acta path. microbiol. scand. **39**:195.
Haldane, J. B. S. & Smith, C. A. B. (1947) Ann. Eugen., Lond. **14**:10.
Harris, H. & Kalmus, H. (1950) Ann. Eugen., Lond. **15**:24.
Lang-Brown, H., Lawler, S. D. & Penrose, L. S. (1953) Ann. Eugen., Lond. **17**:307.
Lawler, S. D. (1954) In: Montalenti, G. & Chiarugi, A., ed. Proceedings of the 9th International Congress of Genetics [Bellagio (Como), 24-31 August 1953], p. 1199. (Issued as a supplementary volume of Caryologia.)
Lawler, S. D. (1958) Brit. med. J. **1**:1538 [Abstract].
Lawler, S. D., Renwick, J. H., Hauge, M., Mosbech, J. & Wildervanck, L. S. (1958) Ann. hum. Genet. **22**:342.
Layrisse, M. & Arends, T. (1956) Nature, Lond. **177**:1083.
Miller, E. B., Rosenfield, R. E. & Vogel, P. (1951) Amer. J. phys. Anthrop. **9**:115.
Mohr, J. (1951) Acta path. microbiol. scand. **29**:339.
Mohr, J. (1954) A study of linkage in man. Munksgaard, Copenhagen.
Mohr, J. (1956) Acta genet. **6**:24.
Morton, N. (1955) Amer. J. hum. Genet. **7**:277.
Morton, N. (1956) Amer. J. hum. Genet. **8**:80.
Morton, N. (1957) Amer. J. hum. Genet. **9**:55.
Mourant, A. E. (1954) The distribution of the human blood groups. Blackwell Scientific Publications, Oxford.
Penrose, L. S. (1953) Ann. Eugen., Lond. **18**:120.
Renwick, J. H. & Lawler, S. D. (1955) Ann. hum. Genet. **19**:312.
Roberts, J. A. Fraser (1959) Brit. med. Bull. **15**:129.
Sanger, R. (1958) Bibl. haemat. **7**:110.
Sanger, R. & Race, R. R. (1958) Heredity **12**:513.
Slizynski, B. M. (1954) Nature, Lond. **174**:310.
Smith, C. A. B. (1953) J. R. statist. Soc. B **15**:153.
Smith, C. A. B. (1957) Ann. hum. Genet. **21**:254.
Smithies, O. & Walker, N. F. (1955) Nature, Lond. **176**:1265.
Tjio, J. H. & Levan, A. (1956) Hereditas, Lund **42**:1.

21 Isolation and characterization of recombination-deficient mutants of Escherichia coli K12

Alvin J. Clark
Ann Dee Margulies*
Department of Bacteriology
University of California
Berkeley, California

Certain features of the process of genetic recombination at the molecular level have recently become evident: (1) Recombination in bacteria and viruses involves the physical interaction of and subsequent inheritance by recombinant progeny of double-stranded elements of DNA derived from two parents.[1-5] (2) The unreplicated recombinant DNA may contain a double-stranded region in which the two complementary strands are derived from different parents.[6] (3) Recombination may involve the removal and resynthesis of small amounts of DNA.[5]

Since it did not seem unlikely that enzymes participate in the events leading to the formation of the completed recombinant DNA structure, one of the authors (A. D. M.) undertook the isolation of mutants in which one or more of the hypothetical recombination enzymes would be defective.

After two mutants had been isolated, the authors learned that each of the above features had been incorporated into a model of recombination devised by Howard-Flanders.[7] His purpose was to show the similarity between the steps in recombination and the steps thought to be involved in the *in vivo* removal from DNA of photoproducts formed by exposure of cells to ultraviolet light and their replacement by undamaged nucleotides.[8,9] The first step in this model of recombination is the breakage of the parental DNA's and synaptic pairing of complementary single-stranded ends of two parental fragments by the formation of hydrogen bonds between complementary sequences of bases. This step has no counterpart in the excision from DNA of photoproducts, the step which initiates the repair of irradiation damage. The subsequent steps of the two processes are formally similar, however, and can be described as follows: (1) In recombination there is degradation of the single strands of terminal regions of the parental fragments which are not involved in the double-stranded region holding the two fragments together. This step may be similar to the step in repair of photodamaged DNA which leads to removal from the DNA of 30 nucleotides for every thymine dimer excised.[9,10] (2) Following the degradation step, a polymerization step occurs during which the gaps in the DNA are filled by newly synthesized single-stranded regions. (3) The final step in both recombination and repair of photodamage is the restoration of the integrity of the phospho-sugar backbone of the DNA by joining the newly synthesized single strands to the extant single strands at the side of the gap opposite the side from which the polymerization began.

The similarity of the models for recombination and the repair of photodamaged DNA has led Howard-Flanders[7] to predict that mutants would occur in which one of the enzymes common to the two processes was defective. Such mutants are expected to possess two phenotypic properties if the mutations are obtained in an F^- strain of *E. coli* K12: (1) inability to form recombinants by conjugation, hence appearing infertile in crosses with Hfr strains, and (2) inability to repair photodamaged DNA, hence appearing very sensitive to the lethal effects of UV irradiation. In this report we wish to describe the isolation and preliminary characterization of two such mutants.

From Proceedings of the National Academy of Sciences **53**:451-459, 1965. Used with permission.

This investigation was supported by USPHS research grant, AI 05371, from the National Institutes of Allergy and Infectious Diseases.

The authors appreciate the assistance of Miss Ann Templin and Miss Haruko Nagaishi in performing some of the experiments described. They also acknowledge the collaboration of Dr. Katherine Brooks in performing the zygotic induction experiments.

*Present address: Department of Microbiology, New York University School of Medicine, New York.

MATERIALS AND METHODS

Strains used. The bacterial strains used in this study are characterized below according to the alleles of relevant nutritional and fermentative genes they carry,[11] their mating type,[11] and their phenotypic response to phages and other lethal agents:[11]

JC-182: *thr+*, *leu+*, *his+*, *pur*–1, *arg+*, *met+*, Thi–, *lac+*, *mal+*, λ^S, P1^S, T6^S, Sm^S, UV^R, D, double male strain. See reference 10.

JC-1020: *thr+*, *leu+*, *his+*, *pur+*, *arg+*, *met+*, Thi–, F-*lac*$^+$/*lac*$^-$, *mal+*, λ^S, P1^S, T6^S, Sm^S, UV?, D, F-*lac* donor derived from strain 200P.

JC-1164: *thr+*, *leu+*, *his+*, *pur+*, *arg+*, *met+*, Thi–, *lac+*, *mal+*, (λ), P1^S, T6^S, Sm^S, UV?, D, a lysogenic derivative of a Hayes Hfr.

JC-411: *thr+*, *leu*–2, *his*–1, *pur+*, *arg*–6, *met*–1, *thi+*, *lac*–1, 4, *mal*–1, λ^R, P1^S, T6^S, Sm^R, UV^R, ND.

JC-1553 and JC-1554: *thr+*, *leu*–2, *his*–1, *pur+*, *arg*–6, *met*–1, *thi+*, lac–1, 4, mal–1, λ^R, P1^S, T6^S, Sm^R, UV^S, ND.

Those strains which carry mutant alleles are either dependent for growth at 37°C on the presence of a nutritional supplement in minimal medium or are unable to utilize particular carbon sources for growth. Ordinarily, the nutritional supplement required can be inferred from the pathway affected by the mutation; however, this is not true of strains carrying *pur*-1. Such mutant strains will utilize adenine as the sole purine source.

Media and mating conditions. All of the media and most of the mating conditions used have been previously described fully by Clark[12] and by Adelberg and Burns.[13]

Mating on plates was accomplished by inoculating strains of one parent onto a "lawn" of the other parent spread on medium selective for recombinants. The inoculum was obtained from growth of nondonor strains on a complex or minimal agar medium either in the form of colonies or heavy confluent growth in patches. The Hfr strain was prepared by washing cells obtained from overnight growth in complex medium with *M*/20 phosphate buffer at pH 7.0. Approximately 2 × 10^9 washed cells of the Hfr strain were spread onto the surface of a minimal medium selective for recombinants. This plate was inoculated by replica plating as was a control plate of selective medium containing no Hfr cells. After the plates had been incubated for 24 and 48 hr they were examined for the presence of recombinant colonies within the areas inoculated by the nondonor strains. Generally, for comparison purposes, strains known to be capable of forming conjugational recombinants were present on every master plate which contained strains suspected of being incapable of forming such recombinants.

Technique of ultraviolet irradiation. Strains were tested for their sensitivity to ultraviolet irradiation by first inoculating a complex medium with cells obtained from confluent growth within patches on a master plate, and then subjecting the inoculated plate to 20 or 30 sec exposure to ultraviolet light. An inoculum was first transferred to a complex medium agar plate by replica plating. Then without prior incubation the freshly inoculated plate was used as a master plate to inoculate two other complex medium agar plates. In this fashion the inoculum was reduced to the point where exposure of one of the latter plates to 20-40 sec of ultraviolet light was sufficient to distinguish UV^S from UV^R strains after incubation of both plates at 37°C for 18 hr had permitted the growth of all surviving cells. All irradiations were carried out at a distance of 25 cm from a Mineralight lamp with an output measured at that distance to be 2.27 ergs/sec/mm^2 in the ultraviolet.

Selection of revertants. Independently isolated revertants were obtained from single colony isolates of mutant strains. Cultures from different colonies were kept on complex medium slants at 4°C after overnight incubation at 37°C. A flask containing fresh liquid complex medium was inoculated with cells from one slant and incubated at 37°C overnight. An aliquot was used to inoculate fresh medium and the remaining cells were harvested by centrifugation. Approximately 1-5 × 10^9 cells were spread on the surface of each of four complex medium agar plates. The plates were then exposed to 20, 30, 40, and 50 sec of ultraviolet irradiation and finally incubated in the dark at 37°C for 18-20 hr. The colonies which appeared were tested for the presence of UV^R cells and cells able to form conjugational recombinants. Usually several serial transfers were required before UV^R revertants were discovered among the survivors.

Mutagenic treatment. Stationary phase cells of JC-411, obtained after overnight growth in complex medium, were collected by centrifugation, washed with 0.1 *M* citrate buffer at pH 5.5, and resuspended in 0.1 *M* citrate buffer supplemented with 50 μg/ml of 1-methyl-3-nitro-1-nitrosoguanidine. They were incubated for 1 hr at 37°C to about 0.1% survival. Appropriate dilutions were then made in phosphate buffer and the cells plated onto minimal glucose medium. The plates were incubated for two days at 37°C and the resulting colonies were screened for mutants.

RESULTS

A multiply marked F- strain of *E. coli* K12, JC-411, was treated with the mutagen 1-methyl-3-nitro-1-nitrosoguanidine and the surviving cells were allowed to grow into colonies on minimal medium. The colonies were then screened for the ability of the cells they contained to produce recombinants when exposed to a population of Hfr cells. The colonies of Leu$^-$ Ade$^+$ F- cells were replicated onto a lawn of Leu$^+$ Ade$^-$ Hfr, JC-182, which had been spread onto a minimal medium selective for Leu$^+$ (Ade$^+$) recombinants. After suitable incubation the plates were examined for the existence of areas which did not contain recombinant colonies although they had been inoculated from an F- colony. Part of the corresponding colony was taken from the master plate, suspended in a small amount of liquid medium, and streaked onto nutrient medium. Two successive single colony isolations were performed testing several colonies at each isolation for their phenotypic characteristics including their infertility when crossed with Hfr JC-182.

From approximately 2000 survivors of exposure to nitrosoguanidine two strains infertile with JC-182 were isolated: JC-1553 and JC-1554. Since their infertility with an *E. coli* Hfr is a trait which would characterize most bacteria of other genera, care was taken to establish the relationship of the isolates with JC-411. Both were examined microscopically under various conditions of growth and were found to be morphologically similar to JC-411. Both were also found to be phenotypically similar to JC-411 with respect to their growth factor requirements, their inability to ferment certain sugars, and their response to streptomycin and certain phages.

Four explanations of the infertility of the mutant strains with Hfr JC-182 may be advanced. (1) Transfer of genetic material from the Hfr to the mutant F- may be impossible because cells of the F- may be incapable of forming effective contacts with cells of the Hfr. (2) Transfer of chromosomal material from Hfr cells to the mutant cells may be prevented even though effective contacts are established between the cells. (3) Genetic material may be destroyed upon its entry into the mutant F-. (4) The mutant F- may be unable to catalyze recombination between the endogenote and exogenote. In order to test these possibilities two experiments were performed. In the first, JC-1020, a streptomycin-sensitive donor strain carrying the F-merogenote F-*lac*, was crossed to the streptomycin-resistant strains JC-411, JC-1553, and JC-1554. Samples of the three mating mixtures were removed after suitable intervals had elapsed from the mixing of the parent strains, and the effective pairs present were disrupted mechanically. In each sample the number of Lac+(Sm^R) merodiploids and Leu+(Sm^R) recombinants were determined. The results are plotted as a function of time in Figure 1. As can be seen from this figure, the kinetics of formation of Lac+(Sm^R) merodiploids is similar when JC-411 and the two mutant strains are used as recipients. The frequency of merodiploids formed is also similar in the three crosses. These facts indicate that the mutant cells participate in the formation of effective contacts and effective pairs and that transferred genetic material is not completely destroyed upon entry into the mutant cells. On the other hand, Figure 1 shows that after two hours of mating the

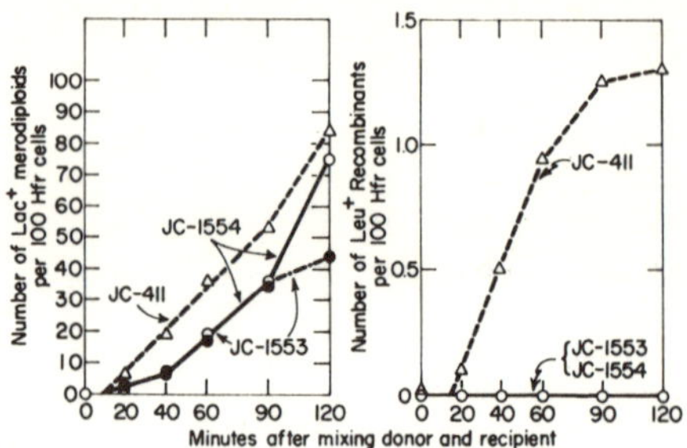

Fig. 1. Results of crossing an F-*lac* donor, JC-1020, with JC-411 and two recombination-deficient mutants JC-1553 and JC-1554. Cells of each of the four strains were grown into log phage in complex medium at 37°C. Five ml of mating mixture were constituted at approximately 1×10^7 cells per ml of JC-1020 and 2×10^8 cells per ml of one of the F^- strains. The 3 mating mixtures were incubated in a 125-ml Erlenmeyer flask without shaking at 37°C. Periodically an aliquot was withdrawn, diluted 1:10 in buffer, and subjected to the shearing action of a Waring Blendor. Appropriate dilutions were then made and aliquots plated on medium selective for Lac^+ (Sm^R) merodiploids and Leu^+ (Sm^R) recombinants.

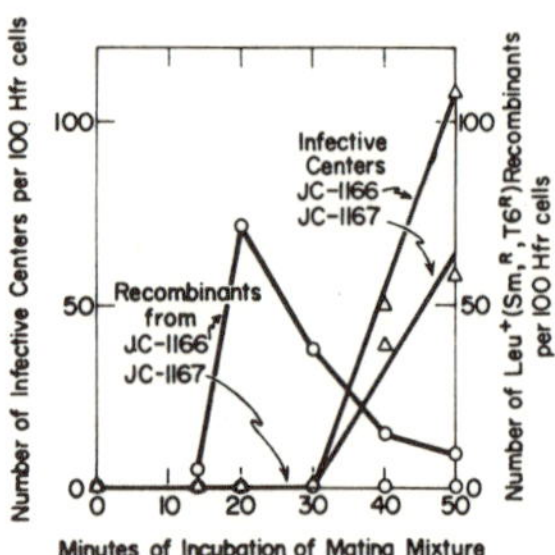

Fig. 2. Cross of a lambda lysogenic Hfr, JC-1164, with nonlysogenic F^- strains JC-1166 and the recombination-deficient JC-1167. Cells of JC-1164 were harvested in log phase of growth by centrifugation at 4°C. They were washed once in complex medium. 0.5 ml of resuspended, washed cells were added to 4.5 ml of a log culture of each F^- strain. The final cell concentration was approximately 1×10^7 Hfr cells per ml and approximately 2×10^8 F^- cells per ml. The mixture was incubated in a 125-ml Erlenmeyer flask without shaking at 37°C for 10 min, and then 0.1 ml was withdrawn and was added to 19.9 ml of fresh complex medium prewarmed to 37°C in a 250-ml flask. At intervals, 0.5 ml of the diluted mating mixture was added to 0.5 ml of a lysate of bacteriophage T6 having a titer of 3×10^9 pfu per ml. The resulting suspension was incubated at 37°C with gentle shaking for 10 min and then was sampled to determine the titer of recombinants and infective centers. A parallel culture in which 0.5 ml of washed, resuspended Hfr cells were added to 4.5 ml of fresh complex broth was treated in the same fashion in order to ascertain the titer of phage produced by spontaneous lysis. The number so obtained (from 2×10^3 to 2×10^5 infective centers per ml) was subtracted from the number of infective centers present per ml of mating mixture in order to calculate the titer of zygotically induced infective centers.

mixtures of JC-1553 and JC-1554 with the donor contain fewer than $^1/_{1000}$ the number of recombinants present in the mating mixture containing JC-411. This substantiates the infertility of the mutant strains detected first with JC-182 and indicates either that chromosomal transfer to the mutants is prevented or else that the mutants suffer from an inability to catalyze recombination.

The second experiment demonstrates that chromosomal markers are transferred to the mutant strains although they are not inherited by recombinants. A cross was performed with a lambda lysogenic donor JC-1164 and the T6-resistant mutants of JC-411 and JC-1553, JC-1166, and JC-1167, respectively. In Figure 2 are shown the results obtained when mating is interrupted at intervals by the addition of T6 phage to samples of the mating mixtures. Leu$^+$ (Sm^R, T6^R) recombinants were formed from JC-1166 beginning about 14 min after mixing the parent strains. They increased in number until about the time the lambda prophage was transferred to the zygotes, and then they decreased in number as zygotic induction took place producing an increase in the number of

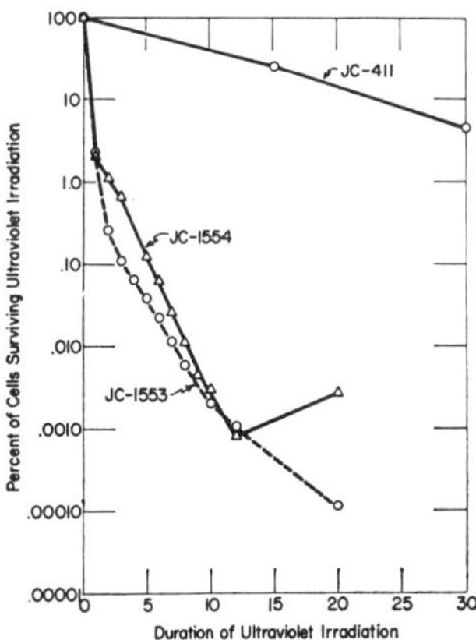

Fig. 3. Sensitivity to ultraviolet irradiation of two recombination-deficient strains JC-1553 and JC-1554 as compared to the relative resistance of JC-411. Cells were prepared for irradiation by inoculating a liquid minimal medium with approximately 10^7 cells per ml obtained from stationary phase cultures in complex medium. The cells were harvested when they reached 5×10^8 per ml and were washed once with *M*/20 phosphate buffer at pH 7. Five-ml aliquots of the cells, resuspended in buffer at 10^8 cells per ml, were placed in glass Petri dishes and were irradiated at a distance of 25 cm from a Mineralight lamp. The samples were kept in the dark until they were diluted and plated on complex medium. All operations were carried out in dim light and the plates were incubated in the dark.

infectious centers. This behavior has been observed by other authors[14] in a similar cross. In the cross of JC-1164 with JC-1167 zygotic induction occurred as the lambda prophage was transferred to zygotes beginning about 30 min after the parents were mixed. Leu$^+$ (Sm^R T6^R) recombinants were not formed in this cross, however, although it is clear from the cross with JC-1166 that *leu*$^+$ preceded the lambda prophage into the zygotes formed. Therefore it seems clear that the mutant JC-1553 and its derivatives are unable to catalyze recombination between endogenote and exogenote. A similar zygotic induction experiment performed with T6^R mutant of JC-1554 shows substantially the same results, thereby demonstrating that the defect in both mutant strains is similar.

Having obtained mutants in which recombination was blocked we were able to determine whether or not their ability to repair photodamage to DNA was also impaired. Aliquots of cultures of JC-411, JC-1553, and JC-1554 were irradiated for different periods of time and the number of survivors determined by plating samples onto a complex medium. All operations were performed in dim light and the plates were incubated in the dark. The results are shown in Figure 3. The fact that after 10 sec of irradiation 40 per cent of the cells of JC-411 but only 0.003 per cent of the cells of JC-1553 and JC-1554 are viable provides clear evidence that the mutant strains are more sensitive to ultraviolet light than is the parent strain.

The mutants JC-1553 and JC-1554 therefore differ from the parent strain in two characteristics, and it then becomes necessary to demonstrate whether one or two mutations are responsible for the mutant phenotype. Two methods of determining this information are available: (1) Back-mutants selected for their reversion to one of the parental phenotypic characteristics may be examined for reversion to the other parental characteristic. (2) Conjugational or transductional recombinants formed by crossing a wild-type donor with the mutant strains may be selected for their inheritance of one of the wild-type traits and then examined for inheritance of the other wild-type trait. The second of these methods depends upon recombination between two mutant genes and their wild-type alleles to produce recombinants usually inheriting only one of the wild-type alleles. In general, this result would distinguish a double mutant from a single mutant; in this case, however, the distinction is not possible. If the mutant strains carried two mutations, only the locus conferring the trait of sensitivity to ultraviolet light would be of selective disadvantage permitting selection for UV^R recombinants. The other mutant locus would presumably prevent recombination, since it is assumed to determine that trait in the mutant strains. Recombination would then be possible only if both wild-type alleles had been transferred to the recipient. This would increase the probability of inheritance by recombinants of both wild-type alleles; in fact the wild-type allele controlling UV^R may not be integrated unless prior integration of the wild-type allele permitting recombination occurs. Thus, recombinational separation of two mutant alleles, one of which prevents recombination, may be impossible.

Because of these considerations the method of reversion was used to determine the number of mutations present in JC-1553 and JC-1554. Two revertants were obtained from among the cells of JC-1553 surviving ultraviolet irradiation in the experiment shown in Figure 3. In similar but independent experiments, another UV^R revertant of JC-1553 and one of JC-1554 were obtained. The properties of these revertants are listed in Table 1; all four have regained either fully or partially the wild-type ability to form conjugational recombinants, as well as resistance to ultraviolet light. In all the experiments performed to obtain the four revertant strains, several thousand survivors of ultraviolet irradiation were tested. Most of these survivors were UV^S and in no case had they recovered wild-type proficiency in conjugational recombinant formation.

DISCUSSION

The term "recombination" when used in the context of bacterial genetics connotes to many either the process of DNA transmission known as conjugation or the formation by conjugation of any progeny which inherit phenotypic traits derived from both parents. It can, however, be used more strictly to denote the series of physical and chemical events which serve to link genes derived from one parental DNA with those derived from another parental DNA.[15] It is in this sense that the word is used to describe the recombination-deficient (Rec-) mutants, JC-1553 and JC-1554, whose isolation is described in this report. These two strains were isolated after mutagenic treatment of their parent culture because they appeared to be unable to form recombinants when crossed with an Hfr strain. They were tested for their ability to engage in the process of zygote formation and were found to form F-*lac* merodiploids in a cross with an F-*lac* donor and to form infective centers by zygotic induction when crossed with an Hfr carrying lambda prophage. These results served to rule out the possibility that the mutants' infertility with an Hfr was a reflection of their inability to engage the donor cells in mating or their acquisition of a new ability to destroy exogenous DNA. It was therefore concluded that the mutants were unable to catalyze at least one of the steps involved in recombination. Upon further examination the mutant strains were found to be much more sensitive than their parent strain to the lethal effects of ultraviolet light. The inference is made that the mutation has affected the ability of irradiated cells to repair photodamage to DNA. Experiments carried out

Table 1. Characteristics of revertants of recombination-deficient mutants JC-1553 and JC-1554

			Results of crosses with F-*lac* donor JC-1020	
Strain	Source of revertant	Survival after 30 sec of UV (%)	Freq. of Lac^+ (Sm^R) merodiploid formation	Freq. of Leu^+ (Sm^R) recombinant formation
JC-411	–	4.1	84*	1.3*
JC-1553	–	3.5×10^{-5}	44	1.1×10^{-4}
JC-1554	–	5.6×10^{-5}	75	1.4×10^{-4}
JC-679	JC-1553	39	108	3.4
JC-680	JC-1553	11	146	2.0
JC-678	JC-1553	8.6	95	1.7
JC-677	JC-1554	0.48	88	0.09

*Number of merodiploids or recombinants per 100 Hfr cells.

Sensitivity to ultraviolet irradiation was measured as described in the legend to Fig. 3. Proficiency in recombination was measured as described in the legend to Fig. 2.

by one of the authors (A. J. C.) in collaboration with Drs. P. Howard-Flanders and R. Boyce at the Yale University Medical School have strengthened this inference. These experiments will be described elsewhere.

Since JC-1553 and JC-1554 were both isolated from the same culture treated with mutagen and show marked similarity in phenotype, they may be siblings. However, the fact that the cells were plated immediately after treatment and that the mutants occurred on different plates renders a sibling relationship unlikely. Consequently they are considered to be independently isolated mutants.

An examination of a number of independently isolated revertants to ultraviolet resistance has lent support to the hypothesis that a single gene mutation is responsible for the UV^S and Rec- traits of the mutants. All revertants isolated had regained not only resistance to ultraviolet irradiation but proficiency in the formation of conjugational recombinants as well. This fact is consistent with the hypothesis that one gene determines an enzyme which catalyzes one of the steps in recombination and one of the steps in the replacement of photoproducts in ultraviolet-damaged DNA. There are, however, other hypotheses which could also account for all the facts reported here: (1) A single suppressor mutation could cause a phenotypic reversion of two independent mutations to Rec$^-$ and UV^S. This possibility is supported by the observation that the revertants show a greater or lower resistance to ultraviolet irradiation than the wild-type strain, JC-411. (2) A mutant protein may cause a modification in the DNA of the mutant cell so that it can neither participate in the process of recombination, nor be repaired after being damaged by ultraviolet light. (3) A single mutation may affect the expression of more than one gene: for example, a mutation in a gene concerned with regulating the operation of genes, some of which participate in recombination and some of which participate in repair. A polarity mutation in an operon containing some genes participating in repair and some participating in recombination would also have the same effect. Tests of each of these hypotheses can be devised and are presently being conducted.

REFERENCES

1. Kellenberger, G., M. L. Zichichi, and J. Weigle, these Proceedings **47**:869 (1961).
2. Meselson, M., and J. J. Weigle, these Proceedings **47**:857 (1961).
3. Ihler, G., and M. Meselson, Virology **21**:7 (1963).
4. Siddiqi, O. H., these Proceedings **49**:589 (1962).
5. Meselson, M., personal communication.
6. Luria, S. E., Ann. Rev. Microbiol. **16**:205 (1962).
7. Howard-Flanders, P., personal communications.
8. Boyce, R. P., and P. Howard-Flanders, these Proceedings **51**:293 (1964).
9. Setlow, R. B., and W. L. Carrier, these Proceedings **51**:226 (1964).
10. Pettijohn, David E., Biophysics Laboratory Report, No. 103 (Stanford University, 1964).
11. Genotypic symbols stand for the genes concerned with the biosynthesis of threonine, *thr;* leucine, *leu;* histidine, *his;* purines, *pur;* arginine, *arg;* methionine, *met;* thiamin, *thi;* and the fermentation of lactose, *lac;* and maltose, *mal.* When used with gene symbols, "+" refers to the wild-type state of the gene. Numbers indicate arbitrary site designations assigned to mutant alleles by Adelberg (personal communication).

    Phenotypic characteristics are indicated by the following abbreviations: Thr, threonine; Leu, leucine; His, histidine; Ade, adenine; Arg, arginine; Met, methionine; Thi, thiamin; Lac, lactose; Mal, maltose; Sm, streptomycin; UV, ultraviolet light; D, donor in conjugation; ND, nondonor in conjugation; "–", "requiring," when used with the abbreviation of an amino acid, and "nonfermenting," when used with the abbreviation of a sugar; "+", nonrequiring and fermenting, when used with the abbreviation of an amino acid and sugar, respectively: *S*, sensitivity to a lethal agent; *R*, resistance to a lethal agent.
12. Clark, A. J., Genetics **48**:105 (1963).
13. Adelberg, E. A., and S. N. Burns, J. Bacteriol. **29**:321 (1960).
14. Jacob, F., and E. L. Wollman, Sexuality and the Genetics of Bacteria (Academic Press, 1961).
15. Schaeffer, P., in The Bacteria, ed. I. C. Gunsalus and R. Y. Stanier (New York: Academic Press, 1964), vol. 5, p. 115.

22 Reciprocal translocations

C. E. Ford
Hilary M. Clegg
Medical Research Council
Radiobiological Research Unit
Harwell, Didcot, Berkshire, England

The term "translocation" was introduced into genetics by Bridges (1923) to connote the transference of a chromosome segment from its normal position to a position in a different chromosome. The first example was inferred from genetic experiments with a mutant stock of *Drosophila melanogaster,* but subsequent cases have usually been identified from observations of chromosome behaviour at meiosis in heterozygotes. Examples are now known in all species that have been subjected to extensive cytogenetic study.

Four types may be distinguished (cf. White, 1954): attachment of a terminal segment of one chromosome to the end of another (simple translocation); exchange of terminal segments between non-homologous chromosomes (reciprocal translocation or interchange); and transference of an intercalary segment to another position either in the same chromosome (shift) or in a different chromosome (insertion). It is the convention to ascribe their origin to chromosome breakage followed by reunion of non-related broken ends. Simple translocation implies joining of a broken end to a normal chromosome end or telomere (Muller, 1932), and it is doubtful whether any true examples occur. Reciprocal translocations require two breaks and rejoins; shift and insertion each require three. On the whole the term "reciprocal translocation" has been favoured by workers with *Drosophila* species and the mouse, the term "interchange" by plant cytogeneticists.

The association of two acrocentric or telocentric chromosomes to form a single metacentric chromosome ("centric fusion" or Robertsonian translocation) might be included as a fifth type of translocation although it does not fall strictly within the definition given in the foregoing paragraph. Furthermore, as the not very satisfactory term "centric fusion" suggests, it may originate not in chromosome breakage and reunion as normally understood but rather as a consequence of abnormal events of a special class occurring at the centromere. Robertsonian translocations are of special interest and importance in human cytogenetics. They have recently been considered by Hamerton (1968) and will not be given further attention here.

In species other than man, most of the known translocations have first been suspected following the recognition of an unexpected reduction in fertility that showed quasi-dominant inheritance, then confirmed by observation of the specific pairing properties of individual chromosome segments at meiosis. In man the primary observation has usually been an unbalanced karyotype in a congenitally malformed propositus. Further investigation has then often revealed an apparently balanced re-arrangement in one of the parents, one chromosome being shorter than normal and another increased by the same amount. Fewest assumptions are required if such a change is interpreted as a reciprocal translocation. As indicated above, simple translocations are rare in other species, if they occur at all, and three-break re-arrangements are less likely than two-break re-arrangements on geometrical grounds (Ford, 1964). Nevertheless, in the absence of evidence of pairing at meiosis, simple translocations and re-arrangements involving three or more breaks cannot be excluded, so interpretation as a reciprocal translocation must remain presumptive.

Lejeune & Berger (1965) have advanced reasons for supposing that some translocations

From British Medical Bulletin **25**:110-114, 1969. Used with permission.

may indeed be three-break re-arrangements (specifically, insertions) and that they may be responsible for some instances in which phenotypically normal carriers and phenotypically abnormal individuals with an apparently identical balanced re-arrangement occur in the same family. They pointed out that crossing-over in an insertion heterozygote between the inserted segment and the normally located homologous segment may give rise to a morphologically similar, but segmentally distinct, chromosome that would cause simultaneous duplication and deficiency (see section 1). An essential condition is that the inserted segment must lie at the same distance from the chromosome end as in its normal location. Lejeune & Berger (1965) have referred to the consequence of such crossing-over as *"aneusomie de recombinaison"*. Additional support for their view has been given by de Grouchy, Aussannaire, Brissaud & Lamy (1966).

Attention will be confined in the following account to presumptive reciprocal translocations. To avoid excessive repetition, the qualification "reciprocal" will usually be omitted.

1. THEORETICAL CONSIDERATIONS

In view of the remarkable universality of the principal events of meiosis in what Whitehouse (1965) has called the "chromosomal kingdom", experience of the basic meiotic behaviour in translocation heterozygotes of other species can confidently be extrapolated to man. It can be expected that each segment of the re-arranged chromosomes will pair at pachytene with the homologous segment of the unchanged chromosomes to give a cross-shaped configuration (Fig. 1). At diplotene, chiasmata may or may not appear in each arm of the cross, independently of the other three, the mean number being positively related to the length of the arm. At diakinesis and first metaphase, the four chromosomes of the complex may therefore remain associated as a ring quadrivalent or as a chain quadrivalent; or they may be separated (in consequence of failure of chiasma formation

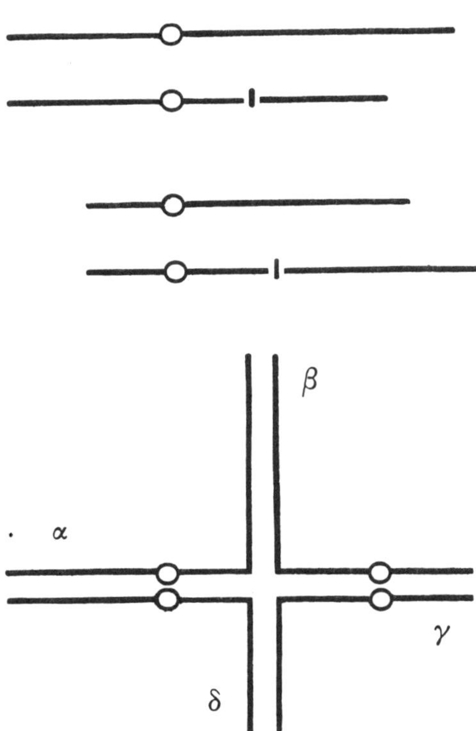

The normal and re-arranged chromosomes (with points of breakage) are shown above and the expected pachytene cross-configuration below. The segments defined by the points of breakage are shown as α, β, γ and δ. Chromosome 2 is αβ; chromosome B is γδ.

Fig. 1. Diagram of a hypothetical reciprocal translocation between the long arm of a No. 2 chromosome and the long arm of a chromosome of the B group.

in two or more arms of the pachytene cross) into trivalent and univalent, two bivalents, bivalent and two univalents, or even four univalents. In the male mouse, where the chiasma frequency is low, failure to form chiasmata in one or more arms of the pachytene cross is common, particularly in arms that are short in comparison with the normal chromosomes (Fig. 2). Few observations have yet been made at meiosis in human translocation heterozygotes but, in view of the much higher chiasma frequency, ring quadrivalents should be common, particularly when the longer chromosomes are involved (Fig. 3; and cf. Lindsten, Fraccaro, Klinger & Zetterqvist, 1965).

At anaphase, a ring association of four chromosomes can disjoin, so that two members go to each pole of the spindle, in three ways. These have been termed alternate, adjacent-1 and adjacent-2 (McClintock, 1945), the two last named being distinguished by the behaviour of the centromeres, homologous centromeres disjoining in adjacent-1 segregations and proceeding to the same pole in adjacent-2 segregations. The final segregational consequences are altered and extended if chiasmata are formed in either or both of the interstitial segments that lie between the points of breakage and the centromeres. Alternate and adjacent-1 disjunctions become equivalent (the normal condition in mouse translocations, since all the chromosomes are acrocentric); and adjacent-2 disjunctions can lead to an additional series of six different meiotic products (as defined by their segmental structure). The full series of ten gametic and zygotic types potentially derivable from a translocation heterozygote following 2:2 disjunction at first anaphase is set out in Table I. If 3:1 and 4:0 disjunctions are included, the number of possible types is increased to 36 (Ford, 1969).

Table I shows that only two of the ten combinations have a balanced segmental constitution. In four there is duplication of one segment and deficiency of another; and in the last four combinations two segments are duplicated and two deficient. These unbalanced combinations are almost invariably lethal to the developing mouse embryo (Carter, Lyon & Phillips, 1955), and the high frequency with which human embryos with corresponding unbalanced genomes are carried to term and born alive, though with varied arrays of

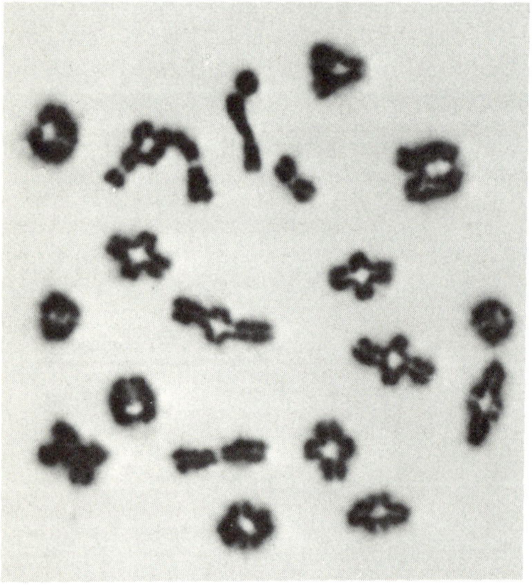

(Retouched print; preparation by Dr E. P. Evans: magnification ×1,900)

The short chromosome frequently appears as a univalent, the other three chromosomes of the complex forming a chain trivalent (III+I). Ring quadrivalents (R IV) are rare

Fig. 2. Chain quadrivalent (Ch IV) at first metaphase in a primary spermatocyte of a mouse heterozygous for the T6 translocation.

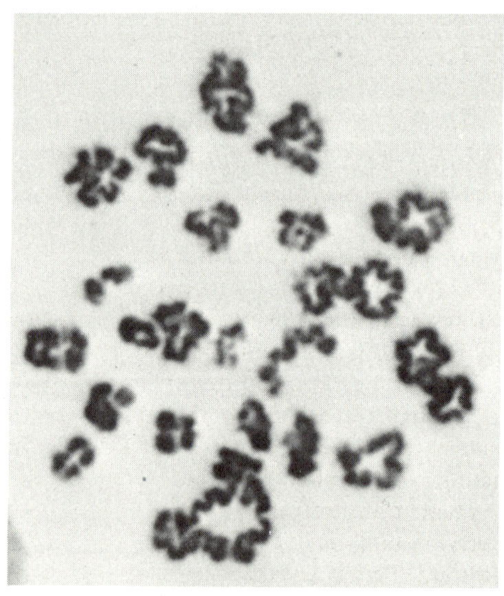

(Retouched print; unpublished case of Dr M. N. Macintyre: magnification ×2,300)

Ring quadrivalents with five or more chiasmata are the common form of association in this case; chain quadrivalents are rare and other types of association have not been seen

Fig. 3. Ring quadrivalent in a primary spermatocyte at first metaphase from a human male heterozygous for a reciprocal translocation t(3p + or q +;Bq−).

Table I. Segregation at meiosis in a heterozygote for the postulated reciprocal translocation shown in Fig. 1

Type of disjunction*	Segmental structure of products	Karotype of corresponding zygote†
Alternate	$\alpha\beta$, $\gamma\delta$	46 (normal)
	$\alpha\delta$, $\gamma\beta$	46,t(2q–;Bq+)
Adjacent-1	$\alpha\beta$, $\gamma\beta$	46,B–,(Bq+)+
	$\alpha\delta$, $\gamma\delta$	46,2–,(2q–)+
Adjacent-2	$\alpha\beta$, $\alpha\delta$	46,(2q–)+,B–
	$\gamma\beta$, $\gamma\delta$	46,2–,(Bq+)+
Adjacent-2 after crossing-over in interstitial segments	$\alpha\beta$, $\alpha\beta$	46,B–,2+
	$\gamma\delta$, $\gamma\delta$	46,2–,B+
	$\alpha\delta$, $\alpha\delta$	46,2–,B–,2(2q–)+
	$\gamma\beta$, $\gamma\beta$	46,2–,B–,2(Bq+)+

*When crossing-over occurs in the interstitial segments, alternate and adjacent-1 disjunctions become geometrically indistinguishable and the first four types of product are expected with equal frequency.

†Arising from syngamy with a normal haploid gamete.

Table II. Ascertainment of propositi

Method of ascertainment	Number of propositi
Child with congenital malformations	118
Woman with history of spontaneous abortion	6
Fortuitous	5
Total	129

congenital malformations, has come as a considerable surprise. However, most reciprocal translocations in man have been ascertained through congenitally malformed propositi (Table II), and this may well go some way towards explaining the apparent difference from findings in the mouse and other species.

It cannot be stressed too strongly that the relative frequencies of the different types of gametes (considered in terms of segmental constitution) are likely to be unique to each individual translocation, notwithstanding the underlying conformity to the rules of meiotic behaviour. This is because the break-points are widely if not entirely randomly distributed over the whole chromosome complement, so that the probability of recurrence of the same set of segment lengths and centromere positions is very low, and because orientation on the spindle (and consequent mode of disjunction) is believed to be dependent largely on the numbers and distribution of chiasmata and the relative positions of the centromeres.

2. SOURCES OF INFORMATION

We wished to identify preseumptive reciprocal translocations in which the chromosomes of the propositus and both parents had been examined. A Medical Literature Analysis and Retrieval System (MEDLARS) search covered the literature up to July 1967. This was supplemented by searches of the titles in *Index Medicus,* the abstracts in *Birth Defects,* the *Human Chromosome Newsletter,* and individual journals held in the Radcliffe Science Library, Oxford, and in the libraries of the Atomic Energy Research Establishment, Harwell, and this Unit. Wherever possible the original articles were consulted and 129 suitable cases were identified. The method of ascertainment of the propositus, and the karyotypes of the propositus and his parents, were recorded. Pedigrees extending over two or more generations were extracted or constructed from data given in the publications concerned. It is probable that some cases that satisfy the criteria have been overlooked. It is also possible that a small number of cases have been included twice through double reporting, but this is unlikely to be a serious source of error and the sample as a whole should be a representative one.

The original references are too numerous to be quoted here but a list of them has been placed on file in the office of the *British Medical Bulletin.*

3. ASCERTAINMENT OF PROPOSITI

Table II shows that the overwhelming majority of propositi of the 129 translocation cases were children with congenital malformations. The almost universal inviability of progeny that bear unbalanced genomes in translocation stocks of mice, *Drosophila* and even plants suggests that, in man, matings involving a heterozygote as one partner might be characterized by an erratic sequence of normal births and spontaneous abortions. Yet only six of the propositi were women with histories of spontaneous abortion. Evidence will be presented in section 6 that, in fact, spontaneous abortion in matings involving human translocation heterozygotes is not strikingly greater than in the general population.

Court-Brown & Smith (p. 74 of this Bulletin) present evidence that the total frequency of structural heterozygotes in the general population may be as high as 0.3%. It is therefore not surprising that some translocation heterozy-

Table III. 129 presumptive reciprocal translocations: distribution of break-points between chromosome groups

Chromosome group	Total relative length of chromosomes in group*	Number of breaks observed†	Number of breaks expected‡	Contribution to χ^2_6 §
A	49.18	32	57.32	11.185
B	25.36	46	29.56	9.143
C	78.06	52	90.98	16.701
D	21.10	52	24.59	33.334
E	18.28	32	21.31	5.362
F	8.50	4	9.90	3.516
G	8.00	25	9.32	26.380
Total	208.48	243	242.98	105.621

*From table of relative lengths of chromosomes given by Penrose (1964). Value for group C includes one and a half times the length of the X chromosome. Value for group G includes half the length of the Y chromosome.

†10 were uncertainly assigned: 6 in group D, 2 in group E and one each in groups C and F. In 15 cases with unbalanced karyotypes only one abnormal chromosome was identified.

‡Calculated in proportion to total relative length of group.

§For $\chi^2_6 = 105.621$, P is negligibly small.

Table IV. Evidence of origin of presumptive reciprocal translocations in 129 propositi

Evidence of origin	Number
Inherited from mother	65
Inherited from father	28
Father and mother both karyotypically normal:	
Propositus not mosaic	30
Propositus a mosaic	6
Total	129

gotes should have been identified by chance in the course of other investigations.

4. PARTICIPATION OF DIFFERENT CHROMOSOME GROUPS

The participation of the different chromosome groups in the 129 translocations is shown in Table III. Counting two breaks per translocation, there should be a total of 258 break-points, but, in 15 cases, only one of the two chromosomes involved in the presumptive translocation was identified. On the assumption that the frequency of breakage is directly proportional to the total relative length of all the chromosomes in the group, the expected distribution of the break-points between groups is also given. This has been calculated from the table of mean relative lengths of individual chromosomes presented by Penrose (1964). Inspection of Table III shows that break-points are much more frequent than expected in groups B, D, E and G and much less frequent than expected in groups A, C and F. A χ^2 test of goodness of fit shows that the probability of obtaining these deviations by chance sampling is negligibly small.

These data do not, of course, provide an indication as to whether the differences are due to primary differences in frequency of breakage between the chromosomes of the different groups, or whether they are to be attributed to selective elimination of certain translocations subsequent to origin. It would be instructive to compare the frequencies shown in Table III with the frequencies of breaks induced by ionizing radiation and other agents known to cause breakage of chromosomes.

It is of interest that two of the break-points were located in the X chromosome and one in the Y chromosome.

5. ORIGIN OF RECIPROCAL TRANSLOCATIONS

Evidence bearing on the origin of the translocations is assembled in Table IV. The explanation of the highly significant difference ($\chi^2_1 = 14.720$, $P < 0.001$) between the numbers of propositi that have received the translocation from their mothers and the numbers that have inherited it from their fathers is not apparent. Moreover this difference appears to conflict with the evidence given in Table V that male and female heterozygotes are almost equally likely to transmit the translocation to their progeny. In any case, since 93 of 129 translocations were inherited from a parent, it may be inferred that each translocation, on the average, is transmitted through about three generations before it is eliminated. This must be regarded as only a very approximate figure, since not only is it probable that the chance of transmission to another generation will vary from one individual translocation to another, but also changes in social conditions, particularly in mean family size, are likely to have a considerable influence.

Of the 36 propositi, both of whose parents had normal karyotypes, six were reported to be normal/translocation mosaics. This figure may be biased on the low side by the fact that if a

Table V. Outcome of 561 pregnancies in matings of presumptive translocation heterozygotes

Sex of heterozygous parent	Number of cases	Livebirth: normal phenotype			Stillbirth or livebirth with abnormal phenotype	Spontaneous abortion
		Translocation heterozygote	Normal karyotype	Chromosomes not investigated		
Male	55	53	30	7	50	33
Female	106	79	56	43	121	89
Total	161	132	86	50	171	122

subject is discovered to be a translocation mosaic the parents are less likely to be investigated, since the translocation could not have been inherited, and subjects whose parents were not investigated were excluded from our sample. On the other hand only one of the 93 heterozygous parents was reported to be a mosaic, whereas, if the proportion of mosaics in the propositi is representative, at least five would be expected. There appears to be a conflict of the data here, but it may be that not all mosaics identified in the course of extensive family investigations have been reported.

In any case the occurrence of normal/translocation mosaics demonstrates that structural changes, like numerical changes, can originate in somatic cells during development and are by no means limited to germ cells. At the present time there do not appear to be any grounds for supposing that chromosomes in either type of cell are the more likely to undergo structural change (other than misdivision of the centromere).

6. GENETICS OF RECIPROCAL TRANSLOCATIONS

An analysis of the outcome of 561 pregnancies in matings in which one partner was heterozygous for a presumptive reciprocal translocation is presented in Table V. The information was extracted from the 89 families for which informative pedigrees were given or could be constructed.

The mean family size of the male heterozygotes is 3.145; of female heterozygotes, 3.660. The significance of the difference has not been tested; in any case there is no striking difference in relative fertility. The reason for the great excess of female heterozygotes that contributed information is not apparent, but it may reflect a tendency of investigators to concentrate on recording the progenies of females in the pedigrees rather than males.

Whatever the mode of disjunction at first meiotic anaphase, "carrier" gametes and normal gametes are expected with equal frequency. The ratio of heterozygotes to normal subjects in the progeny should not, therefore, differ significantly from unity. The data for male and female heterozygotes are homogeneous ($\chi_1^2 = 0.613$, $P > 0.3$) but, when combined, the deviation from the expected 1:1 ratio is highly significant ($\chi_1^2 = 9.706$, $P < 0.01$). It is most unlikely that the survival of heterozygotes should be favoured over normal subjects and the explanation of the deviation may be found in a bias of the investigator towards the examination of the parents of progenies that include abnormal subjects. It is worth notice that, if the subjects whose chromosomes were not investigated should have been all or mostly normal, the deviation from equality would be greatly reduced or even eliminated.

A large, but uncertain, proportion of the two remaining classes were presumptively bearers of unbalanced genomes. A between sexes comparison of the numbers in these two classes with the numbers having a normal phenotype shows that there is no significant heterogeneity ($\chi_2^2 = 1.970$, $P > 0.3$). This suggests that there is no marked difference between the frequencies of the different disjunctional types as between spermatocyte and oocyte meiosis.

It is noteworthy that the frequency of spontaneous abortion is only 21.7%. This does not represent a striking increase over the level in the general population, commonly supposed to be about 15% (see World Health Organization, 1966). The frequency of stillbirths plus deaths within the first postnatal week is about 3% in the populations from which the great majority of the propositi were drawn *(United Nations Demographic Yearbook, 1966)*. If the totals entered in the final two columns are corrected to allow for non-genetic and irrelevant genetic deaths by use of these figures, 3%

and 15% respectively, the ratio of normal to presumptive abnormal zygotes would become 268:196. In view of the very high frequency of chiasmata at meiosis in the human species crossing-over in the interstitial segments should be very frequent if not quite universal, so that a minimum ratio of 1 unbalanced gamete to 1 balanced gamete would be expected to be produced by heterozygotes (see section 1 and Table I). The deviation of the adjusted numbers of abnormal zygotes from equality with normal zygotes, which is very highly significant ($\chi_1^2 = 11.172$, $P < 0.001$), may be a consequence of the failure of some karyotypically unbalanced conceptuses to implant and give clinical signs of pregnancy. Other possibilities, which are considered to be less likely, are under-reporting of abortions and stillbirths and an excess of alternate disjunctions without crossing-over in an interstitial segment.

It is clear from the foregoing analysis that many questions regarding the genetic behaviour of human reciprocal translocations remain to be answered. Their solution would be helped by full unbiassed pedigrees of translocation families with recording, including birth dates, of the progenies of normal subjects as well as heterozygotes, and of males as well as females.

REFERENCES

Bridges, C. B. (1923) Anat. Rec. **24**:426.
Carter, T. C., Lyon, M. F. & Phillips, R. J. S. (1955) J. Genet. **53**:154.
Ford, C. E. (1964) In: Second International Conference on Congenital Malformations, New York City, July 14-19, 1963, p. 22. International Medical Congress, New York.
Ford, C. E. (1969) In: Benirshke, K., ed. Comparative mammalian cytogenetics. Springer-Verlag, New York (In press).
Grouchy, J. de, Aussannaire, M., Brissaud, H. E. & Lamy, M. (1966) Am. J. hum. Genet. **18**:467.
Hamerton, J. L. (1968) Cytogenetics **7**:260.
Lejeune, J. & Berger, R. (1965) Annls. Génét. **8**:21.
Lindsten, J., Fraccaro, M., Klinger, H. P. & Zetterqvist, P. (1965) Cytogenetics **4**:45.
McClintock, B. (1945) Am. J. Bot. **32**:671.
Muller, H. J. (1932) In: Jones, D. F., ed. Proc. VI int. Congr. Genet., Ithaca, 24-30 August 1932, vol. 1, p. 213. Brooklyn Botanic Garden, New York.
Penrose, L. S. (1964) Ann. hum. Genet. **28**:195.
United Nations Demographic Yearbook, 1966 (1967) United Nations, New York.
White, M. J. D. (1954) Animal cytology and evolution, 2nd ed. Cambridge University Press, London.
Whitehouse, H. L. K. (1965) Towards an understanding of the mechanism of heredity. Arnold, London.
World Health Organization (1966) WHO Group on standardization of procedures for chromosome studies in abortion. Bull. Wld. Hlth. Org. **34**:765; and Cytogenetics **5**:361.

chapter 6

Sex linkage and sex determination

SEX CHROMOSOMES

The concept that the karyotype (somatic chromosomal complement) of diploid organisms consists of pairs of morphologically similar units has its exceptions when one considers the sex chromosomes. In some species one finds that the female karyotype contains one more chromosome than the male. In other species both sexes have the same number of chromosomes, but the sex chromosomes are morphologically different in either the male or the female, depending on the species. These different situations regarding the sex chromosomes were described independently in 1905 by Wilson (Ref. 6-1), whose paper is reprinted here, and also by Stevens (Ref. 6-2).

It was quickly surmised that the heterogametic sex (the sex with the odd number of chromosomes or with the morphologically different chromosomes) was the one that determined the sex of the offspring. It was also realized that the general one-to-one ratio between the sexes at birth resembled the results obtained from a Mendelian cross between a heterozygote and a homozygous recessive. In most species, including *Drosophila* and man, it is the male that is the heterogametic sex. Experimental proof of the role of the heterogametic parent in determining the sex of the offspring was provided in 1907 by Correns (Ref. 6-3), who studied dioecious species of the plant *Bryonia.* He found that the pollen produced by the male plants are of two sorts, half male-determining and half female-determining. The eggs produced by the female plant in these species he found to be all alike in respect to sex determination.

SEX LINKAGE

In addition to functioning in sex determination, the sex chromosome designated as "X" carries genes for various traits. Where present, the "Y" chromosome appears to be largely, if not completely, devoid of genes. The first trait in *Drosophila* to be clearly demonstrated as sex-linked was reported by in 1910 by T. H. Morgan (Ref. 6-4), whose paper is included in this chapter. (It will be noted that what Morgan referred to as *sex limited* is now known as *sex-linked*).

The critical characteristic of sex-linked inheritance in *Drosophila,* man, etc. is the absence of father-to-son transmission. This is a necessary result of the fact that the X chromosome of the male is transmitted to none of his sons, yet to all of his daughters. An important exception to this "crisscross" type of inheritance was discovered in *Drosophila* by Bridges in 1916 to be the result of nondisjunction of X chromosomes during oogenesis. This paper was referred to in Chapter 3 (Ref. 3-13). A later paper by L. V. Morgan in 1922 discussed the pattern of inheritance that results from nondisjunction of sex chromosomes (Ref. 6-5).

In man, the classic example of a sex-linked trait is hemophilia, bleeder's disease. Its mode of inheritance was understood by the Jews of the Middle Ages, as seen in their talmudic prohibition against the circumcision of any male child born to a woman whose male relatives were bleeders. The most celebrated cases of hemophilia occurred in the royal families of England, Russia, and Spain as detailed by Haldane in 1939 (Ref. 6-6). Today about 80 sex-linked traits are known in man. Most of them are pathological and caused by recessive genes. These traits include, among others, muscular dystrophy (muscular wasting-away resulting in invalidism), optic atrophy (withering of optic nerve, leading to blindness), nystagmus (uncontrolled rolling of the eyeballs), and anhidrotic ectodermal dysplasia (nonfunctional sweat glands). Some sex-linked traits caused by dominant genes are hypophosphatemic rickets (vitamin D–resistant rickets) and defective tooth enamel, which results in the early wearing-down of teeth to the gums. A recent catalogue of sex-linked and autosomal traits found in man was published by McKusick in 1968 (ref. 6-7).

Although in most species the male is the heterogametic sex, in some organisms it is the female. These include, among others, the moths and the birds. In these instances, the X chromosome of daughters is derived from the male parent as are all their sex-linked traits. The inheritance of pigmentation in moths was found to follow this pattern of sex-linkage by Doncaster and Raynor in 1906 (Ref. 6-8). A similar situation was found for the inheritance of the barred plumage pattern of Plymouth Rock fowls by Goodale in 1909 (Ref. 6-9) and is reported in this chapter.

GENETIC DETERMINATION OF SEX

Whereas some species of plants and most species of higher animals consist of individuals who are either male or female, there are many species that consist of bisexual individuals. In some species all the individuals are hermaphroditic, indicating that one does not always need differences in chromosomes in order to get differences in reproductive systems. In other species, the individuals change from one sex to the other with age or with alterations of the environment. Even in species in which the individual's sex appears to be determined by the sex chromosomes, genetic or environmental factors may modify the sex.

In our discussion of sex chromosomes and sex linkage in *Drosophila,* it was pointed out that the sex of the individual was determined by its sex chromosome combination (i.e., XX is female and XY is male). However, it was reported by Bridges in 1925 (Ref. 6-10) that the effects of the sex chromosomes in sex determination could be altered by changing the ratio of the number of X chromosomes to the number of sets of autosomes in the karyotype of the individual. This became known as the *genic balance theory* of sex determination in *D. melanogaster*. It is not known in how many species the sex of the individual is determined by genic balance. An autosomal gene which, in homozygous condition, alters the sex-determining action of the sex chromosomes in *D. melanogaster* was reported by Sturtevant in 1945 (Ref. 6-11).

In the absence of any information to the contrary, it was assumed by many that sex determination in man was the result of genic balance. However, this erroneous idea was due to the lack of adequate human chromosome preparations. A discovery of significance for an understanding of the mechanism of human sex determination occurred in 1949 when Barr and Bertram (Ref. 6-12) found a deeply staining body in the nuclei of the nerve cells of female cats that was absent from the nerve cells of male cats. This body, called the *Barr body,* was found in the cells of all female mammals but not in the cells of the males. The Barr body proved to be an extremely useful sex indicator in man but was found to have certain notable exceptions. Sterile human females exhibiting the characteristics of Turner's syndrome were found to lack a Barr body in each of their cells, but sterile human males exhibiting the characteristics of Klinefelter's syndrome were found to have a Barr body in each of their cells. These seeming contradictions were resolved through an extension of the work in 1956 by Tjio and Levan (Ref. 6-13). Their paper, which is reprinted here, reported a technique for the preparation of human chromosomes that permitted their separation and study. The resolution of the problem in the case of Turner's syndrome was reported in 1959 by Ford and his co-workers (Ref. 6-14), and that of Klinefelter's syndrome was reported, also in 1959, by Jacobs and Strong (Ref. 6-15). Both of these papers are included in this chapter. As a result of the above investigations, it was concluded that in man, the Y chromosome is absolutely male determining regardless of the number of X chromosomes present in the karyotype. In the absence of a Y chromosome, the individual is female regardless of the composition of her karyotype.

A different method of sex determination is found in some of the species of the insect order Hymenoptera, which includes the ants, bees, sawflies, and wasps. In these species males develop parthenogenetically from unfertilized eggs and have the haploid chromosome number. The females develop from fertilized eggs and have the diploid chromosome number. However, the genetic system in at least one species is somewhat complicated. It was described in 1939 by Whiting (Ref. 6-16). He found that one could obtain diploid males in the parasitic wasp, *Habrobracon juglandis* but that these tended to be inviable. He postulated that the sex of the insect was under the control of a multiple allelic series, so that a combination between any two different alleles produces a female. On the other hand, the presence of either a single allele in the haploid condition or of a homozygote in the diploid condition produces a male. It is not known in how many other species the sex of the individual is determined by the genetic system described above.

The gametes of plants, depending on the species, are produced normally either on the same plant (monoecious) or on different plants (dioecious). One dioecious plant whose sex-determining mechanism has been well studied is *Melandrium.* It was reported in 1948 by Westergaard (Ref. 6-17) that the sex-determining mechanism resembles that of man. Male plants of the species are XY in genotype and the Y chromosome is absolutely essential for the development of a male reproductive system. Another dioecious plant, asparagus, has a somewhat different sex-determining mechanism. This was described in 1943 by Rick and Hanna (Ref. 6-18), whose paper is reprinted in this chapter. An extensive summary of sex determination in the flowering plants was prepared in 1940 by Allen (Ref. 6-19).

The last type of genetic determination of sex that we shall consider is the one which operates in the colon bacillus, *Escherichia coli.* The sex of the bacterial cell is controlled by an extrachromosomal cytoplasmic factor. This was discovered in 1953 by Hayes (Ref. 6-20). The sex factor takes one of many possible forms, of which we shall consider only two: F^+ (malelike) and F^- (femalelike). When F^+ and F^- cells conjugate, the male cell will transfer an F^+ factor to the female cell, converting the latter into an F^+ cell. The original male cell remains F^+ and both cells can transform other F^- cells into F^+-containing cells.

Our consideration of the genetic determination of sex shows that it can take many forms. These include a genic balance system, a dependency on the presence of a specific chromosome, a dependency on a single gene, or an extrachromosomal cytoplasmic factor. In some cases there may be an interaction of two or more of the above sex-determining mechanisms in the development of a reproductive system. Further research will undoubtedly reveal other genetic mechanisms of sex determination.

ENVIRONMENTAL DETERMINATION OF SEX

There are many examples of environmental determination of sex. One such instance involves the snail, *Crepidula,* and was reported in 1936 by Coe (Ref. 6-21). He found that all young specimens are males. If raised in isolation, the male develops into a female. If the young animal is attached to a female, it will remain as a male. If a large number of young males are attached to one another, some of the males become female. An individual that becomes a female remains in that state.

Another instance illustrating environmental control over sexual development was described in 1917 by Lillie (Ref. 6-22) and concerned the occurrence of twin calves in cattle. When the calves are of different sexes, the female member, called a *freemartin,* is sterile and has female external genitalia but male internal organs. The male twin is usually normal. It was hypothesized that the formation of the freemartin was due to the transfer of male sex hormones to the developing female and a consequent suppression of the formation of internal female organs. This would indicate that, in cattle, the male hormone is produced earlier during development than the female hormone.

The report of an experiment affecting reproductive system development in the mosquito was made in 1961 by Horsfall and Anderson (Ref. 6-23) and is the final paper reprinted in this chapter.

REFERENCES

Sex chromosomes

6-1. Wilson, E. B. 1905. The chromosomes in relation to the determination of sex in insects. Science **22**:500-502.

6-2. Stevens, N. M. 1905. Studies in spermatogenesis with especial reference to the "accessory chromosome." Carneg. Inst. Wash. Publ. No. 36. Washington, D. C.

6-3. Correns, C. 1907. Die Bestimmung und Vererbung des Geschlechtes. Gebrüder Borntraeger, Berlin.

Sex linkage

6-4. Morgan, T. H. 1910. Sex limited inheritance in *Drosophila.* Science **32**:120-122.

6-5. Morgan, L. V. 1922. Non-criss-cross inheritance in *Drosophila melanogaster.* Biol. Bull. **42**:267-274.

6-6. Haldane, J. B. S. 1939. Royal blood. Living Age **35**:626-631.

6-7. McKusick, V. A. 1968. Mendelian inheritance in man. 2nd ed. Johns Hopkins Press, Baltimore.

6-8. Doncaster, L., and Raynor, G. H. 1906. On breeding experiments with Lepidoptera. Proc. Zool. Soc. Lond. **1**:125-133.

6-9. Goodale, H. D. 1909. Sex and its relation to the barring factor in poultry. Science **29**:1004-1005.

Genetic determination of sex

6-10. Bridges, C. B. 1925. Sex in relation to chromosomes and genes. Am. Nat. **59**:127-137.

6-11. Sturtevant, A. H. 1945. A gene in *Drosophila melanogaster* that transforms females into males. Genetics **30**:297-299.

6-12. Barr, M. L., and Bertram, E. G. 1949. A morphological distinction between neurones of the male and female, and the behavior of the nucleolar satellite during accelerated nucleoprotein synthesis. Nature **163**:676-677.

6-13. Tjio, J. H., and Levan, A. 1956. The chromosome number of man. Hereditas **42**:1-6.

6-14. Ford, C. E., Jones, K. W., Polani, P. E., de Almeida, J. C., and Briggs, J. H. 1959. A sex chromosome anomaly in a case of gonadal dysgenesis (Turner's syndrome). Lancet **1**:711-713.

6-15. Jacobs, P. A., and Strong, J. A. 1959. A case of human intersexuality having a possible XXY sex determining mechanism. Nature **183**:302-303.

6-16. Whiting, P. W. 1939. Multiple alleles in sex determination of *Habrobracon.* J. Morphol. **66**:323-355.

6-17. Westergaard, M. 1948. The relation between chromosome constitution and sex in the offspring of triploid *Melandrium.* Hereditas **34**: 257-279.

6-18. Rick, C. M., and Hanna, G. C. 1943. Determination of sex in *Asparagus officinalis* L. Am. J. Bot. **30**:711-714.

6-19. Allen, C. E. 1940. The genotypic basis of sex-expression in Angiosperms. Bot. Rev. **6**:227-300.

6-20. Hayes, W. 1953. Observations on a transmissible agent determining sexual differentiation in *Bact. coli.* J. Gen. Microbiol. **8**:72-88.

Environmental determination of sex

6-21. Coe, W. R. 1936. Sexual phases in *Crepidula.* J. Exptl. Zool. **72**:455-477.

6-22. Lillie, F. R. 1917. The freemartin: a study of the action of sex hormones in the fetal life of cattle. J. Exptl. Zool. **23**:371-452.

6-23. Horsfall, W. R., and Anderson, J. F. 1961. Suppression of male characteristics of mosquitoes by thermal means. Science **133**:1830.

23 The chromosomes in relation to the determination of sex in insects

Edmund B. Wilson
Zoological Laboratory
Columbia University
New York, New York

Material procured during the past summer demonstrates with great clearness that the sexes of Hemiptera show constant and characteristic differences in the chromosome groups, which are of such a nature as to leave no doubt that a definite connection of some kind between the chromosomes and the determination of sex exists in these animals. These differences are of two types. In one of these, the cells of the female possess one more chromosome than those of the male; in the other, both sexes possess the same number of chromosomes, but one of the chromosomes in the male is much smaller than the corresponding one in the female (which is in agreement with the observations of Stevens on the beetle *Tenebrio*). These types may conveniently be designated as *A* and *B*, respectively. The essential facts have been determined in three genera of each type, namely, (type *A*) *Protenor belfragei, Anasa tristis* and *Alydus pilosulus,* and (type *B*) *Lygoeus turcicus, Euschistus fissilis* and *Coenus delius.* The chromosome groups have been examined in the dividing oogonia and ovarian follicle cells of the female and in the dividing spermatogonia and investing cells of the testis in case of the male.

Type *A* includes those forms in which (as has been known since Henking's paper of 1890 on *Pyrrochoris*) the spermatozoa are of two classes, one of which contains one more chromosome (the so-called 'accessory' or heterotropic chromosome) than the other. In this type the somatic number of chromosomes in the female is an even one, while the somatic number in the male is one less (hence an odd number) the actual numbers being in *Protenor* and *Alydus* ♀ 14, ♂ 13, and in *Anasa* ♀ 22, ♂ 21. A study of the chromosome groups in the two sexes brings out the following additional facts. In the cells of the female all the chromosomes may be arranged two by two to form pairs, each consisting of two chromosomes of equal size, as is most obvious in the beautiful chromosome groups of *Protenor,* where the size differences of the chromosomes are very marked. In the male all the chromosomes may be thus symmetrically paired with the exception of one which is without a mate. This chromosome is the 'accessory' or heterotropic one; and it is a consequence of its unpaired character that it passes into only one half of the spermatozoa.

In type *B* all of the spermatozoa contain the same number of chromosomes (half the somatic number in both sexes), but they are, nevertheless, of two classes, one of which contains a large and one a small 'idiochromosome.' Both sexes have the same somatic number of chromosomes (fourteen in the three examples mentioned above), but differ as follows: In the cells of the female (oogonia and follicle-cells) all of the chromosomes may, as in type *A,* be arranged two by two in equal pairs, and a small idiochromosome is not present. In the cells of the male all but two may be thus equally paired. These two are the unequal idiochromosomes, and during the maturation process they are so distributed that the small one passes into one half of the spermatozoa, the large one into the other half.

These facts admit, I believe, of but one interpretation. Since all of the chromosomes in the female (oogonia) may be symmetrically paired, there can be no doubt that synapsis in this sex gives rise to the reduced number of symmetrical bivalents, and that consequently all of the eggs receive the same number of chromosomes. This number (eleven in *Anasa,* seven in *Protenor* or *Alydus*) is the same as that present in those spermatozoa that contain the 'accessory' chromosome. It is evident that both forms of spermatozoa are functional, and that in type *A* females are produced from eggs fertilized by spermatozoa that contain the 'accessory' chromosome, while males are produced from eggs fertilized by spermatozoa that lack this chromosome (the reverse of the conjecture made by McClung). Thus if *n* be the

From Science **22:**500-502, 1905.

somatic number in the female $n/2$ is the number in all of the matured eggs, $n/2$ the number in one half of the spermatozoa (namely, those that contain the 'accessory'), and $n/2-1$ the number in the other half. Accordingly:

In fertilization

$$\text{Egg } \frac{n}{2} + \text{spermatozoon } \frac{n}{2} = n \text{ (female).}$$

$$\text{Egg } \frac{n}{2} + \text{spermatozoon } \frac{n}{2}-1 = n - 1 \text{ (male).}$$

The validity of this interpretation is completely established by the case of *Protenor,* where, as was first shown by Montgomery, the 'accessory' is at every period unmistakably recognizable by its great size. The spermatogonial divisions invariably show but one such large chromosome, while an equal pair of exactly similar chromosomes appear in the oogonial divisions. One of these in the female must have been derived in fertilization from the egg-nucleus, the other (obviously the 'accessory') from the sperm-nucleus. It is evident, therefore, that all of the matured eggs must before fertilization contain a chromosome that is the maternal mate of the 'accessory' of the male, and that females are produced from eggs fertilized by spermatozoa that contain a similar group (*i.e.,* those containing the 'accessory'). The presence of but one large chromosome (the 'accessory') in the somatic nuclei of the male can only mean that males arise from eggs fertilized by spermatozoa that lack such a chromosome, and that the single 'accessory' of the male is derived in fertilization from the egg-nucleus.

In type *B* all of the eggs must contain a chromosome corresponding to the large idiochromosome of the male. Upon fertilization by a spermatozoon containing the large idiochromosome a female is produced, while fertilization by a spermatozoon containing the small one produces a male.

The two types distinguished above may readily be reduced to one; for if the small idiochromosome of type *B* be supposed to disappear, the phenomena become identical with those in type *A.* There can be little doubt that such has been the actual origin of the latter type, and that the 'accessory' chromosome was originally a large idiochromosome, its smaller mate having vanished. The unpaired character of the 'accessory' chromosome thus finds a complete explanation, and its behavior loses its apparently anomalous character.

The foregoing facts irresistibly lead to the conclusion that a causal connection of some kind exists between the chromosomes and the determination of sex; and at first thought they naturally suggest the conclusion that the idiochromosomes and heterotropic chromosomes are actually sex determinants, as was conjectured by McClung in case of the 'accessory' chromosome. Analysis will show, however, that great, if not insuperable, difficulties are encountered by any form of the assumption that these chromosomes are specifically male or female sex determinants. It is more probable, for reasons that will be set forth hereafter, that the difference between eggs and spermatozoa is primarily due to differences of degree or intensity, rather than of kind, in the activity of the chromosome groups in the two sexes; and we may here find a clue to a general theory of sex determination that will accord with the facts observed in Hemiptera. A significant fact that bears on this question is that in both types the two sexes differ in respect to the behavior of the idiochromosomes or 'accessory' chromosomes during the synaptic and growth periods, these chromosomes assuming in the male the form of condensed chromosome nucleoli, while in the female they remain, like the other chromosomes, in a diffused condition. This indicates that during these periods these chromosomes play a more active part in the metabolism of the cell in the female than in the male. The primary factor in the differentiation of the germ cells may, therefore, be a matter of metabolism, perhaps one of growth.

24 Sex limited inheritance in Drosophila

T. H. Morgan†
Woods Hole, Massachusetts

In a pedigree culture of *Drosophila* which had been running for nearly a year through a considerable number of generations, a male appeared with white eyes. The normal flies have brilliant red eyes.

The white-eyed male, bred to his red-eyed sisters, produced 1,237 red-eyed offspring, (F_1), and 3 white-eyed males. The occurrence of these three white-eyed males (F_1) (due evidently to further sporting) will, in the present communication, be ignored.

The F_1 hybrids, inbred, produced:

2,459 red-eyed females,
1,011 red-eyed males,
782 white-eyed males.

No white-eyed females appeared. The new character showed itself therefore to be sex limited in the sense that it was transmitted only to the grandsons. But that the character is not incompatible with femaleness is shown by the following experiment.

The white-eyed male (mutant) was later crossed with some of his daughters (F_1), and produced:

129 red-eyed females,
132 red-eyed males,
88 white-eyed females,
86 white-eyed males.

The results show that the new character, white eyes, can be carried over to the females by a suitable cross, and is in consequence in this sense not limited to one sex. It will be noted that the four classes of individuals occur in approximately equal numbers (25 per cent.).

AN HYPOTHESIS TO ACCOUNT FOR THE RESULTS

The results just described can be accounted for by the following hypothesis. Assume that all of the spermatozoa of the white-eyed male carry the "factor" for white eyes "W"; that half of the spermatozoa carry a sex factor "X" the other half lack it, *i.e.*, the male is heterozygous for sex. Thus the symbol for the male is "WWX," and for his two kinds of spermatozoa WX–W.

Assume that all of the eggs of the red-eyed female carry the red-eyed "factor" R; and that all of the eggs (after reduction) carry one X, each, the symbol for the red-eyed female will be therefore RRXX and that for her eggs will be RX–RX.

When the white-eyed male (sport) is crossed with his red-eyed sisters, the following combinations result:

WX–W (male)
RX–RX (female)

RWXX (50%)–RWX (50%)
Red female Red male

When these F_1 individuals are mated, the following table shows the expected combinations that result:

RX–WX (F_1 female)
RX–W (F_1 male)

RRXX–	RWXX–	RWX–	WWX
(25%)	(25%)	(25%)	(25%)
Red female	Red female	Red male	White male

It will be seen from the last formulae that the outcome is Mendelian in the sense that there are three reds to one white. But it is also apparent that all of the whites are confined to the male sex.

It will also be noted that there are two classes of red females–one pure RRXX and one hybrid RWXX–but only one class of red males (RWX). This point will be taken up later. In order to obtain these results it is necessary to assume, as in the last scheme, that, when the two classes of the spermatozoa are formed in the F_1 red male (RWX), R and X go together–otherwise the results will not follow (with the symbolism here used). This all-important point can not be fully discussed in this communication.

The hypothesis just utilized to explain these results first obtained can be tested in several ways.

From Science **32**:120-122, 1910.
†Deceased.

Verification of hypothesis

First verification. If the symbol for the white male is WWX, and for the white female WWXX, the germ cells will be WX–W (male) and WX–WX (female), respectively. Mated, these individuals should give

WX–W (male)
WX–WX (female)

WWXX (50%)–WWX (50%)
White female White male

All of the offspring should be white, and male and female in equal numbers; this in fact is the case.

Second verification. As stated, there should be two classes of females in the F_2 generation, namely, RRXX and RWXX. This can be tested by pairing individual females with white males. In the one instance (RRXX) all the offspring should be red–

RX–RX (female)
WX–W (male)

RWXX–RWX

and in the other instance (RWXX) there should be four classes of individuals in equal numbers, thus:

RX–WX (female)
WX–W (male)

RWXX–WWXX–RWX–WWX

Tests of the F_2 red females show in fact that these two classes exist.

Third verification. The red F_1 females should all be RWXX, and should give with any white male the four combinations last described. Such in fact is found to be the case.

Fourth verification. The red F_1 males (RWX) should also be heterozygous. Crossed with white females (WWXX) all the female offspring should be red-eyed, and all the male offspring white-eyed, thus:

RX–W (red male)
WX–WX (white female)

RWXX–WWX

Here again the anticipation was verified, for all of the females were red-eyed and all of the males were white-eyed.

CROSSING THE NEW TYPE WITH WILD MALES AND FEMALES

A most surprising fact appeared when a white-eyed female was paired to a wild, red-eyed male, *i. e.*, to an individual of an unrelated stock. The anticipation was that wild males and females alike carry the factor for red eyes, but the experiments showed that all wild males are heterozygous for red eyes, and that all the wild females are homozygous. Thus when the white-eyed female is crossed with a wild red-eyed male, all of the female offspring are red-eyed, and all of the male offspring white-eyed. The results can be accounted for on the assumption that the wild male is RWX. Thus:

RX–W (red male)
WX–WX (white female)

RWXX (50%)–WWX (50%)

The converse cross between a white-eyed male RWX and a wild, red-eyed female shows that the wild female is homozygous both for X and for red eyes. Thus:

WX–W (white male)
RX–RX (red female)

RWXX (50%)–RWX (50%)

The results give, in fact, only red males and females in equal numbers.

GENERAL CONCLUSIONS

The most important consideration from these results is that in every point they furnish the converse evidence from that given by Abraxas as worked out by Punnett and Raynor. The two cases supplement each other in every way, and it is significant to note in this connection that in nature only females of the sport *Abraxas lacticolor* occur, while in *Drosophila* I have obtained only the male sport. Significant, too, is the fact that analysis of the result shows that the wild female *Abraxas grossulariata* is heterozygous for color and sex, while in *Drosophila* it is the male that is heterozygous for these two characters.

Since the wild males (RWX) are heterozygous for red eyes, and the female (RXRX) homozygous, it seems probable that the sport arose from a change in a single egg of such a sort that instead of being RX (after reduction) the red factor dropped out, so that RX became WX or simply OX. If this view is correct it follows that the mutation took place in the egg of a female from which a male was produced by combination with the sperm carrying no X, no R (or W in our formulae). In other words, if the formula for the eggs of the normal female is RX–RX, then the formula for the particular egg that sported will be WX; *i. e.*, one R dropped out of the egg leaving it WX (or no R and one X), which may be written OX. This egg we assume was fertilized by a male-producing sperm. The formula for the two classes of spermatozoa is RX–O. The latter, O, is the male-producing sperm, which combining with the egg OX (see above) gives OOX (or WWX), which is the formula for the white-eyed male mutant.

The transfer of the new character (white eyes) to the female (by crossing a white-eyed male, OOX to a heterozygous female (F_1)) can therefore be expressed as follows:

OX–O (white male)
RX–OX (F_1 female)

RXOX–	RXO–	OOXX–	OOX
Red female	Red male	White female	White male

It now becomes evident why we found it necessary to assume a coupling of R and X in one of the spermatozoa of the red-eyed F_1 hybrid (RXO). The fact is that this R and X are combined, and have never existed apart.

It has been assumed that the white-eyed mutant arose by a male-producing sperm (O) fertilizing an egg (OX) that had mutated. It may be asked what would have been the result if a female-producing sperm (RX) had fertilized this egg (OX)? Evidently a heterozygous female RXOX would arise, which, fertilized later by any normal male (RX–O) would produce in the next generation pure red females RRXX, red heterozygous females RXOX, red males RXO, and white males OOX (25 per cent.). As yet I have found no evidence that white-eyed sports occur in such numbers. Selective fertilization may be involved in the answer to this question.

25 Sex and its relation to the barring factor in poultry

H. D. Goodale

W. J. Spillman[1] has suggested that the barring factor and sex in poultry are correlated in such a way that the female is always heterozygous in respect to sex and also barring when present. The male, on the other hand, is always homozygous in respect to sex and may be either homozygous or heterozygous in respect to barring. I have recently performed the following experiments, which bear directly on this point and confirm his theoretical deductions.

Experiment 1. A Buff Rock male (non-barred) bred to Barred Rock females give, in F_1, barred males and non-barred (blacks or buff) females.

Experiment 2. A Barred Rock male bred to Buff Rock females (non-barred) or a Rhode Island Red female (non-barred) gives, in F_1, all barred birds in both sexes.

Experiment 3. A Buff Rock male bred to F_1 females (non-barred) from experiment 1 gives, in F_2, chicks which do not show the down pattern characteristic of chicks from barred parents, thus indicating an entire absence of barring in F_2.

These experiments may be formulated thus: Using B = barring factor, b = its absence; F = the female sex factor, f = its absence or the male sex factor. Assume that B and F can not occur in the same gamete (Spillman). Then,

Experiment 1 becomes $bf \cdot bf \times Bf \cdot bF = Bf \cdot bf + bf \cdot bF$.

Experiment 2 becomes $Bf \cdot Bf \times bf \cdot bF = Bf \cdot bf + Bf \cdot bF$.

Experiment 3 becomes $bf \cdot bf \times bf \cdot bF = bf \cdot bf + bf \cdot bF$.

Other crosses giving similar results are:

Experiment 4. A White Rock male (carrying barring as a cryptomere) mated with Brown Leghorn females gives in F_1 both sexes barred.

Experiment 5. The reciprocal cross, viz., a Brown Leghorn male mated with White Rock females gives black chicks and chicks having a down pattern like that of Barred Rock chicks. These chicks, however, are yet too young to enable a determination of their sex.

Experiment 6. A White Rock male (carrying barring) bred to Buff Rock females gives, in F_1, both sexes barred.

Experiment 7. From the reciprocal cross I have only two birds as yet, both barred males.

Experiment 8. One of the F_1 barred males from experiment 7 mated with a Buff Rock female gives, in F_2, barred and non-barred chicks, which are still too young to permit of their sex being determined.

While my results appear to confirm Spillman's suggestion, I wish to point out that experiment 3, rather than experiment 2, 4, 6 or 8, furnishes us the true test of his suggestion, for the reason that the presence of the F factor may simply prevent the B factor from becoming visible under certain conditions. In some experiments, at any rate, I find that the presence of the F factor operates to modify barring, making it appear obscure and blurred as compared with males from the same parents. On the other hand, we may refer this obscuring of barring to some other cause, perhaps the heterozygous nature of the female.

The details of these experiments are reserved for a later paper.

Since the above note went to *Science* some F_2 chicks in experiment 3 have reached the stage at which barred chicks usually exhibit distinct barring in their first feathers. Such barring is absent in these F_2 chicks.

From Science **29:**1004-1005, 1909.

[1] Am. Nat., Vol. **XLII.,** No. 50, 1908.

26 The chromosome number of man

Joe Hin Tjio*
Albert Levan
Estación Experimental de Aula Dei
Zaragoza, Spain
and
Cancer Chromosome Laboratory
Institute of Genetics
Lund, Sweden

While staying last summer at the Sloan-Kettering Institute, New York, one of us tried out some modifications of Hsu's technique (1952) on various human tissue cultures carried in serial *in vitro* cultivation at that institute. The results were promising inasmuch as some fairly satisfactory chromosome analyses were obtained in cultures both of tissues of normal origin and of tumours (Levan, 1956).

Later on both authors, working in cooperation at Lund, have tried still further to improve the technique. We had access to tissue cultures of human embryonic lung fibroblasts, grown in bovine amniotic fluid; these were very kindly supplied to us by Dr. Rune Grubb of the Virus Laboratory, Institute of Bacteriology, Lund. All cultures were primary explants taken from human embryos obtained after legal abortions. The embryos were 10-25 cm in length. The chromosomes were studied a few days after the *in vitro* explantation had been made.

In our opinion the hypotonic pre-treatment introduced by Hsu, although a very significant improvement especially for spreading the chromosomes, has a tendency to make the chromosome outlines somewhat blurred and vague. We consequently tried to abbreviate the hypotonic treatment to a minimum, hoping to induce the scattering of the chromosomes without unfavourable effects on the chromosome surface. Pre-treatment with hypotonic solution for only one or two minutes gave good results. In addition, we gave a colchicine dose to the culture medium 12-20 hours before fixation, making the medium 50×10^{-9} mol/l for the drug. The colchicine effected a considerable accumulation of mitoses and a varying degree of chromosome contraction. Fixation followed in 60 % acetic acid, twice exchanged in order to wash out the salts left from the culture medium and from the hypotonic solution that would otherwise have caused precipitation with the orcein. Ordinary squash preparations were made in 1 % acetic orcein. For chromosome counts the squashing was made very mild in order to keep the chromosomes in the metaphase groups. For idiogram studies a more thorough squashing was preferable. In many cases single cells were squashed under the microscope by a slight pressure of a needle. In such cases it was directly observed that no chromosomes escaped.

THE CHROMOSOME NUMBER

With the technique used exact counts could be made in a great number of cells. Figs. 1*a* and *b* represent typical samples of the appearance of the chromosomes at early metaphase *(a)* and full metaphase *(b)*, showing the ease with which the counting could be made. In Table 1 the numbers of counts made from the four embryos studied are recorded.

We were surprised to find that the chromosome number 46 predominated in the tissue cultures from all four embryos, only single

Table 1. Number of exact chromosome counts made

Embryo No.	Number of cultures	Number of counts
1	5	15
2	10	98
3	3	119
4	4	29
Total	22	261

From Hereditas **42**:1-6, 1956. Used with permission.

We wish to express our sincere thanks to the Swedish Cancer Society for financial support of this investigation, and to Dr. Rune Grubb for supplying us with tissue cultures.

*Present address: National Institutes of Health, Bethesda, Maryland.

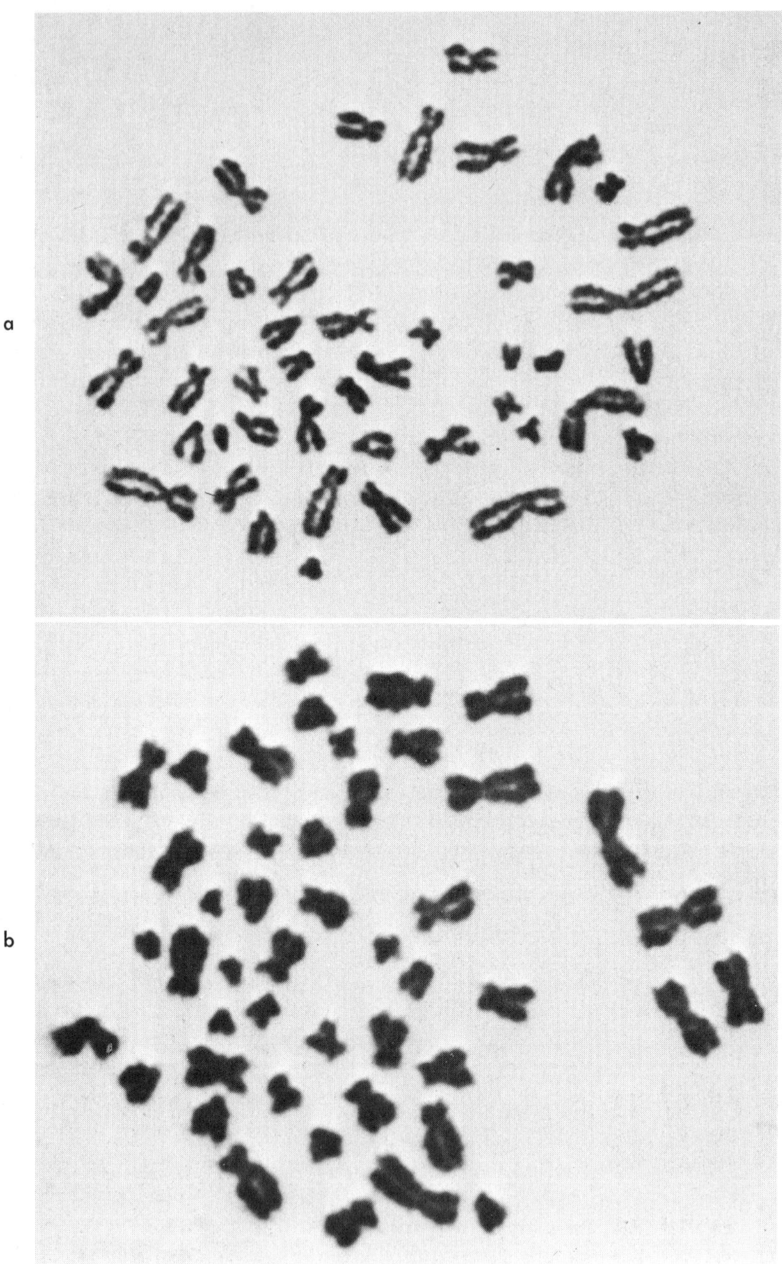

Fig. 1. Colchicine-metaphases of human embryonic lung fibroblasts grown *in vitro.* **a,** early metaphase; **b,** full metaphase. The two cells are from embryos 2 and 3 (Table 1), respectively.—X 2300.

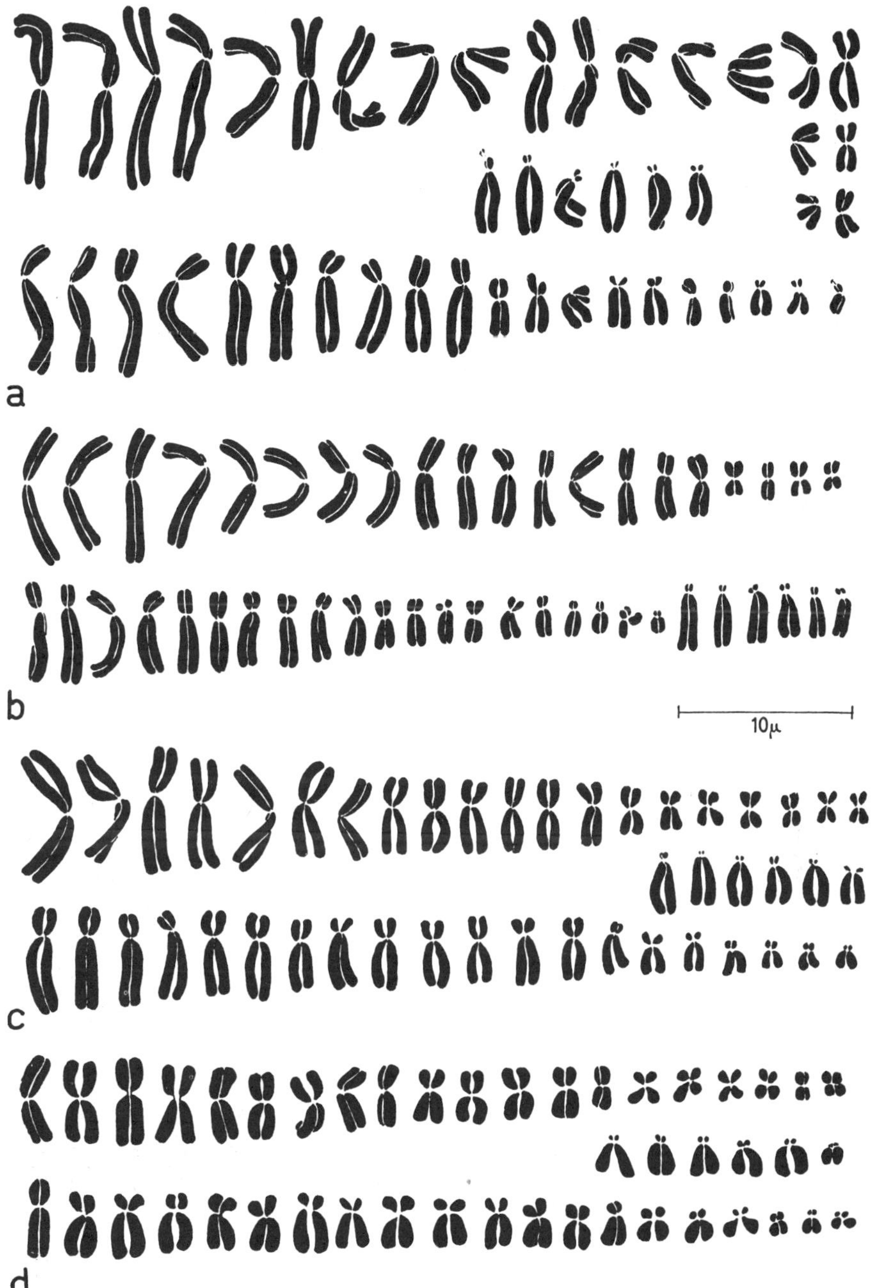

Fig. 2. Four idiogram analyses of human embryonic lung fibroblasts grown *in vitro.* The chromosomes have been grouped in three classes: M (top row), S (bottom row), and T (in between, except in **b,** where T is at the end of the S row). Within each class the chromosomes have been roughly arranged in diminishing order of size.—X 2400.

cases deviating from this number. Lower numbers were frequent, of course, but always in cells that seemed damaged. These were consequently disregarded just as the solitary chromosomes and the groups with but a few chromosomes, which were frequent. In some doubtful cases the numbers 47 and 48 were counted (in four cases not included in the table). This may be due to one or two solitary chromosomes having been pressed into a 46-chromosome plate at the squashing. It is also possible that deviating numbers may originate through nondisjunction, thus representing a real chromosome number variation in the living tissue. This kind of variation will probably increase as a consequence of the change in environment for the tissue involved in the *in vitro* explantation. Hsu (1952) reports a certain degree of such variation in his primary cultures. Levan (1956), studying long-carried serial subcultures, found hypotriploid stemline numbers in two of them, and a near-diploid number in a third culture. In this culture one cell with 48 chromosomes was analysed. Naturally, at that time, this was thought to represent the normal diploid number.

CHROMOSOME MORPHOLOGY

Some data on the chromosome morphology of the 46 human chromosomes will be communicated here. The detailed idiogram analysis will be postponed, however, until we are able to study individuals of known sex, the sex of the present embryos being unknown. The comparative study of germline chromosomes in spermatogonial mitoses constitutes an urgent supplement to the present work.

In Fig. 2 four cells are analysed ranging from late prophase *(a)* to late c-metaphase *(d)*. The chromosomes of metaphases with moderate colchicine contraction vary in length between 1 and 8 μ (Fig. 2*b*), but the entire range of variation of Fig. 2 is from 1 to 11 μ. The chromosome morphology is roughly concordant with the observations of earlier workers, as, for instance, the idiogram of Hsu (1952). The chromosomes may be divided into three groups: M chromosomes (median-submedian centromere; index long arm : short arm 1-1,9), S chromosomes (subterminal centromere; arm index 2-4,9), and T chromosomes (nearly terminal centromere; arm index 5 or more.

The M and S chromosomes are present in about equal numbers (twenty of each), while six T chromosomes are found. The classification of the three groups is arbitrary, of course, since gradual transitions of arm indices occur between the three groups. Certain submedian M chromosomes are hard to distinguish from some of the S chromosomes, and the most asymmetric S chromosomes approach the T group.

The chromosomes are easily arranged in pairs, but only certain of these pairs are individually distinguishable. Thus, the M chromosomes include the three longest pairs, which can always be identified. The two longest pairs are different: the second having a decidedly more asymmetric location of its centromere. The two or three smallest M pairs are also recognizable. Between the three longest and the three shortest pairs there are four intermediate pairs that cannot be individually recognized.

The S chromosomes are hardly identifiable, since they form a series of gradually decreasing length. The largest pair, however, is characteristic. Certain chromosomes were seen to have a small satellite on their short arms. Secondary constrictions, too, have been observed now and then, so that it may be hoped that the detailed morphologic study will lead to the identification of more chromosome pairs. The T chromosomes are recognizable; they constitute three pairs of middle-sized chromosomes. Unlike the mouse chromosomes, the human T chromosomes evidently have a small shorter arm.

CONCLUSION

The almost exclusive occurrence of the chromosome number 46 in one somatic tissue derived from four individual human embryos is a very unexpected finding. To assume a regular mechanism for the exclusion of two chromosomes from the idiogram at the formation of a certain tissue is unlikely, even if this assumption cannot be entirely dismissed at this stage of inquiry. Our experience from one somatic tissue in mice and rats, *viz.*, regenerating liver, speaks against this assumption. The exact diploid chromosome set was always found in regenerating liver.

After the conclusion had been drawn that the tissue studied by us had 46 as chromosome number, Dr. Eva Hansen-Melander kindly informed us that during last spring she had studied, in cooperation with Drs. Yngve Melander and Stig Kullander, the chromosomes of liver mitoses in aborted human embryos. This study, however, was temporarily discontinued because the workers were unable to find all the 48 human chromosomes in their material; as a matter of fact, the number 46 was repeatedly counted in their slides. We have seen potomicrographs of liver prophases from this study, clearly showing 46 chromosomes. These findings suggest that 46 may be the correct chromosome number for human liver tissue, too.

With previously used technique it has been extremely difficult to make counts in human material. Even with the great progress involved in Hsu's method exact counts seem difficult, judging from the photomicrographs published (Hsu, 1952 and elsewhere). For instance, we think that the excellent photomicrograph of Hsu published in Darlington's book (1953, facing p. 288) is more in agreement with the chromosome number 46 than 48, and the same is true of many of the photomicrographs of human chromosomes previously published.

Before a renewed, careful control has been made of the chromosome number in spermatogonial mitoses of man we do not wish to generalize our present findings into a statement that the chromosome number of man is $2n=46$, but it is hard to avoid the conclusion that this would be the most natural explanation of our observations.

SUMMARY

The chromosomes were studied in primary tissue cultures of human lung fibroblasts explanted from four individual embryos. In all of them the chromosome number 46 was encountered, instead of the expected number 48. Since among 265 mitoses counted all except 4 showed the number 46, this number is characteristic of the tissue studied. The possible bearing of this result on the chromosome number of man is discussed.

LITERATURE CITED

Darlington, C. D. 1953. The facts of life. – London, 467 pp.

Hsu, T. C. 1952. Mammalian chromosomes *in vitro.* – The karyotype of man. – J. Hered. **43**:167-172.

Levan, A. 1956. Chromosome studies on some human tumors and tissues of normal origin, grown *in vivo* and *in vitro* at the Sloan-Kettering Institute. – Cancer (in the press).

27 A sex-chromosome anomaly in a case of gonadal dysgenesis (Turner's syndrome)

C. E. Ford
K. W. Jones
Medical Research Council Radiobiological Research Unit
Atomic Energy Research Establishment
Harwell, Berkshire, England
P. E. Polani
J. C. de Almeida
J. H. Briggs
Guy's Hospital
London SE 1, England

Gonadal dysgenesis (ovarian agenesis, gonadal dysplasia) is a clinical syndrome usually presenting as a failure of secondary sex characteristics at puberty in girls whose gonads are absent or rudimentary. It is often associated with other congenital malformations such as small stature, digital anomalies, and, more rarely, webbed neck, congenital heart-disease, renal anomalies, intellectual subnormality, and other developmental errors. The more extreme expressions are often referred to as Turner's syndrome.

A considerable proportion of patients with gonadal dysgenesis are chromatin-negative (Décourt, Sasso, Chiorboli, and Fernandes 1954, Polani, Hunter, and Lennox 1954, Wilkins, Grumbach, and Van Wyck 1954), although chromatin-negativity (Barr and Bertram 1949) is an invariable feature of normal males. One possible explanation is that gonadal dysgenesis in man is due to castration while an embryo, since the experimental castration of embryonic rabbits results in the production of animals of female phenotype irrespective of the genetic sex-constitution of the embryo (Jost 1947). However, in abnormal individuals chromatin negativity or positivity ("nuclear sexing") may not necessarily indicate true chromosomal sex (*Lancet* 1956, Polani, Lessof, and Bishop 1956). An alternative explanation for the findings in gonadal dysgenesis might be abnormal sex differentiation following anomalous sex determination in the zygote.

Two approaches to the problem of certainly identifying the sex chromosomes present in Turner's syndrome suggested themselves: direct cytological observation, and the study of colour-blindness, which is a sex-linked recessive character and an X-chromosome marker. The results obtained by the second method agreed with the simple interpretation of the "nuclear sexing" results: chromatin-negative patients with gonadal dysgenesis seemed to have only one X chromosome (Polani et al. 1956). The presence or absence of the Y chromosome could not be determined and it was thought likely that the patients had an XY sex-chromosome constitution, although the possibility that they might be XO was also considered. Danon and Sachs (1957) also suggested that some patients with gonadal dysgenesis might have an XO sex-chromosome constitution, but that other patients might be examples of somatic mosaicism in respect of their sex-chromosome constitution. A study of the blood-groups of three patients with gonadal dysgenesis (Platt and Stratton 1956) supplied evidence that these individuals were not haploid—i.e., were not XO merely because all their chromosomes were unpaired.

Technical developments have recently made it possible to obtain accurate information regarding the somatic chromosomes of human patients, either in bone-marrow cells briefly incubated in vitro (Ford, Jacobs, and Lajtha 1958) or in cells from tissue cultures (Tjio and Puck 1958). In consequence the normal number of human chromosomes and their normal morphology are now reasonably well known.

From The Lancet 1:711-713, 1959. Used with permission.

We wish to thank Dr. P. M. F. Bishop for permission to study a patient under his care. We acknowledge the skilful technical assistance of Mr. G. D. Breckon, Miss P. A. Moore, and Miss S. R. Wakefield.

The subject of this report is a chromatin-negative case of Turner's syndrome whose bone-marrow cells proved to contain 45 chromosomes only, instead of the normal number of 46, and whose sex-chromosomal constitution is determined to be XO.

CASE-REPORT

The patient presented at the age of 14 with a short stature, primary amenorrhoea, and absence of secondary sex characteristics. In addition she was backward at school.

Family history. Parents healthy. Father 5 ft. 4 in., mother 5 ft. 2 in. A maternal aunt, who died at the age of 21, was dwarfed and had had only two scanty periods; she was known to have pernicious anaemia. Two brothers 6 and 10 years old and one sister aged 9 were all healthy.

Personal history. Maternal health good during pregnancy, and delivery normal. Birth weight 5 lb. 4 oz. Early development normal.

On examination, height 51 in., lower segment 25 in., arm-span 50 in. Weight 4 st. 13 lb.

There was slight facial asymmetry, low implantation of the ears, and a small chin. There was a high arched palate, a short broad neck without webbing, slight funnel deformity of the chest, cubitus valgus, pes cavus, and digital deformities. Cardiovascular system normal. Blood-pressure 110/70. Normal femoral pulses. There was no evidence of puberty.

Investigations

Examination of skin biopsy and blood smear showed a chromatin-negative pattern.

Follicle-stimulating hormone positive to 32 mouse units 17-ketosteroids 10.8 mg. per day.

Radiographically chest, heart, intravenous pyelography normal. Radiological assessment of bone age corresponded to the chronological age.

No defect in colour-vision (Ishihara) in patient or parents.

The marrow cells, obtained by a routine marrow puncture, were suspended in a mixture of glucose-saline and serum from the patient herself, and were sent to Harwell for cytological processing. After incubation the cells were exposed to colchicine for one hour, then fixed and stained by the Feulgen procedure. Squash preparations were made and the chromosomes were studied in cells arrested in the metaphase of mitosis by the action of the colchicine.

The chromosomes were counted in 102 cells of which 99 cells were found to have 45 chromosomes only. The remaining 3 cells contained fewer than 45 chromosomes and previous experience suggests that the deficiency is likely to be a consequence of damage to the cells during the making of the preparations. 14 cells were selected for detailed study. In every one of them 4 small acrocentric chromosomes were present, as in a normal female: in a normal male there are 5 of these chromosomes, one being the Y chromosome. All the selected cells also contained 15 medium-length metacentric chromosomes, as in a normal male—a normal female having 16 which include the two X chromosomes (Ford et al. 1958). These observations of themselves strongly suggest that the chromosome constitution is XO.

The individual recognition of the X and Y chromosomes may be a matter of some difficulty. However Tjio and Puck (1958) assert that X and Y chromsomes can be recognised individually in their preparations made from tissue cultures. We agree that the Y chromosome can be distinguished in favourable cells of normal males, but we have not yet been able to identify the X chromosome (or chromosomes) unequivocally in bone-marrow preparations. Nevertheless in many of the selected cells of the present patient it was possible to make a reasonably satisfactory classification of the chromosomes into 22 pairs and one odd chromosome. A photograph of one of these cells is reproduced in Fig. 1. In Fig. 2 the chromosomes from the same cell are shown arranged in pairs. Suspicion that the odd chromosome is the X is inevitable, but this chromosome and the two members of pair 6 are very similar in length and arm-ratio and their true relationships remain uncertain. The probability that one of the three is X is strengthened by the good agreement of their proportions with those of undoubted X-chromosomes in X-Y bivalents at metaphase in primary spermatocytes (Ford and Hamerton 1956). Experience of numerical chromosomal abnormalities in animals and plants (Swanson 1957) suggests that it is very improbable that a human individual who has only 45 chromosomes as the result of the loss of a large or medium-sized autosome would be viable, but that an XO zygote might well develop to maturity. We therefore conclude that the sex-chromosome constitution of the patient is XO.

DISCUSSION

These observations are of interest not only with reference to Turner's syndrome. Here is an individual who is female anatomically and psychologically, whose cells are "male" as judged by nuclear sexing, and whose chromosomes are neither normally male nor normally

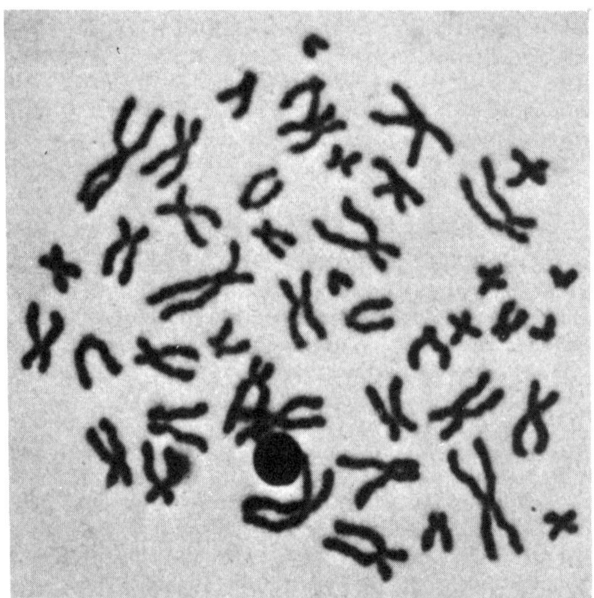

Fig. 1. Chromosomes (45) of the patient with Turner's syndrome discussed in the text. Colchicine-arrested metaphase in a bone-marrow cell. Feulgen squash preparation (X 2200). The round black body is probably an oil droplet.

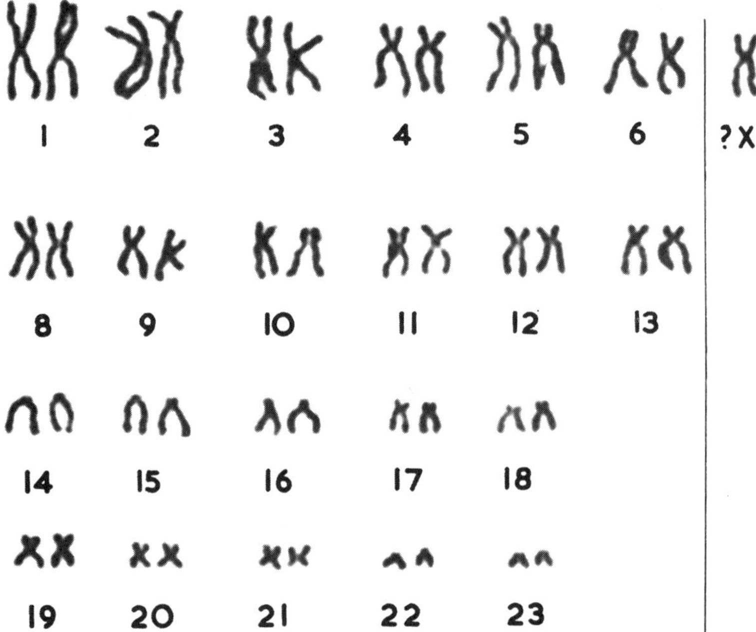

Fig. 2. Chromosomes from the cell shown in Fig. 1 arranged in pairs (X 2200).

female. However, as judged by her chromosomes she has no male component, but half a normal female component, and there seems no justification for considering her to be really male in any sense. It must therefore be accepted that chromatin negativity does not necessarily imply maleness and it would probably be best if the phrase "nuclear sexing" were dropped from the vocabulary and the more accurate if less striking terms, chromatin negativity or positivity, were always used instead. The very real clinical reasons for doing this have already been stressed.

An explanation of the origin of the sex chromosome anomaly in gonadal dysgenesis can be sought in the process of non-disjunction, best known as an abnormality of oogenesis in *Drosophila melanogaster* (Morgan, Bridges, and Sturtevant 1925). Non-disjunction of the sex-chromosomes in the female fly implies the migration of two X chromosomes to one pole of the spindle during one of the two meiotic anaphases. Thus the ovum comes to contain either the haploid number of autosomes plus *two* X chromosomes, or only the haploid number of autosomes *without* X chromosomes. Fertilisation of an ovum of the latter type by a Y-bearing sperm results in a non-viable YO zygote; fertilisation by an X-bearing sperm yields an XO zygote which develops into a sterile male. Our findings suggest that in man an XO zygote develops into a sterile "agonadal" individual whose phenotype is female.

Fertilisation of the other type of abnormal ovum (XX) by an X-bearing sperm gives, in *Drosophila,* an XXX zygote with poor viability, but which occasionally survives pupation and then emerges as a fly with accentuated female secondary sexual characteristics, technically called a "super-female"; fertilisation by a Y-bearing sperm yields an XXY zygote which develops into a fertile female. In man the XXX state is as yet unknown, but evidence that the XXY individual appears as a chromatin-positive case of Klinefelter's syndrome has been presented (Ford, Polani, Briggs, and Bishop 1959, Jacobs and Strong 1959.) The evidence in favour of the occurrence of non-disjunction in man is thereby strengthened.

Non-disjunction has so far been considered as occurring during oogenesis only. Family colour-vision studies in cases of gonadal dysgenesis suggest that it may also occur during spermatogenesis. If an O ovum is fertilised by an X sperm and the resultant individual has a major red-green colour-vision defect, the father should also show the colour-vision defect. But in four families where patients with gonadal dysgenesis (chromatin-negative) have a major red-green colour-vision defect this was not the case (Lenz 1957, Stewart 1958 personal communication; and two of our families, see Bishop et al. 1959). It would appear that in these patients the X-chromosome with the anomalous colour-vision gene was not of paternal origin but was derived from a heterozygous (carrier) mother. These patients have not been examined cytologically, but, if they are XO, it will follow that they developed from zygotes arising from the fertilisation of normal X-bearing ova by sperm carrying neither X nor Y. Such sperm would arise as a result of non-disjunction during spermatogenesis. Evidence that this is by no means unlikely is provided by the observation that X and Y chromosomes are sometimes unpaired at metaphase in first spermatocytes (Ford and Hamerton 1956). Both chromosomes would then be expected to migrate to the same pole in approximately 50% of the ensuing anaphases.

In conclusion it should be emphasised that the XO patient should not be referred to as an instance of "sex-reversal", as a "chromosomal male", or as a "genetic male": she is a female, with an abnormal genotype.

REFERENCES

Barr, M. L., Bertram, E. G. (1949) Nature, Lond. **163**:676.

Bishop, P. M. F., Lessof, M. H., Polani, P. E. (1959) Memoir no. 7. Society for Endocrinology, London. Ed.: C. R. Austin (in the press).

Danon, M., Sachs, L. (1957) Lancet **ii**:20.

Décourt, L., Sasso, W. da S., Chiorboli, E., Fernandes, J. M. (1954) Rev. Assoc. med. Brazil **1**:203.

Ford, C. E., Hammerton, J. L. (1956) Nature, Lond. **178**:1020.

——Jacobs, P. A., Lajtha, L. G. (1958) *ibid.* **181**:1565.

——Polani, P. E., Briggs, J. H., Bishop, P. M. F. (1959) *ibid.* (in the press).

Jacobs, P. A., Strong, J. A. (1959) *ibid.* **183**:302.

Jost, A. (1947) C. R. Soc. Biol., Paris **141**:126.

Lancet (1956) **2**:127.

Lenz, W. (1957) Acta Geneticae Medicae et Gemellologiae **6**:231.

Morgan, T. H., Bridges, C. B., Sturtevant, A. H. (1925) Bibliog. Genetica **2**:1.

Platt, R., Stratton, F. (1956) Lancet **ii**:120.

Polani, P. E., Hunter, W. F., Lennox, B. (1954) *ibid.* **ii**:120.

——Lessof, M. H., Bishop, P. M. F. (1956) *ibid.* **ii**:118.

Stewart, J. S. S. (1958) Personal communication.

Swanson, C. P. (1957) Cytology and Cytogenetics. (Prentice Hall, Englewood Cliffs, N.J.).

Tjio, J. H., Puck, T. T. (1958) Proc. Nat. Acad. Sci, **12**:1229.

Wilkins, L., Grumbach, M. M., Van Wyck, J. J. (1954) J. clin. Endocrin. **14**:1270.

28 A case of human intersexuality having a possible XXY sex-determining mechanism

Patricia A. Jacobs*
Dr. J. A. Strong
Medical Research Council Group for Research on the General Effects of Radiation
Department for Endocrine and Metabolic Diseases
Western General Hospital
University of Edinburgh
Edinburgh, Scotland

Recent improvements in techniques for the examination of human somatic chromosomes have made possible the study of the chromosome complement of human intersexes; consequently, it is now practicable to investigate the relationship in these cases between sex as determined by direct chromosome study, and sex as inferred from the study of 'nuclear sex chromatin' of the type described by Barr and Bertram[1]. This report is concerned with one of a series of patients with gonadal dysgenesis who are under investigation, and the particular feature of interest is the occurrence of 47 somatic chromosomes in contrast with the normal number of 46 in man.

In recent years the diploid chromosome number of 46 has been recorded in a large number of instances. In addition to the 60 cases cited in a previous publication[2] we have recorded a diploid number of 46 in bone marrow preparations from a further 40 European subjects. Kodani, on the other hand, has recently published results of counts made on testicular material from 36 Japanese and 8 American white males[3,4]. He claims that in 16 Japanese and 1 American there was a diploid number of 48; in 2 Japanese a diploid number of 47, and that in the remaining 13 Japanese and 7 whites the number was 46. He suggests that 46 is the basic diploid chromosome number for man, but in some instances there are additional "supernumerary chromosomes". The occurrence of this type of supernumerary chromosome, however, has not been reported previously among the vertebrates and awaits confirmation by other workers.

Our patient, an apparent male aged twenty-four, was presented as a case of gonadal dysgenesis with gynaecomastia and small testes associated with poor facial hair-growth and a high-pitched voice. Biopsy examination of testicular tissue showed the seminiferous tubules to be extremely hyalinized and atrophic, and also an apparent increase in the number of interstitial cells. Chromosome studies were attempted on part of this material, but no spermatogonial mitotic or meiotic divisions were seen. Smears made from both the buccal mucosa and the blood were examined by Dr. B. Lennox of the Department of Pathology, Western Infirmary, Glasgow, and found to demonstrate typical female morphology with regard to their nuclear sex chromatin.

Material obtained by sternal marrow puncture was used for investigating the somatic chromosomes. The technique used for culturing the material in the presence of colchicine and for making squash preparations has already been described[2].

The chromosomes were counted in 44 cells in metaphase and the results are shown in Table 1.

The majority of the cells contained 47 chromosomes, and in all those cells where the chromosomes were well fixed and spread, the count was undoubtedly 47 (Fig. 1). The apparent variation is in all probability due to technical errors. Fragments of cells containing chromosomes may become lost during the squashing process so that counts lower than the diploid number are obtained; and occasionally chromosomes split at the centromere and individual chromatids may be counted as

From Nature **183:**302-303, 1959. Used with permission.

We would like to thank Dr. B. Lennox of the Department of Pathology, Western Infirmary, Glasgow, for checking the nuclear sex of the preparations of buccal mucosa and blood and Miss M. Brunton for technical assistance.

*Present address: University of California at San Diego, La Jolla, California.

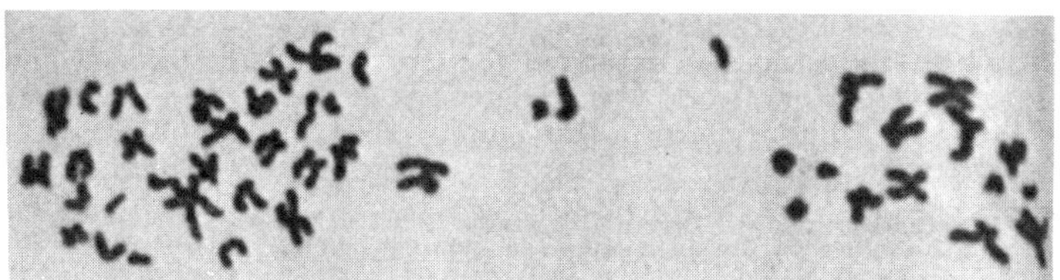

Fig. 1. Metaphase plate showing 47 chromosomes.

Table 1

Chromosome No.	45	46	47	48	49
No. of cells	2	7	29	5	1

Table 2

Father	Chromosome No.	44	45	46	47	48	49
32 cells counted	No. of cells	2	3	26	–	–	1
Mother	Chromosome No.	44	45	46	47	48	49
39 cells counted	No. of cells	1	3	33	2	–	–

chromosomes, giving a count higher than the diploid number[2].

A study of the chromosome morphology in 8 cells of a suitably high standard showed that each of these had a normal male complement with the *Y* chromosome present, but that there was also an extra chromosome having a sub-median centromere occurring in the medium size range. In the normal male there are 15 chromosomes in this range, and in the female 16, all having sub-terminal or sub-median centromeres[2]. Owing to the slight variations in their size and morphology these chromosomes have proved difficult to pair, and it is in this category that the *X* chromosome is to be found.

There are strong grounds, both observational and genetic[5,6], for believing that human beings with chromatin-positive nuclei are genetic females having two *X* chromosomes. The fact that this patient is chromatin-positive and has an additional chromosome within the same size range as the *X*, as well as an apparently normal *Y*, makes it seem likely that he has the genetic constitution *XXY*. The possibility cannot be excluded, however, that the additional chromosome is an autosome carrying feminizing genes.

The presence of the extra chromosome might have been due to one or other of the parents having 47 chromosomes, and, therefore, chromosome studies were made on marrow specimens from both parents. Both were found to have a diploid number of 46 (Table 2), and analysis of cells of suitable quality showed the morphology of the chromosomes to be normal.

The occurrence of the extra chromosome therefore may be due to non-disjunction at either mitosis or meiosis during gametogenesis in one or other parent. Alternatively, it may be due to non-disjunction occurring during the patient's very early embryological development, in which case there is a possibility that the patient may be a mosaic. Unfortunately, it is not possible with the techniques at present available to examine the chromosomes of tissues arising from different germ layers of the embryo.

A further report of this and other cases will follow.

REFERENCES

1. Barr, M. L., and Bertram, E. G., Nature **163:**676 (1949).
2. Ford, C. E., Jacobs, P. A., and Lajtha, L. G., Nature **181:**1565 (1958).
3. Kodani, M., Proc. U.S. Nat. Acad. Sci. **43:**285 (1957).
4. Kodani, M., Amer. J. Human Genetics **10:**125 (1958).
5. Grumbach, M. M., and Barr, M. L., "Recent Progress in Hormone Research" **14:**255 (1958).
6. Polani, P. E., Bishop, P. M. F., Lennox, B., Ferguson-Smith, M. A., Stewart, J. S. S., and Prader, A., Nature **182:**1092 (1958).

29 Determination of sex in Asparagus officinalis L.

Charles M. Rick*
G. C. Hanna
Division of Truck Crops
University of California
Davis, California

Asparagus officinalis L., the garden Asparagus, is normally a dioecious species. Rudimentary organs of the opposite sex appear in both staminate and pistillate flowers but rarely develop to a functional state. Staminate and pistillate plants are represented in approximately equal numbers according to Flory (1932), Robbins and Jones (1925), Shoji and Nakamura (1928), and unpublished data of the writers. In this and other investigations (Flory, 1932; Kamo, 1929) attempts have been made to distinguish the sex-determining pair of chromosomes cytologically, but no heteromorphic pair has been observed. Furthermore, up to the present time no other critical evidence has been presented to establish the genetic basis of sex determination in this species.

Wide variation in the development of pistils in staminate flowers has been frequently observed (Lewitsky, 1925; and others). Although usually abortive, such pistils may very rarely function to produce parthenocarpic or seed-bearing fruits. Seeds of this origin are often developmentally subnormal as testified by their smaller size and poorly developed integuments. Many of them, nevertheless, will germinate and produce viable seedlings. Evidence of this hermaphroditic development may be readily observed late in the growing season when any berries developed on staminate plants are quite conspicuous by virtue of their bright red color.

Self-pollination is probably responsible for the production of berries by these occasional perfect flowers. The typical staminate corolla is narrowly campanulate. When the anthers dehisce they form a contiguous mass of pollen which almost completely obstructs the corolla above the abortive pistil. The stigma of a functional pistil would be placed immediately below this pollen mass or would project into it. Thus it is difficult to conceive of pollination of a perfect flower by foreign pollen.

Berries produced on 61 staminate plants were collected in the fall of 1939; seeds were extracted and planted the following year. Of the plants thus produced the following frequency of sexes was observed: 155 males (78.3 ± 3.0 per cent) and 43 females (21.7 ± 3.0 per cent). Four additional plants did not flower. This segregation of sexes in the progeny of what were, in effect, male selfings or male-by-male crosses indicated the male sex as heterogametic. The close approximation to a 3:1 ratio suggested that sex in *Asparagus* is inherited as a simple Mendelian factor, maleness being dominant and homozygous males being equally as viable as heterozygous males. The staminate plants in this progeny were then examined for characters by which the homozygous and heterozygous types might be identified. In this preliminary survey no such characters were found.

In order to test this proposed basis of sex determination, 25 males comprising seven families (that is, originating from berries on seven staminate plants) were crossed with supposedly normal pistillate plants. The pistillate parents used were considered normal as to sex determination, since the group in which they appeared exhibited an approximately even sex ratio. As an additional check on the sex-transmitting nature of females of this group, nine of them were crossed with males of the same group.

All except 0.7 per cent of the offspring of these test and control crosses have flowered. Observations on sex ratio and other characters are presented according to families in Table 1. These results fully support the hypothesis that two types of males—homozygous and heterozygous—were being tested, for eight of the test progenies were composed exclusively of males and seventeen showed segregations of males and

From American Journal of Botany **30:**711-714, 1943. Used with permission.

*Present address: Professor and Geneticist, Department of Vegetable Crops, University of California, Davis, California.

Table 1. Summary of progeny sex ratios and measurements of male parents

Progeny number	Male parent	Males	Females	Undetermined	Per cent male	Measurements of male parent: Length of pistil	Diameter of pistil	Pollen abortion
210	Control	23	16	..	59.0	...	...	...
214	Control	15	23	..	38.5	...	...	...
215	Control	18	15	1	54.6	...	...	...
229	Control	17	23	..	42.5	...	...	...
230	Control	13	9	..	59.1	...	...	...
231	Control	18	21	..	46.2	...	...	...
232	Control	19	13	1	59.4	...	...	...
233	Control	18	18	..	50.0	...	...	...
236	Control	7	7	1	50.0	...	...	...
202	1493-7	33	..	..	100.0	...	...	...
204	3742-3	25	..	1	100.0	a	a	a
205	3636-2	10	..	..	100.0	1.35 mm.	0.83 mm.	8.9%
206	1493-9	39	..	..	100.0	1.31	0.84	10.9
207	4358-7	29	..	..	100.0	1.51	0.84	4.7
219	4538-1	10	..	..	100.0	1.40	0.92	82.3
225	1493-14	36	..	..	100.0	1.49	0.79	10.9
237	3712-5	44	..	..	100.0	1.32	0.87	15.2
201	1493-8	10	11	..	47.6	1.60	1.12	17.1
203	3636-4	13	5	..	72.3	...	...	...
208	3742-2	21	19	..	52.5	1.33	0.96	7.9
209	1493-2	40	37	..	51.9	...	...	...
212	4358-2	22	17	..	56.4	...	...	...
213	3033-3	17	22	..	43.6	1.12	0.92	20.6
216	3636-3	22	17	..	56.4	...	...	...
217	1493-11	22	9	1	70.9	1.36	0.80	14.5
220	1493-5	6	8	..	42.8	1.32	0.83	10.0
221	4358-4	9	13	2	40.9	1.49	0.87	4.2
222	4358-6	20	15	..	57.2	1.31	0.91	4.2
223	1493-6	16	15	..	51.7	...	...	...
224	1493-1	35	5	..	87.5	1.36	0.80	14.9
226	1493-4	18	17	1	51.4	1.55	0.81	5.5
227	1493-10	20	17	..	54.1	...	...	...
228	1493-12	11	5	..	68.8	1.44	0.83	6.8
235	3033-5	8	8	..	50.0	1.45	0.89	7.1

[a]No measurements indicate failure of plant to survive until the time of measurements.

females, in most cases in an approximate 1:1 ratio. The ratio of one homozygous male to two heterozygous males obtained also supports the assumption previously mentioned that maleness is inherited as a single dominant factor. The normal sex ratio, therefore, is maintained in *Asparagus* as in the majority of dioecious plants by what is essentially a back-cross segregation of a single gene or gene group, the male sex being heterogametic.

Table 2 summarizes an analysis of sex ratio in the segregating families of test and control crosses. For the control crosses the fit to a 1:1 ratio is satisfactory, indicating a normal sex determination of the females used in these crosses. In the test cross families, however, the analysis indicates a significant deviation from the expected ratio as well as a significant heterogeneity between families. According to the separate χ^2 tests of each family, two families—namely, #217 and #224—deviate significantly at the two per cent level. When these two deviating families are omitted from the analysis (cf. Table 2) a reasonably close fit is obtained and no heterogeneity is indicated. The fit is also satisfactory if only one family, #224, is eliminated; both families were omitted, however, for reasons to be mentioned later. Furthermore, the distribution of sex ratios in the remaining test cross families corresponds closely to that of the control cross families.

It is noteworthy that the two families, #217 and #224, eliminated from the foregoing χ^2 tests agree in their deviation toward an excess of males. These two families, moreover, have a common grandparent, #1493, a male which produced seed. The probability that any two families taken at random would show an excess

Table 2. Chi-square analysis of segregating families

	Test crosses			Control crosses		
	x^2	d.f.	P	x^2	d.f.	P
All families:						
Total	37.733	15	<0.01	6.321	9	0.7-0.8
Deviation	6.333	1	0.01-0.02	0.030	1	0.8-0.9
Difference (Heterogeneity)	31.400	14	<0.01	6.291	8	0.5-0.7
All families except 217 & 224:						
Total	9.781	13	0.7-0.8	....	..	......
Deviation	0.444	1	0.5-0.7	....	..	......
Difference (Heterogeneity)	9.337	12	0.5-0.7	....	..	......

of males and would trace to this single grandparent is not significantly low, being 0.15; nevertheless, this evidence suggests an inherited basis—possibly partial sex linkage of some lethal factor—for the production of excess males. It also warrants, to some extent, the omissions made in the foregoing tests. The nature of this tendency for the production of excess males is being tested by experiments now in progress.

After the genetic difference between homozygous and heterozygous males had been established, these tested male individuals were re-examined for any morphological character by which they might be identified. Any differences might be expected to manifest themselves in characters by which females differ from normal heterozygous males. No differences in size or shape of flower are apparent between the two classes of males. The length of abortive pistil is slightly greater in homozygous males, and diameter of pistil is slightly greater in heterozygous males, but neither of these differences approaches statistical significance, not to mention their uselessness as key characters for identification of the two types of males.

Sterility is characteristic of supersex types in *Drosophila melanogaster* (Bridges, 1925), although this condition may be related to chromosomal unbalance rather than to unbalance of the sex determining genes *per se.* In the present experiment pollen sterility in the homozygous group (cf. Table 1) is somewhat higher, but the difference again is not significant and lacks identification value.

Obviously enough, every morphological character has not been tested in this search for differences, but up to the present time the writers have failed to detect any means other than the genetic test which will reliably identify homozygous males. The highly heterozygous nature of cultivated *Asparagus,* consequent to its dioecious condition and to the systems of variety maintainance to which it has been subjected in its horticultural history, might mask small differences. Until closely inbred material is obtained, such differences, if they exist, cannot be demonstrated.

To the extent that sex expression has been investigated in this work, maleness behaves as a completely dominant character in *Asparagus.* A similar example in angiosperms is furnished by *Mercurialis annua* L., in which species Gabe (1939) discovered homozygous males in an investigation, the results of which very closely parallel those of the present experiment. In *Mercurialis* heterozygous and homozygous males were indistinguishable except for differences in color and dehiscence of anthers, which apparently attended the much greater pollen sterility of the homozygous males.

The equal viability of homozygous and heterozygous males possibly points to a genetically active Y-chromosome or male allelomorph-bearing chromosome in *Asparagus* and *Mercurialis* in contrast to the frequently inert Y-chromosome in animals. Thus the dioecious condition is apparently much more superficial in these plants than it is generally in animals, probably in keeping with the comparatively lower level of evolution of the dioecious condition in plants. Although seemingly less likely, the alternative explanation of low genetic activity of both the X and Y chromosomes must be admitted as compatible with these findings.

A point in regard to the seed-producing males merits attention. The proportions of staminate plants producing seed in various pedigrees in two years as shown in Table 3

Table 3. Proportion of staminate plants producing seed in various families of known pedigree

Pedigree	1939	1940
All progenies not sired by #312	1.90±0.39	1.24±0.32
All progenies sired by #312. . .	11.4 ±1.5	3.79±0.93
Progeny of #484 × #312	1.6 ±1.1	0.00
Progeny of #437 × #312	34.5 ±5.9	6.0 ±2.9

suggest an influence of heredity as well as environment upon the appearance of perfect flowers. As evidence of genetic control, the progenies of plant #312 included a significantly higher proportion of males producing seed than the progenies of all other males examined. In fact, two-thirds of males producing seed in 1939 were sired by this one plant. The percentages for two individual progenies also included in Table 3 illustrate the extremes of the wide fluctuation in various progenies of #312. The nature of the genetic control involved cannot be established with the available data, but it seems likely that at least two genes influence this character. The staminate plants derived from perfect flowers are still too small to allow an estimate of their seed-producing capacities. A mere comparison of the percentages in 1939 and 1940 is sufficient to indicate a very pronounced environmental effect. This seasonal influence seems to be greater on genetically more productive males.

Functional perfect flowers appear but very rarely even on those staminate plants which display the strongest tendencies of this type of hermaphroditism. In spite of the tremendous number of flowers produced by single staminate plants, only 204 viable seeds were produced by 1656 mature staminate plants in 1939. Thus a staminate plant yielding a few seeds can scarcely be regarded as a true hermaphrodite. It is rather a male with slight female tendencies—a condition influenced by genetic modifiers as indicated in the preceding discussion. Plants which would more closely approximate the concept of a true hermaphrodite—that is, having a closer balance of expression of the two sexes—have been reported as rare individuals by Norton (1913), but have not been found in the work at this station.

DISCUSSION AND APPLICATION

Flory (1932) proposed the male sex to be heterogametic in *Asparagus officinalis* on the grounds that the heterogametic sex would be expected to display more variation in the development of sexual parts in the direction of the other sex. He had observed rare hermaphroditic flowers only on staminate plants and a greater development of pistils in staminate flowers than of stamens in pistillate flowers. The size of pistils in flowers of homozygous males should be considered on the basis of this contention. The mean size of pistil in homozygous males would be expected to fall under that of heterozygous males, but, as pointed out previously, this is not true, differences obtained for pistil length and diameter being respectively greater and smaller in homozygous males, and neither difference being significant. One might also expect a reduction in variation of pistil development in homozygous males. This expectation is realized, for the inter-plant standard deviations for pistil length, 0.086 mm., and of diameter, 0.044 mm., for the homozygous males are smaller than the respective values, 0.132 mm. and 0.094 mm., for the heterozygous males. Only in the case of diameters, however, is the difference significant to the five per cent level. Thus the present survey contributes evidence, admittedly slight, to support the hypothesis that the heterogametic sex in plants is the more variable in its expression of sex.

According to Robbins and Jones (1928) and others, staminate *Asparagus* plants outyield pistillate plants by about 25 per cent. Although staminate plants produce a greater total yield, the female shoot is larger and has a more desirable appearance. Yet, for the sake of higher yields alone, a method which would ensure all-male plantings might be desirable.

The present experiment suggests a technique by means of which all-male populations could be conveniently produced. The planting for seed production should be located at a sufficient distance from other plantings to preclude cross-pollination with them. This planting would include normal female plants selected from a seedling nursery the preceding season and homozygous males obtained and identified by genetic test as previously described. A much smaller number of males than females should be planted—only a sufficient number to ensure complete pollination of all females. If this planting were then allowed to set berries by insect pollination, it would yield seed which would give rise to staminate plants only. This method would also give control of the product to the seed producer, since the progeny, being entirely male, could not be propagated further by seed.

SUMMARY

Plants grown from seed produced by hermaphroditic flower development on staminate plants exhibited a ratio of approximately three males to one female. One-third of the male progeny were homozygous for sex-transmission, yielding only male offspring when crossed to normal females; the remainder proved to be heterozygous. In all comparisons attempted, homozygous males could not be distinguished from heterozygous males except by means of progeny tests. The male sex is thus established as heterogametic in *Asparagus officinalis,* and sex is inherited as a simple Mendelian factor, maleness being dominant.

Evidence is presented to indicate genetic as well as environmental control of the tendency of staminate plants to produce berries.

A possible application of these findings in the production of all-male populations is discussed.

LITERATURE CITED

Bridges, C. B. 1925. Sex in relation to chromosomes and genes. Amer. Nat. **59**:127-137.

Flory, W. S., Jr. 1932. Genetic and cytological investigations on *Asparagus officinalis* L. Genetics **17**:432-467.

Gabe, D. R. 1939. Inheritance of sex in *Mercurialis annua* in relation to cytoplasmic theory of sex in inheritance. C. R. Acad. Sci. U. R. S. S. **23**:478-481.

Kamo, I. 1929. Einige Beobachtungen über die Chromosomen von *Asparagus officinalis* L. Bot. Mag. Tokyo **43**:127-133.

Lewitsky, G. 1925. On the phenomenon of abortion in the organs of reproduction of *Asparagus officinalis* L. Bull. Appl. Bot. & Pl. Breed. **14**:133-141. (Russ. with Engl. sum.)

Norton, J. B. 1913. Methods used in breeding Asparagus for disease resistance. U. S. Dept. Agric. Bur. Pl. Ind. Bull. **263**:1-60.

Robbins, W. W., and H. A. Jones. 1925. Secondary sex characters in *Asparagus officinalis* L. Hilgardia **1**:183-202.

———, and ———. 1928. Further studies on sex in Asparagus. Proc. Amer. Soc. Hort. Sci. **25**:13-16.

Shoji, T., and T. Nakamura. 1928. On the dioecism of garden Asparagus (*Asparagus officinalis* L.) Jap. Jour. Bot. **4**:125-151.

30 Suppression of male characteristics of mosquitoes by thermal means

William R. Horsfall
John F. Anderson
Department of Entomology
University of Illinois
Urbana, Illinois

Abstract. *Dimorphism in* Aedes stimulans, *a northern floodwater mosquito, may be decreased possibly to obliteration by exposing larvae for most of their lives to abnormally high temperature. Determiners for maleness fail to express themselves when larvae are exposed to a temperature of 29°C throughout their lives. Not only are male characteristics eliminated, but normal female ones such as ovaries, spermathecae, and cerci develop. The resultant adult is structurally a female. Forms showing characteristics of both sexes occur when the number of days of exposure to 29°C is lessened.*

The sexes of mosquitoes differ in appearance in a number of easily recognizable ways, as has been well summarized by Snodgrass *(1)*. Anteriorly, the appendages of the head are distinctive for each sex. The antennae and palpi of males are strikingly more hirsute than their counterparts in females. On the other hand, mouth parts are reduced from the female complement of seven to two functional appendages. Caudally, the male has an elaborate set of copulatory appendages, while the female has none. A pair of flaplike cerci marks the caudal portion of the female externally. Internally, the males have testes, vasa deferentia, seminal vesicles, and a bilobed accessory gland. The female has ovaries, oviducts, spermathecae, and a small saclike accessory gland. Significant changes in these structures have been brought about by treating larvae to abnormally high temperature.

Mosquitoes showing external characteristics of both sexes have been collected in different parts of the world. Such anomalies, called intersexes by Kitzmiller *(2)*, number less than 40. Most of them have come from northern latitudes, and in some instances two or more individuals have been collected from the same vicinity at approximately the same time. The facts of location and repetitive occurrence suggest possible genetic or environmental causes for the anomalies. Unfortunately, specimens have been collected so infrequently that little more can be inferred.

Aedes stimulans, a snow-pool mosquito common to Canada and northern latitudes of conterminous United States, has been induced to express marked intersexual tendencies under laboratory conditions. The two sides of the responding genetic males are affected alike, and no unilateral responses have been elicited. When a uniformly mixed population of larvae (Table 1) is separated into two lots immediately after hatching and is exposed in one instance to a continuous temperature of 24°C and in another to one of 29°C, marked differences between the resulting males always occur. Genetically intended males from larvae reared at 24°C develop antennae, palpi, mouth parts, external genitalia, accessory glands, seminal vesicles, vasa deferentia, and testes that are normal in appearance and function. Larvae of genetically intended males when reared at 29°C without exception are like females in all respects except for slight differences in palpi. Internally, the anomalous males have ovaries, oviducts, and spermathecae, and they lack testes, vasa deferentia, seminal vesicles, and bilobed accessory glands. The ovaries have globular egg chambers indistinguishable from those of young genetic females (see *3*).

Abnormally high temperature exerts its modifying effect on larvae of potential males according to the duration of exposure (Table 1). High temperature applied late (last 3 days) in larval life produces no structural defects but prevents rotation of genitalia to the copulatory position. An extension of exposure to high

From Science **133:**1830, June 9, 1961.

Funds for support of this work were in part from a grant made by the Graduate College. Reinhart Brust has given much help in this work and our gratitude is expressed to him.

Table 1. Effect of high-rearing temperature on sexual characteristics of genetically intended males of *Aedes stimulans* in the laboratory (M and F, normal male and female; a.m. and a.f., abnormal male and female; +, present; −, absent)

Imaginal parts		Sex at serial-rearing temperatures of 24° and 29°C						
	Days at 24° C:	8	5	4	3	2	1	0
	Days at 29° C:	0	3	4	5	6	7	8
		Internal parts						
Gonads		M	M	M	M	F	F	F
Tubes to gonads		M	M	M	M	M,F	F	F
Sperms		+	+	+	−	−	−	−
Spermathecae		−	−	−	−	+,−	+	+
Accessory gland		M	M	M	M	M,F	F	F
		Caudal parts						
Parameres		M	M	M	a.m.	a.m.	−	−
Phallosome		M	M	M	M	−	−	−
Genitalia position		M	a.m.	a.m.	a.m.	F	F	F
Cerci		−	−	−	a.f.	F	F	F
		Cephalic parts						
Antennae		M	M	M	F	F	F	F
Palpi		M	M	a.m.	a.m.	a.m.	a.m.	a.f.
Mouth parts		M	M	M	F	F	F	F

temperature to include the last 6 days of larval life causes a series of changes that produces intersexes. Larvae exposed to 29°C for 7 days or more grow into apparent females, some of which have been inseminated by normal males. Insemination was determined by examining the spermathecae under a compound microscope for the presence of sperms.

Larvae that bear female determiners are unaffected by a temperature of 29°C. They give rise to females that are normal in appearance, copulate readily by the artificial means described by McDaniel and Horsfall *(4)*, feed on blood, and develop eggs in a normal manner.

REFERENCES AND NOTES

1. R. E. Snodgrass, Smithsonian Inst. Publs. Misc. Collections **139**(8):1 (1959).
2. J. B. Kitzmiller, Rev. brasil. malariol. e doenças trop. **5**:285 (1953).
3. R. F. Harwood and W. R. Horsfall, Ann. Entomol. Soc. Am. **50**:555 (1957).
4. I. N. McDaniel and W. R. Horsfall, Science **125**:745 (1957).

chapter 7

Chromosome numbers

Each species is characterized by a particular number of chromosomes. In sexually reproducing forms, there are actually two basic numbers: (1) diploid (2n), which is found in the somatic cells; and (2) haploid (n), which is found in the germ cells. In asexually reproducing species, only one characteristic number is present; this can be either a diploid or a haploid number, depending on the evolutionary origin of the species. As is true of most generalizations, exceptions to the concept of a single diploid or haploid number of chromosomes for a given species exist. As discussed in Chapter 6, in some insect species the female contains one more chromosome in her somatic cells than does the male. In other insect species, the female is diploid, but the male is haploid.

A karyotype can either be composed of whole sets of chromosomes and designated as *euploid* or contain an incomplete set of chromosomes and then be designated as *aneuploid.* Euploid karyotypes include haploid (n), diploid (2n), triploid (3n), tetraploid (4n), etc. Any condition that involves a multiple of the *n* number is called *polyploidy.* In contrast to a euploid, an aneuploid karyotype includes such variations in chromosome complement as monosomic (2n−1); double-monosomic (2n−1−1); nullisomic (2n−2), when both lost chromosomes are homologues; trisomic (2n+1); double-trisomic (2n+1+1); tetrasomic (2n+2), when both extra chromosomes are homologues; etc.

ANEUPLOIDY

An early study of aneuploidy, which involved all the chromosomes of a species, was made by Blakeslee and Belling in 1924 (Ref. 7-1). They worked with jimsonweed, which has twelve pairs of chromosomes in its diploid complement. Through experimental breeding, they were able to obtain twelve different trisomics: each had one of the chromosomes of the normal set present in triplicate, and each was recognizable by its appearance and distinguishable from the other trisomics.

Attempts to study changes in human karyotypes were hampered for a long time by lack of a proper technique for the preparation of chromosomes. As pointed out in Chapter 6, the problem was solved by Tjio and Levan in 1956 (Ref. 6-13). Human chromosomes can be arranged in groups based on size and position of centromere. The resulting classification of chromosomes is given in Table 7-1.

With the advancement in techniques for studying human chromosomes, a concerted effort was made to discover whether any known human defects were associated with aneuploidy. The first such association found was Down's syndrome (mongolism). This discovery was made by Lejeune and his co-workers in 1959 and was referred to in Chapter 3 (Ref. 3-14). Individuals with Down's syndrome are characterized by mental retardation, short stature, peculiarity of palm prints, and congenital malformations, especially of the heart. An analysis of their karyotype shows that they are trisomic for chromosome 21. Some individuals with Down's syndrome have reproduced. A review of such cases was prepared by Thompson in 1961 (Ref. 7-2) and is reprinted in this chapter. A bibliography of mongolism (Down's syndrome), covering the literature up to September, 1964, was prepared by Stiles and Isoun in 1966 (Ref. 7-3).

Other types of autosomal aneuploidy have also been found. One of them, Patau's syndrome, is considered to be a trisomy for chromosome 13, 14, or 15 and is characterized by deformities of the eyes, lips, palate, brain, and heart. Almost half of the affected infants die within one month, about three fourths are dead by six months, and only one in every five survives to the end of the first year. Survival beyond early infancy, always with severe mental defect, sometimes occurs. Another type of autosomal aneuploidy, Edward's syndrome, involves trisomy for chromosome 17 or 18 and is characterized by a narrow and long skull, small mouth and jaw, cleft palate, and defects of the brain and heart. About one third of the affected infants die within one month, half of them within two, three fourths within three months, and about 90% within a year. Those infants that survive to the later stages are

Table 7-1. Human chromosome analysis by groups*

Group	Size and centromere position	Ideogram number	Number in diploid cell
A or I	Large; metacentric/submetacentric	1-3	6
B or II	Large; submetacentric	4, 5	4
C or III	Medium; submetacentric	6-12 and X	15 (male) or 16 (female)
D or IV	Medium; acrocentric	13-15	6
E or V	Small; metacentric/submetacentric	16-18	6
F or VI	Small; metacentric	19, 20	4
G or VII	Smallest; acrocentric	21, 22, and Y	5 (male) or 4 (female)

*From Levine, L. 1969. Biology of the gene. The C. V. Mosby Co., St. Louis.

grossly defective. A review of the above types of autosomal aneuploids, as well as others, was prepared by Polani in 1969 (Ref. 7-4).

Aneuploidy is not restricted to the autosomes but can also involve the sex chromosomes. In the case of human females exhibiting Turner's syndrome, it was found that these individuals were monosomic for their X chromosome. They are characterized by a complete lack of development of ovaries, an underdevelopment of breasts, short stature, webbed neck, and mental retardation. The chromosomal basis of Turner's syndrome was reported, as discussed in Chapter 6, by Ford and his co-workers in 1959 (Ref. 6-14). In the case of human males exhibiting Klinefelter's syndrome, it was found that these individuals had one or more X chromosomes added to their normal male XY karyotype. These males are sterile because of a defective development of their testes. In addition, most of these males are mentally retarded. The chromosomal basis of Klinefelter's syndrome was reported, as discussed in Chapter 6, by Jacobs and Strong in 1959 (Ref. 6-15). Sex chromosome aneuploidy involving the Y chromosome was discovered by Sandberg and his co-workers in 1961 (Ref. 7-5), and their paper is included in this collection. The possible association of aggressive behavior with sex chromosome aneuploidy will be considered in Chapter 12.

A complication in karyotype analysis occurs when all the cells of the body do not contain the same chromosome complement. This occurs when the individual is either a mosaic or a chimaera. A review of such cases in man was prepared in 1969 by Ford (Ref. 7-6), whose paper is reprinted in this chapter.

One of the benign side effects of the different types of human aneuploidy is the characteristic change that each causes in the dermatoglyphics of the affected individual. *Dermatoglyphics* is a collective name for all the skin patterns of the fingers, toes, palms, and soles. The relationship of karyotype to dermatoglyphics was reviewed in 1963 by Penrose (Ref. 7-7), whose paper is included in this chapter.

POLYPLOIDY

Depending on the source of its chromosomes, a polyploid is classified either as an autopolyploid or an allopolyploid. *Autopolyploids* arise either from single, asexually reproducing individuals or from individuals in an interbreeding population. All the sets of chromosomes of an autopolyploid are homologous. *Allopolyploids* arise from hybrids between species. The chromosome of the species-hybrid are normally not homologous. However, doubling the number of chromosomes of the species-hybrid does provide each chromosome with a homologue.

In nature, one finds both autopolyploids and allopolyploids. The establishment of polyploidy in an asexually reproducing population depends strictly on the competitive advantage that the particular polyploid may have over the parental type. However, in sexually reproducing, cross-fertilizing forms, the reproductive process itself presents great difficulties for the survival of a polyploid regardless of its competitive superiority. This is because most polyploids will occur as a single tetraploid individual in a population. The tetraploid will have to mate or cross with one of the surrounding diploids, thus forming triploid offspring. Triploids are viable and experience no difficulties in cell division and growth, since the pairing of chromosomes is not involved in mitosis. However, at meiosis, one finds that for the autotriploids three sets of homologous chromosomes are present whereas for allotriploids two sets of homologous chromosomes are present. In both cases, there will be a normal segregation of two of the sets of homologous chromosomes, and the extra chromosomes will be distributed to the resultant cells in random fashion. Most of the gametes will therefore not have a balanced complement of chromosomes. This will result in aneuploid offspring, which tend to be nonviable. It is apparent that the establishment of polyploidy in an interbreeding population must carry with it some radical modification of the life cycle. This, in fact, is the case. Most polyploids, plant as well as animal, exhibit some form of parthenogenesis. These individuals have become essentially asexual in their mode of reproduction. Without meiosis, extra sets of chromosomes present no hazard for reproduction.

The first well-demonstrated case of autopolyploidy was reported by Lutz in 1907 (Ref. 7-8). The discovery of polyploids led to a proposal by Winge in 1917 (Ref. 7-9) that polyploidy was a means by which new species with greater chromosome numbers may originate. An experimental demonstration of the origin of a new species through allopolyploidy was provided in 1925 by Clausen and Goodspeed (Ref. 7-10), whose paper is included in this collection. It is interesting to note that the species of wheat used in the making of bread is thought to be an allotetraploid between some ancient type of wheat and some form of wild goat grass. The probable steps involved in the origin of modern wheat was experimentally demonstrated by McFadden and Sears in 1946 (Ref. 7-11). A detailed discussion of polyploidy in plants was published by Stebbins in 1950 (Ref. 7-12).

Polyploidy is less frequent in animals than it is in plants, although an increasing number of animal species have been found either to contain polyploid populations or to be polyploids of some characteristic generic chromosome number. Until recently, most of the known cases of animal polyploidy occurred in the invertebrates. The first well-studied species was the brine shrimp, *Artemia salina,* which was found to contain (1) normal diploid bisexual populations; (2) parthenogenetic diploid populations; (3) parthenogenetic triploid populations; and (4) parthenogenetic tetraploid populations. The overwhelming number of offspring produced in the parthenogenetic populations are females. The existence of a parthenogenetic diploid population led to the hypothesis that the evolution of polyploidy in the brine shrimp consisted of two stages: the first stage to involve the development of a parthenogenetic diploid population from one or more diploid individuals and the second stage to involve the origin of parthenogenetic polyploids from parthenogenetic diploids. It is difficult to evaluate this hypothesis because not all species containing polyploid populations include a parthenogenetic diploid group. Polyploidy has also been found in the bagworm moth, *Solenobia triquetrella;* the sow bug, *Trichoniscus elizabethae;* the weevil, *Otiorrhynchus scaber;* and others. Bungenberg de Jong in 1958 (Ref. 7-13) reviewed the literature on the occurrence of polyploidy in these and other species.

More recently an increasing number of investigators have turned their attention to the vertebrates in search of naturally occurring polyploid groups. Uzzell in 1963 (Ref. 7-14) reported the discovery of populations of salamanders related to *Ambystoma jeffersonianum* that were female and triploid and intermingled with diploid populations of the same and other related species. The triploid females mate with diploid males, but their offspring are always triploid and female. A study of the mechanism by which this occurs was reported by Macgregor and Uzzell in 1964 (Ref. 7-15). They found that early in oogenesis, the chromosomes of triploid females undergo an extra division and become hexaploid. These oogonia then undergo meiosis and produce triploid ova. Sperm are necessary to stimulate the eggs to develop, but they do not contribute chromosomes to the triploid egg nucleus. It is not known whether these triploids are autopolyploids or allopolyploids.

A study of parthenogenetic species of the whiptail lizards (genus *Cnemidophorus*) by Pennock in 1965 (Ref. 7-16) revealed that at least three of the species are triploid and a fourth species, diploid. The members of the parthenogenetic species are all female. The mechanism by which triploidy is maintained in these lizards is unknown, and it is also unknown whether these species are autopolyploids or allopolyploids.

The examples of vertebrate polyploids discussed above involved parthenogenetic populations. However, other vertebrate polyploid populations have been found that are bisexual and exhibit a normal pattern of sexual reproduction. These populations have been found among the frogs and toads (order Anura). A report of such a polyploid population was made by Wasserman in 1970 (Ref. 7-17).

In humans, complete triploidy has thus far been discovered only in aborted embryos. It would appear that triploidy results in too many malformations to permit a complete development and birth of the individual. However, a number of cases of diploid/triploid mosaics have been found, and these are viable. The first such human mosaic reported was a boy who was described by Book and Santesson in 1960 (Ref. 7-18). A case of a girl with triploid cells was reported by Ellis and co-workers in 1963 (Ref. 7-19). As stated above, complete triploidy in humans appears to be lethal. The first such case was reported in 1961 by Penrose and Delhanty (Ref. 7-20), whose paper is the final article reprinted in this chapter. A paper that summarizes a good deal of the information on triploidy in man was published by Edwards and co-workers in 1967 (Ref. 7-21).

One of the questions asked by early investigators of polyploidy was whether there was any evolutionary advantage to polyploidy. The most apparent advantage is an increase in the amount of genetic material that is available for mutation and selection and, hence, for evolutionary divergence from the parental stock. This additional genetic material could permit the polyploids to adapt to habitats that were unfavorable for the diploid type. It is also apparent that the asexual reproduction which usually accompanies polyploidy permits the migration and multiplication of the polyploid type in a new area even as a single, isolated individual. A discussion of the geographic distribution of polyploids as compared to diploids was prepared by Vandel in 1940 (Ref. 7-22) for animals and by Löve and Löve in 1943 (Ref. 7-23) for plants.

REFERENCES

Aneuploidy

7-1. Blakeslee, A. F., and Belling, J. 1924. Chromosomal mutations in the Jimson weed, *Datura stramonium.* J. Heredity **15:**195-206.

7-2. Thompson, M. W. 1961. Reproduction in two female mongols. Can. J. Genet. Cytol. **3:**351-354.

7-3. Stiles, K. A., and Isoun, M. F. 1966. Bibliography of mongolism. Acta Genet. Med. Gemellol. **15:**247-272.

7-4. Polani, P. E. 1969. Autosomal imbalance and its syndromes, excluding Down's. Brit. Med. Bull. **25:**81-93.

7-5. Sandberg, A. A., Koepf, G. F., Ishihara, T., and Hauschka, T. S. 1961. An XYY human male. Lancet **ii:**488-489.

7-6. Ford, C. E. 1969. Mosaics and chimaeras. Brit. Med. Bull. **25:**104-109.

7-7. Penrose, L. S. 1963. Finger-prints, palms and chromosomes. Nature **197:**933-938.

Polyploidy

7-8. Lutz, A. M. 1907. A preliminary note on the chromosomes of *Oenothera lamarckiana* and one of its mutants, *O. gigas.* Science **26:**151-152.

7-9. Winge, O. 1917. The chromosomes—their numbers and general importance. Comp. Rend. Trav. Lab. Carlsberg **13:**131-275.

7-10. Clausen, R. E., and Goodspeed, T. H. 1925. Interspecific hybridization in *Nicotiana.* II. A tetraploid *glutinosa-tabacum* hybrid. An experimental verification of Winge's hypothesis. Genetics **10:**278-284.

7-11. McFadden, E. S., and Sears, E. R. 1946. The origin of *Triticum spelta* and its free-threshing hexaploid relatives. J. Heredity **37:**107-116.

7-12. Stebbins, G. L. 1950. Variation and evolution in plants. Columbia University Press, New York.

7-13. Bungenberg de Jong, C. M. 1958. Polyploidy in animals. Bibliographica Genetica **17:**111-228.

7-14. Uzzell, T. M., Jr. 1963. Natural triploidy in salamanders related to *Ambystoma jeffersonianum.* Science **139:**113-115.

7-15. Macgregor, H. C., and Uzzell, T. M., Jr. 1964. Gynogenesis in salamanders related to *Ambystoma jeffersonianum.* Science **143:**1043-1045.

7-16. Pennock, L. A. 1965. Triploidy in parthenogenetic species of the Teiid lizard, genus *Cnemidophorus.* Science **149:**539-540.

7-17. Wasserman, A. O. 1970. Polyploidy in the common tree toad *Hyla versicolor* Le Conte. Science **167:**385-386.

7-18. Böök, J. A., and Santesson, B. 1960. Malformation syndrome in man associated with triploidy. Lancet **i:**858-859.

7-19. Ellis, J. R., Marshall, R., Normand, I. C. S., and Penrose, L. S. 1963. A girl with triploid cells. Nature **198:**411.

7-20. Penrose, L. S., and Delhanty, J. D. A. 1961. Triploid cell cultures from a macerated foetus. Lancet **i:**1261-1262.

7-21. Edwards, J. H., Yuncken, C., Rushton, D. I., Richards, S., and Mittwoch, U. 1967. Three cases of triploidy in man. Cytogenetics **6:**81-104.

7-22. Vandel, A. 1940. La parthénogenèse géographique. IV. Polyploidie et distribution géographique. Bull. Biol. France et Belg. **74:**94-100.

7-23. Löve, A., and Löve, D. 1943. The significance of differences in the distribution of diploids and polyploids. Hereditas **29:**145-163.

31 Reproduction in two female mongols

Margaret W. Thompson
Department of Paediatrics
University of Alberta
Edmonton, Alberta, Canada

In view of the current interest in reproduction in mongols, two cases of reproduction in female mongols are briefly reported.

CASE REPORTS

Case 1. A female mongol aged 21 residing in a private home for mental defectives produced a liveborn male child in 1946. There appears to have been reason to believe that the father was also a mongol, but this cannot be confirmed. The mother was later admitted to a provincial mental institute and died of tuberculosis at the age of 25. The diagnosis of mongolism was made by qualified psychiatrists and seems unequivocal. Photographs show her to have been typically mongoloid in appearance, with short stature, mongoloid facial features, short stubby hands, and a very wide gap between the first and second toes. A marked convergent squint was present. She was the child of normal parents and had four normal sibs. The maternal age at her birth was 40 years.

This mongol's child became a public ward at birth, and because his unfavorable background was considered to make him an unfit prospect for adoption he was maintained in foster homes until 10 years of age. At that time, in view of his normal development and satisfactory school progress, he was given in adoption. He now resides in a rural area of the province, where he continues to exhibit normal development and intelligence. He has not been available for chromosome studies or dermal pattern analysis.

Case 2. A female mongol aged 22 was admitted to hospital in Edmonton in April 1960 for hysterotomy, termination of pregnancy, and sterilization. Two years earlier she had been delivered of a stillborn macerated female foetus at approximately the 28th week of development, the pregnancy having been unsuspected until the time of delivery.

The diagnosis of mongolism in this patient was made on the usual clinical grounds. She was short in stature, typically mongoloid in appearance, and had an apical systolic heart murmur. Her dermal pattern index of +6.6 was high in the mongoloid range (Walker, 1958). She had an arch tibial pattern in the hallucal area of each sole, a high axial triradius on each palm, and an ulnar loop pattern on each finger. The third interdigital pattern was not present. She had double flexion creases on both palms, her fifth fingers were only slightly incurved, and there was no obvious gap between her first and second toes. Chromosome studies have not yet been undertaken.

Following termination of this patient's second pregnancy, the foetus was available for examination. It was a male, at approximately the 20th week of development (crown-rump length 180 mm.; weight 442 gm.). Even upon casual examination it appeared obviously mongoloid, because of the markedly incurved fifth digits of the hands, the presence of a borderline type of single flexion crease on one hand, and the gap between the first and second toe of each foot. The facial features also suggested mongolism; typical epicanthal folds were present, and the ears had the square, crushed appearance commonly seen in newborn mongols. The dermal pattern index of +6.9, like that of the mother, was high in the mongoloid range. There was an arch tibial pattern in the hallucal area of each sole, a high axial triradius on each palm, and an ulnar loop on each finger. The third interdigital pattern was present on the left palm, absent on the right.

An attempt was made to culture kidney tissue from the foetus for chromosome analysis, but unfortunately this was unsuccessful.

The family history is essentially negative, but it is of interest that the mongol patient was the second-born child of a 22-year-old mother, who

From Canadian Journal of Genetics and Cytology 3:351-354, 1961. Used with permission.

The author wishes to thank Dr. James Calder, Dr. D. C. Ritchie, and officials of the Child Welfare Branch and the Department of Public Health of the Government of Alberta for permission to report these cases. Financial support through National Health Grant # 608-13-12 has made the investigation possible.

since then has had five other children including a normal son born when she was 42 years of age. The patient's mother was one of 11 sibs and the father one of 14. No other case of mongolism is known in this large kindred.

DISCUSSION

Although mongols usually are sterile, this report brings to 11 the number of confirmed cases of female mongols who have reproduced. Six mongol mothers have produced a total of 7 normal children including a monozygotic twin pair (Forssman and Thysell, 1957; Forssman, Lehmann, and Thysell, 1961; Levan and Hsu, 1959 and 1960; Mullins *et al.,* 1960; Sawyer, 1949; Sawyer and Shafter, 1957; Thuline and Priest, 1961, twin case; and our Case 1). Four other mongol mothers have produced a total of 5 mongol children (Hanhart, 1960, and Hanhart *et al.,* 1961, two mongol sibs; LeLong *et al.,* 1949; Rehn and Thomas, 1957, and Stiles, 1958; and our Case 2). One child (Schlaug, 1957) is reported to have shown certain stigmata of mongolism. Chromosome studies have been reported for the case described by Forssman *et al.,* one of Hanhart's two cases, the case of Levan and Hsu, and the monozygotic twin pair observed by Thuline and Priest *(loc. cit.);* in all these cases the mongol mothers and children have had 47 chromosomes with trisomy for a small acrocentric chromosome (probably chromosome # 21), and the normal children have had a normal karyotype.

The original demonstration by Lejeune and Turpin (1959) that mongolism is associated with trisomy for one of the small acrocentric chromosomes has now been amply confirmed. It has also become clear that translocation in a previous generation, involving the mongolism chromosome, can give rise to "functional trisomy", with a chromosome number of 46 but with essentially triple representation of chromosome # 21 (Polani *et al.,* 1960). Translocation mongols are clinically typical but can readily be distinguished from standard trisomics or normals by karyotype analysis. Their occurrence appears to be independent of maternal age. Their frequency is not known, but is certainly not higher than 10 per cent of all mongols. A third possible type of mongol is the mosaic mongol, whose tissues are made up partly of 47-chromosome and partly of normal 46-chromosome cells (as suggested by Jacobs *et al.,* 1959).

Any type of mongol could, if fertile, produce an entirely normal child. In the standard trisomic mongol, ova with 23 and 24 chromosomes should be formed in equal numbers, and, if viability of the two types of ovum is approximately equal, the risk of mongolism in the child should be approximately 50 per cent.

There is as yet no direct evidence concerning the possible types of gametes of translocation mongols and their relative frequencies. Mosaic mongols could produce a proportion of mongol children if the mosaicism involved all or part of the gonads. In view of the rarity of these types as compared with the more usual standard trisomic mongols, they are omitted from consideration here.

Although the total reported number of offspring of mongols is still far too small for statistical analysis, the data appear to be concordant with the theoretical expectation of a 50 per cent risk of mongolism in the child of a mongol mother. Mongols differ in this way from XXX females, who, although occasionally fertile (Barr and Carr, 1960), have not been reported to produce trisomic offspring.

In cases similar to our Case 1, clinical examination supported by karyotype analysis could promptly determine whether the child was a suitable candidate for adoption. If the child had a normal 46-chromosome or the translocation-bearing 45 chromosome karyotype, he could be considered entirely normal, and long pre-adoption custodial care could be avoided. On the other hand, a standard trisomic or translocation-bearing 46 chromosome karyotype would confirm that the child was a mongol.

SUMMARY

Two cases of pregnancy in female mongols are described. One of these mongols produced a normal male; the other produced a mongol foetus. A total of 11 cases of mongols who have become pregnant have now been reported. Seven normal and five mongol offspring have been produced by 10 of these mothers in 11 pregnancies. The data are in accordance with the theoretical expectation of a 50 per cent risk of mongolism in the offspring of mongols.

REFERENCES

Barr, M. L., and Carr, D. H. 1960. Sex chromatin, sex chromosomes, and sex anomalies. Can. Med. Assoc. J. **83**:979-986.

Forssman, H., and Thysell, T. 1957. A woman with mongolism and her child. Am. J. Mental Deficiency **62**:500-503.

Forssman, H., Lehmann, O., and Thysell, T. 1961. Reproduction in mongolism: chromosome studies and re-examination of a child. Am. J. Mental Deficiency **65**:495-498.

Hanhart, E. 1960. Mongoloid idiocy in mother and two children from an incestuous relationship. Acta Genet. Med. Gemellol. **9**:112-130.

Hanhart, E., Delhanty, Joy D. A., and Penrose, L. S. 1961. Trisomy in mother and child. Lancet **i**:403.

Jacobs, P. A., Baikie, A. G., Brown, W. M. Court, and Strong, J. A. 1959. The somatic chromosomes in mongolism. Lancet **i**:710.

Lejeune, J., Gautier, M., and Turpin, R. 1959. Etudes des chromosomes somatiques de neuf enfants mongoliens. Compt. Rend. Acad. Sci., Paris **248**: 1721-1722.

LeLong, M., Borniche, P., Kreisler, L., and Baudy, R. 1949. Mongolien issu de mère mongolienne. Arch. franc. pédiat. **6**:231-238.

Levan, A., and Hsu, T. C. 1959. The human idiogram. Hereditas **45**:665-674.

Levan, A., and Hsu, T. C. 1960. The chromosomes of a mongoloid female, mother of a normal boy. Hereditas **46**:770-772.

Mullins, D. H., Estrada, W. R., and Gready, T. G. 1960. Pregnancy in an adult mongoloid female: report of a case. Obstet. and Gynecol. **15**:781-783.

Polani, P. E., Briggs, J. H., Ford, C. E., Clarke, C. M., and Berg, J. M. 1960. A mongol girl with 46 chromosomes. Lancet **i**:721-724.

Rehn, A. T., and Thomas, E., Jr. 1957. Family history of a mongoloid girl who bore a mongoloid child. Am. J. Mental Deficiency **62**:496-499.

Sawyer, G. M. 1949. Case report: reproduction in a mongoloid. Am. J. Mental Deficiency **54**:204-206.

Sawyer, G. M., and Shafter, A. J. 1957. Reproduction in a mongoloid: a follow-up. Am. J. Mental Deficiency **61**:793-795.

Schlaug, R. 1957. A mongolian mother and her child: a case report. Acta Genet. Statist. Med. **7**:533-540.

Stiles, K. A. 1958. Reproduction in a mongoloid imbecile. Proc. Intern. Congr. Genet. 10th Congr. Montreal **2**:276-277.

Thuline, H. C., and Priest, Jean H. 1961. Pregnancy in a 14-year-old mongoloid. Lancet **i**:1115-1116.

Walker, N. F. May, 1958. The use of dermal configurations in the diagnosis of mongolism. Pediat. Clin. North America. pp. 531-543.

32 # An XYY human male

A. A. Sandberg
G. F. Koepf
T. Ishihara
T. S. Hauschka
The Medical Foundation of Buffalo
and
Roswell Park Memorial Institute
Buffalo, New York

Four abnormal sex-chromosome combinations may result from faulty separation of the sex-chromosomes during oogenesis and/or spermatogenesis: XO, XXY, XXX, and XYY. To date, only the first three conditions (and the rare XXXX, XXYY, and XXXXY) have been described and confirmed. We wish to present preliminary findings on a probable XYY male.

A forty-four-year-old white man of average intelligence and without physical defects, despite the 47 chromosomes found in his marrow and blood, started work at seventeen after two years of high school. He has been married twice; 6 children (1 died at age 3 as a "blue baby") resulted from the first marriage, and 2 from the second. Each of his wives gave birth to at least one abnormal child. An eighteen-year-old daughter of the first marriage has had amenorrhoea, no breast development and, at operation, was found to have no internal sex organs. The nuclei of her buccal mucosa were chromatin-positive and her marrow cells were shown to have 46 chromosomes with a normal XX constitution. A twenty-two-month-old daughter from the second marriage is a typical 47-chromosome mongoloid. Also one miscarriage occurred in each marriage. All other children are apparently living and well. Since they have been anonymously adopted we have not been able to study their chromosomes.

From The Lancet **ii**:488-489, 1961. Used with permission.

Their father claims to have normal libido, and results of examination were unremarkable. He has somewhat large facial features, is obese and weighs 287 lb., and has a neurodermatitis, an umbilical hernia, and, in the left mandible, has had a cystic lesion for many years. The buccal-mucosa cells were chromatin-negative.

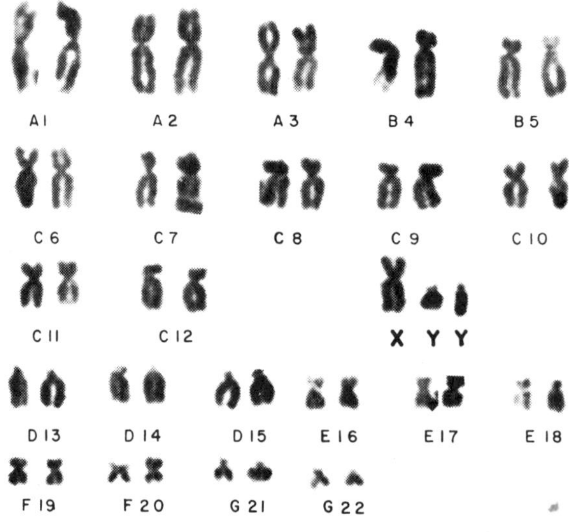

Karyotype of father.

Fig. 1

Table 1. Chromosome constitution

Source of cells	Cells with chromosome number of 45	46	47	48	Total cells counted
Blood	0	3	37	1	41
Blood	0	1	26	2	29
Marrow	1	8	68	0	77

His chromosome constitution was determined on three occasions. Two separate blood samples were cultured by the procedure of Moorhead et al.[1] and metaphases in freshly aspirated sternal and iliac crest marrow were analysed after performing the "direct squash" technique.[2] Nearly 90% of the metaphases contained 47 chromosomes (see accompanying table). In its arm ratio and size, the extra chromosome shown in the karyotype of the figure consistently resembles the Y more than the small autosomes G21 and 22, judged according to the criteria of the Human Chromosome Study Group.[3] Absence of the severe mental and physical stigmata regularly associated with autosomal trisomy is further evidence against the latter interpretation. The relatively small gene content of the Y is in keeping with the absence of the congenital defects that characterise other types of sex-chromosome imbalance. This case may therefore be considered as an XYY male.

A more detailed study of the patient's skin and chromosome data for other members of his family will be published later.

[1] Moorhead, P. S., Nowell, P. C., Mellman, W. J., Battips, D. M., Hungerford, D. A. Exp. Cell. Res. 1960, **20**:613.

[2] Sandberg, A. A., Ishihara, T., Miwa, T., Hauschka, T. S. Cancer Res. 1961, **21**:678.

[3] Hered. 1960, **51**:214.

33 Mosaics and chimaeras

C. E. Ford
Medical Research Council
Radiobiological Research Unit
Harwell, Didcot, Berkshire, England

The words "mosaic" and "chimaera" have a long history of use in biology and have sometimes been employed synonymously. In human cytogenetics both connote subjects with cells of two or more chromosomally different kinds. General definitions of the two terms given by Anderson, Billingham, Lampkin & Medawar (1951) will be adopted here. According to these authors, a chimaera ". . . is an organism whose cells derive from two or more distinct zygote lineages . . .", whereas a mosaic ". . . is formed of the cells of a single zygote lineage." This simple theoretical distinction eases the problem of classification, but a preliminary discussion of its implications is required.

Anticipating later more detailed consideration, two groups of human chimaeras have been distinguished, one group thought to arise from rare instances of two separate acts of syngamy in one ovum, the other from infrequent cases of intermixture or exchange of cells between individuals of independent zygotic origin. Mosaics, on the other hand, must be supposed to originate as consequences of irregularities during the cell cycle. Regardless of the nature of the chromosomal differences, however, the component cell lines of mosaics will have an underlying genetic identity, whereas in chimaeras they will be genetically distinct. On the evidence of karyotypic analysis alone then, it is formally possible for a subject with chromosomally distinct cell lines to be either a mosaic or a chimaera. It follows that chimaerism can be proved only by genetic evidence and that mosaicism (in individual cases) cannot be proved at all. Nevertheless the very great majority of human individuals with two or more chromosomally distinct cell lines must be mosaics, since otherwise it would be necessary to postulate the coincidence of two rare events to account for their origin. As a working rule, therefore, mosaicism may be assumed unless there are special reasons for suspecting that the subject may be a chimaera.

1. DIAGNOSIS OF MOSAICS

The cytogenetic evidence for the diagnosis of mosaicism is in most cases sufficiently striking for acceptance without question. It is only when the number of cells recorded for one cell line is low that uncertainty arises. The obvious criterion to adopt is that there should be more cells of the presumptive independent cell line in the sample studied than can reasonably be attributed to chance (Court-Brown, Jacobs & Doll, 1960). Unfortunately this criterion cannot be applied with rigour. It is true that the probabilities of obtaining given sets of observations by chance can be calculated by use of the binomial distribution; it is the assumptions involved that are open to question, as the following hypothetical example illustrates.

Suppose a sample of 20 cells from a child with Down's syndrome consists of 17 cells with a count of 47 chromosomes including 5 of group G, and 3 cells with 46 chromosomes, 2 of them with 4 in group G, and 1 with 5 in group G but lacking a chromosome from one of the other groups. If hypomodal cells originate by random losses of chromosomes (whether through mitotic error or preparative artifact is immaterial), the probability that the observed group of hypomodal cells originated by chance independent events is given by the sum of the last two terms of the binomial expansion $(42/47 + 5/47)^3$. This value is 0.041, or 1 in 25. Even if this were the sole basis for a judgment, the probability level is not low enough for a confident diagnosis of mosaicism and the

From British Medical Bulletin **25**:104-109, 1969. Used with permission.

I am deeply indebted to Dr. Race and Dr. Sanger for permission to read, in proof, chapter 23 of the fifth edition (1968) of their invaluable book, *Blood groups in man.*

obvious course would be to increase the number of observations. Had all three hypomodal cells lacked a chromosome of the G group, the associated probability would have been $(5/47)^3$, or approximately 0.001. This figure is below the level that can reasonably be attributed to chance. However, it rests on the assumption that losses of chromosomes are random, which is almost certainly not true. If, as is probable, the loss of the smaller chromosomes is favoured, the corresponding frequency to be entered in the binomial would be greater than 5/47; and if, for purposes of illustration, this is set at 0.4, the probability of finding 3 out of 3 hypomodal cells all lacking a chromosome of group G would become $0.4^3 = 0.064$, making the assumption of mosaicism questionable. Even if the mean frequency of hypomodal cells lacking a G chromosome were known for normal subjects, calculations based on it could still only be regarded as indicative, since the true frequency would doubtless differ from case to case owing to factors unique to the individual and to variations in preparative technique. Despite their limitations, such calculations can be useful. The topic has been discussed by Harnden (1964), who gives examples.

A different question, but one that can be answered explicitly, is the probability that a random sample of cells of a given size from a mixed population of cells shall not include at least one cell of a cell line that is present in given frequency. The probability required is $(1 - p)^n$ where p is the frequency of the cell line in question. This is shown in Table I for samples of 10, 20 and 50 cells and for frequencies 0.5-0.05. It can be inferred that, if the cell line comprises only 5% of the cells in the population, it would pass unsuspected approximately 6 times in 10 trials if the sample size were 10 cells, approximately once in 3 trials if the sample were increased to 20, and approximately once in 12 trials if the sample were of 50 cells.

Mention must also be made of the fact that lymphocytes of the constitution 45,X increase in frequency in the blood of both normal males and normal females as they get older (Court-Brown, Buckton, Jacobs, Tough, Kuenssberg & Knox, 1966). If this were not taken into consideration it could conceivably lead to spurious identification of normal subjects as mosaics with a 45,X cell line. The point has been discussed by Court-Brown (1967) who presents a table (Table 1.9, page 15) giving the numbers of presumptive 45,X cells that would be expected by chance in exceptional ($P = 0.001$) samples of cells from women of three age-groups.

Table I. The probability that a random sample of cells from a mosaic subject with two cell lines will not include at least one cell of the minor cell line

Frequency of minor component	Number of cells in sample		
	10	20	50
0.5	0.001	<0.001	<0.001
0.4	0.006	<0.001	<0.001
0.3	0.028	0.001	<0.001
0.2	0.107	0.012	<0.001
0.1	0.349	0.122	0.005
0.05	0.598	0.358	0.077

2. SEX-CHROMOSOME MOSAICS

Since the first presumptive case of mosaicism involving the sex chromosomes was described in 1959 it has become evident that mosaics are not only remarkably frequent among subjects with abnormalities of the sex chromosomes, but there is also a great variety of different

Table II. Numerical mosaics classified according to the number of cell lines and the minimum number of abnormal mitotic events that must be postulated to account for them

Two cell lines (One event)		Three cell lines (One event)	Two cell lines (Two events)	Three cell lines (Two events)
XO/XX	XY/XYY	XO/XX/XXX	XO/XXXY	XO/XY/XXY
XO/XY	XX/XYYY	XO/XY/XYY		XO/XXY/XXYY
XO/XXX	XY/XXYY			XO/XX/XY
XO/XYY	XXY/XXXY			XX/XY/XXY
XX/XXX	XXY/XXYY			XX/XXY/XXYYY
XX/XXY	XXX/XXXX			
XY/XXY	XXXX/XXXXX			
	XXXY/XXXXY			

types involving either simple numerical differences or structurally altered X and Y chromosomes, or both. To give one example, of the 262 cases described by Court-Brown, Harnden, Jacobs, Maclean & Mantle (1964), 49 were mosaics. Table II gives a list of types of mosaics with cell lines that differ only in the number and kind of sex chromosomes they contain. This list is certainly incomplete but will serve to illustrate the variety that exists. It has been assembled mainly from papers published during the years 1960 to 1967 inclusive, in the following journals: *American Journal of Human Genetics, Annals of Human Genetics, Cytogenetics, Journal of Medical Genetics* and *The Lancet.* Other sources were: de la Chapelle, 1962; Lindsten, 1963; Court-Brown *et al.* 1964; and Ricci, Dallapiccola, Ventimiglia, Tiepolo & Fraccaro, 1968.

It has generally been supposed that errors at anaphase, non-disjunction or lagging on the spindle, are responsible for the origin of the variant cell lines in numerical mosaics. There is no reason to doubt these assumptions, though neither abnormality has yet been directly demonstrated in a human cell. Some mosaics, such as the XO/XX type, can be explained by lagging, leading to the loss of a single chromosome from one or both daughter cells. In others, such as the XO/XXX type, it is obvious that non-disjunction, leading to complementary loss and gain of a chromosome in the daughter cells, is a more plausible assumption. It is convenient to classify numerical mosaics according to the number of cell lines identified and to the minimum number of abnormal events that must be assumed to account for them. For example, in the XO/XXXY case, none of the possible intermediate cell lines could have given rise to the 2 observed lines as the result of a single abnormal event; 2 is the minimum. The explanation requiring fewest assumptions is that the zygote was XXY, that 2 successive non-disjunctional events gave rise to 4 cell lines, XO, XY, XXY and XXXY, and that the 2 intermediate cell lines either did not contribute to the founder cells of the observed cell population or did so in such small numbers that they were undetected in the sample of cells studied.

As an extreme example of numerical mosaicism, a case of Klinefelter's syndrome studied by Wahrman & Gersht (1966) may be quoted. Karyotypes were prepared from 107 cells in preparations from 3 blood cultures, and identifications were made as follows: XX (13 cells), XY (16 cells), XXY (67 cells), XXXY (8 cells), XXYY (2 cells), and XYY (1 cell). Notwithstanding the complexity, if the 1 XYY cell is ignored the observations can be accounted for by assuming that the zygote constitution was XXY and that 2 non-disjunctional events took place. If the single XYY cell represents a sixth cell line a third disjunctional error must be postulated.

First thoughts suggest that the abnormal events must occur very early in development indeed for a resultant cell line to be detectable in the absence of strongly unequal proliferation; and in the case of mosaics like XO/XXX there is an immediate presumption that non-disjunction had occurred at the first cleavage division. This inference of abnormality during one of the very first divisions of the cleaving egg becomes open to question when it is remembered that in most cases the evidence of mosaicism is provided by a single and possibly unrepresentative group of cells—the lymphocytes that respond in culture to stimulation by phytohaemagglutinin. Even when supporting chromosomal evidence is drawn from other sources, such as dermal fibroblasts and bone-marrow, the situation is not fundamentally changed, since the cells concerned are all derived from a single one of the embryonic germ layers, the mesoderm. All that can be said, therefore, is that the abnormal event must have occurred either in the single cell that served as progenitor of all mesodermal cells, or in one of its near progeny.

Whatever the exact point in development at which a variant cell line originates, other factors are also likely to influence its frequency in any given tissue or organ. There is first the partition of cells between the future trophoblast and the inner cell mass of the blastocyst. Second, there is the diversion of an unknown proportion of cells from the one-time inner cell mass to contribute to the development of the foetal membranes. Third, there is the partition of cells between sites during organogeny. And, continuing over the whole period, there are possible influences of cell migration and differential proliferation.

Mosaics with one or more cell lines containing a structurally abnormal X chromosome are common among subjects with Turner's syndrome (ovarian dysgenesis). Combining the results of four extensive series, 28 of 163 cases were found to be structural mosaics of various kinds (de la Chapelle, 1962; Lindsten, 1963; Court-Brown *et al.* 1964; Ferguson-Smith, Alexander, Bowen, Goodman, Kaufmann, Jones & Heller, 1964). The most remarkable features of these, and similar cases reported elsewhere, are that an XO cell line is nearly always present, but only very rarely a normal one. The principal kinds are:

XO/XXqi	(long-arm isochromosome)
XO/XXpi	(short-arm isochromosome)
XO/XXq–	(long-arm deletion)
XO/XXp–	(short-arm deletion)
XO/XXq+	(long-arm extension)
XO/XXr	(ring chromosome).

The almost invariable presence of an XO cell line suggests very strongly that a single abnormality must be responsible for both X monosomy and the structural change. A plausible explanation in the special case of the isochromosome mosaics is provided by the primary assumption of misdivision of the centromere of an X chromosome at meiosis, leading to incorporation of one arm as a telocentric chromosome, first in a gamete nucleus, then in the fusion nucleus of a zygote, followed finally by failure of centromere division at the first cleavage mitosis. This would result in an isochromosome that can only go into one daughter cell, necessarily leaving the other monosomic.

In view of the well-known propensity of ring chromosomes to undergo non-disjunction it is not surprising to find an associated XO cell line and also, as in the case reported by Lindsten (1963), an XXrXr line as well. No explanation can be offered for the presence of the XO line in the remaining three classes.

Further information on X-chromosome mosaics and a discussion of karotype-phenotype relations is given by Ferguson-Smith *et al.* (1964). Mosaics involving structural changes of the Y chromosome are mentioned by Jacobs, p. 94 of this Bulletin.

3. AUTOSOME MOSAICS

Mosaics involving the autosomes have been reported very infrequently by comparison with sex-chromosome mosaics. In three cytogenetic surveys of cases of Down's syndrome only 10 mosaics were identified in a total of 430 cases studied (Chitham & MacIver, 1965; Richards, Stewart, Sylvester & Jasiewicz, 1965; Mikkelsen, 1967). Other types of autosome mosaic appear to be equally rare and reports of only 15 cases were found in a search of the same sources as were used to compile Table II. This apparent difference between the sex chromosomes and the autosomes could be primary in the sense of a greatly reduced likelihood of involvement in disjunctional errors, or secondary through the operation of selective forces on the mixed-cell population, or partly one and partly the other. It is also possible that differences of ascertainment may have had an influence on the numbers of mosaics recorded.

Variation between individual chromosomes in liability to undergo disjunctional errors may be real but must be dismissed from further consideration at present, owing to lack of information. Selective proliferation, however, at least as far as autosomal differences are concerned, may be expected as a consequence of associated differences of cell genotype and has been demonstrated in several cases. The most striking is that of a child born with many of the features of trisomy 17-18. A blood culture set up on the day of birth showed that the child was a 46,XX/47,XX,17+ or 18+ mosaic with approximately 90% of trisomic cells. Five further cultures, however, set up between 10 and 16 months later, contained only normal 46,XX cells; the trisomic cells had apparently all been replaced (La Marche, Heisler & Kronemer, 1967).

A longitudinal study of normal/trisomy-G-mosaic subjects has been reported by Taylor (1968). Cultures of blood from 8 children were studied over periods up to 4 years. In 3 of the children there were pronounced increases in the proportions of normal cells, in one there was a progressive but relatively small increase of normal cells from an initially high proportion (86%), in two the proportions fluctuated and no real trend was apparent, and, surprisingly, in two cases there was an apparent increase of trisomic cells.

These and other similar results raise the possibility that a karyotypically normal cell might be produced by a disjunctional error in a non-mosaic subject with an unbalanced karyotype, and, by selective proliferation, give rise to a cell line of detectable magnitude. This may be the explanation of observations reported by Thompson, Melnyk & Hecht (1967). They studied testicular preparations from a male with an apparently non-mosaic 47,XYY karyotype, as determined by blood culture. Although they examined 22 spermatogonial and 155 spermatocytal metaphases they were unable to find a single cell with two Y chromosomes, the great majority being consistent with a normal 46,XY karyotype.

A more extensive consideration of selective proliferation is given by Ford (1964). For other autosome mosaics, see the paper by Polani, p. 81 of this Bulletin.

4. MOSAIC MONOZYGOTIC TWINS

A surprising result of investigations in human cytogenetics has been the discovery of undoubted monozygotic twins with markedly different phenotypes. Several examples are now known, of which the most remarkable are two brother-sister pairs (Turpin, Lejeune, Lafourcade, Chigot & Salmon, 1961; Edwards, Dent & Kahn, 1966). The sister in the first pair was a typical case of Turner's syndrome with a 45,X karyotype; the brother was below average height (165 cm.) but otherwise a phenotypically and karyotypically normal male. The sister in the second pair exhibited several stigmata of Turner's syndrome, but was not a severe case; she was a 45,X/46,XY mosaic. Unexpectedly, cells of the constitution 45,X only were identified in her brother. Since he is of normal physique, though rather short, and has normal testes, there is a very strong presumption that he is also a 45,X/46,XY mosaic, despite the failure to identify cells with the normal male karyotype. The evidence for monozygotic origin in both cases includes complete concordance in respect of blood-group and other genetic markers.

It is appropriate at this point to refer to the striking differences in phenotype that can be presented by single-born 45,XO/46,XY mosaics. They can range from typical cases of Turner's syndrome (Court-Brown *et al.* 1964) to phenotypically normal though infertile males (Kjessler, 1966). Doubtless, this great variation is a consequence of the chances of distribution of cells of the two lines during embryogenesis, particularly into the gonadal rudiments.

Other examples of phenotypically distinct, monozygotic twins are the normal boy and his mongol brother, reported by Lejeune, Lafourcade, Schärer, de Wolff, Salmon, Haines & Turpin (1962), and the twin girls with 45,X/46,XX constitutions, one normal, the other seriously abnormal, reported by Mikkelsen, Frøland & Ellebjerg (1963).

5. CHIMAERAS

It hardly needs to be said that knowledge of human chimaeras has come primarily from blood grouping, later supported by typing of other genetic markers. However, in many cases chromosome investigations have provided confirmation of conclusions already drawn and an independent estimate of the proportions of the different cell lines in the blood. Although this has applied as yet only to 46,XX/46,XY combinations, the segregation of an autosomal marker feature could well provide useful information in some combinations with identical sex chromosomes.

It has already been said that two groups of human chimaeras have been distinguished, one originating through two separate acts of syngamy, the other from the association of cells ultimately derived from two independent zygotes. The two groups can themselves be subdivided and a classification is given in Table III embodying 9 types, including 3 that are theoretically possible but as yet unsupported by published evidence.

Four types involve dispermy and lead to consequences that are potentially distinguishable by differences in the partition of genetic markers between the different cell lines. The markers in the two haploid nuclei contributed by the mother may have a partly complementary relationship to one another; those in the two sperm from the father would necessarily be independent. The genetic situation is analysed in Table III. The letters *A, a,* and *B, b* jointly represent alleles at all the marker loci at which the mother is heterozygous: *A, a* those that segregated at the first meiotic division and *B, b* those that segregated at the second meiotic division. If the second meiotic division is suppressed, or if the nucleus of the first polar body takes part (types 1 and 2), the segregation that normally occurs at the second division would take place subsequent to syngamy so that four (type 1), or three (type 2) genetically distinct cell lines would result and, chance exclusion apart, would be represented in the embryo.

Evidence interpreted as favouring dispermy was first provided by Gartler, Waxman & Giblett (1962), who found two distinct populations of red cells in a female child in whom 46,XX and 46,XY cell lines had already been identified by chromosome analysis. The evidence of genetic markers required two contributions from the father but only one from the mother, although she was heterozygous at four loci that contributed information. The probability of identity at all four loci if two

Table III. A classification of chimaeras

Type	Mechanism	Genotypes of maternal contributions*	Number of independent paternal contributions	Number of genetically distinct cell lines	Means of distinction	Example
	Two separate acts of syngamy	(1) (2)				
1	Dispermy: suppression of second meiotic division	$\frac{A\ B}{A\ b}$ $\frac{a\ B}{a\ b}$	2	4	Identification of cell genotypes	Not found
2	Dispermy: participation of ovum nucleus and nucleus of first polar body	$\frac{A\ B}{A\ b}$ aB or ab or $\frac{a\ B}{a\ b}$ AB or Ab	2	3	Identification of cell genotypes	Not found
3	Dispermy: participation of ovum nucleus and nucleus of second polar body	$A\ B$ Ab or $a\ B$ ab	2	2	Identification of cell genotypes	See Zuelzer, Beattie & Reisman (1964)
4	Dispermy: participation of two haploid nuclei, daughters of the ovum nucleus	Identical	2	2	Identification of cell genotypes	See Gartler, Waxman & Giblett (1962)
5	Fusion of one daughter of zygote nucleus with nucleus of second polar body	$A\ B$ Ab or $a\ B$ ab	1	2	Chromosome counts: 2n† and 3n cell lines	Schmid & Vischer (1967)
	Contributions from two independent zygotes					
6	Early fusion of two embryos	Unrelated	2	2	Single birth. Intimate mixture of cell lines, potentially in all tissues	Not found
7	Placental cross-circulation between dizygotic twins	Unrelated	2	2	Dizygotic twin. One cell line effectively confined to organs of lymphomyeloid complex	Woodruff, Fox, Buckton & Jacobs (1962)
8	Maternal-foetal transplacental exchange	Unrelated	2	2	Single birth. Genetic identity of one cell line with mother's cells	See Taylor & Polani (1965)
9	Artificial: transfusion or grafting	Unrelated	2	2	Medical records	Naiman, Punnett, Destiné & Lischner (1966)

*See text, section 5, for explanation of symbols.
†n represents the haploid, or gametic, number of chromosomes.

independent female pronuclei had been involved is one in 16, though it could be less if one or more of the loci were located close to the centromere and the two female nuclei concerned had been the normal pronucleus and the nucleus of the second polar body (type 3). The authors considered that their evidence favoured origin according to type 4, but they pointed out that other mechanisms were not excluded (see below). The child came to medical notice for repair of an enlarged clitoris and was then observed to have one eye brown and one hazel. Laparotomy was performed and revealed a normal ovary on the left, an ovotestis on the right.

A second example came to attention when

two types of cell were detected at a blood-grouping laboratory and subsequent examination of cultured blood revealed a mixture of 46,XX and 46,XY cells (Zuelzer, Beattie & Reisman, 1964). This time the evidence of genetic markers pointed to the participation of two different maternal nuclei as well as two different sperm. The subject is a male of racially mixed origin and his skin is a mixture of lighter and darker areas.

The last case was presumed by the authors to be an example of type 3, though again other possibilities are not excluded. It would require an exceptionally fortunate combination of complementary heterozygosity at marker loci in father and mother to be able to say more about an individual case. Of the four types of dispermic combination listed, only type 4 is capable of formal proof and, since it requires the combination of one theoretically possible, but unreported, abnormality (suppression of the first meiotic division) with another that is very rare, it may be unrealistic to postulate it at all. However, it is possible that a coincidence of rare events may be detected by cytogenetic analysis, as the following example shows.

Examination of a culture of blood from a child with multiple congenital abnormalities revealed two cell lines, one with 48 and the other with 71 chromosomes (Schmid & Vischer, 1967). A combination of autoradiography and direct morphological analysis showed the constitutions of the two lines to be 48,XXYY and 71,XXXYY respectively. The most plausible explanation, unlikely as it may seem, is that an abnormal 25,XYY sperm fertilized a normal 23,X ovum, that the fusion nucleus divided once and that one of the two products fused again with the nucleus of the second polar body. In another child with diploid and triploid cell lines, evidence was obtained from blood grouping that pointed to normal fertilization and the inclusion in the triploid line of an additional, genotypically distinct, maternal haploid genome (Ellis, Marshall, Normand & Penrose, 1963). These two cases are considered to be examples of type 5: other explanations would require the postulation of still more unlikely events.

Of the four ways listed in Table III in which cells derived from two independent zygotes may be combined in a chimaeric subject, two are established, one is uncertain and the fourth remains as a possibility. The last-named (type 6) is the early fusion of two separate embryos to give a single product, a process that can be considered the converse of monozygotic twinning. It has been achieved artificially in the mouse (Tarkowski, 1961; Mintz, 1964) and promises to be of great analytical value (Mintz & Baker, 1967; Mystkowska & Tarkowski, 1968). The experimental production of these "egg-fusion chimaeras" involves the removal of the zona pellucida from 8-cell embryos by mechanical or enzymic means and the subsequent attachment of the groups of naked blastomeres in pairs. It may be questioned whether an analogous process could occur in vivo, but on present genetic evidence it cannot be excluded as a formal possibility. Although the two cell lines would be different in both hypothetical origin and genetic relationship from those of chimaeras of types 3 and 4, distinction by the segregation pattern of a limited number of genetic markers would not be possible in the individual case. This means that genetic evidence of two paternal contributions is not a decisive indication of dispermy in the strict sense.

The type of chimaerism at once best known and most securely established is that attributed to exchange of cells between dizygotic twins in utero through placental vascular connexions (type 7). Chimaeras of this kind were first identified by Owen (1945) in his classic investigation of blood groups in cattle twins and are now known to occur in a number of other mammalian species. Race & Sanger (1968) list eight human pairs. Blood cultures from three heterosexual pairs have been examined and the proportions of 46,XX and 46,XY cells estimated. In one pair there was a close similarity of proportions between sister and brother; in a second pair there was a clear difference, although 46,XY cells were dominant in both; in the third pair 46,XX cells were overwhelmingly dominant in the sister, 46,XY cells overwhelmingly dominant in the brother. Although there is no means of cross-identifying red-cell types with chromosomal types, there is a very good presumptive correlation between the two sets of data for each pair of twins. This suggests that there may be a relationship between the progenitors of red blood cells and the progenitors of the lymphocytes that respond in culture to the action of phytohaemagglutinin.

Taylor & Polani (1965) reported that they had examined the chromosomes in a blood culture from a 12-week aborted male foetus and had found 10% of 46,XX cells, although there was no sign of a twin. A similar proportion of 46,XX cells was stated to have been found also in the thymus and the suggestion was made that the 46,XX cells had originated from the mother by transplacental bleeding. The case is therefore a possible example of type 8. This interpretation may get some support from the fact that of the mother's eleven previous pregnancies, seven had terminated in spontaneous abortion; this suggests that normal uterine function had been disturbed. However, chimaerism of any one of types 1-6 is not excluded. Evidence of genetic identity of one cell line with the mother's cells will be necessary for demonstration of type-8 chimaerism.

Evidence for chimaerism of type 9 needs no labouring. The example chosen is a case where intra-uterine transfusion of blood to a foetus was followed, after birth, by the identification of cells in cultured blood derived from a donor. Cells of the donor's type were also found in thymic tissue obtained at necropsy (Naiman, Punnett, Destiné & Lischner, 1966). It is of particular interest that identification of the two cell lines was achieved through differences in the size of the Y chromosome. The child was a male. His father and two paternal uncles had an exceptionally long Y chromosome, a male donor had a very short Y chromosome, and cells with both these unusual Y chromosomes were found in the child. The child showed signs of a graft-versus-host reaction and died at the age of three months.

REFERENCES

Anderson, D., Billingham, R. E., Lampkin, G. H. & Medawar, P. B. (1951) Heredity, Lond. **5**:379.

Chapelle, A. de la (1962) Acta endocr., Copenh. **40,** Suppl. No. 65.

Chitham, R. G. & MacIver, E. (1965) Ann. hum. Genet. **28**:309.

Court-Brown, W. M. (1967) Human population cytogenetics. North-Holland, Amsterdam.

Court-Brown, W. M., Buckton, K. E., Jacobs, P. A., Tough, I. M., Kuenssberg, E. V. & Knox, J. D. E. (1966) Chromosome studies on adults. (Eugenics Laboratory Memoirs, No. 42.) Cambridge University Press, London.

Court-Brown, W. M., Harnden, D. G., Jacobs, P. A., Maclean, N. & Mantle, D. J. (1964) Spec. Rep. Ser. med. Res. Coun. No. 305.

Court-Brown, W. M., Jacobs, P. A. & Doll, R. (1960) Lancet **1**:160.

Edwards, J. H., Dent, T. & Kahn, J. (1966) J. med. Genet. **3**:117.

Ellis, J. R., Marshall, R., Normand, I. C. S. & Penrose, L. S. (1963) Nature, Lond. **198**:411.

Ferguson-Smith, M. A., Alexander, D. S., Bowen, P., Goodman, R. M., Kaufmann, B. N., Jones, H. W., Jr. & Heller, R. H. (1964) Cytogenetics **3**:355.

Ford, C. E. (1964) In: Harris, R. J. C., ed. Cytogenetics of cells in culture, p. 27. (Symposia of the International Society for Cell Biology, vol. 3.) Academic Press, New York.

Gartler, S. M., Waxman, S. H. & Giblett, E. (1962) Proc. natn. Acad. Sci. U.S.A. **48**:332.

Harnden, D. G. (1964) In: Second International Conference on Congenital Malformations, New York City, July 14-19, 1963, p. 43. International Medical Congress, New York.

Kjessler, B. (1966) Karyotype, meiosis and spermatogenesis in a sample of men attending an infertility clinic. Karger, Basel.

La Marche, P. H., Heisler, A. B. & Kronemer, N. S. (1967) Rhode Isl. med. J. **50**:184.

Lejeune, J., Lafourcade, J., Schärer, K., Wolff, E. de, Salmon, C., Haines, M. & Turpin, R. (1962) C. r. hebd. Acad. Séanc. Sci., Paris **254**:4404.

Lindsten, J. (1963) The nature and origin of X chromosome aberrations in Turner's syndrome. Almqvist & Wiksell, Stockholm.

Mikkelsen, M. (1967) Ann. hum. Genet. **31**:51.

Mikkelsen, M., Frøland, A. & Ellebjerg, J. (1963) Cytogenetics **2**:86.

Mintz, B. (1964) J. exp. Zool. **157**:273.

Mintz, B. & Baker, W. W. (1967) Proc. natn. Acad. Sci. U.S.A. **58**:592.

Mystkowska, E. T. & Tarkowski, A. K. (1968) J. Embryol. exp. Morph. **20**:33.

Naiman, J. L., Punnett, H. H., Destiné, M. L. & Lischner, H. W. (1966) Lancet **2**:590 [Letter].

Owen, R. D. (1945) Science, N.Y. **102**:400.

Race, R. R. & Sanger, R. (1968) Blood groups in man, 5th ed. Blackwell Scientific Publications, Oxford.

Ricci, N., Dallapiccola, B., Ventimiglia, B., Tiepolo, L. & Fraccaro, M. (1968) Cytogenetics **7**:249.

Richards, B. W., Stewart, A., Sylvester, P. E. & Jasiewicz, V. (1965) J. ment. Defic. Res. **9**:245.

Schmid, W. & Vischer, D. (1967) Cytogenetics **6**:145.

Tarkowski, A. K. (1961) Nature, Lond. **190**:857.

Taylor, A. I. (1968) Nature, Lond. **219**:1028.

Taylor, A. I. & Polani, P. E. (1965) Lancet **1**:1226 [Letter].

Thompson, H., Melnyk, J. & Hecht, F. (1967) Lancet **2**:831 [Letter].

Turpin, R., Lejeune, J., Lafourcade, J., Chigot, P. L. & Salmon, C. (1961) C. r. hebd. Séanc. Acad. Sci., Paris **252**:2945.

Wahrman, J. & Gersht, N. (1966) Israel J. med. Sci. **2**:123.

Woodruff, M. F. A., Fox, M., Buckton, K. A. & Jacobs, P. A. (1962) Lancet **1**:192.

Zuelzer, W. W., Beattie, K. M. & Reisman, L. E. (1964) Am. J. hum. Genet. **16**:38.

34 Finger-prints, palms and chromosomes

Prof. L. S. Penrose*
Galton Laboratory
University College
London, England

HISTORICAL NOTES

The analysis of finger-prints from the point of view of personal identification is well known. Early examples can be taken from Chinese documents. In modern times the method is used by police and immigration officials. The principle involved is that the ridges formed by the raised apertures of sweat glands have unique detailed formations which have arisen almost by pure chance. The large-scale pattern of ridges, however, is much less fortuitous in origin. There are anatomical regularities[1]. There are also normal variations, which represent hereditary differences between members of separate populations, members of the same population or members of the same family.

The contrast between general pattern and detail is made fairly clear by the study of identical twins. The general configurations of ridges agree, as hereditary traits should, on the hands of such pairs, and there are usually only slight differences in arrangement; but the details do not agree and are individually specific.

All configurations are laid down permanently at a very early period of development, during the third month of foetal life. Best known are those on the finger-tips.

DESCRIPTION OF TRIRADII, MAIN LINES AND CREASES

The systematic classification of ridge patterns, which can be used as a preliminary in personal identification, was first proposed by Galton[2]. It depended on the recognition of 'deltas', now known as 'triradii'. These are natural centres, and the radiant ridges form the main lines of the patterns or configurations.

A triradius can be defined as the meeting point of three spokes which demarcate three regions, each containing a system of almost parallel ridges. One triradius on a finger-tip always accompanies a loop pattern (Fig. 1) and two produce a whorl. In the absence of any triradius, the pattern is known as an open field, arc or arch.

On the palm there are normally four triradii, one at the base of each finger, called *a, b, c* and *d,* and another, known as *t* near the base of the fourth metacarpal bone or at some point on its axis.

Fig. 1. Loop pattern on a finger: the arrows mark continuations of the three ridge lines which radiate from the delta or triradius. The white line joining the delta to the pattern core crosses 13 ridges.

From Nature **197**:933-938, 1963. Used with permission.

Substance of a lecture delivered at University College, London, on January 15.

I thank a great number of colleagues who have helped me to obtain material for dermatoglyphic studies. In particular I thank Prof. P. E. Polani, Dr. J. Lindsten and Dr. M. Fraccaro for many hand-prints from cases of Turner's syndrome, Dr. J. H. Edwards, Dr. C. E. Blank, Dr. W. J. Mellman, Dr. O. J. Miller and Dr. A. Shapiro for obtaining prints from patients with other chromosomal aberrations, Dr. Sarah B. Holt for making available to me the results of her measurements, and Mr. A. J. Lee for his drawings.

*Present address: Kennedy-Galton Centre, Harperbury Hospital, St. Albans, England.

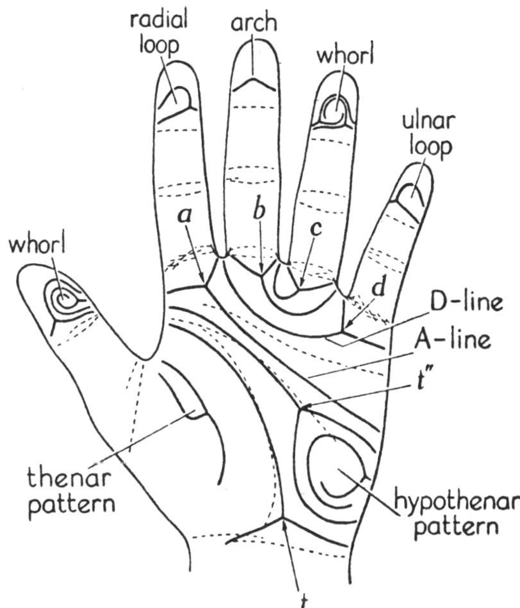

Fig. 2. Dermatoglyphics on palms and finger-tips showing nomenclature.

By starting at the triradii, the configurations of the main lines of the dermal ridges can be marked out (Fig. 2). The patterns produced on different hands fall into natural groups and the most prominent lines are named, for example, *A* and *D*.

At this point it may be well to emphasize that the dermal ridge main lines are of an entirely different character from the flexion creases which have such an attraction for fortune-tellers. The flexion lines on the palm, traditionally named after the head, the heart and life, and the creases on fingers certainly show interesting variations, but the ancient interpretations have to be revised.

PHYSIOLOGICAL VARIATIONS

These large-scale patterns of lines and creases are all strongly, but not exclusively, influenced by heredity. On the corresponding hands of identical twins, dissimilarities can be attributed to environmental modifications; they differ about as much from one another as do normally the left and right hands of a single person.

Just as the sexes differ in stature and other physical traits, so also are there characteristic peculiarities of dermal ridges to be found in males and females by careful analysis. All these variations can be considered to be physiological, like the fact that female stature is less than male stature; their presence sometimes obscures pathological changes in pattern.

TESTS OF INHERITANCE OF DERMATOGLYPHIC TRAITS

Analytical investigations of hereditary influence can best be made by quantitative methods which involve counting ridges between specified points or measuring angles. One well-established result is that obtained by measuring the size or 'intensity' of the finger-tip patterns. The measurement is made by counting the ridges crossed by a straight line drawn from the triradius to the core of the pattern[3]. An arch scores zero and a loop 12, on the average: usually whorls score still higher, about 19 (only the larger of the two numbers obtained in a whorl is recorded). The total score for all ten fingers averages about 145 in males and 127 in females. It is a trait which, like stature, is determined by many genes with a combined additive action. This is demonstrated by analysis of families, in the traditional manner, using correlation coefficients to measure hereditary likeness. The characteristic value, in a random population, for parents and their children and for sib pairs is 0.5; and this is very closely approached by observations on the total finger-ridge count[4]. Equality of inheritance from father and mother to sons and daughters excludes sex-chromosomal influence.

Other traits, such as the position of *D*-line exit, the third interdigital pattern[5], the number of ridges between *a* and *b,* the presence or absence of thenar[6] patterns and the maximal *atd* angle[7], can all be shown to have strong genetical determination; but these features are more capricious than the total finger pattern quantity and more influenced by early foetal environment. Some of them, like angular measurements, can alter with age.

DISTORTIONS CAUSED BY GROWTH ABNORMALITIES

Any early disturbance of growth which affects the hands or the feet is liable to cause distortions of the natural dermatoglyphic patterns. The classical example is polydactyly, in which an extra digit is often associated with an extra triradius. Conversely, octrodactyly[8], or loss of digits, and also syndactyly, fusion of digits, leads to loss of triradii or alterations of arrangements which may be very extensive (Fig. 3). These distortions can be correctly said to be

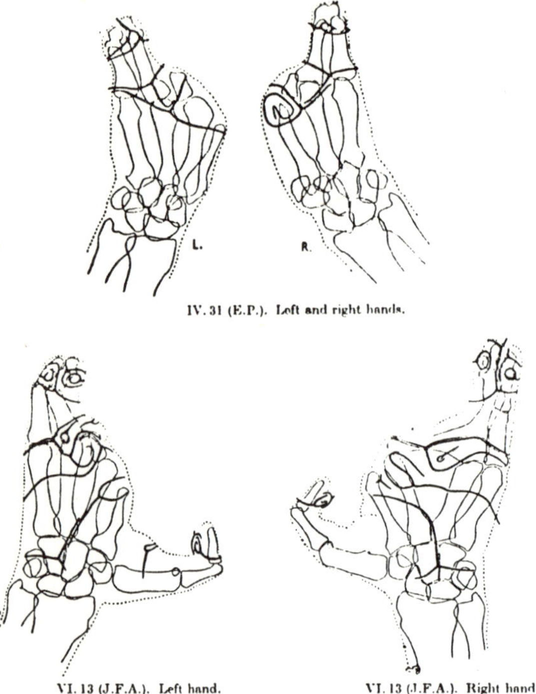

Fig. 3. Dermatoglyphic distortion caused by a single gene for ectrodactyly.

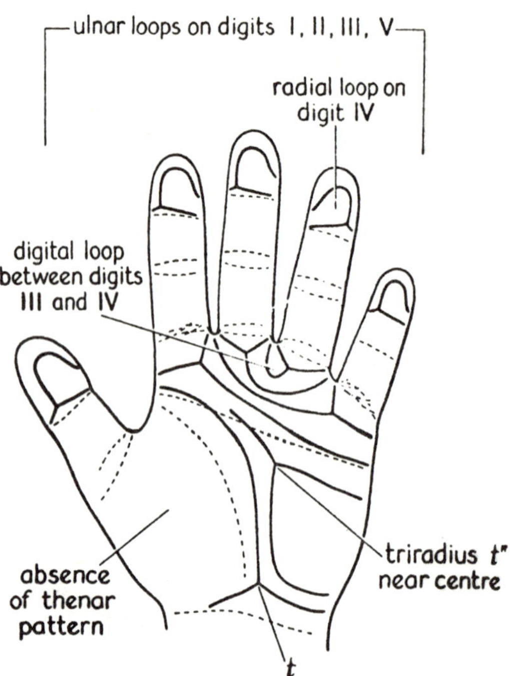

Fig. 4. Stereotype pattern found in mongolism.

caused by abnormal genes, but they are secondary effects; the underlying developmental disturbances are the more direct results of abnormal gene action. Some distortions are quite obscure in origin, such as the very rare vertical alignment found in some palms and usually connected with digital deformity.

DISTORTIONS CAUSED BY CHROMOSOMAL ABERRATIONS

Many curious and unexpected distortions of pattern on palms and soles occur in connexion with chromosomal aberrations. The common types indeed present highly characteristic peculiarities. I will describe them in turn.

(1) *Mongolism.* This condition, in which there is a multiplicity of anomalies, was first recognized nearly a hundred years ago by Down. It has, of course, nothing to do with Mongolian peoples specifically. Recently its origin in chromosomal aberration has been proved. The hands were first examined dermatoglyphically by Cummins[9] in 1936. Before this the flexion creases had interested clinicians, and the transverse palmar crease, or simian line, played an important part in diagnosis. According to the students of cheiromancy, this crease is equivalent to the fusions of the lines of the head and the heart; it is a sign of concentrative mental power. Other alterations in flexion lines of mongols include a marked crease between first and second toes and occasional absence of the distal creases on the fifth fingers[10]: these have passed unnoticed by cheiromancers.

On the hands, one noticeable feature of the mongol stereotype is the strong tendency for every finger to possess a loop pattern rather than a whorl or an arch (Fig. 4). Moreover, the distribution of radial loops on digits IV and V in mongolism is unusual as compared with normal hands. The total finger-ridge count is less variable than among normal people[11]. Very characteristic is the central position, on the palm, of the triradius t which sub-tends an angle of about 80° with the points a and d (and in this position it is called t'') instead of the normal 45°. Hypothenar patterns are common and thenar patterns rare. A distal loop is nearly always present between b and c (the third interdigital region). The ridges themselves are often poorly formed.

On the hallucal area of the sole, patterns are infrequent and an open field is typical[12].

One of the recent important discoveries in human cytology is that more than one genetical cell type can take part in the formation of the

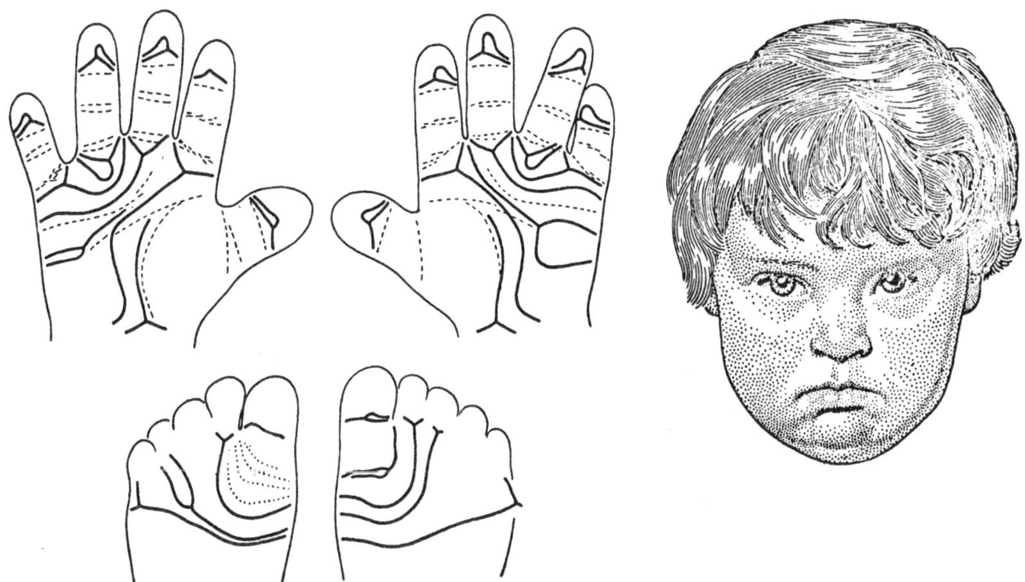

Fig. 5. Main lines on prints of palms and soles from a mosaic mongol female child whose cells are about 1/3 trisomic and 2/3 normal. Note the open field on left hallucal area (scale for hands and feet doubled).

same individual. A person in whom this occurs is called a mosaic. A person can be a partial mongol in that only a proportion of his cells are abnormal. These abnormal cells are usually distributed fairly evenly over the whole body and thoroughly intermingled. It has, for many years, been observed that some people exist who have a few characteristics of mongolism but that they are otherwise perfectly normal. There are also mentally sub-normal patients who are incomplete mongols. Some of these people are in fact mongol-normal mosaics in varying degrees. Their dermatoglyphs are interesting in that they show some of the features of the mongol stereotype (Fig. 5)[13]. The degree of distortion is not closely related to the proportion of abnormal cells, as demonstrated in cultures, nor to the level of mental development attained.

(2) *Trisomy 13.* An entirely different group of peculiarities characterizes other trisomic conditions. Those connected with trisomy of chromosome No. 13 (it may be No. 14 or No. 15—we are not quite sure) are striking. Deformities of the eyes, lips, palate and ears are part of the clinical picture. Polydactyly may be present and this itself is likely to upset the dermatoglyphic pattern. The stereotype includes a very distal axial triradius, the position of which, as measured by the angle *atd,* is distorted more than it is in mongolism and could sometimes be called t''' (Fig. 6). The extra

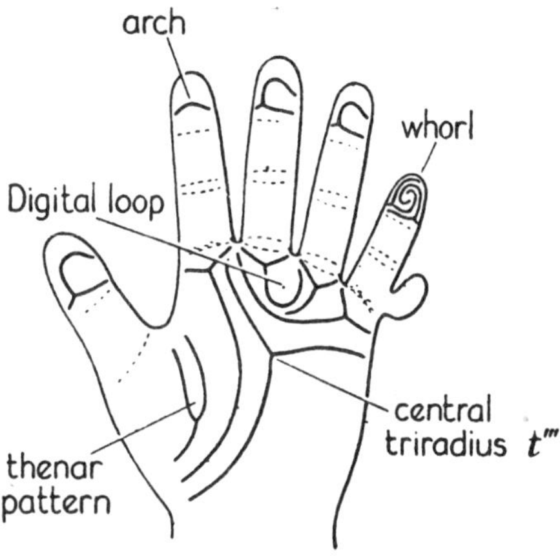

Fig. 6. Palmar pattern stereotype found in trisomy 13.

patterns, however, tend to be in the thenar, not the hypothenar, region. Unlike the mongols, their fingers show both arches and whorls as well as loops (Fig. 7). On the soles of the feet there are few triradii and peculiar loops or open fields are the rule[14]. These observations are obtained with considerable difficulty since most of these cases fail to survive more than a few weeks and the hand-printing even of normal new-born infants is far from easy.

(3) *Trisomy 17 or 18.* The third kind of well-established autosomal aberration concerns trisomy of either chromosome No. 17 or No.

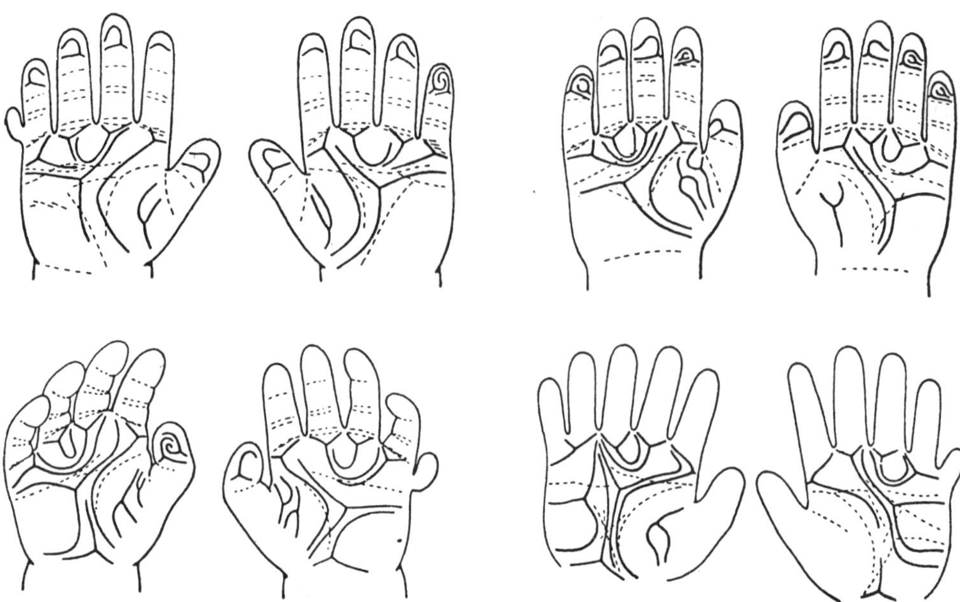

Fig. 7. Pairs of hands in cases of trisomy 13 (scale doubled).

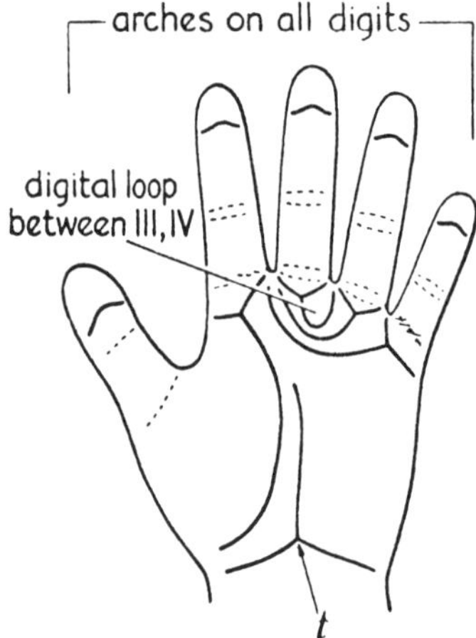

Fig. 8. Pattern stereotype found in trisomy 17 or 18.

18, it is uncertain which. The head is long, ears are 'low set' and the chin is small. Again there is a stereotyped distortion of dermal ridge patterns along with a group of physical peculiarities (Fig. 8). The hands are specially difficult to examine because the fingers are held permanently flexed but with the distal joint extended. There is absence of the distal flexion creases on most fingers and the tips only rarely have any triradii on them and nearly all the patterns are arches[14]. In contrast to the mongols and trisomics No. 13, the palms show no obvious abnormality. The hallucal areas of the soles, however, have somewhat unusual configurations with fairly wide open fields and there are arches on the toes.

(4) *Turner and Klinefelter types.* When we come to sex chromosomes there is a wealth of material for study, and some surprising results have been derived. First of all, Turner's syndrome, in which there are only 45 chromosomes in each cell nucleus because there is one X and no Y present, shows various peculiarities. The patients are females with short stature and are liable to have webbing of the neck and slight limb deformities. There is no very stereotyped dermatoglyphic pattern but a general tendency to certain types of arrangement (Fig. 9). The pattern intensity on the finger-tips is high, large whorls are common and the total ridge number is, on the average, greater than normal. On the palms, the axial triradius is found nearer the centre than is usual in controls: on the average it is in the position t' (Fig. 10). Thenar patterns are not uncommon. The upper region of the palm is often peculiar in that triradius b tends to move towards the ulnar side of the hand. Correspondingly, the distance between a and b and the mean number of ridges between them is increased.

Mosaics composed of Turner and normal cells show the same features as in Turner's syndrome and, also, other conditions, where one of the X chromosomes in a female is grossly aberrant morphologically as, for example, when it forms a ring and has suffered deletion.

One might suppose that a person who has too many X chromosomes, instead of too few, would show the opposite characteristics of the Turner's syndrome. The triple-X females, however, show very little deviation from the normal dermatoglyphs. The tendency for high ridge scores to occur on finger-tips in a Turner is replaced by a normal, and perhaps a slightly lowered, count. Males, with additional X chromosomes giving rise to the syndrome of Klinefelter, are tall rather than dwarfed and they have, in association, rather long hands and feet. Finger-prints show a slight general tendency towards patterns with low counts and arches are quite frequently observed. The axial triradius, on the average, is low down in the palm, probably a little lower than in normal hands, especially when there are more than two X chromosomes[15], as occasionally happens (Fig. 11); there is, however, no obvious stereotype connected with the possession of too many X chromosomes. There is insufficient information to say what effect is produced by having too many Y chromosomes.

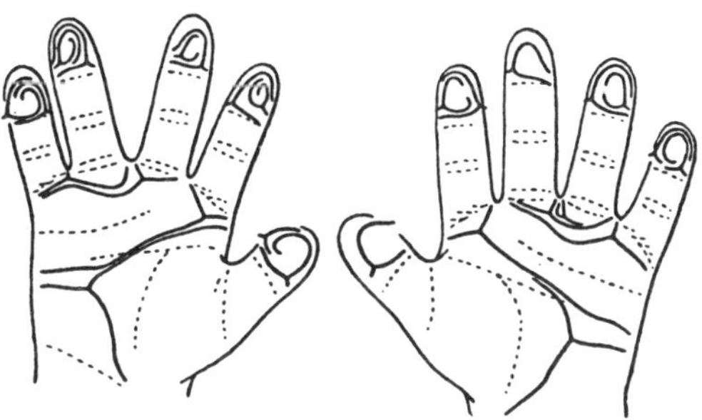

Fig. 9. Typical dermatoglyphics in Turner's syndrome.

(5) *Miscellaneous anomalies.* Many miscellaneous examples are known where chromosomal aberrations have been found and where enthusiasts have printed the hands and feet in the hope of finding characteristic variations. Sometimes observers are lucky and sometimes not. The key factor seems to be to what extent the chromosomal change induces a disorder of

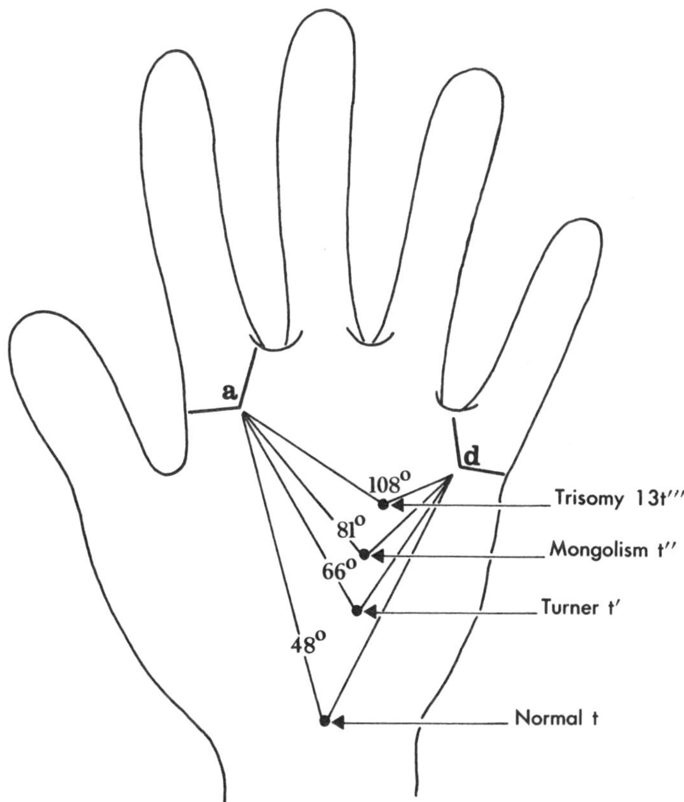

Fig. 10. Mean position of most distal triradius *t* in children up to 4 years of age.

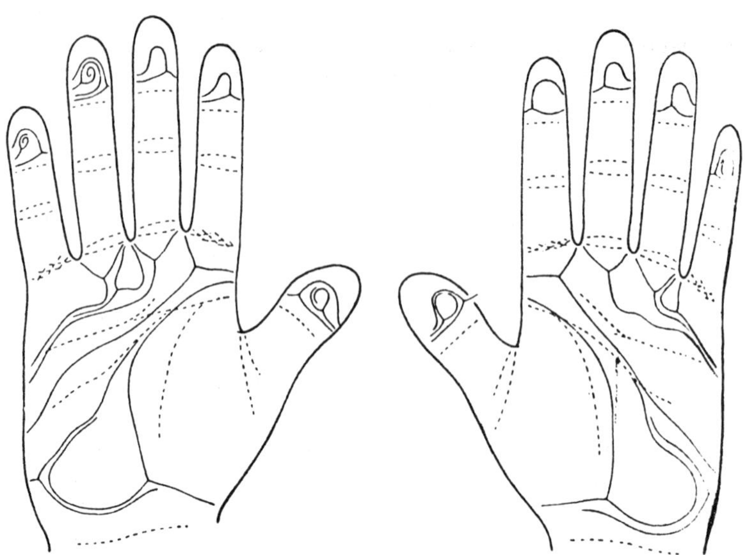

Fig. 11. Example of dermatoglyphics in a male with *XXXXY* sex chromosomes.

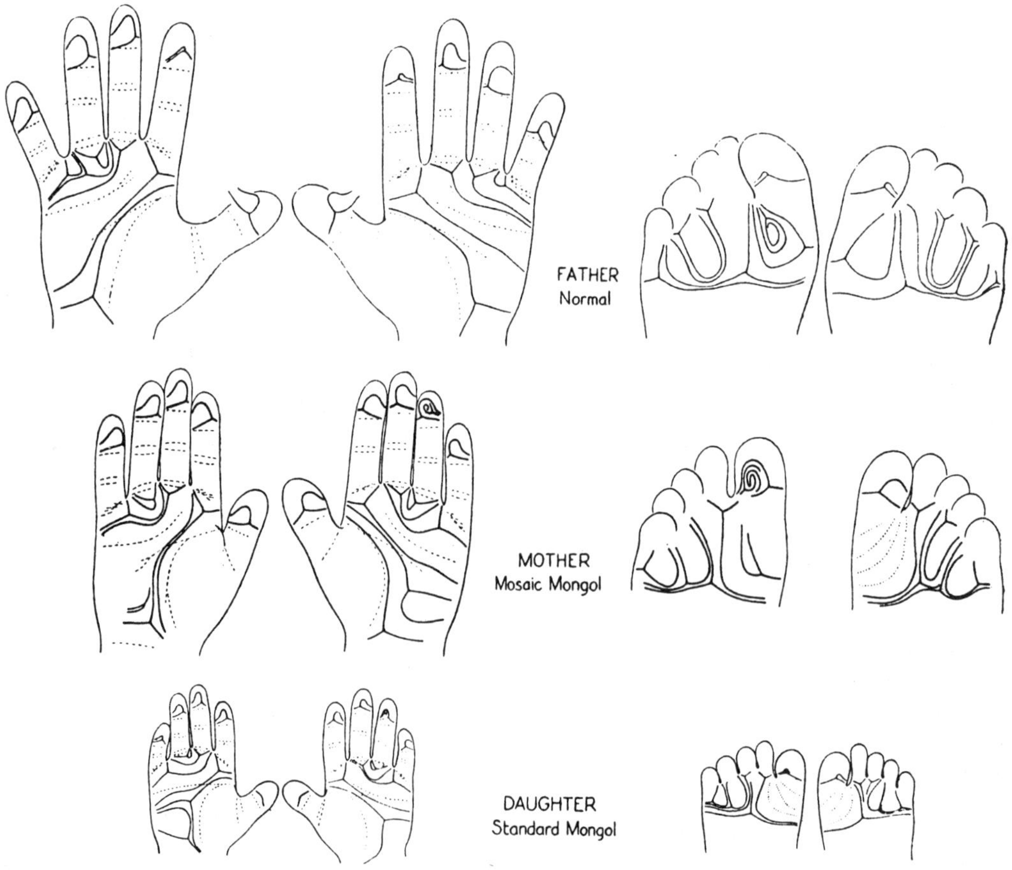

FATHER
Normal

MOTHER
Mosaic Mongol

DAUGHTER
Standard Mongol

Fig. 12. Palms and soles in a family with mosaic mongol mother and standard mongol daughters.

limb growth for, as we have seen, any such change is likely to distort the normal patterns. Since many of the aberrations observed are unique, the evaluation of the findings is difficult. Usually the most useful information obtained is that the patient's dermal ridge patterns do, or do not, resemble a known stereotype associated with known trisomics. The observation may then be used to aid the cytologist in sorting out which chromosome is involved in a particular aberration. For example, if the hands of a patient trisomic for a small chromosome do not show any traits of mongolism, there is a strong likelihood that the chromosome concerned is not the usual one, for example, No. 22, not No. 21.

GENETICAL STUDIES ON DISTORTIONS

Somewhat more definite results can be obtained when groups of close relatives are studied. The correspondence between a chromosomal aberration and a dermatoglyphic pattern can be followed through more than one generation. Here are some examples.

(1) *Mongolism with mosaicism.* It was possible, over a period of years, to examine the inheritance of the position of the axial triradius in relation to mongolism before anything could be known about its chromosomal origin. It became clear that mothers of mongols, especially if they had more than one affected child, were more likely than control females to have centrally placed axial triradii[7]. No explanation then seemed satisfactory. However, if the mother should, as is now known to happen, suffer from an aberration involving chromosome No. 21, this could both tend to cause mongolism in her children and produce some features of the mongol stereotype on her own hands and feet (Fig. 12). The best example of this occurs when the mother is herself a mongol or a mosaic mongol[16].

(2) *Mongolism with translocation.* Another type of situation occurs when the mother transmits what is called a translocated chromosome, caused by breakage followed by the rejoining of the wrong ends. In one section of a certain family, every individual who carried an abnormal chromosome after it had arisen by translocation also had a central axial triradius like the two mongols who occurred among its members[17].

(3) *Other unbalanced-autosomal complements.* Equally striking are some cases of abnormal children with aberrations of unusual kinds affecting the larger chromosomes and whose hands showed remarkable dermatoglyphic patterns which included strong thenar configurations[18]. Unbalanced in chromosomal complement they showed, as in mongolism, evidence of growth disturbances and mental retardation. In two families the father, though physically and mentally normal, carried the same translocation as the abnormal children in a compensated form; in each case, his palms exhibited less marked but similar peculiarities to those of his offspring.

CONCLUSION: INFORMATION ABOUT LOCATION OF GENES FROM DERMATOGLYPHS

It would be natural to infer, from observations like those on the families just described, that the gene for this or that dermatoglyphic pattern was located on the aberrant chromosome concerned. One must, however, be wary of all such inferences unless supported by family investigation. Some people have rashly supposed that, because Turner patients are short and Klinefelters are tall, the genes for stature must occupy loci on the *X* chromosome. Previous observations on the inheritance of stature, however, have shown it to be autosomal, not sex-linked. Exactly the same reasoning precludes us from attributing the observed changes in finger-ridge and *a-b* ridge counts in Turner's syndrome to the *X* chromosome genes. Similarly, the stereotypic peculiarities in trisomics, like triple-21 (mongolism) and triple-13, must not be attributed perfunctorily to genes located on No. 21 or No. 13, respectively. The problem of gene location involves extensive long-term investigations. In the meantime let us take pleasure in the fact that even the obscure study of dermatoglyphic patterns is, in spite of its inherent difficulties, at last beginning to play a significant part in the ever-widening field of human genetics.

REFERENCES

1. Cummins, H., and Midlo, C., Finger Prints, Palms and Soles (Philadelphia: The Blakiston Co., 1943).
2. Galton, F., Finger Prints (London, Macmillan and Co., 1892).
3. Bonnevie, K., J. Genet. **15:**1 (1924).
4. Holt, S. B., Brit. Med. Bull. **17:**247 (1961).
5. Fang, T. C., J. Ment. Sci. **96:**780 (1950).
6. Weninger, M., Mitt. anthrop. Ges. Wien **55:**182 (1935).
7. Penrose, L. S., Ann. Hum. Genet. (Lond.) **19:**10 (1954).
8. MacKenzie, H. J., and Penrose, L. S., Ann. Eugen. (Lond.) **16:**88 (1951).

9. Cummins, H., Anat. Rec. **6**:11 (1936).
10. Penrose, L. S., Lancet **ii**:585 (1931).
11. Holt, S. B., Ann. Eugen. (Lond.) **15**:355 (1950).
12. Ford Walker, N., J. Pediat. **50**:19, 27 (1957).
13. Clarke, C., Edwards, J. H., and Smallpeice, V., Lancet **ii**:1028 (1961).
14. Uchida, I. A., Patau, K., and Smith, D. W., Amer. J. Hum. Genet. **14**:345 (1962).
15. Miller, O. J., Breg, W. R., Schmickel, R. D., and Tretter, W., Lancet **ii**:78 (1961).
16. Blank, C. E., Gemmell, E., Casey, M. D., and Lord, M., Brit. Med. J. **ii**:378 (1962).
17. Penrose, L. S., and Delhanty, J. D. A., Ann. Hum. Genet. (Lond.) **25**:243 (1961).
18. Edwards, J. H., Fraccaro, M., Davies, P., and Young, R. B., Ann. Hum. Genet. (Lond.) **26**:163 (1962).

35 Interspecific hybridization in Nicotiana. II. A tetraploid glutinosa-tabacum hybrid, and experimental verification of Winge's hypothesis

R. E. Clausen†
T. H. Goodspeed†
University of California
Berkeley, California

Numerous investigations have shown that in plants the chromosome numbers of the species of a genus are often in arithmetical progression. Excellent examples are afforded by Triticum, $n = 7$, 14 and 21; by Rosa, $n = 7$, 14, 21, 28; and by Chrysanthemum, $n = 9$, 18, 27, 36 and 45. Appeal has often been made to tetraploidy as a method of increase in chromosome number, but Winge (1917) has pointed out that successive doubling of chromosome number would give rise to geometrical rather than arithmetical series, and has suggested as an alternative hypothesis interspecific hybridization followed by doubling of chromosome number. The process suggested by Winge would establish tetraploid interspecific hybrids having $2\ (n_1 + n_2)$ chromosomes, where n_1 and n_2 represent the haploid numbers of the parent species; and since such forms are essentially homozygous diploids, they may reasonably be expected to be fertile and constant. Winge was unable, however, to present any experimental evidence in support of his hypothesis. In this article a preliminary account is given of a tetraploid hybrid of *Nicotiana glutinosa* and *N. tabacum,* and attention is invited to *Primula kewensis,* apparently a tetraploid *P. floribunda-verticillata* hybrid, which together offer the necessary experimental verification of the hypothesis.

The two species of Nicotiana employed in the present investigations are very distinctly different. *N. glutinosa* is particularly characterized by its villose pubescence; its distinctly petioled, cordate leaves; its bilabiate flowers; and its sparsely branched, racemous inflorescence. It is one of the most distinct species of the genus. *N. tabacum* is familiar to almost everyone as the tobacco of commerce. Descriptions and more complete illustrations of the forms employed in the experiments described below may be found in Setchell's (1912) account of the genus. The chromosome number of *N. glutinosa* is $n = 12$ (Goodspeed 1923), of *N. tabacum,* $n = 24$ (White 1913, Goodspeed 1923).

The two species have frequently been crossed. Reciprocal hybrids may be obtained, although hybridization is attended with some difficulty. Usually only a few viable seeds are produced in a capsule; in our experience an average of about ten or twelve. The F_1 hybrids are weak in germination and development, but they grow on to maturity. This, in brief, is the behavior which has been noted for all varieties of *N. tabacum* which have been tested, except one, the variety "Cuba" (cf. Goodspeed 1915). From "Cuba" ♀ X *glutinosa* ♂ full capsules of seed are obtained. These seeds are of the same order of viability as pure seed of the species; the seedlings are vigorous; the hybrid plants develop to a height approximately equal to that of "Cuba"; and they branch profusely. Gärtner (1849) apparently also observed marked differences in the vigor of F_1 *glutinosa-tabacum* hybrids, when different *tabacum* varieties were employed; but as Focke (1881) points out, it is difficult to know how to judge these results because of the numerous discrepancies in Gärtner's account of his observations. Despite these differences in vigor, the F_1 hybrids are always intermediate in appearance and they are apparently completely sterile. Numerous attempts to secure seed by backcrossing to the parental species under a variety of conditions have failed, and no seed has been found in open-pollinated capsules.

In 1922 from a single capsule of *glutinosa* ♀ X *tabacum* var. *purpurea* ♂, three plants were secured, which were grown under the garden number, 22062. Two of these plants were obvious hybrids. They were both small plants, about two feet in height, with few, slender branches and small leaves. The flowers exhibited a strong tendency towards the bilabiate shape of *N. glutinosa,* but the color was

From Genetics **10**:278-284, 1925. Used with permission.

†Deceased.

carmine, like that of *purpurea.* The leaves showed distinct evidences of *glutinosa* in their cordate shape. One of these plants was partially fertile, the other completely sterile. No other differences were noticed at that time. The third plant was very strikingly different from the other two; in fact, if our notes and memory may be depended upon, it was identical with the *purpurea* haploid which was obtained later (cf. Clausen and Mann 1923). Unfortunately, we lost it during the winter of 1922-1923 because of unfavorable greenhouse conditions.

A number of flowers on the single partially fertile plant were hand-pollinated and gave selfed seed without difficulty. The capsules were harvested separately. In the season of 1923, 155 plants were obtained from one of these capsules. These, however, were set out in the field late in the season, and they did not mature. In 1924 a culture of 65 plants was grown under the garden number, 24123. For purposes of comparison there was available at the same time a culture, 24192, of 15 F_1 plants of *purpurea* ♀ X *glutinosa* ♂. Much to our surprise, with one exception, the 65 F_2 plants of 24123 were uniform and almost identical with the F_1 plants of 24192. There were, however, important minor differences, which were found constantly to characterize the two populations. The F_1 plants were completely sterile. They set no seed on open-pollination, and twenty-five attempts at back-pollination with each of the parental species failed to give seed. Under these circumstances, capsules were retained for as long as three weeks, but in no case did they reach maturity. The plants of the F_2 population were reasonably fertile, and uniform in this respect. Large plump capsules were obtained from open-pollinated and hand-pollinated flowers and also from crosses with the parent species. These capsules contained a fair quantity of seed, but not so much as capsules of normal species.

The plants of the two populations exhibited a close correspondence in morphological characters. Both populations were very uniform. The F_2 plants had slightly, but constantly, larger flowers than F_1 plants, and the anthers were conspicuously larger. F_2 plants produced abundant pollen, most of the grains of which were normal in appearance: F_1 plants produced scanty pollen, consisting entirely of shrivelled empty grains. F_1 plants averaged about two feet in height, F_2 plants about a foot and a half. Despite the difference in height, which was probably due to their unfavorable start in the flats, the general impression given by F_2 plants was that of a slight enlargement to scale of characters of F_1, aside from those features obviously connected with the difference in fertility.

One F_2 plant stood out from the rest by reason of its remarkable robustness. This plant eventually attained a height of six feet, and produced numerous stout branches. Despite the difference in size, however, the general morphological characters were those of the other plants of the population on an enlarged scale. Vegetative characters were proprotionately enlarged, flower size only slightly. The plants of this population were all very weak as seedlings, and they grew very feebly during the time they were in flats. It is believed that the general small size of the plants in the population was due to this stunting during their early growth and that 24123P55, the robust individual, merely by some fortunate chance overcame this difficulty.

The uniformity of F_2 and its close resemblance to F_1 immediately suggested the need for cytological examination. Excellent aceto-carmine smears of pollen mother cells were easily secured; the stage of development of anthers containing them in proper condition being rather later than is usually the case. Examinations were made of material from several plants, including the robust plant described above, which gave exactly the same results as the others. The general impression of the cytological figures was one of regularity of meiotic division rather different from the irregular distribution seen in normal F_1 *glutinosa-tabacum* hybrids. Numerous counts of first-metaphase figures showed 36 bivalents. In a few instances it was possible to count both metaphase plates in the second division, and to determine that each contained 36 chromosomes. There was of course some doubt as to the exact count in a number of figures, but only to the extent of one or two chromosomes. There were minor irregularities in distribution, evidenced by precocious splitting, lagging, and microcyte formation; but these features, while noticed, were not studied in detail. There is no doubt that the chromosome number of the plants of this population was uniformly $n = 36$, $2n = 72$.

Pollen-mother-cell heterotypic anaphase conditions in the sterile F_1 *glutinosa-tabacum* hybrid and in the tetraploid hybrid are illustrated in Figures 1 and 2, while Figure 3 shows two homotypic metaphase plates of the tetra-

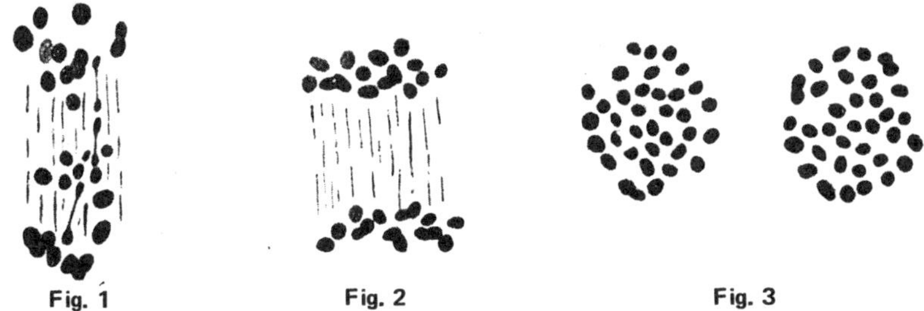
Fig. 1 Fig. 2 Fig. 3

Fig. 1. Portion of an anaphase of the normal F_1 *glutinosa-tabacum* hybrid.
Fig. 2. The same of the tetraploid hybrid.
Fig. 3. Homotypic metaphase, polar view, of the tetraploid hybrid.

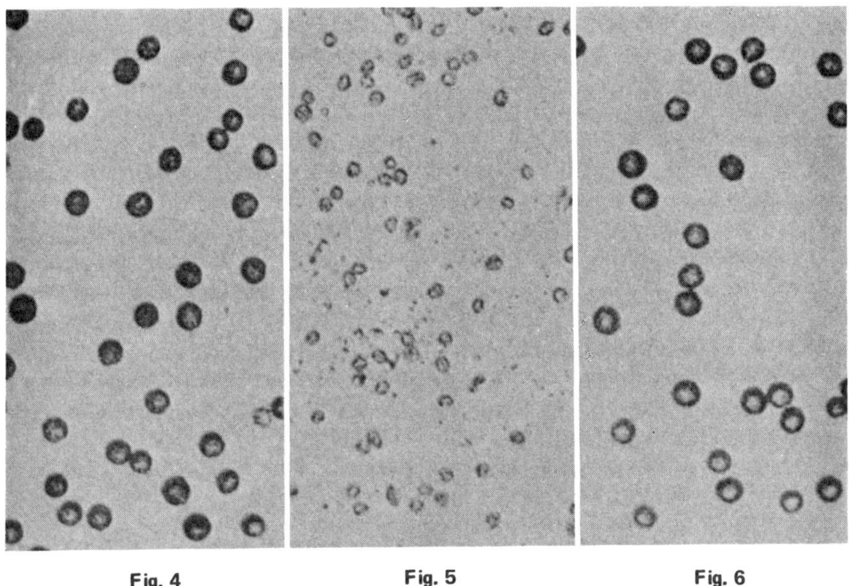
Fig. 4 Fig. 5 Fig. 6

Fig. 4. Portion of a photomicrograph of pollen of the tetraploid *glutinosa-tabacum* hybrid, 24123P55.
Fig. 5. Of the normal F_1, 24192P4.
Fig. 6. Of another plant of the tetraploid hybrid, 24123P58. The preparations were stained in aceto-carmine and the photomicrographs were taken at the same magnification.

ploid, each polar view containing 36 chromosomes. Figure 1 was drawn from fixed material, the other two figures from aceto-carmine preparations. As will be noted (Figure 1), the behavior of the bivalent and univalent chromosomes closely parallels that found in the F_1 *tabacum-sylvestris* hybrid elsewhere described (Goodspeed 1923). The bivalent partners are approaching the poles while the univalents are in the equatorial zone, either dividing or preparing to divide. In the tetraploid, on the other hand, there appear to be no univalent chromosomes and the bivalent partners move in regular fashion to the poles. No attempt is made in either Figure 1 or 2 to represent the full chromosome complement.

Pollen conditions are illustrated in Figures 4 and 6 for the tetraploid hybrid and in Figure 5 for the normal sterile F_1. These figures are reproduced from photomicrographs of pollen preparations stained with aceto-carmine. As will be noted, the pollen of the normal hybrid consists exclusively of shrivelled grains devoid of contents. The pollen of the tetraploid hybrid consists mostly of large grains apparently normal in protoplasmic contents. Measurements were made of pollen grains of the tetraploid hybrid and of its *tabacum* parent, but unfortunately no pollen of *glutinosa* was available at the time measurements were made. The average diameter of pollen grains in the tetraploid hybrid was found to be 46.3 microns, of *tabacum*, 36.7. The volumes are therefore, in the ratio of approximately 2:1 (106:49).

The cytological findings supply an obvious explanation for the uniformity and constancy of this hybrid. Since in *glutinosa*, $n = 12$, and in *tabacum*, $n = 24$, the F_1 hybrid normally has 36 chromosomes. This was undoubtedly the case in the sterile F_1 described above. The original fertile F_1 plant, 22066P2, must have arisen from a doubling of the chromosome number immediately or soon after fertilization, by which a tetraploid hybrid with 36 pairs of chromosomes was produced. Such a plant may be represented by the chromosomal formula, 12 GG+24 TT; and, if *glutinosa* and *tabacum* homologues pair regularly, fertility and constancy follow as a matter of course, for every gamete would then contain 12 *glutinosa* and 24 *tabacum* chromosomes. If this explanation is correct, an interspecific hybrid may be expected to become fertile and constant by simple doubling of its chromosome number.

These observations naturally recall the case of *Primula kewensis*, the much discussed hybrid of *P. floribunda* with *P. verticillata*. According to accounts of its origin as described by Miss Digby (1912) and by the Misses Pellew and Durham (1916), the original hybrid was sterile; but it eventually produced a fertile bud-sport which gave rise immediately to the fertile, comparatively constant form now known as *P. kewensis*. Miss Digby found that the chromosome numbers of *P. floribunda* and *P. verticillata* were both $n = 9$ and $2n = 18$, that the sterile hybrid had 18 chromosomes and the fertile *P. kewensis*, 36. *P. kewensis*, therefore, is evidently a tetraploid hybrid; and as Winkler (1920) and Renner (1924) suggest, this fact probably accounts for its genetic behavior. If it contains 9 pairs each of chromosomes of *P. floribunda* and *P. verticillata*, and homologues of each species pair regularly, the situation is exactly the same as that described in the tetraploid Nicotiana. It seems more reasonable to adopt this explanation of its chromosome number, since it accounts so well for the genetic results thus far obtained with it, rather than that of transverse fission suggested by Farmer and Digby (1914) on the basis of chromosome measurements, which has been accepted by Gates (1924) in his recent discussion of polyploidy.

The confirmation of Winge's hypothesis afforded by the instances described above extends only to establishment of the tetraploid chromosome condition, and not to the method of origin described by him. The establishment of the condition is evidently a mutational event, analogous to that which occurs in the establishment of the tetraploid condition in pure species. The tetraploid hybrid condition may, however, arise in a variety of ways: (1) by doubling of chromosome number immediately subsequent to fertilization; (2) by bud-variation in an F_1 interspecific hybrid; (3) by crossing together tetraploid representatives of two different species; and (4) by irregular distribution of chromosomes in an interspecific hybrid in which the chromosomes do not pair in meiosis, as suggested by Collins and Mann (1923). It may be possible, therefore, that tetraploid hybrids have the significance in the origin of new chromosome numbers ascribed to them by Winge.

LITERATURE CITED

Clausen, R. E., and Mann, Margaret C., 1924. Inheritance in *Nicotiana tabacum*. V. The occurrence of haploid plants in interspecific progenies. Proc. Nation. Acad. Sci. **10:**121-124.

Collins, J. L., and Mann, Margaret C., 1923. Interspecific hybrids in Crepis. II. A preliminary report on the results of hybridizing *Crepis setosa* Hall. with *C. capillaris* (L.) Wallr. and with *C. biennis* L. Genetics **8:**212-232.

Digby, L., 1912. The cytology of *Primula kewensis* and of other related Primula hybrids. Annals of Bot. **36:**357-388.

Farmer, J. B., and Digby, L., 1914. On dimensions of chromosomes considered in relation to phylogeny. Phil. Trans. Roy. Soc. B **205:**1-25.

Focke, W. O., 1881. Die Pflanzen-Mischlinge. 569 pp. Berlin: Borntraeger.

Gates, R. R., 1924. Polyploidy. British Jour. Exp. Biol. **1:**153-182.

Gärtner, C. F., 1849. Versuche und Beobachtungen über die Bastarderzeugung im Pflanzenreich. 790 pp. Stuttgart: The Author.

Goodspeed, T. H., 1915. Parthenogenesis, parthenocarpy and phenospermy in Nicotiana. Univ. California Publ. Bot. **5:**249-272.

1923 A preliminary note on the cytology of Nicotiana species and hybrids. Svensk Bot. Tid. **17:**472-478.

Pellew, Caroline, and Durham, Florence M., 1916. The genetic behavior of the hybrid *Primula kewensis*, and of its allies. Jour. Genetics **5:**159-182.

Renner, O., 1924. Vererbung bei Artbastarden. Zeit. indukt. Abstamm. u. Vererb. **33:**317-347.

Setchell, W. A., 1912. Studies in Nicotiana. I. Univ. California Publ. Bot. **5:**1-86.

White, O. E., 1913. The bearing of teratological development in Nicotiana on theories of heredity. Amer. Nat. **47:**206-228.

Winge, O., 1917. The chromosomes. Their numbers and general significance. Compt. Rend. Lab. Carlsberg **13:**131-275.

Winkler, H., 1920. Verbreitung und Ursache der Parthenogenesis im Pflanzen und Tierreiche. 231 pp. Jena: Gustav Fischer.

36 Triploid cell cultures from a macerated foetus

L. S. Penrose*
Joy D. A. Delhanty
Galton Laboratory
University College
London, England

Although the first accurate observations on normal human chromosomes (Tjio and Levan 1956) were made with cell cultures taken from foetal tissues, the opportunities for investigating the chromosomes in cases of natural abortion have not so far been greatly exploited. We here describe the results of culturing cells from foetal remnants in a case of missed abortion.

The *patient,* whose offspring was studied, is a healthy woman aged 44. She had one previous pregnancy at the age of 41, which resulted in a normal boy. The second pregnancy began during the third week of August, 1960 (last menstrual period, Aug. 8). The xenopus test was positive on a specimen taken on Oct. 3. About a week later a very slight bloodstained discharge was noted. Eight weeks afterwards (Nov. 30) the uterus was found to be definitely too small to be compatible with a normal pregnancy of twelve weeks' duration or more, and missed abortion was suspected. Curettage was performed on Jan. 2, 1961. A routine pathological report, by Dr. N. F. C. Gowing, indicated the presence of living and dead placenta, showing normal vascularisation. The patient's recovery was uneventful.

A sterile *specimen* from the uterine contents, including the supposedly foetal material (in glucose saline) was kindly provided by Mr. Clifford Simmons. The specimen consisted of an amorphous spongy mass, about 15 cc. in volume. When spread out, it was found to contain some completely disorganised tissues which appeared to be of foetal origin but which were inseparable from the main mass of degenerated placenta. There was evidence of calcification. In the centre was a yellowish narrow firm tubular structure, not freely detachable, which was thought to be foetal gut. A small bladder with a delicate transparent wall, attached to a short stalk and full of clear fluid, was believed to be the yolk sac.

Two *cultures,* A and B, were taken from the gut area, and one, C, from the yolk-sac wall. The explants, each about 4 c.mm. in volume, were divided into smaller pieces for culture. The cultures were set up on Jan. 4, the procedure closely following that described by Harnden (1960). Growth was, at first, rapid in cultures B and C; culture A grew more slowly. Slides for cytological examination were obtained on Jan. 23. Good preparations were difficult to obtain, especially in B and C, because of the tendency of cells to aggregate in clumps in a manner characteristic of cultured epithelial cells, and mitoses were not plentiful. Later the cultures continued to grow, but very slowly.

Investigation of the *karyotype* in A showed that all cells contained 46 chromosomes, as in a normal female. Cultures B and C, however, revealed a triploid condition, with 69 chromosomes in nearly all unbroken cells, though a few were hexaploid, with 138 chromosomes. Analysis of the triploid cells clearly revealed 7 small acrocentric chromosomes, one of which could reasonably be supposed to be a Y. As the figure shows, there can be little doubt about the interpretation of these cells as true examples of a triploid condition in which one set of 23 chromosomes was derived from a sperm, and two sets from an ovum. The fact that in each trio (and especially the larger ones) two chromosomes seem to be morphologically very similar, supports this view. Probably the cause of the triploidy was fertilisation of an unreduced ovum which had undergone the first, but not the second, meiotic division.

Two cytological investigations were made to throw light on the origin of these karotypes. The cells in a culture of the patient's peripheral blood, made by Miss Ruth Marshall, contained a normal female chromosome complement; so the cells which grew in culture A could have been endometrial in origin. To test the possibility that foetal membranes might normally be triploid, a culture was taken from the amniotic sac of a normal male foetus at four months. This was found to be diploid.

Triploidy in mice, rabbits, and rats is known to occur spontaneously (Austin 1960); it can also be produced artificially in mice by temperature shock, or by treating ova with colchicine before fertilisation (Beatty 1951). Such

From The Lancet **i:**1261-1262, 1961. Used with permission.

We are grateful to the Medical Research Council for a grant to J. D. A. D., and to the United States Public Health Service for a research grant (RG-6984).

*Present address: Kennedy-Galton Centre, Harperbury Hospital, St. Albans, England.

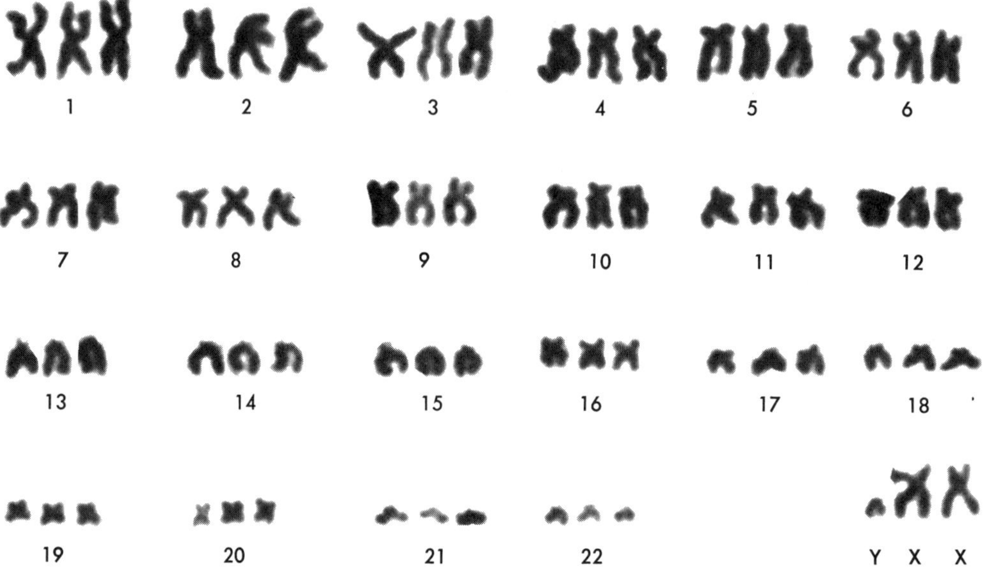

Triploid cell from yolk sac, culture C, with chromosome analysis.

embryos rarely complete their development. Mosaic or chimaeric animals—heteroploids with both triploid and diploid cells—have also been found. In man there is the example of a malformed 2-year-old boy reported recently (Böök and Santesson 1960, Böök 1961). Here, 84% of the cells from skin biopsy, and 49% of cells cultured from connective tissue, but none of the cells in peripheral blood or bone-marrow, showed a triploid condition almost exactly like that shown in the figure. The normal cells were diploid male. It is possible that the foetus we studied may also have been heteroploid, though this seems unlikely because none of the cultures was mosaic.

Two points arise from this investigation. Triploidy following fertilisation of unreduced ova might cause abortion more often than has been formerly supposed. Successful culturing of material from a macerated foetus, nearly two months after it had ceased to develop, is unexpected but encouraging for future studies of spontaneous abortion.

REFERENCES

Austin, C. R. (1960) F. cell. comp. Physiol. **56,** suppl. 1, p. 1.
Beatty, R. A. (1951) Anim. Breed. Abstr. **19**:283.
Böök, J. A. (1961) Personal communication.
____ Santesson, B. (1960) Lancet **i**:858.
Harnden, D. G. (1960) Brit. F. exp. Path. **41**:31.
Tjio, J. H., Levan, A. (1956) Hereditas **42**:1.

chapter 8

Extrachromosomal inheritance

Chromosomal, or, as it is also called, Mendelian, inheritance is characterized by the fact that genes from the male and female parents contribute equally to the genetic constitution of the progeny. Thus in Mendelian inheritance, reciprocal crosses between parents of different homozygous genotypes will yield offspring of identical genotype and phenotype. In *extrachromosomal* inheritance, male and female parents do not make equal contributions to the genetic constitution of the progeny; hence reciprocal crosses yield different results. Investigations of extrachromosomal inheritance have been complicated by the fact that the hereditary factors may be contained in (1) normal cell constituents (e.g., plastids, mitochondria), (2) infective particles that may be symbiotic microorganisms (e.g., kappa, mu in *Paramecium aurelia, Treponema* in *Drosophila willistoni*), (3) episomes that can be either chromosomal or extrachromosomal (e.g., phage lambda (λ) in *Escherichia coli*), or (4) as yet undetected constituents of the cell (e.g., nonchromosomal factors in *Chlamydomonas reinhardi*).

NORMAL CELL CONSTITUENTS

The first genetic trait to be clearly demonstrated as extrachromosomal was color in *plastids.* This was reported in 1909 by Correns (Ref. 8-1), who worked with the four-o'clock plant, *Mirabilis jalapa.* He found that the leaves of the plants were green, white, or a mottled pattern of green and white. The flowers on green-leaved branches produced only green seedlings, and flowers from the white-leaved branches produced only white seedlings, regardless of the genotype and phenotype of the pollen parent. Flowers from the variegated branches yielded a mixture of green, white, and variegated plants in widely varying ratios.

An analysis of the color-producing factor showed that in the cells of the leaves are self-reproducing cytoplasmic structures, plastids, which may contain chlorophyll (i.e., chloroplasts) or may not (i.e., leukoplasts). Chloroplasts may lose their chlorophyll and become leukoplasts. Green-colored cells have at least some chloroplasts in their cytoplasm. White-colored cells contain only leukoplasts. The division of the cytoplasm at each mitosis is such that both resultant cells contain plastids. However, the cells need not receive the same number of plastids, nor need they receive some of every type of plastid present in the original cell. The inheritance of color in the four-o'clock plant was found to be determined by the nature of the plastids in the egg cell. The pollen cell contains very little cytoplasm and is, in most cases, completely devoid of plastids. This mode of inheritance of color has been shown to occur in many plants, including barley, maize, and rice.

If plastids are self-duplicating structures that possess their own hereditary factors, they should contain a complete genetic system. Such a system would have to provide for DNA replication, DNA-mRNA transcription, and mRNA-protein translation. A review of the evidence that plastids and another type of cytoplasmic structure, *mitochondria,* do in fact contain complete genetic systems was prepared by Gibor and Granick in 1964 (Ref. 8-2) and is included in this chapter.

In a number of cases, it has been found that chromosomal genes can either produce or stimulate the production of plastid mutations. A well-studied case in maize was reported by Rhoades in 1943 (Ref. 8-3).

INFECTIVE PARTICLES

In certain stocks of *Paramecium aurelia* a number of hereditary characteristics are found that are extrachromosomal in nature and appear to be caused by the presence of infective particles in the cytoplasm of the organism. The first such trait was discovered by Sonneborn in 1938 (Ref. 8-4). He had placed individuals from different stocks in the same culture vessel. In some combinations, he observed that individuals from one stock (later called *killers*) survived but individuals from the other stock (later called *sensitives*) died. His later research demonstrated that killer and sensitive paramecia could conjugate and if they were separated immediately thereafter, no killing occurred. He further found that the killer factor, called *kappa,* could be transmitted from a killer to a sensitive individual during conjugation provided that in addition to an exchange of micronuclei, an exchange of cytoplasm also occurred. It was subsequently discovered that kappa was maintained in an individual only if the organism was either homozygous or heterozygous for a dominant chromosomal gene, designated *K*. The killer factor, kappa, was found to be associated with minute particles present only in the cytoplasm of killer paramecia. Approximately half the kappa particles are transmitted to each daughter cell at mitosis, and the full population level is restored by self-duplication of the kappa particles before the next fission. If, as a result of autogamy, a killer paramecium of *Kk* genotype becomes *kk*, the kappa particles are maintained at their characteristic number for at least eight generations and then suddenly disappear from the cytoplasm of the cell. The kappa particle was found to contain DNA and RNA and appears to be either a large virus or *Rickettsia* capable of infecting, and living as a symbiont in, killer-type paramecia. A comprehensive review of the work on kappa was prepared by Sonneborn in 1959 (Ref. 8-5).

Another type of killer strain in *P. aurelia*, called *mate-killer*, was discovered by Siegel in 1952 (Ref. 8-6). In this case, the cytoplasmic particle associated with the killer trait was called *mu*. The maintenance of mu in the cytoplasm of an individual requires that the organism be either homozygous or heterozygous for one of two dominant alleles, labeled M_1 and M_2. However, it was reported by Gibson and Beale in 1962 (Ref. 8-7) that the disappearance of mu from the cytoplasm of an $m_1m_1m_2m_2$ individual was apparently controlled by the presence of still another cytoplasmic particle, which they called a *metagon*. The metagon particle consists of RNA and is thought to be mRNA of either M_1 or M_2. Later experiments showed that both mu and metagons could be transferred from *P. aurelia* to *Didinium nasutum* when the latter feeds on the former. From the evidence reported by Gibson and Sonneborn in 1964 (Ref. 8-8), it would also appear that the metagons transferred from *Paramecium* increase in number while in *Didinium*. A general review of infective particles in paramecia and the role of metagons in the maintenance of mu was prepared by Sonneborn in 1965 (Ref. 8-9).

Infective particles that act as extrachromosomal hereditary units have been found in various organisms. In a number of species of *Drosophila*, an occasional female is discovered that gives rise to progenies predominantly or exclusively female. This condition has been called *sex ratio*. It was found that the sex-ratio factor is not only transmitted from mother to daughter but can also be transferred from a sex-ratio female to an unaffected one. An experiment demonstrating the transfer of the sex-ratio factor from one female to another was reported in 1957 by Malogolowkin and Poulson (Ref. 8-10), whose paper is included in this chapter. The nature of the sex-ratio agent was discovered in 1961 by Poulson and Sakaguchi (Ref. 8-11), whose paper is reprinted in this chapter. Later studies showed that the sex-ratio factor could also be transferred from affected females of one species to unaffected females of another species. The success of such interspecific transfers and the expression of the transferred sex-ratio factor in the new species, however, were found to vary, as shown by Ikeda in 1965 (Ref. 8-12).

EPISOMES

The extrachromosomal genetic systems we have considered thus far are always found to be physically distinct from the cell's chromosomal genes. However, certain hereditary particles can exist in the cell interchangeably, either as extrachromosomal or as chromosomal elements. These particles were named *episomes* by Jacob and Wollman in 1958 (Ref. 8-13). Their existence was hypothesized to explain certain observations on virus-infected bacteria. It had been observed by many workers that not all bacterial cells infected with viruses would be lysed. In some instances, the virus would enter the cell and seem to disappear. In these cases, the cells would remain intact, grow, and divide for many generations, but among their descendants an occasional cell would break open and yield a crop of new virus particles. A bacterial cell that harbors a virus and is capable of being lysed by it is called a *lysogenic* bacterium. The virus that enters and remains latent within a lysogenic bacterium is called a *temperate* virus. While in its latent stage, the temperate virus is called a *provirus* or a *prophage*. A virus that multiplies rapidly within a bacterial cell and eventually kills the bacterium is called a *virulent* virus, and the doomed host is called a *nonlysogenic* bacterium, with regard to the particular virus.

It was discovered by Lwoff and his co-workers in 1950 (Ref. 8-14) that if the descendants of a lysogenic bacterium were exposed to UV light, virus production was initiated inside virtually every cell of the culture, with concomitant lysis of the affected cells. This demonstrated that each of the descendants of a lysogenic bacterium contained at least one latent virus. It was realized that the association of a latent virus and a lysogenic bacterium cannot be of a passive nature, since it was impossible to free lysogenic strains from the phage they carry, by heat, anti-phage serum neutralization, or single-colony purification. These methods normally kill or remove virus particles located in a bacterial cell's cytoplasm. It seemed evident that the disappearance of the invading temperate virus within the lysogenic bacterium and its subsequent distribution to descendant cells could be explained if the DNA of the virus became incorporated in the bacterial chromosome. An experiment demonstrating that the latent virus was located in the bacterial chromosome was reported in 1953 by Lederberg and Lederberg (Ref. 8-15), whose article is reprinted in this chapter. A later paper by Jacob in 1955 (Ref. 8-16) reported the transduction of lysogeny. It has become customary to refer to the prophage as the *integrated* stage of the viral life cycle, since at this time the virus is incorporated into the bacterial chromosome. The rapid reproductive phase of the virus, which occurs in the bacterial cell's cytoplasm, is referred to as the *autonomous* stage of the viral life cycle. A general review of lysogeny was prepared by Lwoff in 1953 (Ref. 8-17). A detailed discussion of extrachromosomal inheritance in bacteria was prepared by Novick in 1969 (Ref. 8-18).

UNDETECTED CELL CONSTITUENTS

A number of traits in various organisms follow a uniparental mode of inheritance but cannot be associated with any cell constituents yet known. One of the best studied examples of this situation involves the green alga, *Chlamydomonas reinhardi.* Members of this species can reproduce asexually by mitotic cell division and can also engage in a form of sexual reproduction in which cells pair, fuse, and form zygotes. This organism is characterized by two mating types that have been designated mt^+ and mt^-. The diploid zygote must then be mt^+mt^- with respect to mating type. An analysis of the cells resulting from meiosis of the zygote shows a 1:1 ratio of mt^+:mt^-. Mating type therefore behaves as a trait that is determined by a Mendelian gene. However, other traits follow a different mode of inheritance. It was observed by Sager in 1954 (Ref. 8-19) that in most cases mutant cells that were resistant to streptomycin would pass on this trait through a sexual phase only if these cells were mt^+. If the cells were mt^-, in most instances the trait for streptomycin resistance appeared to be either lost or destroyed in the zygote. A large number of traits whose transmission is under the control of the mating-type gene were subsequently discovered. A paper by Sager and Ramanis in 1963 (Ref. 8-20) discussing the evidence for the particulate nature of these genes in *Chlamydomonas* is included in this chapter. An experiment exploring the linkage relationship of some of these extrachromosomal genes was reported by Gillham and Fifer in 1968 (Ref. 8-21). An analytic review of uniparental inheritance in *Chlamydomonas reinhardi* was prepared by Gillham in 1969 (Ref. 8-22).

Extrachromosomal or non-Mendelian inheritance undoubtedly occurs in many, if not all, species. In many organisms, man included, the pattern of mating and the long life cycle have made it very difficult to detect any indication of the existence of this type of genetic system. However, as research continues, the number of species that exhibit non-Mendelian inheritance of some of their traits increases. As with other aspects of genetics, we have no evidence to indicate that this particular mode of inheritance is restricted to a few species. The existence of this second genetic system in organisms will require that we expand our concept of the genotype of the individual and, in the case of man, will add a new dimension to the search for the genetic bases of inherited diseases.

REFERENCES

Normal cell constituents

8-1. Correns, C. 1909. Vererbungsversuche mit blass (gelb) grünen und buntblättrigen Sippen bei *Mirabilis jalapa, Urtica pilulifera* und *Lunaria annua.* Zeit. Induk. Abst. u. Vererbung. **1**:291-329.

8-2. Gibor, A., and Granick, S. 1964. Plastids and mitochondria: inheritable systems. Science **145**:890-897.

8-3. Rhoades, M. M. 1943. Genic induction of an inherited cytoplasmic difference. Proc. Natl. Acad. Sci. U.S.A. **29**:327-329.

Infective particles

8-4. Sonneborn, T. M. 1938. Mating types, toxic interactions and heredity in *Paramecium aurelia.* Science **88**:503.

8-5. Sonneborn, T. M. 1959. Kappa and related

particles in *Paramecium.* Advances in Virus Research **6:**229-356.

8-6. Siegel, R. W. 1952. The genetic analysis of mate-killing in *P. aurelia.* Genetics **37:**625-626.

8-7. Gibson, I., and Beale, G. H. 1962. The mechanism whereby the genes M_1 and M_2 in *Paramecium aurelia,* stock 540, control growth of the mate-killer (mu) particles. Genet. Res., Cambridge **3:**24-46.

8-8. Gibson, I., and Sonneborn, T. M. 1964. Is the metagon an m-RNA in *Paramecium* and a virus in *Didinium?* Proc. Natl. Acad. Sci. U.S.A. **52:**869-876.

8-9. Sonneborn, T. M. 1965. The metagon: RNA and cytoplasmic inheritance. Am. Nat. **99:**279-307.

8-10. Malogolowkin, C., and Poulson, D. F. 1957. Infective transfer of maternally inherited abnormal sex-ratio in *Drosophila willistoni.* Science **126:**32.

8-11. Poulson, D. F., and Sakaguchi, B. 1961. Nature of the "sex-ratio" agent in *Drosophila.* Science **133:**1489-1490.

8-12. Ikeda, H. 1965. Interspecific transfer of the "sex-ratio" agent of *Drosophila willistoni* in *Drosophila bifasciata* and *Drosophila melanogaster.* Science **147:**1147-1148.

Episomes

8-13. Jacob, F., and Wollman, E. L. 1958. Les épisomes, éléments génétiques ajoutés. Compt. Rend. Acad. Sci., Paris **247:**154-156.

8-14. Lwoff, A., Siminovitch, L., and Kjeldgaard, N. 1950. Induction de la production de bactériophages chez une bactérie lysogène. Ann. Inst. Pasteur **79:**815-859.

8-15. Lederberg, E. M., and Lederberg, J. 1953. Genetic studies of lysogenicity in *Escherichia coli.* Genetics **38:**51-64.

8-16. Jacob, F. 1955. Transduction of lysogeny in *Escherichia coli.* Virology **1:**207-220.

8-17. Lwoff, A. 1953. Lysogeny. Bacteriol. Rev. **17:**269-337.

8-18. Novick, R. P. 1969. Extrachromosomal inheritance in bacteria. Bacteriol. Rev. **33:**210-235.

Undetected cell constituents

8-19. Sager, R. 1954. Mendelian and non-Mendelian inheritance of streptomycin resistance in *Chlamydomonas.* Proc. Natl. Acad. Sci., U.S.A. **40:**356-363.

8-20. Sager, R., and Ramanis, Z. 1963. The particulate nature of nonchromosomal genes in *Chlamydomonas.* Proc. Natl. Acad. Sci., U.S.A. **50:**260-268.

8-21. Gillham, N. W., and Fifer, W. 1968. Recombination of nonchromosomal mutations: a three-point cross in the green alga *Chlamydomonas reinhardi.* Science **162:**683-684.

8-22. Gillham, N. W. 1969. Uniparental inheritance in *Chlamydomonas reinhardi.* Am. Nat. **103:**355-388.

37 Plastids and mitochondria: inheritable systems

Do plastids and mitochondria contain a chromosome which controls their multiplication and development?

A. Gibor*
S. Granick
Rockefeller Institute
New York, New York

Two important types of cytoplasmic organelles are the plastids of plant cells, which function in photosynthesis, and the mitochondria of both plant and animal cells, which function in oxidative respiration. Within the last few years new information has become available which supports the hypothesis that these organelles do not arise *de novo* but that plastids arise from preexisting plastids and mitochondria arise from preexisting mitochondria. The original evidence reviewed by Granick *(1)* included observations on the division of plastids in algae and genetic studies of chloroplast inheritance in variegated plants and *Oenothera.* Recent studies indicate that (i) the plastids and mitochondria contain DNA and RNA; (ii) the organelles are self-duplicating bodies that do not arise *de novo;* (iii) the DNA represents a multigenic hereditary system which is not derived from the nucleus; (iv) the multigenic system of an organelle is responsible, in part, for the specific biochemical properties of the organelle; and (v) the organelles are controlled by an adaptive mechanism which in the case of plastids is responsive to light and in the case of mitochondria is responsive to O_2.

In this article we review the evidence for the foregoing statements. We consider especially the data provided by the genetic systems of the plastids of *Euglena* and the mitochondria of yeast. A comparison of the properties of these two genetic systems illustrates their fundamental similarities (Table 1). For reasons not known, these systems are exceptionally mutable, a property, however, which has made possible the recognition of their multigenic components.

EVIDENCE FOR NUCLEIC ACIDS IN PLASTIDS

In order to establish that plastids are semiautonomous units with their own hereditary apparatus, it is necessary to show that (i) they contain DNA and RNA; (ii) the DNA replicates in the organelle; (iii) the DNA functions to make messenger RNA; and (iv) the messenger RNA codes specifically for certain proteins or enzymes of the organelle.

Experiments to test for these properties are of various kinds, and some of the tests are more conclusive than others. However, taken together, the data strongly support the view that these organelles contain an autonomous DNA genetic apparatus.

DNA content. Reported findings of traces of DNA in plastids have been questioned because of the possibility of contamination with nuclear DNA. Gibor and Izawa *(2)* overcame this difficulty in analyses of *Acetabularia* chloroplasts. They succeeded in growing this large unicellular alga free from bacterial contamination. At the right stage of growth the single large nucleus at the base of the plant was cut off. The chloroplasts were squeezed out from the enucleated portion of the cell and purified by differential centrifugation, and their DNA content was determined before and after treatment with deoxyribonuclease. The chloroplasts were found to contain 0.2 microgram of DNA per milligram of protein, or approximately 1×10^{-16} gram of DNA per plastid. This DNA content is similar to that found in the vaccinia virus and in the even-numbered T bacteriophages *(3)* and would be sufficient for

From Science **145**:890-897, August 28, 1964.

We wish to acknowledge the critical review of the manuscript by Dr. Miriam Jacob, and the advice and criticisms of Dr. E. Tatum. The work of our laboratory which is presented here and the preparation of this article were supported by NSF research grant G-14418.

*Present address: University of California at Santa Barbara, California.

Table 1. Properties of the cytoplasmic mutations in *Euglena* and yeast

Euglena	Yeast
Phenomenon	
Bleaching; impaired photosynthetic apparatus	"Petit" colony; impaired oxidative respiration
Anatomical changes	
Proplastids fail to differentiate on exposure to light; no chloroplast lamellae develop *(50-52)*	Promitochondria fail to differentiate on exposure to air; no cristae membranes develop *(49)*
Mutation rate (spontaneous)	
1%; higher in unstable clones *(50)*	1%; higher in unstable clones *(53)*
Mutation rate (induced)	
Up to 100% *(50, 54, 55)*	Up to 100% *(39, 53)*
Agents that induce the mutations	
1) Ultraviolet light, 260 mμ maximum, photoreactivable *(10)* 2) Elevated temperature during growth *(55)* 3) Basic drugs (streptomycin) *(54)*	1) Ultraviolet light, 260 mμ maximum, photoreactivable *(41)* 2) Elevated temperature during growth *(53)* 3) Basic drugs (acridines) *(53)*
Variability of mutants	
Differences in carotenoid content and in porphyrin-synthetic abilities Less defective plastids are larger *(50, 51)*	Suppressive and neutral types; differences in degree of suppressiveness Suppressives can give rise to neutrals *(56)*
Inheritance	
Cytoplasmic, by ultraviolet microbeam *(13)*	Cytoplasmic, by genetic analysis of crosses *(39)*

the coding of several hundred genes. Baltus and Brachet *(4)* also recently reported the presence of DNA in chloroplasts of enucleated *Acetabularia* which were apparently free of bacteria.

By density-gradient ultracentrifugation the DNA from the green plants was found to be composed of a major and minor (or satellite) component, by Chun, Vaughan, and Rich *(5)*, Sager and Ishida *(6)*, and Leff, Mandel, Epstein, and Schiff *(7)*. It has been suggested by a number of workers that the satellite DNA of plants is actually the DNA of the chloroplasts. When purified chloroplast preparations were examined, as much as 50 percent of their DNA was found to be of the satellite type. Such analyses suggested that the satellite DNA was derived from the chloroplasts. A second satellite DNA which was sometimes seen may have been derived from the mitochondria.

Available data suggest that not only *Acetabularia* chloroplasts but also the chloroplasts of higher plants and of *Euglena* contain approximately 10^{-15} to 10^{-16} gram of DNA. Chun *et al.* *(5)* found that the DNA of the chloroplasts of spinach and beets was about 1 percent of the total DNA. If the number of plastids in a leaf parenchyma cell is assumed to be 100, the DNA per plastid would be 0.01 percent of the nuclear DNA. The amount of DNA in a diploid nucleus of a bean plant was estimated, by Rasch, Swift, and Klein *(8)*, to be 18×10^{-12} gram; thus, a plastid of a higher plant would contain 10^{-15} to 10^{-16} gram of DNA. In *Euglena,* the total amount of DNA per cell is 2.5 to 4.3×10^{-12} gram *(9)*. Leff *et al.* *(7)* estimated that in *Euglena* the satellite DNA comprises 4 percent of the total DNA. Lyman *et al.* *(10)* determined that a *Euglena* cell contains 30 ultraviolet-sensitive replicating units which control the development of chloroplasts. Thus, each replicating unit contains approximately 4×10^{-15} gram of DNA.

Electron microscope studies provide independent support for the conclusion that DNA is made in the plastids. Ris and Plaut *(11)* were the first to observe fibrils 25 to 30 angstroms thick in the chloroplasts of *Chlamydomonas;* the fibrils were removed by deoxyribonuclease digestion. Similar DNA fibrils have been found in Swiss chard proplastdis by Kislev, Swift, and Bogorad *(12)*.

Replication of DNA in the organelle. Evidence that DNA is made in the plastids, rather than derived from nuclear DNA, is indirect. It depends on showing that mutations of the plastids in *Euglena* can be produced by ultraviolet irradiation of the cytoplasm even though the nucleus is shielded from irradiation. The ultraviolet light causes irreversible bleaching of the plastids, and the irradiated *Euglena* individuals in successive generations never recover the ability to green. Lyman, Epstein, and Schiff

Table 2. Enzyme systems in the differentiation of the organelles in *Euglena:* differentiation of proplastids to chloroplasts by exposure to light *(50, 58, 59)*

Structure or function involved	Normal		Bleached	
	Dark	Light	Dark	Light
Organized lamellae	–	+	–	–
Chlorophylls	–	+	–	Porphyrins*
Photosynthetic enzymes	–	+	–	+†
Carotenoids	Trace	+	–	+*
Paramylum synthesis	+	+	+	+
Plastic multiplication	+	+	+	+

*In some strains *(50)*. †In Astasia *(59)* the presence of ribulose-diphosphate-carboxylase, a photosynthetic enzyme, was demonstrated.

(10) found that the action spectrum for bleaching had a maximum at 260 mμ, indicative of an effect of ultraviolet light on nucleic acids. To decide whether the nucleic acids responsible for these mutations were localized in the cytoplasm or nucleus, we *(13)* used a microbeam of ultraviolet light to irradiate the cytoplasm alone while shielding the nucleus, or to irradiate the nucleus alone while shielding the cytoplasm. Only irradiation of the cytoplasm caused irreversible mutation of the plastids. The affected plastids then multiplied as tiny proplastids from generation to generation. They had lost the ability to differentiate into chloroplasts. If the DNA units of the cytoplasm which were sensitive to ultraviolet light had been formed by the nucleus, then one might expect that new DNA units, formed by the nucleus, would replace the damaged DNA units of the cytoplasm and thus prevent bleaching. Because the nonirradiated nucleus did not "cure" the bleached cells, it is inferred that the DNA units of the cytoplasm, presumably in the plastids, did not originate from the nucleus. The experiments of Lyman *et al. (10)* indicated that about 30 hits on a *Euglena* cell were required for bleaching. Since each cell contains about ten plastids, this result would be compatible with the hypothesis that there are 30 replicate DNA units, three per plastid, each of which must undergo mutation before bleaching can occur. Brawerman *(14)* found that *Euglena* plastids contained DNA with a unique base ratio; the adenine and thymine pair accounted for 80 percent of the total bases. The high thymine content may explain in part the sensitivity of the plastids to ultraviolet irradiation, since the damaging effect of such irradiation on DNA may result from the dimerization of the thymine molecules *(15)*.

A number of earlier studies also provide evidence that the inheritable factors, which we now equate with DNA, reside in the plastids. In the same cell—that is, under the influence of the same nucleus and cytoplasm—one may have chloroplasts that are different from each other. For example, van Wisselingh *(16)* observed that in *Spirogyra triformis* some cells contained normal chloroplasts with pyrenoids and, in addition, a chloroplast which lacked pyrenoids. In subsequent cell divisions the latter, abnormal chloroplast persisted in the cells in the presence of the normal chloroplasts. If a mutation that affected pyrenoid development had occurred in the nucleus or cytoplasm, it should have affected all the plastids alike.

Other examples include Bauer's *(17)* classic work on variegated *Pelargonium zonale* var. Albomarginata, Schwemmle's *(18)* interspecies crosses of *Oenothera,* and Michaelis's *(19)* studies on *Epilobium.* These are genetic studies in which advantage is taken of the fact of maternal inheritance—that is, of the fact that plastids of the male gamete are not transmitted to the zygote during fertilization. The results of a cross in which maternal inheritance is involved are as follows. If the female gamete contains only damaged or bleached, colorless plastids, the new plant will be colorless. If the female gamete contains normal plastids, the new plants will be green. The new cells that arise are the products of a female nucleus, male nucleus, female cytoplasm, male cytoplasm, and female plastids, but not of male plastids. Whether the new plant will be colorless or green thus depends on the inheritable units of the plastids themselves.

In a recent report, Bell and Muhlethaler *(20)* reach conclusions diametrically opposed to those cited above. These investigators propose that, in the fertilized egg of the fern *Pteridium equilinum*, the plastids and mitochondria originate *de novo* from outfoldings of the nuclear membrane. (This view is not in agreement with the genetic evidence of maternal inheritance of the defective plastids in higher plants.)

Evidence that the DNA in plastids replicates to make more DNA is also suggested by radioautographic studies. The plastids in *Spirogyra* cells *(21)* and in *Euglena (22)* were found to incorporate tritiated thymidine. In *Euglena* the labeled thymidine was removed from the plastids only by treatment with deoxyribonuclease. Since, as discussed above, the DNA of the plastids appears not to be derived from the nucleus, its synthesis probably occurs within the plastids.

DNA as template for RNA. The presence of RNA in plastids is well supported by experiment *(1)*. Here we cite only the most recent work. Biggins and Park *(23)* found that defatted spinach chloroplasts, which were isolated by Behren's nonaqueous method, contained 7 percent RNA. Lyttleton *(24)* isolated ribosomes which contained 45 percent RNA from spinach chloroplasts. With the electron microscope, ribosome like granules, digestible by ribonuclease, were observed in corn proplastids by Jacobson, Swift, and Bogorad *(25)*, and in spinach chloroplasts by Murakami *(26)*.

Evidence that the DNA of the plastids serves as a template for RNA is indirect. Kirk *(27)* reported the incorporation of labeled precursors of RNA by isolated bean chloroplasts. Treatment with actinomycin D reduced the amount of the label incorporated. This drug blocks the synthesis of RNA which is dependent on a DNA template. Similarly, Gibor and Izawa *(28)* found that the rate of incorporation of labeled uridine into the RNA of isolated chloroplasts of *Acetabularia* was reduced in the presence of actinomycin D. This suggests that DNA-dependent RNA synthesis occurs in these plastids. Brawerman *et al. (29)* found that a new kind of RNA was synthesized when *Euglena* cells grown in the dark were exposed to light. The new RNA contained a lower concentration of guanosine and cytosine than the RNA of the dark-grown cells. The satellite DNA of green *Euglena (7)* also had a lower content of guanosine and cytosine relative to adenine and thymine than the DNA of the nucleus had. All these observations suggest that the RNA species which is synthesized in response to light is coded by the satellite DNA.

Although protein synthesis in isolated plastids is suggested by the incorporation of radioactive amino acids *(29, 30)*, no net synthesis of specific protein of the plastids has been demonstrated as yet.

EVIDENCE FOR NUCLEIC ACIDS IN MITOCHONDRIA

The RNA content of mitochondria is low compared to that of the plastids. Lindberg and Ernster *(31)* suggested that a value of about 0.5 percent of the dry weight was a reasonable estimate for the RNA of mitochondria, and Rendi *(32)* obtained a value of 1.2 percent for the ratio (by weight) of RNA to protein in liver mitochondria.

Kroon *(33)* has reported that protein synthesis by isolated liver mitochondria was inhibited in the presence of actinomycin D. This finding suggests that there must be active synthesis of RNA on a DNA template for protein synthesis to occur in mitochondria.

Direct chemical evidence for the presence of DNA in mitochondria is still lacking, probably because of the small amounts of DNA present and the difficulty of avoiding contamination with nuclear DNA during isolation procedures. With the electron microscope, Nass and Nass *(34)* observed the presence of fibrils 25 to 100 angstroms thick which could be removed through digestion by deoxyribonuclease. The fibrils occupied a clear area in the fixed mitochondrion, similar to the clear area occupied by the DNA fibers of a bacterium. A long mitochondrion had two of these areas, as if the mitochondrion had not yet undergone fission. The studies of Chevremont *(35)* also suggested that DNA is formed by mitochondria. Chevremont found that chick fibroblasts, when grown in vitro for 1 day at 16°C or when treated with deoxyribonuclease II, accumulated DNA in their mitochondria. The DNA was demonstrated by Feulgen staining, by deoxyribonuclease digestion, and by radioautography after the incorporation of tritiated thymidine.

The presence of DNA in mitochondria is also shown by studies on the kinetoplasts of hemoflagellate protozoa. Steinert, Firket, and Steinert *(36)* demonstrated that the kinetoplasts gave a positive Feulgen reaction which disappeared when cells were pretreated with deoxyribonuclease. They thus concluded that the kinetoplasts contained DNA. Rapid incorporation of tritium-labeled thymidine into the kinetoplasts suggested that DNA is synthesized within them *(36)*. Electron microscope studies by Steinert on *Trypanosoma (37)* and by Rudzinska, D'Alesandro, and Trager on *Leishmania (38)* revealed that the kinetoplast is a specialized mitochondrion. It was found that the kinetoplast enlarges during metamorphosis from the parasitic leishmania form to the free living leptomonad form and develops into a single convoluted mitochondrion with typical cristae which carry out oxidative respiration.

The kinetoplast may represent a special mitochondrion in which the DNA is condensed in one segment of the organelle. In the development of the large respiring kinetoplast, the DNA may or may not become distributed into the branching mitochondrial arms of this structure.

In support of the conclusion that DNA is present in mitochondria one must also consider the genetic evidence provided by the outstanding studies of Ephrussi and his co-workers *(39)*. By genetic crosses of certain yeast strains, they showed that one kind of "respiratory deficiency" was the result of mutations in the inheritable factors which reside in the cytoplasm. These mutations were not "cured" by the normal nucleus, and therefore one could conclude that the inheritable factors for respiratory deficiency were not derived from the nucleus. By electron microscope studies, the respiratory deficiency was correlated with an absence of cristae in the mitochondria *(40)*. Respiratory deficiency could be induced in yeast by ultraviolet irradiation *(41)*. The action spectrum for this induction had a maximum at 260 mμ, suggesting that nucleic acids were the targets being irradiated. It is therefore inferred that the hereditary factors for respiratory deficiency are localized in the mitochondria and that these factors probably contain DNA.

In animals some instances have been found in which mitochondria are inherited only through the egg, not through the sperm. These examples of maternal inheritance may prove useful for investigating the genetics of mitochondrial DNA, perhaps in tissue culture. Bourne *(42)* has cited the following examples. When the fertilized egg of the bat divides, the male mitochondria sometimes pass into only one of the first two cells. In echinids the male mitochondria were traced into only one cell of a 32-cell embryo. In Nereis the middle piece of the sperm which carries the mitochondrial material does not enter the egg at all. The individual in this case contains its mother's and father's nuclear chromosomes, but only its mother's mitochondria.

SELF-DUPLICATION OF PLASTIDS AND MITOCHONDRIA

Chloroplasts and mitochondria may be considered to represent fully differentiated stages, respectively, of proplastids and promitochondria. Both the proplastids and the promitochondria are characterized morphologically as vesicles, approximately 1 micron in diameter, surrounded by a double membrane. During differentiation the inner membrane invaginates at various points to give rise to tubules that become the cristae of the mitochondria; in the plastids, they are transformed to chloroplast discs. It has often been suggested that the proplastids and promitochondria are identical *(43)*, but the concept of a particular DNA for each is incompatible with this hypothesis. The high sensitivity of the plastids of *Euglena* to various mutagens and the much lower sensitivity of the mitochondria of the same cells (Fig. 1) illustrate the difference between the DNA of the plastids and of the mitochondria.

The plastids and mitochondria, whether differentiated or not, divide by fission. One may expect that the DNA unit of the organelle will replicate before fission takes place, so that each daughter organelle will contain an identical DNA unit. Electron microscope studies of

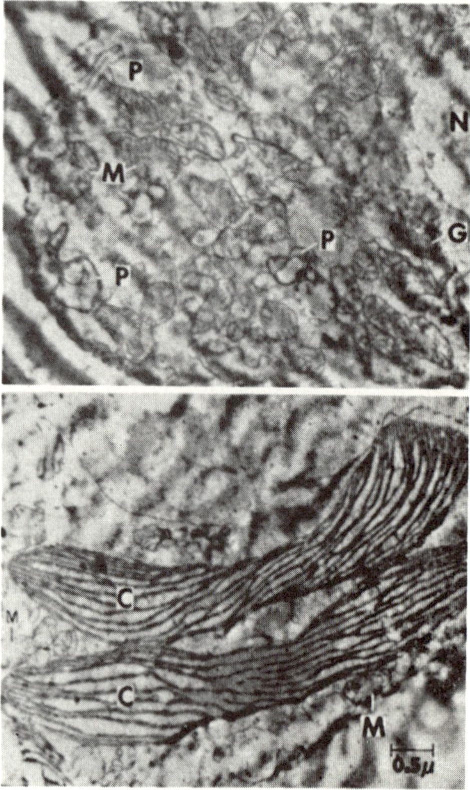

Fig. 1. (Top) Normal mitochondria and mutated plastids of an untraviolet-bleached *Euglena;* (bottom) normal mitochondria and chloroplasts of a green *Euglena. C,* Chloroplast; *G,* golgi body; *M,* mitochondria; *N,* nucleus; *P,* plastid. [From Granick *et al.* *(51)*].

the organelles in fission may eventually provide visual evidence of such multiplication. The problem becomes complicated if organelles contain not single but, rather, multiple DNA units. For example, the organelle may increase in size but not undergo fission even if the DNA has replicated. An analogous case is seen in some bacteria where the pinching off of the cells is delayed and long bacterial threads are formed which contain many DNA units. It is possible that a situation of this kind may occur in algae, which have huge chloroplasts. For example, the long strap-shaped chloroplasts of *Oedogonium* continue to grow without separating into individual chloroplasts. The high DNA content of the single chloroplast of *Chlamydomonas (6)* may also represent multiple DNA units. Similarly, the high DNA content of the kinetoplast may represent a multiple DNA unit of the mitochondrion.

The question of the origin and individuality of the mitochondrion is less clear than the analogous question of the origin of the plastid. The small size of the mitochondrion, its plasticity, and its lack of pigmentation have led to diverse interpretations. A clear example of the origin of a mitochondrion from a preexisting one was discovered by Manton during her electron microscope studies of the tiny unicellular marine alga *Micromonas (44)*. This organism contains one nucleus, one chloroplast, and one mitochondrion; the three divide synchronously at the time of cell division. Further evidence that mitochondria arise from preexisting ones is provided by the experiments of Luck *(45)* on a choline-deficient strain of *Neurospora.* By means of radioactive choline, the mitochondria of the cells were pulse-labeled in an early logarithmic phase of growth. The labeled lecithin formed was tightly bound to the mitochondria and did not turn over during this growth period. When the mitochondria increased in number, all were found to contain labeled lecithin. The results were compatible with the hypothesis that the new mitochondria did not arise *de novo* but arose through growth and division of the originally labeled organelles.

However, a number of observations on mitochondria are not readily explained. For example, both fission and fusion of mitochondria have been observed with the phase microscope *(46)*, but it has not been established by electron microscopy whether the fusion between two mitochondria represents merely temporary adhesion or the disappearance of the four membranes at the point of contact. If fusion does occur, then the possibility of genetic recombination should be investigated. The unexpected finding by Fletcher and Sanadi *(47)* that liver mitochondria have a half-life of about 10 days—an estimate based on the rate of disappearance of three incorporated isotopes—indicates that there is much to learn about the life history of these organelles.

The spontaneous origin of yeast mitochondria was suggested by Linane, Vitols, and Nowland *(48)*, on the basis of an electron microscope study of anaerobically grown yeast in which no mitochondria could be identified. Yotsuyanagi *(49)*, however, found small, faintly staining promitochondria in anaerobic yeast. These structures developed into mitochondria within several hours after exposure to oxygen.

Evidence for a multigenic system in *Euglena* plastids and yeast mitochondria. The hereditary factors in *Euglena* plastids and yeast mitochondria are remarkably similar with respect to a number of properties (Tables 1 and 2). These similarities suggest the hypothesis that more than one hereditary factor or gene, including a regulator control mechanism, resides in each of these organelles, and that most of the mutations bring about a defective differentiation of proplastids in *Euglena* and of promitochondria in yeast. A discussion of the properties listed in Table 1 follows.

Evidence for proplastids and promitochondria in mutated cells. In the bleached *Euglena* cell, green chloroplasts do not develop and photosynthesis cannot take place. In respiratory deficiency of yeast, oxidative respiration, a process associated with mitochondria, does not occur. It has been suggested that these metabolic changes result from a permanent loss of the respective organelles; this, however, is not the case (Figs. 1 and 2). In *Euglena,* it has been established cytochemically *(50)* and with the electron microscope *(51, 52)* that the plastids do not disappear but are present as tiny proplastids. These cannot differentiate to form organized chlorophyll-bearing lamellae, nor can they form the associated enzymes required for photosynthesis. Likewise, in the respiratory deficiency of yeast (Fig. 2), Yotsuyanagi *(40)* has found that mitochondira do not disappear but are present in a "promitochondrial" state—that is, a state in which they lack cristae membranes and some of the enzymes of the electron transport chain and are incapable of using O_2 for oxidative respiration.

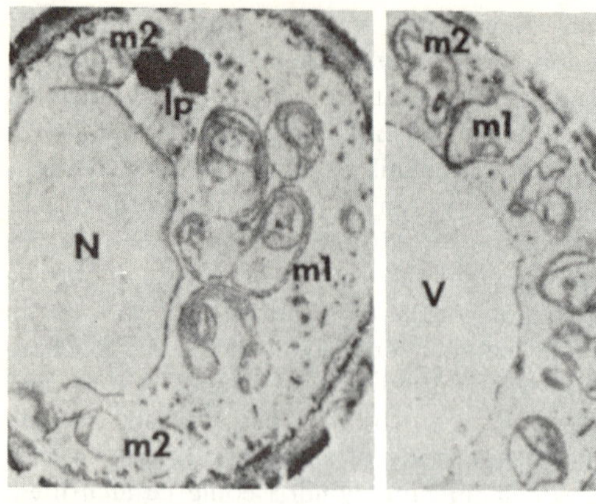

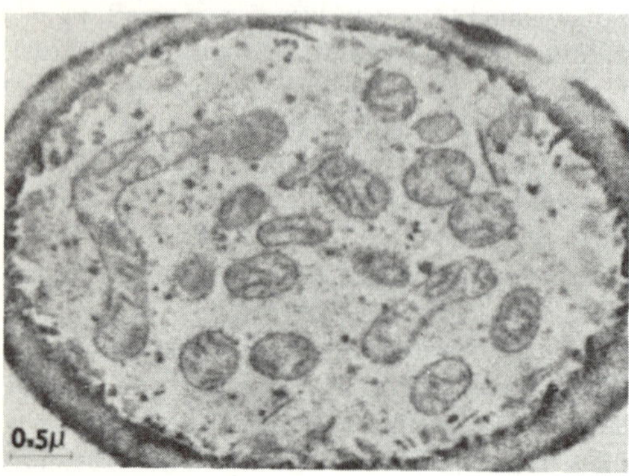

Fig. 2. Mitochondria of (top) a "petit" mutant of yeast, and of (bottom) a normal yeast. *m1* and *m2*, Mitochondria; *N*, nucleus; *V*, vacuole. [From Yotsuyanagi *(40)*].

Mutation rate. The rate of mutation to the bleached or to the respiratory-deficient state, respectively, was found to be very high. The frequency of occurrence of spontaneous mutations could be as high as 1 percent. This frequency could be increased to 100 percent by feeble irradiation with ultraviolet light. The action-spectrum maximum in both cases was at 260 mμ, which is the absorption maximum for nucleic acids *(10, 41)*. The mutation-inducing effect of ultraviolet light could be reversed by light of longer wavelength—that is, by photoreactivation. The mutation rate could also be increased by growing these organisms at elevated temperatures or by treatment with basic drugs such as streptomycin or acridine dyes *(53-55)*. Such a high mutation rate has not been observed for nuclear genes.

Variability of mutants. The plastids of different bleached strains of *Euglena* were found to vary in their biochemical properties *(50)*. Colonies derived from single cells bred true. The mutant strains differed in their ability to form carotenoids and in the enzymes of the porphyrin biosynthetic chain. Morphologic differences were also observed with the electron microscope; the less defective plastids usually appeared to be larger in size. These results, together with the results of the irradiation experiments described above, suggested that a number of cytoplasmic genes, presumably in the plastids, had undergone mutation.

Variations in the genetic properties of different mitochondrial-gene mutants have been observed in yeast. Ephrussi *et al.* *(56)* found two types of respiratory-deficient cells, suppressive and neutral. When neutral respiratory-deficient cells were crossed with cells containing normal mitochondria the progeny had normal respiration, as if the normal mitochondria had replaced or perhaps "cured" the defective neutral promitochondria. Strains derived from crossing suppressive respiratory-deficient cells and cells containing normal mitochondria had inhibited respiration. The suppressive cells could also mutate to the neutral form. No simple interpretation is possible as yet to explain the apparently "dominant" and "recessive" types of the mitochondrial mutations. Electron microscope studies on the fate of the different types of mitochondria in a zygote should help to clarify these points.

AGENTS OF ENVIRONMENTAL CONTROL OVER GENES OF THE ORGANELLES

In both types of organelles, differentiation is controlled by specific environmental agents. For the *Euglena* plastids, the agent is light; for the yeast mitochondria it is O_2.

Differentiation in *Euglena* plastids. When grown in the dark the cell contains tiny proplastids 1 to 2 microns in diameter, faintly yellow, and associated with paramylum grains (the carbohydrate-reserve polymer of euglenoids). The proplastids reproduce by fission; synchronized division of proplastids as teardrop-shaped bodies pulling apart was observed at the time of cell division *(51)*. When the cells are exposed to light the proplastids enlarge, develop chloroplast lamellae and the associated pigments and enzymes of photosynthesis, and become functional chloroplasts. The location in the cell of the light-absorbing pigment system which triggers the differentiation is not known. It has been suggested that, in higher plants, phytochrome is the pigment *(57)*.

Table 3. Enzyme systems in the differentiation of promitochondria to mitochondria by exposure to oxygen *(40, 53)*

Structure of function involved	Normal		Mutated	
	Anaerobic	Aerobic	Anaerobic	Aerobic
Organized cristae	−	+	−	−
Cytochrome oxidase	−	+	−	−
Succinic cytochrome *c*-reductase	−	+	−	−
DPN cytochrome *c*-reductase	−	+	−	−
Cytochrome *c*	Trace	+	Trace	+
Cytochrome *c*-peroxidase	−	+	−	+
Catalase	Trace	+	Trace	+

The undifferentiated proplastids of normal cells grown in the absence of light are similar to the plastids found in bleached *Euglena* mutants (Table 2). The mutated plastids, however, cannot differentiate even when they are exposed to light. Any mutation that inhibits either adequate protein synthesis for the formation of lamellae or adequate pigment synthesis will prevent chloroplast formation. A number of bleached *Euglena* mutants have been observed to respond to light by synthesizing more carotenoids or by increasing the rate at which they convert δ-aminolevulinic acid to porphyrins. This suggests that in such mutants the mutation is probably not in the control mechanism that initiates the differentiation.

Differentiation in yeast mitochondria. When normal yeast cells are grown in the absence of O_2 they contain a few tiny promitochondria, less than 1 micron in diameter. These promitochondria possess few internal lamellae or membranes, have only a trace of cytochrome *c*, lack cytochrome oxidase and cytochrome *b*, and cannot respire O_2. When such cells are transferred to an atmosphere containing O_2 (see Table 3), the mitochondria become numerous by the end of the logarithmic phase of growth, have well-organized cristae, contain the enzymes of the electron transport system, and respire O_2. Evidently there is a mechanism triggered by O_2 that induces the differentiation of mitochondria to a form that can use the O_2 for oxidative metabolism. The nature of the triggering mechanism and its location in the cell are not yet known.

The promitochondria in the normal yeast cell grown in the absence of O_2 resemble those found in the respiratory-deficient mutants of yeast. The control mechanism for differentiation need not be damaged to cause respiratory deficiency. For example, when respiratory-deficient cytoplasmic mutants, grown in the absence of O_2, were then grown in air, it was observed that increased amounts of cytochrome *c*, cytochrome *c*-peroxidase, and catalase were synthesized. Morphologically, the exposure of respiratory-deficient cells to O_2 caused an increase in density of the outer membranes, together with enlargement of the promitochondria *(49)*.

SIGNIFICANCE OF ORGANELLES IN FUNCTION AND HEREDITY

What is the significance of each cell having multiple numbers of organelles, each with its own DNA unit? One possibility is that the mitochondrial energy-releasing factory and the photosynthesizing factory are thus maintained inviolate, through having many replicates with their own specialized protein-synthesizing equipment. Multiple numbers of organelles would therefore provide greater phylogenetic stability. On the other hand, certain mutations could occur independently in each organelle DNA unit, and these mutations could be carried along so that, when drastic environmental changes occurred, there could be selection for the most suitable organelles. Thus, a multiple number of mutated organelles per cell could provide for more rapid evolutionary change.

WORKING HYPOTHESIS

On the basis of the facts and inferences that have been presented, and in analogy with schemes that have been used in bacterial genetics, we suggest the following model (Fig. 3) and working hypothesis for the genetic mechanisms of plastids and mitochondria.

Each organelle contains a DNA unit, or multiples of these units. A unit is sufficiently long to code for several hundred different proteins. The DNA unit may perhaps be a circular chromosome, analogous to the circular chromosome of some phages and bacteria.

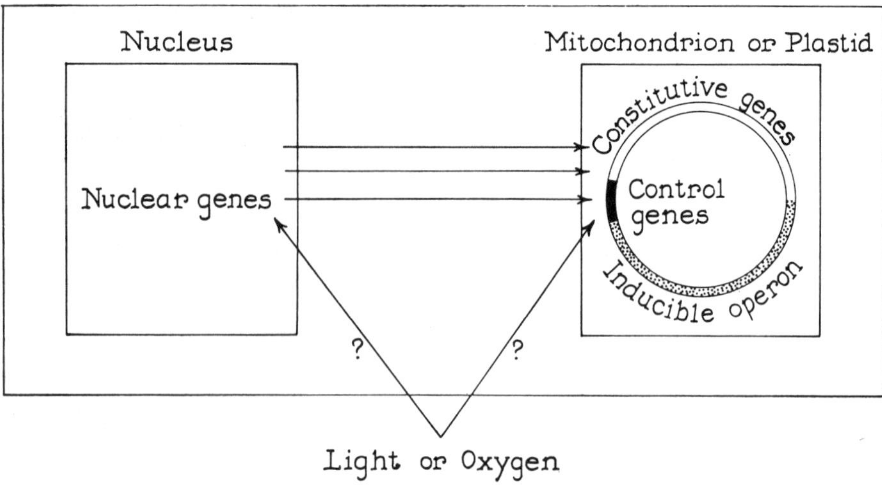

Fig. 3. Diagram of the hypothetical spherical chromosome of a cytoplasmic organelle and factors controlling its activity.

Some of the genes function as constitutive genes, while others are inducible. The constitutive genes are responsible for duplication of the organelles as well as for other biochemical processes, such as starch synthesis in the plastids.

Differentiation from proplastids to chloroplasts or from promitochondria to mitochondria is inducible and is controlled by a regulator-type gene of an operon unit. In *Euglena* and yeast it is the genes within this operon which easily undergo mutation, thus blocking the normal differentiation of the organelles. It is not yet known whether the induction by light or O_2 occurs by way of a direct effect on the genes of the organelle or whether the effect is mediated by activation of nuclear genes.

In addition to the control of the inducible genes of the organelle by environmental factors, products of nuclear genes are known to affect, directly or indirectly, the expression of the genes of the organelle. There are numerous nuclear genes in higher plants which affect the normal development of the plastids. The mutation of certain nuclear genes of yeast cause the suppression of mitochondrial differentiation. Nuclear genes could affect the organelles, for example, by limiting the supply of essential metabolites such as the pyridine nucleotides, or they could, more specifically, affect the organelles by the production of specific inhibitors or activators that would act on certain key genes of the organelle.

SUMMARY AND CONCLUSION

Evidence has been cited which indicates that RNA and DNA are present in plastids and mitochondria. A multigenic apparatus in the plastid is deduced from the properties of bleached *Euglena* strains. Control mechanisms are present for the differentiation of proplastids to chloroplasts in *Euglena* and in higher plants, and for the differentiation of promitochondria to mitochondria in yeast. An operon-regulator mechanism for this control is suggested. A comparison of the hereditary cytoplasmic units of *Euglena* plastids and yeast mitochondria indicates great similarities in their properties. Because of these similarities in two unrelated organisms, we suggest that a DNA unit which is self-duplicating and which serves as a code for RNA is the basic hereditary unit of each plastid and mitochondrion. Much work must be done if this reasonable hypothesis is to be converted into well-founded theory.

Some pressing problems await solution. We do not understand the nonrandom distribution of plastids in the mitotic divisions of variegated plants. A related unresolved problem is that of maternal inheritance, in which nonrandom segregation of cytoplasmic organelles occurs after fertilization, causing elimination of the organelles which are contributed by the male parent. How different are the gene components of one plastid in a cell from the gene components of other plastids in the same cell, and how do we test for these differences? Can gene exchange or recombination occur between organelles within the same cell? The answers to these questions may have to await development of more sophisticated techniques, such as the ability to transplant these organelles between different cells or to culture cellular organelles in vitro.

REFERENCES AND NOTES

1. S. Granick, "The plastids; their morphological and chemical differentiation," in Cytodifferentiation and Macromolecular Synthesis, M. Locke, Ed. (Academic Press. New York, 1963), p. 144; ———, "Plastid structure, development and inheritance," in Encyclopedia of Plant Physiology, W. Ruhland, Ed. (Springer, Berlin, 1955), vol. 1, p. 507.
2. A. Gibor and M. Izawa, Proc. Natl. Acad. Sci. U.S. **50:**1164 (1963).
3. A. C. Allison and D. C. Burke, J. Gen. Microbiol. **27:**181 (1962).
4. E. Baltus and J. Brachet, Biochim. Biophys. Acta **76:**490 (1963).
5. E. H. L. Chun, N. H. Vaughan, Jr., A. Rich, "The isolation and characterization of DNA associated with chloroplast preparations," J. Mol. Biol. **7:**130 (1963).
6. R. Sager and M. R. Ishida, "Chloroplast DNA in *Chlamydomonas*" Proc. Natl. Acad. Sci. U.S. **50:**725 (1963).
7. J. Leff, M. Mandel, H. T. Epstein, J. A. Schiff, "DNA satellites from cells of green and aplastidic algae," Biochem. Biophys. Res. Commun. **13:**126 (1963).
8. E. Rasch, H. Swift, R. M. Klein, "Nucleoprotein changes in plant-tumor growth," J. Biophys. Biochem. Cytol. **6:**11 (1959).
9. R. H. Neff, "Volume nucleic acid and nitrogen content of strains of green and colorless *Euglena gracilis* and of *Astasia longa*," J. Protozool. **7:**69 (1960); D. E. Buetow and B. H. Levedahl, "Decline in cellular content of RNA, protein and dry weight during the logarithmic growth of *Euglena gracilis*," J. Gen. Microbiol. **28:**579 (1962).
10. H. Lyman, H. T. Epstein, J. A. Schiff, Biochim. Biophys. Acta **50:**301 (1961).
11. H. Ris and W. Plaut, J. Cell Biol. **13:**383 (1962).
12. H. Kislev, H. Swift, L. Bogorad, unpublished.
13. A. Gibor and S. Granick, J. Cell Biol. **15:**599 (1962).
14. G. Brawerman, personal communication.
15. J. K. Setlow, Photochem. Photobiol. **3:**393 (1963).
16. C. van Wisselingh, "Uber variabilitat und Erblichkeit," Z. Induktive Abstammungs-Vererbungslehre **22:**65 (1920).
17. E. Bauer, "Einfuhring in die Vererbungslehre" (Borntrager, Berlin, ed. 2, 1930), p. 431.
18. J. Schwemmle, "Genetische und zytologische Untersuchungen an Eu-Oenatheren," Z. Induktive Abstammungs-Vererbungslehre **75:**358 (1938).
19. P. Michaelis, "Cytoplasmic inheritance in *Epilobium* and its theoretical significance," Advan. Genetics **6:**287 (1954).
20. P. R. Bell and K. Muhlethaler, "The degeneration and reappearance of mitochondria in the egg cells of a plant," J. Cell Biol. **20:**235 (1964).
21. C. R. Stocking and E. M. Gifford, Biochem. Biophys. Res. Commun. **1:**159 (1959).
22. L. Sagan and S. Scher, "Evidence for cytoplasmic DNA in *Euglena gracilis,*" J. Protozool. **8,** suppl., 20 (1961).
23. J. Biggins and R. B. Park, Bio-Organic Chem. Quart. Rept. U.C.R.L. 9900 (1962), p. 39.
24. J. W. Lyttleton, Exptl. Cell Res. **26:**312 (1962).
25. A. B. Jacobson, H. Swift, L. Bogorad, "Cytochemical studies concerning the occurrence and distribution of RNA in plastids of Zea-mays," J. Cell Biol. **17:**557 (1963).
26. S. Murakami, Exptl. Cell Res. **32:**398 (1963).
27. J. T. O. Kirk, "DNA-dependent RNA syntehsis in chloroplast preparations," Biochem. Biophys. Res. Commun. **14:**393 (1964).
28. A. Giber and M. Izawa, unpublished experiments.
29. G. Brawerman, A. O. Pogo, E. Chargaff, "Induced formation of ribonucleic acids and plastid protein in *Euglena gracilis* under the influence of light," Biochim. Biophys. Acta **55:**326 (1962).
30. A. A. App and A. Jagendorf, "Incorporation of labelled amino acids by chloroplast ribosomes," *ibid.* **76:**286 (1963); J. Eisenstadt and G. Brawerman, "Incorporation of amino acids into protein of chloroplasts and chloroplast ribosomes of Euglena," *ibid.*, p. 319.
31. O. Lindberg and L. Ernster, "Chemistry and physiology of mitochondria and microsomes," in Protoplasmatologia, L. V. Heilbrun and F. Weber, Eds. (Springer, Vienna, 1954), vol. 3, pt. 4.
32. R. Rendi, "On the occurrence of intramitochondrial RNA particles," Exptl. Cell Res. **17:**585 (1959).
33. A. M. Kroon, "Inhibitors of mitochondrial protein synthesis," Biochim. Biophys. Acta **76:**585 (1959).
34. M. M. K. Nass and S. Nass, "Intramitochondrial fibers with DNA characteristics. I, Fixation and electron staining reactions," J. Cell Biol. **19:**593 (1963).
35. M. Chevremont, "Cytoplasmic DNA," in Cell Growth and Cell Division, R. J. C. Harris, Ed. (Academic Press, New York, 1963), p. 323.
36. G. Steinert, H. Firket, M. Steinert, "Synthèse d'acide desoxyribonucleique dans le corps parabasal de *Trypanosoma mega,*" Exptl. Cell. Res. **15:**632 (1958).
37. M. Steinert, "Mitochondria associated with the kineto-nucleus of *Trypanosoma mega,*" J. Biochem. Biophys. Cytol. **8:**542 (1960).
38. M. M. Rudzinska, P. A. D'Alesandro, W. Trager, "The fine structure of *Leishmania donovani* and the role of the kinetoplast in the leishmania leptomonad transformation," J. Protozool., in press.
39. B. Ephrussi, "Nucleo-Cytoplasmic Relations in Microorganisms" (Oxford Univ. Press, New York, 1953).
40. Y. Yotsuyanagi, "Etudes sur le chondriome de la levure I and II," J. Ultrastructure Res. **7:**121 (1962).
41. C. Raut and W. L. Simpson, "The effect of x-rays and of ultra-violet light of different wave lengths

on the production of cytochrome deficient yeast," Arch. Biochem. Biophys. **57**:218 (1955).

42. G. H. Bourne, "Mitochondria and the golgi complex," in Cytology and Cell Physiology, G. H. Bourne, Ed. (Oxford Univ. Press, New York, ed. 2, 1951) p. 234.
43. E. H. Newcomer, "Concerning the duality of the mitochondria and the validity of the osmiophilic platelets in plants, Am. J. Botany **33**:684 (1946).
44. I. Manton, "Electron microscopical observations on a very small flagellate: the problem of *Chromulina pusilla* Butcher," J. Marine Biol. Assoc. U.K. **38**:319 (1959).
45. D. J. L. Luck, "Formation of mitochondria in *Neurospora crassa,*" J. Cell Biol. **16**:483 (1963).
46. S. G. Wildman, T. Hongladarom, S. I. Honda, "Chloroplasts and mitochondria in living plant cells," Science **138**:434 (1962).
47. M. S. Fletcher and D. R. Sanadi, "Turnover of rat-liver mitochondria," Biochim. Biophys. Acta **51**:356 (1961).
48. A. W. Linnane, E. Vitols, P. G. Nowland, J. Cell Biol. **13**:345 (1962).
49. Y. Yotsuyanagi, personal communications, 1964.
50. A. Gibor and S. Granick, J. Protozool. **9**:327 (1962).
51. S. Granick, J. L. Granick, A. Gibor, unpublished results.
52. L. G. Moriber, B. Hershenov, S. Aaronson, B. Barsky, "Teratological chloroplast structures in *Euglena gracilis* permanently bleached by exogenous physical and chemical agents," J. Protozool. **10**:80 (1963).
53. S. Nagai, N. Yanagishima, H. Nagai, "Advances in the study of respiration-deficient mutation in yeast and other microorganisms," Bacteriol. Rev. **25**:404 (1961).
54. L. Provasoli, S. Hutner, A. Shatz, "Streptomycin-induced chlorophyll-less races of *Euglena,*" Proc. Soc. Exptl. Biol. Med. **69**:279 (1948).
55. E. G. Pringsheim and O. Pringsheim, "Experimental elimination of chromatophores and eye-spot in *Euglena gracilis*," New Phytologist **51**:65 (1952).
56. B. Ephrussi, H. D. Hottinguer, H. Roman, Proc. Natl. Acad. Sci. U.S. **41**:1065 (1955).
57. H. I. Virgin, A. Kahn, D. Van Wettstein, "The physiology of chlorophyll formation in relation to structural changes in chloroplasts," Photochem. Photobiol. **2**:83 (1963).
58. R. M. Smillie, "Formation and function of soluble proteins in chloroplasts," Can. J. Botany **41**:123 (1963).
59. R. C. Fuller and M. Gibbs, "Intracellular and phylogenetic distribution of ribulose 1,5-diphosphate carboxylase and D-glyceraldehyde-3-phosphate dehydrogenase," Plant Physiol. **34**:324 (1959).
60. P. P. Slonimski, "Adaptation respiratoire: développement du système hemoprotéique induit par l'oxygène," Proc. Intern. Congr. Biochem. 3rd Brussels (1955), p. 242; H. Chantrenne, "Formation induite de cytochrome peroxydase chez la levure," Biochim. Biophys. Acta **14**:157 (1954).

38 Infective transfer of maternally inherited abnormal sex-ratio in Drosophila willistoni

Chana Malogolowkin*
D. F. Poulson
Department of Zoology
Columbia University
New York, New York
Faculdade Nacional de Filosofia
Rio de Janeiro, Brazil
and
Osborn Zoological Laboratory
Yale University
New Haven, Connecticut

Deviations from the normal 1/1 ratio of sexes are known in natural populations of several species of *Drosophila.* These deviations usually take the form of production of unisexual progenies which consist mainly or exclusively of daughters. In *D. obscura, D. pseudoobscura, D. persimilis,* and *D. azteca,* this condition is inherited through the X-chromosome *(1).* On the other hand, Cavalcanti *(2),* Magni *(3),* and Carson *(4)* discovered in *D. prosaltans, D. bifasciata,* and *D. borealis,* respectively, deviations from the normal sex-ratio which appear to be inherited through the cytoplasm. Females of certain strains produce progenies that consist mainly or only of daughters, regardless of which males they are crossed to, and this condition is transmitted, in turn, to all of their female offspring. This cytoplasmically inherited "sex-ratio" condition resembles, in many ways, the oversensitivity to CO_2 that was studied by l'Heritier and his school *(5).* Recently, B. Spassky observed that a single female of *D. willistoni,* from Jamaica, and a single female of *D. paulistorum,* from Sierra Nevada de Santa Marta, Colombia, produced nearly unisexual female progenies and that this peculiarity was inherited by their offspring. Spassky has very generously given these stocks to one of us (C.M.) for study.

The "sex-ratio" condition of *D. willistoni* has been examined in some detail. Females from the "sex-ratio" strain produce nothing but daughters in outcrosses to males from most of the strains which have been tested in this respect. However, outcrosses to males from three strains collected at Recife, Brázil, from one strain from the island of Saint Lucia, West Indies, from one strain from Costa Rica, and from a laboratory strain that contains the second chromosome mutants Star, Hooked, abbreviated, and brown, produce intermediate or normal (1 ♀/1 ♂) sex-ratios after one or more generations of crossing and backcrossing. Thus, the "sex-ratio" condition is not transmitted through the usual chromosomal inheritance, but it is not independent of chromosomal genes *(6).*

Eggs deposited by "sex-ratio" females fall into two readily distinguishable classes when they are dechorionated about 2 to 4 hours after deposition. Approximately half of the eggs begin to show translucent areas, both anteriorly and posteriorly, following formation of the blastoderm. These eggs show no further normal development and yield no larvae. Presumably, they represent dying male zygotes. A fraction of eggs which appear normal in early stages produce embryos which fail to hatch and darken markedly between 24 and 36 hours after being laid. Although the sex of these embryos is not yet certain, it seems probable that they are female.

To test the possibility of the transfer of the "sex-ratio" condition to normal females, early abnormal eggs from "sex-ratio" females were punctured (about 3 to 6 hours after deposition) with a micropipette. Ooplasm was taken into the pipette and injected into the abdomens of young virgin females from the Recife strain. Uninjected females from this strain give, with great regularity, a normal 1/1 ratio of the sexes.

From Science **126**:32, July 5, 1957. Used with permission.

*Present address: P. O. Box 1064, Tel-Aviv, Israel.

The injected females were then mated to males of their own strain and transferred, at 2-day intervals, to a fresh culture medium. Eggs were collected in this way until the end of the life of each of the injected females. In most of the cases the broods from each of the females for the first 2 weeks of egg production yielded normal proportions of males and females. However, at the end of this period, five out of the 16 females began to produce mainly daughters, and finally they produced daughters exclusively. One of the females showed a ratio of 2 ♀/1 ♂ from the beginning and, at the end of the first 2 weeks, began to produce only females. Daughters of the injected females derived from the successive broods of eggs were then tested by mating to brothers or to males from the normal Recife strain. In the two most thoroughly tested cases of broods from the later period, when only females were being produced, 17 daughters of one injected female all showed "sex-ratio" in their progeny, 11 giving no males at all; the others, only a few males. Twelve daughters of the other injected female all gave "sex-ratio" progeny; among them were five that gave no males at all. In all cases the progenies were sufficiently large to leave no doubt of the presence of the "sex-ratio" condition in these flies. Subsequently, the F_2 daughters have produced "sex-ratio" progenies; hence, it is clear that the original infection has now been transmitted through three generations. Stocks of these new "sex-ratio" strains are now being maintained.

Examination of the eggs of the new "sex-ratio" females shows the same abnormalities that were encountered in the original "sex-ratio" strain of Jamaican origin *(7)*.

A series of controls was carried out, along with the "sex-ratio" injections. For these, ooplasm of unfertilized 3- to 6-hour eggs from virgin females of the Recife strain was injected into young virgin females of this strain, by means of the same procedures that were followed in the experimental series. Broods from eggs laid at 2-day intervals were raised, and the sex ratio was determined. In none of the 15 females of this control were there any significant deviations from the normal ratio of 1/1, even after the 2-week period in which the experimental series showed the striking changes that have been described in this report.

It is therefore clearly demonstrated that the "sex-ratio" condition in *Drosophila willistoni* can be transferred to normal females, and that it is essentially infectious in nature.

REFERENCES AND NOTES

1. S. Gershenson, Genetics **13**:488 (1928); A. H. Sturtevant and Th. Dobzhansky, *ibid.* **21**:473 (1936); Th. Dobzhansky, Scientia Genet. (Turin) **1**:67 (1939).
2. A. G. L. Cavalcanti and D. N. Falcão, Caryologia **6**:1233 (1954).
3. G. E. Magni, Ist. Lombardo di Scienze e Lettere **LXXV**:3 (1952); Caryologia **6**:1213 (1954).
4. H. L. Carson, Drosophila Information Service **30**:109 (1956).
5. P. l'Heritier, Cold Spring Harbor Symposia **16**:99 (1951).
6. A detailed report of this inheritance is in preparation.
7. A report of the developmental disturbances and the cytology of the "sex-ratio" eggs and ovaries is in preparation.

39 Nature of "sex-ratio" agent in Drosophila

D. F. Poulson*
B. Sakaguchi
Department of Zoology
Yale University
New Haven, Connecticut

Abstract. *Several lines of evidence implicate small spirochetes, presumably treponemata, as etiologic agents in the production of the maternally transmitted "sex ratio" condition (SR) in* Drosophila nebulosa, *in* D. willistoni, *and in strains of* D. melanogaster *into which the SR condition has been artificially transferred. The presence of treponemata in the hemolymph of adult females of these species is completely correlated with the production of unisexual progenies and like this condition is dependent on the genotype of the host and of the infectious agent.*

A condition of unisexual (or near unisexual) progenies in *Drosophila* known as "sex-ratio" (SR) has been intensively studied in a number of species by several investigators *(1)* who showed it to be maternally transmitted and dependent on an agent variously interpreted as a plasma gene, cytoplasmic particulate, or virus *(2)*. The pattern of transmission has been established in *D. bifasciata, D. equinoxialis, D. nebulosa, D. paulistorum, D. prosaltans,* and *D. willistoni (3)*, and in all except the first it has been shown that the stability and persistence of the condition is also dependent on the nuclear genotype of the flies. In *D. bifasciata* and *D. equinoxialis* this condition has been shown to be temperature sensitive and subject to thermic cure *(4)*. The unisexual progenies are a consequence of disturbances of development in male zygotes which lead to 50 percent egg mortality *(5)*. The pattern of disturbance and the stage of onset are strongly influenced by the genotype of the zygote, and there is evidence that female as well as male zygotes may be affected *(6)*.

From Science **133**:1489-1490, May 12, 1961.

The research was supported by grant No. 6017 from the National Science Foundation.

*Present address: Department of Biology, Yale University.

That the "sex-ratio" agent is of an infectious nature (even though the condition is not contagious) was demonstrated in experiments in which the condition was transferred by injection of ooplasmic materials from SR strains into previously normal strains of *D. willistoni* and *D. equinoxialis (7)*. In the course of such experimental infections the condition makes its appearance after an incubation period of 10 to 12 days in many or all of the infected females. In some instances the condition may only make its appearance in later generations after a period of latent transmission *(8)*. A thorough study of the course of several such infections showed that the otherwise normal daughters of injected females may transmit a sporadic and nonspecific zygote lethality which can be carried through more than 20 generations of their descendants *(9)*.

A study of the distribution of the "sex-ratio" agent in tissues and organs of SR adults of *D. willistoni* demonstrated its presence in ovary, fat body, flight muscle, and in exceptionally high concentration in hemolymph *(10)*. The agent was also found in high concentration in the hemolymph of the rare surviving sons of SR females in this species. Further, the agent may be present in latent form in flies of apparently normal strains of *D. willistoni,* as shown by infections produced from injections of extracts of such flies into other females of the same strain. Attempts to separate and concentrate the infective agent of *D. willistoni* by centrifugation and ultrafiltration of extracts of flies showed that the activity (as measured by frequency of infections) of supernatant fluids from whole-fly homogenates is not reduced on passage through a Millipore filter of pore size 0.3 μ, but is cut to about one-third by passage through a Millipore filter of pore size 100 mμ *(11)*.

Microscopic examination of the hemolymph of females of *D. nebulosa* and *D. willistoni* giving strictly unisexual progenies shows the regular presence of many very fine filaments which are more numerous in older than in newly emerged flies. These filaments, which are absent from the blood of normal females of both species, are in constant motion in fresh preparations of hemolymph mounted in Crown immersion oil. They are visible only with dark-field or phase-contrast microscopy and in thin preparations can be seen to be of the order of 0.1 to 0.2 μ in diameter and to average 4 to 5 μ in length, although occasional individuals may be 8 to 10 μ in length. In favorable freshly mounted preparations, waves of sinusoidal or helicoidal movement are seen to pass along the length of the filaments, giving them a regular spiral appearance. In such preparations the filaments remain visible and active up to 48 hours at 25°C, although they become increasingly granular and fuzzy-edged, and evidences of spiralizing vanish. During this time there is an increase in the number of minute granules free in the hemolymph. At the end of a week's time the hemolymph preparations contain principally such small granules, evidently derived from the filaments. The granules appear to have a minute, almost invisible, tail or flagellum and maintain an active movement distinguishable from Brownian movement. Similar small active granules are numerous in the blood of rare surviving sons of SR females. Occasional granules of this sort are encountered in the hemolymph of most normal females and males.

The most satisfactory permanent preparations have been obtained by fixation of small drops or smears of hemolymph in formaldehyde vapor followed by one or another of the standard staining procedures *(12)*. The filaments stain a strong purplish-red with Giemsa, are intensely colored by Fast Green employed as a basic protein stain, and are moderately colored by Azure B under conditions of ribonucleic acid staining. The filaments are Gram-negative, but give indications of slight Feulgen positivity. They blacken readily with Fontana's silver method. Their characteristic appearance in Giemsa preparations is seen in Fig. 1.

The filaments are sensitive to penicillin and rapidly disappear from the hemolymph of SR females injected with penicillin G. However, they reappear after a few days in the absence of further injections of penicillin. Normal hosts injected with penicillin-treated SR blood pro-

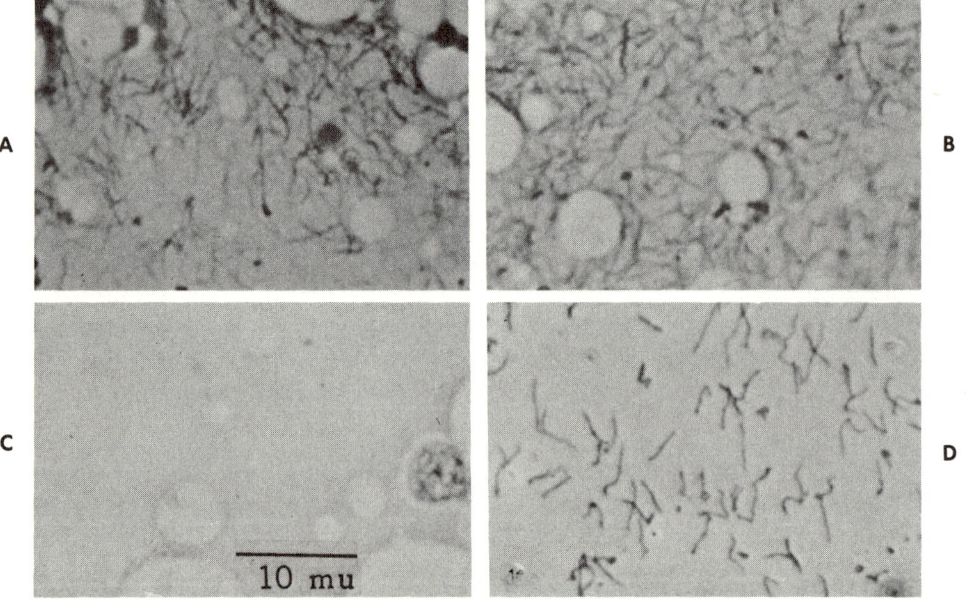

Fig. 1. Hemolymph smears from SR and normal females. Giemsa stain, phase contrast, scale for all photos is given in **C. A** and **B**, ***Drosophila nebulosa,*** SR strain PV-45, fixed in formaldehyde vapor, showing numerous spirochetes. **C,** ***D. nebulosa,*** normal strain PV-59, no spirochetes; clear areas indicate fat droplets; small blood cell nucleus at right. **D,** ***Drosophila melanogaster,*** SR strain (XXY) derived by transfer of SR from the SR B-3 strain of ***D. willistoni;*** fresh Giemsa preparation without fixation; spirochetes somewhat swollen and well stained.

duced no SR progeny, while controls gave 100 percent SR offspring. Filaments were completely absent in the former and invariably present in the latter. The filaments have all the characteristics of small spirochetes and resemble those of the genus *Treponema* more closely than any other described forms *(13)*.

Further evidence of the correlation between the "sex-ratio" condition and the presence of treponemata derives from experiments in which the condition was transferred from *Drosophila willistoni* into *D. melanogaster (14)*. There, as in the original form, all females producing unisexual progenies possess the treponemata in their hemolymph. Where loss of the unisexual condition has occurred the spirochetes are lacking. The stability of the condition in *D. melanogaster* is dependent on genotype and, in some instances at least, on temperature.

Although apparently different from a larger spirochete described by Chatton in 1912 from *D. confusa* and named by him *Treponema drosophilae (15)*, the rare occurrence of giant forms in the blood of very old flies (especially in *D. nebulosa*) does raise the question of a possible relationship. Since it has been demonstrated that the "sex-ratio" agent, and thus spirochetes, may be carried in latent condition, the whole subject of interrelationships of spirochetes and *Drosophila* deserves thorough investigation.

It is concluded that the "sex-ratio" condition in *Drosophila* species is a consequence of a very special, stabilized relationship between spirochete and host in which the genotypes of both are of prime importance.

REFERENCES AND NOTES

1. G. E. Magni, Ist. Lombardo Sci. Lettere **75**:3 (1952); A. G. L. Cavalcanti and D. N. Falcão, Caryologia **6**:1213 (1954); A. G. L. Calvalcanti, D. N. Falcão, L. E. Castro, Univ. Brasil Publ. fac. nacl. fil. Ser. cient. No. 1 (1958); H. L. Carson, Drosophila Inform. Serv. **30**:109 (1956); C. Malogolowkin and D. F. Poulson, Science **126**:32 (1957); C. Malogolowkin, Genetics **43**:274 (1958).
2. A. G. L. Calvalcanti, D. N. Falcão, L. E. Castro, *op. cit.* (1958); G. E. Magni, Atti 2nd Riun. Assoc. Genet. Ital. **26**:1 (1956); *ibid.* **27**:1 (1957); C. Malogolowkin and D. F. Poulson, Science **126**:32 (1957); D. F. Poulson and C. Malogolowkin, Proc. Intern. Congr. Zool. 15th Congr. London (1959), p. 200.
3. G. E. Magni, Ist. Lombardo Sci. Lettere **75**:3 (1952); Atti 2nd Rivn. Assoc. Genet. Ital. **26**:1 (1956); A. G. L. Cavalcanti, D. N. Falcão, L. E. Castro, Am. Naturalist **91**:321 (1957); Univ. Brazil Publ. fac. nacl. fil. Ser. cient. No. 1 (1958); C. Malogolowkin, Genetics **43**:274 (1958); C. Malogolowkin, Am. Naturalist **93**:365 (1959); D. F. Poulson, unpublished data on *Drosophila nebulosa* from strains kindly provided by H. L. Carson of Washington University, St. Louis, Mo.
4. G. E. Magni, Caryologia **6**:1213 (1954); C. Malogolowkin, Am. Naturalist **93**:365 (1959).
5. C. Malogolowkin and D. F. Poulson, Science **126**:32 (1957); C. Malogolowkin, D. F. Poulson, E. Y. Wright, Genetics **44**:59 (1959).
6. D. F. Poulson and S. J. Counce, Anat. Record **134**:625 (1959); unpublished data.
7. C. Malogolowkin, D. F. Poulson, E. Y. Wright, Genetics **44**:59 (1959); C. Malogolowkin, Am. Naturalist **93**:365 (1959).
8. C. Malogolowkin, D. F. Poulson, E. Y. Wright, Genetics **44**:59 (1959); D. F. Poulson, unpublished data.
9. D. F. Poulson, unpublished data.
10. B. Sakaguchi and D. F. Poulson, Ann. Rept. Natl. Inst. Genet. Japan **10**:27-28 (1959).
11. ———, unpublished data.
12. R. D. Lillie, Histopathologic Technic and Practical Histochemistry (Blakiston, New York, 1954), p. 399; R. E. Campbell and P. D. Rosahn, Yale J. Biol. and Med. **22**:527 (1950).
13. Bergey's Manual of Determinative Bacteriology (Williams and Wilkins, Baltimore, ed. 7, 1957), pp. 896-907; T. B. Turner and D. H. Hollander, "Biology of the Treponematoses," World Health Organization Monogr. Ser. No. 35 (1957).
14. B. Sakaguchi and D. F. Poulson, Anat. Record **138**:381 (1960); D. F. Poulson and B. Sakaguchi, *ibid.* **138**:376 (1960); C. Malogolowkin, by personal communication, informs us that the SR strain of *Drosophila equinoxialis* carries treponemata indistinguishable from those reported here.
15. E. Chatton, Compt. rend. soc. biol. **73**:212 (1912).

40 Genetic studies of lysogenicity in Escherichia coli

Esther M. Lederberg*
Joshua Lederberg
Department of Genetics
University of Wisconsin
Madison, Wisconsin

Recent research on *Escherichia coli* phages has outlined the biology of viruses that promptly lyse their bacterial hosts (Delbrück 1950). In addition to the progressive parasitic relationship that these studies have analyzed, many phage-bacterium complexes persist in a more enduring symbiosis, lysogenicity. The experiments to be described in this paper were designed to probe two related questions: how is the virus of a lysogenic bacterium transmitted in vegetative and sexual reproduction? and how is a symbiotic complex established following infection by the virus, as an alternative to the parasitization and lysis of the host bacterium? Complementary problems, especially concerning the growth and release of virus in lysogenic bacteria have received more emphasis from other workers (Bertani 1951; Lwoff and Gutmann 1950; Weigle and Delbrück 1951).

Our interest in lysogenicity was provoked by the discovery that *E. coli* strain K-12 was lysogenic. On two occasions, mixtures of certain mutant stocks appeared to be contaminated with bacteriophage. The plaques were unusual in showing turbid centers, suggesting those figured by Burnet and Lush (1936). It soon became apparent that practically all K-12 cultures carried this latent phage. The novelty consisted of two exceptional mutant substrains, W-435 and W-518 which were sensitive to the phage, now referred to as λ. These two strains had been maintained in our stocks as nonfermenting mutants for lactose (Lac_3^-) and galactose ($Gal_4^- Lac_1^-$), respectively, isolated from ultraviolet-treated suspensions. Both cultures are derived from 58-161, a methionine-requiring auxotroph previously used in many recombination experiments (Tatum 1945; Tatum and Lederberg 1947). The lysogenicity of strain K-12 had remained unsuspected despite its maintenance for over 25 years and close study as the subject of mutation and recombination experiments since 1944. However, the only objective criterion of a lysogenic symbiosis is the lysis of another sensitive strain that functions as an indicator. Thus, in the absence of an appropriate conjunction of strains the virus carried by the K-12 subline would remain undetected. Because of the low frequency of sensitive strains, such opportunities are rare. The development of crossing techniques in strain K-12 has allowed the virus to be studied as a genetic factor. Intercrosses among strains differing with respect to λ and the development of lysogenic from sensitive strains are the main subjects of this report.

TERMINOLOGY

Although the adoption of a fixed terminology would be premature, for convenience, a few terms will be defined for use in this account. Lysogenicity will be understood as the regular and persistent transmission of virus potentiality during the multiplication of a bacterium, without overt lysis. When tested directly with the phage, a bacterial culture is *sensitive* (lysed) or *resistant* (not lysed). When tested with a sensitive indicator strain, the bacteria are *lysogenic* (carriers of λ) if the indicator is lysed, or *nonlysogenic* if not. Bacteria that are resistant to λ but nonlysogenic are termed *immune.* The virus as transmitted in lysogenic bacteria will be referred to as *latent virus.*

MATERIALS AND METHODS

Preparation of free phage

Suspensions of λ were first obtained from filtrates of 6–8-hour bacterial cultures devel-

From Genetics **38**:51-64, 1953. Used with permission.

Paper No. 490 of the Department of Genetics, University of Wisconsin. The investigation was supported in part by a research grant (E72-C3) from the National Microbiological Institute of the National Institutes of Health, Public Health Service.

*Present address: Department of Medical Microbiology, Stanford University Medical School, Palo Alto, California.

oped from mixed inocula of λ-sensitive and λ-lysogenic strains in nutrient broth. Thereafter, further batches were prepared by growing λ-sensitive cells with virus according to the usual methods (Adams 1950). Lysis in broth is indicated by decreased turbidity rather than marked clearing. A convenient method for obtaining high titer λ directly from lysogenic bacteria has been developed by Weigle and Delbrück (1951) from the methods described by Lwoff *et al.* (1950). A lysogenic strain grown in a yeast-extract broth is subjected to a dose of ultraviolet irradiation which kills only a small fraction of a genetically comparable λ-free strain. After 40–50 minutes incubation in yeast-extract broth the majority of lysogenic cells lyse with a burst of about 100 phage particles each. Virus titrations were made by established methods (Adams 1950; Delbrück 1950). All lysates were sterilized by filtration through nine- or fourteen-pound test Mandler candles.

Some pertinent physical and morphological characteristics of λ have been described by other investigators (Weigle and Delbrück 1951).

Media

The media recommended for observing phage-bacterium interactions are less useful with the λ system because of the presence of bacterial survivors (which prove to be either resistant or sensitive) in the plaques. The lysed areas are, however, accentuated by their discoloration on an eosin–methylene blue agar medium without the fermentable sugar customarily added (EMB base, Lederberg 1950). Plaques from free λ suspensions were counted on TSA (tryptone saline agar, Weigle and Delbrück 1951). It was sometimes supplemented with ten percent citrated bovine blood to test the release of hemolysins during bacterial lysis by phage (Schiff and Bornstein 1940). A positive reaction is the clearing of the blood at the zone where sensitive bacteria are exposed to λ or to lysogenic bacteria. It must be cautioned, however, that occasional cultures are normally hemolytic, perhaps owing to a high rate of spontaneous lysis.

Scoring for sensitivity and lysogenicity

Susceptibility to λ is tested by streaking a phage suspension across a dry EMB agar plate with a broad loop. A small loopful of the cells to be tested is then streaked at right angles to the phage. To test for lysogenicity, the bacteria are similarly streaked against a sensitive indicator. As a precautionary measure, the tested cells are also deposited at a control spot. As shown in figure 1, positive tests consist of the interruption in the continuity of growth of the indicator, or plaques and discoloration at the conjunction of phage with sensitive bacteria. The technique of replica plating (Lederberg and Lederberg 1952) facilitates large-scale tests of lysogenicity. Instead of individual tests on bacterial colonies from a plate, these are transferred en masse by means of velveteen

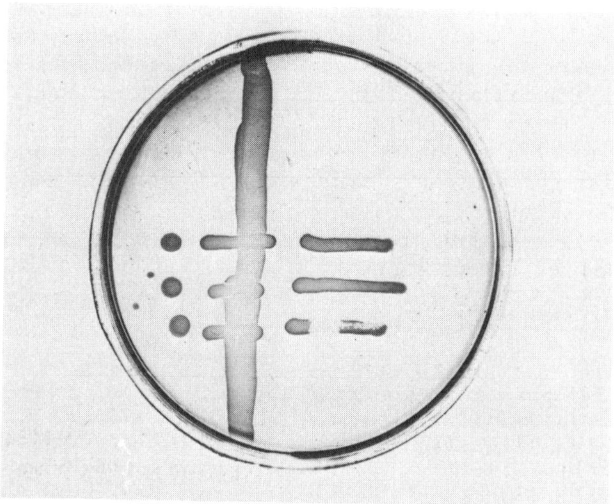

Fig. 1. Reactions of sensitive, lysogenic and immune. Extreme left: control spots. Left center: cross-streaks against sensitive indicator bacteria. Right: cross-streaks against λ. From top to bottom $Lp_1{}^s$, $Lp_1{}^+$, and $Lp_1{}^r$ respectively.

fabric to a TSA plate previously layered with 10 ml of TSA seeded with about 10^8 indicator cells. On the replica, each lysogenic colony is surrounded by a zone of lysis.

Crosses

Crosses are carried out by plating washed cultures differing in nutritional characters on minimal agar (Tatum and Lederberg 1947; Lederberg *et al.* 1951) or with added streptomycin where streptomycin sensitive (S^s) prototrophs are crossed with resistant (S^r) diauxotrophs (Lederberg 1951a). The resulting progeny are picked and purified by streaking on a complete medium, and from this, single colonies are isolated for further characterization of segregating markers, including lysogenicity.

Selection of lysogenic and nonlysogenic cultures

Lysogenic bacteria may be routinely isolated from turbid plaques on sensitive bacteria plated with λ or from the residual growth after mixed inoculation of sensitive bacteria and phage in broth on agar. Successive single colony purifications result in stably lysogenic isolates free of extraneous λ and sensitive bacteria. By isolating one lysogenic derivative from a series of single plaques, it was demonstrated that the transfer of λ from well-marked lysogenic to previously sensitive stocks occurs without any alteration of the known genetic markers of the new host other than its reaction to λ.

The isolation of nonlysogenic (immune or sensitive) types from lysogenic bacteria is less predictable but they have been obtained by the following procedure. EMB or blood agar plates were spread with 10^8 cells and exposed to ultraviolet light so that about 100 to 200 colonies survived. New types have been sporadically detected either by testing large numbers of normal-appearing colonies or by the partially lysed appearance of phage-contaminated sensitive colonies. These consist of lysogenic and sensitive sectors, and free λ. Immune mutants have arisen from sensitive bacteria after selection with phage. The various occurrences of nonlysogenic derivatives are listed in table 1.

RESULTS

Intercrosses of various phenotypes

Crosses among the sensitive, lysogenic, and immune strains are all fully fertile. They have been repeated many times with the following

Table 1. Principal stocks used in lysogenicity studies

Strain number	Source strain	History	Genotype[1]
		Sensitive $(Lp_1{}^s)$	
W-435	58-161	UV (ultraviolet)	$M^-Lac_3{}^-$
W-518	Y-87	UV	$M^-Lac_1{}^-Gal_4{}^-$
W-1267		W-518 × W-588, f-1 segregant	$T^-L^-Lac_1{}^-Gal_4{}^-$
W-1485	K-12	UV; blood agar	wild type sensitive
W-1486	W-811	plating with streptomycin	$M^-Lac_1{}^-Gal_4{}^-S^r$
W-1487	W-1405	plating with streptomycin	$T^-L^-Lac_1{}^-Gal_4{}^-S^r$
W-1502	W-1245	spontaneous variation	M^-
W-1503	W-1296	spontaneous variation	T^-L^-
W-1655	58-161	UV	M^-
W-1872	K-12	UV	wild type sensitive
		Immune-1 and -2 $(Lp_1{}^r$ and $Lp_2{}^r)$	
W-1027	Y-70	UV	$T^-L^-Lac_1{}^-Lp_1{}^r Lp_2{}^s$
W-1924	W-518	selection with λ	$M^-Lac_1{}^-Gal_4{}^-Lp_1{}^r Lp_2{}^s$
W-1248	W-518	selection with λ	$M^-Lac_1{}^-Gal_4{}^-Lp_1{}^s Lp_2{}^r$
W-1603	W-1177	UV	T^-L^-etc., $Lp_1{}^s Lp_2{}^r$
W-1245	W-478	UV	M^-; unstable immune
W-1296	W-588	UV	T^-L^-; unstable immune
		Lysogenic $(Lp_1{}^+$: $Lp_2{}^r$ or $Lp_2{}^s)$	
58-161		standard parent	$M^-Lp_2{}^s$
W-1177		multiple marker parent	$T^-L^-Lac_1{}^-Mal_1{}^-Xyl^-Gal_5{}^-S^r Lp_2{}^r$
W-811	W-518	infection with λ	$M^-Lac_1{}^-Gal_4{}^-Lp_2{}^s$
W-1439	W-811	selection with λ-2	$M^-Lac_1{}^-Gal_4{}^-Lp_2{}^r$

[1] The significance of the genotypic symbols, and further details of ancestry of many stocks are given in Lederberg *et al.* (1951) and Lederberg (1952).

qualitative results, based on 200 or more tests for each cross.

1. Lysogenic X lysogenic: all progeny lysogenic.
2. Sensitive X sensitive: all progeny sensitive.
3. Sensitive X lysogenic: the progeny segregate into sensitive and lysogenic, with ratios depending on the nutritional markers of the parents.

The total number of tests of crosses 1 and 2 is actually much larger, for exceptional progeny would have been apparent upon inspection of similar crosses conducted for other purposes.

Since only the parental types are found in cross 3, it might be inferred that lysogenic differs from sensitive only by one factor, the presence of the λ. However, the consideration of λ as a cytoplasmic factor leads to a possible paradox: when λ is contributed by just one parent, it segregates among the progeny, but it always appears when contributed by both parents. It should be emphasized that the same segregation ratios for other markers have been obtained regardless of the presence or absence of λ in the parents. No evidence has been found to date for the functioning of λ as a gamete or other sexual form (cf. Hayes 1952; Lederberg, Cavalli and Lederberg 1952).

Further crosses involving two immune parents gave the following results:

4. Immune-1 X sensitive: parental only.
5. Immune-2 X sensitive: parental only.
6. Immune-1 X lysogenic: parental only.
7. Immune-2 X lysogenic: parental and sensitive.
8. Immune-1 X immune-2: parental and sensitive.

The appearance of a sensitive recombination class in cross 8 implicates two loci in resistance to λ. Sensitive can be described as $Lp_1{}^s$ $Lp_2{}^s$, immune-1 as $Lp_1{}^r$ $Lp_2{}^s$ and immune-2 as $Lp_1{}^s$ $Lp_2{}^r$. From the result of cross 6, in contrast to cross 7, lysogenicity is also determined at the Lp_1 (latent phage) locus. Evidence for two kinds of lysogenic, $Lp_2{}^s$ (those so far discussed) and $Lp_2{}^r$, respectively, will be presented in another section.

Occasional sensitive progeny would have been anticipated in cross 6 on the hypothesis that lysogenic is genotypically equivalent to sensitive, and differs only by the presence of cytoplasmic λ, but were not found. The independent segregation of λ (cross 3) and of the genetic factor Lp_1 (cross 4) would have resulted in some λ-free recombinants sensitive to the virus. The results of all these crosses hinted at a primarily "chromosomal" determination of lysogenicity.

Linkage behavior of lysogenicity

The concept of an Lp_1 locus was strengthened by the outcome of linkage tests in which various markers were segregating. A loose linkage of Lp_1 to *Xyl* and to *S* was indicated in preliminary crosses with a multiple marker stock. However, Lp_2 was also segregating, thus doubling the number of genotypic classes, and perhaps confusing the issue. The closest linkage of Lp_1 thus far found has been to Gal_4, as shown in table 2. As it happens, this is the distinctive marker of W-518, in which λ-sensitivity was first noticed.

The linkage of Lp_1 with Gal_4 has been verified by crosses with various combinations of lysogenic stocks resynthesized from sensitive auxotrophs. Some of the latter were newly developed from W-1485, a λ-sensitive directly derived from strain K-12. There can be little doubt, therefore, that the segregating factor is directly associated with lysogenicity. The linked segregations justify the assignment of a new allele, $Lp_1{}^+$, characteristic of lysogenicity. The result indicated for cross 6 points to this as a third allele at the same locus as the contrasting $Lp_1{}^r$ (immune-1) and $Lp_1{}^s$ (sensitive).

Segregation of λ from diploids

Heterozygotes selected as Lac^+/Lac^- or Gal^+/Gal^- were obtained and shown to be segregating for a number of other factors (Lederberg 1949), but these selections were either λ-sensitive or λ-lysogenic. Similar results were obtained in immune, *Het* crosses. It was thought, however, that the λ-determinant might be hemizygous in these diploids, like the *Mal* and *S* factors previously studied (Lederberg *et al.* 1951). This difficulty has been circumvented by the use of diploid X haploid crosses, in which the segmental elimination (of *Mal* and *S*) apparently does not occur. A lysogenic diploid parent (T^- L^- Gal^+ Lac^+ Mal^+/Lac^- Mal^-) was crossed with a sensitive, haploid auxotroph (M^- $Gal_4{}^-$ Lac^- Mal^+) on minimal agar. The resulting prototrophs were almost all diploid, and several were identified as lysogenic, but segregating Gal^+/Gal^- as well as other factors. As shown in table 3, presence vs. absence of λ segregated in the same coupling as shown by the parents: Gal^+ lysogenic/Gal^- nonlysogenic.

Table 2. Linked segregation of Gal_4 and Lp_1 among prototrophic recombinants

		Parents			Prototroph recombinants: $M^+T^+L^+$ . . .			
	A	$M^-T^+L^+$	×	$M^+T^-L^-$	$Gal^+ Lp^+$	$Gal^+ Lp^s$	$Gal^- Lp^+$	$Gal^- Lp^s$
1		Gal^+Lp^s	×	Gal^-Lp^+	1	83	90	2
2		Gal^+Lp^+	×	Gal^-Lp^s	33	1	3	41
3		Gal^-Lp^s	×	Gal^+Lp^+	55	0	5	53
4		Gal^-Lp^+	×	Gal^+Lp^s	1	42	44	1
	B	$M^-H^+L_2^+$	×	$M^+H^-L_2^-$				
1		Gal^-Lp^+	×	Gal^+Lp^s	0	34	40	1
	C	$M^-M_2^+G^+$	×	$M^+M_2^-G^-$				
1		$Gal^-Lp_1^+$	×	Gal^+Lp^s	0	40	39	1
2		Gal^-Lp^s	×	Gal^+Lp^+	64	2	1	67

The crosses were conducted on EMS galactose medium, from which approximately equal numbers of Gal^+ and Gal^- prototrophs were picked for further test. Similar results were obtained when the proportion of Gal^+ and Gal^- was not thus fixed, as on non-indicator glucose minimal agar, but the preponderance of one parental type among the prototrophs limited the usefulness of unselected isolates for linkage tests. The $H^+L_2^-$ and $M_2^-G^-$ parents indicated in B and C are histidine-leucine and methionine-glycine auxotrophs, respectively, recently derived from W-1485. All parents in these crosses were Lp_2^s, but V_1, Lac_1, and S were segregating in their usual patterns.

Table 3. Segregation of Gal_4 and Lp_1 from heterozygous diploids

	$Gal^+ Lp^+$	Gal^+Lp^s	Gal^-Lp^+	Gal^-Lp^s
H-295	36	1	1	39 (19 Lp_2^s)
H-297	29	0	0	11 (3 Lp_2^s)

Segregant (pure) Gal^+ and Gal^- colonies were picked from EMB galactose agar at random, and tested for susceptibility to λ and λ-2, and for lysogenicity. The phenotypically λ-sensitive (Lp_2^s) moiety of the Lp_1^s segregants is shown in parentheses. Almost all of the Lp_1^+ were Lp_2^r.

Unfortunately, this diploid is also segregating Lp_2, so that the nonlysogenic segregants include immune-2 as well as λ-sensitive.

The linkage and segregation evidence shows that a chromosomal factor is altered when a cell becomes lysogenic. In addition, a cytoplasmic factor (λ itself) may be postulated, but genetic evidence for it is entirely inconclusive. Two possible interpretations may be considered: 1) The virus or provirus occupies a definite niche on the chromosome, near Gal_4. Lysogenicity results from the cellular or even chromosomal fixation of the latent virus. 2) The chromosomal factor is a gene, Lp_1^s, which mutates spontaneously to an allele Lp_1^+ that potentiates a symbiotic relationship of λ in the bacterial cytoplasm. On this hypothesis, the role of λ in the induction of lysogenicity is confined to the selection of the pre-adaptive mutation, Lp_1^+. A similar dilemma in the determination of the killer trait in *Paramecium aurelia* has been resolved in terms similar to the second interpretation (Sonneborn 1950), although the first was originally favored (Sonneborn 1945). Its substantiation for lysogenicity would require the recognition of the possible genotypes: Lp_1^s no-λ (presumably the sensitives); Lp_1^s infected with λ (presumably lethal); Lp_1^+ with lambda (the lysogenic); and a new combination, Lp_1^+ no-λ. This last type, genetically pre-conditioned for lysogenicity, would presumably be recognized as an apparently immune form that would promptly absorb λ to become lysogenic. It has not yet been identified among immune stocks of K-12, or immune progeny collected from a variety of strain intercrosses.

Mechanism of infection

When λ-sensitive bacteria are plated with λ, survival ratios in the range of ten to fifty percent are usually encountered. Many of the survivors are apparently lysogenic. The hypothesis of spontaneous variation at the Lp_1 locus would be untenable if, as these facts appear to show *prima facie,* several percent of sensitive bacteria became lysogenic under the direct influence of the virus. Only preliminary experiments have been done on this aspect of the problem, with results that are not yet conclusive. A striking feature of platings of diluted bacteria-virus mixtures of varying relative multiplicity has been the development of contaminated colonies, similar to those figured by Boyd (1951). These colonies displayed a very characteristic appearance on EMB agar. They were often delayed in their development, lagging a few hours behind their neighbors, and later show either a central "necrosis" or plaquing, or often single or multiple pericentric plaques.

When the contaminated colonies were restreaked, they typically gave rise to a mixture of contaminated, sensitive and lysogenic colonies.

Many of the latter are only apparently lysogenic, for they include sensitive bacteria as shown by serial restreaking of single colonies. It is not unlikely (though not yet proven) that contaminated colonies may arise from single infected cells. If this is the case, the determination of lysis versus lysogenicity is effected during the development of a contaminated clone, and there would be greater opportunity for genetic variation and natural selection. On the other hand, if a fair proportion of infected cells are actually converted directly into lysogenics, it would be concluded that λ itself induces or fixes the mutation from $Lp_1{}^s$ to $Lp_1{}^+$.

Virus and host mutations

Following irradiation of a type lysogenic, a self-lysed colony was noted from which a distinctive virus was isolated. This virus, λ-2, differs from λ in its ability to destroy $Lp_1{}^+$ bacteria. Attempts to develop a symbiosis of λ-2 with each of a variety of bacterial stocks have been unsuccessful. Its relationship to λ as a "host range mutant" is supported by the concurrent development of resistance to λ with mutations from sensitivity to resistance to λ-2. Several recurrences of λ-2 have been detected in lysed colonies after ultraviolet irradiation, and in λ stocks grown on sensitive cells. It has not, however, been observed in routine bacterial cultures, although it would presumably have been conspicuous. This is in contrast to the rapid accumulation of comparable virus mutants in cultures of the lysogenic staphylococci studied by Burnet and Lush (1936).

Immune bacterial mutants have been observed among survivors of both irradiated lysogenic cultures and sensitive cultures exposed to the viruses. Immune-1 has occurred very infrequently, and is resistant to (and nonlysogenic for) λ, but sensitive to λ-2. Immune-2 is resistant to both phages, showing neither lysis nor the development of lysogenicity. As already mentioned, different loci, Lp_1 and Lp_2, appear to be involved. Although immune-2, $Lp_1{}^s\ Lp_2{}^r$, does not respond to free λ, selection for resistance to λ-2 in a lysogenic stock gives the genotype $Lp_1{}^+\ Lp_2{}^r$ which remains lysogenic for λ. Crosses of such lysogenics with sensitive ($Lp_1{}^+\ Lp_2{}^r \times Lp_1{}^s\ Lp_2{}^s$) gave all four of the expected types: immune-2 ($Lp_1{}^s\ Lp_2{}^r$) and type lysogenic ($Lp_1{}^+\ Lp_2{}^s$), in addition to the parents. Current stocks of K-12 are mixed populations with respect to Lp_2. It is not surprising, therefore, that several mutant derivatives, notably W-1177 extensively used in crossing experiments, are already $Lp_2{}^r$. Two λ-immune selections have been found, both sensitive to λ-2, which were unstable and frequently engendered λ-sensitive colonies. Tests for allelism with $Lp_1{}^r$ were inconclusive owing to this instability.

Other mutants of the virus have been sought, but only plaque variants not readily scored were observed. Resistance to λ and λ-2 is concomitant with resistance to p-14, a phage isolated from sewage. Morphologically, the plaques of p-14 are intermediate between those of λ and λ-2, with turbid centers associated with a spurious or unstable lysogenicity which persisted in slow-growing isolates at 30° and was rapidly lost at 37°. Despite its initial promise as a selective agent for other bacterial mutations related to λ, p-14 did not elicit any otherwise unrecognized types.

A "weakly lysogenic" bacterium was recovered after ultraviolet irradiation of a typical lysogenic form. When inoculated with the indicator strain, the variant induced very few plaques, so that it was not readily distinguished from immune nonlysogenic forms. When the virus was transferred from the weakly lysogenic form to sensitives normal lysogenicity ensued. This suggests that reduced lysogenicity was a property of the host rather than of the phage. It was conceivable, however, that the plaques of free virus represent reverse-mutants from a virus population that otherwise remains entirely latent within the infected variant bacterium. To eliminate this possibility, sensitive recombinants from crosses of the weak lysogenic with sensitive were infected individually with type λ. Both types of lysogenicity were expected on the hypothesis of bacterial mutation, and this was actually observed. A modifier locus is thus revealed, but its relationships with other factors have not been explored.

Another intermediate reaction type was isolated from plates spread with 10^8 bacteria and λ-2. Most of the survivors were fully resistant to both λ and λ-2, but some exhibited a partial resistance to λ and λ-2, which was reflected in overgrowth of cross-streaks and reduced efficiency of plating and plaque size for both viruses, similar to the expression of V_1p (partial resistance to T_1, Lederberg 1951b; Wahl and Blum-Emerique 1952). λ-lysogenic derivatives were prepared which were still semi-resistant to λ-2. The mutation thus involved either a third allele, Lp_2p, at the Lp_2 locus or mutation at another locus.

In view of speculation concerning the dispersion of lytic phages into genetic subunits during intracellular growth, the possibility that fragments of λ might persist in apparently nonlysogenic cells was considered. The reconstitution of lytically active λ from components carried in different nonlysogenic recombinants or variants would be relevant evidence. However, such a recurrence of phage from appropriate mixtures and crosses has hitherto not been demonstrated.

Disinfection

Two lysogenic streptomycin-sensitive (S^s) cultures plated on streptomycin agar have been observed to yield large numbers of resistant (S^r) mutant colonies which showed the characteristically mottled margins of phage attack. These colonies gave rise to S^r λ-sensitive isolates. Reconstruction experiments with these mutants or their re-infected derivatives failed to establish any foundation for either a selective advantage or a specific inductive effect of streptomycin to explain the accumulation of λ-sensitive. By indirect selection (Lederberg and Lederberg 1952), it was possible to extract the S^r components, and show their λ-sensitive character without exposing them to streptomycin. The λ-sensitive and S^r characters were not distinguishable from mutations previously isolated in single steps. No explanation for this remarkable association can be offered.

Systematic attempts were made to remove λ from lysogenic bacteria by a number of other methods. As none were successful, details will be omitted. The treatments that were tried included cultivation at limiting temperatures and pH ranges (as originally suggested by d'Herelle 1926), and exposure to antibiotics and antiviral chemicals, including streptomycin, aureomycin, chloromycetin, Phosphine GNR, 2-nitro-5-aminoacridine, citrate ion, cobaltous ion, and desoxypyridoxine. A serious limitation to this type of investigation is the inadequacy of earlier methods of detecting disinfected variants, if they occur infrequently. Replica plating should help to surmount this problem, but was not available at the time of these experiments.

Almost all of our original λ-sensitive stocks in strain K-12 have been noticed following exposure to treatment with ultraviolet light. Inasmuch as this agent, under certain conditions, preferentially kills lysogenic cells by inducing lysis (Weigle and Delbrück 1951), it cannot be concluded whether a selective or inductive (disinfective) action is involved.

Lysogenicity and other *E. coli* strains

The λ reaction of about 2000 strains under investigation for intercrossability has been routinely tested. No recurrence of λ itself has been identified, but five new strains are sensitive to λ and λ-2. One apparently unstable immune strain gave rise to sensitive subtypes, which, however, could not be made lysogenic on K-12 line indicators for either virus. All of the new sensitive lines, including NTCC 123 (Cavalli and Heslot 1949) are fertile with K-12, suggesting a statistical correlation of λ receptors with compatibility. Most of the 50 or so interfertile strains that have been screened are, however, immune to λ.

Although a large proportion of the strains tested produced an antibiotic or colicin (Fredericq 1948) active on K-12, less than one percent were lysogenic. The lysogenic cultures (which include, for example, the Waksman strain used in biochemical genetic studies, Davis 1950) carry what appear to be quite distinctive phages, judging from plaque type and resistance patterns. Two of the new latent phages have been successfully transferred to the K-12 line. Triply lysogenic K-12 strains were maintained without any overt effects on the λ system or other characters of the bacteria. The genetic determination of lysogenicity for other phages may differ from that of λ, however, in so far as clear-cut segregation for them was not observed in crosses or from diploids also segregating λ.

DISCUSSION

This work was initiated in the expectation that λ would behave as an extranuclear factor, and might indeed provide a favorable model system for studies of cytoplasmic heredity. Phenotypic changes associated with the transfer of λ have, so far as known, been confined to the direct consequences of virus infection. For example, lysogenic bacteria are more susceptible to ultraviolet light, owing to the "induction" of the latent phage and lysis of the bacterium (Weigle and Delbrück 1951). In other systems, latent viruses have been shown to determine the pattern of susceptibility to other viruses, the "lysotype" (Nicolle and Hamon 1951; Williams-Smith 1948; Anderson 1951), by a mutual exclusion effect. With one dubious exception, no phages that would differentiate λ-sensitive from λ-lysogenic were found in tests of some thousands of coliphage plaques from sewage. In principle, however, a virus-symbiosis might be detected in terms of the intercellular transfer of a genetically active agent not readily recognizable as a lytic phage (Lominski 1938).

This view of λ may have to be qualified in view of the genetic tests discussed in this paper. No genetic evidence of λ as a cytoplasmic agent was found. In the most critical tests, segregation from heterozygote diploids, lysogenicity behaved precisely as if it were controlled by a nuclear factor, linked to other segregating factors. This result provides strong support for the "provirus" concept of the symbiosis. The segregation of uninfected, virus-sensitive haploids from a lysogenic diploid is not readily compatible with the presence of free, mature virus in the latter. It is not, however, conclusive against a cytoplasmic provirus. The segregation of lysogenicity/sensitivity may reflect the overriding control by a segregating nuclear factor which is concerned with the maintenance of the pro-λ. The mutational origin of this segregating factor is, however, still in question.

It should not be assumed that these results can be generalized to other lysogenic symbioses. In *Salmonella typhimurium*, Boyd (1951) has shown that the multiplicity of infection is an important element in the determination of lysogenicity. This would leave little room for bacterial variability, but a closer analysis of the incidents immediately related to the development of lysogenic cells might reveal a situation more comparable to that in *E. coli.* In preliminary studies of the transmission of other viruses, transferred to K-12 from other lysogenic strains, diploids lysogenic for two phages showed segregation for λ but not for the second phage. The apparent difference with respect to nuclear determination may be a consequence of the antiquity of the association of K-12 with λ in contrast to the newly introduced phages.

It may be noteworthy that λ has not recurred in extensive samplings of other *E. coli* strains and of sewage. The occurrence of λ-sensitive isolates has already been mentioned. It is rather striking that all five of these isolates should be cross-fertile, compared to the four to five percent of the whole population. Whether this speaks for a close genetic relationship or for the closer attention given these lines cannot be said. It should be emphasized that all of the evidence argues against any functional relationship between lysogenicity and sexual fertility. The most decisive point, perhaps, is that nonlysogenic crosses are as fertile as crosses involving one or both lysogenic parents, both within strain K-12, and as between strains.

The biological significance of the lysogenic symbiosis is attested to not only by the behavior of individual examples, but by its prevalence in many groups of bacteria. Burnet (1945) and others have emphasized the biological advantages to the parasite as well as the host of symbiotic adaptation. In addition, the virus genotype represents an additional reservoir of genetic material subject to adaptive variation. This adaptation will often lead to an amelioration of the pathogenic effects of the virus. One can imagine a situation in which a virus remains trapped within a host that it never lyses. A bacterial mutation for weak lysogenicity illustrates this trend, and it has perhaps been realized in Lominski's (1938) experiments. The extreme case would however restrict the migration of the virus to other genotypes, as well as our ability to recognize it as a virus. It is conceivable that the immune-1 ($Lp_1{}^r$) mutation represents such a bound virus, although free λ has not recurred even in its crosses to λ-sensitive.

The most prominent mutation of λ has, according to this picture, only short-term evolutionary advantages. The virulent mutant, λ-2 will rapidly destroy lysogenic bacteria, and thus displace λ from viral populations. The exhaustion of sensitive hosts will, however, limit its long-term survival. The early literature on bacteriophage contains many references to the so-called spontaneous generation of bacteriophage in bacterial cultures. While some of these reports are possibly founded on technical faults, probably most of them represent instances of the mutation of virus in lysogenic bacteria not recognized as such. If it were not for the availability of an indicator strain for λ, the occurrence of lysis due to λ-2 in platings of ultraviolet irradiated K-12 would have passed either for a contamination or such "spontaneous generation."

For technical reasons, the phages of the T series acting on *E. coli* B have received considerable attention. These phages have been used for such work precisely because they are atypical in their prompt destruction of sensitive bacteria, high efficiency of plating, the limited number of secondary resistants, and clear plaques. A plating of sewage with indicator bacteria shows at a glance that phages of this kind are relatively infrequent. Although the analysis of phage-bacterium relationships on a logically sound, particulate basis has demanded systems with these technical properties, it would be a fallacy to generalize too hastily on virus biology from the study of a restricted set of materials.

SUMMARY

Escherichia coli strain K-12 carries a symbiotic phage, λ. This phage was discovered only by the occurrence of "mutant" substrains sensitive to λ, and serving as indicators for it. In addition to the lysogenic (carrier) and sensitive bacteria, two immune types ("1" and "2") were found. These are defined as resistant but nonlysogenic.

The various types have been intercrossed to elucidate the genetic basis of lysogenicity. The crosses lysogenic X lysogenic; immune-1 X immune-1 and sensitive X sensitive have yielded only the parental class. Similarly, only the two parental classes were found in lysogenic X sensitive; lysogenic X immune-1; and sensitive X immune-1. The segregation of lysogenicity has been confirmed by the synthesis of diploid stocks heterozygous for lysogenicity, which behaves as a factor linked to Gal_4 (galactose fermentation). Genetic evidence of the transmission of λ as a cytoplasmic factor was not found. A locus for latent phage, Lp_1, which controls the maintenance of λ, or to which λ is bound is postulated. The detailed role of λ in the alteration of the Lp_1 locus that is associated with the resynthesis of lysogenic from sensitive has not been clarified.

Mutation of λ to a more virulent mutant λ-2 has been observed. λ-2 lyses λ-lysogenic as well as λ-sensitive bacteria. Immune-2 confers resistance both to λ and to λ-2. It does not, however, interfere with the maintenance of λ in bacteria already lysogenic. It is genetically separable from immune-1. A few additional *E. coli* stocks sensitive to λ, or lysogenic for other phages, have been found. In an extensive survey, λ itself has not recurred.

LITERATURE CITED

Adams, M. H., 1950. Methods of study of bacterial viruses. Methods in Medical Research **2**:1-73.

Anderson, E. S., 1951. The significance of Vi-phage types F1 and F2 of *Salmonella typhi.* J. Hyg. **49**:458-470.

Bertani, G., 1951. Studies on lysogenesis. I. The mode of phage liberation by lysogenic *Escherichia coli.* J. Bact. **62**:293-300.

Boyd, J. S. K., 1951. Observations on the relationship of symbiotic and lytic bacteriophage. J. Path. Bact. **63**:445-457.

Burnet, F. M., 1945. Virus as Organism. 134 pp. Cambridge, Mass: Harvard University Press.

Burnet, F. M., and D. Lush, 1936. Induced lysogenicity and mutation of bacteriophage within lysogenic bacteria. Austr. J. Exp. Biol. Med. Sci. **14**:27-38.

Cavalli, L. L., and H. Heslot, 1949. Recombination in bacteria: outcrossing *Escherichia coli* K-12. Nature **164**:1057-1058.

Davis, B. D., 1950. Studies on nutritionally deficient bacterial mutants isolated by means of penicillin. Experientia **6**:41-50.

Delbrück, M. (ed.), 1950. Viruses 1950. Pasadena: California Institute of Technology.

D'Herelle, F. D., 1926. The Bacteriophage and Its Behavior. Transl. G. H. Smith. Baltimore: Williams and Wilkins. See p. 237.

Fredericq, P., 1948. Actions antibiotiques réciproques chez les Enterobactériaceae. Rév. belge path. med. exp. **19** (Supplement 4):1-107.

Hayes, W., 1952. Recombination in *Bact. coli* K 12: unidirectional transfer of genetic material. Nature **169**:118.

Lederberg, E. M., 1952. Allelic relationships and reverse mutation in *Escherichia coli.* Genetics **37**:469-483.

Lederberg, J., 1949. Aberrant heterozygotes in *Escherichia coli.* Proc. Nat. Acad. Sci. **35**:178-184.

1950. Isolation and characterization of biochemical mutants of bacteria. Methods in Medical Research **3**:5-22.

1951a. Prevalence of *Escherichia coli* strains exhibiting genetic recombination. Science **114**:68-69.

1951b. Genetic studies with bacteria. Genetics in the 20th Century. viii + 634 pp. New York: Macmillan.

Lederberg, J., L. L. Cavalli, and E. M. Lederberg, 1952. Sex compatibility in *Escherichia coli.* Genetics **37**:720-730.

Lederberg, J., and E. M. Lederberg, 1952. Replica plating and indirect selection of bacterial mutants. J. Bact. **63**:399-406.

Lederberg, J., E. M. Lederberg, N. D. Zinder, and E. R. Lively, 1951. Recombination analysis of bacterial heredity. Cold Spring Harbor Symp. Quant. Biol. **16**:414-443.

Lominski, I., 1938. Souches lysogènes nonactives (cryptolysogènes). Compt. Rend. Soc. Biol. **129**: 151-153.

Lwoff, A., and A. Gutman, 1950. Recherches sur un *Bacillus megaterium* lysogène. Ann. Inst. Pasteur **78**:711-739.

Lwoff, A., L. Siminovitch, and N. Kjelgaard, 1950. Induction de production de phage dans une bactérie lysogène. Ann. Inst. Pasteur **79**:815-858.

Nicolle, P., and Y. Hamon, 1951. Recherches sur les facteurs qui conditionnent l'appertenance des bacilles paratyphiques B aux différents types bacteriophagiques de Felix et Callow. III. Nouvelles études sur les transformations de types par l'action des bactériophages extraits des bacilles lysogènes. Ann. Inst. Pasteur **81**:614-630.

Schiff, F., and S. Bornstein, 1940. Hemolytic effect of typhoid cultures in combination with pure lines of bacteriophage. J. Immunol. **39**:361-367.

Sonneborn, T. M., 1945. The dependence of the physiological action of a gene on a primer and the relation of primer to gene. Amer. Nat. **79**:318-339.

1950. The cytoplasm in heredity. Heredity **4**:11-36.

Tatum, E. L., 1945. X-ray induced mutant strains of *Escherichia coli*. Proc. Nat. Acad. Sci. **31**:215-219.

Tatum, E. L., and J. Lederberg, 1947. Gene recombination in the bacterium *Escherichia coli*. J. Bact. **53**:673-684.

Weigle, J. J., and M. Delbrück, 1951. Mutual exclusion between an infecting phage and a carried phage. J. Bact. **62**:301-318.

Williams-Smith, H., 1948. Investigations on the typing of staphylococci by means of bacteriophage. J. Hyg. **46**:74-81, 82-89.

Wahl, R., and L. Blum-Emerique, 1952. Les bactéries semi-résistantes au bactériophage. Ann. Inst. Pasteur **82**:29-43.

41 The particulate nature of nonchromosomal genes in Chlamydomonas

Ruth Sager*
Zenta Ramanis
Department of Zoology
Columbia University
New York, New York

The existence of stable genetic determinants not carried on chromosomes and occurring in a wide variety of organisms is well documented and has been summarized in a number of reviews.[1] In most of these studies, the evidence is derived from formal genetic analysis, demonstrating that the pattern of segregation of the mutant phenotype among the progeny of crosses cannot be attributed to chromosomal segregation, whether normal or aberrant. These findings have aroused much speculation, the essence of which can be summarized in terms of two alternative hypotheses: (1) that the mutant phenotypes result from mutations of primary genetic information carried elsewhere than on a chromosome;[1,2] or (2) that the mutant phenotypes result from metabolic events which shift a series of interlocking reactions from one steady-state level to another, producing an altered phenotype not resulting from any change in primary genetic information.[3]

These alternative views are similar to those that faced Mendel and other analysts of heredity at the turn of the century: Are genetic determinants particulate, discrete units? Do they persist generation after generation whether expressed or not? The triumph of Mendelian analysis resulted from the recovery of pure parental genotypes among the progeny of F_1 hybrids in which genetic factors from the two parents were mixed. Thus, it was shown that alternative factors from the two parents could coexist in the same organism and subsequently segregate out unaltered in later generations. Recessive genes (unexpressed) were no less stable and persistent than their dominant alleles. Mendel dealt with unit factors, one per haploid genome, but his method is applicable as well to factors present in several replicates, provided that segregation occurs.

The finding that streptomycin is mutagenic for nonchromosomal genetic determinants[4] has provided a wide range of new mutant strains for analysis. In this paper we report genetic studies of segregation and recombination with a number of these strains of the alga *Chlamydomonas reinhardi,* demonstrating the particulate nature of the nonchromosomal factors under observation. Our results provide evidence for the existence of a class of genetic determinants which are nonchromosomal in location, stable in replication and transmission whether expressed or not, represented by alternative allelic states in wild-type and mutant cells, and influencing a wide range of cellular traits. It is proposed that this class of genetic determinants be called NC genes.

The nonchromosomal (NC) genes described in this report are transmitted primarily in a uniparental manner in crosses, as previously reported:[5,6] only one of the two mating types, mt^+, regularly transmits its NC genes to all of the progeny, the corresponding NC complement of the mt^- parent usually being lost and not reappearing in subsequent generations. In such a system, the analysis of segregation and recombination would seem to be precluded. However, as previously noted,[7] exceptional zygotes are found in which NC genes from both parents persist. By selecting for these exceptional zygotes it has proved possible to study NC gene segregation and recombination.

MATERIALS AND METHODS

The parental strains, media, and methods of genetic analysis employed in this study have been previously described, as have the chromosomal gene pairs *act-r*/*act-s* (actidione-resistance and -sensitivity), *ms-r*/*ms-s* (methionine sulfoximine-resistance and -sensitivity), mt^+/mt^- (mating type), and the NC gene for streptomycin dependence, *sd.*[6] The new nonchromosomal mutants described in this paper were obtained by growth of a streptomycin-resistant strain, *sr-500,*

From Proceedings of the National Academy of Sciences **50**:260-268, 1963. Used with permission.

Supported by grant E-1605 of the U.S. Public Health Service and institutional grant IN-4 of the American Cancer Society.

*Present address: Professor of Biological Sciences, Hunter College, New York, New York.

on streptomycin as previously described.[4] The streptomycin-sensitive strains are denoted *ss.* The three unlinked chromosomal gene pairs (*act-r/act-s, ms-r/ms-s,* and mt^+/mt^-) segregated in all crosses to be described, and served to identify the four meiotic products (i.e., the zoöspores) and their clonal descendants.

Selection for exceptional zygotes was carried out by plating aliquot suspensions of zygotes on differential media. Gametes of the two parental strains (mt^+ and mt^-) were mixed under conditions allowing rapid fusion of mating pairs with a zygote yield of about 90%. Zygote suspensions were plated on various media at desired dilutions, and allowed to mature for one week in the dark. Residual vegetative cells were killed by exposure to chloroform vapor for 30 seconds. Germination of zygotes was induced by light. Each resulting colony was derived from a single zygote and contained all of the progeny types, each of which multiplied as a clone within the colony. Progeny analysis was performed by suspending individual zygote colonies in liquid and replating the cells at dilution to give about 200 colonies per plate. Progeny types were identified by replica plating and restreaking when required. Tetrad analysis, the hand-dissection of the four products of meiosis from the germinating zygote, was performed as previously described.[6]

Selection of exceptional zygotes. In previous studies[5-7] the survival of the *sr-500* determinant from the mt^- parent in the progeny of up to 10% of the zygotes had been noted in crosses of the type: *ss* mt^+ X *sr* mt^-. This observation raised the possibility that, in some fraction of zygotes in all crosses, NC genes from both parents might be transmitted to the progeny, thus providing material for an analysis of their segregation patterns.

To explore this possibility, initially we used the NC gene pair *sd/ss.* Exceptional zygotes could be selected in reciprocal crosses, *sd* mt^+ X *ss* mt^- and *ss* mt^+ X *sd* mt^-, as shown in Table 1. When zygotes of cross 1 (*sd* mt^+ X *sr* mt^-. This observation raised the possibility selecting for the *ss* factor, 0.07% of the zygotes germinated and produced colonies, in contrast to full survival on streptomycin-agar plates. Similarly, in cross 2, full survival occurred in the absence of streptomycin, and 0.08% of the zygotes formed colonies in the presence of the drug. The absence of exceptional zygotes in control crosses 3 and 4, *sd* X *sd* and *ss* X *ss,* performed with aliquots of the same suspensions used for crosses 1 and 2, makes it unlikely that mutants in the parental mt^+ populations were the source of the exceptions. Further evidence against this possibility is provided by studies of exceptional zygotes in two factor crosses, described below.

Table 1. Recovery of zygote colonies in crosses of streptomycin-dependent and -sensitive strains

Cross	Per cent zygote colonies	
	Streptomycin-agar	Minimal-agar
(1) *sd* mt^+ X *ss* mt^-	100	0.07
(2) *ss* mt^+ X *sd* mt^-	0.08	100
(3) *sd* mt^+ X *sd* mt^-	100	<0.0001
(4) *ss* mt^+ X *ss* mt^-	<0.0001	100

PROGENY OF EXCEPTIONAL ZYGOTES

The genetic constitution of the progeny from exceptional zygote colonies was analyzed by cloning and replica plating. Twenty exceptional zygotes of cross 1 were studied. Both *ss* and *sd* pure clones were recovered from each zygote colony, indicating that both of the parental NC alleles had been transmitted to the progeny.

Presumably, the presence of the *ss* allele made possible germination and colony formation under the selective conditions. As previously described,[6] in *sd* mt^+ X *ss* mt^- crosses, the *sd* zoöspores from nonexceptional zygotes cannot undergo even one division in the absence of streptomycin. In the exceptional zygotes, however, pure *sd* clones were found after many doublings in the absence of streptomycin in the zygote colonies. The explanation for this behavior, analyzed below, lies in the fact that the segregation of *sd/ss* occurs in the postmeiotic divisions. The four initial haploid products usually contain both *sd* and *ss* factors; these *sd/ss* cells are phenotypically streptomycin-resistant.

The exceptional zygote colonies from cross 1 were found to contain eight genetically different progeny classes; each of the four chromosomal segregant types consist both of *sd* and of *ss* subclones. Thus, *ss* and *sd* must have segregated from each other at postmeiotic divisions. The *ss* clones were in excess because of selection against *sd* in the absence of streptomycin. In cross 2, exceptional zygotes were selected by plating on streptomycin-agar, thus killing any *ss* progeny which might have been present. As will be shown below, both *ss* and *sd* determinants can be recovered in two-factor crosses, by using the ac^+/ac^- pair to select exceptional zygotes.

Exceptional zygotes have also been selected in crosses involving other NC mutant genes, recently isolated after streptomycin-induced mutagenesis.[4] A series of auxotrophic NC mutants were each crossed to wild type (mt^-), and the resulting zygotes plated both on supplemented and on minimal media. The results, listed in Table 2, show the inability of the majority of zygotes in these crosses to produce zygote colonies in the absence of the growth factor required by the mt^+ parent. This behavior is in contrast to that of chromosomal genes, exemplified by *ac-16,* an acetate-requiring mutant which, in crosses with ac^+ mt^-, produces zygote colonies equally well in the presence and absence of acetate.

Table 2. Recovery of zygote colonies in crosses of various SM-induced mutant strains *(mt⁺)* X wild-type *(mt⁻)*

Mutant strain	Per cent zygote colonies	
	Supplemented-agar	Minimal-agar
ac^-16 (ac^-, chromosomal gene mutant)	100	100
S-3 (ac^-)	100	0.10
S-5 (ac^-)	100	0.05
S-12 (ac^-)	100	0.02
S-22 (ac^-)	100	0.05
S-216-1-3 (grows well, then dies on C, poor but viable on M)	+	−
S-266-3-5 (auxotrophic: grows on C minus acetate)	+	−
S-293-1-1 (auxotrophic: grows on C minus acetate)	+	−
S-397-2-1 (auxotrophic: grows on C minus acetate)	+	−
S-473 (ac^-; tiny on all media)	+	−
S-340-1 (ac^-; tiny on all media)	+	−
S-388-1-3 (ac^-; semilethal)	+	−
S-408-2-1 (ac^-; very leaky)	+	−
S-421-1-3 (ac^-; nonleaky)	+	−

C = minimal medium plus 0.1% YE, 0.1% casamino acids, 0.001% each purine and pyrimidine, 0.1% sodium acetate.

Table 3. Tetrad and progeny analysis of strain S-12 *(ac^-)*

Cross	Progeny	No. of zygotes analyzed
(A) Tetrad analysis of unselected zygotes		
S-12 *($ac^- mt^+$ act-r ms-r)* X *$ac^+ mt^-$ act-s ms-s*	all *ac^-* 2 *mt^+*: 2 *mt^-* 2 *act-r*: 2 *act-s* 2 *ms-r*: 2 *ms-s*	14
6961a (F_1) *$ac^- mt^+$ act-s ms-r* X *$ac^+ mt^-$ act-r ms-s*	all *ac^-* 2 *mt^+*: 2 *mt^-* 2 *act-r*: 2 *act-s* 2 *ms-r*: 2 *ms-s*	14
(B) Progeny analysis of exceptional zygotes		
6958c (F_1): *$ac^- mt^-$ act-s ms-r sr-500* X *$ac^+ mt^+$ act-r ms-s ss*	8 genotypes, each zoöspore clone segregating *ac^+/ac^-* subclones	7
	segregating *ac^+/ac^-*, but incomplete sets	48
*6978a (F_2): *$ac^- mt^+$act-r ms-s ss* X *$ac^+ mt^-$ act-s ms-r sd*	16 genotypes, each zoöspore clone segregating *ac^+/ac^-* and *sd/ss* subclones	6
	segregating *ac^+/ac^-* and *sd/ss*, but incomplete sets	7
*6978c (F_2): *$ac^- mt^-$ act-s ms-r ss* X *$ac^+ mt^+$ act-r ms-s sd*	16 genotypes, each zoöspore clone segregating *ac^+/ac^-* and *sd/ss* subclones	25
	segregating as above, but incomplete sets	29

*Exceptional zygotes were selected on streptomycin-agar.

Tetrad analysis of zygotes containing chromosomal as well as NC markers (Table 3), germinated on supplemented media, showed normal segregation of chromosomal gene markers, but all progeny were uniformly auxotrophic like the mt^+ parent. Progeny analysis of exceptional zygotes from the same crosses, however, showed the presence of both the wild-type and mutant NC alleles for auxotrophy among the progeny, segregating after meiosis as in the *sd/ss* system, and giving rise to eight genetically different progeny classes instead of four.

The complete results of one cross of this type are given in Table 3. Tetrad analysis of 14 unselected zygotes germinated on acetate-agar showed that all of the progeny were uniformly ac^- like the mt^+ parent, although they were segregating normally for the unlinked pairs of chromosomal genes, *act-r/act-s* and *ms-r/ms-s*. Similarly, in the F_1 backcross generation, (ac^- mt^+ X ac^+ mt^-), the progeny were entirely ac^-, although the chromosomal markers showed normal segregation.

The zygote colonies recovered by selection of exceptional zygotes were found to contain both ac^+ and ac^- clones. Progeny analysis revealed the presence of eight genetically different classes in both the F_1 and F_2 backcross generations, thus demonstrating the persistent occurrence of postmeiotic segregation of the ac^+/ac^- NC gene pair.

SEGREGATION AND RECOMBINATION IN TWO-FACTOR CROSSES

Questions raised by the results just cited for one-factor crosses include the following: (1) Are more than one pair of NC alleles preserved in the same exceptional zygotes? (2) Do different pairs of NC alleles segregate independently? (3) At which mitotic division does postmeiotic segregation of NC alleles occur? (4) What segregation ratios are found with NC genes? These questions were investigated with a pair of reciprocal two-factor crosses.

Cross A:
sd ac^+ mt^+ *act-r ms-s* X *ss* ac^- mt^- *act-s ms-r*
Cross B:
ss ac^- mt^+ *act-r ms-s* X *sd* ac^+ mt^- *act-s ms-r*

The ac^- strains used in these crosses were F_1's derived from S-12 (Table 3).

(1) In cross A, one can select exceptional zygotes by plating on acetate-agar without streptomycin, thus selecting for the *ss* NC allele of the mt^- parent, and permitting equal growth of ac^+ and ac^- containing cells. The progeny can then be scored for segregation both of the *sd/ss* pair and of the ac^+/ac^- pair. The *sd* determinant is not lost when *sd/ss* cells are plated in the absence of streptomycin. However, as soon as segregation occurs, the resulting *sd* cells are at a selective disadvantage in the absence of the drug.

In cross B, three different selective media are available: acetate-streptomycin agar, which selects for *sd;* minimal-agar which selects for ac^+; and minimal-streptomycin-agar, which selects for both *sd* and ac^+. If the zygote colonies, which arise on either type of streptomycin-agar, are transferred as microcolonies to streptomycin-free medium, *ss* clones can be recovered from *sd/ss* cells which had not already segregated.

The fraction of exceptional zygotes arising in these crosses is in the same range (0.02-0.1%) as that found in one-factor crosses (Tables 1 and 2). Furthermore, selecting simultaneously for two NC genes did not reduce the yield to the extent expected if their probabilities of survival in the zygote were independent. All of the zygotes analyzed in these crosses were segregating both for the ac^+/ac^- pair and for the *sd/ss* pair, except for one zygote from cross B selected on minimal-streptomycin-agar which was entirely ac^+ though segregating for *sd/ss.*

These data indicate that a fraction of zygotes exists in which NC genes from both parents are preserved. It seems likely that the entire class of NC genes, not only those under observation, survive in the exceptional zygotes and are transmitted to all the progeny. These results also provide evidence that exceptional zygotes do not arise by mutation of the particular NC genes being studied.

(2) Analysis of progeny from crosses A and B showed that the *sd/ss* and the ac^+/ac^- NC gene pairs segregated independently in postmeiotic divisions. Sixteen genetically different meiotic products were recovered from 25 exceptional zygotes of cross A and 6 of cross B. In addition, 56 incomplete sets were recovered, in which no bias was detected toward any particular recombinant type.

Reciprocal two-factor crosses involving ac^+/ac^- and the NC gene *sr-500* instead of *sd* have also been studied. In the cross ac^-sr mt^+ X ac^+ss mt^-, zygotes were plated on minimal-agar, and 6 exceptional zygote colonies were recovered. One of them contained all 16 possible recombinant types; the other 5 were incomplete but showed no bias. In the cross ac^+ss mt^+ X ac^-sr mt^-, selected on streptomycin-agar, all progeny were *sr* as a result of the selection medium, but the 10 zygote colonies analyzed were all segregating for ac^+/ac^- and contained 8 genetically different classes. Here, too, *act-r/act-s* and *ms-r/ms-s* were employed to mark the 4 products of meiosis.

(3) The times of segregation of the NC gene pairs were determined by dissecting exceptional zygote colonies from crosses A and B at early stages after germination and classifying the progeny clones. Cross A was analyzed at the 4-16 cell stage (1-2 doublings of zoöspores) and both crosses were analyzed at the 50-100 cell stage (4-5 doublings). Since the zoöspores do not divide with full synchrony after hatching, the 4 tetratypes are not equally represented in each zygote microcolony. Nonetheless, the data presented in Table 4 were pooled because the segregation of NC genes was found to be unbiased by chromosomal constitution.

The results summarized in Table 4 demonstrate that most of the segregation of NC gene pairs occurs in the interval between 2 and 5

Table 4. Time of segregation of NC genes in progeny of crosses A and B

Type of segregation			Per cent of progeny of each segregation type		
	ac^+/ac^-	sd/ss	4-16 cell (A)	50-100 cell (A)	50-100 cell (B)
(1)	+	+	6.0	62.0	42.0
(2)	+	−	6.9	23.0	46.6
(3)	−	+	2.2	6.0	2.0
(4)	−	−	85.0	8.8	9.4
Number of progeny tested			360	272	238
Number of zygotes sampled			65	12	13

doublings of the progeny clones after meiosis. Among the zygote colonies assayed at the 4-16 cell stage, 85% of the progeny had not yet segregated for either NC gene pair, whereas at the 50-100 cell stage, only about 9% of the progeny in both crosses A and B were still of this type. At the later time, in both crosses, the *ac* pair had segregated in over 85% of the progeny. However, the *sd/ss* pair had segregated to a much lesser extent: 68% in cross A and 44% in cross B. Thus, it seems evident that the 2 NC gene pairs segregate, on the average, at different times. It is less clear whether the difference between crosses A and B in segregation time for *sd/ss* is also significant.

(4) The ratio of ac^+ and of ac^- clones found among the segregated progeny of crosses A and B was 268 ac^+:271 ac^-. The two types were recovered with equal frequency in both crosses. The allelic segregation ratios for ac^+/ac^- could be determined with fair accuracy since acetate-medium supports growth at similar rates of both ac^+ and ac^- clones. However, no medium could be devised for growth both of *ss* and *sd* clones, and consequently it was not possible to determine segregation ratios for this allelic pair.

RECOVERY OF NEW PHENOTYPES FROM CROSSES

In addition to the parental phenotypes which reappear in the majority of progeny, a few new types have been isolated. Extreme ac^- strains have been found among the progeny of cross A, characterized by total inability to grow on minimal medium, a more stringent requirement than that of the parental ac^-. A range of new types have been identified by their altered repsonse to streptomycin. Of particular interest is the question whether these new types represent quantitative shifts in the number of determinants, thus showing a dosage effect, or qualitative changes resulting from mutation or recombination. Crosses designed to probe this question are now in progress.

DISCUSSION

In this paper we have described the segregation of nonchromosomal genes by selecting for analysis a class of zygotes in which the nonchromosomal determinants from both parents are transmitted to the progeny. In these exceptional zygotes, the segregation of nonchromosomal determinants occurs after meiosis, among the clonal descendants of each zoöspore (primary meiotic product). By the use of two-factor crosses, in which the segregation of two pairs of nonchromosomal determinants (*sd/ss* and ac^+/ac^-) could be followed simultaneously, it was demonstrated that (1) both pairs were transmitted in the same exceptional zygotes; (2) the pairs segregated independently, giving rise to 16 genetically different classes of progeny per zygote; (3) on the average, the ac^+/ac^- pairs segregated at an earlier mitotic division than did the *sd/ss* pair, over 85 per cent of the progeny having segregated for ac^+/ac^-, but only about 50 per cent having segregated for *sd/ss* by the 50-100 cell stage (4-5 doublings of the zoöspores); and (4) on the average, the ac^+/ac^- segregation was 1:1 among progeny clones.

What are exceptional zygotes? In wild-type *Chlamydomonas,* the mating-type locus appears to control the selective elimination or inactivation of NC genes from the mt^- parent in the zygote at some time between zygote formation and meiosis. Evidence that zygotes from the cross *sd* mt^- X *ss* mt^+ become sensitive to streptomycin before meiosis[6] indicates that the loss (or inactivation) occurs well before meiosis. We have recently found that a heat treatment immediately after zygote formation can increase the yield of exceptional zygotes about 10-fold, suggesting that the fate of NC genes is under enzymic control. Presumably, the exceptional zygotes represent a class in which this control system has failed.

The occasional appearance of streptomycin-resistant progeny in crosses of *ss* $mt^+ \times$ *sr* mt^- was noted in the original description of this system[5] and in subsequent reports.[6,7] Further analysis was postponed until, with the availability of more NC mutants, segregation could be investigated in reciprocal two-factor crosses. In the present study, the frequency of exceptional zygotes recovered by selective techniques, was in the range of 0.02-0.1%. In our earlier work, we observed frequencies of 4-10 per cent, and recently Gillham,[8] studying the same *ss* $mt^+ \times$ *sr* mt^- cross, has reported frequencies of 3-13%. Perhaps the *sr* determinant is less sensitive to mating-type inactivation than are *ac* or *sd*. These preliminary results suggest that the mechanism of NC gene elimination in the zygote may be amenable to analysis.

Postmeiotic segregation. The stability of the NC phenotypes during vegetative growth[5,6] suggests that a mechanism exists which ensures continuity of NC genes in replication and transmission. The requisite regularity could result if each NC factor were present in many replicates, distributed at random in cell division, or in a few replicates, distributed in an orderly fashion. In our present experiments, the segregation of ac^+/ac^- in the first few postmeiotic divisions suggests that the number of segregating units is low. Gillham's preliminary observations of *sr*/*ss* segregation[8] also suggest that few units are involved. This evidence would not preclude the existence of many replicates, but does indicate that they would have to segregate as a group (analogous, perhaps, to a multistranded chromosome). One might ask why postmeiotic segregation occurs at all. Perhaps the haploid cells cannot cope with the doubled number of NC genes.

The picture is less clear with respect to the *sd*/*ss* pair, which segregates on the average later than does the *ac* pair. Perhaps there are more segregating replicates of this NC gene than of the others. We are attempting to clarify this question with crosses in which several different NC mutant genes, including ac^-, *sd*, and *sr*, are segregating simultaneously.

CONCLUSION

Our findings provide strong evidence that the nonchromosomal determinants are carriers of primary genetic information. The alternative hypothesis, that nonchromosomal heredity merely reflects the existence of various cytoplasmic states which alter the expression of chromosomal genes,[3] has been excluded, so far as analysis at the cellular level permits, by the following lines of evidence: (1) postmeiotic segregation of NC gene pairs in clones which are already haploid for the chromosome complement; (2) retention of identity of NC alleles while present together in the diploid zygote and in the haploid zoöspore; (3) survival of *ss* and ac^- determinants in cells which are phenotypically streptomycin-resistant and acetate-independent; (4) stability of NC mutant genes in vegetative growth and in meiosis.

Thus, the NC genes share with chromosomal genes the properties of stability, mutability, maintenance of identity in heterozygotes, segregation of alleles, and the classical dichotomy between genotype and phenotype. It is hoped that, with this conceptual clarification and with the availability of many new NC mutant strains, the pressing questions of molecular identification and functional role of NC genes will be open to fruitful analysis.

SUMMARY

A method has been developed for analysis of segregation and recombination of nonchromosomal (NC) genes in crosses. As shown in one-factor crosses involving several different NC mutant genes, and in reciprocal two-factor crosses with the gene pairs ac^+/ac^- and *sd*/*ss*, the NC genes segregate in postmeiotic divisions. The NC mutant genes are considered to be particulate carriers of primary genetic information.

REFERENCES

1. Rhoades, M. M., Encycl. Plant Physiol. **1:**19 (1955); Caspari, E., Advances in Genet. **2:**2 (1948); Beale, G. H., Sci. Progr. **49:**17 (1961).
2. Sager, R., Science **132:**1458 (1960).
3. Delbrück, M., in Unités Biologiques Douées de Continuité Génétique (Paris, 1949), p. 33; Pollock, M. R., in Adaptation in Microorganisms, Proc. Soc. Gen. Microbiol., 3rd Symposium (Cambridge, 1953), p. 150; Beale, G. H., Genetics of *Paramecium aurelia* (Cambridge, 1954); Sonneborn, T. M., these Proceedings **46:**149 (1960).
4. Sager, R., these Proceedings **48:**2018 (1962); Sager, R., and Y. Tsubo, Archiv Mikrobiol. **42:**159 (1962).
5. Sager, R., these Proceedings **40:**356 (1954).
6. Sager, R., and Y. Tsubo, Z. Vererbungs. **92:**430 (1961).
7. Sager, R., Genetics **40:**594 (1955).
8. Gillham, N. W., Genetics **48:**431 (1963).

chapter 9

Gene concepts

Our earlier discussions of the gene examined its chemical composition and code, its relative position in a chromosome, and its basic functions of replication and transcription. It will be recalled that the sole difference between alleles of a gene lies in their nucleotide sequences. In this chapter we shall consider how such differences could have arisen. Our study of crossing-over showed that genes occupy a particular position in the chromosome. It left unanalyzed where the gene begins and where it ends and whether it is internally subdivisible. Our discussion of gene functions left unspecified the mechanisms for initiating and terminating its actions. We shall now examine the gene from these three points of view: (1) as a mutable unit, (2) as a recombinational unit, and (3) as a functional unit. We ought to anticipate that the gene may not show the same characteristics when considered in these three different aspects.

MUTABLE UNIT

A gene can be changed by altering either the number or the types of its nucleotides. Alterations in number can involve either *addition* or *deletion* of one or more nucleotides. Either of these changes will not only result in a change in codon at the point of addition or deletion of the nucleotides but also will usually cause radical changes in the rest of the message transcribed from the gene. A drastically altered message can easily result in the production of an inactive protein and can lead to the death of the cell. Alteration in the types of nucleotides that make up a gene involves a *replacement* of one nucleotide by another. This will affect only the codon that includes the replaced nucleotide and will result in a protein with a single amino-acid substitution. The above three types of nucleotide change set the lower limit of the gene as a mutable unit at a single nucleotide. This does not mean that all mutations are the result of single-nucleotide changes. It does indicate that a mutation may be produced by that small a change in nucleotide sequence. Changes of this type are referred to as *point mutations.*

It is important to consider how the above three types of nucleotide changes can occur. With reference to a replacement of one nucleotide by another, Watson and Crick in 1953 (Ref. 1-12) suggested a mechanism for spontaneous mutation based upon their model of DNA structure. They hypothesized that, during DNA replication, a tautomeric shift of a hydrogen atom in a particular DNA base would permit either an adenine to pair with a cytosine or a guanine to pair with a thymine. After another cycle of DNA synthesis, during which the tautomers returned to their normal configuration, a DNA molecule would be formed in which an A-T nucleotide pair had been replaced by a G-C pair. Such a change—involving the replacement of one purine in a polynucleotide chain by another purine and, correspondingly, the replacement of one pyrimidine in the complementary chain by another pyrimidine—is called a *transition.* It is not known whether tautomeric shift in base structure is a frequent cause of spontaneous mutations. A great deal of research on transition-type mutation has involved the use of compounds that are analogues of the purine and pyrimidine bases in DNA. These compounds are incorporated into DNA instead of the normal bases, and in so far as these base analogues are capable of existing in tautomeric states, they can cause mutations. An example of this type of experiment was reported by Freese in 1959 (Ref. 9-1).

In analyzing the experimental results involving mutagenic compounds, it is assumed that two compounds cause the same type of mutation (e.g., transition) if one compound can be used to revert the mutations produced by the other. Thus, such varied compounds as 2-aminopurine, 5-bromouracil, and hydroxylamine are all considered to cause the same type of mutation, since each can, in general, revert the mutations produced by the others. They are all believed to cause transitions, since they are base analogues that exhibit a high frequency of tautomerism. When compounds do not revert one another's mutations, it is assumed that they produce their respective mutations by different molecular mechanisms. A study of the revertability of T4 phage mutations by various mutagens was reported by Freese in 1961 (Ref. 9-2).

Another type of base-pair substitution involves the replacement of a purine by a pyrimidine in one DNA chain and the reverse replacement in the complementary chain. This type of mutation is called a *transversion.* Transversions appear to occur less frequently than transitions and are thought to require the temporary removal of a base from a DNA chain with its subsequent replacement by a complementary type of base. Research on transversions has centered on the removal of bases from DNA by various experimental techniques and an analysis of the revertability of the induced mutations by base analogues and other mutagens. The results of such a study are included in Ref. 9-2.

Mutations that involve an addition or deletion of nucleotide pairs from a DNA molecule have been obtained experimentally by the use of the acridine dye, proflavin. As mentioned earlier, such mutations usually lead to nonfunctional proteins. These mutations cannot be reverted by the use of base analogues, but they can be reverted by proflavin. It has been found that most spontaneous T4 phage mutations, but not those of other organisms, are of the addition-deletion type. A study of the effects of acridine-induced mutations on protein production was reported by Terzaghi and his co-workers in 1966 (Ref. 9-3). The mechanism by which acridine dyes induce mutations has not been settled. It was reported by Lerman in 1961 (Ref. 9-4) that proflavin greatly increases the viscosity of DNA. In addition, he found changes in the x-ray diffraction pattern of the DNA that led him to hypothesize that acridines can be sandwiched between adjacent base-pairs, forcing them 6.8 A. apart rather than 3.4 A. If this happens during replication in one chain of the DNA but not in the other, it could lead to the addition or subtraction of a base. It is not known whether there are any normally occurring compounds in the cell that could produce this effect. A general discussion of the chemical production of mutations was published by Auerbach in 1967 (Ref. 9-5) and is reprinted in this chapter.

In addition to the use of chemicals, radiations can be used as mutagenic agents. The discovery that x-rays could induce mutations was reported by Muller in 1927 (Ref. 9-6). Relatively little is known about the mechanism by which ionizing radiations (x-rays and gamma rays; alpha and beta rays; electrons, neutrons, protons) produce mutations. It is known that the ions formed when these radiations pass through a substance must undergo chemical reactions to neutralize their charge and reach a stable configuration. It is during these chemical reactions that the mutagenic effects of ionizing radiations are thought to take place. A known effect of ionizing radiations is the breaking of the sugar-phosphate backbone of the polynucleotide strands. This can result in the various types of chromosomal rearrangements discussed in Chapter 5.

Another type of radiation that can cause mutations is ultraviolet (UV) light. It is a nonionizing type of radiation. Its main effect is to link adjacent thymine bases on the same DNA nucleotide strand and thus form "thymine dimers." Dimerization interferes with the proper base-pairing of the thymines with adenines during DNA replication and is probably the mechanism by which UV produces mutations. However, mutagenesis through the use of UV is complicated by the existence of two mechanisms that can prevent the production of mutations. A series of experiments dealing with UV-induced mutations and the repair mechanisms that prevent their production were reported by Witkin in 1966 (Ref. 9-7) and is reprinted in this chapter. A general review of the molecular mechanisms involved in the production of the various types of mutations was published by Freese in 1963 (Ref. 9-8).

RECOMBINATIONAL UNIT

Early experiments involving crossing-over indicated the relative position of a gene in a chromosome but yielded no information on the internal structure of a gene. The first attempts to obtain recombinants between alleles of a particular gene were unsuccessful, and it was thought by some that functional allelism implied structural allelism. However, Oliver in 1940 (Ref. 9-9) reported a reversion to wild type associated with crossing-over between alleles. Many other instances of recombinations formed by members of allelic series were subsequently discovered, and the term *pseudoalleles* was applied to all these cases. Pseudoallelism indicated that the ultimate unit of recombination was much smaller than the functional gene. However, the actual size of the smallest unit of recombination remained largely unknown until research with microorganisms revealed the complexities of intragenic structure.

Yanofsky in 1963 (Ref. 9-10) reviewed various studies on the fine structure of the gene that transcribes the enzyme tryptophan synthetase of *Escherichia coli.* This enzyme is one of many that are involved in the production of the amino acid tryptophan. Recombinations between mutant strains were obtained through transduction, as discussed in Chapter 2. Some recombinations were found that appear to represent recombinational events between adjacent nucleotides in the same coding unit.

Experiments have also been conducted to examine the fine structure of the gene in higher organisms. A series of such studies were reported by Chovnick and his co-workers in 1964 (Ref. 9-11). They studied the rosy *(ry)* gene, which is one of the genes that control the synthesis of the enzyme xanthine dehydrogenase (XDH) in *Drosophila melanogaster.* This enzyme is involved in the production of eye pigment. Mutations at this locus affect the formation of XDH and result in a reddish brown rather than a normal red-colored eye. It was found that recombination within the *ry* gene can occur and that the resolving power of this type of study, with regard to the fine structure of the gene, approaches that of microbial systems.

The experiments both with microbial systems and higher organisms indicate that it should be possible to obtain recombination between adjacent nucleotide pairs. This would demonstrate that the smallest divisible segment of a gene is a single nucleotide. It would mean that the smallest limit of a gene both as a mutable unit and as a recombinational unit is a single nucleotide.

FUNCTIONAL UNIT

When a bacterial virus is mixed with a culture of bacteria in melted nutrient agar and poured into a petri dish, the agar will solidify. Most of the agar will become cloudy as a result of the dense growth of bacteria. Where the virus has infected a bacterium, there will be a clear *plaque* due to the successive invasions and lyses of bacteria adjacent to the point of infection. The morphology of the plaque (size, nature of its edges, etc.) is characteristic for each type of phage. Mutations can occur in the viruses that affect plaque morphology. In the bacteriophage T4 a large number of such mutations are known that produce a particular effect. The viruses that contain these mutations are called *r* mutants. The various *r* mutants lyse the bacterial cells, and all produce large plaques with sharp edges, whereas the wild type, r^+, virus produces smaller plaques with irregular edges. Depending on the region of the T4 chromosome in which the mutation is located, the mutants have been labeled *rI, rII,* or *rIII.* All the *r* mutants produce the characteristic large, sharp-edged plaque when *E. coli* strain B is used as a host. However, when *E. coli* strain K12 (λ) is used as a host, only mutants *rI* and *rIII* form plaques. Although the mutant *rII* can enter K12 (λ) cells, it normally cannot reproduce and lyse the cells. Phage with the wild-type allele r^+, form their smaller, irregular-edged plaques with both strain B and strain K12 (λ). The interactions of the various viruses and the different bacterial strains are shown in Table 1.

The T4*r* locus that has been most extensively analyzed is T4*rII.* One of the techniques that has been extremely important in analyzing the structure of the T4*rII* locus is that of mixed infections. It has been found that if K12(λ) bacteria are infected simultaneously with both *rII* and r^+ viruses, phage of both genotypes will enter the cells and reproduce. As noted above, *rII* mutants by themselves normally cannot reproduce in K12(λ) bacteria, although they can enter the cells. It is apparent that the r^+ virus produces some necessary substance that the *rII* phage can utilize for its own reproduction but cannot, itself, produce. The mixed infection provides the equivalent of the diploid state for the virus genome. With this technique it is possible to test for both functional and structural allelism between different *rII* mutants. If the mutations are functional alleles, no viral reproduction will occur. If the mutations are located in different genes, they will comple-

Table 1. Types of plaques formed by T4 phages in *E. coli* strains B and K12(λ)*

Phage T4 genotype	Plaques formed on *E. coli* strain	
	B	K12(λ)
rI	r	r
rII	r	None
rIII	r	r^+
r^+	r^+	r^+

*From Levine, L., 1969. Biology of the gene. The C. V. Mosby Co., St. Louis.

ment each other, and both mutants will be able to reproduce. Much of our knowledge of the structure of the T4*rII* locus comes from the work of Benzer, whose paper in 1955 (Ref. 9-12) is included in this chapter. He found that the *rII* locus is divisible into two functionally separable segments, each of which is called a *cistron.* A cistron is a section of the DNA molecule that specifies the composition of a particular polypeptide chain. In effect, a cistron is a gene, although operationally it is convenient to describe a gene in terms of the final protein end product or phenotype and to state that the gene consists of either one or more cistrons. For phage T4, it is customary to consider the entire *rII* locus as a gene that controls plaque morphology and to refer to the two functional segments of the *rII* locus as cistrons A and B.

REGULATION OF GENE ACTION

Studies on the regulation of gene action have centered largely on *E. coli* and its enzymes. Most important have been those enzymes that are produced only when there is need for them. A group of such enzymes are associated with the "lactose" *(lac)* region of the *E. coli* chromosome. These enzymes are involved in the hydrolysis of lactose into galactose and glucose and are synthesized only when lactose is present in the medium. Enzymes that are produced only when their substrates are available are called *inducible* enzymes.

The *lac* region of the *E. coli* chromosome controls the synthesis of three enzymes: β-galactosidase, which catalyses the hydrolysis of lactose into galactose and glucose; β-galactoside permease, which is responsible for the uptake and concentration of lactose in the cell; and galactoside-transacetylase, an enzyme whose function in the cell is not known. Mutations affecting the structure of these three enzymes have been found. It will be recalled that we discussed in Chapter 2 how conjugation between *E. coli* cells can result in a temporary diploid state and in recombinations of genes from donor and recipient chromosomes. With the above materials and techniques it was found that the genes specifying the structures of β-galactosidase (*z* gene), permease (*y* gene), and transacetylase (*ac* gene) map in adjacent loci of the bacterial chromosome.

Evidence that the functioning of these three genes was under the control of some other chromosomal locus was provided by Lederberg in 1951 (Ref. 9-13). It was known that wild-type *E. coli* could not use neolactose (a β-galactoside of altrose) as a carbon source. A mutant *E. coli* was isolated that could use this sugar. The enzyme produced by the mutant that utilized neolactose as a substrate was β-galactosidase. It was also found that the mutant produced the enzyme continuously or, as it is usually termed, *constitutively.* It seemed evident that for wild-type *E. coli,* neolactose is a substrate but not an inducer of β-galactosidase. It also appeared evident that the mutant *E. coli* had a perfectly normal *z* gene but that a mutation had occurred in some "regulator" type gene that determines not the structure of the enzyme but the conditions necessary for its synthesis. The regulator gene was labeled the *i* (for "inducibility") locus. The wild-type allele, i^+, causes the production of all three *lac* enzymes only when lactose is present (i.e., inducibly), whereas the mutant allele, i^-, causes the continuous production of the three enzymes regardless of the presence or absence of lactose (i.e., constitutively). Subsequent recombination experiments showed that the *i* locus was located close to, but not contiguous with, the *z* gene.

Analysis of the mechanism by which the *i* gene controls enzyme synthesis was reported by Pardee and his co-workers in 1959 (Ref. 9-14). They arranged conjugations between *E. coli* cells that brought together in the same cell one *lac* region that was $i^+z^+y^-$ and another *lac* region that was $i^-z^-y^+$. Under these conditions, all the enzymes were produced inducibly. This experiment demonstrated that i^+ was dominant to i^-. It seemed evident that whatever controlled the functioning of the structural genes was transmissable through the cytoplasm and that the i^+ gene had produced a *repressor* substance that prevented the y^+ gene from functioning despite its association with the i^- gene. The repressor substance was isolated by Gilbert and Muller-Hill in 1966 (Ref. 9-15) and was found to be protein, indicating that the i^+ gene functions through the production of a normal gene end product.

Regulation of the structural genes of the *lac* region proved to have additional complications. It was reported by Jacob and Monod in 1959 (Ref. 9-16) that amongst cells demonstrated to be i^+, mutants arose that produced β-galactosidase constitutively. To explain these results, they hypothesized the existence of another locus which they called the "operator" *(o)* gene. The wild-type *E. coli* was thought to contain the o^+ allele, which was the attachment site for the repressor of the i^+ gene. When the o^+ was freed of the repressor through some unknown interaction between the repressor and the inducer, lactose, transcription of the *lac* structural genes could occur. The i^+ constitutive mutant was considered to contain the o^c allele, which in some then unknown fashion prevented the repressor from attaching itself to the operator locus. Without the repressor attached to the operator, the *lac* structural genes transcribed their messages continuously. The operator alleles were found, through recombination experiments, to be very close to the *z* gene. It was also found that an *o* allele was only effective in controlling the functioning of the structural genes adjacent to it on the same chromosome (i.e., in the "cis" position). An *o* allele cannot affect the functioning of the structural genes of some other chromosome in the cell (i.e., in the "trans" position).

The above experiments led Jacob and Monod in 1961 (Ref. 9-17) to propose their "operon" hypothesis. They postualted that the gene as an operational unit consists of an operator gene and one or more structural genes. This linked combination, which they called the *operon,* was postulated to be under the control of a repressor whose synthesis would be determined by a regulator gene that did not necessarily have to be linked to the operon. A recent review of this theory was prepared by Beckwith in 1967 (Ref. 9-18) and is the last paper reprinted in this chapter. (It will be noted that in 1967 it was thought, as seen in Figs. 1 and 4 of Ref. 9-18, that the promoter gene was located between the operator and the *z* gene.)

It is not known whether the operational gene of higher organisms is in the form of an operon that is under the control of a regulator gene. However, a control system of gene action that has certain resemblances to the type found in *E. coli* has been known in maize for some time. A gene called *Dissociation (Ds)* was discovered, which affected the functioning and the mutability of the structural genes adjacent to it. However, the *Ds* gene can only function if another gene called *Activator (Ac)* is also present somewhere in the maize genome. One unusual feature of this system is that the *Ac* gene, when by itself, can also affect the functioning and mutability of structural genes adjacent to it. Another unusual feature is the fact that both the *Ds* and *Ac* genes are capable of undergoing changes in location either within the same chromosome or between chromosomes. In their new locations, they produce their characteristic effects. A summary of the experiments involved in the maize study was published by McClintock in 1956 (Ref. 9-19). A comparison of the control systems in maize and in bacteria was published by McClintock in 1961 (Ref. 9-20).

REFERENCES

Mutable unit

9-1. Freese, E. 1959. The specific mutagenic effect of base analogues on phage T4. J. Mol. Biol. **1**:87-105.

9-2. Freese, E. 1961. The molecular mechanisms of mutations. Proc. 5th Intern. Congr. Biochem., Moscow **1**:204-229.

9-3. Terzaghi, E., Okada, Y., Streisinger, G., Emrich, M. I., and Tsugita, A. 1966. Change of a sequence of amino acids in phage T4 lysozyme by acridine-induced mutations. Proc. Natl. Acad. Sci. U.S.A. **56**:500-507.

9-4. Lerman, L. 1961. Structural considerations in the interaction of DNA and acridines. J. Mol. Biol. **3**:18-30.

9-5. Auerbach, C. 1967. The chemical production of mutations. Science **158**:1141-1147.

9-6. Muller, H. J. 1927. Artificial transmutation of the gene. Science **66**:84-87.

9-7. Witkin, E. 1966. Radiation-induced mutations and their repair. Science **152**:1345-1353.

9-8. Freese, E. 1963. Molecular mechanism of mutations, pp. 207-269. *In* J. H. Taylor [ed.] Molecular genetics, Part I. Academic Press, Inc., New York.

Recombinational unit

9-9. Oliver, C. P. 1940. A reversion to wild-type associated with crossing-over in *Drosophila melanogaster.* Proc. Natl. Acad. Sci. U.S.A. **26**:452-454.

9-10. Yanofsky, C. 1963. Amino acid replacements associated with mutation and recombination in the A gene and their relationship to in vitro coding data. Cold Spring Harbor Symp. Quant. Biol. **28**:581-588.

9-11. Chovnick, A., Schalet, A., Kernaghan, R. P., and Krauss, M. 1964. The rosy cistron in *Drosophila melanogaster* genetic fine structure analysis. Genetics **50**:1245-1259.

Functional unit

9-12. Benzer, S. 1955. Fine structure of a genetic region in bacteriophage. Proc. Natl. Acad. Sci. U.S.A. **41**:344-354.

Regulation of gene action

9-13. Lederberg, J. 1951. Genetic studies with bacteria, pp. 263-289. *In* L. C. Dunn [ed.] Genetics in the 20th century. The Macmillan Co., New York.

9-14. Pardee, A. B., Jacob, F., and Monod, J. 1959. The genetic control and cytoplasmic expression of "inducibility" in the synthesis of β-galactosidase by *E. coli.* J. Mol. Biol. **1**:165-178.

9-15. Gilbert, W., and Muller-Hill, B. 1966. Isolation of the lac repressor. Proc. Natl. Acad. Sci. U.S.A. **56**:1891-1898.

9-16. Jacob, F., and Monod, J. 1959. Gènes de structure et gènes de régulation dans la biosynthèse des protéines. Compt. Rend. Acad. Sci. Paris **249**:1282-1284.

9-17. Jacob, F., and Monod, J. 1961. Genetic regulatory mechanisms in the synthesis of proteins. J. Mol. Biol. **3**:318-356.

9-18. Beckwith, J. R. 1967. Regulation of the lac operon. Science **156**:597-604.

9-19. McClintock, B. 1956. Intranuclear systems controlling gene action and mutation. Brookhaven Symp. Biol. **8**:58-74.

9-20. McClintock, B. 1961. Some parallels between gene control systems in maize and in bacteria. Am. Nat. **95**:265-277.

42 The chemical production of mutations

The effect of chemical mutagens on cells and their genetic material is discussed

Charlotte Auerbach
Department of Genetics
University of Edinburgh
Edinburgh, Scotland

Twenty years ago, *Science* published an article with the above title *(1)*. A few years earlier, the first potent chemical mutagens had been discovered, and this discovery started a vigorous and astonishingly successful search for more substances with mutagenic ability. The hopes which my colleagues and I set on the new field of chemical mutation research were expressed as follows. "If, as we assume, a mutation is a chemical process, then knowledge of the reagents capable of initiating this process should throw light not only on the reaction itself but also on the nature of the gene, the other partner in the reaction. Moreover, it could be hoped that among chemical mutagens there might be some with particular affinities for individual genes. Detection of such substances not only would be of high theoretical interest but would open up the long sought-for way to the production of directed mutations." It is interesting to look back and see how far these hopes of 20 years ago have been fulfilled.

REACTIONS BETWEEN MUTAGENS AND GENES

The chemical nature of the gene has not been elucidated by research on mutation but in entirely different ways. On the contrary, mutation research now starts from the presumption that "the gene, the other partner in the reaction" consists of DNA, or—in some viruses—of RNA. The DNA molecule is a duplex structure, consisting of two sugar-phosphate strands which are helically wound round each other and which carry attached sequences of the four nucleotide bases adenine, guanine, thymine, and cytosine. The two strands are held together by hydrogen bonds between opposite bases; since, for steric reasons, the purine adenine is always opposite the pyrimidine thymine and the purine guanine is opposite the pyrimidine cytosine, the whole structure is internally complementary. At replication, the two strands separate, and each constructs a new complementary strand. The genetic information is coded by the sequence of bases, and, if this has been changed by mutation, the same principle of complementarity that governed replication of the original sequence now leads to perpetuation of the mutated one. It appears that this structure of the genetic material is common to all living species, from viruses to man. Among viruses there are some exceptions, but they have retained the principles of coding by base sequence and of replication by complementarity. In some viruses, DNA is single-stranded when it is not engaged in replication, and in some it has been replaced by RNA. The RNA molecule is single-stranded except at replication; three of its nucleotide bases are the same as those in DNA, but thymine has been replaced by uracil. When speculating about the action of mutagens, we no longer ask whether they react with DNA, but how they react with it. We shall see that, as these questions were answered for a series of mutagens, specificities of reaction appeared at the level of nucleotides and nucleotide sequences and furnished valuable clues for the deciphering of the genetic code. Thus, chemical mutagens have, after all, proved important analytical tools for the study of the genetic material, but at a level of chemical structure that was still quite unsuspected 20 years ago.

Knowledge of the structure of DNA furnishes a framework for the classification of mutations. The least drastic alteration—which, however, may have drastic consequences for the organism—is replacement of one nucleotide

From Science **158**:1141-1147, December 1, 1967.

Dr. Auerbach holds a personal chair in the Department of Genetics at the University of Edinburgh and is a Fellow of the Royal Society of London.

base by another. The most frequently observed base changes are "transitions," in which a purine has been replaced by another purine (adenine by guanine or vice versa) or a pyrimidine by a pyrimidine (thymine by cytosine or vice versa). Much rarer are "transversions," in which a purine has been replaced by a pyrimidine or vice versa. Since the genetic code is read in triplets of bases, each triplet coding for one amino acid, mutations due to base changes—whether transitions or transversions—usually result in a protein in which one amino acid has been replaced by another. In which way this will affect the organism depends on the type of amino acid change and on the position in the protein in which it occurred. Occasionally, a base change results in a so-called "nonsense triplet" which, instead of coding for an amino acid, codes for premature termination of the growing polypeptide chain, so that only incomplete protein molecules can be formed. These mutations lead to complete loss of gene function. A mutant gene that arose through a base change may be reversed by another base change that restores the original code. Such mutations are called "reverse" mutations; they restore a gene function that has become altered or abolished by the first mutation. Since reverse mutations require a very precise chemical change, they are very rare even after mutagenic treatment and can be studied profitably only in material that permits work on very large numbers, such as bacteria or bacteriophages.

A different class of mutations is due to deletions of one or more bases from the sequence of a gene. Since the code is read in base triplets from one end of the gene to the other, deletions of even one base will alter every triplet from the missing base to the end of the gene, and this in turn will alter every one of the corresponding amino acids. Insertion of an extra base has the same effect. Such mutations are often called "reading frame shifts"; they result in loss of gene function. They cannot be reversed by base changes, but, if a deletion and an insertion occur sufficiently close together, the original reading frame will be restored after the second change. If the portion between the two changes is relatively unimportant, this may result in a sufficiently normal protein to pass as a reversion.

Changes in genetic information that take place within the confines of a gene and affect the action of this gene only are called "gene mutations." On a grosser scale, we have to consider changes that affect the order of the genes on the chromosomes. In this article, I shall use the term "chromosome" for all linear structures that carry hereditary information in any organism whatsoever. This terminology is simple and suffices for our purposes, but it obscures the fact that in all organisms beyond the evolutionary level of bacteria the chromosomes are complex structures, containing not only DNA—often several replicas of it—but also proteins and RNA. The role of this complexity in the production of mutations is likely to be important, but hardly anything is known about it. Within the chromosome, there may be deletions of whole genes or sequences of genes. This can happen when a chromosome is broken into several fragments, one of which gets lost when the broken ends rejoin. Rejoining of chromosome fragments in the wrong order may also result in other types of "chromosome rearrangement," for example, a "translocation" in which two broken chromosomes have exchanged pieces. Some of these chromosome rearrangements resemble gene mutations in their effects on the organism and in their mode of hereditary transmission.

THE DISCOVERY OF CHEMICAL MUTAGENS

The first agent shown to produce mutations was x-irradiation, but the search for chemical mutagens had started already before this discovery and was continued after it. Although some of the substances tested in the 1930's, for example, iodide and coppersulphate, seemed to have weakly mutagenic effects on the fly *Drosophila,* no clear positive results were obtained before the early years of World War II when, independently, the mutagenic action of mustard gas was discovered in Edinburgh and that of urethane was found in Germany. These first successes provided a strong stimulus for the testing of more substances, many of which proved mutagenic. By now, we know a large number of chemical mutagens, belonging to a variety of chemical classes. Very different principles have been used in the selection of chemicals for testing. Some of the most powerful mutagens were discovered on the basis of what we now know to be the wrong concept of the gene. Mutation research offers good cautionary examples against the belief that a successful experiment necessarily proves the hypothesis that inspired it. In the following, I shall briefly review the most important classes of mutagenic chemicals.

ALKYLATING AGENTS

This class contains some of the most potent mutagens, including mustard gas. The rationale for testing mustard gas was the pharmacological similarity between mustard-gas burns and x-ray burns, coupled with the knowledge that x-rays cause damage to chromosomes and genes. While this speculation was fully vindicated by the results, pharmacological observations are not always a reliable guide to the detection of chemical mutagens. This became clear very early when lewisite, like mustard gas, a potent vesicant war gas, proved quite ineffective in mutation tests. All the same, tests of pharmacologically active substances for mutagenicity have retained their value as part of the program for protecting man and his domestic animals and plants against genetic damage. Occasionally, this may lead to the discovery of a new group of strong mutagens. This happened with the pyrrolozidine alkaloids (for example, heliotrine) which were tested because they produce liver damage in sheep that ingest them in ragwort. The antibiotic streptonigrin has produced mutations in fungi and chromosome breaks and rearrangements in cells of plants and mammals. In mutation tests on *Drosophila,* streptomycin and the insecticide DDT were ineffective, and carcinogenic hydrocarbons gave, at best, doubtful results. It must be kept in mind, however, that negative results of mutation tests can rarely be considered as final. A chemical that fails to yield mutations in a particular type of experiment may be mutagenic under different conditions or for a different organism or cell type. Examples for this will be found in this article.

Chemically, mustard gas is dichloroethyl sulphide $S(CH_2CH_2Cl)_2$; the related and equally mutagenic nitrogen mustard "NH2" has the formula $CH_3N(CH_2CH_2Cl)_2$. These and other "mustards" owe their biological activity to their chloroethyl groups; they act by alkylating of biologically important macromolecules. In addition to mustards, many other compounds, in particular epoxides, ethylene imines, and alkylmethanesulphonates, have alkylating abilities, and many of them are mutagenic. Indeed, the correlation between alkylating and mutagenic abilities is so strong that a tendency has arisen to attribute alkylating reactions to mutagens whose mode of action is not yet understood, such as a number of nitroso compounds and the pyrrolizidine alkaloids mentioned above.

In vitro, alkylation affects preferentially the guanine in DNA, and this appears to be so also in vivo, although in a bacteriophage with single-stranded DNA all four bases were attacked *(2).* Alkylation of guanine is thought to produce mutations mainly through the tendency of alkylated guanine to pair erroneously with thymine instead of cytosine. At the next replication, thymine will pair correctly with adenine, and the final result will be a transition from a guanine-cytosine to an adenine-thymine pair at the site of mutation. It has also been suggested that mutations may arise through the relative ease with which alkylated guanine detaches from the DNA backbone, leaving an "apurinic gap," which might be filled by a wrong base. This could lead to transversions as well as transitions, but so far there has been no clear evidence for the production of transversions by alkylation. The assumption that alkylation usually changes guanine-cytosine into adenine-thymine but only rarely adenine-thymine into guanine-cytosine agrees well with the fact that alkylating agents are not usually able to revert the mutations that they themselves have produced, while other mutagens—able to change adenine-thymine into guanine-cytosine (see below)—may do so.

In addition to gene mutations, alkylating agents also produce deletions and other types of chromosome rearrangement. Indeed, their genetical effects are so strikingly similar to those of x-rays that the term "radiomimetic substances" is often applied to them. Differences, however, also exist, and these are of special interest for an analysis of the mutation process. Two major differences from x-rays were found early and seem to be characteristic of all alkylating agents, possibly of most chemical mutagens. One is a relative shortage of chromosome rearrangements compared with gene mutations. This is found for all alkylating agents, but its magnitude varies from a moderately high number of rearrangements after treatment with ethylene imines to their almost complete absence after treatment with diethylsulphate. For mustard gas it has been shown that the shortage of rearrangements is not due to a shortage of chromosome breaks but to an inhibition of the process by which the broken ends rejoin into new arrangements. How far this factor contributes to the shortage of rearrangements after treatment with other alkylating agents has not been established.

The inhibition of reunion between chromosome fragments is mainly a consequence of the second and more fundamental peculiarity of alkylating agents and, possibly, of most other chemical mutagens. This is a tendency for injuries to the genetic material not to result directly in chromosome breakage and mutation but to remain latent over a period that may extend over many cell cycles. Since the formation of a chromosome rearrangement requires the simultaneous presence in the same cell of two chromosome breaks, while single unjoined breaks usually result in cell death, a potential rearrangement will be lost every time when the two breaks that might have given rise to it open in different cell cycles.

Although the phenomenon of delayed chromosome breakage and mutation was first observed and studied in *Drosophila,* it can be illustrated more easily by an example from experiments on fission yeast, *Schizosaccharomyces pombe,* in which mutations from red to white colony color can be scored. When cells of the red strain are exposed to the alkylating agent ethylmethanesulphonate and subsequently plated out, the majority grow again into red colonies, but a minority of mutated ones grow into white colonies. In addition, there are "mosaic" colonies that are partly white, partly red. In the early days of mutation research, this observation alone would have suggested that some mutations to white must have arisen with a delay of at least one division. This argument has become invalid with our realization of the duplex nature of DNA. If a chemical change affects only one strand of DNA—and the majority are likely to do so—the first division should result in one mutated and one nonmutated cell, and these should grow into a mosaic colony. Decisive proof for the delayed occurrence of mutations after treatment with ethylmethanesulphonate was obtained when cells from mosaic colonies were respread. As would be expected, the majority grew into either wholly red or wholly white colonies, but in most experiments some grew again into mosaics, and these, in turn, yielded some mosaics when respread. Moreover, although the mutations to white involved a number of different genes, in any particular line of mosaics it was always the same gene that gave rise to delayed mutations. The simplest way of describing these observations, and similar ones obtained in other systems and with other mutagens, is to say that treatment with alkylating agents, in addition to producing mutations immediately, may cause instabilities of individual genes that continue to give rise to mutations. Moreover, since in these lines each mosaic must have started with an instability, and since several cells from the same mosaic may again grow into mosaics, the instability must be able to replicate as such. No satisfactory explanation of these instabilities has so far been put forward. Their nature, in particular their ability to replicate in the unstable state, is difficult to fit into a molecular explanation at the level of DNA unless, as has been suggested very recently, an apurinic gap may replicate as such *(3)*. Probably, different types of instability arise in different ways; this applies in particular to instabilities for gene mutations on the one hand, delayed chromosome breakage on the other.

Many alkylating mutagens are carcinostatic, and some have been found to be carcinogenic. This triad of effects is produced also by other agents, notably urethane and x-rays. The correlation between carcinogenicity and mutagenicity is obscure; it has been used in support of the somatic mutation theory of cancer. The carcinostatic action of mutagens is doubtless connected with their ability to break chromosomes; for chromosome breakage preferentially kills dividing cells such as are present in malignant tissue. Among alkylating agents, only those with two or more functional (for example, chloroethyl) groups are carcinostatic, while related ones with but one functional group are not. Yet many such agents not only produce high frequencies of gene mutations but are also efficient chromosome breakers. A clue to this discrepancy between oncological and genetical observations may be found in recent experiments on *Drosophila* in which pairs of closely related monofunctional and polyfunctional epoxides and ethylene imines were compared *(4)*. When treated spermatozoa were utilized on the day following treatment, the ratios between chromosome rearrangements and gene mutations caused by mono- and polyfunctional members of the same pair were the same. This, however, was drastically changed when the treated spermatozoa were first stored for six or more days in the seminal receptacles of untreated females. During storage, the frequency of chromosome breaks and rearrangements, but not that of mutations, increased up to 15-fold after treatment with the polyfunctional compounds, while it did not change at all after treatment with the monofunctional ones. The superiority of polyfunctional compounds in cancer therapy may, therefore, be related to the fact that treatment allows the full effect of storage to manifest itself.

URETHANE

Like x-rays and many alkylating agents, urethane is mutagenic, carcinostatic, and carcinogenic. In flowering plants, as well as in *Drosophila,* it produces chromosome breaks and rearrangements; in *Drosophila,* the breaks produced by urethane resemble breaks caused by x-rays, at least to the extent that fragments produced by the two treatments given in succession combine as freely with each other as do fragments produced by only one of them. In *Drosophila,* urethane also produced gene mutations which appeared to be unconnected with chromosome breakage. Urethane is, however, a very "spotty" mutagen, acting strongly in some organisms and not at all in others. The fungus *Neurospora* was entirely recalcitrant to its action, even when the whole of the genome was tested for mutations and deletions by a special technique *(5).* Possibly, this organism specificity of urethane as mutagen is related to its even more striking organism specificity as a carcinogen. It produces lung cancers in some rodent species but not in others; the active principle in this case appears to be a metabolite of urethane which is produced in mice but not in guinea pigs *(6).*

PHENOLS

These are similarly "spotty" in their mutagenic effects. A number of phenols, for example, pyrogallol and hydroquinone, produced chromosome fragmentations in plants, although few rearrangements were formed. In *Drosophila,* exposure of the explanted and subsequently reimplanted larval ovary to phenol produced high frequencies of mutations (or small deletions) in some experiments and none in others, and, in spite of prolonged and determined efforts, the conditions for successful application could not be elucidated. These results have gained new significance through the finding of an increased frequency of chromosome rearrangements in lymphocytes of men that had been exposed to ambient benzene *(7).*

FORMALDEHYDE

When formaldehyde is mixed into the food of *Drosophila,* it may act as a strong mutagen. The conditions for its action are, however, more stringent than for any other mutagen. Although, as shown by labeling experiments, formaldehyde penetrates into all germ cells of larvae and adults, mutations occur exclusively during one part of the cell cycle of one germ cell stage in one developmental phase of one sex, namely in early larval spermatocytes. Moreover, the frequency of mutations depends on the nutritional status of the larvae: any conditions that slow down development, including an excessive dose of formaldehyde, decrease the frequency of mutations, and under very poor growth conditions no mutations at all are recovered. On synthetic media, adenosine riboside is an indispensable adjuvant; whether it is involved in the production of a secondary product with mutagenic ability or whether it aids in the release of free formaldehyde from a reversibly bound form or acts in some other way could not so far be decided. Studies on the distribution of mutations along the chromosome have led to the conclusion that mutations are produced during the time of DNA replication and occur in close neighborhood to the point of replication. Casein that has been treated with formaldehyde is also mutagenic; this has raised some doubt about the advisability of feeding breeding pigs skim milk sterilized with formalin. Cytologically, formaldehyde produces few gross chromosome rearrangements but many small deletions and repeats, that is, duplications of small chromosome regions in tandem or reverse. Since duplications are generally assumed to have played an important part in evolution, it is of interest to see that they can be produced by a compound that is closely related to normal metabolic processes.

ORGANIC PEROXIDES AND IRRADIATED MEDIUM

Formaldehyde can also produce mutations when applied in aqueous solution to microorganisms or *Drosophila* spermatozoa. Applied in this way it is a very weak mutagen whose effectiveness, however, can be greatly enhanced by the addition of hydrogen peroxide, which by itself is hardly mutagenic. This suggests that a mutagenically active peroxide is formed and that small amounts of this substance can arise through reaction of formaldehyde with metabolically produced hydrogen peroxide. Indeed, the addition compound of formaldehyde and hydrogen peroxide, as well as other organic peroxides, are fairly good mutagens. They form a link with radiation mutagenesis, for peroxides have been made responsible for the mutagenic action of bacterial medium that has been exposed to heavy irradiation with ultraviolet light or ionizing radiation. In recent years, the sterilization of human food with very heavy x-ray doses has given concern about the possible introduction of mutagens into human consumption. Plant chromosomes can, indeed, be broken by heavily irradiated fruit juices or sugar solutions or by growing of the plants on irradiated potato medium *(8)*. Mutation tests on *Drosophila,* on the other hand, have given negative results in some investigations and slightly, though significantly, positive ones in others *(9)*. The bearing of these findings on human affairs is doubtful. Experiments on mice might give clearer evidence, although even extrapolation from mice to man is hazardous when one is dealing with slight genetical effects of chemicals that have been introduced to the germ cells by way of the food.

INORGANIC SALTS

A number of inorganic salts produce chromosome breaks and rearrangements in plant cells and mutations in bacteria. In particular, manganese chloride is an excellent mutagen for some bacterial strains, but its action depends strongly on ancillary conditions such as the presence of other salts before treatment. In fungi, attempts to produce mutations with $MnCl_2$ have been unsuccessful. It seems likely that the genetical effects of inorganic salts are not due to direct reactions with DNA, but to the creation of cellular conditions that favor the occurrence of "spontaneous" mutations and chromosome breaks.

PURINE DERIVATIVES

Certain alkylating agents, as well as formaldehyde, were first tested because of their known reaction with proteins. This was natural at a time when most geneticists believed that the specificity of the gene resided in its protein moiety. Yet even then, it was thought possible that reaction with the DNA moiety might produce mutations, and a number of purines and pyrimidines were tested for mutagenicity. While the results with pyrimidines were negative, various purines were found to produce chromosome breaks in plants and mutations in fungi and bacteria. Special interest was aroused by the mutagenic action of caffeine because of the large amounts of it that civilized man consumes in tea or coffee. Tests on mice that had been given the highest tolerated dose of caffeine in their drinking water gave no evidence for the production of either chromosome breakage or mutation, but chromosome breaks, although no rearrangements, were found in human cell cultures that had been exposed to solutions of caffeine *(10)*. Experiments on *Drosophila* gave contradictory results, but the latest evidence indicates that feeding or injection of caffeine has a weak mutagenic effect *(11)*. As in the case of food sterilized with radiation, the application to human affairs is doubtful and hazardous. An interesting feature of the experiments on bacteria was the finding that adenosine riboside, which, as mentioned above, is required for the mutagenic action of formaldehyde, abolished that of caffeine and related purines. Moreover, it greatly reduced the frequency of spontaneously occurring mutations, suggesting that a proportion of them is due to ingested or metabolically produced mutagenic purines. The frequency of radiation-induced mutations was not reduced by adenosine riboside or related "antimutagens."

SUBSTANCES TESTED BECAUSE OF THEIR REACTIVITY WITH DNA

Base analogs. With the recognition of DNA as the essential genetic material, the search for mutagens became directed towards substances that are known to react with DNA or may be presumed to do so. The first of these to be tested were analogs of the purine and pyrimidine bases in DNA. It was thought that these analogs, by being mistakenly incorporated into DNA, might misdirect the subsequent incorporation of the natural bases at the time of replication, leading to transitions from adenine-thymine to guanine-cytosine or from guanine-cytosine to adenine-thymine. Indeed, the pyrimidine 5-bromouracil which, under appropriate conditions, may replace most of the thymine in the DNA of bacteria and viruses, proved to be an excellent mutagen; so, however, did the adenine analog 2-aminopurine of which only traces are incorporated. Thus, although incorporation into DNA almost certainly is a prerequisite for the mutagenic action of base analogs, their mutagenic efficiency does not necessarily depend on the chemically detectable degree of incorporation. Since then, evidence has been obtained for assuming that 5-bromouracil, when incorporated instead of thymine, is only weakly mutagenic, presumably because its pairing preference for adenine is almost as high as that of thymine. However, in the rare and chemically not yet detected instances in which it is incorporated instead of cytosine, this same strong pairing preference for adenine leads to the eventual replacement of a guanine-cytosine pair by an adenine-thymine pair. It is in agreement with this assumption—and, in fact, has been used in formulating it—that mutations which are easily reversed by alkylating agents are also easily reversed by 5-bromouracil. 2-Aminopurine, on the contrary, preferentially reverts mutations that do not respond to alkylating agents, including those produced by them. This is taken as evidence that 2-aminopurine usually acts by incorporating instead of adenine, pairing erroneously with cytosine, and finally changing an adenine-thymine pair into a guanine-cytosine pair. Base analogs have also proved mutagenic in some fungi, and they are able to break chromosomes in human cell cultures. The use of one of them, iododeoxyuridine, for the treatment of herpes lesions in the cornea has caused some concern for possible genetical consequences on the patient; but the chance that this strictly localized treatment will allow appreciable amounts to penetrate to the gonads seems negligible.

Acridines. These have long been known to react with nucleic acids, and instances of chromosome breakage and mutation by substances such as acridine orange or proflavine have been reported repeatedly in the past. In most of these experiments, however, visible light had not been excluded, and the effects might have been due to the so-called photodynamic action of the dyes, that is, their ability to sensitize other molecules to the action of visible light. A different and more interesting mechanism of mutagenesis by acridines was suggested by the finding that in vitro acridine molecules intercalate between the nucleotides of DNA. If this should happen also in vivo, replication of the affected strand would be disturbed, and this, it was thought, might lead to the insertion or deletion of a base in the normal sequence. In other words, mutations produced by acridines should be reading-frame shifts and should be characterized by three properties: (i) complete absence of gene function, (ii) lack of reversibility by agents producing only base changes and (iii) ability of being at least partially reversed by other reading-frame shifts of the complementary type that had occurred close enough to the first one to give a functional protein. Experiments with bacteriophage confirmed all these predictions: acridine-induced mutations resulted in completely inactive genes; they could not be reversed by base analogs, which are assumed to cause mutations exclusively by transitions; they could be reversed, or partially so, by additional acridine mutations in their neighborhood. The final proof for the hypothesis was brought by amino acid analysis of the enzyme lysozyme in a certain type of bacteriophage *(12).* The enzyme was completely absent from two strains with acridine-induced mutations, and was partially active in a strain combining these two mutations. As predicted, the enzyme in the doubly mutant strain had a short stretch of faulty amino acids between the sites of the two mutations but was normal beyond it. Moreover, the faulty amino acids were of just the types expected from a reading-frame shift applied to the original base sequence. Acridines, sometimes in the form of a so-called "acridine mustard" with a mustard as well as an acridine moiety, have produced mutations also in fungi and *Drosophila.* There is evidence that these mutations, too, are due to reading-frame shifts.

Nitrous acid and hydroxylamine. Nitrous acid was tested many years ago because of its known reactions with proteins. These experiments gave suggestive results, but they were not followed up. More recently, new experiments were stimulated by the fact that nitrous acid in vitro deaminates three of the bases of DNA and RNA: adenine, guanine, and cytosine. Deamination of adenine produces hypoxanthine, whose pairing properties resemble those of guanine, so that the process in vivo might lead to a transition from adenine to guanine (in double-stranded DNA from adenine-thymine to guanine-cytosine). Deamination of guanine yields xanthine, whose pairing properties resemble those of guanine, so that no mutational change is expected. Finally, deamination of cytosine yields uracil. In RNA, this change results in an immediate change of code; in DNA it might be expected to do so after replication, because uracil tends to pair with adenine rather than with guanine. Genetic experiments, first on tobacco mosaic virus (containing RNA), then on bacteria and fungi, showed that nitrous acid is an excellent mutagen, provided the *p*H is kept low. The results with tobacco mosaic virus were of special importance because of their contribution to the decoding of the nucleotide base triplets. Coding triplets were derived from the correlation between the chemically predicted base changes and the observed amino acid changes in the mutant proteins, and they agreed remarkably well with those derived in other ways. In addition to base changes, nitrous acid can produce deletions; this may be related to its ability to form crosslinks between the strands of DNA. Nitrous acid was the first agent to cause mutations when applied to free nucleic acid in vitro, either to the RNA of tobacco mosaic virus or to the DNA of bacteria. The induced mutations could be detected because RNA, like the virus from which it has been isolated, infects tobacco leaves, while bacterial DNA can enter other bacterial cells and become integrated into their chromosomes by a process called transformation.

Hydroxylamine, too, can produce mutations in free bacterial DNA. In vitro, it reacts preferentially with cytosine. The *p*H dependence of this reaction parallels that of mutation frequency in extracellularly treated bacteriophage, suggesting that mutations are produced exclusively by reaction with cytosine, presumably followed by a transition from guanine-cytosine to adenine-thymine. This high chemical specificity has made hydroxylamine into a standard mutagen for deriving the chemical specificities of others. Thus, the fact that 5-bromouracil tends to revert the same mutations that are also reverted by hydroxylamine, while 2-aminopurine rarely does so, is the mainstay for the assumption that the former preferentially changes guanine-cytosine into adenine-thymine; the latter, changes adenine-thymine into guanine-cytosine. It is, however, important to realize that this simple picture is valid only for bacteriophage treated outside the bacterial cell. Applied to phage inside its host cell, hydroxylamine produces mutations by acting on all four bases *(2)*. This should be taken as a warning against the frequently made assumption that the genes of cellular organisms, which are unavoidably treated inside their "host cell," will necessarily undergo the same reactions with mutagens as does DNA treated in the test tube or bacteriophage treated outside its host.

DIRECTED MUTATION AND THE MUTATION PROCESS

What has become of the hope that chemical mutagens might be a tool for producing directed mutations by reacting selectively with certain genes? Since the vast majority of randomly produced mutations are harmful, means for directing mutation into desirable channels would be of immense importance for "mutation breeding." At a time when we conceived of genes as complex nucleoprotein molecules with highly specific overall composition and structure, the hope of finding selectively acting chemicals did not seem unreasonable. Nowadays, when we conceive of genes as linear sequences made up of the same four nucleotides, it seems a forlorn hope. It is true that, in the nucleotide bases, mutations do not occur at random but tend to attack preferential sites, whose positions within a given gene depend on the mutagen. The nature of these "hot spots" is obscure. They cannot be due to specific reactions between a given chemical and one out of four nucleotides, every one of which occurs many times within the gene. They have been tentatively attributed to the effect of neighboring nucleotides on the chemical reactivity of a given base. Since, however, the same short nucleotide sequence that produces a hot spot in one gene is likely to recur in many or most other genes, the phenomenon of hot spots, however interesting in its own right, holds out no promise for the production of directed mutations.

Does this mean that we have to give up all hope of achieving at least a modicum of control over the direction of mutation? I do not think it does. This defeatist conclusion could be drawn only if numbers and types of mutation were determined wholly by the reaction between DNA and mutagen, and this is most certainly not so. It is true that a change in the information carried by DNA is a necessary condition for mutation, but it is not a sufficient one. It is preceded as well as followed by secondary steps, and these act as so many sieves determining whether a change in DNA will take place and whether, once it has taken place, it will give rise to an observable mutation, that is, to a population of cells with a new type of genetic information. It is at the level of these sieves that specificities may be expected, and have already been found *(13)*.

The sieves that precede the reaction between mutagen and DNA are concerned with the chemical changes that a mutagen may undergo before reaching the gene, and with the accessibility of the gene. They are influenced by strain, cell type, and metabolic state and probably depend on the degree of coiling of chromosomes and chromosome regions and on the amount and type of the chromosomal components other than DNA. A possibility that has been considered but not yet adequately tested is that active genes are more accessible to certain mutagens than repressed ones. If this were true, it would offer a means of selecting genes for mutagen attack. There is also the more remote possibility that one might maneuver a mutagenic group, say a chloroethyl or ethylene imine group, into close neighborhood of repressed genes by attaching it to a repressor substance. This should give a complementary response pattern to the above, and the pattern should be similar if similar mutagenic groups were attached to the same type of repressor substance.

The sieves following the reaction between mutagen and gene determine which of the changes in DNA will eventually appear as observable mutations. The first sieve is repair which, in one form or another, seems capable of reversing chemical changes in DNA after treatment with most or all mutagens. Repair processes are presently under intensive study *(14)*. They involve enzymes, and their efficiency depends on the time available between the production of the chemical change and the next replication of DNA. The latter factor certainly is influenced by mutagenic treatment, the former probably is so in many cases. This is one of several ways in which a mutagen may act as a screening agent for the potential mutations that it has itself induced. There has as yet been no systematic search for repair processes that are specific to certain genes; research on *Neurospora* and bacteria suggests that they may exist.

Once the mutational change in DNA has become stabilized it has to be transcribed by messenger RNA. This particular sieve may be clogged by substances like fluorouracil, which specifically inhibit the manifestation of certain mutational changes *(15)*. Then follows a series of sieves concerned with translation; they include all processes by which a mutated cell is formed under the influence of the new messenger RNA. When the mutation leads to the formation of a new enzyme or other protein, ribosomes and transfer RNA are involved; other steps have to be carried through for the formation of a new type of transfer, or ribosomal, RNA. Again, many factors including the mutagen itself may affect the action of these sieves, and some may do so specifically. Thus, streptomycin and neomycin specifically prevent the manifestation of certain types of mutation by misdirecting translation on the ribosomes *(16)*. Finally, the mutated cell has to grow into a population of mutant cells. In mutation experiments on microorganisms, this last sieve has often to be passed in competition with a vast majority of nonmutant cells. Moreover, in the most frequently used type of experiment, the screening for mutations that render the mutant cells resistant to conditions by which the nonmutant ones are killed, there is a race between death and the completion of the mutation process, and all sieves have to be passed within a strict time limit.

For many years, the remarkable successes in the molecular analysis of mutagenesis at the DNA level have channeled most mutation research into this line of approach, and a study of mutation as biological rather than chemical process has hardly begun. There are, however, already a fair number of cases which point to the importance of cellular events for making or marring potential mutations. In some of them, the observed effects were directed ones in the sense that the proportion between different types of mutation could be profoundly altered by conditions such as temperature, *p*H, visible light, plating medium, type and dose of mutagen, treatment before or after with mutagenic or nonmutagenic chemicals, or the introduction of new genes into the genetic background of those to be screened for mutation *(17)*. In my opinion, it is along these lines that progress toward a direction of mutation is to be expected. Moreover, this approach will help us to understand one of the most interesting biological processes, by which a change in the information carried in DNA leads to the emergence of a population of cells with altered hereditary properties.

SUMMARY

Since the discovery of the first potent mutagens over 20 years ago, progress in mutation research has been rapid. Many new mutagens, belonging to a variety of chemical classes, have been discovered, and for some of them the reaction with DNA in vitro has been established. It seems that the findings of these chemical investigations usually also apply to viruses which are treated outside the cell. This has made chemical mutagens into an important tool for the analysis of the genetic code. When DNA is treated inside the cell, its reactions would not be expected to be always identical with those observed in vitro; in one case they have, indeed, been found to be different.

A chemical change in DNA is a necessary but not a sufficient condition for the production of an observable mutation. Intercalated between this primary change and the emergence of a population of cells with a new hereditary property is a whole series of cellular events, including a variety of repair mechanisms, transcription and translation of the new information, and growth of the mutant cell into a mutant population, often in the face of severe competition from nonmutant cells. These events act as so many sieves that screen out a proportion of potential mutations for realization. The study of mutation as cellular process has hardly begun, but it already shows the importance of these cellular events for the numbers and types of mutation produced. In addition to its theoretical interest, this approach is the only one likely to lead to the production of directed mutations for "mutation breeding."

REFERENCES

1. C. Auerbach, J. M. Robson, J. G. Carr, Science **105**:243 (1947).
2. I. Tessman, R. K. Poddar, S. Kumar, J. Mol. Biol. **9**:352 (1964); I. Tessman, H. Ishiwa, S. Kumar, Science **148**:507 (1965).
3. J. D. Karkas and E. Chargaff, Proc. Nat. Acad. Sci. U.S. **56**:1241 (1966).
4. W. A. F. Watson, Z. Vererbungsl. **95**:374 (1964); Mutation Res. **3**:455 (1966).
5. C. Auerbach, Microbial Genet. Bull. **17**:5 (1960).
6. S. Rogers, J. Nat. Cancer Inst. **15**:1675 (1955).
7. I. M. Tough and W. M. Court Brown, Lancet **1965-II**:684 (1965).
8. M. S. Swaminathan, V. L. Chopra, S. Bhaskaran, Radiat. Res. **16**:182 (1962); V. L. Chopra, A. T. Natarajan, M. S. Swaminathan, Radiat. Bot. **3**:1 (1963); V. L. Chopra and M. S. Swaminathan, Naturwissenschaften **50**:374 (1963); F. C. Steward, R. D. Holsten, M. Sugii, Nature **213**:178 (1967).
9. V. L. Chopra, Nature **208**:699 (1965); O. S. Reddi, G. M. Reddy, J. J. Rao, D. N. Ebenezer, M. S. Rao, *ibid.*, p. 702; R. R. Rinehart and F. J. Ratty, Genetics **52**:1119 (1965).
10. W. Ostertag, E. Duisberg, M. Stürmann, Mutation Res. **2**:293 (1965); W. Ostertag, *ibid.* **3**:249 (1966).
11. W. Ostertag and J. Haake, Z. Vererbungsl. **98**:299 (1966).
12. E. Terzaghi, Y. Okada, G. Streisinger, J. Emrich, M. Inouye, A. Tsugita, Proc. Nat. Acad. Sci. U.S. **56**:500 (1966).
13. C. Auerbach, Soviet Genet. **2**:1 (1966).
14. E. M. Witkin, Science **152**:1345 (1966).
15. S. P. Champe and S. Benzer, Proc. Nat. Acad. Sci. U.S. **48**:532 (1962).
16. J. Davies, W. Gilbert and L. Gorini, *ibid.* **51**:883 (1964); L. Gorini and E. Kataja, *ibid.*, p. 487.
17. V. L. Chopra, Mutation Res. **4**:382 (1967); C. Auerbach and D. Ramsay, *ibid.*, p. 508; B. J. Kilbey, in press.

BIBLIOGRAPHY

C. Auerbach, Biol. Rev. **24**:355 (1949).

————, Radiat. Res. **9**:33 (1958).

————, in Nat. Acad. Sci.-Nat. Res. Council Publ. 891 (1961), p. 120.

E. Freese, in Molecular Genetics, J. H. Taylor, Ed. (Academic Press, London, 1963), part 1, pp. 207-269.

E. Freese and E. B. Freese, Radiat. Res. Suppl. **6**:97 (1966).

B. A. Kihlman, Actions of Chemicals on Dividing Cells (Prentice-Hall, Englewood Cliffs, N.J., 1966).

D. R. Krieg, in Progress in Nucleic Acid Research, J. N. Davidson and W. E. Cohn, Eds. (Academic Press, New York, 1963), vol. 2, pp. 125-168.

A. Loveless, Genetic and Allied Effects of Alkylating Agents (Butterworth, London, 1966).

43 Radiation-induced mutations and their repair

Bacteria reduce the mutagenic effects of ultraviolet light by repairing DNA damaged by the radiation

Evelyn M. Witkin
Department of Medicine
Downstate Medical Center
New York, New York

Ultraviolet light kills bacteria, and it also induces mutations among the survivors. For many years, radiobiologists have suspected that both effects start with photochemical changes in the nucleic acids of the exposed cells *(1)*. Recently, they have succeeded in identifying specific photoproducts, formed in the DNA of irradiated bacteria, that contribute to the bactericidal effect of ultraviolet light, and have begun to understand how bacteria sometimes repair potentially lethal radiation damage (for recent reviews, see *2-4*). This article explores the roles played by newly discovered products of ultraviolet irradiation and by the mechanisms whereby such damage is repaired in the induction of mutations.

The first photochemical lesion found in the DNA of irradiated bacteria was the thymine dimer *(5)*. Ultraviolet light produces thymine dimers mainly by linking adjacent thymine bases in the same strand of DNA, via carbon-to-carbon bonds. (Normally, of course, the purine and pyrimidine bases in a single strand of DNA are connected only to the sugar-phosphate "backbone," and not to each other.) Other pyrimidine dimers (cytosine-cytosine and cytosine-thymine) are also formed in irradiated DNA, but probably less efficiently than dimers of thymine *(2, 6, 4)*. Dimers containing thymine block DNA replication in vitro *(7)* and in vivo *(8)* and are responsible for an important fraction of the lethal effects of low doses of ultraviolet light in some strains of bacteria *(8)*.

From Science **152:**1345-1353, June 3, 1966.

Research supported by grant A1-01240 from NIAID and conducted with the able assistance of N. A. Sicurella. I thank Dr. L. Gorini for supplying the "conditional streptomycin-dependent" strains, Dr. R. Hill for strain WP2s, and Dr. S. Kondo for the ultraviolet-sensitive phr^+ and phr^- strains (see *22*).

Pyrimidine dimers are subject to repair in the bacterium *Escherichia coli.* They may be eliminated from the DNA of irradiated cells in one of two known ways. The first requires exposure, after irradiation, to an intense source of visible light (the most effective wavelengths being those around 4000 angstroms), a treatment known to reverse or "photoreactivate" many of the biological effects of ultraviolet light *(9)*. In one kind of photoreactivation, pyrimidine dimers are split, *in situ (10),* by a light-dependent "photoreactivating enzyme" found in *E. coli* and yeast *(11)*. Alternatively, in the dark, pyrimidine dimers are removed from the DNA, apparently by the excision of short, single-stranded segments that include the dimers and a small number of neighboring nucleotides *(12)*. Subsequent steps in the dark repair of pyrimidine dimers are not known in detail, but the gap left by excision is presumably filled by "repair synthesis," with the nucleotide sequence opposite the excised segment serving as template. Evidence for a "patching" mechanism following excision has been described *(13)*.

These advances in our knowledge of the effects of ultraviolet light on DNA, and of the ways in which cells minimize radiation damage by repair, raise many questions about mutagenesis. Do pyrimidine dimers cause mutations? Are some ultraviolet-induced mutations caused by photoproducts that are demonstrably different from pyrimidine dimers? Does repair of radiation damage reduce the chance of induced mutation, or are most induced mutations caused by "mistakes" in the repair process itself? Is repair of radiation damage equally efficient in all genes? Is there any specificity in the kinds of products of ultraviolet irradiation that cause mutations in different genes? Direct biochemical methods are not very helpful in answering such questions, since an induced mutation is a rare and elusive event. The experiments described here make use, instead, of mutant strains that differ from each other in their ability to effect particular kinds of repair of radiation damage. Comparing mutagenesis in pairs of such strains has yielded some preliminary answers to the questions posed.

MATERIAL AND METHODS

The experiments, except where noted, were done with strains WP2, a tryptophan-requiring derivative of *Escherichia coli* B/r, and $WP2_s$, an ultraviolet-sensitive mutant strain derived from WP2 *(14)* and also known as WP2 (hcr^-). Methods of growing cultures, preparing them for irradiation, and exposing them to ultraviolet light and photoreactivating light were all as previously described *(15)*, unless otherwise specified. In experiments in which ultraviolet light doses of less than 100 ergs per square millimeter were used, the dose rate of the lamp was reduced about 20-fold by masking all but a 2.5-centimeter central section of the bulb in aluminum foil *(16)*.

Three kinds of genetic changes were investigated: mutations from streptomycin-sensitivity to streptomycin-resistance; from ability to inability to ferment lactose; and from auxotrophy (growth-factor requirement) to prototrophy (no growth-factor requirement). Streptomycin-resistant mutants were selected as previously described *(17)*, except that streptomycin was added to the plates after 6 hours of incubation, and colony counts were made after 3 days. Mutants which were unable to ferment lactose were detected by inspection of colonies on Bacto eosin-methylene blue indicator agar after 2 days of incubation. Only mutants that formed colorless colonies (or sectors of colonies) on these plates and failed to grow when lactose was the sole carbon source were scored as "lactose-negative." Mutations from auxotrophy to prototrophy were selected on "semi-enriched minimal" (SEM) agar *(15)*, a minimal agar containing 5 percent nutrient broth. Survival was determined on the same media used to detect the induced mutations, except that streptomycin was omitted from the plates used to assay survival in experiments with streptomycin-resistance. All mutation frequencies are corrected for spontaneous mutations.

MUTATIONS AND EXCISION

Strain WP2, like its parent B/r, is about as resistant to ultraviolet light as any strain of *E. coli*, and possesses, in active form, all genes known to promote repair of radiation damage. Strain $WP2_s$, which arose by mutation from WP2, is about 20 times more sensitive to ultraviolet light and lacks measurable ability to excise thymine-containing dimers from its DNA *(14)*. The finding that many strains of *E. coli*, differing widely in their sensitivity to ultraviolet light, all have the same number of thymine-containing dimers produced in their DNA by the same ultraviolet dose *(8)* supports the assumption that the same ultraviolet dose produces the same amount of damage in the two strains and that the only relevant difference between the strains is their excision ability. Thus the sensitivity of strain $WP2_s$ can be ascribed to reduced repair of radiation damage, resulting from the failure of normal excision. A comparison of mutagenesis in WP2 and $WP2_s$ should tell us whether the excision of damaged segments of the DNA affects the induction of mutations. Hill's finding *(14)* that $WP2_s$ is much more susceptible than WP2 to the induction of mutations to tryptophan-independence indicates that the loss of detectable excision ability greatly increases the probability of this kind of mutation.

Figure 1 shows survival and the frequency of induced mutations resulting in independence of tryptophan in strains WP2 and $WP2_s$, after various doses of ultraviolet light. The results confirm those of Hill in that they show that mutations of this kind are produced abundantly in the sensitive strain at doses far below those required to induce comparable numbers in the resistant strain.

Figure 2 shows survival and the frequency of mutations resulting in resistance to streptomycin, in the sensitive and resistant strains, as a function of ultraviolet dose. Strain $WP2_s$ is much more susceptible than WP2 to the induction of mutations to streptomycin-resistance, the frequencies obtained at very low doses actually exceeding those observed at any dose in the resistant strain.

Table 1 shows the frequency of mutations resulting in failure to ferment lactose induced in the sensitive and resistant strains by ultraviolet light. These mutations, too, are produced in strain $WP2_s$ at a dose too low to show a significant mutagenic effect in the resistant strain.

The higher mutation frequencies found in the sensitive strain $WP2_s$, in association with reduced excision capability, indicate that

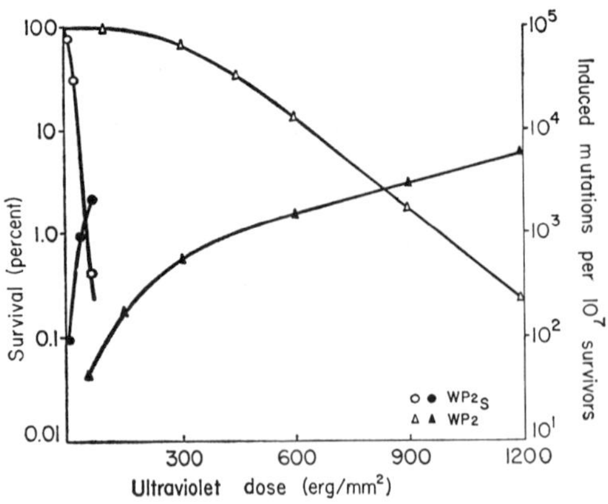

Fig. 1. Survival and frequency of induced mutations that result in independence of tryptophan in *Escherichia coli* strains WP2 and WP2s, after ultraviolet irradiation. Open symbols, survival; closed symbols, mutation frequency. Each point is the average of four experiments.

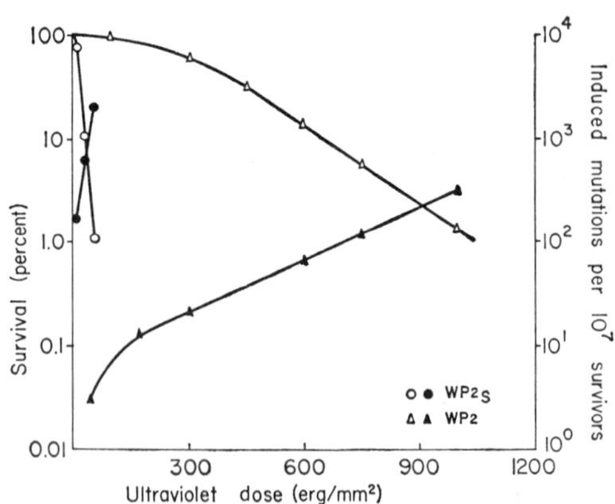

Fig. 2. Survival and frequency of induced mutations that result in resistance to streptomycin in strains WP2 and WP2s, after ultraviolet irradiation. Open symbols, survival; closed symbols, mutation frequency. Each point is the average of four experiments.

Table 1. Frequencies of mutations induced by ultraviolet light that result in inability to ferment lactose in *Escherichia coli* strains WP2 and $WP2_S$

Treatment after irradiation	Ultraviolet dose (erg/mm²)	Survival (%)	No. of colonies screened	No. of mutants: Whole colony	No. of mutants: Sectored colony	No. of induced mutants per 10^6 survivors
			Strain WP2			
None	0	100	26,588	2	0	
None	60	100	32,294	3	0	0
None	1200	0.92	46,312	12	17	540
			Strain $WP2_S$			
None	0	100	24,438	1	1	
None	60	1.2	55,917	15	23	589
PRL*	60	100	48,935	8	28	653

*PRL, photoreactivating light; 10-minute exposure to visible light immediately after irradiation with ultraviolet light.

photoproducts of the kind that are potentially excisable can play an essential part in the induction of all three types of mutation. If only nonexcisable photoproducts were involved in mutagenesis, differences in excision ability would be irrelevant, and the frequency of induced mutations should be the same, at a given dose, in the sensitive and resistant strains. Since mutation frequencies are lower when excision is normal, we may conclude also that mutations are less likely to occur as "mistakes" in the course of normal repair than as errors due to the failure or derangement of normal repair itself. If there is no excision at all in the sensitive strain, the mutations observed in $WP2_S$ must be caused, directly or indirectly, by unexcised photochemical lesions. If the sensitive strain is actually able to excise radiation damage, but only at a rate too low to be detected, it is possible that the mutations in $WP2_S$ are caused by errors specific to abnormal repair following extremely slow excision. In either case, normal excision greatly reduces the probability of mutation. Thus at least 99 percent of the mutations that would otherwise result from ultraviolet irradiation are prevented from occurring, in the resistant strain, by repair processes that include excision. This conclusion depends upon the comparison of mutation frequencies at the same ultraviolet dose in sensitive and resistant strains, and therefore applies only to the very low doses (less than 100 erg/mm²) at which mutagenesis can be studied in the sensitive strain. Other evidence

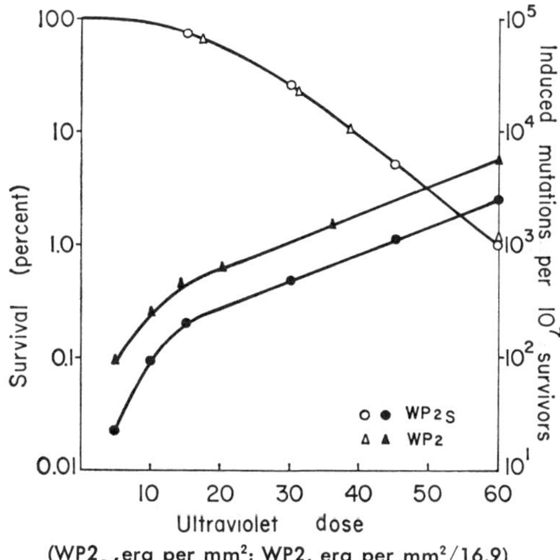

Fig. 3. Frequency of induced mutations that result in independence of tryptophan in strains WP2 and $WP2_S$, after treatment with ultraviolet irradiation at doses resulting in equal survival. Open symbols, survival; closed symbols, mutation frequency. Each point is the average of four experiments. Doses used for strain WP2 were 16.9 times those shown on the abscissa.

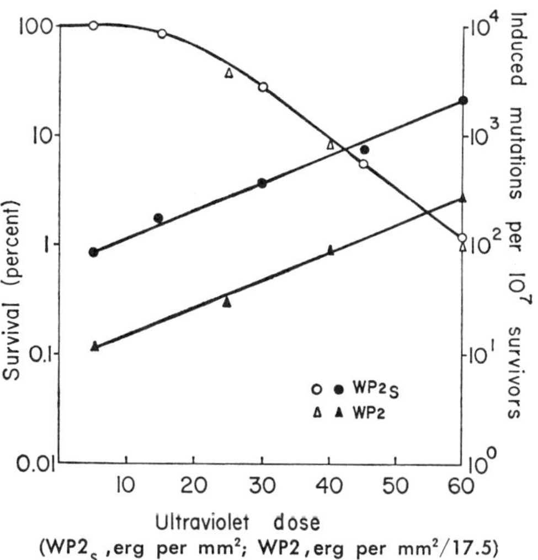

Fig. 4. Frequency of induced mutations that result in resistance to streptomycin in strains WP2 and $WP2_S$, after treatment with ultraviolet irradiation at doses resulting in equal survival. Open symbols, survival; closed symbols, mutation frequency. Each point is the average of four experiments. Doses used for strain WP2 were 17.5 times those shown on the abscissa.

(18), however, suggests that mutation frequencies are greatly reduced by repair of ultraviolet damage at much higher doses, as well.

It is important to know whether the loss of detectable excision ability, in the sensitive strain, increases sensitivity to the lethal and mutagenic effects of ultraviolet light equally. If so, the mutation frequencies obtained in the sensitive and resistant strains, at doses having equivalent effects on survival, should be the same. As shown in Figs. 3 and 4, the survival curves of the two strains may be superimposed if appropriate scales of ultraviolet dose are used, the resistant strain requiring about 17 times as much radiation as the sensitive strain to produce the same lethal effect. At doses of ultraviolet light equated for survival, mutations to tryptophan-independence (Fig. 3) are about twice as frequent in the resistant strain *(19)*. If the relative sensitivity of the two strains is expressed as the difference in the amount of radiation required to produce equal effects, $WP2_S$ is 17 times as sensitive as WP2 to the lethal effect of ultraviolet light, but only about ten times as sensitive to the induction of such mutations.

A different result is obtained with mutations to streptomycin-resistance, as shown in Fig. 4. In this case, doses equated for survival produce almost ten times as many mutations in the sensitive strain, and thus it appears that susceptibility to the induction of this kind of mutation is increased much more, in strain $WP2_S$, than is sensitivity to the lethal effect. The loss of detectable excision ability is clearly not accompanied by equal increases in the sensitivity to lethal and mutagenic effects, or by equal increases in the sensitivity to the induction of different mutations. Survival is probably influenced by the production and repair of potentially lethal radiation damage anywhere in the DNA, whereas the frequency of a particular kind of mutation is likely to be affected only by the fate of photoproducts produced in a particular gene or small group of genes. Unequally increased sensitivity to killing and to the induction of various mutations, in a strain owing its sensitivity to reduced excision ability, could mean (i) that repair processes involving excision are more efficient in some parts of the DNA than in others, or (ii) that the amount of irreparable damage varies in different regions of the DNA. For example, the lower yield of mutations leading to streptomycin-resistance induced in the resistant strain, by doses equated for survival, could indicate either that repair is much more efficient in the streptomycin locus than in the DNA as a whole, or that this locus contains an unusually small residue of irreparable ultraviolet damage, or both.

MUTATIONS AND PYRIMIDINE DIMERS

Potentially excisable products of ultraviolet irradiation are involved in the induction of mutations in strain WP2$_S$. Since the specificity of excision is not known, the foregoing experiments do not tell us whether the premutational photoproducts are pyrimidine dimers or some other kinds of ultraviolet damage that may also be excisable. One way to find out is to study the photoreversibility of the mutations induced in the sensitive strain. Photoreversibility alone is not an adequate criterion for the involvement of pyrimidine dimers in an ultraviolet-initiated effect, since, in addition to the light-dependent splitting of pyrimidine dimers by "photoreactivating enzyme," there are indirect mechanisms of photoreversal that do not require this enzyme and that probably do not operate by monomerization of pyrimidine dimers *(20)*. Photoreversibility of mutations by the direct, enzymatic mechanism, however, would indicate that pyrimidine dimers are involved in the induction of the mutations, since dimer-splitting appears to be the only direct photoreversal mechanism of which *E. coli* is capable *(4)*.

Figures 5 and 6 show that about 90 percent of the mutations to tryptophan-independence and to streptomycin-resistance, respectively, that are induced in the sensitive strain WP2$_S$ are eliminated by exposure to visible light. No significant part of this reversal can be ascribed to indirect effects of the light, since (i) wavelengths known to produce such effects were excluded by a filter *(21)*; (ii) no reversal of mutations was obtained if the exposure to visible light *preceded* irradiation with ultraviolet light (a condition that permits indirect photoreversal, but not enzymatic dimer-splitting) *(20)*; and (iii) induced mutations resulting in prototrophy or streptomycin-resistance were not reversed by exposure to visible light in another ultraviolet-sensitive strain that lacked the activity of the photoreactivating enzyme and presumably was unable to split pyrimidine dimers *(22)*. It seems certain that the photoreversibility of the mutations to streptomycin-resistance and to tryptophan-independence induced in strain WP2$_S$ is of the direct, enzymatic kind in which pyrimidine dimers are split. Pyrimidine dimers, therefore, must participate in the induction of at least 90 percent of these mutations.

In contrast to these results, induced mutations resulting in inability to ferment lactose, as shown in the last two lines of Table 1, are not

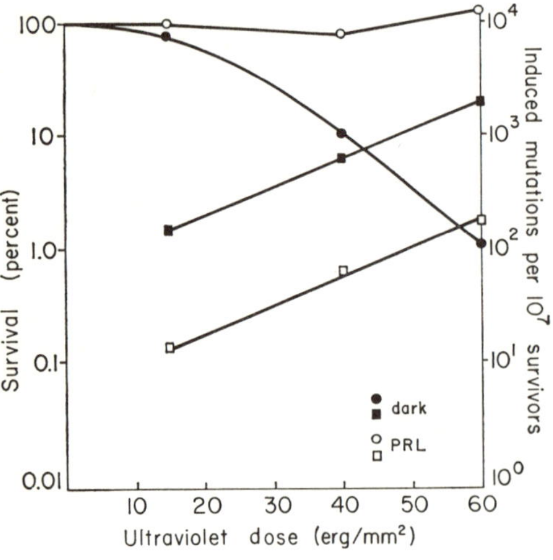

Fig. 5. Reversal by visible light, of potentially lethal damage and induced mutations that result in independence of tryptophan in strain WP2$_S$. Circles, survival; squares, mutation frequency. PRL = 10 minutes' exposure to photoreactivating light immediately after ultraviolet irradiation. Each point is the average of three experiments.

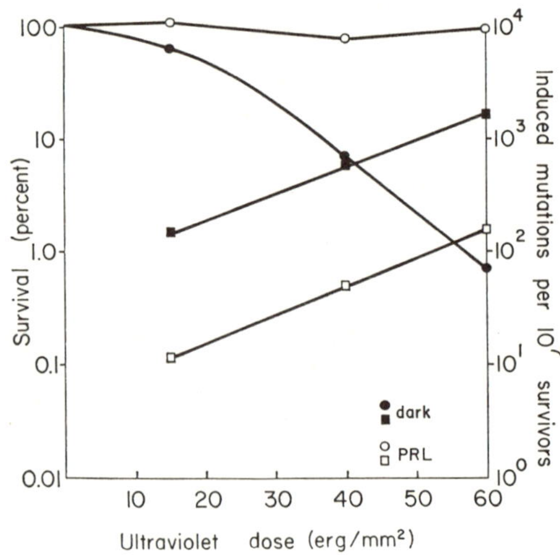

Fig. 6. Reversal by visible light, of potentially lethal damage and of induced mutations that result in resistance to streptomycin in strain WP2$_S$. Circles, survival; squares, mutation frequency. PRL = 10 minutes' exposure to photoreactivating light immediately after ultraviolet irradiation. Each point is the average of three experiments.

photoreversible to any significant extent. The frequency of lactose-negative mutants in $WP2_s$ is just as high among cells treated with photoreactivating light as among controls kept in the dark, despite the hundred-fold increase in the number of survivors due to the photoreversal of potentially lethal damage. Products of ultraviolet irradiation that lead to lactose-negative mutations are excisable but are not reversible by visible light. This indicates either that they are not pyrimidine dimers or that, if they are, conditions specific to the loci at which these mutations arise prevent the monomerization of pyrimidine dimers by "photoreactivating enzyme." Nonphotoreversibility of certain mutations resulting in fermentation inability has been reported previously *(23)*.

MUTATIONS IN RESISTANT STRAINS

Is the residual, unrepaired damage that gives rise to mutations in resistant strains qualitatively different from the larger amount of damage that is effectively repaired? Or does it comprise a random sampling of similar damage and merely reflect the limitations of the repair systems? Since pyrimidine dimers are involved in the induction of most mutations resulting in tryptophan-independence and streptomycin-resistance in the sensitive strain $WP2_s$, these questions really ask whether the same is true of the mutations induced in the resistant strain.

For ultraviolet-induced mutations to streptomycin-resistance, in the radiation-resistant strain B/r, about 90 percent of the mutations obtained in the dark are eliminated by treatment with light, and this reversal does not occur at all in strains lacking photoreactivating enzyme *(24)*. Products of ultraviolet irradiation that are involved in the induction of mutations to streptomycin-resistance, in both sensitive and resistant strains, are subject to direct, enzymatic photoreversal, and may therefore be described as pyrimidine dimers. Both fractions of the premutational damage (the fraction effectively repaired in resistant strains, and the fraction not repaired) include pyrimidine dimers.

No such simple situation exists for mutations to tryptophan-independence induced in strain WP2, however. These mutations belong to a class (broadly described as mutations from auxotrophy to prototrophy) that has been intensively investigated in ultraviolet-resistant derivatives of strain B/r. Although the induction of most of the mutations to tryptophan-independence obtained in the sensitive strain $WP2_s$ (corresponding to those normally repaired in resistant strains) involves pyrimidine dimers, the same kind of mutation produced by higher doses in resistant strains appears to originate from products of ultraviolet irradiation that have unusual properties. These mutations are irreversibly eliminated by a process that has been interpreted as enzymatic dark repair *(18, 25)*, if conditions immediately after ultraviolet irradiation are unfavorable for protein synthesis, and mutations are obtained in large numbers only if conditions after irradiation permit active protein synthesis *(26)*. Since the same treatments that eliminate mutations to prototrophy (amino acid deprivation, chloramphenicol inhibition) fail to affect survival and the frequency of mutations to streptomycin-resistance *(17)*, the premutational photoproducts leading to induced prototrophy must differ, somehow, from those responsible for death and streptomycin-resistance. Not only does the specificity of the conditions promoting its dark repair set this class of mutations apart, but so does its photoreversibility. Mutations to prototrophy are photoreversible in resistant strains *(27)* as well as in sensitive strains. In the sensitive strain $WP2_s$, all this reversibility is accounted for by direct, enzymatic dimer-splitting, as shown above. In resistant strains, on the contrary, induced prototrophy remains fully photoreversible when photoreactivating enzyme and dimer-splitting ability are eliminated by mutation *(24, 28)*. Even in strains possessing the ability to split pyrimidine dimers, most of the photoreversal of induced prototrophy can be shown to be of the indirect type, since it can be prevented by treatments applied after the exposure to reactivating light, and hence after any dimer-splitting that may contribute to the photoreversal is completed *(15)*. The absence of dimer-splitting as a direct component in their photoreversal, as well as the specificity of their dark repair, has led to the hypothesis *(28)* that the photoproducts responsible for the mutations to prototrophy observed in resistant strains are not pyrimidine dimers (or at least not pyrimidine dimers having normal reparability).

SUPPRESSORS AND REPAIR

Not all auxotrophic strains respond to ultraviolet light by producing mutations to prototrophy that have unusual repair properties. Among auxotrophic substrains of *E. coli* B/r, isolated at random, only about 20 to 30 percent given relatively high yields (at least 1 per 10^5 survivors) of induced mutations to prototrophy that can be greatly reduced by omission of amino acids from the plating medium used after irradiation, and that may hence by considered subject to the kind of dark repair that is blocked by protein synthesis. In all auxotrophic strains that do respond in this way, including strain WP2, most of the ultraviolet-induced prototrophs owe their independence of growth factors to suppressor mutations, each capable of correcting a large number of mutational defects scattered around the genetic map *(29)*. Suppression of this kind is believed to operate through alterations in the specificity of the genetic code *(30)*. The unusual repair properties of the ultraviolet-induced suppressor mutations may reflect singularities associated with genes coding one or more elements of genetic "translation" (activating enzymes, transfer RNA's, ribosomes). The singularity could be primary (unique bases or base sequences leading to unique ultraviolet photoproducts) or secondary (unusual physical state of the DNA at the suppressor loci that affects reparability of radiation damage).

Gorini and Kataja *(31)* have described "conditional streptomycin-dependent" auxotrophs, in which the auxotrophic defect is corrected by streptomycin, and have also shown that the same defect is correctable by suppressor mutations *(32)*. Since streptomycin affects ribosomes, thereby altering the specificity of incorporation of amino acids in vitro *(33)*, it is likely that genetic suppressors of defects suppressible by streptomycin also change the reading of the code. In a group of 30 conditional streptomycin-dependent auxotrophs tested for their response to ultraviolet light, all but one yielded high frequencies of ultraviolet-induced suppressor mutations that are subject to "amino-acid-sensitive" dark repair (that is, they are obtained only on media enriched with a pool of amino acids). Suppressibility by this kind of mutation is highly correlated with suppressibility by streptomycin among auxotrophic strains selected for their response to streptomycin. These findings support the view that ultraviolet-induced suppressors exhibiting amino-acid-sensitive dark reparability originate in genes that code activating enzymes, transfer RNA's, or ribosomal components, and that their unusual repair properties may be related to singularities at one or more of these loci.

SUPPRESSORS AND EXCISION

In resistant strains, suppressor mutations induced by ultraviolet light are obtained in large numbers only if conditions after irradiation favor protein synthesis. If protein synthesis is briefly inhibited after irradiation (for instance, by amino acid deprivation or treatment with chloramphenicol), the potential mutations are irreversibly lost. Actually, only a small fraction of the premutational ultraviolet damage capable of leading to suppressor mutations is subject to this amino-acid-sensitive dark repair. Most of the premutational photoproducts are effectively repaired, in resistant strains, whether the medium contains amino acids or not. This can be inferred from the much higher yield of mutations in the sensitive strain, which lacks detectable excision ability, as well as from the greatly increased mutation frequencies obtained when repair, in resistant strains, is reduced phenotypically by treatment, after irradiation, with such agents as caffeine *(34)* or acriflavine *(18)*. Since neither the survival nor the frequency of induced mutations to streptomycin-resistance is affected by deprivation of amino acids after irradiation *(17)*, neither seems to depend upon photoproducts that are subject to dark repair only under conditions of inhibited protein synthesis. Such photoproducts, as far as we know, are associated exclusively with the genes in which suppressor mutations originate, where they constitute a small, but demonstrable, fraction of the total premutational ultraviolet damage.

The only dark-repair mechanism known in any detail includes, as an essential step, the excision of radiation damage from the DNA. Does the amino-acid-sensitive dark repair of suppressor mutations also involve excision? Several striking parallels exist between the excision of thymine-containing dimers and the loss of suppressor mutations that occurs when protein synthesis is inhibited after irradiation: both processes show kinetics that are exponential and dose-dependent *(34, 35)*; both require energy *(26, 4)*; both occur in the presence of chloramphenicol *(26, 4)*; and both are blocked by acriflavine *(15, 35)*. The most convincing evidence that the amino-acid-sensitive dark repair of suppressor mutations involves excision comes from mutant strains deliberately selected for their loss of the ability to carry out "mutation frequency decline," as the disappearance of suppressor mutations when protein synthesis is inhibited is sometimes called. These "mfd^-" mutants produce high yields of induced suppressor mutations even when plating is preceded by a period of amino acid deprivation or chloramphenicol inhibition as shown in Fig. 7, in which an mfd^- mutant is compared with its mfd^+ parent strain *(36)*.

The frequency of induced suppressor mutations ultimately obtained drops rapidly as the mfd^+ strain is incubated in minimal medium or chloramphenicol, but is reduced only slightly in the mfd^- mutant. Although the ultraviolet-sensitivity of the mfd^- strain is the same as that of the resistant strain from which it was derived, the mfd^- strain excises thymine-containing dimers at a rate markedly lower than normal *(37)*. Mutants unable to carry out mutation frequency decline have been isolated repeatedly, and all recover from the inhibition of DNA synthesis following irradiation more slowly than normal. This fact suggests that a reduction in the rate of excision of radiation damage invariably accompanies loss of the ability to effect amino-acid-sensitive dark repair of suppressor mutations. Although other explanations are possible, this association of the two effects in a single mutation is most simply explained by the assumption that excision of thymine-containing dimers and mutation frequency decline are mediated by the same enzyme. Figure 8 shows that strain WP2$_S$, the sensitive strain lacking detectable excision ability, is also unable to carry out mutation frequency decline. In this strain, the frequency of induced suppressor mutations (to tryptophan-independence) is scarcely reduced by conditions that promote mutation frequency decline in the resistant strain WP2, after a dose equivalent with respect to survival. This might also seem to support the hypothesis that excision is an essential step in the amino-acid-sensitive dark repair of suppressors, but the association of the two effects is less meaningful in the sensitive strain than in the resistant mfd^- mutant. In sensitive strains, the mutations observed arise primarily from damage that is usually repaired, in resistant strains, whether amino acids are present or not. This fraction of the damage may simply not be subject to mutation frequency decline. The reduced excision rate observed in the ultraviolet-resistant mfd^- mutants, however, does support the idea that mutation frequency decline involves the excision of a fraction of the premutational damage, produced in certain suppressor loci, that can be excised only if protein synthesis is inhibited after irradiation.

If the amino-acid-sensitive dark repair of suppressor mutations does include excision,

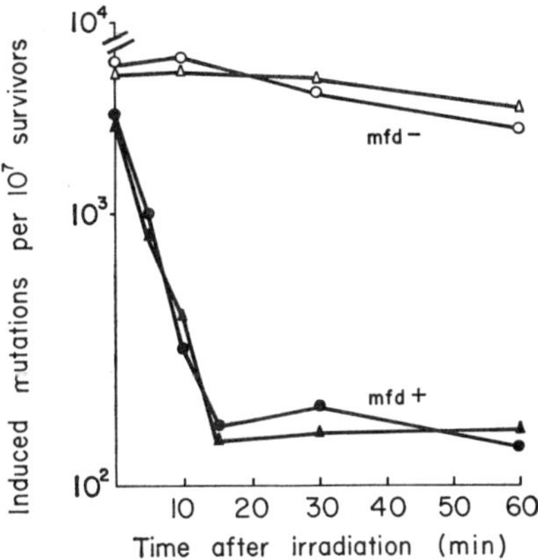

Fig. 7. "Mutation frequency decline" in strains WU36-10 (*mfd*$^+$) and WU36-10-45 (*mfd*$^-$). Saline suspension of each strain was irradiated with ultraviolet light; diluted 10^{-1} in minimal medium (circles) or nutrient broth containing 25 units of chloramphenicol per milliliter (triangles); incubated times indicated on ordinate, then plated (0.2 ml undiluted) on amino-acid-enriched agar to determine frequency of mutations resulting in independence of tyrosine. Ultraviolet dose: 600 erg/mm^2. Titer before irradiation: *mfd*$^+$, 3.3 × 10^8/ml; *mfd*$^-$, 3.5 × 10^8/ml. Survival: *mfd*$^+$, 11.1 percent; *mfd*$^-$, 12.6 percent. Each point is the average of two experiments.

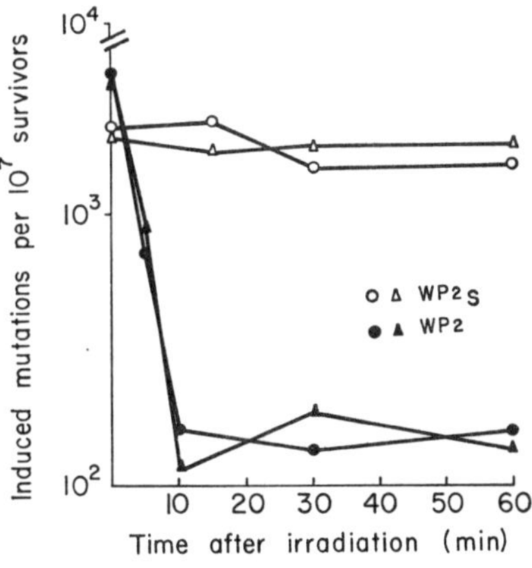

Fig. 8. "Mutation frequency decline" in strains WP2 and WP2$_S$. Procedure as described in legend of Fig. 7; suspensions were plated on semi-enriched minimal agar to determine frequency of mutations resulting in independence of tryptophan. Ultraviolet dose: WP2, 850 erg/mm^2; WP2$_S$, 50 erg/mm^2. Survival: WP2, 5.8 percent; WP2$_S$, 7.5 percent. Each point is the average of two experiments.

why does it occur only if protein synthesis is inhibited immediately after exposure to ultraviolet light? One possibility is that only under these conditions can the unique photoproducts at the suppressor loci bind excision enzyme. Protein synthesis after irradiation may trigger a change in the state of the DNA as a whole, or of the suppressor loci specifically, such that the ability of the unique photoproducts to compete successfully with pyrimidine dimers for available excision enzyme is reduced. If the suppressor genes code elements of translation, one consequence of inhibiting protein synthesis might be the repression of the suppressor loci. It is possible that the recognition of radiation damage by excision enzyme, and hence the reparability of the damage, may be more efficient in the repressed state than in the active state of a particular gene.

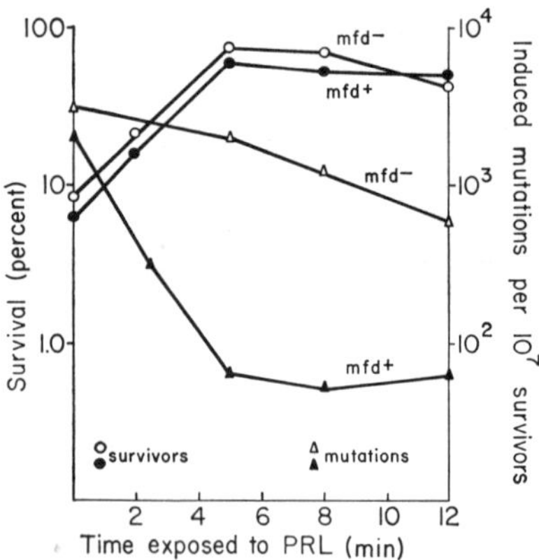

Fig. 9. Efficiency of photoreversal of potentially lethal damage and induced mutations in strains WU36-10 (*mfd*$^+$) and WU36-10-45 (*mfd*$^-$). Ultraviolet dose: 600 erg/mm^2. PRL = photoreactivating light, exposure immediately after ultraviolet irradiation.

PHOTOREVERSAL OF SUPPRESSORS

Suppressor mutations in resistant strains are unusual in their photoreversibility, as well as in their dark repair. Treatments known to inhibit amino-acid-sensitive dark repair (caffeine, acriflavine) can prevent most of the photoreversal of these mutations, even if the treatments are administered after exposure to the photoreactivating light. Since dimers are split at the instant of exposure to visible light, most of the photoreversal of suppressor mutations (even in strains having photoreactivating enzyme) cannot be direct, enzymatic dimer-splitting. The indirect nature of the photoreversal of suppressor mutations, in resistant strains, is further indicated by the photoreversibility of these mutations in an *mfd*$^-$ strain, selected for loss of the ability to carry out amino-acid-sensitive dark repair. As shown in Fig. 9, the photoreversibility of the potentially lethal effect of ultraviolet light is unchanged, in this strain, an indication that the bacteria are able to split pyrimidine dimers normally. However, the photoreversal of suppressor mutations is much less efficient in this strain than in the *mfd*$^+$ parent strain, from which it differs in lacking the ability to carry out mutation frequency decline. The *mfd*$^-$ mutation duplicates the phenotypic effect of acriflavine *(15)*, which also inhibits amino-acid-sensitive dark repair of suppressor mutations and reduces the efficiency of their photoreversal. These results are readily explained by assuming that most of the photoreversal of suppressor mutations, in resistant strains, is really cryptic dark repair. The reactivating light appears to act indirectly, in this case, by creating metabolic conditions that favor mutation frequency decline. The *mfd*$^-$ mutants thus provide additional evidence that suppressor mutations, in resistant strains, arise primarily from photoproducts that are distinguishable, by their reparability, from typical pyrimidine dimers.

SUMMARY

Mutagenesis is compared in an ultraviolet-resistant strain of *E. coli* (WP2) and an ultraviolet-sensitive derivative (WP2_S) that lacks detectable ability to excise thymine-containing dimers from its DNA. High frequencies of induced mutations resulting in resistance to streptomycin, independence of tryptophan, and inability to ferment lactose are obtained in the sensitive strain after ultraviolet doses too low to induce significant numbers of such mutations in the resistant strain. It is concluded that potentially excisable photoproducts participate in the induction of all three kinds of mutations; that at least 99 percent of the mutations observed in the sensitive strain are prevented in the resistant strain by repair that includes excision; and that mutations are less likely to occur as "mistakes" in the course of normal repair than as the result of the failure or derangement of normal repair.

At ultraviolet doses having equivalent effects on survival, the sensitive strain exhibits only about half as many mutations to tryptophan-independence, but about ten times as many mutations to streptomycin-resistance, as the resistant strain. This indicates either that repair processes involving excision are not equally efficient in all genes or that some genes contain larger residues of irreparable damage than others, or both.

In the sensitive strain, about 90 percent of the induced mutations to tryptophan-independence and to streptomycin-resistance can be reversed by "direct" enzymatic photoreactivation, that is, by the splitting of pyrimidine dimers. In the same strain, mutations to the inability to ferment lactose are not photoreversible. It is concluded that pyrimidine dimers are involved in the induction of at least 90 percent of the mutations to tryptophan-independence and streptomycin-resistance observed in the sensitive strain, but that pyrimidine dimers (or at least pyrimidine dimers subject to enzymatic splitting) are not involved in the induction of mutations that result in inability to ferment lactose.

In resistant strains, the ultraviolet-induced mutations arise from a fraction of the photoproducts that is not effectively repaired despite the optimal repair capability of these strains. In the case of streptomycin-resistance, pyrimidine dimers are included in this fraction and participate in the induction of at least 90 percent of the mutations observed in the resistant strain, as shown by their direct photoreversibility. Mutations to prototrophy, in resistant strains, are photoreversible, but mainly by a mechanism not directly involving splitting of pyrimidine dimers. They are repaired in the dark, but only if protein synthesis immediately after irradiation is inhibited (a condition not required for the dark repair of thymine-containing dimers). Although pyrimidine dimers participate in the induction of mutations to prototrophy in the sensitive strain $WP2_S$, most of those observed in resistant strains originate from a residue of photoproducts distinguishable from typical pyrimidine dimers by their pattern of reparability.

Mutations to prototrophy that have unusual repair properties (indirect photoreversal, dark repair blocked by conditions favoring protein synthesis) are invariably suppressor mutations, each capable of correcting many scattered auxotrophic defects, and therefore probably originate in genes coding elements of translations (activating enzymes, transfer RNA's, ribosomes). The unique pattern of reparability exhibited by these mutations may be associated with singularities of these loci.

The dark repair of suppressor mutations, in resistant strains, that occurs only if protein synthesis after irradiation is briefly inhibited ("mutation frequency decline") probably involves excision. This inference rests mainly on the reduced rate of dimer excision found in a mutant (mfd^-) strain selected for the loss of the ability to repair suppressor mutations under conditions normally promoting "mutation frequency decline." It is proposed that suppressor mutations, in resistant strains, arise primarily from photoproducts that can compete successfully with pyrmidine dimers for excision enzyme only if protein synthesis after ultraviolet irradiation is inhibited.

REFERENCES AND NOTES

1. F. L. Gates, J. Gen. Physiol. **14**:31 (1930); R. W. Kaplan, Z. Naturforsch. **7b**:291 (1952).
2. A. Wacker, in Progress in Nucleic Acid Research, J. N. Davidson and W. E. Cohn, Eds. (Academic Press, New York, 1963), p. 369.
3. A. D. McLaren and D. Shugar, Photochemistry of Proteins and Nucleic Acids (Pergamon, Oxford, 1964); K. C. Smith, Radiation Res., in press.
4. J. K. Setlow, in Current Topics in Radiation Research, M. Ebert and A. Howard, Eds. (North-Holland, Amsterdam, in press), vol. 2.
5. A. Wacker, H. Dellweg, D. Jacherts, J. Mol. Biol. **4**:410 (1962).
6. R. B. Setlow, W. L. Carrier, F. J. Bollum, Abstracts, Biophysical Society (1965), p. 111.
7. F. J. Bollum and R. B. Setlow, Biochim. Biophys. Acta **68**:599 (1963).
8. R. B. Setlow, P. A. Swenson, W. L. Carrier, Science **142**:1464 (1963).
9. A. Kelner, J. Bacteriol. **58**:511 (1949).
10. D. L. Wulff and C. S. Rupert, Biochem. Biophys. Res. Commun. **7**:237 (1962).
11. S. H. Goodgal, C. S. Rupert, R. M. Herriott, in The Chemical Basis of Heredity, W. D. McElroy and B. Glass, Eds. (Johns Hopkins Press, Baltimore, 1957).
12. R. B. Setlow and W. L. Carrier, Proc. Nat. Acad. Sci. U.S. **51**:226 (1964); P. R. Boyce and P. Howard-Flanders, *ibid.*, p. 293.
13. D. Pettijohn and P. Hanawalt, J. Mol. Biol. **9**:395 (1964).
14. R. F. Hill, Photochem. Photobiol. **4**:563 (1965).
15. E. M. Witkin, Proc. Nat. Acad. Sci. U.S. **50**:425 (1963).
16. Modification of method described by Hill *(14)*.
17. E. M. Witkin and E. C. Theil, Proc. Nat. Acad. Sci. U.S. **46**:226 (1960).
18. E. M. Witkin, J. Cellular Comp. Physiol. **58** (Suppl. 1):135 (1961).
19. Hill *(14)* describes a similar result, but considers the difference statistically insignificant; in these experiments, the difference is unquestionably significant.
20. J. Jagger and R. S. Stafford, Biophys. J. **5**:75 (1965).
21. In these experiments, a Corning filter No. 3060, which cuts off light of wavelengths below 3800 Å,

was placed between the photoreactivating light source and the tubes to be illuminated.

22. Strains R-03 and 002, obtained from Dr. S. Kondo, were used in these experiments. Both strains require arginine and are sensitive to ultraviolet, with a survival and frequency of induced prototrophy similar to that of $WP2_S$; R-03 is *phr*$^+$, 002 is *phr*$^-$. Result for mutations from *arg*$^-$ to *arg*$^+$ are those obtained by Kondo (personal communication); results for streptomycin-resistance observed by me. Both strains were derived from strain H/r30 *(28)*.
23. H. B. Newcombe and H. A. Whitehead, J. Bacteriol. **61**:243 (1951).
24. E. M. Witkin, N. A. Sicurella, G. M. Bennett, Proc. Nat. Acad. Sci. U.S. **50**:1055 (1963).
25. M. Lieb, Z. Vererbungslehre **92**:419 (1961).
26. E. M. Witkin, Cold Spring Harbor Symp. Quant. Biol. **21**:123 (1956).
27. C. O. Doudney and F. H. Haas, Genetics **45**:1481 (1960).
28. E. M. Witkin, Mutation Res. **1**:22 (1964).
29. ———, Genetics **48**:916 (1963) (abstract).
30. S. Brody and C. Yanofsky, Proc. Nat. Acad. Sci. U.S. **50**:9 (1963); M. G. Weigert and A. Garen, J. Mol. Biol. **12**:448 (1965); A. O. W. Stretton and S. Brenner, *ibid.*, p. 456.
31. L. Gorini and E. Kataja, Proc. Nat. Acad. Sci. U.S. **51**:487 (1964).
32. ———, *ibid.*, p. 995.
33. J. Davies, W. Gilbert, L. Gorini, *ibid.*, p. 883 (1964).
34. E. M. Witkin, in Proc. Intern. Congr. Genet. 10th, Montreal, 1958 (1959), vol. 1, p. 280.
35. R. B. Setlow, J. Cell. Comp. Physiol. **64**:Supp. 1, 51 (1964).
36. The parent strain (*mfd*$^+$) is WU36-10, a diauxotroph requiring tyrosine and leucine, derived from WU36 *(15)*. The *mfd*$^-$ mutant has the same requirements. In Figs. 7 and 9, mutations to tyrosine-independence are followed on SEM agar supplemented with 20 μg of leucine per milliliter.
37. J. Setlow, personal communication.

44 Fine structure of a genetic region in bacteriophage

Seymour Benzer*
Biophysical Laboratory
Departments of Biological Sciences and Physics
Purdue University
Lafayette, Indiana

This paper describes a functionally related region in the genetic material of a bacteriophage that is finely subdivisible by mutation and by genetic recombination. The group of mutants resembles similar cases which have been observed in many organisms, usually designated as "psuedo-alleles." (See reviews by Lewis[1] and Pontecorvo.[2]) Such cases are of special interest for their bearing on the structure and function of genetic determinants.

The phenomenon of genetic recombination provides a powerful tool for separating mutations and discerning their positions along a chromosome. When it comes to very closely neighboring mutations, a difficulty arises, since the closer two mutations lie to one another, the smaller is the probability that recombination between them will occur. Therefore, failure to observe recombinant types among a finite number of progeny ordinarily does not justify the conclusion that the two mutations are inseparable but can only place an upper limit on the linkage distance between them. A high degree of resolution requires the examination of very many progeny. This can best be achieved if there is available a selective feature for the detection of small proportions of recombinants.

Such a feature is offered by the case of the rII mutants of T4 bacteriophage described in this paper. The wild-type phage produces plaques on either of two bacterial hosts, B or K, while a mutant of the rII group produces plaques only on B. Therefore, if a cross is made between two different rII mutants, any wild-type recombinants which arise, even in proportions as low as 10^{-8}, can be detected by plating on K.

This great sensitivity prompts the question of how closely the attainable resolution approaches the molecular limits of the genetic material. From the experiments of Hershey and Chase,[3] it appears practically certain that the genetic information of phage is carried in its DNA. The amount of DNA in a particle of phage T2 has been determined by Hershey, Dixon, and Chase[4] to be 4×10^5 nucleotides. The amount for T4 is similar.[5] If we accept the model of DNA structure proposed by Watson and Crick,[6] consisting of two paired nucleotide chains, this corresponds to a total length of DNA per T4 particle of 2×10^5 nucleotide pairs. We wish to translate linkage distances, as derived from genetic recombination experiments, into molecular units. This cannot be done very precisely at present. It is not known whether all the DNA in a phage particle is indispensable genetic material. Nor is it known whether a phage "chromosome" (i.e., the physical counterpart of a linkage group identified by genetic means) is composed of a single (duplex) DNA fiber or whether genetic recombination is equally probable in all chromosomal regions. For the purpose of a rough calculation, however, these notions will be assumed to be true. Thus we place the total linkage map of T4 in correspondence with 2×10^5 nucleotide pairs of DNA. The total known length of the three linkage groups[7] in phage T4 amounts to some 100 units (one unit = 1 per cent recombination in a standard cross). In addition, there is evidence[8] for roughly another 100 units of length connecting two of the groups. Therefore, if we assume 200 recombination units to correspond to 2×10^5 nucleotide pairs, the recombination per nucleotide pair is 10^{-3} per cent. That is to say, given two phage mutants

---

From Proceedings of the National Academy of Sciences **41**:344-354, 1955. Used with permission.

I am much indebted to A. D. Hershey and A. H. Doermann for stocks of their genetically mapped mutants, to Sydney Brenner and David Krieg for stimulating discussion, and to Max Delbrück for his invaluable moderating influence.

Supported by a grant-in-aid from the American Cancer Society upon recommendation of the Committee on Growth of the National Research Council.

*Present address: California Institute of Technology, Pasadena, California.

whose mutations are localized in their chromosomes at sites only one nucleotide pair apart, a cross between these mutants should give rise to a progeny population in which one particle in 10^5 results from recombination *between* the mutations (provided, of course, that recombination is possible between adjacent nucleotide pairs). This computation is an exceedingly rough one and is only intended to indicate the order of magnitude of the scale factor. Some preliminary results are here presented of a program designed to extend genetic studies to the molecular (nucleotide) level.

r Mutants

The wild-type phages T2, T4, and T6 produce small plaques with rough edges when plated on strain B of *Escherichia coli.* From sectors of clearing in these plaques, mutants can be readily isolated which produce large, sharp-edged plaques (Hershey[9]). These mutants have been designated "r" for rapid lysis; they differ from the wild type by a failure to cause "lysis inhibition" on strain B (Doermann[10]). The wild type has a selective advantage over r mutants when the two types grow together on B. The genetics of r mutants was studied by Hershey and Rotman,[11] who found three regions in the linkage map of T2 in which various mutations causing the r phenotype were located, including one large "cluster" of mutants which were shown to be genetically distinct from one another. The genetic study of T4 by Doermann and Hill[7] showed r regions corresponding to two of those in T2. T6 also has at least two such r regions.

Table 1. Phenotypes (plaque morphology) of T4 wild and rII mutant plated on various hosts

	Host strain		
	E. coli B	*E. coli* K12S	*E. coli* K12S(λ)
T4 wild type	Wild	Wild	Wild
T4 rII mutant	r type	Wild	. .

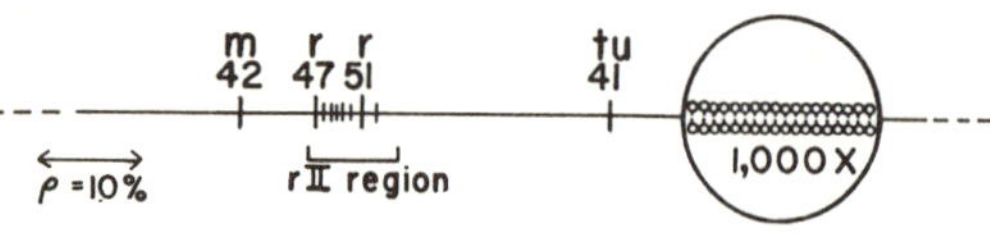

Fig. 1. Partial linkage map of T4 (Doermann), indicating the location of the rII region. *m* and *tu* designate "minute plaque" and "turbid plaque" mutations. The circular inset shows, diagrammatically, the corresponding dimensions of the DNA chain magnified 1,000 diameters.

The rII group

For all three phages, T2, T4, and T6, the r mutants can be separated into groups on the basis of their behavior on strains other than B. This paper will be concerned only with one group, which will be called the "rII group." Mutants of the rII group are distinguished from those of other groups, and from wild type, by a failure to produce plaques on certain lysogenic strains[12] of *E. coli* which carry phage λ. As shown in Table 1, a mutant of the rII group produces r-type plaques on strain B, wild-type plaques on strain K12S (nonlysogenic strain sensitive to λ), and no plaques on K12S (λ) (derived from K12S by lysogenization with λ). The wild-type phage produces similar plaques on all three strains. In the case of T4, with which we shall be concerned in this paper, the efficiencies of plating are approximately equal on the three strains, except, of course, for rII on K12S (λ). The three bacterial strains will be here designated as "B," "S," and "K."

Approximately two-thirds of the independently arising r mutants isolated on B are of the rII type. This group includes the "cluster" of r mutants of T2 described by Hershey and Rotman and the r47 and r51 mutants described by Doermann and Hill in the corresponding map region of T4 but does not include r mutants located outside that region. Similarly, all newly isolated mutants showing the rII character have turned out to fall within the same region, as indicated in Figure 1.

The properties of the rII group are especially favorable for detailed genetic study. An rII mutant has three different phenotypes on the three host strains (Table 1): (1) altered plaque morphology on B, (2) indistinguishable from wild type on S, and (3) unable to produce plaques on K. These properties are all useful. By virtue of their altered plaque type on B, r mutants are readily isolated, and those of the rII group are identified by testing on K. Where it is desired to avoid a selective disadvantage compared with wild type, e.g., in measuring mutation rates, S can be used as a nondiscriminating host. The failure of rII mutants to plage on K enables one to detect very small proportions of wild-type particles due to reversion or due to recombination between different rII mutants.

Fate of rII mutants in K

Wild-type and rII mutants adsorb equally well to strains S and K. Whereas the wild type provokes lysis and liberation of a burst of progeny on both strains, the rII mutant grows normally only on S. Infection of K with an rII mutant provokes very little (and/or very late) lysis, although all infected cells are killed. The block in growth of rII mutant is associated with the presence of the carried phage λ. The reason for this association is unknown.

Quantitative differences in phenotype

While all rII mutants show the same phenotypic effect of poor multiplication on K, they differ in the degree of this effect. A certain proportion of K infected with rII actually liberates some progeny, which can be detected by plating the infected cells on B. The fraction of infected cells yielding progeny defines a "transmission coefficient" characteristic of the mutant. The transmission coefficient is insensitive to the multiplicity of infection but depends strongly upon the physiological state of the bacteria (K) and upon temperature. Under given conditions, however, the coefficient can be used as a comparative index of degree of phenotypic effect, a "leaky" mutant having a high coefficient. As can be seen in Table 2, a wide range of values is found.

Plaques on K

Some rII mutants produce no plaques on K, even when as many as 10^8 particles (as measured by plaque count on B) of a stock are plated. Other rII mutants, however, produce various proportions of plaques on K. When the plaques appearing on K are picked and retested, they fall into three categories: (1) a type which, like the original mutant, produces very few plaques on K and r-type plaques on B; (2) a type which produces plaques (often smaller than wild type) on K with good efficiency but r-type plaques on B; and (3) a type indistinguishable from the original wild. These three types are understood to be due to the following: (1) "leaking" effects, i.e., ability of the mutant to grow slightly on K, so that there is a chance for a few visible plaques to form; (2) a mutation which partially undoes the effect of the rII mutation, so that multiplication in K is possible, but the full wild phenotype is not achieved; and (3) apparent reverse mutation, which may or may not be genuine, to the original wild type.

The proportion of each type occurring in a stock is characteristic and reproducible for a particular rII mutant but differs enormously from one rII mutant to another. There is no evident correlation in the rates of occurrence of the three types.

Table 2. Properties of T4 mutants of the rII group*

Mutant number	Map position	Transmission coefficient	Reversion index (units of 10^{-6})
r47	0	0.03	<0.01
r104	1.3	.91	<1
r101	2.3	.03	4.5
r103	2.9	.02	<0.2
r105	3.4	.02	1.8
r106	4.9	.55	<1
r51	6.7	.02	170
r102	8.3	.02	<0.01

*Three parameters are given for each mutant. The map position is computed from the sum of the nearest intervals shown in Figure 2 and is given in percentage recombination units, taking the position of r47 as zero. The "transmission coefficient" is a measure of phenotypic effect determined by infecting bacteria K with the mutant in question and is given as the fraction of such infected cells yielding plaques on strain B. The "reversion index" is the average fraction of wild-type particles arising in lysates of the mutant grown from a small inoculum on a nonselective host.

Reversion rates of rII mutants

Reversion of r mutants to a form indistinguishable from wild type was demonstrated by Hershey,[9] who made use of the selective advantage of wild type on B to enrich its proportion in serial transfers. Given the inability of rII mutants to produce plaques on K, such reversions are easily detected, even in very small proportion. An index to the frequency of reversion of a particular rII mutant can be obtained by preparing a lysate from a small inoculum (about 100 particles, say, so that there is very little chance of introducing a wild-type particle present in the stock). If S is used as the host, both rII mutant and any reversions which arise can multiply with little selection, as shown by control mixtures. The average fraction of wild-type particles present in several lysates is an index which can be shown to be roughly proportional to the probability of reversion per duplication of the rII mutant. Under the conditions of measurement the index is of the order of 10-20 times the probability of reversion per duplication. The plaques appearing on K must be tested by picking and replating on B. This eliminates the "spurious" plaques produced by partial reversions and by leaky mutants, which show up as r type on B. As may be seen in Table 2, the reversion indices for rII mutants vary over a very wide range. One mutant has been found which reverts 10 times more frequently than r51, so that the reversion rates cover a known range of over 10^5-fold.

It has not been proved that these apparent reversions constitute a genuine return to the original wild type. However, the possibility of suppressor mutations distant from the site of the rII mutation has been ruled out by backcrosses to the original wild type. Krieg[13] found very few, if any, r-type recombinants in backcrosses of several reversions, localizing the reverse changes to within a few tenths of a per cent linkage distance from the original rII mutations. One case of "partial reversion" has also been tested by backcrossing, and failure to observe rII-type recombinants localized the "partial reverse mutation" to within the rII region.

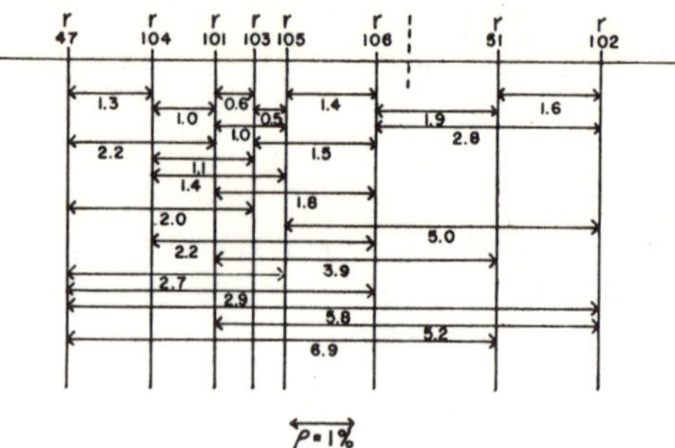

Fig. 2. Larger-scale map of eight rII mutants, including Doermann's r47 and r51. Newly isolated mutants are numbered starting with *101*. The recombination value (in per cent) for each cross is obtained by plating the progeny on K and on B and doubling the ratio of plaque count on K to count on B.

Mapping of the rII region

A cross between two rII mutants is made by infecting a culture of B with equal multiplicities (three per bacterium) of each type. The yield after lysis contains the two parental types and, if the parents are genetically distinct, two recombinant types, the double mutant and wild type. In the average yield from many cells, the recombinant types occur in equal numbers.[11] In all cases thus far tested, double rII mutants, like single mutants, do not produce plaques on K. On the assumption that this is generally true, the proportion of recombinants in the yield can be measured simply by doubling the ratio of the plaque count on K (which registers only the wild recombinant) to the count on B (which registers all types). The percentage of wild type thus measured agrees well with a direct count of plaque types on B.

In this way, a series of six rII mutants of T4 (the first six isolated—not selected in any way) have been crossed with each other and with r47 and r51 (kindly supplied by A. H. Doermann) in 23 of the 28 possible pairs. The results of these crosses are given in Figure 2 and are compatible with the indicated seriation of the mutants. The distances are only roughly additive; there is some systematic deviation in the sense that a long distance tends to be smaller than the sum of its component shorter ones. Part of this discrepancy is accounted for by the Visconti-Delbrück correction for multiple rounds of mating.[14] Reversion rates were small enough to be negligible in these crosses. Thus, while all rII mutants in this set fall into a small portion of the phage linkage map, it is possible to seriate them unambiguously, and their positions *within* the region are well scattered.

Tests for pseudo-allelism

The functional relatedness of two closely linked mutations causing similar defects may be tested by constructing diploid heterozygotes containing the two mutations in different configurations.[1,2] The *cis* form, with both mutations in one chromosome, usually behaves as wild type, since the second chromosome supplies an intact functional unit (or units). However, the *trans* form, containing one of the mutations in each chromosome, may or may not produce the wild phenotype. If it does, it is concluded that the two mutations in question are located in separate functional units.

In applying this test to the rII mutants, the diploid heterozygote can be simulated by a mixed infection with two kinds of phage. The rII phenotype is a failure to lyse K, whereas the wild phenotype is to cause lysis. If K is mixedly infected with wild type and rII mutant, the cells lyse, liberating both types of phage. Thus the presence of wild type in the cell supplies the function which is defective in rII type, and the rII mutation can be considered "recessive." Although it has not yet been tested, the *cis* configuration of double rII mutant plus wild type is also presumed to produce lysis in all cases. The *trans* configuration is obtained by infecting K with the pair of rII mutants in question. This is found to give lysis or not, depending upon which rII mutants compose the

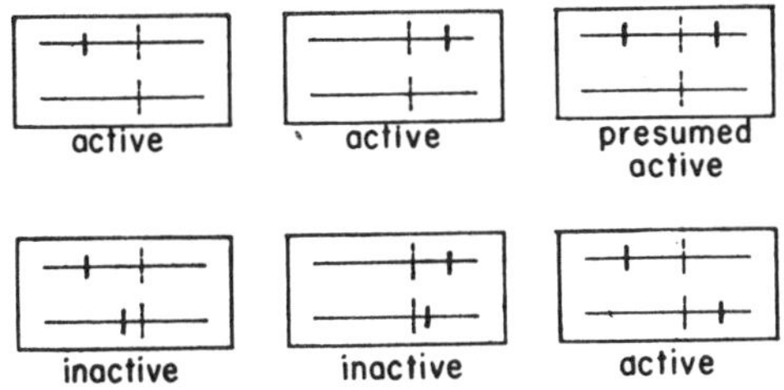

Fig. 3. Summary of tests for "position-effect pseudo-allelism" of rII mutants. Each diagram represents a diploid heterozygote as simulated by mixed infection of a bacterium (K) with two types of phage containing the indicated mutations. *Active* means extensive lysis of the mixedly infected cells; *inactive* means very little lysis. The dotted line represents a dividing point in the rII region, the position of which is defined by these results.

pair. The results are summarized by the dotted line in Figure 2, indicating a division of the rII region into two segments. If both mutants belong to the same segment, mixed infection of K gives the mutant phenotype (very few cells lyse). If the two mutants belong to different segments, extensive lysis occurs with liberation of both infecting types (and recombinants). These results are summarized in Figure 3. Thus, on the basis of this test, the two segments of the rII region correspond to independent functional units.

Actually, for mixed infection of K with two (nonleaky) mutants of the *same* segment, a very small proportion of the cells do lyse and liberate wild recombinants, that proportion increasing with the linkage distance between the mutations. For two rII mutants separated by 1 per cent linkage distance (measured by a standard cross on B) the proportion of mixedly infected K yielding any wild particles is about 0.2 per cent.

This value has bearing upon the effect upon K/B values of the heterozygous phage particles which arise in a cross between two rII mutants on B. In such a cross between closely linked rII mutants, the progeny should include about 2 per cent of particles containing a *trans* configuration heterozygous piece.[15] When one of these is plated on K, there is a certain chance that a wild recombinant may form in the first cycle of infection, leading to production of a plaque. If it is assumed that these are no more likely to do so than a mixed infection of K with two complete mutant particles, it can be concluded that the effect of these heterozygous particles upon the count on K is negligible, provided that both rII mutants belong to the same segment. For mutants in different segments, however, the "efficiency" of the heterozygous particles should be much greater, and recombination values measured by the K/B method should run considerably higher than the true values. The recombination values in Figure 2 for crosses which transgress the segmental divide are probably subject to some correction for this reason.

Rough mapping by spot test

If a stock of either of two rII mutants is plated on K, no plaques arise; but if both are plated together, some bacteria become infected by both mutants and, if this leads to the occurrence of wild-type recombinants, plaques are produced. If the two mutants are such that wild recombinants cannot arise between them (e.g., if they contain identical mutations), no plaques appear. A given rII mutant may thus be tested against several others on a single plate by first seeding the plate with K plus the mutant in question (in the usual soft agar top layer) and then spotting with drops containing the other rII mutants.

Inspection of such a plate immediately places the unknown mutant in the proper segment, since spotting any mutant of segment A against any mutant of segment B gives a very clear spot, due to the extensive lysis of mixedly infected bacteria. However, for a pair of mutations belonging to the same segment, plaques are produced only by the relatively few mixedly infected bacteria which give rise to wild recombinants. The greater the linkage distance between the mutations, the larger the number of plaques that appear in the spot. A group of mutants of the same segment may thus be seriated by seeding one plate with each and spotting with all the others. Given a previously seriated group, a new mutant can thus be quickly located within the group. This method works best for mutants which are stable (i.e., low reversion rate) and nonleaky, so that large numbers of phage particles can be plated. Reversions or pronounced leaking effects obviously cause an obscuring background.

This test has been applied to a large group of stable, nonleaky rII mutants. Their approximate locations as deduced from these tests are shown in Figure 4. Some of the mutants showed anomalies which made it impossible to locate them as members of a series. They gave very little recombination with any of the mutants located within a certain span, while behaving normally with respect to mutants located outside that span. They are indicated in Figure 4 by horizontal lines extending over the span.

Spot tests on numerous other mutants have shown that mutants of varied reversion rates,

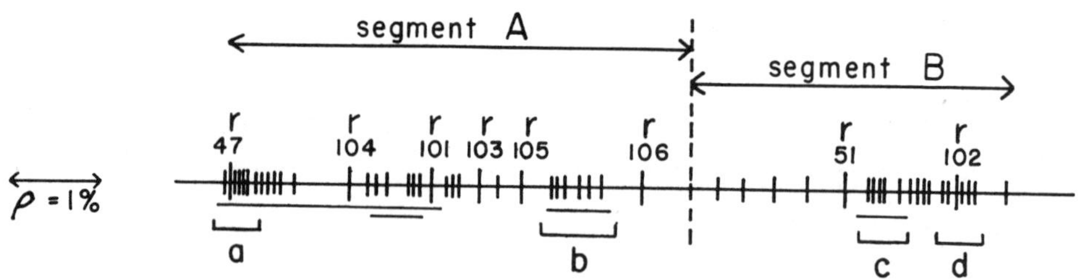

Fig. 4. Preliminary locations of various rII mutants, based upon spot tests.

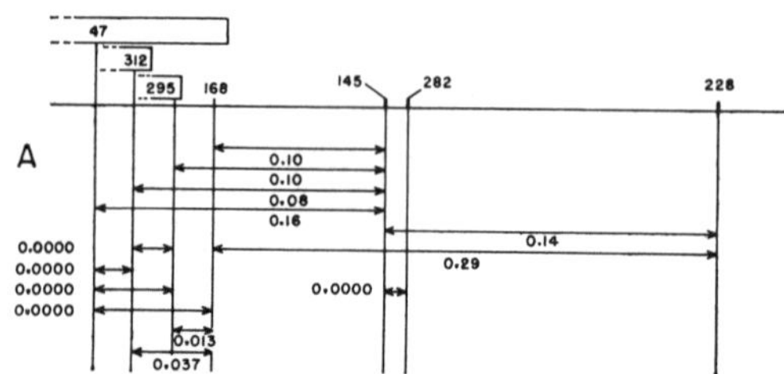
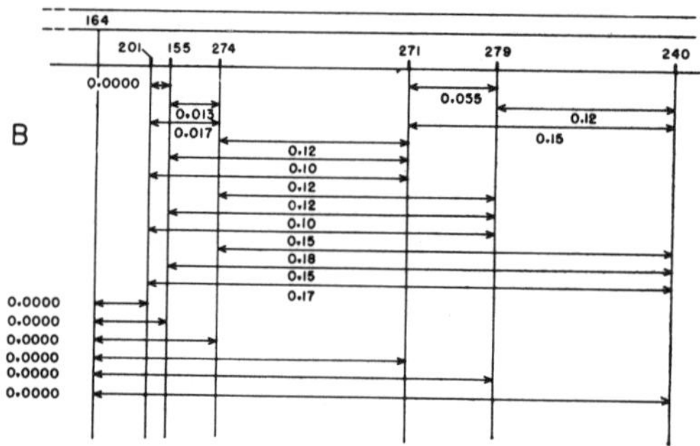
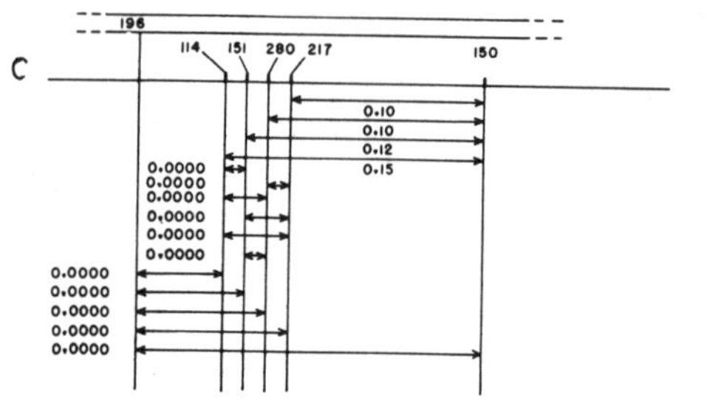
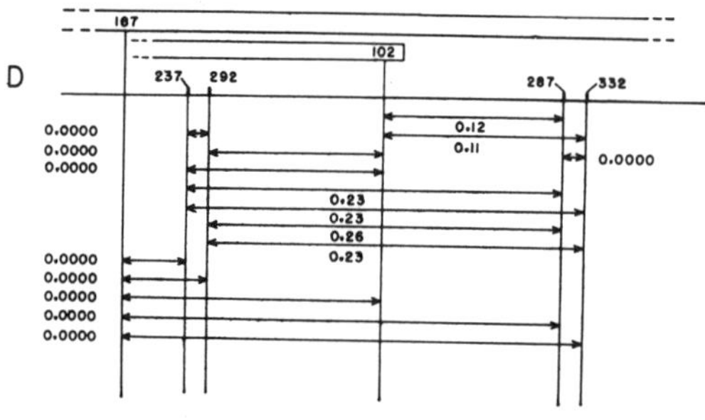

Fig. 5. Maps of microclusters.

transmission coefficients, and rates of "partial reversion" occur at scattered positions in both segments.

Mapping of "microclusters"

The spot test enables us to pick out "microclusters," i.e., groups of very closely neighboring mutations. Four such groups selected for further study are indicated in Figure 4, and the results of mapping them are given in Figure 5. While some intervals show reasonably good additivity properties, there are some mutants which give violently anomalous results. Thus in microcluster *a,* r47 gives no wild recombinants (i.e., less than 1 in 10^6) with any of the other three mutants, but two pairs of the three do show recombination. These results can be understood if it is assumed that each mutation extends over a certain length of the chromosome, and production of wild type requires recombination within the space *between* those lengths. According to this interpretation, the mutations would cover the lengths indicated by the bars in Figure 5. These anomalies resemble those observed in the spot tests, only they are more limited in span.

This observation raises the question of whether there exist true "point" mutations (i.e., involving an alteration of only one nucleotide pair) or whether all mutations involve more or less long pieces of the chromosome. It must be remembered that the mutants used in these experiments were selected for extreme stability against reversion. This procedure would be expected to enrich the proportion of mutants containing gross chromosomal alterations. So far as is known, the anomalous cases observed could equally well be imagined to be due to double (i.e., two near-by "point") mutations, inversions, or deletions of the wild-type chromosome. In continuing these experiments, it would seem well advised to employ only mutants for which some reversion is observed.

DISCUSSION

The set of rII mutants defines a bounded region of a linkage group in which mutations may occur at various locations, all the mutations leading to qualitatively similar phenotypic effects. The rII region would seem, therefore, to be functionally connected, so that mutations arising anywhere within the region affect the same phenotype. This effect is expressed, in case strain B is the host, by failure to produce lysis inhibition; in case S is the host, by no consequence; and in case K is the host, by inability to multiply normally. The failure of an rII mutant to mature in K can be overcome by the presence of a wild-type phage in the same cell. This could be understood if the function of the region in the wild-type "chromosome" were to control the production of a substance or substances needed for reproduction of this phage in K cells.

The phenotypic test for "pseudo-allelism" leads to the division of the region into two functionally distinguishable segments. These could be imagined to affect two necessary sequential events or could go to make up a single substance the two parts of which must be unblemished in order for the substance to be fully active. For example, each segment might control the production of a specific polypeptide chain, the two chains later being combined to form an enzyme. While it is not known whether this sort of picture is applicable, a model of this kind is capable of describing the observed properties of the rII mutants. The map position of a mutation would localize a change in the region (and also in the "enzyme" molecule), the reversion rate would characterize the type of change involved in the genetic material, and the degree of phenotypic effect would be an expression of the degree of resultant change in the activity of the enzyme. A "leaky" mutant would be one where this latter effect was small. While no obvious correlation has yet been observed among these three parameters of rII mutants, one may well show up upon more exhaustive study.

"Clustering" of similar mutants separable by crossing-over has been observed for several characters in phage by Doermann and Hill and appears to represent the rule. This may well be the rule in all organisms, simply because functional genetic units are composed of smaller recombinational and mutational elements. One would expect to see this effect more readily in phage because the probability of recombination per unit of hereditary material is much greater than for higher organisms.

By extension of these experiments to still more closely linked mutations, one may hope to characterize, in molecular terms, the sizes of the ultimate units of genetic recombination, mutation, and "function." Our preliminary results suggest that the chromosomal elements separable by recombination are not larger than the order of a dozen nucleotide pairs (as calculated from the smallest non-zero recombination value) and that mutations involve variable lengths which may extend over hundreds of nucleotide pairs.

In order to characterize a unit of "function," it is necessary to define what function is meant. The entire rII region is unitary in the sense that mutations anywhere within the region cause the rII phenotype. On the basis of phenotype tests of *trans* configuration heterozygotes, this region can be subdivided into two functionally separable segments, each of which is estimated to contain of the order of 4×10^3 nucleotide pairs. If one assumes that each segment has the "function" of specifying the sequence of amino acids in a polypeptide chain, then the specification of each individual amino acid can as well be considered a unitary function. It would seem feasible, with this system, to extend genetic studies even to the level of the latter functional elements.

SUMMARY

It has been discovered that the mutations in the rII region of phage T4 have a characteristic in common which sets them apart from the mutations in all other parts of the map. This characteristic is a host-range reduction, namely, a failure to produce plaques on a host (K) lysogenic for phage λ. The mutant phage particles adsorb to and kill K, but normal lysis and phage release do not occur.

All mutants with this property are located within a sharply defined portion of the phage linkage map. Within that region, however, their locations are widely scattered. An unambiguous seriation of the mutants, with roughly additive distances, can be accomplished, except for certain anomalous cases.

The simultaneous presence of a wild-type phage particle in K enables the multiplication of rII mutants to proceed, apparently by supplying a function in which the mutant is deficient. A heterozygous diploid in the *trans* configuration is simulated by a mixed infection of K with two mutant types. The application of the phenotype test to pairs of rII mutant leads to the division of the region into two functionally separable segments.

Spontaneous reversion to wild-type had been observed for most of these mutants. It remains to be seen whether these are genuine reversions. Each mutant reverts at a characteristic rate, but the rates for different mutants differ enormously. Partial reversions to intermediate types are also observed.

The mutants differ greatly in degree of residual ability to grow on K. There is no evident correlation between map position, reversion rate, and degree of residual activity of the various mutants.

The selective feature of K for wild-type recombinants offers the possibility of extending the recombination studies to an analysis of the fine details of the region.

Preliminary studies of this type indicate that the units of recombination are not larger than the order of one dozen nucleotide pairs and that mutations may involve various lengths of "chromosome."

REFERENCES

1. E. B. Lewis, Cold Spring Harbor Symposia Quant. Biol. **16**:159-174, 1951.
2. G. Pontecorvo, Advances in Enzymol. **13**:121-149, 1952.
3. A. D. Hershey and M. Chase, J. Gen. Physiol. **36**:39-56, 1952.
4. A. D. Hershey, J. Dixon, and M. Chase, J. Gen. Physiol. **36**:777-789, 1953.
5. E. K. Volkin, personal communication.
6. J. D. Watson and F. H. C. Crick, Cold Spring Harbor Symposia Quant. Biol. **18**:123-131, 1953.
7. A. H. Doermann and M. B. Hill, Genetics **38**:79-90, 1953.
8. G. Streisinger and V. Bruce, personal communication.
9. A. D. Hershey, Genetics **31**:620-640, 1946.
10. A. H. Doermann, J. Bacteriol. **55**:257-276, 1948.
11. A. D. Hershey and R. Rotman, Genetics **34**:44-71, 1949.
12. E. M. Lederberg and J. Lederberg, Genetics **38**:51-64, 1953.
13. D. Krieg, personal communication.
14. N. Visconti and M. Delbrück, Genetics **38**:5-33, 1953.
15. A. D. Hershey and M. Chase, Cold Spring Harbor Symposia Quant. Biol. **16**:471-479, 1951; C. Levinthal, Genetics **39**:169-184, 1954.

45 Regulation of the lac operon

Recent studies on the regulation of lactose metabolism in Escherichia coli support the operon model

Jonathan R. Beckwith
Department of Bacteriology and Immunology
Harvard Medical School
Boston, Massachusetts

The mechanism of regulation of gene expression is today one of the most actively studied problems in molecular biology, in good part as a result of the pioneering work of Jacob and Monod on the control of the genes involved in lactose metabolism in the bacterium *Escherichia coli.* Since 1961, when Jacob and Monod first proposed the operon model for gene regulation *(1),* a number of alternative suggestions have been published for the ways in which genes are controlled *(2-5),* some of them radically different from the Jacob-Monod model.

The lactose *(lac)* system is still the best-studied example of gene regulation. In the years since 1961, there has been a considerable amount of new information, both genetic and biochemical, on the *lac* operon. On the basis of the most recent information, which will be discussed in this article, it appears that the original formulation of the operon, in most of its aspects, is still the simplest model fitting all the known facts about the *lac* system *(6).*

LACTOSE METABOLISM IN ESCHERICHIA COLI

The initial steps in the metabolism of lactose by *E. coli* involve two protein components: (i) a membrane-bound protein [M-protein *(7)* or permease *(8)*] which is probably responsible for both the transport of lactose into the bacterial cell and for its concentration therein; and (ii) the enzyme β-galactosidase which catalyzes the hydrolysis of lactose within the cell to glucose and galactose. The structure of these two proteins is determined by two chromosomal genes, *y* for the permease and *z* for β-galactosidase. In wild-type strains of *E. coli* grown on almost any carbon source but lactose, the activities of these genes are repressed, their products being found in only very small amounts. However, growth on lactose as sole carbon source, or addition to the growth medium of various compounds structurally related to lactose, results in the induction of gene expression, with an increase in the amounts of these proteins of as much as 1000 times. Under these optimum conditions, β-galactosidase represents approximately 3 percent of the total protein of the cell. Induction also results in the appearance of another enzymic activity, thiogalactoside transacetylase *(9),* corresponding to a third structural gene, *a.* Although the *a* gene is regulated in parallel with the *z* and *y* genes, the enzyme plays no essential role in lactose metabolism *(10).* These three structural genes of the lactose system lie next to one another on the chromosome, mapping in the order *zya* (Fig. 1).

THE MODEL

The Jacob-Monod operon model of control, with some additions and modifications resulting from recent work on the *lac* system (Fig. 1) may be described as follows. In the absence of any regulation, the expression of the three *lac* structural genes involves two steps. First, the information from these genes is transcribed into a single RNA (messenger RNA) molecule. The information from each gene-copy within the RNA is then translated by the protein-synthesizing machinery into the structure of the three protein products. The synthesis of this mRNA molecule is initiated at the promoter site, *p,* which is adjacent to, or part of the *z* gene. The structure of the promoter, or of a site in the

From Science **156:**597-604, May 5, 1967.

I thank Dr. B. Davis, Dr. W. Epstein, Dr. J. Scaife and Dr. E. Signer for discussions and criticisms of the manuscript. Supported by PHS research career program award GM-9027, PHS research grant GM-13017, and by NSF grant GB-5763.

NO INDUCER PRESENT

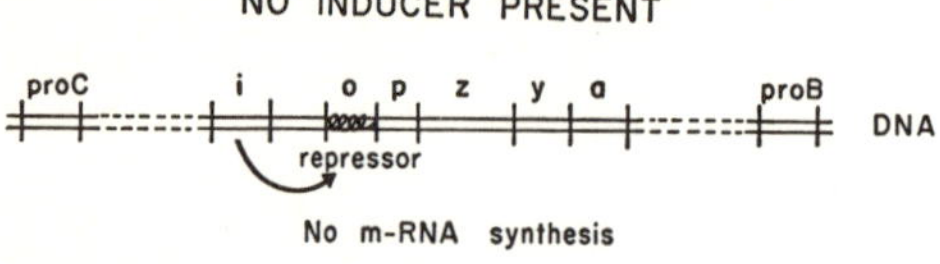

INDUCER PRESENT

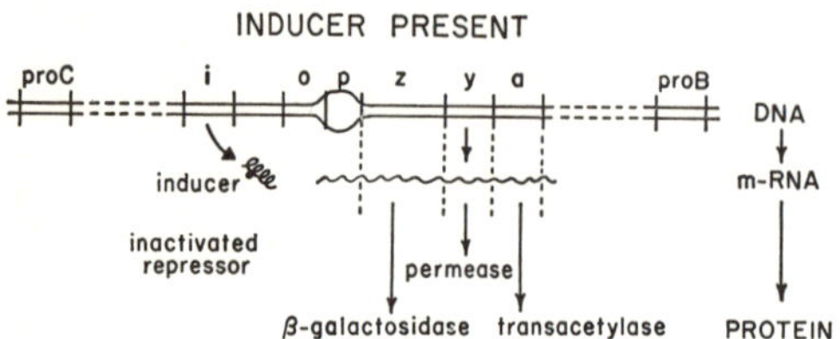

Fig. 1. The control of expression of the *lac* operon. The *lac* region and two of the markers surrounding it (genes involved in proline biosynthesis) on the chromosome are represented as regions of double-stranded DNA. The picture of the operator and promoter in this figure is only one of many ways of visualizing their interaction. The promoter is shown serving as an initiation site for mRNA synthesis only when its two DNA strands are not held together by hydrogen bonding in the "closed" configuration. The repressor binds to and closes the operator, resulting in a closing of the promoter. An inducer alters the repressor so that it can no longer interact with the operator in this way. Pro C and pro B indicate two genes involved in the biosynthesis of proline.

same region, determines the rate at which the message is transcribed from these genes under conditions of maximal expression. The control of the expression of the *lac* structural genes is effected by a repressor molecule, the protein product of the closely linked *i* gene. The repressor acts on the DNA to inhibit the transcription of these genes. The combination of the repressor with the operator site, *o*, which is adjacent to *p*, in some way inhibits the initiation process. Compounds that cause induction of the expression of the *lac* genes act by either destroying or altering the repressor-operator complex, thus allowing initiation of mRNA synthesis at the promoter.

A group of genes whose activity is coordinated by an operator is termed an operon. According to this model, the operator is not essential for operon activity, but rather serves as a controlling site superimposed on a functioning unit.

MAPPING OF THE ELEMENTS OF THE LAC OPERON

Jacob and Monod have presented evidence from three-factor crosses demonstrating the gene order *iozy (11)*. In addition, strains which carry deletions extending into the *lac* genes from outside either end of the *lac* operon have been isolated, providing further confirmation for the order *iozya* (see Fig. 4) *(10, 21)*. Class-I deletions remove the *i* gene and leave the operator and structural genes intact, while class-IVa deletions remove the *a* gene but leave all other sites in the operon intact.

TRANSLATION OF THE LAC OPERON

The translation process. To discuss certain aspects of *lac* operon control, I must first describe what is known about the mode of translation of mRNA information into protein. The mRNA copies of genes are thought to be translated by the following process *(13)*. Ribosomes attach to one end of the message (corresponding to the amino-terminal end of the protein), and, in conjunction with the other components of the protein-synthesizing machinery, begin to move along the message as the peptide bonds are formed. New ribosomes continually attach to this end of the message and proceed in this way, so that at any one time the mRNA will carry many ribosomes along its length. This complex of ribosomes and mRNA is called a polysome. At the end of the gene-copy in the mRNA, the translation machinery meets a codon that signals termination and release of the completed polypeptide chain.

As a result of mutation, chain-terminating codons may be introduced into the gene at various points preceding the normal site of chain termination. The two well-studied chain-terminating codons arising by mutation are the amber and ochre codons, UAG and UAA (for uracil, adenine, and guanine) respectively *(14, 15)*. Chain termination by an amber mutation (and probably by an ochre mutation also) results in a quantitative release of an NH_2-terminal fragment of the protein coded for by a particular gene. The length of the protein fragment depends upon the distance of the mutation from the beginning of the gene (corresponding to the NH_2-terminal end of the protein).

Another type of mutation that interferes with translation is the frame-shift mutation *(16)*. These mutations, through the addition or removal of one or more base pairs from a gene, cause the reading of an incorrect sequence of codons in the mRNA gene-copy.

The operon messenger RNA. There is still no direct evidence that all the information from

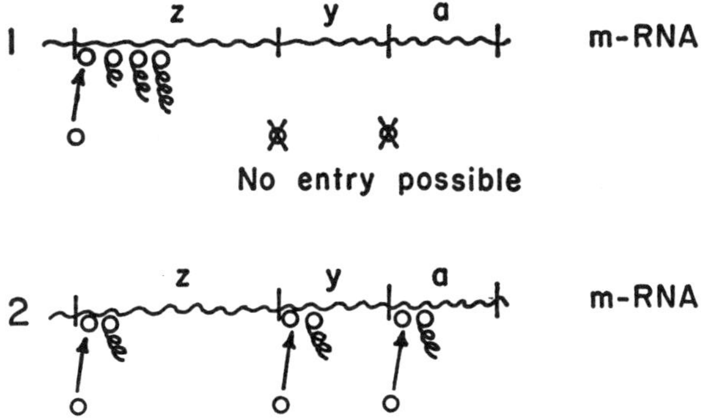

Fig. 2. Two possible models for translation of the *lac* operon mRNA. The circles represent ribosomes, and the squiggly lines attached to them represent the growing polypeptide chain.

the structural genes of the *lac* operon is contained in a single piece of mRNA. However, studies on polarity (described later) do suggest that this is so. Also, Kiho and Rich *(17)* have presented evidence that an amber mutation prematurely terminating translation in the *y* gene affects the size of the polysome on which β-galactosidase is made. If the *y* and *z* gene-copies were on different mRNA molecules, no effect should have been observed. We shall assume that there is a single operon mRNA.

The amino-terminal end of the *z* gene. A knowledge of the direction of translation of the structural genes of the operon is important in interpreting the various experiments on operon expression. The direction of translation indicates the direction of transcription of the operon, since it has been shown by biochemical and genetic experiments that both these processes begin at the 5′ end of the RNA molecules *(18)*. Studies by Fowler and Zabin *(19)* on three chain-terminating mutants of the *z* gene *(20)* indicate that translation is initiated at the proximal *(21)* end of the *z* gene-copy in the mRNA. These mutants, one of which lies very close to the *y* gene, map at the distal end of the *z* gene.

Each mutant makes a large amount of a protein that is immunologically similar to β-galactosidase. If this distal end of the *z* gene corresponded to the NH_2-terminal end of the protein, translation should have begun at this end, and, in these mutants, terminated within a short distance, releasing a small polypeptide fragment. The finding of large amounts of a large protein molecule is thus very strong evidence that translation begins at the proximal end of the *z* gene, and that in these mutants, most of the *z* polypeptide chain is made before the chain-terminating triplet is read and the protein released. As mentioned earlier, this conclusion also leads to the further conclusion that operon mRNA synthesis is initiated at this end of the operon. Even stronger evidence for the operator-proximal end of a gene corresponding to the NH_2-terminal end of the protein comes from studies on the tryptophan operon by Yanofsky and co-workers *(22)*.

POLAR MUTANTS

Two different models for the translation of the gene-copies in an operon mRNA have been considered (Fig. 2). (i) A ribosome can enter onto the mRNA at only one end of the molecule *(2)*; in the *lac* operon, this end would correspond to the NH_2-terminal end of the *z* gene. Then, during translation, the ribosome proceeds down the message and must complete passage through the *z* gene in order to start translation of *y* and *a*. (ii) Ribosomes can enter independently at the starting points for all three gene-copies in the mRNA, *z*, *y*, and *a*, without any requirement for having translated a previous gene-copy *(23)*.

There is no conclusive evidence to distinguish between the two models. However, any model for the translation of the *lac* operon mRNA must take into account the class of mutants known as polar mutants. Polar mutations in the *z* gene of the *lac* operon are usually point mutations that not only abolish *z* gene activity, but, in addition, reduce or abolish the expression of the *y* and *a* genes *(24, 25)*. Such mutations in the *y* gene affect the activity of the *a* gene but not of the *z* gene. Although many polar mutants are not well characterized, a large percentage are amber and ochre mutations *(23)*. Since these chain-terminating muta-

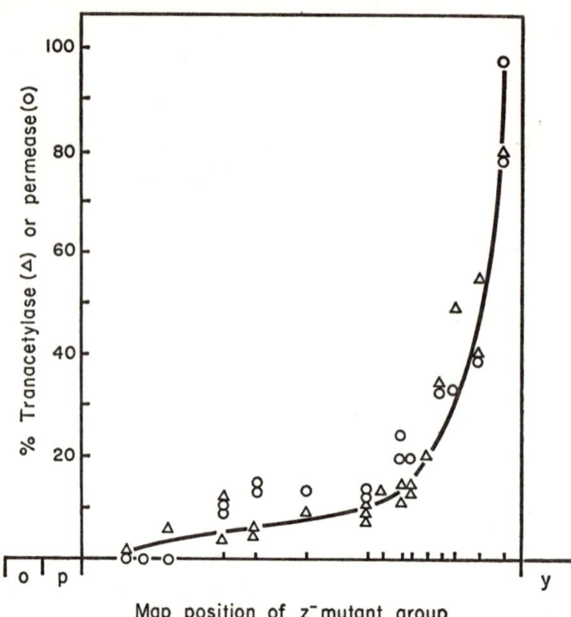

Fig. 3. The gradient of polarity of chain-terminating mutants in the *z* gene. This figure is copied from Newton *et al.* *(23).* The permease activity (○) and transacetylase activity (△) of many z^- amber and ochre mutants are plotted against the position within the *z* gene. These positions are only rough approximations of true location on the genetic map location. However, the order of the mutants is unambiguous.

tions exert their effects on translation, the polarity must be due primarily to an interference with the translation of the operon mRNA.

The polarity of a chain-terminating mutation in the *z* gene depends upon its position within the gene (Fig. 3) *(23).* Amber and ochre mutants mapping toward the distal end of the *z* gene are not very strongly polar, whereas those chain-terminating mutants mapping in a proximal segment of the *z* gene do not make any detectable amounts of the *y* and *a* gene products. The extent of polarity of a chain-terminating mutation in *z* is determined by the distance of the mutation from the boundary between *z* and *y*, and not by its proximity to the operator. Thus, when an extremely polar mutation, mapping at the proximal end of the *z* gene, is combined with a succeeding deletion in the gene, effectively moving the mutation closer to the *y* gene, the polarity effects are markedly reduced *(23, 26, 27).* The distance from the operator end of the *z* gene to this mutation has not changed. This finding shows that, even if translation of the operon is terminated by a chain-terminating mutation very shortly after it begins, the rest of the operon can still be expressed if this mutation is very close to the *y* gene.

Polar mutations in the *z* gene not only reduce the amounts of the proteins, permease and transacetylase, but also reduce the amount of *lac* mRNA present in the cell *(28).* Again, the reduction of the amount of *lac* mRNA is correlated with the position of the mutation in the *z* gene. Extracts of induced cells carrying one of the extremely polar mutations at the proximal end of the *z* gene contain no detectable amounts of *lac* mRNA; cells harboring z^- polar mutations closer to the *y* gene contain detectable, but still reduced, levels of *lac* mRNA. There is some indication that in these mutants there are normal amounts of the fragment of *lac* mRNA corresponding to the proximal segment of the *z* gene preceding the mutation, but that complete operon mRNA molecules are present only in reduced amounts. Thus, as a result of the interference by polar mutations with the translation process, a polarity effect on *lac* mRNA levels also results.

The results of studies on polar mutants fit both of the models for operon mRNA translation considered above. If the ribosomes can only enter at the beginning of *z* (model 1), then, after encountering an amber or ochre mutation in the *z* gene-copy, ribosomes would continue to pass down the message. However, since these ribosomes are no longer engaged in translation, there is an increased probability of their falling off the mRNA. The longer the distance to be traversed after the site of chain-termination, the lower the probability that the ribosome will reach the *y* gene-copy and there initiate translation. Recent studies of Malamy *(29)* on the strong polarity effects of what appear to be frame-shift mutations in the distal segment of the *z* gene suggest that ribosomes after encountering chain-terminating mutations do continue moving along the mRNA. He finds that double mutants, in which the "frame-shift" is preceded by an ochre or amber mutation in *z*, are still just as polar as the original frame-shift mutant. If ribosomes fell off the mRNA at chain-terminating codons, they should never encounter the frame-shift, and the polarity effects should be reversed. However, it should be noted that apparent frame-shift mutations in one gene of the histidine operon in *Salmonella typhimurium* do not have the strong polarity effects seen in similar mutants in the *lac* operon *(30)*. Confirmation of the proposal that the extreme polarity effects of these mutants in the *lac* operon are due to their frame-shift character would be strong supporting evidence that ribosomes continue their movement after chain-terminating codons.

If ribosomes cannot continue along an mRNA after a chain-terminating codon, then, in order to explain why most amber and ochre mutants allow some expression of *y* and *a*, we must admit that ribosomes can enter the *y* and *a* gene-copies independently of the completion of passage of the *z* gene-copy (model 2). The polarity effects of chain-terminating mutants can be accounted for by one of the following additional hypotheses.

1) Initially, the operon mRNA sticks to the DNA. As ribosomes move down the mRNA during translation and get closer to the beginning of the *y* gene-copy, there is an increasing chance that they will cause release of this part of the mRNA, thus freeing it for ribosome entry *(23)*. Premature chain-termination in the *z* gene and falling off of ribosomes would make it more likely that the mRNA will stick to the DNA.

2) The introduction of a chain-terminating mutation in the *z* gene, leaving an untranslated portion of the mRNA, results in a more rapid destruction of the distal segment of the mRNA *(15, 31)*. The amount of destruction could depend upon the segment of mRNA which is not engaged in translation.

3) The secondary structure of the RNA surrounding the ribosome entry points for *y* and *a* is such that ribosomes cannot enter unless there has been enough progress of ribosomes on the *z* gene-copy to disrupt this structure *(32)*. Premature chain-termination in the *z* gene and falling off of ribosomes would reduce the probability of exposing the entry site on *y*.

Although all of these hypotheses must include some explanation for the decreased amounts of *lac* mRNA in polar mutants, it is not a very important consideration in formulating the models. If this decrease in mRNA is not directly predicted by the model for polarity, then the suggestion can always be added that, under the particular conditions, the mRNA is destroyed very rapidly. If, of course, it is proved that the mRNA is not rapidly destroyed in these mutants, then certain models of polarity must be discarded.

Certain important findings do come out of the studies on polarity. First, the conclusion that the amount of *lac* mRNA in a cell somehow depends upon the translation of the operon has led to new speculations on mechanisms of repression of operon activity involving effects on translation. Second, it can be concluded that translation of an early part of the *z* message is not necessary for the release of the operon mRNA from the DNA, except in so far as this region is distant from the beginning of the *y* part of the mRNA. Stent *(3)*, Yanofsky *(33)*, and others have suggested that there is a critical region at the beginning of the first gene-copy in an operon mRNA, and this region must be translated in order to release the entire mRNA from the DNA. In the *lac* operon, this model cannot account for the extreme polar mutants.

SEQUENTIAL EXPRESSION OF THE LAC GENES

When the kinetics of expression of the *lac* operon are followed after the addition of inducer to a culture of *E. coli*, β-galactosidase activity begins to increase a minute or so before transacetylase activity rises *(34, 35)*. Since the appearance of an enzymic activity requires transcription, translation, and assembly of a protein into the correct configuration, any or all of these processes could be responsible for this lag. Experiments designed to determine the basis of this lag do show that there is a measurable time lapse between the transcription of the *z* gene and of the *a* gene *(35, 36)*. However, it is not clear yet whether this delay is responsible for the lag in appearance of transacetylase activity. In any case, the results showing the sequence of transcription of the *lac* genes are another indication that transcription begins at the operator end of the operon.

UNBALANCED TRANSLATION OF LAC GENES

Zabin has shown that the molar amounts of transacetylase synthesized are five or more times lower than the molar amounts of β-galactosidase *(37)*. One explanation for this finding derives from the studies on polar mutants which show that the distance between a chain-termination event in one gene and the initiation of translation in the next determines how much of the next protein is made. It is possible that, after the chain-terminating triplet in either *z* or *y*, there is a "dead space" of untranslated nucleotide sequence, the length of which determines the amount of transacetylase made.

THE OPERATOR

I have described evidence that the transcription of the *lac* operon starts at the operator end. But is transcription actually initiated at the operator itself? In the original operon hypothesis, it was proposed that the operator has three functions. (i) It forms the first part of the *z* gene; (ii) it is the site of repressor action; and (iii) it is the site of initiation of transcription. The second function is confirmed by the existence of operator-constitutive mutants (O^c) *(1)*, which have partly lost sensitivity to repressor. As a result, these mutants make large amounts of the *lac* enzymes in the absence of inducer, but can still be induced to make more. The nature of the operator is indicated by the finding that O^c mutations cause constitutivity only for the *lac* genes which are linked to that operator. Thus, in a partial diploid strain carrying two copies of the *lac* region, one of which is O^c, the genes attached to the O^+ operator are still normally repressible, while those linked to the O^c operator are still partially constitutive.

There is reason to believe that all O^c's are deletions *(38)*. First, certain of the O^c's can be shown to be deletions, since they remove the *i* gene also (as discussed later) *(38)*. Second, the frequency of O^c's is not increased by mutagens which only cause base substitutions, but it is increased by treatment with x-rays, which does cause deletions *(38)*. Third, no suppressible O^c's have yet been found *(39)*. With another operon there is also mutagenic evidence that O^c's are always deletions *(40)*.

The third function for the operator was based on the existence of a second class of mutants ("O^0") mapping at the beginning of the *z* gene, which permanently shut off the *z, y,* and *a* genes *(1)*. However, it was subsequently shown that complete activity is restored to the *y* and *a* genes by deletion of the "O^0" mutant site *(41)*. Thus, this region of the operon is not necessary for operon expression and cannot be the site of initiation of mRNA transcription for the operon. In addition, it was shown that "O^0" mutations do not, in fact, lie in the operator region, but are only extreme examples of the polar mutants *(15, 23, 31, 41)*. Therefore, the term "O^0" is a misnomer, since these mutants have nothing to do with operator control. As a result of these studies and further work indicating that O^c mutations do not affect the properties of β-galactosidase *(38, 41, 42)*, it has been concluded that the operator lies outside the first structural gene of the operon.

THE PROMOTER

The following discussion summarizes evidence which suggests that transcription begins not at the operator, but at the adjacent promoter site. It is an elaboration of the argument for the promoter presented by Jacob, Ullman, and Monod *(38)*. This concept of operon function is based on the properties of mutants in which various components of the *lac* operon are deleted. Starting with strains in which the *lac* operon has been inactivated by certain mutations, it is possible to select O^c mutations that render the operon partially or fully constitutive *(38)*. In addition, by selecting for O^c mutations under conditions where only *y* gene function is required, O^c's are found which are z^-y^+ *(38, 41)*. In all cases examined, O^c mutations of this latter type (Fig. 4, class III) are a result of deletions which cover a proximal segment of the *z* gene, the operator, and the *i* gene, and presumably fuse the intact *y* and *a* genes to another operon or gene. In most cases, the gene or genes to which *lac* has become fused have not been identified. However, two sets of deletions of this type have been isolated in which *lac* has been fused to known operons. In one case, the *y* and *a* genes have come under the control of an operator for two genes involved in purine biosynthesis *(43)*,

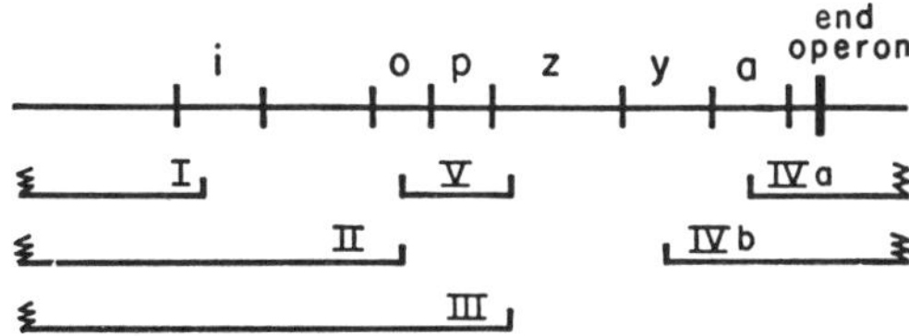

Fig. 4. Deletions of the *lac* operon.

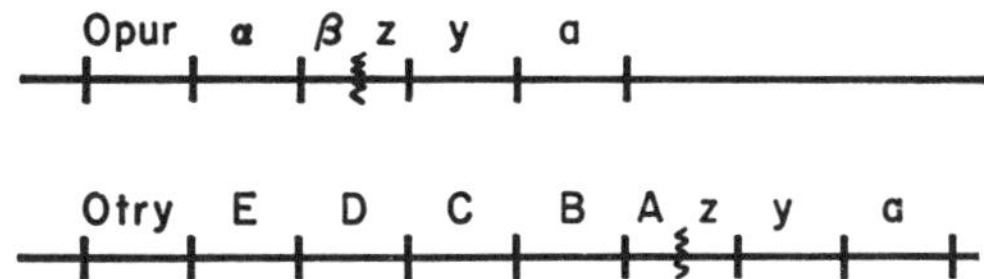

Fig. 5. Fused *lac* operons. The *site* of fusion of the *lac* and *try* operons was determined by E. R. Signer *(53)*.

and, in the other, the *lac* genes have come under the control of the operon determining tryptophan biosynthesis (Fig. 5) *(12)*. In these cases, derepression of the tryptophan *(try)* or purine *(pur)* operons leads to a corresponding derepression of the remaining *lac* genes. In addition, the presence of purines or of tryptophan in the growth medium of the two types of strains results in strong repression of the synthesis of permease and transacetylase.

These results show that when a deletion removes the *lac* operator and a proximal segment of the *z* gene and fuses the *y* and *a* genes to an operon in the repressed state, the *y* and *a* genes are not expressed. Thus, it is not possible to restore *y* and *a* function to an inactivated *lac* operon merely by deleting the proximal end of the operon without regard to where the other end of the deletion lies. On the contrary, it seems likely that, to allow expression of the *y* and *a* genes, the deletions must connect these genes to another functioning gene or operon. Further evidence that this is the case comes from the failure to find, among a large number of O^c mutations, deletions covering part of the *z* gene which do not extend past the *i* gene. If it were possible to connect the *y* and *a* genes to any region of the DNA and allow function, one would have expected to find among the O^c-z^- deletions some in which the deletion either ended in the operator or in whatever space might exist between *i* and *o* (Fig. 4, class V). These results suggest that the *lac* operon has an associated site which is essential for expression and that removal of this site, without the substitution of a new one, inactivates the operon. Apparently, deletions of class V inactivate the operon, since they are not found. Therefore, this essential site, in the case of the *lac* operon, must lie between *o* and *z*. On the basis of this argument, Jacob, Ullman, and Monod *(38)* have proposed this location for the promoter site.

SECOND FUNCTION FOR THE PROMOTER

The studies on deletions of the *lac* operon also provide information on the site or sites which determine the rate of expression of the operon under conditions of full induction. All deletions of class III (Fig. 4) result in a marked reduction in the amount of expression of the remaining *y* and *a* genes. In most cases, the activities of permease and transacetylase in these strains are only 10 percent or less of the activities seen in fully induced wild-type strains *(38, 41, 43)*. Several explanations can account for these findings.

1) The *lac* operon may ordinarily be transcribed at a very fast rate compared to that of most other genes and operons, either because of repression of the other genes acting on transcription, or because of lower intrinsic potentials for transcription of the other genes. According to this proposal, the *y* and *a* genes are always likely to function at a lower rate when connected to a different gene or operon.

2) The *lac* operon mRNA is translated at a very fast rate compared to that of other genes or operons. This argument concerning the rate of translation is analogous to the first argument.

3) The deletions fusing *z* to some other gene create a sequence of nucleotides which results in polarity effects, thus reducing the expression of *y* and *a*. For example, the deletion may create a frame-shift in the operon or a nonsense mutation at the site of fusion. This explanation could account for low *y* and *a* in some cases, but it seems very unlikely that every deletion creates just the correct conditions for such an effect.

Another set of deletions was isolated which removed only part of the operator but again deleted the *i* gene (class II) *(38)*, thus connecting the *lac* operon, through its operator, to some nearby segment of the chromosome. Deletions of class II are far more frequent than those of class III. None of these class II deletions completely abolishes operator function, as measured by its sensitivity to repressor. In the presence of the wild-type allele of the *i* gene, these O^c mutants are still partly repressible. But, the striking property of these deletions, in contrast to those of class III, is that they all produce maximally exactly the same amounts (100 percent) of *lac* gene products as the wild-type strain. Thus, since extensive deletion of the operator does not reduce the maximal rate of operon expression, the operator *(44)* cannot be the site which sets the maximal potential for expression of the *lac* operon.

How can we explain the difference between deletions of classes II and III? Since we have concluded that a site *p* lying between *o* and *z* is essential for expression of the operon, we may now suggest that this site (or a site in the same region) is also involved in determining the maximum rate of this expression *(45)*. Deletions of class III remove this site and connect the *lac* operon to some other gene or operon with a promoter which functions at a lower rate. Deletions of class II, in contrast, do not delete the promoter, and so the site determining the maximum rate of expression of the operon is left fully functional. In addition, there are no sites at the distal end of the operon determining the maximum level of operon expression, since class IVa and IVb deletions have no effect on the expression of the intact proximal genes *(12)*.

REGULATION OF OPERON TRANSCRIPTION

The information obtained from study of these deletions is critical in analyzing the various mechanisms proposed for regulation and expression of the operon. These arguments can be reduced to the question of whether or not the operator (or some site previous to it) is the starting point of mRNA synthesis. I now consider these two possibilities for operon regulation and their implications.

1) The operator, in addition to carrying the information determining the sensitivity to repressor, is also the site of initiation of mRNA synthesis for the operon. Since we have concluded that the operator does not set the maximum level of operon expression, it follows from this model that the rate of initiation of mRNA synthesis is not the rate-limiting factor in the setting of this maximum. In other words, all genes or segments of the DNA to which the operon has become attached by deletions of class III are transcribed at the same rate as the *lac* operon is normally (when fully induced). The variation in maximal levels of gene expression is always a result of different rates of translation. An implication of this model is that *E. coli* has not evolved a mechanism for setting different maximal rates of RNA synthesis for different genes and operons. Although there is no strong evidence against this possibility, it is rather unattractive.

2) The only function of the operator is that of being the repressor-sensitive site. The promoter site is the initiation point for mRNA synthesis, and its structure determines how much transcription takes place in the absence of repression. According to this model, in contrast to the first possibility, different rates of mRNA synthesis can be set by different sequences of nucleotides (promoters) on the DNA.

In view of the implications of model 1, I favor this latter picture of transcription of the *lac* operon. Since the promoter would be the site of initiation of mRNA synthesis, according to this model, the operator cannot be transcribed as part of the operon mRNA. Therefore, the operon mRNA does not contain a repressor-recognition site. Then, in order to affect the expression of the *lac* structural genes, the repressor must recognize the operator on the chromosome. If repression does take place on the chromosome, the repressor could either directly repress the initiation of synthesis of operon mRNA or interact with the mRNA-DNA complex (or even a ribosome-RNA-DNA complex) to prevent further synthesis of operon mRNA.

It is very likely that a promoter region plays some part in determining the activity of the *lac* operon. It should be possible, therefore, to isolate mutants of the promoter which alter this maximal level. Mutants have been found which reduce the maximum expression of the *lac* operon and which have other properties expected of promoter mutants *(45)*. These are mutants which, when fully induced, make only 5 to 10 percent of the normal fully induced levels of all three gene products. The mutants are not connected with the *i-o* control system, do not appear to be polar mutants, and map in the same region as the promoter.

THE REPRESSOR

The existence of two classes of mutations affecting *lac* regulation and defining the *i* locus, led to the concept of the *i* gene product as a repressor molecule which interacts both with the operator and the inducer *(1)*. Strains carrying i^- mutations which abolish activity of the *i* gene result in a maximal synthesis of the *lac* proteins, even in the absence of inducer. Moreover, in partial diploid strains carrying both an i^- mutation and the wild-type i^+ allele of the *i* gene, the operon becomes repressible *(1, 46)*. Thus, the *i* gene appears to be the structural gene for a diffusible product which is responsible for the repression of the operon. The i^s (super-repressed) mutations, in contrast, result in noninducibility of the *lac* operon *(24, 47)*, in spite of the presence of unaltered operator, promoter, and structural genes. These mutants are thought to produce a repressor which has no affinity for inducer, so that the operon is permanently shut off. However, as pointed out by Brenner *(5)*, on the basis of only these two classes of mutations one can devise somewhat more complex pictures of *i* gene action; for example, the *i* gene product is not itself the repressor but is an enzyme catalyzing the synthesis of a small molecule which is part of the repressor. However, if there were another molecule involved in repression, one would have expected to find mutations in another locus resulting in constitutivity. No such mutants have been found. Thus, the original Jacob-Monod idea of *i* gene action is still the simplest.

There is now both genetic and chemical evidence that the *i* gene ultimately directs the synthesis of a protein molecule. The genetic evidence is that there exist suppressible amber mutants (i^-) of the *i* gene *(39, 48)*. Since it is known that these mutants affect translation, the *i* gene messenger must be translated into protein. Recently, Gilbert and Muller-Hill *(49)* have isolated the *i* gene product and have shown that it is at least partially composed of protein. The isolation was achieved by purification of a fraction of *E. coli* protein which binds the inducer IPTG (isopropyl-β-D-thiogalactoside) with the expected affinity constant. The identification of this protein as the *i* gene product was established first by the demonstration that this protein could not be detected by this technique in i^-, i^s and *i*-deletion strains; more importantly, the affinity constant of this protein for IPTG was altered in a strain carrying an *i* mutation leading to an increased efficiency of IPTG as an inducer.

OLD THEORIES AND FUTURE EXPERIMENTS

I have discussed the genetic work on the *lac* operon and its implications in terms of the Jacob-Monod operon hypothesis. As mentioned earlier, several alternative models of control have been proposed. In addition to the Jacob-Monod model, in which the operator is not transcribed into the operon mRNA, the following possibilities have been suggested. The first possibility is a model in which the operator is transcribed into the operon mRNA, but is not translated. For example, (i) the repressor binds to the mRNA copy of the operator, inhibiting the initiation of protein synthesis (at the promoter?), which, in turn, is necessary for continuing synthesis and release of the mRNA from the DNA. In this model, repression during translation inhibits the transcription process *(2, 3)*. (ii) The operon mRNA is synthesized in constant amounts, but the repressor inhibits initiation of protein synthesis and thus causes a very rapid destruction of the mRNA *(15, 31)*. (iii) The operon mRNA is made in constant amounts, but the repressor is a ribonuclease which destroys the mRNA *(5, 31)*.

The second possibility includes models in which the operator is transcribed into the operon mRNA and then must be translated in order to allow translation of the structural genes of the operon *(2-4)*. Again, the failure to translate may result either in the mRNA's sticking to the DNA or in rapid destruction of the mRNA.

All of these models take into account the experiments of Attardi *et al.* *(28)* which show that repression results in a disappearance of *lac* mRNA from the cell. This finding, like the similar finding with the strong polar mutants, must be considered in formulating a model for *lac* operon regulation, but it does not severely limit the number of possibilities.

The thesis of this article has been that the current knowledge of the *lac* operon suggests that the operator is not transcribed into the operon mRNA and that, therefore, the Jacob-Monod suggestion for operon control is most likely to be correct for this system. Some of the other models listed are difficult to reconcile with some of the evidence discussed. One of the most striking of the recent findings is that most and probably all O^c's are deletions but still retain some sensitivity to repressor. Although this finding makes any picture of the repressor-operator interaction somewhat difficult to visualize, it makes particularly unattractive models in which the operator is translated into protein or is the initiation site for operon protein synthesis. One would expect deletions of the operator to have drastic effects on operon functioning in such models, and this does not appear to be so.

Final proof of one model or another will probably have to come from biochemical experiments on operon functioning. It is possible to set up a system in vitro in which *lac* mRNA is made, with RNA polymerase and DNA preparations that contain a high proportion of *lac* genes *(50)*. In such experiments, other mRNA species are made also, but these can be eliminated by annealing with DNA preparations from appropriate strains. Then, the amount of *lac* mRNA made can be estimated. Using this system, Gilbert and Muller-Hill *(51)* are attempting to ascertain whether the *lac* repressor protein will inhibit synthesis of *lac* mRNA. In the same system, we are attempting to see whether the potential promoter mutants affect *lac* mRNA synthesis in vitro.

OTHER REGULATORY SYSTEMS

Of the regulatory systems that have been studied in detail, the *lac* operon appears to be one of the simplest. Although the operon model can account for all the information concerning *lac*, in some other systems the control is clearly more complex, and even the basic control mechanism may be entirely different *(52)*. However, in none of these cases is there yet any strong evidence against the operon model.

REFERENCES AND NOTES

1. F. Jacob and J. Monod, J. Mol. Biol. **3**:318 (1961).
2. B. N. Ames and P. E. Hartman, Cold Spring Harbor Symp. Quant. Biol. **28**:349 (1963).
3. G. Stent, Science **144**:816 (1964).
4. W. Maas and E. McFall, Ann. Rev. Microbiol. **18**:95 (1964).
5. S. Brenner, Brit. Med. Bull. **21**:244 (1965).
6. Both Jacob [Science **152**:1470 (1966)] and Monod [*ibid.* **154**:475 (1966)] have in their Nobel laureate lectures discussed many historical and recent aspects of the *lac* operon and its regulation.
7. C. F. Fox and E. P. Kennedy, Proc. Nat. Acad. Sci. U.S. **54**:891 (1965).
8. H. V. Rickenberg, G. N. Cohen, G. Buttin, J. Monod, Ann. Inst. Pasteur **91**:829 (1956).
9. I. Zabin, A. Kepes, J. Monod, J. Biol. Chem. **237**:253 (1962).
10. C. F. Fox, J. R. Beckwith, W. Epstein, E. R. Signer, J. Mol. Biol. **19**:576 (1966).
11. F. Jacob and J. Monod, Biochem. Biophys. Res. Commun. **18**:693 (1965).
12. J. R. Beckwith, E. R. Signer, W. Epstein, Cold Spring Harbor Symp. Quant. Biol. **31**:393 (1967).
13. W. Gilbert, J. Mol. Biol. **6**:374 (1963); A. Gierer, *ibid.*, p. 148; J. R. Warner, P. Knopf, A. Rich, Proc. Nat. Acad. Sci. U.S. **49**:122 (1963); F. O. Wettstein, T. Staehelin, H. Noll, Nature **197**:430 (1963); H. M. Dintzis, Proc. Nat. Acad. Sci. U.S. **48**:247 (1961).
14. S. Benzer and S. P. Champe, Proc. Nat. Acad. Sci. U.S. **48**:1114 (1962); A. Garen and O. Siddiqi, *ibid.*, p. 1121; A. S. Sarabhai, A. O. W. Stretton, S. Brenner, A. Bolle, Nature **201**:13 (1964); S. Brenner, A. O. W. Stretton, S. Kaplan, *ibid.* **206**:994 (1965); M. G. Weigert and A. Garen, *ibid.*, p. 992.
15. S. Brenner and J. R. Beckwith, J. Mol. Biol. **13**:629 (1965).
16. F. H. C. Crick, L. Barnett, S. Brenner, R. J. Watts-Tobin, Nature **192**:1227 (1961).
17. Y. Kiho and A. Rich, Proc. Nat. Acad. Sci. U.S. **54**:1751 (1965).
18. R. E. Thach, M. A. Cecere, T. A. Sundarajan, P. Doty, *ibid.*, p. 1167; M. Salas, M. A. Smith, W. M. Stanley, A. J. Wahba, S. Ochoa, J. Biol. Chem. **240**:3988 (1965); G. Streisinger, Y. Okada, E. Terzaghi, J. Emrich, A. Tsugita, M. Inouye, Cold Spring Harbor Symp. Quant. Biol., in press; A. Goldstein, J. Kirschbaum, A. Roman, Proc. Nat. Acad. Sci. U.S. **54**:1669 (1965); V. Maitra and J. Hurwitz, *ibid.*, p. 815; H. Bremer, M. W. Konrad, K. Gaines, G. S. Stent, J. Mol. Biol. **13**:540 (1965); J. R. Guest and C. Yanofsky, Nature **210**:799 (1966).
19. A. V. Fowler and I. Zabin, Science **154**:1027 (1966).
20. Studies on complementation with z^- mutants suggest the possibility that the *z* gene is composed of three or more genes specifying distinct polypeptide chains [A. Ullman, D. Perrin, F. Jacob, J. Monod, J. Mol. Biol. **12**:918 (1965)]. Although this question has not been conclusively resolved, most of the evidence indicates that the *z* gene directs the synthesis of only *one* polypeptide chain [J. L. Brown, S. Koorajian, J. Katze, I. Zabin, J. Biol. Chem. **241**:2826 (1966); G. Craven, personal communication; D. Fan, personal communication].
21. In this paper, the proximal end of a gene or of the operon refers to that end which is closest to the operator. A distal end is that farthest from the operator.
22. C. Yanofsky, B. C. Carlton, J. R. Guest, D. R. Helinski, U. Henning, Proc. Nat. Acad. Sci. U.S. **51**:266 (1964).
23. W. A. Newton, J. R. Beckwith, D. Zipser, S. Brenner, J. Mol. Biol. **14**:290 (1965).
24. F. Jacob and J. Monod, Cold Spring Harbor Symp. Quant. Biol. **26**:193 (1961).
25. N. C. Franklin and S. E. Luria, Virology **15**:299 (1961).
26. W. A. Newton, Cold Spring Harbor Symp. Quant. Biol. **31**:181 (1967).
27. D. Zipser and W. A. Newton, J. Mol. Biol., in press.
28. G. S. Attardi, S. Naono, J. Rouviere, F. Jacob, F. Gros, Cold Spring Harbor Symp. Quant. Biol. **28**:363 (1963); G. Contesse, S. Naono, F. Gros, Compt. Rend. **263**:1007 (1966).
29. M. Malamy, Cold Spring Harbor Symp. Quant. Biol. **31**:89 (1967).

30. R. G. Martin, D. F. Silbert, D. W. E. Smith, H. J. Whitfield, J. Mol. Biol. **21**:357 (1966).
31. J. R. Beckwith, Abhandl. Deutsch. Akad. Wiss. Berlin Kl. Med. **4**:119 (1964).
32. M. R. Capecchi, J. Mol. Biol. **21**:173 (1966); G. N. Gussin, *ibid.*, p. 435.
33. C. Yanofsky and J. Ito, *ibid.*, p. 313.
34. D. H. Alpers and G. M. Tomkins, Proc. Nat. Acad. Sci. U.S. **53**:797 (1965).
35. A. Kepes, in preparation.
36. D. H. Alpers and G. M. Tomkins, J. Biol. Chem. **241**:4434 (1966).
37. I. Zabin, Cold Spring Harbor Symp. Quant. Biol. **28**:431 (1963); W. Epstein, personal communication; D. Perrin, personal communication.
38. F. Jacob, A. Ullman, J. Monod, Compt. Rend. **258**:3125 (1964).
39. S. Bourgeois, M. Cohn, L. E. Orgel, J. Mol. Biol. **14**:300 (1965); S. Bourgeois, personal communication.
40. T. Ramakrishnan and E. A. Adelberg, J. Bacteriol. **87**:566 (1964).
41. J. R. Beckwith, J. Mol. Biol. **8**:427 (1964).
42. E. Steers, G. R. Craven, C. B. Anfinsen, Proc. Nat. Acad. Sci. U.S. **54**:1174 (1965).
43. F. Jacob, A. Ullman, J. Monod, J. Mol. Biol. **13**:704 (1965).
44. One might consider that since no 100-percent O^c's have been found, a right-hand extremity of the operator is involved in determining the rate of operon expression. This argument is not very different from the one discussed in that rather than one site with two distinguishable regions there are two sites.
45. J. Scaife and J. R. Beckwith, Cold Spring Harbor Symp. Quant. Biol. **31**:403 (1967).
46. A. B. Pardee, F. Jacob, J. Monod, J. Mol. Biol. **1**:165 (1959).
47. C. Willson, D. Perrin, M. Cohn, F. Jacob, J. Monod, *ibid.*, **8**:582 (1964).
48. B. Muller-Hill, *ibid.* **15**:374 (1966).
49. W. Gilbert and B. Muller-Hill, Proc. Nat. Acad. Sci. U.S. **56**:1891 (1966).
50. High-frequency transducing lysates for the *lac* genes have been isolated with the restricted transducing phage $\phi 80$ [J. R. Beckwith and E. R. Signer, J. Mol. Biol. **19**:254 (1966)]. Isolation of DNA from phage lysates results in a DNA preparation which contains genetic material, the *lac* region, and a small portion of the chromosome surrounding *lac.*
51. W. Gilbert and B. Muller-Hill, personal communication.
52. E. Engelsberg, J. Irr, J. Power, N. Lee, J. Bacteriol. **90**:946 (1965) (arabinose metabolism); H. Echols, A. Garen, S. Garen, A. Torriani, J. Mol. Biol. **3**:425 (1961); A. Garen and N. Otsuji, *ibid.* **8**:841 (1964) (alkaline phosphatase); M. Schwartz, personal communication (maltose metabolism).
53. E. R. Signer, unpublished data.

chapter 10

Genes and metabolism

Although the immediate functions of a gene appear to be limited to replication and transcription, the net effect of these actions is gene control of the organism's metabolism. This results from the fact that transcription leads to translation, which yields both structural and enzymatic proteins. These in turn determine cell metabolism. An understanding of the genetic control of metabolism rests on a study of the factors that affect transcription and translation and of the consequences to the organism of changes in the types and amounts of different proteins produced.

FACTORS AFFECTING TRANSCRIPTION

In transcription, RNA-polymerase catalyzes the synthesis of an mRNA from a DNA template. An agent that affects transcription is *actinomycin D.* This antibiotic preferentially inhibits the RNA polymerase reaction only when DNA containing guanine is involved. No inhibition occurs when a synthetic DNA polymer that contains only adenine, thymine, and cytosine is used. A report by Trakatellis and his co-workers in 1964 (Ref. 10-1) illustrates the effectiveness of various amounts of actinomycin D per 100 grams of mouse body weight on RNA synthesis.

Actinomycin D does not represent a normally occurring compound for limiting transcription. However, a group of naturally occurring transcription inhibitors are proteins called *histones.* Histones are localized in the nuclei of higher organisms where they are complexed with DNA, forming the chromosomes of the individual. The strong interaction of histones with DNA led Stedman and Stedman in 1950 (Ref. 10-2) to hypothesize that the histones of the chromosomes in higher organisms play a role in controlling gene action. An experimental demonstration of the role of histones in transcription was provided in 1963 by Bonner and his co-workers (Ref. 10-3), whose paper is reprinted in this chapter.

FACTORS AFFECTING TRANSLATION

Translation is the process by which amino acids are assembled into a polypeptide chain in an order specified by the sequence of codons in an mRNA. It is to be anticipated that environmental factors will affect the translation process. In addition, however, factors affecting the translation processes could include the amino acids themselves, the tRNA's involved in the transportation of the amino acids, and the ribosomes on which the polypeptide chains are formed.

The possible role of *amino acids* in affecting translation has been examined in a number of studies. A paper that reviews the earlier work and reports additional information on this topic was published by Grunberger and his co-workers in 1969 (Ref. 10-4) and is included in this collection.

In considering the role of *tRNA* in translation, one must take into account the fact that tRNA has two functions. The first function is to "recognize" the proper amino acid, and the second function is to "recognize" the proper codon on the mRNA being translated. Each of these functions requires a separate recognition site somewhere on the tRNA molecule, and each recognition site is subject to alteration by mutation in that portion of the DNA molecule from which the tRNA is transcribed. A mutation that results in a wrong amino acid being incorporated into a protein is called a *missense* mutation. Although a missense mutation causes the production of a different protein from that specified by the DNA, the mutation does not, strictly speaking, result in an alteration of the translation process. When such a mutation leads to the formation of a functional enzyme (wild type) in a mutant strain that normally produces an inactive protein, the mutation is called a *missense suppressor.* Experimental evidence was provided by Gupta and Khorana in 1966 (Ref. 10-5) that the incorporation of glycine instead of cysteine into a cell-free amino acid–incorporating system prepared from *Escherichia coli* was due to a change in the amino acid–recognizing function of tRNA. A study of a missense suppressor that had an altered mRNA codon-recognition site was reported by Carbon and his co-workers in 1966 (Ref. 10-6).

A mutation in which the resulting triplet does

not code for any amino acid is called a *nonsense* mutation. This type of mutation directly affects the translation process, since it results in the premature termination of the nascent polypeptide chain. There are three nonsense triplets: UAG or amber, UAA or ochre, and UGA. Several mutants are known in *E. coli* that read UAG as an amino acid codon and thus allow continued translation of the message. These mutants are called *nonsense suppressors.* The mutations involved must affect the mRNA codon-recognition site of the tRNA molecule. An example of this type of mutation was reported by Capecchi and Gussin in 1965 (Ref. 10-7).

Another factor that could affect the translation process would be a change in the *ribosome* that would result in a misreading of an mRNA codon. An example of this was reported by Davies and his co-workers in 1964 (Ref. 10-8) and is included in this chapter.

GENE END PRODUCTS

As mentioned earlier in this chapter, the end products of genes are either structural or enzymatic proteins, and it is through the production of these types of proteins that genes control metabolism. An excellent example of the genetic control of the production of a structural protein can be seen in *hemoglobin,* which is responsible for the oxygen-carrying capacity of the individual's blood. Any reduction in the amount of oxygen available will severely affect all biochemical reactions in the cells of an organism and may even prove fatal. Mutations affecting hemoglobin production in man may involve a substitution of one amino acid by another (e.g., sickle cell anemia) or may involve the production of a reduced number of hemoglobin molecules per erythrocyte (e.g., thalassemia).

In the case of sickle cell anemia, evidence was presented by Neel in 1949 (Ref. 10-9), indicating that the disease occurred in individuals who were homozygous for a recessive autosomal gene. This paper was followed by a demonstration by Pauling and his co-workers in 1949 (Ref. 10-10) that sickle cell hemoglobin differs in electrophoretic mobility from normal hemoglobin. A chemical analysis of the globins of normal and sickle cell anemia hemoglobin by Ingram in 1957 (Ref. 10-11) showed that out of nearly 300 amino acids in the alpha and beta chains of the two proteins, only one is different. One of the glutamic acid residues in the beta chains of normal hemoglobin is replaced by a valine residue in sickle cell anemia hemoglobin.

With regard to thalassemia, a paper by Weatherall in 1969 (Ref. 10-12), which reviews our knowledge of the various forms of the disease and their possible etiologies, is reprinted in this chapter.

A general review of the role of genes in human hemoglobin production was published by Baglioni in 1963 (Ref. 10-13).

There have been many investigations of the genetic control of enzyme production. An example of these involves *tryptophan synthetase,* one of the enzymes that mediate the production of the amino acid tryptophan. Tryptophan synthetase has been studied mostly in *E. coli.* The enzyme is composed of two protein subunits, called components A and B. Mutants have been discovered that produce inactive forms of one or the other of these protein subunits. Yanofsky and his co-workers reported in 1961 (Ref. 10-14) on the effects of mutation on the composition of the A subunit of the enzyme. Further research resulted in a paper by Yanofsky and his co-workers in 1967 (Ref. 10-15), which is included in this chapter.

METABOLIC PATHWAYS AND ENZYME BLOCKS

Gene end products, namely proteins, control cell metabolism through the role they play in various biochemical pathways. Each chemical reaction in a given series is mediated by an enzyme. The effectiveness of the enzyme will determine whether the metabolic pathway will be completed normally, shunted into some alternate pathway, or blocked completely. One of the better-studied metabolic pathways is that of phenylalanine and tyrosine in man. Human beings cannot manufacture phenylalanine or tyrosine, and a complete lack of these amino acids in the diet will result in the death of the individual. As will be seen in the figure on p. 308, phenylalanine and tyrosine are involved in a widely diverging series of biochemical reactions.

An example of a genetically controlled completely-blocked metabolic pathway, that occurs in some human beings, was reported by Garrod in 1902 (Ref. 3-3). The disease was *alcaptonuria,* which is caused by a defect in a liver enzyme, homogentisic acid oxidase. When this enzyme is defective, large amounts of homogentisic acid are excreted in the urine, which turns black on exposure to the air. A

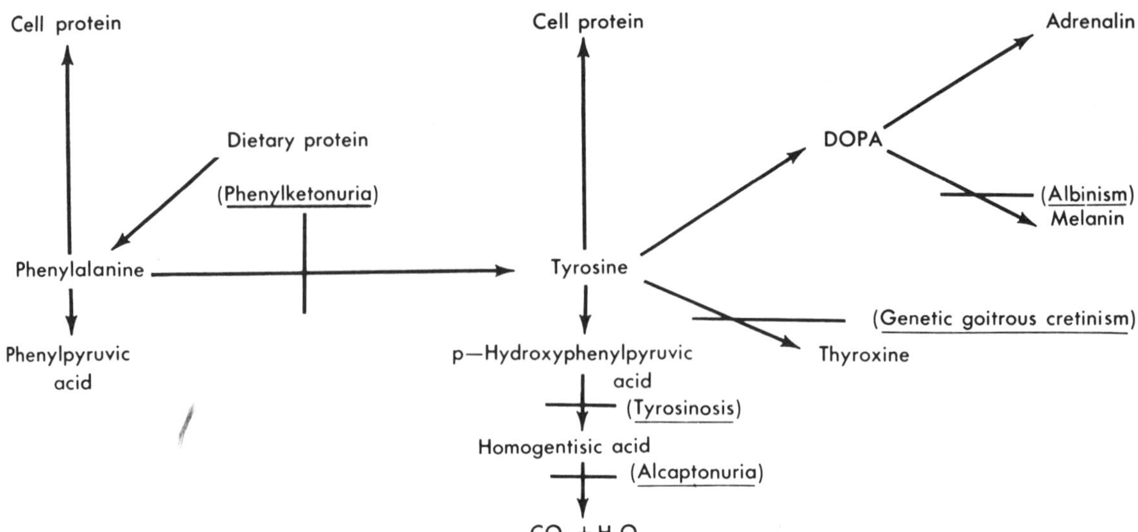

Phenylalanine and tyrosine metabolism in man. (From Levine, L.: Biology of the gene, St. Louis, 1969, The C. V. Mosby Co.)

review of this and other heritable biochemical diseases was published by Garrod in 1909 (Ref. 10-16). A later discovery of a human genetic disease, involving an alternate pathway of the phenylalanine-tyrosine sequence, was made by Følling in 1934 and reviewed by Følling and his co-workers in 1944 (Ref. 10-17). The disease was *phenylketonuria,* which is caused by a defect in a liver enzyme, phenylalanine hydroxylase. When this enzyme is defective, phenylalanine is not oxidized to tyrosine but is converted to phenylpyruvic acid. People with this enzyme deficiency are feebleminded. Unfortunately, although the relationship of enzyme and disease was generally acknowledged in the cases of alcaptonuria and phenylketonuria, it was not realized that the genes caused the diseases by directing the production of defective enzymes. The effective uniting of biochemistry and genetics was accomplished by an entirely different route.

Early experiments that led to an understanding of gene-enzyme relationships involved eye-color mutants in *Drosophila.* Normal *Drosophila* eye color is red and is produced by a combination of brown and red pigments. Some mutations cause the absence of the normal brown pigment, and the resulting eye color is bright red (e.g. vermilion, cinnabar). It was reported by Beadle and Ephrussi in 1937 (Ref. 10-18) that transplants of eye tissue from vermilion into cinnabar larvae resulted in the transplants' developing wild-type eye color. However, transplants of eye tissue from cinnabar into vermilion larvae resulted in the transplants' developing cinnabar eye color. It was hypothesized that vermilion and cinnabar represented two successive steps in the production of brown eye pigments. Under this hypothesis, host cinnabar larvae are able to produce the compound that the wild-type allele of vermilion would form. This compound would then diffuse into the cells of the vermilion eye-tissue transplant. Since all other genes in the vermilion eye-tissue transplant are wild type, the cells of the transplant can metabolize the compound supplied by the cinnabar host tissue, along the normal brown-pigment pathway. However, in the reverse transplantation, the host vermilion larvae do not produce any intermediate compound that the cinnabar eye-tissue transplant can metabolize for the production of brown pigment. It was evident from these results that the eye-color mutations had resulted in enzyme blocks at specific points in a synthetic pathway.

The work on *Drosophila* stimulated further research on biochemical genetics. However, it remained for experiments on the mold *Neurospora crassa* to put the study of biochemical genetics on a firm basis. This was accomplished in 1941 by Beadle and Tatum (Ref. 10-19), whose paper is included in this chapter.

DRUG METABOLISM IN MAN

An environmental factor of ever-increasing importance in affecting human metabolic pathways has been introduced by man himself. This is the widespread use of therapeutic drugs in modern medicine. A large number of discoveries have been made of individuals whose metabolism is adversely affected by the drugs. In many instances the sensitivity to the administered drug has been found to be inherited. From studies of these situations has developed the field of *pharmacogenetics.*

One example of an adverse drug reaction that is genetically determined involves *primaquine,* a compound which has been used in the treatment of malaria. It was reported by Cordes in 1926 (Ref. 10-20) that some individuals who took primaquine exhibited hemolytic anemia. It was later demonstrated by Carson and co-workers in 1956 (Ref. 10-21) that primaquine-sensitive individuals have a deficiency of the enzyme glucose-6-phosphate dehydrogenase (G-6-PD). A deficiency of G-6-PD results in an abnormality in the direct oxidation of glucose in the erythrocytes of affected individuals. The relationship of reduced glycolysis efficiency and erythrocyte fragility has not been satisfactorily explained. Primaquine sensitivity, hence G-6-PD deficiency, is inherited as a sex-linked recessive trait. It has also been shown, as reviewed by Tarlov and his coworkers in 1962 (Ref. 10-22), that a hemolytic crisis in persons with G-6-PD deficiency follows the ingestion of fava beans, naphthalene mothballs (i.e., accidentally by children), sulfanilamide, sulfoxone (antileprosy drug), nitrofuratoin (urinary antiseptic), and others. G-6-PD deficiency thus appears to be the basis for a number of drug-induced hemolytic anemias. In the absence of these drugs, the affected individuals appear to have normal health.

A sometimes fatal drug reaction has come to light with the introduction of barbiturates in modern medicine and particularly with their use in anesthesia. The reaction occurs in people who suffer from a hereditary disease called *porphyria,* which is caused by a dominant gene. These people exhibit an excess formation and liberation of porphyrins in the body. A review of this phenomenon was published in 1969 by Dean (Ref. 10-23), whose paper is the last article reprinted in this chapter.

A book that treats extensively the field of pharmacogenetics was published by Kalow in 1962 (Ref. 10-24).

REFERENCES

Factors affecting transcription

10-1. Trakatellis, A. C., Axelrod, A. E., and Montjar, M. 1964. Actinomycin *D* and messenger-RNA turnover. Nature **203:**1134-1136.

10-2. Stedman, E., and Stedman, E. 1950. Cell specificity of histones. Nature **166:**780-781.

10-3. Bonner, J., Huang, R. C., and Gilden, R. V. 1963. Chromosomally directed protein synthesis. Proc. Natl. Acad. Sci. U. S. A. **50:**893-900.

Factors affecting translation

10-4. Grunberger, D., Weinstein, I. B., and Jacobson, K. B. 1969. Codon recognition by enzymatically mischarged valine transfer ribonucleic acid. Science **166:**1635-1637.

10-5. Gupta, N. K., and Khorana, H. G. 1966. Missense suppression of the tryptophan synthetase A-protein mutant A78. Proc. Natl. Acad. Sci. U. S. A. **56:**772-779.

10-6. Carbon, J., Berg, P., and Yanofsky, C. 1966. Studies of missense suppression of the tryptophan synthetase A-protein mutant A-36. Proc. Natl. Acad. Sci. U. S. A. **56:**764-771.

10-7. Capecchi, M. R., and Gussin, G. N. 1965. Suppression in vitro: Identification of a serine sRNA as a "nonsense" suppressor. Science **149:**417-422.

10-8. Davies, J., Gilbert, W., and Gorini, L. 1964. Streptomycin, suppression, and the code. Proc. Natl. Acad. Sci. U. S. A. **51:**883-890.

Gene end products

10-9. Neel, J. V. 1949. The inheritance of sickle cell anemia. Science **110:**64-66.

10-10. Pauling, L., Itano, H. A., Singer, S. J., and Wells, I. C. 1949. Sickle cell anemia, a molecular disease. Science **110:**543-548.

10-11. Ingram, V. M. 1957. Gene mutations in normal haemoglobin: the chemical difference between normal and sickle cell haemoglobin. Nature **180:**326-328.

10-12. Weatherall, D. J. 1969. The genetics of the thalassaemias. Brit. Med. Bull. **25:**24-29.

10-13. Baglioni, C. 1963. Correlations between genetics and chemistry of human hemoglobins. pp. 405-475. *In* J. H. Taylor [ed.] Molecular genetics. Part I. Academic Press, Inc., New York.

10-14. Yanofsky, C., Helinski, D. R., and Maling, B. D. 1961. The effects of mutation on the composition of the A protein of *E. coli* tryptophan synthetase. Cold Spring Harbor Symp. Quant. Biol. **26:**11-24.

10-15. Yanofsky, C., Drapeau, G. R., Guest, J. R., and Carlton, B. C. 1967. The complete amino acid sequence of the tryptophan synthetase A protein (α subunit) and its colinear relationship with the genetic map of the A gene. Proc. Natl. Acad. Sci. U. S. A. **57:**296-298.

Metabolic pathways and enzyme blocks

10-16. Garrod, A. E. 1909. Inborn errors of metabolism. Oxford University Press, Inc., New York.

10-17. Føolling, A., Mohr, O. L., and Rund, L. 1944. Oligophrenia phenylpyrouvica, a recessive syndrome in man. Vid.-Akad. Skr. I. Mat.-Naturv. Klasse. **13:**5-44.

10-18. Beadle, G. W., and Ephrussi, B. 1937. Development of eye colors in *Drosophila;* diffusable substances and their interrelations. Genetics **22:**76-86.

10-19. Beadle, G. W., and Tatum, E. L. 1941. Genetic

control of biochemical reactions in *Neurospora.* Proc. Natl. Acad. Sci. U. S. A. **27**:499-506.

Drug metabolism in man

10-20. Cordes, W. 1926. Experiences with plasmochin in malaria (preliminary reports). Ann. Rept. 15th, United Fruit Company, Med. Dept., pp. 66-71.

10-21. Carson, P. E., Flanagan, C. L., Ickes, C. E., and Alving, A. S. 1956. Enzymatic deficiency in primaquine sensitive erythrocytes. Science **124**:484-485.

10-22. Tarlov, A. R., Brewer, G. J., Carson, P. E., and Alving, A. S. 1962. Primaquine sensitivity. Glucose-6-phosphate dehydrogenase deficiency: An inborn error of metabolism of medical and biological significance. Arch. Int. Med. **109**:209-234.

10-23. Dean, G. 1969. The porphyrias. Brit. Med. Bull. **25**:48-51.

10-24. Kalow, W. 1962. Pharmacogenetics. Heredity and the response to drugs. W. B. Saunders Co., Philadelphia.

46 Chromosomally directed protein synthesis

James Bonner
Ru-chih C. Huang
Ray V. Gilden*
Division of Biology
California Institute of Biology
Pasadena, California

The different kinds of specialized cells of a higher organism differ among themselves in the kinds of enzymes and other proteins which they contain. Nonetheless, each possesses in the chromosomes of its nucleus the complete genomal DNA containing information concerning the manufacture of all of the kinds of enzymes contained in all of the kinds of cells of that creature. It is clear, therefore, that there is in the nucleus some mechanism for the control of genetic activity, a mechanism responsible for that orderly repression and derepression of gene activity which makes possible development. The present work concerns the control of genetic activity.

Genetic activity consists in the direction of DNA-dependent RNA synthesis, the RNA thus formed containing in coded form the message appropriate for the ribosomal synthesis of a particular protein. For the study of the control of genetic activity, we first establish an *in vitro* system for the conduct of chromosome-dependent RNA synthesis. We then couple our chromosomal RNA-generating system to a messenger RNA-dependent ribosomal protein synthesis system.

With our coupled system established we next focus attention upon a particular gene and its product-protein, in our case the gene of the pea plant responsible for the production of the reserve globulin of pea cotyledons. This protein is synthesized in developing pea cotyledons, and is not synthesized by such other pea plant tissues as buds or roots. It will be shown that a specific protein, pea seed globulin, is synthesized by our coupled system in response to the presence of cotyledon chromatin, and is not synthesized in response to the presence of pea bud chromatin. Thus, the control of genetic activity characteristic of the living cell is preserved in the isolated chromatin. It will be further shown that such control is exerted by the histone component of the chromosome.

MATERIALS AND METHODS

Preparation of chromatin. Chromatin was prepared by the methods of Huang and Bonner.[1] The organ or tissue (in this work, of pea plants, var. Alaska is first ground in a blendor, in a grinding medium composed of sucrose, 0.25 *M;* tris, pH 8, 0.05 *M;* $MgCl_2$ 0.001 *M.* The homogenate after filtration through cheesecloth and miracloth is centrifuged at 4,000 × *g* for 30 min and the supernatant discarded. The crude chromatin, which contains 90-95 per cent of the tissue DNA, is scraped from the underlying starch pellet, resuspended in 0.05 *M* tris, pH 8, and repelleted for 15 min at 10,000 × *g.* This step is repeated twice. The crude chromatin is then layered on 1.9 *M* sucrose and centrifuged at 22,000 rpm for 120 min in the SW-25 Spinco head. The pellet, resuspended in dilute saline citrate (0.016 *M*), constitutes the purified chromatin which has been freed of contaminating protein by this procedure. In the case of chromatin of pea buds, we have found it advantageous to use the crude chromatin dialyzed against dilute saline citrate without further purification.

RNA polymerase. RNA polymerase was prepared from log phase cells of *E. coli* strain B by the method of Chamberlin and Berg[2] and purification of the enzyme carried through ammonium sulfate fractionation to their fraction 3. The freshly made enzyme at this stage possessed a specific activity of approximately 160 mμm AMP incorporated into RNA/10 min/mg enzyme protein, a dependency ratio (activity in presence of added DNA/activity in absence of added DNA) of 40- to 100-fold and a 280/260 ratio of 1.2 to 1.5.

Preparation of ribosomes. The *E. coli* ribosomal system was prepared from cells ground with alumina according to the procedure of Nirenberg and Matthaei.[3] The resulting homogenate was centrifuged at 30,000 × *g* for 60 min and the supernatant recentrifuged for a further 20 min. This supernatant was then incubated for 45 min at 37°C in the presence of a

From Proceedings of the National Academy of Sciences **50:**893-900, 1963. Used with permission.

Report of work supported by the U.S. Public Health Service, grants RG-3977, RG-5143, and A-3102.

We wish to acknowledge the fine technical assistance of Stella Tan, Ludia Brown, and Abbas Soleimani, as well as the continuing helpful counsel of Professor Paul O. P. Ts'o.

*Present address: Wistar Institute, Philadelphia, Pa.

complete protein synthesis reaction mixture, including 19 amino acids, ATP, and phosphoenolpyruvate and its kinase. Ribosomes were then pelleted from the preincubation mixture, resuspended, repelleted, and finally resuspended and exhaustively dialyzed. The supernatant from the initial ribosomal pellet, containing activating and other soluble enzymes, as well as *E. coli* transfer RNA, was exhaustively dialyzed and then freed of *E. coli* DNA and its RNA polymerase by 18 hr of centrifugation at 105,000 × *g*. We have found the optimum composition of the ribosomal system to consist of, per 50 μg ribosomal protein, 70 μg protein of supernatant fraction supported by 20 mμm of RNA generated by the *E. coli* polymerase system.

Analytical procedures. Analysis of RNA, DNA, and protein were done by the methods standard to our laboratory.[1,4] Determination of radioactivity incorporated into RNA or protein was accomplished by 5 × repeated washing of the sample with cold 10% TCA, after which it was dissolved in 0.2 *N* NH_4OH, and pipetted on to and dried on a glass planchet for counting.

Immunochemical detection of pea seed globulin. Authentic pea seed globulin was prepared from mature pea seeds, var. Alaska, by the method of Danielsson.[5] Potent antisera were developed in chickens by repeated immunization with the globulin over a period of six months. A single lot of pooled antiserum was used for all of the work here reported. The amount of antiserum required to maximally precipitate a given amount of globulin (slight antibody excess) was determined by standard quantitative precipitin methods and used thereafter. Synthesis of globulin both *in vivo* and by the *in vitro* ribosomal system was detected as follows: to the reaction mixture containing soluble C^{14}-labeled proteins, 0.5 mg of authentic carrier globulin (except in the case of pea cotyledons *in vivo*, which contain endogenous globulin) was added the appropriate amount of antiserum protein. Precipitate formation was then allowed to develop maximally (1 hr at 37°C). The precipitated complex, containing approximately 2.5 mg of protein (ratio of pea globulin to antibody, 1:3), was centrifuged off and washed by five successive complete resuspensions and recentrifugations in cold 0.15 *M* NaCl. The complex was then dissolved in 0.2 *N* NH_4OH and counted.

EXPERIMENTAL RESULTS

Isolated chromatin possesses the ability to conduct DNA-dependent RNA synthesis, using as substrate the four riboside triphosphates,[6] and contains bound chromosomal RNA polymerase.[1] However, chromatin, pretreated at 60°C for 5 min to inactivate its endogenous polymerase and other enzymes so that it can serve only as template, can support RNA synthesis by the purified polymerase of *E. coli*, as is shown in Table 1. In addition, the RNA synthesis supported by chromatin in response to added *E. coli* RNA polymerase consists of molecules which are largely free, that is, do not remain bound to and sediment with the chromatin, as does RNA synthesized by chromosomal RNA polymerase.[7]

Table 1. Activity of chromatin of developing pea cotyledons and of pea DNA in support of RNA synthesis by *E. coli* RNA polymerase

System*	RNA synthesis (μμm nucleotide/ 10 min)
125 μg deproteinized DNA: *coli* polymerase present	2,220†
37.5 μg deproteinized DNA: *coli* polymerase present	2,030
12.5 μg deproteinized DNA: *coli* polymerase present	740
125 μg DNA in form of native chromosomal nucleohistone: *coli* polymerase present	25
125 μg DNA as whole chromatin: *coli* polymerase present	760
125 μg DNA as whole chromatin heated at 60°C for 5 min: *coli* polymerase present	940
125 μg DNA as whole chromatin: no added *coli* polymerase	17

*Reaction mixture includes tris, pH 8, 20 μm; $MnCl_2$, 0.5 μm; $MgCl_2$, 2μm; 8-C^{14}-ATP, 0.13 μm, 1.5 μc/μm; GTP, CTP, and UTP each 0.2 μm; β-mercaptoethanol, 6 μm; and chromatin or DNA and *E. coli* polymerase (approx. 10 μg) as indicated, all in 0.5 ml final volume. Incubation at 37°C.

†Incorporation by polymerase alone subtracted.

The data of Table 1 show not only that whole chromatin, in this case of developing pea cotyledons, supports RNA synthesis by *E. coli* RNA polymerase, but that, in addition, the nucleohistone component of such chromatin prepared by the methods of Bonner and Huang[4] and in which DNA and histone are complexed in very nearly equivalent amounts is almost totally inactive in the support of RNA synthesis. Since this nucleohistone component (in which the histone greatly stabilizes the DNA against melting[4]) includes the bulk of the DNA of pea cotyledon chromatin, only a small fraction of the DNA of such chromatin would appear to be present in an active state. The data of Table 1 show in fact that under conditions in which DNA is limiting, the DNA of pea cotyledon chromatin is as active in the support of RNA synthesis as about one-tenth as much deproteinized whole genomal pea DNA.

Chromatin-dependent RNA synthesis will next be coupled to a messenger RNA-dependent ribosomal system. For this purpose ribo-

Table 2. Dependency of ribosomal protein synthesis system on exogenous messenger RNA as affected by preincubation of ribosomal system in complete protein synthesis reaction mixture

System*	C^{14}-Leucine incorporation into protein (μμm/60 min/mg ribosomal protein) Preincubated	Not preincubated
Ribosomal system alone	88	251
Ribosomal system + pea DNA + RNA polymerase	844	754

*Reaction mixture contains tris, pH 8, 50 μm; MgAc$_2$, 2 μm; MnCl$_2$, 0.5 μm, KCl, 20 μm; β-mercaptoethanol, 3 μm; 18 amino acids, each 0.02 μm; C^{14}-leucine, 6.8 μc/μm, 0.075 μm; ATP(K), 1 μm; GTP, CTP, UTP each 0.1 μm; 50 μg ribosomal protein; 69 μg 105,000 × *g* supernatant protein; pea DNA, 100 μg; and approx. 20 μg *E. coli* RNA polymerase, all in total volume of 0.3 ml. Incubation at 37°C.

Table 3. Dependence of protein synthesis by *E. coli* ribosomal system on pea and *E. coli* RNA polymerase.

Additions to ribosomal system*	C^{14}-Leucine incorporation into protein (μμm/mg ribosomal protein/ 30 min)
Pea DNA, 125 μg; RNA polymerase	1,158
T2 DNA, 125 μg; polymerase	828
T4 DNA, 125 μg; polymerase	756
Polymerase; no DNA	100
Pea DNA, 125 μg; no polymerase	102
No DNA; no polymerase	104

*Reaction mixture includes components specified in Table 2.

somes are borrowed from *E. coli.*[8] The mode of their preparation, described under *Materials and Methods,* makes the system highly dependent upon exogenous messenger RNA. Toward the achievement of this end, preincubation of the crude system in a complete protein synthesis reaction mixture contributes importantly, as is shown in Table 2. Protein synthesis by the ribosomal system is highly dependent, as shown in Table 3, upon both added DNA and upon added RNA polymerase. Pea plant DNA is as effective in this function as is that of bacteriophage T2 or T4. The protein synthesized by the ribosomal system in response to RNA formed

Table 4. Support of ribosomal protein synthesis by chromatin-dependent RNA synthesis*

System†	C^{14}-Leucine incorporation into protein (cpm/30 min) Two-step expt.	One-step expt.
Cotyledon chromatin; RNA polymerase: ribosomal system	505	721
Cotyledon chromatin; ribosomal system: no polymerase	87	133
Bud chromatin + RNA polymerase: ribosomal system	182	924
Bud chromatin; ribosomal system: no polymerase	122	110

*In the two-step experiment, RNA is generated by chromatin for 30 min, the chromatin removed by centrifugation, and the ribosomal system added to supernatant. In the one-step experiment, all ingredients are present simultaneously.

†Reaction mixture includes components specified in Table 2 except for C^{14}-leucine. In this case, uniformly labeled leucine, 0.02 μm, 131 μc/μm, was used. Chromatin containing 250 μg DNA used in each case.

by the DNA-dependent RNA generating system is largely soluble, i.e., is released from the ribosomes. Of the protein synthesized, 65-90 per cent (av. 75 per cent) remains in solution after removal of the ribosomes by centrifugation.

We now couple chromatin-dependent RNA synthesis to messenger RNA-dependent protein synthesis. This may be accomplished in two steps, as is shown by the data of Table 4. In the first step chromatin and RNA polymerase synthesize RNA. The chromatin is then centrifuged off at 16,000 × *g* and the ribosomal system added to the RNA-containing supernatant. As the data of Table 4 also show, however, it is more effective to simply combine chromatin, polymerase, and ribosomal system, and thus to directly couple RNA synthesis to protein synthesis. This is particularly true for the case of protein synthesis supported by bud chromatin.

We next focus upon the particular gene (in fact, genes[5]) and its product-protein, pea seed globulin, which we have selected for study. We consider first the extent to which globulin is synthesized by varied tissues of the plant. For the experiment of Table 5, 1- to 10-gm samples of each of a variety of pea plant tissues were excised, and each incubated aerobically with a small volume of buffer containing C^{14}-leucine and penicillin. After 2 hr of incubation, the tissue samples were ground (Virtis homogenizer), filtered through miracloth, and centrifuged for 120 min in the no. 40 Spinco head at 40,000 rpm. The clear supernatants were then freed of C^{14}-leucine by dialysis (four successive changes of 0.1 *M* tris buffer, pH 8, 48 hr). To each supernatant representing the soluble proteins of that tissue, authentic carrier globulin and the appropriate amount of high-titer chicken antiserum were added. The specifically precipitated globulin was then centrifuged off, washed as described under *Materials and Methods,* as counted, as were aliquots of the whole supernatant. The data of Table 5 concern then the proportion which globulin-antiserum reacting material, known hereafter as globulin, constitutes of the total protein synthesized by the particular tissue during the experimental period. As is to be expected, globulin constitutes an important protein product of developing pea cotyledons, making up 4-10 per cent of total protein synthesized, the exact amount depending upon the exact stage at which the developing cotyledons are harvested.[5] The flower, too, produces a significant amount of globulin. For such organs as apical buds and root, however, the radioactivity of the globulin-antiserum precipitate constitutes only 0.1-0.2 per cent of total radioactive protein synthesized. A portion, or all, of even this amount may be due to nonspecific contamination of the carrier-precipitate by soluble, labeled but nonglobulin proteins. It may be concluded that pea apical buds synthesize little or no globulin.

For the study of globulin synthesis by the *E. coli* ribosomal system in response to chromosomally generated RNA, incubation mixtures were made on a sufficiently large scale to incorporate 5,000-50,000 cpm of C^{14}-leucine into protein. At the end of the incubation period, the cooled reaction mixtures were centrifuged 120 min in the no. 40 Spinco head at 40,000 rpm for removal of chromatin and ribosomes. The soluble proteins of the supernatant were dialyzed as outlined above. From each such reaction mixture of dialyzed soluble proteins added, authentic globulin was precipitated by the appropriate amount of antibody, just as with the *in vitro* reaction mixtures. The data of the example of Table 6 show that globulin is synthesized by the ribosomal system in response to the presence of chromatin of developing cotyledons. The chromatin of pea buds, although it supports protein synthesis equally as well as does that of pea cotyledons, does not cause the formation of any globulin detectable above the level of the background.

Table 5. Synthesis of pea seed globulin *in vivo* by varied organs of the pea plant

	C^{14}-Leucine incorporation into protein		
Organ	Total soluble protein (cpm)	Globulin-antibody precipitate (cpm)	Globulin/total protein (%)
Flower	17,500	510	2.9
Cotyledon	7,100	337	4.7
Older cotyledon	9,200	862	9.3
Roots	17,150	31	0.18
Apical bud	11,600	14	0.12
Apical bud	27,600	42	0.15

Tissues were separately incubated in L-leucine (1 μc/ml, 19 μc/μm) for 2 hr at 25°C, in the presence of penicillin, 6 μg/ml. Soluble protein obtained by dialysis of 105,000 × *g* supernatant of tissue homogenate.

Table 6. Synthesis of pea seed globulin by ribosomal system of *E. coli* in response to varied templates for RNA synthesis

	C^{14}-Leucine incorporation into protein[a]		
Template for RNA synthesis	Total soluble protein (cpm)	Globulin/antibody precipitate (cpm)	Globulin/total protein (%)
Apical bud chromatin[b]	15,650	16	0.10
Apical bud chromatin[c]	41,200	54	0.13
Cotyledon chromatin[c]	23,600	341	1.45
Cotyledon chromatin[c]	14,000	226	1.61
DNA of bud chromatin[d]	15,200	60	0.40
DNA of bud chromatin[d]	14,200	72	0.51
DNA of cotyledon chromatin[d]	5,600	22	0.39
DNA of cotyledon chromatin[d]	50,100	157	0.31

[a]Reaction mixtures included components of Table 2 but on scale 15-40× greater.

[b]Chromatin not subjected to 60°C pretreatment.

[c]Chromatin pretreated at 60°C, 5 min, in dilute saline citrate, 0.015 *M*.

[d]DNA prepared by deproteinization of purified chromatin.

Although the chromatin of pea buds does not support the synthesis of globulin, the gene or genes responsible for such synthesis is of course present in it. Table 6 includes data from experiments in which protein synthesis by the ribosomal system was supported by the presence of DNA prepared by deproteinization of pea bud or of pea cotyledon chromatin. The amount of radioactivity in the globulin complex prepared from such reaction mixtures, approximately 0.4 per cent of total protein formed, is appreciably greater than that formed either by the pea-bud-chromatin-primed ribosomal system or by pea buds *in vivo*.

DISCUSSION

We acknowledge that the immunochemical methods here used for the separation and determination of pea seed globulin are not rigorous in the sense that we have not chemically established that the product protein is, in fact, globulin. This is a task for the future. In the present work we are, however, concerned with comparison between the amounts of anti-globulin-reacting material formed in the living cell and in the chromosomally supported ribosomal system, and in the similar comparison between the products of ribosomal systems supported by different chromatins. Used in this manner the immunochemical assay of globulin would be a rigorous one were it not for the background: the small amount of activity associated with the complex when carrier globulin is precipitated with its antibody from protein mixtures containing little or no globulin, as in the case of the proteins of pea buds, pea roots, etc. Whether this background represents nonspecific absorption or a small amount of synthesis of globulin or other cross-reacting material, we do not know. The differences in proportion of globulin formed as between systems supported by cotyledon chromatin, pea bud chromatin, and deproteinized DNA from pea bud or cotyledon chromatin are in any case well beyond the range of this background.

The protein produced by the ribosomal system in response to pea whole-genomal DNA includes a substantial proportion of globulin, in the examples of Table 6, 4 or 5 times above the background. It may be that the genes for synthesis of globulin constitute a larger-than-average proportion of the pea genome. Our immunochemical methods of globulin detection are, however, inadequate to establish this point rigorously.

The results presented above show not only that messenger RNA-dependent protein synthesis can be supported by RNA formed by chromatin-dependent RNA synthesis, but also that during the isolation of chromatin there is preserved to a considerable degree the control of genetic activity characteristic of the particular tissue in life. Cells in which globulin synthesis is repressed yield chromatin which does not support globulin synthesis in our coupled system, and vice versa. It is true, however, that protein synthesis as supported by chromatin of developing pea cotyledons *in vitro* does not result in a proportion of globulin as high as is characteristic of pea cotyledons *in vivo*. This may be due to some derepression of normally repressed genes during the preparation of pea cotyledon chromatin. It may result also from some fractionation of the genome during chromatin preparation, although as is shown in Table 6, the DNA of cotyledon chromatin is approximately as effective as that of pea buds in support of globulin synthesis. It is clear, however, that the repressed state of the genes responsible for globulin synthesis is preserved during the preparation of pea bud chromatin. What agent is responsible for such repression? It has been shown that removal of protein from pea bud chromatin derepresses the genes for globulin synthesis. The principal protein of chromatin is histone, although other proteins are also present. That it is removal of histone which is responsible for derepression is indicated by the facts (Table 1; see also refs. 1, 4, and 9) that the nucleohistone component of chromatin, in which DNA is fully complexed with histone, is inert in support of RNA synthesis and that ability of chromatin to support RNA synthesis is greatly increased by the specific removal of histone.[1] The histone component of chromatin is, then, a repressor of genetic activity.

Even though we now see that the histone of the chromosome is one agent responsible for genetic repression, the principal questions concerning control of genetic activity remain unanswered. How does nature turn genes off and on—put on or take off histones? How is programming of genetic activity exerted? We cannot yet answer these questions. We do, however, have a system with which such matters can be studied. It appears probable, too, that the present type of system in which the well-standardized, readily preparable components of *E. coli*, such as ribosomes and polymerase, are used as tools for the investigation of chromosomal activity may be of general utility in the study of development in higher organisms.

SUMMARY

The ability of chromatin, isolated from varied tissues of the pea plant, to support DNA-dependent RNA synthesis is enhanced in the presence of the RNA polymerase of *E. coli.* We have coupled such chromatin-dependent RNA synthesis to a messenger RNA-dependent ribosomal protein synthesis system, the latter also derived from *E. coli.* The material synthesized by the ribosomal system under the direction of chromatin isolated from developing pea cotyledons includes a protein characteristic of such cotyledons, the pea seed reserve globulin. Chromatin of pea buds, which do not synthesize pea seed globulin *in vivo,* does not support the synthesis of such globulin by the isolated ribosomal system. Hence the control of genetic activity characteristic of the living cell is, to an appreciable extent, preserved in the isolated chromatin. This control is exerted by the histone of the chromosome. Thus, the removal of histone from pea bud chromatin, in which the genes for globulin synthesis are repressed, yields DNA which supports globulin synthesis.

REFERENCES

1. Huang, R. C. C., and J. Bonner, these Proceedings **48:**1216 (1962).
2. Chamberlin, M., and P. Berg, these Proceedings **48:**81 (1962).
3. Nirenberg, M., and J. H. Matthaei, these Proceedings **47:**1588 (1962).
4. Bonner, J., and R. C. C. Huang, J. Mol. Biol. **3:**169 (1963).
5. Danielsson, C. E., Biochem. J. **44:**387 (1949). Globulin prepared by this method consists of two related major components, legumin and vicilin, present in approximately equal amounts. Our antiserum contains antibodies to both components.
6. Huang, R. C. C., N. Maheshwari, and J. Bonner, Biochem. Biophys. Res. Comm. **3:**689 (1960).
7. Bonner, J., R. C. C. Huang, and N. Maheshwari, these Proceedings **47:**1548 (1961).
8. Ribosomal systems from *E. coli* which are dependent upon a DNA-dependent RNA generating system have also been described by Wood, W., and P. Berg, these Proceedings **48:**94 (1962); by Furth, J., F. Kahan, and J. Hurwitz, Biochem. Biophys. Res. Comm. **9:**337 (1962); and by Nirenberg, M., and J. H. Matthaei, Biochem. Biophys. Res. Comm. **9:**339 (1962). The effect of preincubation in increasing the messenger RNA-dependency of the system has been described by Nirenberg and Matthaei, and by Ning, C., and A. Stevens, J. Mol. Biol. **5:**660 (1962).
9. Huang, R. C. C., J. Bonner, and K. Murray, J. Mol. Biol., in press.

47 Codon recognition by enzymatically mischarged valine transfer ribonucleic acid

D. Grunberger
I. B. Weinstein
K. B. Jacobson
Institute of Cancer Research
Department of Medicine
College of Physicians and Surgeons of Columbia University
New York, New York
and
Biology Division
Oak Ridge National Laboratory
Oak Ridge, Tennessee

Abstract. *Direct evidence for the adaptor hypothesis has been obtained by examining the codon recognition of a purified* Escherichia coli *valine transfer ribonucleic acid which was enzymatically mischarged with phenylalanine labeled with carbon-14 by reaction with purified phenylalanyl-transfer ribonucleic acid synthetase from* Neurospora crassa. *The mischarged transfer ribonucleic acid recognized the valine codons but failed to recognize the phenylalanine codon when tested in trinucleotide-directed ribosomal binding assay.*

The adaptor hypothesis of Crick *(1)* states that the amino acid sequence of a protein is determined, during the course of the translation and protein synthesis, by the alignment of aminoacyl transfer ribonucleic acids (tRNA) at corresponding nucleotide triplets (codons) in messenger RNA (mRNA). The specificity of the translation mechanisms depends on base pairing between a nucleotide region of tRNA, the anticodon, and nucleotides in the codon. The hypothesis predicts that the amino acid bound to tRNA does not participate in this recognition. The first evidence for the passive role of the amino acid was obtained by Chapeville *et al. (2)*; they showed that [^{14}C]cysteine, while attached to its tRNA (tRNA^{Cys}), *(3)*, was converted by catalytic reduction with Raney nickel to [^{14}C]alanine. When the resulting complex, [^{14}C]alanyl-tRNA^{Cys}, was tested in a protein synthesizing system, radioactivity was incorporated under the direction of poly UG, a polymer which normally stimulates the incorporation of cysteine but not alanine *(2)*. Similarly, von Ehrenstein *et al. (4)* found that the chemically altered aminoacyl-tRNA delivered [^{14}C]alanine, in place of cysteine, in a peptide of hemoglobin.

We used a different system to confirm the results of Chapeville *et al. (2)*. Our approach was made possible by the discovery *(5)* that a phenylalanyl-tRNA synthetase of *Neurospora crassa* can mischarge phenylalanine onto two *Escherichia coli* tRNA's which have been identified as tRNA^{Val} and tRNA^{Ala} *(6)*. Earlier studies had indicated that the phenylalanyl-tRNA^{Ala} responded to poly UCG in the same manner as alanyl-tRNA^{Ala} of *N. crassa* and *E. coli (6, 7)*. We have asked whether valine or phenylalanine codons are recognized when the mischarged tRNA is tested in a codon-directed ribosome binding assay.

We used *E. coli* tRNA^{Val} (acceptance of 1100 pmole/O.D. unit, measured at 260 nm) purified by reversed-phase chromatography *(8)*. *Neurospora crassa* phenylalanyl-tRNA synthetase (peak C), one of three chromatographically separable, cytoplasmic phenylalanyl-tRNA synthetases, was purified as described by Kull and Jacobson *(9)*. The aminoacylation of *E. coli* tRNA^{Val} by *N. crassa* phenylalanyl-tRNA synthetase was performed according to Ritter *et al. (10)*. The reaction system contained (in 1 ml): 50 m*M* potassium-cacodylate buffer, *p*H 6.3; 0.5 μmole of ATP; 7.5 μmole of $MgCl_2$; 2.7 nmole of [^{14}C]phenylalanine [817

From Science **166**:1635-1637, December 26, 1969.

Supported by NIH grant R10 CA-02332, and by the U.S. Atomic Energy Commission under contract with the Union Carbide Corporation. I.B.W. is a Career Scientist of the Health Research Council of the City of New York (I-190). We thank Dr. J. F. Weiss for the *E. coli* tRNA^{Val} and Miss G. Carvajal for technical assistance.

Table 1. The incubation mixture (0.05 ml) contained 0.05*M* tris-acetate (*p*H 7.2), 0.05*M* KCl, 0.03*M* magnesium acetate, and 2.1 O.D. units (260 nm) of *E. coli* ribosomes. Trinucleotides [0.1 O.D. units (260 nm)] and [^{14}C] aminoacyl-tRNA's were added as indicated. Incubation was carried out at 24°C for 20 minutes. The reaction mixture was diluted with 3 ml of buffer and filtered through Millipore filters: the radioactivity was measured in a Tricarb liquid scintillation counter.

	[^{14}C] Valyl-tRNA^{Val} (5.5 pmole)		[^{14}C] Phenylalanyl-tRNA^{Val} (4.8 pmole)		[^{14}C] Phenylalanyl-tRNA (7.4 pmole)	
Trinucleotides	pmole	Δpmole	pmole	Δpmole	pmole	Δpmole
None	0.298		0.301		0.223	
GUC	0.300	0.002	0.298	-0.003		
GUU	0.511	0.213	0.598	0.297		
GUA	1.005	0.707	1.203	0.902		
GUG	1.259	0.961	1.315	1.014		
UUU	0.287	-0.011	0.291	-0.010	0.516	0.293

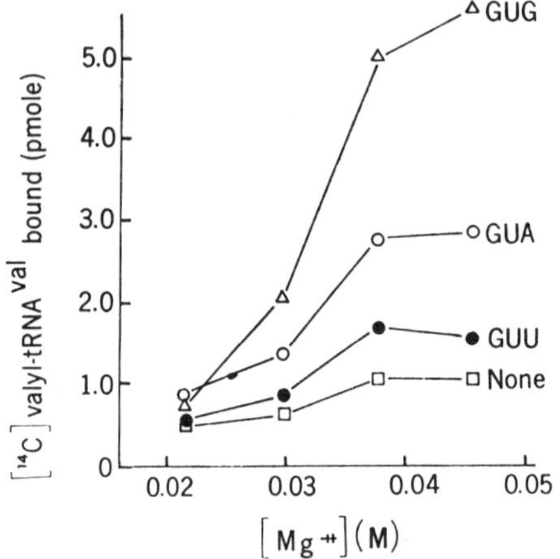

Fig. 1. Effect of Mg^{2+} concentration on the binding of [^{14}C] valyl-tRNA^{Val} to ribosomes in the presence of trinucleotides. The incubation mixture contained 27.7 pmole of [^{14}C] valyl-tRNA^{Val}, 0.1 O.D. unit (260 nm) of the indicated trinucleotides, and the specific concentration of magnesium acetate. The remaining conditions were as described in Table 1.

disintegrations per minute (dpm)/pmole; and 70 pmole of *E. coli* tRNA^{Val}. Aminoacylation of *E. coli* tRNA^{Val} with [^{14}C] valine and of unfractionated *E. coli* tRNA with [^{14}C] phenylalanine were performed with a crude mixture of *E. coli* aminoacyl-tRNA synthetases *(11)*. *Escherichia coli* ribosomes were prepared from *E. coli* B (General Biochemicals) as described *(12)*. The trinucleotides GUU, GUC, GUA, and GUG were prepared enzymatically with ribonuclease T_1 from guanosine 2′,3′-cyclic phosphate and appropriate dinucleotide phosphates *(13)*. (UUU was a product of Miles Laboratories.) The binding of aminoacyl-tRNA to ribosomes was assayed by the procedure of Nirenberg and Leder *(14)*.

Since the codon assignments *(15)* for valine are GUU, GUC, GUA, and GUG, and for phenylalanine are UUU and UUC, we tested the binding of [^{14}C] valyl-tRNA^{Val} and [^{14}C] phenylalanyl-tRNA^{Val} to ribosomes in the presence of these triplets. Both [^{14}C] valyl-tRNA^{Val} and [^{14}C] phenylalanyl-tRNA^{Val} recognized GUG and GUA and to a lesser extent GUU (Table 1). The binding of the [^{14}C] phenylalanyl-tRNA^{Val} to these valine codons was quantitatively similar to that obtained with [^{14}C] valyl-tRNA^{Val}. At the same time, *E. coli* tRNA^{Val}, whether charged with valine or phenylalanine, failed to recognize the phenylalanine codon UUU. In contrast, the [^{14}C] phenylalanyl-tRNA, synthesized in a homologous system containing both *E. coli* aminoacyl-tRNA synthetase and unfractionated *E. coli* tRNA, gave the expected positive response to the UUU codon (Table 1, last column). These codon response data are consistent with studies of Jacobson *(16)* demonstrating that phenylalanyl-tRNA^{Val} and phenylalanyl-tRNA^{Ala} do not allow phenylalanine to enter into normal positions of the hemoglobin chain when tested in a subcellular reticulocyte system. Other results indicated that [^{14}C] phenylalanyl-tRNA$^{Val+Ala}$ failed to support the incorporation of phenylalanine into protein in the presence of poly U and ribosomes *(17)*.

The response of a highly purified *E. coli* valyl-tRNA^{Val} to the codons GUG, GUA, and GUU is of interest because this pattern of codon recognition is not predicated in the wobble hypothesis of Crick *(18)*. Similar results, however, have been obtained in studies with the major *E. coli* valine tRNA *(19, 20)*, a yeast valine tRNA *(21)*, and a rat liver valine tRNA *(22)*. Since the binding assay is strongly dependent on Mg^{2+}, we examined the effect of Mg^{2+} concentration on the ability of *E. coli* valyl-$tRNA^{Val}$ to recognize these codons (Fig. 1). At a Mg^{2+} concentration of 0.02*M*, significant stimulation of binding occurred with GUG and GUA, but not with GUU. The stimulatory effect of GUU became apparent only at higher Mg^{2+} concentrations. These results could explain a possible discrepancy with the data of Yaniv and Barrell *(23)*, who used 0.02*M* Mg^{2+} in their binding assays *(24)*. However, Nishimura *(17)* found Val-$tRNA^{Val}$ bound to the GUU-ribosome complex at 0.02*M* Mg^{2+}. Although the primary sequence of *E. coli* $tRNA^{Val}$ has been elucidated *(23)*, the nucleotide in the 5′-position of the anticodon region has not been fully characterized. Studies suggest that it is a uridine derivative with an additional residue, presumably a carboxyl group *(23, 25)*. Its specific function may be recognition of both A and G, and to a lesser degree U, in the third position of valine codons, not only in *E. coli (19)*, but perhaps also in a valine tRNA of yeast *(21)* and rat liver *(22)*.

These results provide direct evidence that an amino acid, once attached to tRNA, does not itself participate in codon recognition. The decisive role in codon recognition is played by the precise structure of the tRNA.

REFERENCES AND NOTES

1. F. H. C. Crick, Symp. Soc. Exp. Biol. **12**:138 (1958).
2. F. Chapeville, F. Lipmann, G. von Ehrenstein, B. Weisblum, W. J. Ray, Jr., L. B. Benzer, Proc. Nat. Acad. Sci. U.S. **48**:1086 (1962).
3. Abbreviations are: A, adenosine; C, cytidine; G, guanosine; U, uridine; cys, cysteine; ala, alanine; val, valine; poly U, poly uridylic acid; poly UG, poly uridylic-guanylic acid; O.D., optical density; and tris, tris(hydroxymethyl)-aminomethane.
4. G. von Ehrenstein, B. Weisblum, S. Benzer, Proc. Nat. Acad. Sci. U.S. **49**:669 (1963).
5. W. E. Barnett and K. B. Jacobson, *ibid.* **51**:642 (1964).
6. W. E. Barnett and J. L. Epler, Cold Spring Harbor Symp. Quant. Biol. **31**:549 (1966).
7. J. L. Epler, W. E. Barnett, K. B. Jacobson, Fed. Proc. **24**:408 (1965).
8. J. F. Weiss, R. L. Pearson, A. D. Kelmers, Biochem. J. **7**:3479 (1968).
9. F. J. Kull and K. B. Jacobson, Proc. Nat. Acad, Sci. U.S. **62**:1137 (1969).
10. P. O. Ritter, F. J. Kull, K. B. Jacobson, Biochim. Biophys. Acta **179**:524 (1969).
11. V. Z. Holten and K. B. Jacobson, Arch. Biochem. Biophys. **129**:283 (1969).
12. D. Grunberger, L. Meissner, A. Holý, F. Šorm, Collect. Czech. Chem. Commun. **32**:2625 (1967).
13. D. Grunberger, A. Holý, F. Šorm, *ibid.* **33**:286 (1968).
14. M. Nirenberg and P. Leder, Science **145**:1399 (1964).
15. ———, M. Bernfield, R. Brimacombe, J. Trupin, F. Rottman, C. O'Neal, Proc. Nat. Acad. Sci. U.S. **53**:1161 (1965).
16. K. B. Jacobson, Cold Spring Harbor Symp. Quant. Biol. **31**:719 (1966).
17. S. Nishimura, personal communication.
18. F. H. C. Crick, J. Mol. Biol. **19**:548 (1966).
19. D. A. Kellogg, B. P. Doctor, J. E. Loebel, M. W. Nirenberg, Proc. Nat. Acad. Sci. U.S. **55**:912 (1966).
20. K. Oda, F. Kimura, F. Harada, S. Nishimura, Biochim. Biophys. Acta **179**:97 (1969).
21. A. D. Mirzabekov, D. Grunberger, A. I. Krutilina, A. Holý, A. A. Bayev, F. Šorm, *ibid.* **176**:75 (1968).
22. S. Nishimura and I. B. Weinstein, Biochemistry **8**:832 (1969).
23. M. Yaniv and B. G. Barrell, Nature **222**:278 (1969).
24. M. Yaniv, personal communication.
25. F. Harada, F. Kimura, S. Nishimura, Biochim. Biophys. Acta **182**:590 (1969).

48 Streptomycin, suppression, and the code

Julian Davies*
Walter Gilbert
Luigi Gorini
Department of Bacteriology and Immunology
Harvard Medical School
Boston, Massachusetts
and
Jefferson Laboratory of Physics
Harvard University
Cambridge, Massachusetts

We have found that an external agent, streptomycin, can upset the genetic code, producing specific misreadings during *in vitro* polypeptide synthesis. This interference is at the level of the ribosome-messenger RNA-sRNA complex, for a modification in the ribosome makes the *in vitro* system insensitive to this effect.

Streptomycin[1] is a bacteriocidal agent. It is a basic molecule that can bind strongly to nucleic acids.[2] It interferes with and finally blocks protein synthesis while permitting continued RNA and DNA synthesis.[3] The mechanism of its killing is not known, but the existence of single mutations to high-level resistance suggests a unitary cause, a single, vital point of attack. Spotts and Stanier[4] hypothesized that the ribosomes were the sensitive elements, and work with the *in vitro* system, in which the poly U-directed incorporation of phenylalanine was shown to be inhibited by streptomycin,[5,6] further implicated the ribosomes as the site of the shift from sensitivity to resistance. The sensitivity to streptomycin has just been shown by Davies[7] and by Cox, White, and Flaks[8] to reside on the 30s subunit of the 70s ribosome. If these 30s subunits are taken from a sensitive strain, the reconstructed *in vitro* system is sensitive, while if they are derived from a resistant strain, the system is resistant.

THE EFFECTS OF STREPTOMYCIN ON THE SPECIFICITY OF AMINO ACID INCORPORATION

Although the incorporation of phenylalanine into hot TCA-insoluble material is blocked (often by 50-75%) by streptomycin, the incorporation of other amino acids, not normally coded for by poly U, is stimulated. Table 1 shows that, using a purified system, the incorporation of isoleucine (UUA, UAA, CAU),[9] and to a much lesser extent serine (UUC, UCC, AGC) and leucine (UUA, UUC, UUG, UCC), is stimulated in the presence of streptomycin. The same is true for crude extracts (Table 2). Tyrosine (UUA) is not noticeably stimulated, nor are the other amino acids. Asparagine (CAA, CUA, UAA) was examined by damping the incorporation of a C^{14}-chlorella hydrolysate with the cold amino acid, with negative results. The experiments with purified ribosomes were done at the optimum ion concentration for phenylalanine incorporation, 17.5 mM Mg^{++} and 0.19 *M* NH_4^+.

The incorporation of isoleucine and phenylalanine as a function of the magnesium concentration is shown in Figure 1. The ammonium concentration was 0.086 *M* and all twenty amino acids were present. In Figure 1*a* we see that the incorporation of isoleucine is stimu-

From Proceedings of the National Academy of Sciences **51**:883-890, 1964. Used with permission.

This investigation was supported by grants from the American Cancer Society (E-226-B), National Science Foundation (GB-1307), and U.S. Public Health Service (AI-02011-06; GM-09541-03).

Experiments by W. Gilbert were carried out in the Biological Laboratories, Harvard University. We wish to thank Mrs. Anne Nevins and Mrs. Susanne Armour for their technical assistance. We thank Dr. J. D. Watson and Dr. B. D. Davis for their interest and support. We are grateful to Dr. M. Lubin for a gift of poly CA, and Dr. W. T. Sokolski for neomycins B and C.

The following abbreviations are used: Sm = streptomycin; Km = kanamycin; Neo = neomycin; CM = chloramphenicol; DHSm = dihydrostreptomycin; poly U = polyuridylic acid; poly CA = cytidylic acid-adenylic acid copolymer; PEP = phosphoenolpyruvate; PK = phosphoenolpyruvate kinase.

*Present address: Department of Biochemistry, University of Wisconsin, Madison, Wisconsin.

lated by streptomycin by more than a factor of 10 at all magnesium levels, and reaches the same level as the phenylalanine incorporation. The optimum magnesium concentration for this incorporation is very high (about 30 mM at this particular ammonium ion concentration, higher than that for phenylalanine), and at the optimum more isoleucine than phenylalanine is inserted. There is a significant background incorporation of isoleucine in the absence of streptomycin at very high magnesium levels. Figure 1*b* compares the streptomycin-stimulated incorporation of sensitive and resistant ribosomes, using a supernatant from a sensitive strain. The streptomycin-resistant ribosomes are essentially resistant to this miscoding.

The stimulation of isoleucine incorporation by magnesium ions alone is akin to the effects observed by Szer and Ochoa,[10] who have shown that the magnesium optimum for the poly U-directed incorporation of leucine was higher than that for phenylalanine and that there were slight incorporations of isoleucine, serine, and tyrosine in high magnesium, when the amino acids were supplied singly. Our findings differ in that we are following incorporation in the presence of all the amino acids, and we are exploring a still higher (20-60 mM) range of magnesium concentrations.

There is also a monovalent ion effect: if we raise the ammonium ion concentration from 0.086 *M* to 0.19 *M*, much more isoleucine is incorporated in high magnesium. Figure 2*a* shows the streptomycin and magnesium effects in 0.19 *M* NH_4^+ ion. At 30-40 mM Mg^{++}, the code is perturbed sufficiently to put in equal parts of isoleucine and phenylalanine. The additional effect of streptomycin is now only a 2.5-fold stimulation of the isoleucine incorporation near the optimum, but at lower Mg^{++} levels the stimulation is much greater. Again, in Figure 2*b*, we see that resistant ribosomes do not miscode under the influence of streptomycin.

Table 1

C^{14} amino acid	−Poly U	+Poly U	+Sm, poly U
Ala	6	8	8
Arg	8	5	7
Asp	9	8	5
Cys	693	380	396
Glu	24	11	19
Gly	9	8	6
His	30	17	13
Ileu	6	17	340
Leu	17	96	130
Lys	28	21	33
Met	228	102	185
Phe	47	949	452
Pro	10	16	21
Ser	18	11	41
Thr	8	14	7
Try	480	418	482
Tyr	511	396	438
Val	9	13	17

Each amino acid was tested in a 0.050-ml reaction mixture like that for Fig. 2*a*, containing 0.4 μg of each of the 20 amino acids and 0.1 μC of the labeled amino acid, at 0.19 *M* NH_4^+ and 0.0175 *M* Mg^{++}. 0.3 μg poly U and 1 μg Sm were added as indicated. Results are expressed as cpm/reaction mixture. Significant changes produced by the antibiotic are underlined.

THE EFFECT OF OTHER AMINOGLYCOSIDE ANTIBIOTICS

Kanamycin and neomycins B and C are aminoglycoside antibiotics structurally related to streptomycin and appear to have a similar mechanism of action.[1,11] The streptomycin-

Table 2

	−Poly U		+Poly U						
C^{14} amino acid		Sm		Sm	Km	NeoB	NeoC	DHSm	Om
Phe	80	99	1035	1270	1050	735	690	1210	730
Leu	25	24	400	775	763	545	590	740	275
Ileu	13	13	88	925	985	805	895	860	50
Ser	20	30	36	140	545	515	565	104	23
Tyr	255	240	207	198	286	341	335	212	225

The reaction mixtures contained in 0.5 ml, 0.1 *M* Tris, pH 7.8; 0.01 *M* MgAc; 0.075 *M* NH_4Ac; 6×10^{-3} *M* mercaptoethanol; 1×10^{-3} *M* ATP; 5×10^{-3} *M* PEP; 20 μg PK; 3×10^{-5} *M* GTP; 2×10^{-4} *M* of each of 19 L-amino acids minus the C^{14} amino acid; 10^{-5} *M* C^{14} amino acid (approx. 50,000 cpm per reaction mixture); 20 μg of poly U, and 0.025 ml of dialyzed, preincubated *E. coli* crude extract[7] containing *ca.* 150-200 μg ribosomes. Antibiotics (5 μg/0.5 ml) were added before the poly U. Mixtures were incubated at 34°C for 30 min. The cpm per reaction mixture are given.

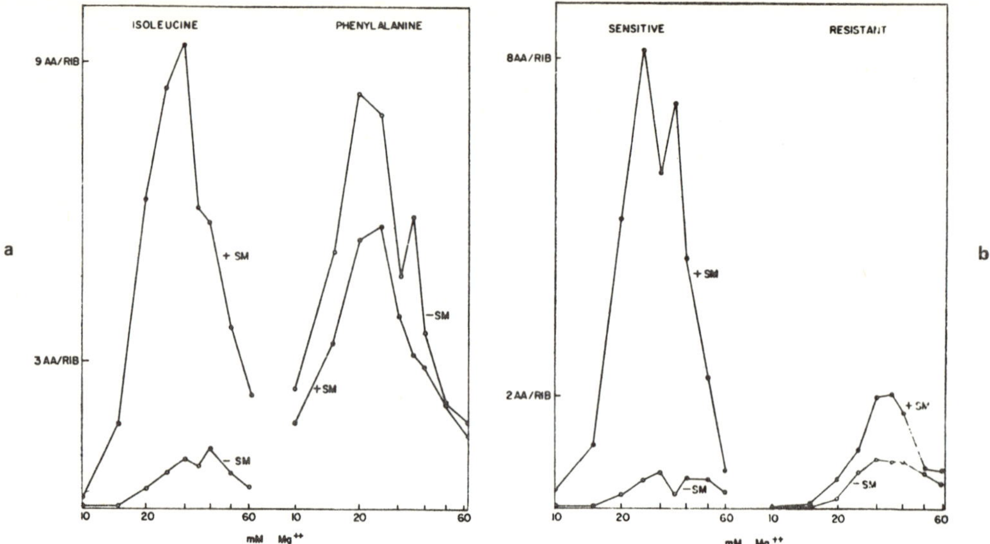

Fig. 1. Streptomycin effect on the incorporation of isoleucine and phenylalanine at 0.086 *M* NH_4^+. 0.050-ml reaction mixtures contained 0.010 ml supernatant fraction from *E. coli* B(Sm^s) (the upper two thirds of a 5-hr, 100,000 *g* centrifugation of a crude extract), 10 μg ribosomes from *E. coli* C600 Sm^s or from a resistant mutant of this strain (the ribosomes were 70s ribosomes isolated on a gradient from an extract preincubated with puromycin (50 μg/ml) and the ATP generating system), 0.1 μg of each of the 20 amino acids and 0.25 μC of the labeled amino acid (0.14 μg of ileu or 0.12 μg of phe), 0.3 μg of poly U, 20 μg ATP, 12 μg GTP, 0.25 μM of PEP, 2 μg PK, 0.10 *M* Tris pH 7.5, 0.086 *M* NH_4Ac, and MgAc as specified. The mixtures were incubated for 90 min at 30°C. The results are given as moles of amino acid/mole of 70s ribosome; the ribosomes are limiting. Streptomycin (20 μg/ml) was added after messenger. **a,** The magnesium dependence of the streptomycin stimulation of isoleucine incorporation and inhibition of phenylalanine incorporation on sensitive ribosomes. **b,** Comparison of the stimulation of isoleucine incorporated on streptomycin-sensitive and -resistant ribosomes.

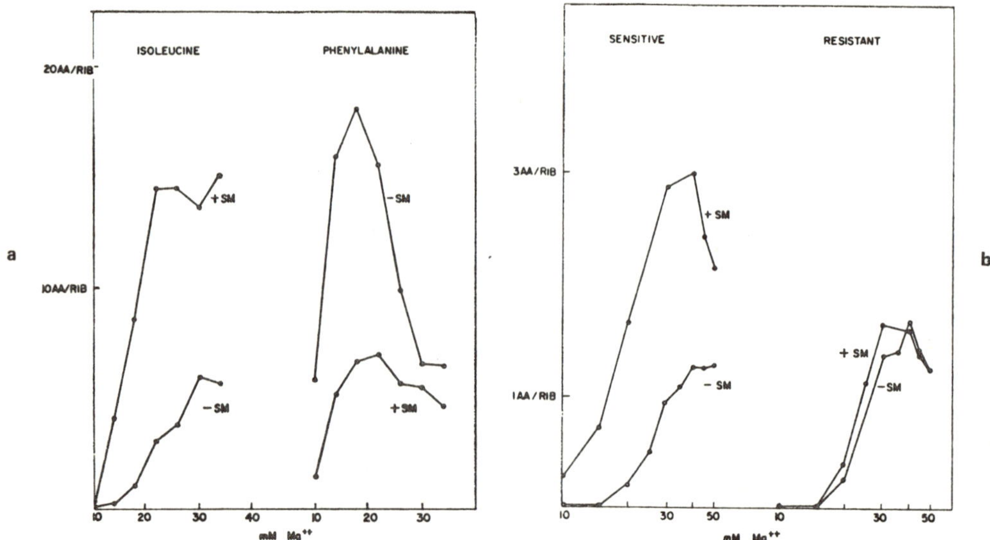

Fig. 2. Streptomycin effect on the incorporation of isoleucine and phenylalanine at 0.19 *M* NH_4^+. The reaction mixtures were as for Fig. 1, but those for **a** contained 0.4 μg of each amino acid plus 0.1 μC of the labeled amino acid, and 10 μg of ribosomes from *E. coli* B. Those for **b** contained 0.1 μg of each amino acid plus 0.1 μC of the labeled amino acid, and 10 μg of ribosomes from *E. coli* C600 Sm^s or Sm^r. Both were at 0.19 *M* NH_4^+ and the specified Mg^{++} concentration. **a,** The stimulation of isoleucine incorporation and the inhibition of phenylalanine incorporation by streptomycin acting on sensitive ribosomes. **b,** Comparison of the stimulation of isoleucine incorporation on streptomycin-sensitive and -resistant ribosomes.

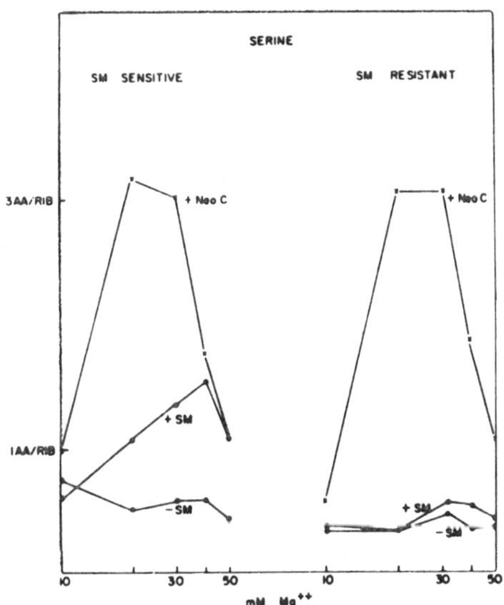

Fig. 3. Effects of streptomycin and neomycin C on serine incorporation with streptomycin-sensitive or -resistant ribosomes. The reaction mixtures were as for Fig. 1, but contained 0.1 μC of C^{14} serine (0.0876 μg) in addition to 0.1 μg of all 20 cold and either 20 μg/ml streptomycin or neomycin C.

Table 3

	−Poly CA		+Poly CA			
C^{14} amino acid	−	Sm	−	Sm	Km	NeoB
Thr	351	340	427	450	710	822
Pro	14	18	111	91	215	196
Hist	44	−	71	105	201	177

Reaction conditions as in Table 2, except that 0.05 ml of dialyzed crude extract was used in each incubation. 20 μg of poly CA was used.

resistant strain used in this investigation was highly sensitive, like the streptomycin-sensitive parent, to the effect of these antibiotics *in vivo*. In the poly U system, kanamycin and neomycins B and C all stimulate the incorporation of isoleucine. In addition, they stimulate tyrosine, leucine, and serine as is shown in Table 2. They are indifferent to whether or not the ribosomes are streptomycin-sensitive or -resistant. One such experiment is shown in Figure 3, where the stimulation of serine incorporation by either streptomycin or neomycin C is compared on streptomycin-sensitive and -resistant ribosomes. We observe a small stimulation with streptomycin on sensitive ribosomes, no effect on resistant ribosomes, while with neomycin C there is a marked, identical stimulation on both. This experiment underscores the specificity of the change from sensitive to resistant ribosomes. We expect that the ribosomes from high-level kanamycin- and neomycin-resistant strains may be resistant to the code shifts induced by their respective antibiotics.

These coding alterations have also been observed with a polymer that does not contain U. We tested the effect of the antibiotics on incorporation directed by poly CA (2:1). This polymer would be expected to code well for proline (CAC, CCC, CUC), to a lesser extent for threonine (ACA, CCA, UCA) and histidine (ACC, AUC), and for aspartic acid (GCA, GUA) and glutamic acid (AAC, UAC).[12] Table 3 shows that neomycin and kanamycin stimulated the incorporation of all three amino acids tested—threonine, proline, and histidine. Streptomycin produced a smaller stimulation of histidine and threonine, while causing an inhibition of proline incorporation. It is not possible to identify specific code changes in this system, but it is obvious that such changes do occur, particularly in the presence of neomycin and kanamycin.

DISCUSSION

The fact that an external agent, acting on the ribosomes, can perturb the code leads to a change in our view of the sources of specificity in protein synthesis. The translation mechanism involves not only the specific hydrogen bonds formed between the sRNA adaptor and the messenger, but also the conformation of the site on the ribosome that holds the sRNA to the messenger: this site must be such as to permit or require only the correct pairing to take place. A modification in this site, for example by the binding of streptomycin, permits a "wrong" sRNA to fit so well against the messenger that a "wrong" amino acid is entered into the polypeptide chain. A further modification, by mutation, changes the structure of the ribosomal site so that the correct sRNA is paired, whether or not streptomycin is present. Still a further modification might make the site require the presence of a streptomycin molecule in order to function correctly; this would constitute a mechanism for the classical streptomycin dependence.

The streptomycin effect gives us confidence that the point of contact between the sRNA and the messenger is on the 30s subunit, as is suggested by messenger binding experiments,[13] although the growing point of the polypeptide chain is probably on the 50s subunit, along with the site that binds the sRNA.[14]

The large perturbations induced by ion shifts, and the results of Szer and Ochoa,[10] raise questions about the validity of the code determined by the *in vitro* system. One would expect that a valid *in vitro* system would display a code insensitive to small changes in the conditions of assay. The ion effects on the code may be due to a direct influence on the secondary structure of the messenger and the sRNA, as is suggested by Szer and Ochoa.[10] Another possibility is that these higher ionic strengths produce a relaxation in the structure of the ribosomes, making them impose less stringent conditions on the pairing between the messenger and sRNA, in analogy to the streptomycin effect.

The phenomenon of intergenic suppression is thought to involve modifications in the translation process. The suppressor genes produce a restoration of enzymatic activity through inducing either an amino acid replacement[15] or a transition from "nonsense" to sense or from the command to "end the chain" to an amino acid.[16] The models that have been proposed for such suppressors involve altered sRNA's or activating enzymes.

Recently, however, Gorini and Kataja[17] have demonstrated the existence of a streptomycin-activated suppressor phenotype, which suggests a new mechanism for suppression. This phenotype describes streptomycin-resistant strains that display suppression only when grown in streptomycin. Only in the presence of streptomycin do these cells make a small amount of functional protein. It is evident that the miscoding property of streptomycin offers an explanation for this suppression. We need only imagine that in some cases the mutation, in the structure of the ribosomes, to streptomycin resistance is not complete, that a residual error frequency, a few parts in a thousand in the presence of streptomycin, remains. This error frequency, involving the specific misreading of a subset of code words, would produce the shift of a number of amino acids. The argument is this: we can interpret our finding of a specific shift, phenylalanine to isoleucine, as a forced misreading of U as A in a specific position in a triplet. We might then expect that we could force U to read as A in that position of any triplet (unless there are sequence effects). This converts 16 triplets into other triplets and produces changes in up to 16 amino acids ("up to" because of degeneracy). Since streptomycin upsets the reading of other bases, still a larger number of replacements is possible. Thus, we might expect the streptomycin-activated suppressor to cause a number of different misreadings of the code, resulting in a variety of amino acid replacements or changes in other coded functions.

The involvement of the ribosomes in the accuracy of the reading, and the interpretation we have put upon the streptomycin-activated suppressor suggests that many suppressors may be modifications in the structure of the ribosomes; that is, the product of the suppressor gene is an altered component of the ribosome, whose incorporation in a complete ribosome makes that ribosome alter the code. Such a model would also explain one characteristic of the suppressor genes: since each suppressor would affect a number of different amino acids, different suppressors would have wide and overlapping spectra of repair. If we wished to provide an explanation of very efficient suppression along these lines, we would require the assumption that the triplets that are being misread by this mechanism are uncommon, as has been suggested for other high suppression mechanisms.[16] The assumption that these triplets are associated with rare sRNA's would then permit the suppressed reading to occur faster than the normal reading and would yield large amounts of the suppressed protein.[18]

We conclude by observing that the gross misreading that we find *in vitro* could be the basic mechanism of streptomycin killing: flooding the cell with nonfunctional proteins would be lethal and would perturb, in an unpredictable way, all other cellular functions. To explain the requirement for growth in the killing phenomenon, we assume that the presence of the messenger prevents the attachment of streptomycin to the ribosomes.[5] During the recycling of ribosomes as the messenger moves through the polyribosome, the ribosomes are free to bind streptomycin. If streptomycin binds irreversibly to the ribosomes when they are exposed, then its action will be irreversible. In order to explain the dominance of sensitivity over resistance[19] we need only observe that an equal mixture of good and bad ribosomes would produce mostly bad proteins because of the multimeric form of many proteins. Such bacteria should not grow in streptomycin but might survive a pulsed exposure, a few good ribosomes enabling the cell to throw off viable daughter cells on subsequent incubation in the absence of the drug. This phenomenon has been shown to occur in Pneumococcus.[20, 21]

SUMMARY

Streptomycin and related antibiotics cause extensive and specific alterations in the coding properties of synthetic polynucleotides *in vitro*. Streptomycin-resistant ribosomes are resistant only to the shift in the code induced by streptomycin. These findings provide evidence that the ribosomes control the accuracy of the reading and may have a role in suppression.

REFERENCES

1. For a general review of streptomycin effects see Davis, B. D., and D. S. Feingold, The Bacteria, ed. I. C. Gunsalus and R. Stanier (New York: Academic Press, 1962), vol. 4, p. 343.
2. Cohen, S. S., J. Biol. Chem. **166**:393 (1946).
3. Anand, N., and B. D. Davis, Nature **185**:22 (1960).
4. Spotts, C. R., and R. Y. Stanier, Nature **192**:633 (1961).
5. Flaks, J. G., E. C. Cox, M. L. Witting, and J. R. White, Biochem. Biophys. Res. Commun. **7**:385 and 390 (1962).
6. Speyer, J. F., P. Lengyel, and C. Basilio, these Proceedings **48**:684 (1962).
7. Davies, J., these Proceedings **51**:659 (1964).
8. Cox, E. C., J. R. White, and J. G. Flaks, these Proceedings **51**:703 (1964).
9. We give currently suggested codons for reference. See Nirenberg, M. W., O. W. Jones, P. Leder, B. F. C. Clark, W. S. Sly, and S. Pestka, in Synthesis and Structure of Macromolecules, Cold Spring Harbor Symposia on Quantitative Biology, vol. 28 (1963), p. 549; and Speyer, J. F., P. Lengyel, C. Basilio, A. J. Wahba, R. S. Gardner, and S. Ochoa, in Synthesis and Structure of Macromolecules, Cold Spring Harbor Symposia on Quantitative Biology, vol. 28 (1963), p. 559.
10. Szer, W., and S. Ochoa, J. Mol. Biol., in press.
11. Feingold, D. S., and B. D. Davis, Biochim. Biophys. Acta **55**:787 (1962).
12. Jones, O. W., and M. W. Nirenberg, these Proceedings **48**:2115 (1962).
13. Okamoto, T., and M. Takanami, Biochim. Biophys. Acta **68**:325 (1963).
14. Gilbert, W., J. Mol. Biol. **6**:389 (1963); Cannon, M., R. Krug, and W. Gilbert, J. Mol. Biol. **7**:360 (1963).
15. Brody, S., and C. Yanofsky, these Proceedings **50**:9 (1963).
16. Garen, A., and O. Siddiqi, these Proceedings **48**:1121 (1962); Benzer, S., and S. Champe, these Proceedings **47**:1025 (1961). These suppressible mutations release fragments in the unsuppressed host (Sarabhai, A. S., A. O. W. Stretton, S. Brenner, and A. Bolle, Nature **201**:13 (1964)).
17. Gorini, L., and E. Kataja, these Proceedings **51**:487 (1964); see also Lederberg, E. M., L. Cavalli-Sforza, and J. Lederberg, these Proceedings **51**:678 (1964).
18. This is the modulation hypothesis of Ames, B. N., and P. E. Hartman, in Synthesis and Structure of Macromolecules, Cold Spring Harbor Symposia on Quantitative Biology, vol. 28 (1963), p. 349.
19. Lederberg, J., J. Bact. **61**:549 (1951).
20. Hotchkiss, R. D., in Enzymes: Units of Biological Structure and Function (New York: Academic Press, 1956), p. 119.
21. Ephrussi-Taylor, H., Nature **196**:748 (1962).

49 The genetics of the thalassaemias

D. J. Weatherall

Department of Medicine
University of Liverpool
Liverpool, England

During recent years considerable progress has been made in working out the genetic control and structure of the human haemoglobins. It is now clear that the inherited disorders of haemoglobin synthesis fall into two main groups. First, there are diseases which result from an inherited structural alteration in one of the globin peptide chains, sickle-cell anaemia for example. The clinical manifestations of these disorders result from the substitution of an amino acid at a critical site in the haemoglobin molecule, so altering its configuration or stability that red-cell survival is shortened or oxygen transportation grossly impaired. The second group of diseases that constitute the haemoglobinopathies are inherited defects of the rate of synthesis of the globin chains. The clinical picture of these disorders, which are known collectively as the thalassaemia syndromes, results from both underproduction of haemoglobin and imbalanced globin-chain synthesis leading to a shortened red-cell survival.

In order to describe the genetic variants of thalassaemia, it is necessary to review briefly the control of normal human haemoglobin synthesis. Unfortunately, although much is known about factors regulating the structure of haemoglobin, there is less information about the control of its rate of synthesis, information which is necessary for a more complete understanding of the pathogenesis of the thalassaemias. The molecular mechanisms of protein synthesis in general, and of haemoglobin in particular, have been the subject of several recent reviews (Arnstein, 1965; Itano, 1965; Weatherall, 1968).

1. STRUCTURE AND GENETIC CONTROL OF NORMAL HUMAN HAEMOGLOBIN

Human adult haemoglobin consists of a major component, haemoglobin A, and a minor component comprising 1.5-3.5% of the total, haemoglobin A_2. In intra-uterine life haemoglobin F is the predominant respiratory pigment, representing about 80% of the total haemoglobin at birth. Synthesis of foetal haemoglobin ceases during the first few months of life, only traces being detectable after 12 months. There is probably an embryonic haemoglobin present up to the 12th week of intra-uterine life.

The gross structure of all the human haemoglobins is similar (Baglioni, 1963). Thus, in each case the molecule consists of two separate pairs of peptide chains, each chain being related to a single haem unit. All the haemoglobins have one pair of chains in common, the α-chains. In haemoglobin A, α-chains are paired with β-chains ($\alpha_2\beta_2$), in haemoglobin A_2 with δ-chains ($\alpha_2\delta_2$), and in haemoglobin F with γ-chains ($\alpha_2\gamma_2$). This pairing of unlike sub-units is essential for the allosteric relationships necessary for normal oxygen transport.

Separate pairs of alleles control the structure of the α-chains, β-chains and δ-chains respectively, and almost certainly the γ-chains are also controlled at a distinct genetic locus. In intra-uterine life, α-chains combine with γ-

From British Medical Bulletin **25**:24-29, 1969. Used with permission.

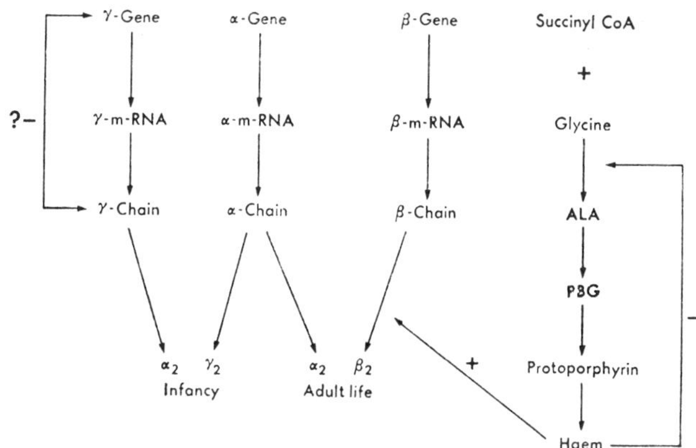

+ indicates stimulation and – signifies inhibition.

ALA: δ-aminolaevulinic acid
PBG: porphobilinogen

It is probable that some kind of feed-back mechanism is responsible for inhibition of m-RNA synthesis, but the mechanism is unknown. It is not clear whether haem is attached to the globin at ribosomal level or on the finished chain, or how haem stimulates globin synthesis

Fig. 1. Genetic control of normal haemoglobin synthesis.

chains to produce haemoglobin F. At term, γ-chain synthesis ceases and β- and δ-chain production is fully activated. α-Chains now combine with β- and δ-chains to produce haemoglobins A and A_2. This scheme is summarized in fig. 1.

The possible control mechanisms involved in this genetic system have been recently reviewed (Itano, 1965; Weatherall, 1968). Haemoglobin production can be controlled at both genetic (nuclear) and cytoplasmic levels. Since the production of RNA ceases in the red-cell precursor at the orthochromatic normoblast stage, control of haemoglobin synthesis in the reticulocyte must be achieved at the cytoplasmic level.

Quite clearly the amount of globin chain synthesized will depend on the amount of the appropriate messenger RNA (m-RNA) which is produced. Nothing is known about the factors which control the rate of mammalian m-RNA production. Since the β- and δ-loci are closely linked (Boyer, Rucknagel, Weatherall & Watson-Williams, 1963), it has been suggested that they form an operon, under the control of a closely linked operator gene. Such a system, based on the model of Jacob & Monod (1961), would provide a mechanism whereby these loci could be activated in the neonatal period and respond to various environmental control factors. There are, however, many objections to this hypothesis (Weatherall, 1968), and the control of haemoglobin synthesis at the nuclear level awaits elucidation.

Control at the cytoplasmic level is more amenable to study, although little progress has been made in this area. The final level of any haemoglobin fraction will depend on the rate at which its constituent chains are assembled on the m-RNA-ribosomal template. This, in turn, depends on the rate of globin-chain initiation, laying down of amino acids and binding in peptide linkage, termination, and association with other chains and haem to form a haemoglobin molecule. The genetic code is degenerate, several codons coding for any one amino acid. The basis for this degeneracy is the existence of several different transfer RNA's (t-RNA's) for any given amino acid. Thus, the availability of different t-RNA's will also be a rate-limiting factor in chain synthesis (Itano, 1965).

The control of haem synthesis is achieved by "feed-back" inhibition, haem depressing a critical enzymic step early in its own synthetic pathway (London, Bruns & Karibian, 1964). Furthermore, haem stimulates globin synthesis, thus allowing a synchronous production of haem and globin. Iron can also inhibit the early steps of haem synthesis and is required for maintaining stability of the aggregates of ribosomes (polysomes) active in globin synthesis.

Thus, haemoglobin synthesis is controlled by

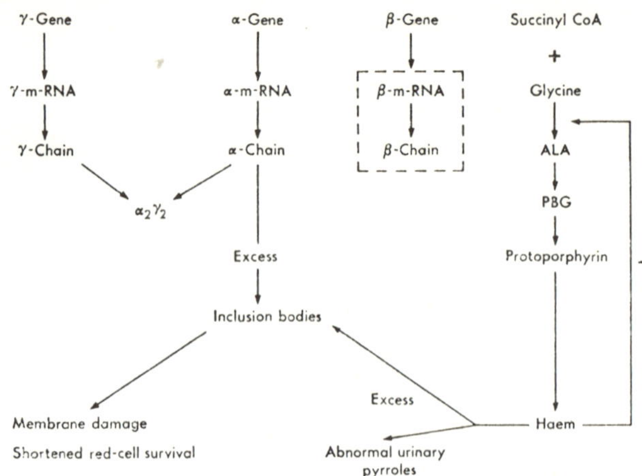

In this situation no β-chains are produced. Excess of haem and α-chain are probably responsible for the haemolytic component in this disorder. A secondary defect in haem synthesis follows a pile-up of haem with inhibition of ALA synthesis

Fig. 2. The break-down of haemoglobin control mechanisms in β-thalassaemia.

the rate of production of m-RNA's for each globin chain, the rate at which each chain is assembled and by the interrelationships between globin, haem and iron. It is against the background of these complex control mechanisms that the thalassaemia syndromes must be examined, both in terms of their pathogenesis and the production of the associated clinical manifestations (fig. 2).

2. A GENETIC CLASSIFICATION OF THE THALASSAEMIAS

From observations of the haemoglobin constitution of patients with structural haemoglobin variants, it was suggested that the thalassaemias might result from inherited abnormalities of the rate of globin synthesis (Itano, 1957, 1965). This idea was extended by Ingram & Stretton (1959), who suggested that there are two main types of thalassaemia: α-thalassaemia, resulting from a reduced rate of α-chain synthesis, and β-thalassaemia, resulting from a defective production of β-chains. This hypothesis has been fully confirmed by more recent studies of the rates of globin-chain production in different types of thalassaemia (Weatherall, Clegg & Naughton, 1965). More recently, family studies have revealed a heterogeneity of both α- and β-thalassaemia (Table I).

The β-thalassaemias are characterized by persistent γ-chain synthesis beyond the neonatal period, resulting in a variable elevation of foetal haemoglobin. In some cases β-chain synthesis is totally defective, only haemoglobins F and A_2 being synthesized. This type of β-thalassaemia runs true within families and is probably a separate genetic entity. In some families with β-thalassaemia, δ-chain synthesis is also defective, this condition being known as δ-β-thalassaemia. The levels of foetal haemoglobin in β-thalassaemia heterozygotes also point to further genetic heterogeneity within this group. Finally, there are a whole series of clinical syndromes resulting from interaction of the β-thalassaemia genes with those for β-chain structural haemoglobin variants (Weatherall, 1965).

A similar heterogeneity exists in the α-thalassaemias. Since defective α-chain synthesis results in depression of both haemoglobins A and F, these disorders find expression in both adult and foetal life. In the foetus a deficiency of α-chains results in an excess production of γ-chains which form γ_4 molecules or haemoglobin Bart's (Ager & Lehmann, 1958). In adult life, deficiency of α-chains results in an excess of β-chains which tetramerize to form β_4 molecules or haemoglobin H. A variety of α-thalassaemias exist which produce α-chain deficiencies of varying severity (Table I).

Finally, there are clinical syndromes which result from interaction of α- or β-thalassaemia with both α- and β-structural haemoglobin variants. These combinations produce the very complex series of genetic disorders which are summarized in Table I.

Table I. The thalassaemia syndromes

Type of thalassaemia	Homozygous state	Heterozygous state
β-Thalassaemia (with haemoglobin A production)	Severe anaemia; high level of haemoglobin F	Increased level of haemoglobin A_2
β-Thalassaemia (with no haemoglobin A production)	Severe anaemia; haemoglobin consists of F and A_2 only	As above
β-Thalassaemia (with no haemoglobin A production)*	Moderate anaemia; haemoglobin consists of F and A_2 only	Increased level of haemoglobin A_2 and high levels of haemoglobin F (5-15% range)
β-δ-Thalassaemia	Severe anaemia; haemoglobin consists of F only	Haemoglobin F in 5-20% range; normal levels of haemoglobin A_2
Haemoglobin Lepore thalassaemia	Severe anaemia; haemoglobin consists of Lepore and F	Normal levels of haemoglobin A_2; haemoglobins F and Lepore present
β-Thalassaemia with β-chain haemoglobin variants	Clinical severity depends on level of haemoglobin A, and proportion of haemoglobin variant. In some cases no haemoglobin A is produced. Most important are S thalassaemia, C thalassaemia and E thalassaemia	
α-Thalassaemia 1†	Death in utero; haemoglobin consists mainly of Bart's	Difficult to detect in adults; haemoglobin Bart's in 5-10% range in infancy
α-Thalassaemia 2†	Not yet recognized	Not detectable in adults; slight elevation of haemoglobin Bart's in infancy
Haemoglobin H disease	–	Probably heterozygous for α-thalassaemia 1 and 2
α-Thalassaemia 1 and 2 with haemoglobin E	–	Severe anaemia; haemoglobins Bart's, F, E and A present
δ-Thalassaemia	–	Reduced haemoglobin A_2 levels
Thalassaemia-like states	Normal levels of haemoglobin A_2 and F, with clinical picture of thalassaemia	

*Probably a distinct form of β-thalassaemia with more haemoglobin F synthesis than in typical β-thalassaemia (see Schokker, Went & Bok (1966)).

†May represent different genetic disorders in different populations.

3. THE β-THALASSAEMIAS

a. True β-thalassaemia (high haemoglobin A_2β-thalassaemia)

The homozygous state for this type of β-thalassaemia produces a clinical picture identical with that first described in 1925 by Cooley & Lee. Anaemia and splenomegaly are noted from the second or third month of life, and ultimately the bone changes, pigmentation and stunting of growth which constitute the well-known clinical picture of Cooley's anaemia are fully developed. There is marked anaemia, the blood film showing hypochromia and great variation in size and shape of the red cells. Many of the red-cell precursors have ragged inclusion bodies in the cytoplasm, which probably consist of precipitated α-chains (Fessas, 1963). The degree of anaemia and morphological abnormality of the red cells in individuals heterozygous for this form of thalassaemia is very variable. In some cases the clinical condition is almost as severe as in the homozygous state, while in other cases there may be no clinical or haematological abnormality.

The haemoglobin pattern in individuals homozygous for this form of thalassaemia consists of a variable increase in haemoglobin F, foetal haemoglobin accounting for 10% to over 90% of the total haemoglobin. The level of haemoglobin A_2 when related to haemoglobin A is markedly elevated. In the heterozygous state there is sometimes a slight increase in foetal haemoglobin and the level of haemoglobin A_2 is elevated. In other families, the clinical picture in the homozygotes is rather mild, while the level of haemoglobin F in the heterozygotes is unusually high (Schokker, Went & Bok, 1966). It seems likely that these types of β-thalassaemia represent separate genetic entities.

b. δ-β-Thalassaemia

This disorder produces a clinical picture similar to that of β-thalassaemia. However, in the homozygous state no β- or δ-chains are synthesized, the haemoglobin consisting entirely of haemoglobin F (Brancati & Baglioni, 1966). Heterozygotes have normal levels of haemoglobin A_2, with levels of haemoglobin F in the 5-20% range.

c. β-Thalassaemia and β-chain haemoglobin variants

Individuals heterozygous for a β-thalassaemia gene and a β-chain structural variant, haemoglobin S for example, have the genotype $\alpha\alpha$ $\beta^{thal.}\beta^{S}$. In cases where some β-chains are produced at the $\beta^{thal.}$-locus, the haemoglobin pattern consists of about 20% haemoglobin A with over 70% haemoglobin S. In cases where no β-chain is synthesized at all owing to the presence of the thalassaemia gene[1], haemoglobin S accounts for most of the haemoglobin, and the pattern is identical to that of sickle-cell anaemia. Thus, the clinical results of carrying genes for both β-thalassaemia and a β-chain structural variant depend on the type of β-thalassaemia and the properties of the particular structural variant.

In Africa, sickle-cell thalassaemia and haemoglobin C thalassaemia are frequently encountered. Sickle-cell thalassaemia with no haemoglobin A production is associated with the clinical picture of sickle-cell anaemia. Those cases in which some haemoglobin A is produced may be associated with very little clinical disability. Similarly, haemoglobin C thalassaemia is a very mild haemolytic disorder, usually producing no clinical disability.

Haemoglobin E thalassaemia is extremely common in South-East Asia. It has been estimated that there are 48,000 patients with this disorder in Thailand alone. The clinical picture is usually similar to that of Cooley's anaemia. The haemoglobin pattern consists of haemoglobins E and F, haemoglobin A not usually being demonstrable. This suggests that the predominant form of β-thalassaemia in South-East Asia is that in which β-chain synthesis is totally deficient.

d. Haemoglobin Lepore thalassaemia

Some individuals with haematological findings indistinguishable from those in heterozygous β-thalassaemia carry about 5-10% of a haemoglobin fraction which migrates in the region of haemoglobin S. This fraction, named Lepore after the family name in which the variant was first noted (Gerald & Diamond, 1958), has been found in individuals of a variety of racial groups, always with the clinical picture of thalassaemia. In the homozygous state no haemoglobin A or A_2 is demonstrable, only haemoglobins F and Lepore being present. The clinical picture is identical with that of severe Cooley's anaemia.

Haemoglobin Lepore has α-chains identical with haemoglobins A and F, but the non-α-chains consist of the *N*-terminal portion of the δ-chain attached to the carboxy-terminal part of the β-chain. It has been suggested that this composite δ-β-chain has arisen from a process of misalignment with non-homologous crossing-over at the closely linked δ- and β-loci (Baglioni, 1962). At least two varieties of haemoglobin Lepore have been reported. In haemoglobin Lepore (Washington) the area of fusion between δ- and β-chains is between the 85th and 115th residues, while in haemoglobin Lepore (Hollandia) the union is closer to the *N*-terminal end of the δ-chain (Barnabas & Muller, 1962).

Presumably the composite δ-β-chain is synthesized at a reduced rate, thus resulting in defective haemoglobin synthesis and the picture of thalassaemia. This suggests that the rate-limiting area of the δ-chain lies within the first four tryptic peptides, since haemoglobin Lepore (Hollandia) contains the 5th β-tryptic peptide and the rest of the β-chain.

e. The chromosomal location of the β-thalassaemia genes

The chemical studies mentioned in the previous sections provide strong evidence for the close linkage of the δ- and β-structural loci. This evidence has been strengthened by the findings in children born of matings between individuals heterozygous for both β-chain and δ-chain haemoglobin variants and normal persons. Of 50 such children reported to date, there was only one rather doubtful case of crossing-over between the β- and δ-loci (see Weatherall, 1967, for references).

The relationship of the β-thalassaemia loci to the β- and δ-structural loci can be assessed by studying children born from matings of normal individuals with those heterozygous for both β-thalassaemia and either β- or δ-chain structural haemoglobin variants. There have been no really well-documented instances of crossing-over between the β-thalassaemia and β-structural loci, while, out of 30 possible chances, there has been at least one, and possibly three, cases of crossing-over between the β-thalassaemia and δ-structural loci.

Quite clearly, far more data are required, but the few critical families studied to date suggest that the β-thalassaemia and δ-structural loci may be separable by linkage studies of this type. In this way, it may be possible to separate the β-thalassaemia locus from that for the β-structural gene.

[1] Sometimes designated the "β^0" gene.

4. THE α-THALASSAEMIAS

The genetics of the α-thalassaemias are not yet fully worked out. This is because there are relatively few α-chain structural variants to act as "markers" of the α-chain locus. In addition, the heterozygous carrier states for α-thalassaemia are extremely difficult to recognize in adult life by either haematological changes or alterations in haemoglobin patterns.

Since, in normal infants, the β-chain locus is fully activated at about term, there will be a short period just after birth during which both β- and γ-chains are competing for available α-chains. Any imbalance of globin-chain production will be particularly noticeable at this time and, in fact, a slight excess of γ-chains is produced in most normal infants, resulting in levels of haemoglobin Bart's (γ_4) which do not exceed 1% of the total haemoglobin (Weatherall, 1963). It is probable that haemoglobin Bart's (γ_4) is produced rather than haemoglobin H (β_4) in situations where there is an excess of β- and γ-chains over α-chains because of the greater affinity of α-chains for β-chains.

For these reasons, the level of haemoglobin Bart's in the neonatal period is probably a useful guide to the degree of α-chain deficiency, and several types of α-thalassaemia can be defined in this way. Infants with almost 100% haemoglobin Bart's do not survive, being stillborn at 34-36 weeks of age or, if they do go to term, dying within a few minutes of birth. Infants with 20-30% haemoglobin Bart's at birth do survive and develop a moderately severe form of thalassaemia associated with similar quantities of haemoglobin H as they grow older. Infants born with levels of haemoglobin Bart's in the 5-10% or 1-2% range are probably carriers for α-thalassaemia genes. In these infants the haemoglobin Bart's disappears by six months of age and is not replaced by haemoglobin H. These conditions have been designated α-thalassaemia 1 and 2 respectively (Wasi, Na-Nakorn & Suingdumrong, 1964).

Each of these varieties of α-thalassaemia will now be considered.

a. The haemoglobin Bart's stillbirth syndrome: ?homozygous α-thalassaemia 1

This syndrome is the cause of much foetal wastage in South-East Asia (Lie-Injo Luan Eng, Lie Hong Ghie, Ager & Lehmann, 1962). Affected infants are usually stillborn at about 34 weeks of age with the clinical picture of severe hydrops foetalis. Occasionally they survive to term but die a few minutes after delivery.

The haemoglobin pattern consists mainly of haemoglobin Bart's with traces of haemoglobin H and a so far unidentified component which migrates just behind haemoglobin Bart's at pH 8.6. Although no α-chains can be demonstrated in many of these infants, a trace of haemoglobin A can occasionally be seen on starch-gel electrophoresis, and α-chains have been isolated by column chromatography (D. J. Weatherall and J. B. Clegg, unpublished observation, 1968).

The parents of these infants show minimal haematological changes and no alteration in their haemoglobin constitution. It is assumed that they are α-thalassaemia carriers and this condition may, in fact, represent the homozygous α-thalassaemia 1 state.

b. Haemoglobin H disease

This condition is characterized by the clinical picture of thalassaemia of intermediate severity associated with 10-30% haemoglobin H (β_4). In addition, excess γ-chains and δ-chains are also usually found in the red cells of these patients. Haemoglobin H is unstable and tends to precipitate in older red cells, resulting in a shortened red-cell survival.

The genetics of this disorder are still far from clear, but it seems likely that it results from the interaction of two α-thalassaemia genes, possibly α-thalassaemias 1 and 2 (Wasi *et al.* 1964; Huehns, 1965). The samples of cord blood of babies born of patients with this disorder nearly always show an increase in haemoglobin Bart's, either in the 1-2% range, or in the 5-10% range. The many instances of direct transmission of haemoglobin H disease from parent to child suggest that either these α-thalassaemia genes are not alleles, or that one of them has an extremely high frequency in populations where haemoglobin H disease is common.

c. Heterozygous α-thalassaemia

The heterozygous α-thalassaemias are very difficult to define in adult life. Furthermore, it is not certain that the level of haemoglobin Bart's in the neonatal period can be used as an index of any one form of the disorder. For instance, it is by no means certain that the α-thalassaemia-1 gene, which is characterized by levels of haemoglobin Bart's in the 5-15% range, is the same entity in South-East Asia and Africa. The hydrops picture is not seen in

Negroes, yet 1-2% of American Negroes carry this amount of haemoglobin Bart's in the neonatal period (Weatherall, 1963). Until careful quantitative data of haemoglobin Bart's levels in samples of cord blood from a variety of racial groups are available, the heterozygous α-thalassaemias cannot be defined further.

α-Chain haemoglobin variants have been found in association with α-thalassaemia in several racial groups. Haemoglobin Q α-thalassaemia is found in Orientals, the haemoglobin consisting of Q, H and Bart's—i.e., no α-chain is produced (Dormandy, Lock & Lehmann, 1961). Haemoglobin I α-thalassaemia was found in a Negro family (Atwater, Schwartz, Erslev, Montgomery & Tocantins, 1960), and in this case there was about 70% haemoglobin I and some haemoglobin A, suggesting that α-chain synthesis was not completely deficient. Clearly the α-thalassaemia genes may, like those for β-thalassaemias, result in total or partial deficiency of chain production.

d. α-Thalassaemia with β-chain haemoglobin variants

The combination of α-thalassaemia 1 and 2 and haemoglobin E produces a well-defined clinical syndrome in South-East Asia (Wasi, Sookanek, Pootrakul, Na-Nakorn & Suingdumrong, 1967). The clinical picture is that of a moderately severe Cooley's anaemia, while the haemoglobin pattern consists of haemoglobins A, E and Bart's with levels of haemoglobin E in the 13-15% range.

It is a curious feature of all the combinations of α-thalassaemia with β-chain haemoglobin variants that the level of β-chain variant is markedly decreased. It appears that, if α-chains are in short supply, normal β-chains are bound as compared with abnormal β-chains.

5. DISTRIBUTION OF THE THALASSAEMIAS

The thalassaemias occur very frequently in the Mediterranean region, the Middle East, parts of India and Pakistan, and the Far East. In addition, sporadic cases occur in practically every racial group.

Detailed information about the types of thalassaemia in different populations is scanty because of the practical difficulties of recognizing heterozygous thalassaemia in the presence of nutritional anaemia and the lack of a method for the identification of α-thalassaemia carriers. A very high incidence of β-thalassaemia has been reported in northern Italy, while in Greece the carrier rate ranges from 6% to 14%. The disease is probably wide spread in the Arab races but less common in the African Negro, occurring with any frequency only in West Africa. Incidences of from 2% to 5% have been reported for β-thalassaemia in Thailand, and it seems likely that the disease will be equally prevalent in other parts of South-East Asia.

α-Thalassaemia, as judged by the presence of haemoglobin Bart's in infancy, is wide spread in South-East Asia, but much less common in the Mediterranean region and Africa (Weatherall, 1965). Other populations have not yet been examined for the presence of haemoglobin Bart's in samples of cord blood.

While evidence suggesting that β-thalassaemia confers resistance to malarial infection has been obtained in Sardinia and New Guinea, the distribution of thalassaemia in Greece does not entirely support this hypothesis. The whole question of the factors which maintain the remarkably high frequency of the thalassaemia gene remains open.

6. RELATIONSHIP BETWEEN ABNORMAL GLOBIN-CHAIN PRODUCTION AND THE CLINICAL PICTURE OF THALASSAEMIA

During the last few years, methods have become available for measuring the rate of globin-chain production in human reticulocytes, and a clear picture of the kinetics of abnormal globin-chain synthesis in the thalassaemias has emerged (Weatherall *et al.* 1965; Bank & Marks, 1966; Huehns & Modell, 1967).

In homozygous β-thalassaemia, α-chain synthesis greatly exceeds that of the β- and γ-chains. Thus, there is a large excess of α-chain produced, β-chain synthesis being either partially or totally defective. Kinetic studies have shown that this excess of α-chain is unstable and rapidly becomes associated with the stromal fraction. It is almost certain that these precipitated α-chains form the inclusion bodies seen in the red-cell precursors in β-thalassaemia. These observations suggest that the anaemia of β-thalassaemia results from the combination of an over-all deficit of haemoglobin synthesis together with the harmful effects of excessive and unstable globin-chain production.

The greatest excess of α-chains will be in those cells in which γ-chain synthesis is least active. The cells with the highest level of foetal haemoglobin do, in fact, have the longest survival in β-thalassaemia homozygotes (Gabuzda, Nathan & Gardner, 1963). In the absence of sufficient γ- or β-chains, the excess α-chains precipitate, forming red-cell inclusions. Some of this material may never leave the bone-marrow, break-down of the haem attached to these chains accounting for the high level of "early labelled" bile and abnormal urinary pyrroles reported in β-thalassaemia. Such inclusion bodies as do reach the peripheral blood are probably removed from the circulation in the spleen, since they are most readily found after splenectomy. How precipitated globin chains produce shortened red-cell survival is uncertain. It is possible that erythrocyte-membrane damage results from mechanical trauma due to the rigidity of the inclusion bodies but binding of membrane sulphydryl groups during precipitation may be equally important. The role of the excess of α-chains in producing the anaemia of β-thalassaemia is emphasized by the good correlation of the ratio of α-chain to β-chain plus γ-chain production with the ^{51}Cr red-cell survival time (Bargellesi, Pontremoli, Menini & Conconi, 1968).

Similar studies have been performed on the cells of patients with haemoglobin H disease (Clegg & Weatherall, 1967). The rate of β-chain synthesis exceeds that of α-chain production by a factor of 1.5 to 3 times. The β-chains rapidly form haemoglobin H, in this state being no longer available to combine with α-chains to make haemoglobin A. Haemoglobin H precipitates as the red cells age, the mechanism of red-cell destruction secondary to precipitation of excess β-chains probably being very similar to that which results from the precipitation of α-chains in β-thalassaemia. Certainly there is evidence of membrane damage as judged by increased permeability to cations in both forms of thalassaemia (Nathan & Gunn, 1966).

It is clear, therefore, that the clinical results of both types of thalassaemia gene are due to the combined effects of defective globin production and globin-chain imbalance with inclusion-body production, membrane damage, and premature red-cell destruction (fig. 2).

7. THE MOLECULAR ABNORMALITY IN THE THALASSAEMIAS

It is now clear that the thalassaemias result from either a partial or total deficit of globin synthesis. In-vitro radioactive experiments have provided no evidence for the production of a grossly abnormal globin chain in any of the thalassaemias in which there is a total deficit of globin. Furthermore, the chemistry of such haemoglobin A as is produced in these disorders is normal. These observations suggest that the underlying defect is a true deficiency of globin chains.

A deficit of globin could arise from either a reduced amount of m-RNA for a particular chain, or from the production of an abnormal m-RNA, such that the mechanism of chain assembly, i.e., initiation, translation and termination, is retarded. Such an abnormality could arise in several ways. Thus, it might follow an alteration in a codon such that chain initiation is ineffective. Alternatively, a t-RNA which is in short supply might be required to insert a given amino acid due to an altered codon anywhere along the m-RNA strand. Finally, the mechanism for chain release might be abnormal.

Recent experimental work on the assembly of globin chains in thalassaemia (Clegg, Weatherall, Na-Nakorn & Wasi, 1968) suggests that, at least in those cases in which some haemoglobin A is produced, the chain-assembly time is normal. These findings suggest that the disease results from a quantitative reduction in m-RNA for the affected chain. This may imply that thalassaemia is, in fact, a "controller" gene disease, but there are several other theoretical ways in which m-RNA synthesis could be retarded. The scanty genetic data on the position of the β-thalassaemia loci would not be incompatible with the "controller" gene hypothesis. Until mammalian m-RNA can be isolated and its rate of production measured, this problem may remain unsolved.

REFERENCES

Ager, J. A. M. & Lehmann, H. (1958) Br. med. J. **1**:929.

Arnstein, H. R. V. (1965) Br. med. Bull. **21**:217.

Atwater, J., Schwartz, I. R., Erslev, A. J., Montgomery, T. L. & Tocantins, L. M. (1960) New Engl. J. Med. **263**:1215.

Baglioni, C. (1962) Proc. natn. Acad. Sci. U.S.A. **48**:1880.

Baglioni, C. (1963) In: Taylor, J. H., ed. Molecular genetics, pt 1, p. 405. Academic Press, London.

Bank, A. & Marks, P. A. (1966) Nature, Lond. **212**:1198.

Bargellesi, A., Pontremoli, S., Menini, C. & Conconi, F. (1968) Eur. J. Biochem. **3**:364.

Barnabas, J. & Muller, C. J. (1962) Nature, Lond. **194**:931.

Boyer, S. H., Rucknagel, D. L., Weatherall, D. J. & Watson-Williams, E. J. (1963) Am. J. hum. Genet. **15**:438.

Brancati, C. & Baglioni, C. (1966) Nature, Lond. **212**:262.

Clegg, J. B. & Weatherall, D. J. (1967) Nature, Lond. **215**:1241.

Clegg, J. B., Weatherall, D. J., Na-Nakorn, S. & Wasi, P. (1968) Nature, Lond. **220**:664.

Cooley, T. B. & Lee, P. (1925) Trans. Am. pediat. Soc. **37**:29.

Dormandy, K. M., Lock, S. P. & Lehmann, H. (1961) Br. med. J. **1**:1582.

Fessas, P. (1963) Blood **21**:21.

Gabuzda, T. G., Nathan, D. G. & Gardner, F. H. (1963) J. clin. Invest. **42**:1678.

Gerald, P. S. & Diamond, L. K. (1958) Blood **13**:61.

Huehns, E. R. (1965) Post-grad. med. J. **41**:718.

Huehns, E. R. & Modell, C. B. (1967) Trans. R. Soc. trop. Med. Hyg. **61**:157.

Ingram, V. M. & Stretton, A. O. W. (1959) Nature, Lond. **184**:1903.

Itano, H. A. (1957) Adv. Protein Chem. **12**:215.

Itano, H. A. (1965) In: Jonxis, J. H. P., ed. Abnormal haemoglobins in Africa: a symposium organized by the Council for International Organizations of Medical Sciences, established under the joint auspices of UNESCO and WHO, p. 3. Blackwell Scientific Publications, Oxford.

Jacob, F. & Monod, J. (1961) J. molec. Biol. **3**:318.

Lie-Injo Luan Eng, Lie Hong Ghie, Ager, J. A. M. & Lehmann, H. (1962) Br. J. Haemat. **8**:1.

London, I. M., Bruns, G. P. & Karibian, D. (1964) Medicine, Baltimore **43**:789.

Nathan, D. G. & Gunn, R. B. (1966) Am. J. Med. **41**:815.

Schokker, R. C., Went, L. N. & Bok, J. (1966) Nature, Lond. **209**:44.

Wasi, P., Na-Nakorn, S. & Suingdumrong, A. (1964) Nature, Lond. **204**:907.

Wasi, P., Sookanek, M., Pootrakul, S., Na-Nakorn, S. & Suingdumrong, A. (1967) Br. med. J. **4**:29.

Weatherall, D. J. (1963) Br. J. Haemat. **9**:265.

Weatherall, D. J. (1965) The thalassaemia syndromes. Blackwell Scientific Publications, Oxford.

Weatherall, D. J. (1967) Prog. med. genet. **5**:8.

Weatherall, D. J. (1968) A. Rev. Med. **19**:217.

Weatherall, D. J., Clegg, J. B. & Naughton, M. A. (1965) Nature, Lond. **208**:1061.

50 The complete amino acid sequence of the tryptophan synthetase A protein (α subunit) and its colinear relationship with the genetic map of the A gene

Charles Yanofsky
Gabriel R. Drapeau*
John R. Guest*
Bruce C. Carlton*
Department of Biological Sciences
Stanford University
Palo Alto, California

Previously we presented findings demonstrating the existence of a colinear relationship between gene structure (the genetic map) and protein structure.[1,2] The altered tryptophan synthetase A proteins produced by a group of mutants of *Escherichia coli* were examined for primary structure changes and each mutant protein was found to differ from the wild-type protein by a change of a single amino acid.[3-6] The colinear relationship was then established by showing that the order of the positions at which these single amino acid changes occurred in the A protein was the same as the order of the respective mutational sites on the genetic map.[1,2] It was also observed in these investigations that distance on the genetic map was reasonably representative of distance in the polypeptide chain.[1,2] Colinearity of gene structure and protein structure has also been convincingly demonstrated as a result of studies with very different experimental material, viz., nonsense mutants[7,8] and frame-shift mutants.[9]

Recently the complete sequence of the 267 amino acid residues in the tryptophan synthetase A protein has been determined,[10] and consequently the relationship between the genetic map of the A gene and the changes in mutationally altered A proteins can be reconsidered in terms of the primary structure of the entire protein. The purpose of this report is to re-examine this relationship.

RESULTS AND DISCUSSION

The amino acid sequence of the A protein, shown in Figure 1, was determined by analysis of fragments derived by treating the protein with various proteolytic enzymes or with cyanogen bromide. The details of the sequence studies will be described elsewhere.[10]

The genetic map of the relevant mutationally altered sites in the A gene is presented in Figure 2. Mutants A38 and A96 do not produce detectable altered A proteins; the map locations of their alterations are included because these sites presently represent the most distant sites in the A gene. The A38 site maps closest to the B gene and to the operator end of the tryptophan operon.[11,12] The other altered sites on the map are the genetic locations of mutational alterations which lead to the single amino acid substitutions in the A protein that are indicated. Of these altered sites, the positions of the A3 and A33 changes are closest to the A38 site. As can be seen in Figure 2, the A3 and A33 mutational alterations lead to amino acid changes at position 48 in the protein, the closest position to the amino-terminal end of the protein at which amino acid changes are observed. The other mutationally altered sites shown correspond to amino acid changes at positions in the protein which are in the same relative order as the respective altered sites on the genetic map, as reported previously.[1,2] The A169 mutational alteration is closest to the A96 site on the genetic map and the affected position in the protein is only 33 residues from the carboxy-terminus of the protein. Thus, for almost the entire length of the map of the A gene, the existing evidence

From Proceedings of the National Academy of Sciences 57:296-298, 1967. Used with permission.

These investigations were supported by grants from the U.S. Public Health Service and the National Science Foundation.

The authors are pleased to acknowledge the assistance of Virginia Horn, Donald Vinicor, Deanna Thorpe, and George Pegelow at various stages of this investigation.

*Present addresses: G. R. D., Department of Microbiology, University of Montreal, Montreal, Quebec, Canada; J. R. G., Department of Microbiology, Sheffield University, Sheffield, England; B. C. C., Department of Biology, Yale University, New Haven, Connecticut.

Met — Gln — Arg — Tyr — Glu — Ser — Leu — Phe — Ala — Gln — Leu — Lys — Glu — Arg — Lys — Glu — Gly — Ala — Phe — Val —
10 20

Pro — Phe — Val — Thr — Leu — Gly — Asp — Pro — Gly — Ile — Glu — Gln — Ser — Leu — Lys — Ile — Asp — Thr — Leu — Ile —
30 40

Glu — Ala — Gly — Ala — Asp — Ala — Leu — Glu — Leu — Gly — Ile — Pro — Phe — Ser — Asp — Pro — Leu — Ala — Asp — Gly —
50 60

Pro — Thr — Ile — Gln — Asn — Ala — Thr — Leu — Arg — Ala — Phe — Ala — Ala — Gly — Val — Thr — Pro — Ala — Gln — Cys —
70 80

Phe — Glu — Met — Leu — Ala — Leu — Ile — Arg — Gln — Lys — His — Pro — Thr — Ile — Pro — Ile — Gly — Leu — Leu — Met —
90 100

Tyr — Ala — Asn — Leu — Val — Phe — Asn — Lys — Gly — Ile — Asp — Glu — Phe — Tyr — Ala — Gln — Cys — Glu — Lys — Val —
110 120

Gly — Val — Asp — Ser — Val — Leu — Val — Ala — Asp — Val — Pro — Val — Gln — Glu — Ser — Ala — Pro — Phe — Arg — Gln —
130 140

Ala — Ala — Leu — Arg — His — Asn — Val — Ala — Pro — Ile — Phe — Ile — Cys — Pro — Pro — Asn — Ala — Asp — Asp — Asp —
150 160

Leu — Leu — Arg — Gln — Ile — Ala — Ser — Tyr — Gly — Arg — Gly — Tyr — Thr — Tyr — Leu — Leu — Ser — Arg — Ala — Gly —
170 180

Val — Thr — Gly — Ala — Glu — Asn — Arg — Ala — Ala — Leu — Pro — Leu — Asn — His — Leu — Val — Ala — Lys — Leu — Lys —
190 200

Glu — Tyr — Asn — Ala — Ala — Pro — Pro — Leu — Gln — Gly — Phe — Gly — Ile — Ser — Ala — Pro — Asp — Gln — Val — Lys —
210 220

Ala — Ala — Ile — Asp — Ala — Gly — Ala — Ala — Gly — Ala — Ile — Ser — Gly — Ser — Ala — Ile — Val — Lys — Ile — Ile —
230 240

Glu — Gln — His — Asn — Ile — Glu — Pro — Glu — Lys — Met — Leu — Ala — Ala — Leu — Lys — Val — Phe — Val — Gln — Pro —
250 260

Met — Lys — Ala — Ala — Thr — Arg — Ser

Fig. 1. Amino acid sequence of the tryptophan synthetase A protein of *E. coli.*[10] The underlined residues are at the positions in the protein at which amino acid changes have occurred in mutants.

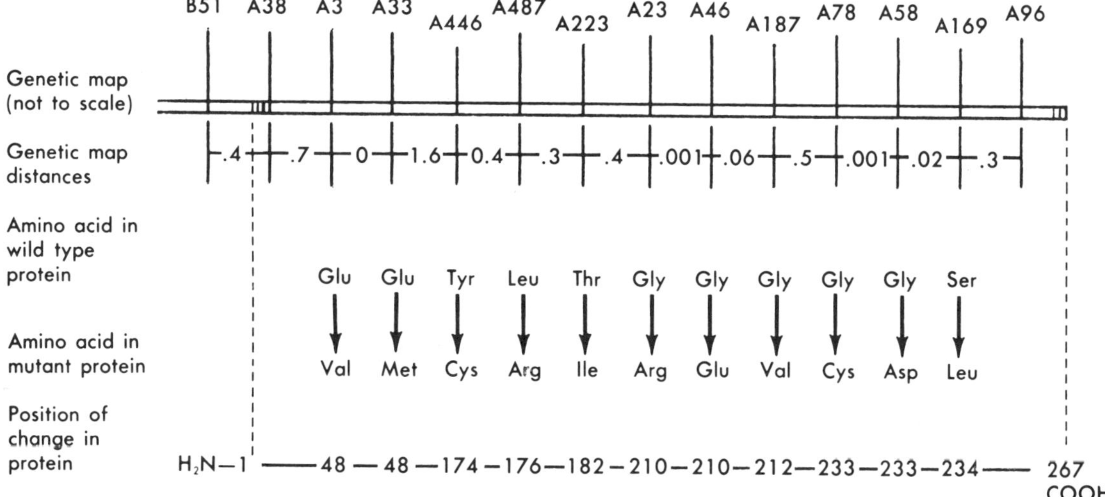

Fig. 2. Genetic map of the A gene and the corresponding amino acid changes in the A protein. The positions of these changes in the amino acid sequence are also indicated. Mutants A3 and A33, and their amino acid replacements, will be described in detail elsewhere.[13]

indicates that it is colinear with the structure of the A protein. The established orientation of the A protein relative to the A gene and consequently to the operator region of the operon, in conjunction with the orientation of the operon on the *E. coli* chromosome,[14] permits the orientation of the amino acid sequence of the A protein relative to the chromosome. The nucleotide sequence corresponding to the A protein runs in a clockwise direction from the region specifying the COOH-terminal end; i.e., the order is *thr-gal*—A gene region specifying the COOH-terminal end of the A protein—A gene region specifying the amino-terminal end of the A protein—*his.*

A representative value relating distance on the genetic map to distance in the polypeptide chain can be calculated by dividing the map distance separating the A3 and A169 sites by the number of amino acid residues in between the positions of the corresponding amino acid changes. The value so obtained is about 0.015 map units per amino acid residue. Using this value we can estimate that the genetic map of the A gene should extend some 0.7 units to the left of the A3 site and about 0.5 units beyond the A169 site. This would give a total length of approximately 4.2 map units for the A gene, and would place the A38 site at or very near the beginning of the A gene.

REFERENCES

1. Yanofsky, C., in Cold Spring Harbor Symposia on Quantitative Biology, vol. 28 (1963), p. 296.
2. Yanofsky, C., B. C. Carlton, J. R. Guest, D. R. Helinski, and U. Henning, these Proceedings **51**:266 (1964).
3. Helinski, D. R., and C. Yanofsky, these Proceedings **48**:173 (1962).
4. Henning, U., and C. Yanofsky, these Proceedings **48**:183 (1962).
5. Guest, J. R., and C. Yanofsky, J. Biol. Chem. **240**:679 (1965).
6. Carlton, B. C., and C. Yanofsky, J. Biol. Chem. **240**:690 (1965).
7. Sarabhai, A. S., A. O. W. Stretton, S. Brenner and A. Bolle, Nature **201**:13 (1964).
8. Fowler, A. V., and I. Zabin, Science **154**:1027 (1966).
9. Streisinger, G., Y. Okada, J. Emrich, J. Newton, A. Tsugita, E. Terzaghi, and M. Inouye, in Cold Spring Harbor Symposia on Quantitative Biology, vol. 31, in press.
10. Guest, J. R., B. C. Carlton, and C. Yanofsky, in preparation; Carlton, B. C., J. R. Guest, and C. Yanofsky, in preparation; Drapeau, G., and C. Yanofsky, in preparation; and Yanofsky, C., J. R. Guest, G. Drapeau, and B. C. Carlton, in preparation.
11. Guest, J. R., and C. Yanofsky, Nature **210**:799 (1966).
12. Matsushiro, A., K. Sato, J. Ito, S. Kida, and F. Imamoto, J. Mol. Biol. **11**:54 (1965).
13. Drapeau, G., W. Brammar, and C. Yanofsky, in preparation.
14. Signer, E. R., J. R. Beckwith, and S. Brenner, J. Mol. Biol. **14**:153 (1965).

51 Genetic control of biochemical reactions in Neurospora

G. W. Beadle*
E. L. Tatum
Biological Department
Stanford University
Palo Alto, California

From the standpoint of physiological genetics the development and functioning of an organism consist essentially of an integrated system of chemical reactions controlled in some manner by genes. It is entirely tenable to suppose that these genes which are themselves a part of the system, control or regulate specific reactions in the system either by acting directly as enzymes or by determining the specificities of enzymes.[1] Since the components of such a system are likely to be interrelated in complex ways, and since the synthesis of the parts of individual genes are presumably dependent on the functioning of other genes, it would appear that there must exist orders of directness of gene control ranging from simple one-to-one relations to relations of great complexity. In investigating the rôles of genes, the physiological geneticist usually attempts to determine the physiological and biochemical bases of already known hereditary traits. This approach, as made in the study of anthocyanin pigments in plants,[2] the fermentation of sugars by yeasts[3] and a number of other instances,[4] has established that many biochemical reactions are in fact controlled in specific ways by specific genes. Furthermore, investigations of this type tend to support the assumption that gene and enzyme specificities are of the same order.[5] There are, however, a number of limitations inherent in this approach. Perhaps the most serious of these is that the investigator must in general confine himself to a study of non-lethal heritable characters. Such characters are likely to involve more or less non-essential so-called "terminal" reactions.[5] The selection of these for genetic study was perhaps responsible for the now rapidly disappearing belief that genes are concerned only with the control of "superficial" characters. A second difficulty, not unrelated to the first, is that the standard approach to the problem implies the use of characters with visible manifestations. Many such characters involve morphological variations, and these are likely to be based on systems of biochemical reactions so complex as to make analysis exceedingly difficult.

Considerations such as those just outlined have led us to investigate the general problem of the genetic control of developmental and metabolic reactions by reversing the ordinary procedure and, instead of attempting to work out the chemical bases of known genetic characters, to set out to determine if and how genes control known biochemical reactions. The ascomycete *Neurospora* offers many advantages for such an approach and is well suited to genetic studies.[6] Accordingly, our program has been built around this organism. The procedure is based on the assumption that x-ray treatment will induce mutations in genes concerned with the control of known specific chemical reactions. If the organism must be able to carry out a certain chemical reaction to survive on a given medium, a mutant unable to do this will obviously be lethal on this medium. Such a mutant can be maintained and studied, however, if it will grow on a medium to which has been added the essential product of the genetically blocked reaction. The experimental procedure based on this reasoning can best be illustrated by considering a hypothetical example. Normal strains of *Neurospora crassa* are able to use sucrose as a carbon source, and are therefore able to carry out the specific and enzymatically controlled reaction involved in the hydrolysis of this sugar. Assuming this reaction to be genetically controlled, it should be possible to induce a gene to mutate to a condition such that the organism could no

From Proceedings of National Academy of Sciences 27:499-506, 1941. Used with permission.

Work supported in part by a grant from the Rockefeller Foundation. The authors are indebted to Doctors B. O. Dodge, C. C. Lindegren and W. S. Malloch for stocks and for advice on techniques, and to Miss Caryl Parker for technical assistance.

*Present address: Professor of Biology, University of Chicago, Chicago, Illinois.

longer carry out sucrose hydrolysis. A strain carrying this mutant would then be unable to grow on a medium containing sucrose as a sole carbon source but should be able to grow on a medium containing some other normally utilizable carbon source. In other words, it should be possible to establish and maintain such a mutant strain on a medium containing glucose and detect its inability to utilize sucrose by transferring it to a sucrose medium.

Essentially similar procedures can be developed for a great many metabolic processes. For example, ability to synthesize growth factors (vitamins), amino acids and other essential substances should be lost through gene mutation if our assumptions are correct. Theoretically, any such metabolic deficiency can be "by-passed" if the substance lacking can be supplied in the medium and can pass cell walls and protoplasmic membranes.

In terms of specific experimental practice, we have devised a procedure in which x-rayed single-spore cultures are established on a so-called "complete" medium, i.e., one containing as many of the normally synthesized constituents of the organism as is practicable. Subsequently these are tested by transferring them to a "minimal" medium, i.e., one requiring the organism to carry on all the essential syntheses of which it is capable. In practice the complete medium is made up of agar, inorganic salts, malt extract, yeast extract and glucose. The minimal medium contains agar (optional), inorganic salts and biotin, and a disaccharide, fat or more complex carbon source. Biotin, the one growth factor that wild type *Neurospora* strains cannot synthesize,[7] is supplied in the form of a commercial concentrate containing 100 micrograms of biotin per cc.[8] Any loss of ability to synthesize an essential substance present in the complete medium and absent in the minimal medium is indicated by a strain growing on the first and failing to grow on the second medium. Such strains are then tested in a systematic manner to determine what substance or substances they are unable to synthesize. These subsequent tests include attempts to grow mutant strains on the minimal medium with (1) known vitamins added, (2) amino acids added or (3) glucose substituted for the more complex carbon source of the minimal medium.

Single ascospore strains are individually derived from perithecia of *N. crassa* and *N. sitophila* x-rayed prior to meiosis. Among approximately 2000 such strains, three mutants have been found that grow essentially normally on the complete medium and scarcely at all on the minimal medium with sucrose as the carbon source. One of these strains *(N. sitophila)* proved to be unable to synthesize vitamin B_6 (pyridoxine). A second strain *(N. sitophila)* turned out to be unable to synthesize vitamin B_1 (thiamine). Additional tests show that this strain is able to synthesize the pyrimidine half of the B_1 molecule but not the thiazole half. If thiazole alone is added to the minimal medium, the strain grows essentially normally. A third strain *(N. crassa)* has been found to be unable to synthesize para-aminobenzoic acid. This mutant strain appears to be entirely normal when grown on the minimal medium to which *p*-aminobenzoic acid has been added. Only in the case of the "pyridoxinless" strain has an analysis of the inheritance of the induced metabolic defect been investigated. For this reason detailed accounts of the thiamine-deficient and *p*-aminobenzoic acid-deficient strains will be deferred.

Qualitative studies indicate clearly that the pyridoxinless mutant, grown on a medium containing one microgram or more of synthetic vitamin B_6 hydrochloride per 25 cc. of medium, closely approaches in rate and characteristics of growth normal strains grown on a similar medium with no B_6. Lower concentrations of B_6 give intermediate growth rates. A preliminary investigation of the quantitative dependence of growth of the mutant on vitamin B_6 in the medium gave the results summarized in table 1. Additional experiments have given results essentially similar but in only approximate quantitative agreement with those of table 1. It is clear that additional study of the details of culture conditions is necessary before rate of weight increase of this mutant can be used as an accurate assay for vitamin B_6.

It has been found that the progression of the frontier of mycelia of *Neurospora* along a horizontal glass culture tube half filled with an agar medium provides a convenient method of investigating the quantitative effects of growth factors. Tubes of about 13 mm. inside diameter and about 40 cm. in length are used. Segments of about 5 cm. at the two ends are turned up at an angle of about 45°. Agar medium is poured in so as to fill the tube about half full and is allowed to set with the main segment of the tube in a horizontal position. The turned up ends of the tube are stoppered with cotton plugs. Inoculations are made at one end of the agar surface and the position of the advancing front recorded at convenient intervals. The frontier formed by the advancing mycelia is remarkably well defined, and there is no difficulty in determining its position to within a millimeter or less. Progression along such tubes is strictly linear with time and the rate is independent of tube length (up to 1.5 meters). The rate is not changed by reducing the inside tube diameter to 9 mm., or by sealing one or both ends. It therefore appears that gas diffusion is in no way limiting in such tubes.

The results of growing the pyridoxinless strain in horizontal tubes in which the agar medium contained varying amounts of B_6 are shown graphically in figures 1 and 2. Rate of progression is clearly a function of vitamin B_6 concentration in the medium.[10] It is likewise evident that there is no significant difference in rate between the mutant supplied with B_6 and the normal strain growing on a medium without this vitamin. These results are consistent with the assumption that the primary physiological difference between pyridoxinless and normal strains is the inability of the former to carry out the synthesis of vitamin B_6. There is certainly more than one step in this synthesis and accordingly the gene differential involved is presumably concerned with only one specific step in the biosynthesis of vitamin B_6.

In order to ascertain the inheritance of the pyridoxinless character, crosses between normal and mutant strains were made. The techniques for hybridization and ascospore isolation have been worked out and described by Dodge, and by Lindegren.[6] The ascospores from 24 asci of the cross were isolated and their positions in the asci recorded. For some unknown reason, most of these failed to germinate. From seven asci, however, one or more spores germinated. These were grown on a medium containing glucose, malt-extract and yeast extract, and in this they all grew normally. The normal and mutant cultures were differentiated by growing them on a B_6 deficient medium. On this medium the mutant cultures grew very little, while the non-mutant ones grew normally. The results are summarized in table 2. It is clear from these rather limited data that this inability to synthesize vitamins B_6 is transmitted as it should be if it were differentiated from normal by a single gene.

The preliminary results summarized above appear to us to indicate that the approach outlined may offer considerable promise as a method of learning more about how genes regulate development and function. For example, it should be possible, by finding a number of mutants unable to carry out a particular step in a given synthesis, to determine whether only one gene is ordinarily concerned with the immediate regulation of a given specific chemical reaction.

It is evident, from the standpoints of biochemistry and physiology, that the method outlined is of value as a technique for discovering additional substances of physiological significance. Since the complete medium used can be made up with yeast extract or with an extract of normal *Neurospora*, it is evident that if, through mutation, there is lost the ability to synthesize an essential substance, a test strain is thereby made available for use in isolating the

Table 1. Growth of pyridoxinless strain of *N. sitophila* on liquid medium containing inorganic salts,[9] 1% sucrose, and 0.004 microgram biotin per cc. Temperature 25°C. Growth period, 6 days from inoculation with conidia

Micrograms B_6 per 25 cc. medium	Strain	Dry weight mycelia, mg.
0	Normal	76.7
0	Pyridoxinless	1.0
0.01	"	4.2
0.03	"	5.7
0.1	"	13.7
0.3	"	25.5
1.0	"	81.1
3.0	"	81.1
10.0	"	65.4
30.0	"	82.4

Table 2. Results of classifying single ascospore cultures from the cross of pyridoxinless and normal *N. sitophila*

Ascus number	1	2	3	4	5	6	7	8
17	–	*pdx*	*pdx*	*pdx*	*N*	*N*	*N*	–
18	–	–	*N*	*N*	–	–	*pdx*	*pdx*
19	–	*pdx*	–	–	–	–	–	*N*
20	–	–	*N*	–	–	–	–	*pdx*
22	–	–	*N*	–	–	–	–	–
23	–	*	*	*	*N*	*N*	*pdx*	*pdx*
24	*N*	*N*	*N*	*N*	*pdx*	*pdx*	*pdx*	*pdx*

N, normal growth on B_6-free medium. *pdx,* slight growth on B_6-free medium. Failure of ascospore germination indicated by dash.

*Spores 2, 3 and 4 isolated but positions confused. Of these, two germinated and both proved to be mutants.

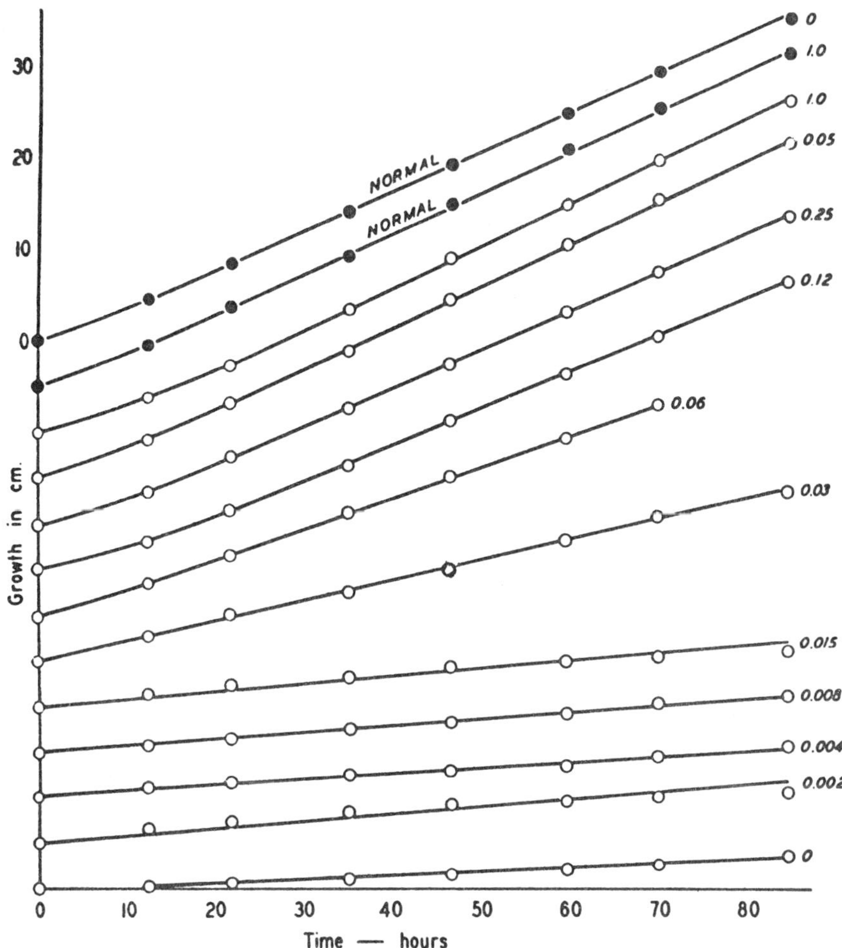

Fig. 1. Growth of normal (top two curves) and pyridoxinless (remaining curves) strains of *Neurospora sitophila* in horizontal tubes. The scale on the ordinate is shifted a fixed amount for each successive curve in the series. The figures at the right of each curve indicate concentration of pyridoxine (B_6) in micrograms per 25 cc. medium.

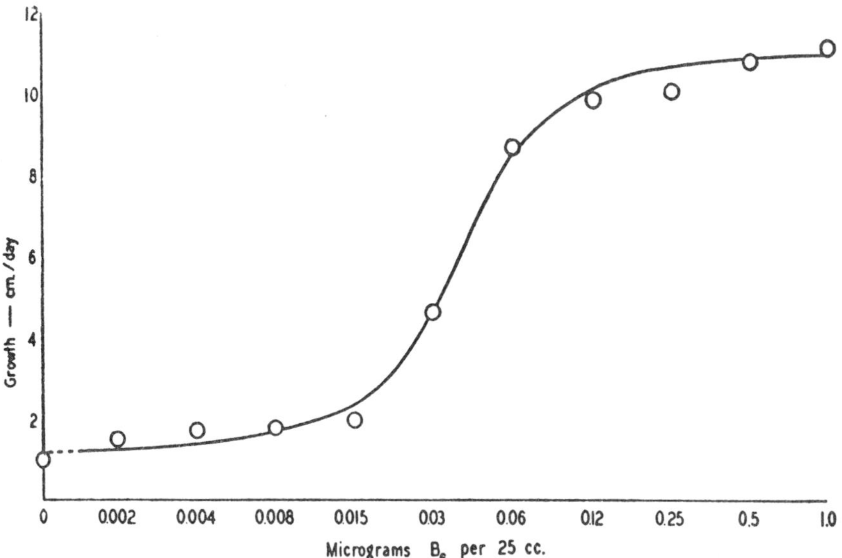

Fig. 2. The relation between growth rate (cm./day) and vitamin B_6 concentration.

substance. It may, of course, be a substance not previously known to be essential for the growth of any organism. Thus we may expect to discover new vitamins, and in the same way, it should be possible to discover additional essential amino acids if such exist. We have, in fact, found a mutant strain that is able to grow on a medium containing Difco yeast extract but unable to grow on any of the synthetic media we have so far tested. Evidently some growth factor present in yeast and as yet unknown to us is essential for *Neurospora.*

SUMMARY

A procedure is outlined by which, using *Neurospora,* one can discover and maintain x-ray induced mutant strains which are characterized by their inability to carry out specific biochemical processes.

Following this method, three mutant strains have been established. In one of these the ability to synthesize vitamin B_6 has been wholly or largely lost. In a second the ability to synthesize the thiazole half of the vitamin B_1 molecule is absent, and in the third para-aminobenzoic acid is not synthesized. It is therefore clear that all of these substances are essential growth factors for *Neurospora.*[11]

Growth of the pyridoxinless mutant (a mutant unable to synthesize vitamin B_6) is a function of the B_6 content of the medium on which it is grown. A method is described for measuring the growth by following linear progression of the mycelia along a horizontal tube half filled with an agar medium.

Inability to synthesize vitamin B_6 is apparently differentiated by a single gene from the ability of the organism to elaborate this essential growth substance.

NOTE: Since the manuscript of this paper was sent to press it has been established that inability to synthesize both thiazole and *p*-aminobenzoic acid are also inherited as though differentiated from normal by single genes.

REFERENCES

1. The possibility that genes may act through the mediation of enzymes has been suggested by several authors. See Troland, L. T., Amer. Nat. **51**:321-350 (1917); Wright, S., Genetics **12**: 530-569 (1927); and Haldane, J. B. S., in Perspectives in Biochemistry, Cambridge Univ. Press, pp. 1-10 (1937), for discussions and references.
2. Onslow, Scott-Moncrieff and others, see review by Lawrence, W. J. C., and Price, J. R., Biol. Rev. **15**:35-58 (1940).
3. Winge, O., and Laustsen, O., Compt. rend. Lab. Carlsberg, Serie physiol. **22**:337-352 (1939).
4. See Goldschmidt, R., Physiological Genetics, McGraw-Hill, pp. 1-375 (1939), and Beadle, G. W., and Tatum, E. L., Amer. Nat. **75**:107-116 (1941) for discussion and references.
5. See Sturtevant, A. H., and Beadle, G. W., An Introduction to Genetics, Saunders, pp. 1-391 (1931), and Beadle, G. W., and Tatum, E. L., loc. cit., footnote 4.
6. Dodge, B. O., Jour. Agric. Res. **35**:289-305 (1927), and Lindegren, C. C., Bull. Torrey Bot. Club **59**:85-102 (1932).
7. In so far as we have carried them, our investigations on the vitamin requirements of *Neurospora* corroborate those of Butler, E. T., Robbins, W. J., and Dodge, B. O., Science **94**:262-263 (1941).
8. The biotin concentrate used was obtained from the S. M. A. Corporation, Chagrin Falls, Ohio.
9. Throughout our work with *Neurospora,* we have used as a salt mixture the one designated number 3 by Fries, N., Symbolae Bot. Upsalienses, Vol. 3, No. 2, 1-188 (1938). This has the following composition: NH_4 tartrate, 5 g.; NH_4NO_3, 1 g.; KH_2PO_4, 1 g.; $MgSO_4 \cdot 7H_2O$, 0.5 g.; NaCl, 0.1 g.; $CaCl_2$, 0.1 g.; $FeCl_3$, 10 drops 1% solution; H_2O, 1 l. The tartrate cannot be used as a carbon source by *Neurospora.*
10. It is planned to investigate further the possibility of using the growth of *Neurospora* strains in the described tubes as a basis of vitamin assay, but it should be emphasized that such additional investigation is essential in order to determine the reproducibility and reliability of the method.
11. The inference that the three vitamins mentioned are essential for the growth of normal strains is supported by the fact that an extract of the normal strain will serve as a source of vitamin for each of the mutant strains.

52 The porphyrias

Geoffrey Dean*
Eastern Cape Provincial Hospital
Port Elizabeth, Union of South Africa

Porphyrins formed by the union of four pyrrole rings are wide spread in vegetable and animal life. Chlorophyll, which traps the energy of the sun, is a compound of porphyrin and magnesium; the haem of haemoglobin is a compound of porphyrin and iron. Porphyrins have the remarkable property of showing a brilliant pink fluorescence in ultraviolet light.

The porphyrias are disorders involving the metabolism of porphyrins and porphyrin precursors. They may be symptomless, may cause cutaneous lesions, or may be responsible for an acute illness (acute porphyria), with profound psychological, neurological and abdominal symptoms. Although they were originally described as "inborn errors of metabolism" (Garrod, 1923), it is now appreciated that certain disorders of porphyrin metabolism may be acquired.

1. PORPHYRIA VARIEGATA

On emigrating to South Africa in 1947 and starting practice in Port Elizabeth, in eastern Cape Province, I was called in consultation to see a number of patients, usually young women, who had become emotionally disturbed after a thiopentone anaesthetic or after taking barbiturates. They complained of abdominal and muscle pain, had attacks of vomiting and a rapid pulse and showed marked muscle weakness. The first three patients I saw with this condition developed a peripheral neuritis and died. Their urine was dark and gave a brilliant pink fluorescence when examined in ultraviolet light with a Wood's filter. I realized that these patients had died from acute porphyria.

The first of these patients was a young nurse, van Rooyen. I talked with other members of her family. Her father had a sensitive skin on the back of the hands that abraded easily when it was knocked; a number of scars from previous sores could be seen (see Plate I: A). He told me it was the "van Rooyen skin" and that three of his brothers, and also his father and grandfather, had a similar skin. The patient's greatgrandfather was Gerrit Renier van Rooyen, who was born in 1814, and I decided to trace all his living descendants. There were 574 of them and eventually I managed to trace all those who were living and tested their urine, and later their stool, for increased porphyrin excretion; 74 in this family group were found to have porphyria and at least 16 had died from acute porphyria (Dean & Barnes, 1955). There was a high concentration of porphyrin in the liver but not in the bone-marrow of those who had inherited porphyria and there was nearly always a high excretion of coproporphyrin and protoporphyrin in the stool (see Table I).

This South African porphyria was evidently a Mendelian dominant hepatic porphyria (fig. 1) that could cause the exposed skin to abrade and blister easily and could also cause acute illness—acute porphyria—if certain drugs, particularly barbiturates and sulphonamides, were administered. Because the South African type of porphyria could present in a variety of ways, with skin lesions, with an acute attack, with either, neither or both, Barnes and I called it porphyria variegata (Dean & Barnes, 1959).

Porphyria variegata is so common in South Africa for the same reason that the name Botha or van der Merwe is common. One million of the three million white people in South Africa hold 40 family names and derive these names from 40 original free burghers. Of the family groups of porphyrics which I have studied, I

From British Medical Bulletin **25**:48-51, 1969. Used with permission.

*Present address: Director of Medico-Social Research, The Medico-Social Research Board, An Bord Taighde Pobal-Liachta, Customs House, Dublin, Republic of Ireland.

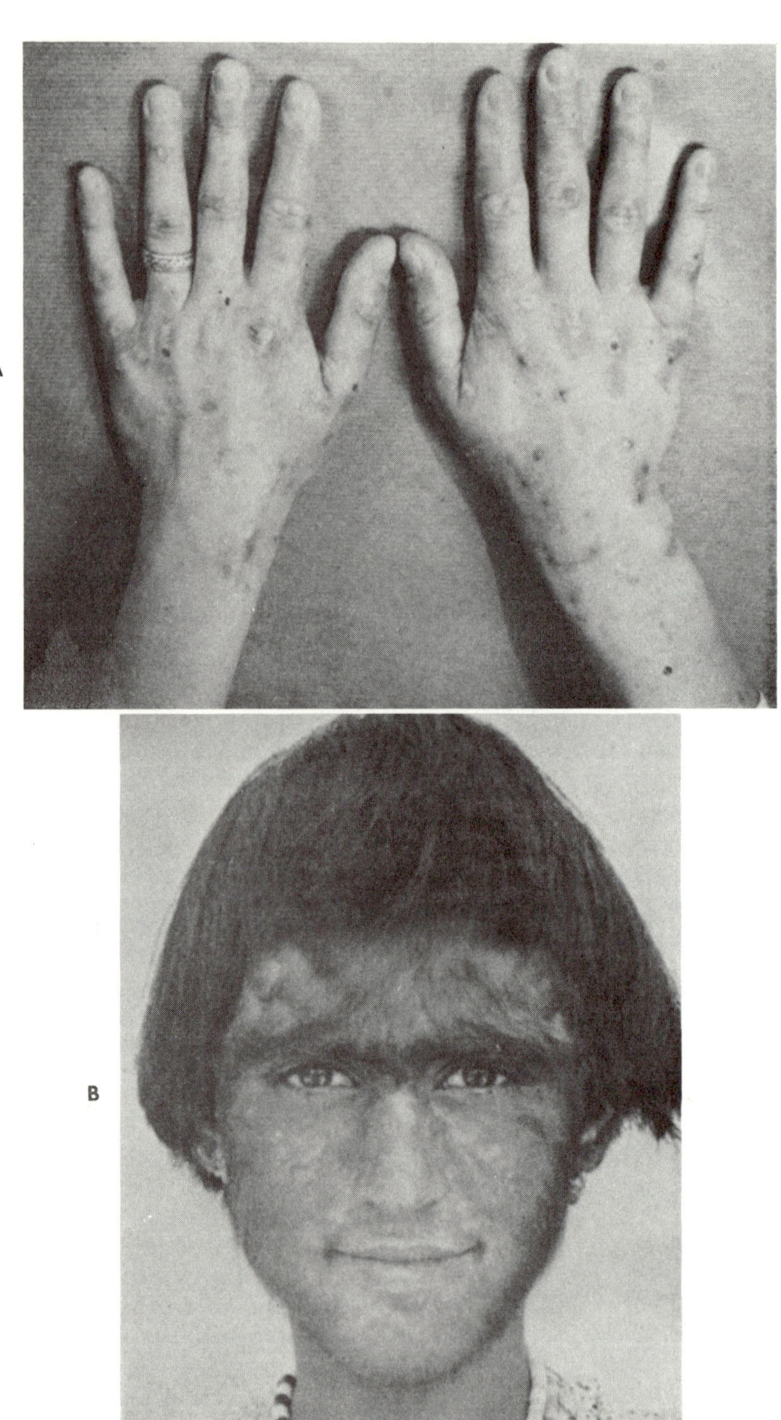

Plate I. A, Skin lesions in porphyria variegata. **B,** Hirsutism of Turkish child with toxic porphyria due to eating bread made from seed wheat treated with hexachlorobenzene. (**A** from Dean (1965) by permission of South African Journal of Medicine.)

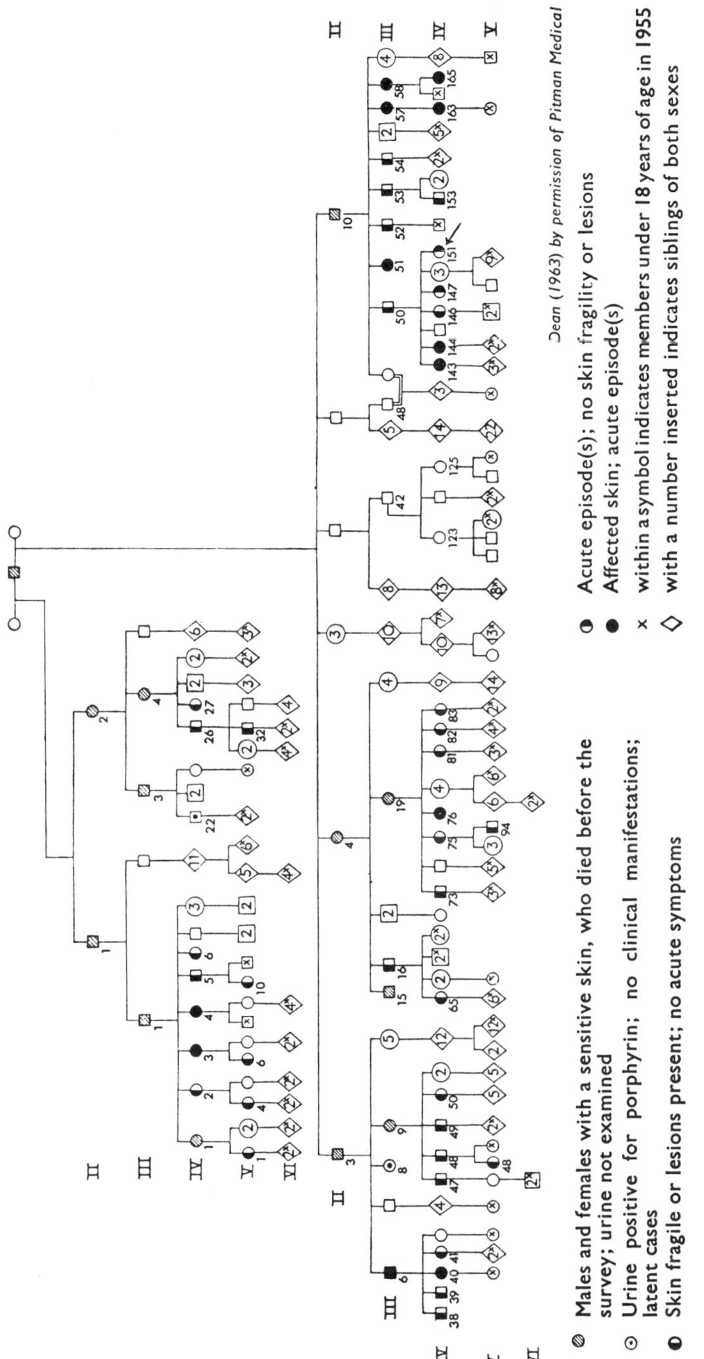

Fig. 1. Genealogical tree of G. R. van Rooyen, born 1814.

Table I. Characteristic findings in three different forms of porphyria

	Porphyria variegata ("protocoproporphyria") (South African type)	Intermittent acute porphyria ("pyrroloporphyria") (Swedish type)	Coproporphyria
Skin	Often sensitive	Not sensitive	Not sensitive
Stool porphyrin	Usually increased (protoporphyrin more so than coproporphyrin)	Normal or nearly normal	Usually increased (coproporphyrin more so than protoporphyrin)
Urinary porphyrin	Slightly increased (uroporphyrin more so than coproporphyrin)	Normal or slightly increased	Slightly increased (coproporphyrin more so than uroporphyrin)
Watson-Schwartz test	Positive only during acute attack	Positive in adults	Positive only during acute attack

succeeded in tracing the first 32 to Gerrit Jansz (Gerrit the son of Jan–they had no surnames in those days–who came from Deventer) who married Ariaantje Jacobs, one of the eight orphans sent out by the Lords Seventeen from the orphanage at Rotterdam on the ship "China" to be wives for the first free burghers–they were all married on arrival at the Cape (Dean, 1963).

Today there are about 9,000 white South Africans who have inherited porphyria variegata in the 12th to 16th generation. The incidence is highest in eastern Cape Provinces as the early porphyrics were trek Boers who trekked east from Cape Town with their cattle and sheep. In Port Elizabeth the incidence of porphyria variegata among the white population is 1 in 250. For all white South Africans the estimated incidence is 1 in 400 (see Dean (1960)). Porphyria variegata is also fairly common among the coloured people of South Africa.

2. INTERMITTENT ACUTE PORPHYRIA (PYRROLOPORPHYRIA)

Waldenström (1937) described a form of Mendelian dominant porphyria in Sweden which he called intermittent acute porphyria. I went to Sweden to study this type of porphyria and found that acute attacks occurred after certain drugs, as in the South African type, but that the skin did not abrade unduly. In the Swedish type in adults, Ehrlich's aldehyde reagent caused a purple coloration of the urine, as shown by the Watson-Schwartz test (Watson & Schwartz, 1941), both in acute attacks and also in the quiescent phase, whereas in the South African type the Watson-Schwartz test was positive only during an acute attack. In intermittent acute porphyria there is little or no increase in the stool porphyrin (see Table I).

3. COPROPORPHYRIA

Recently I have seen, in a Portuguese family from Lourenço Marques, yet a third type of Mendelian dominant hepatic porphyria that can cause acute attacks after barbiturates. In this type of porphyria there is a large amount of porphyrin in the stools, but it is mostly coproporphyrin (Dean, 1969). An account of this type of porphyria has been well described by Goldberg and colleagues, who called the disorder "hereditary coproporphyria" (Berger & Goldberg, 1955; Goldberg, Rimington & Lochhead, 1967). In coproporphyria the Watson-Schwartz test is positive only during acute attacks. In the quiescent phase, there is usually an increase in the coproporphyrin and uroporphyrin in the urine, but it may not be marked. Skin lesions do not usually occur (see Table I).

4. ERYTHROPOIETIC PORPHYRIAS

Besides the hepatic porphyrias, there are two rare forms of erythropoietic porphyrias in which the disorder of metabolism occurs in the bone-marrow: congenital or erythropoietic porphyria inherited as a Mendelian recessive, and erythropoietic protoporphyria inherited as a Mendelian dominant, characteristic. In erythropoietic or congenital porphyria, the skin lesions occur in infancy and there is also anaemia, splenomegaly and pink staining of teeth and bone (Fischer, Hilmer, Lindner & Putzer, 1925). Fourie and Rimington described this type of porphyria in cattle, in which it is known as "pink tooth disease" (Fourie, 1936; Rimington, 1936). Congenital porphyria is characterized by fluorescence of the normoblasts of the marrow in ultraviolet light. Erythropoietic protoporphyria causes an urticaria-like rash (see Haeger-Aronsen (1962)).

5. SYMPTOMATIC PORPHYRIA

There is also a hepatic form of porphyria that is not, as far as we know, inherited and which is associated with a gross disturbance of liver function, with abnormal liver-function tests. This type of symptomatic porphyria is sometimes called "porphyria cutanea tarda symptomatica" and is usually associated with the excessive use of alcohol (Brunsting, 1954). In South Africa it is common in the Bantu, especially in those addicted to a home-brewed alcoholic drink, *skokiaan* (Barnes, 1955), but surprisingly it is uncommon among the Africans of Nigeria who drink palm wine. Symptomatic cutaneous porphyria, usually associated with alcoholism, also occurs occasionally among white South Africans.

In symptomatic porphyria there is a high excretion of urinary porphyrin, and the urine is dark in colour. There is little or no increase in porphyrin in the stools. The exposed skin blisters and abrades easily, but acute attacks do not follow the taking of barbiturates.

From 1957 onwards a most unusual outbreak of symptomatic porphyria occurred in the eastern part of Turkey (Cam, 1959). It particularly affected children; and over 5,000 children had the cutaneous lesions of porphyria, not only sores and blisters but also darkening of the skin and marked hairiness, so that they often looked like monkeys and were called the monkey children (see Plate I:B).

This outbreak was caused by eating bread made from seed wheat provided by the United States of America, which had been treated with hexachlorobenzene to prevent destruction in the ground by the fungus *Tilletia tritici*. It was only when a dye was added to the seed wheat, so that bread made from it became blue in colour, that the outbreak was stopped. Many of these children still have disturbed liver function and, in the summer, the cutaneous lesions of porphyria.

6. THE SYNDROME OF ACUTE PORPHYRIA

The acute illness, acute porphyria, with emotional disturbance, abdominal and muscular pain and, finally, generalized peripheral neuritis and paralysis, may be inadvertently precipitated in intermittent acute porphyria, porphyria variegata and coproporphyria by certain drugs, particularly barbiturates and sulphonamides. Other less commonly used drugs can also precipitate attacks, for instance chloroquine diphosphate (Aralen diphosphate), griseofulvin and the sulphones. Women who have inherited these types of porphyria may suffer from minor abdominal pains, especially in the premenstrual period and during pregnancy, when they are especially liable to be given a barbiturate sedative or to have an exploratory laparotomy under an anaesthetic such as thiopentone. The oral oestrogen-progesterone contraceptives may precipitate an attack of acute porphyria, in acute intermittent porphyria, and a hepatotoxic reaction, with a great increase in skin blistering due to a high level of circulating porphyrin, in porphyria variegata (Dean, 1965).

The attack starts with severe pain in the abdomen, back or limbs. The patient is usually very emotional. She does not lie still and cries easily. Hallucinations may occur. Vomiting is frequent and there is constipation, increasing the risk that a misdiagnosis of intestinal obstruction will be made. There are often muscle twitches and sometimes convulsions. The pulse is usually rapid and the blood pressure may be raised. There is often a marked loss of salt in the urine. In the early stages there is no muscular weakness, but over the course of a few days the tendon reflexes disappear and muscle weakness becomes apparent. The voice may be weakened and the patient then talks in a whisper because the muscles of respiration are affected, and she is in grave danger of death, although with good attention most patients nowadays recover. The time for recovery depends to a great extent on whether or not there is paralysis; if there is severe peripheral neuritis, recovery usually takes 6-9 months. In porphyria variegata the recovery is complete, but in intermittent acute porphyria some degree of paralysis may persist for years. A depressive psychosis sometimes follows an acute attack.

In porphyria variegata, the tendency for the skin to abrade and blister easily may become more pronounced during an acute attack and persist for a long time after the acute attack is over.

a. Laboratory findings

During an acute attack the urine is coloured, often like port wine, by the greatly increased porphyrin. If the urine is left to stand in sunlight, it becomes darker and may become the colour of one of the cola drinks. The presence of porphyrin can be confirmed by the characteristic spectroscopic absorption bands and by the pink fluorescence in ultraviolet light (using a Wood's filter).

Increased porphobilinogen in the urine is detected by the Watson-Schwartz test by the use of Ehrlich's aldehyde reagent. (To 2 ml. of urine are added 2 ml. of Ehrlich's aldehyde reagent and 4 ml. of a saturated solution of sodium acetate. The mixture is shaken. A purple colour results if either porphobilinogen or urobilinogen is present, but urobilinogen is soluble in chloroform and may be distinguished by chloroform extraction.) In prophyria variegata and in coproporphyria, the Watson-Schwartz test will quickly show whether or not the patient is in an acute attack, because it is positive only during an acute attack. In intermittent acute porphyria, the Watson-Schwartz test is positive during an acute attack and, in adults, it is generally positive even when there are no symptoms. Examination of the blood and cerebrospinal fluid may show a marked fall in serum sodium, chloride, calcium, potassium and magnesium during an acute attack, especially in porphyria variegata.

Intermittent acute porphyria, porphyria variegata and coproporphyria can be diagnosed before an acute attack has been inadvertently precipitated. In intermittent acute porphyria (pyrroloporphyria), the Watson-Schwartz test for porphobilinogen is usually positive in adults and is the simplest screening test, but it is of no value in detecting quiescent porphyria variegata or coproporphyria. In order to make sure that all the relatives, including the children, who have inherited the gene for pyrroloporphyria are identified, quantitative analysis of porphobilinogen and δ-aminolaevulinic acid can be undertaken and slight increases above the normal range can in this way be detected (Mauzerall & Granick, 1958).

A slight increase in urinary porphyrin excretion can be detected by adding 1 ml. of a solvent consisting of equal parts of amyl alcohol, glacial acetic acid and ether to 10 ml. of urine. The mixture is shaken, and the solvent will float to the top in 15 minutes. On examination in ultraviolet light, the solvent at the top of the urine will show a purple fluorescence if porphyrin is increased. It must be remembered that increased urinary porphyrin can also occur in other disorders, such as lead poisoning.

Examination of the faeces for increased porphyrin is the quickest and simplest way of screening for porphyria variegata and coproporphyria. The faecal porphyrin in this disorder is usually very high both in the quiescent phase and during an acute attack. The clinician can screen the stool for high content of porphyrin by dissolving a small fragment of faeces, generally easily obtained on a finger-stall, in the ether solvent described above. If a marked excess of porphyrin or chlorophyll is present, the solution will show a brilliant pink fluorescence in ultraviolet light, even when diluted several times. Chlorophyll can be separated from coproporphyrin or protoporphyrin by adding 2 ml. of 1.5 normal hydrochloric acid to the solution. After the mixture is shaken and allowed to stand, the porphyrin will pass into the acid solution at the bottom of the test tube (Dean, 1956). For confirmation, the quantitative methods of Holti and Rimington should be used (see Holti, Rimington, Tate & Thomas (1958)). The diagnosis of porphyria variegata depends on a high faecal porphyrin, often combined with a slightly raised urinary porphyrin, and either a personal or family history of a tendency for the exposed skin to abrade and blister easily. The high excretion of coproporphyrin relative to protoporphyrin in the stool, or to uroporphyrin in the urine, distinguishes coproporphyria from porphyria variegata. A high faecal porphyrin can also occur on occasion in other conditions, such as carcinoma of the stomach.

If a patient has intermittent acute porphyria (pyrroloporphyria), porphyria variegata or hereditary coproporphyria, other members of the family are likely to have inherited the disorder. Diagnosis includes a detailed study of the family, so that those who have inherited the gene can be warned of the danger of certain commonly prescribed drugs.

b. Treatment

Prevention of attacks of acute porphyria is the most important part of treatment. Those who inherited porphyria variegata in South Africa and intermittent acute porphyria in Sweden increased and multiplied in the past as rapidly as the rest of the population, before the introduction of certain modern drugs, and this is strong evidence that acute attacks of porphyria causing the death of the patient must have been very uncommon before the beginning of this century. In my experience, attacks of acute porphyria in porphyria variegata have always followed the use of barbiturates or sulphonamides. Oral contraceptives sometimes aggravate the skin lesions and there are a few uncommonly used drugs that occasionally upset porphyrics. Nevertheless in daily medical practice the drugs to be avoided in intermittent acute porphyria, porphyria variegata and coproporphyria are the barbiturates, especially the pentobarbitones such as thiopentone, and the sulphonamides.

Acute porphyria, whether it has been precipitated in porphyria variegata, intermittent acute porphyria, or coproporphyria, is characterized by abdominal and muscle pain, emotional disturbance, often vomiting and constipation and a fast pulse. The Watson-Schwartz test on the urine will be positive during the acute attack.

In an acute attack the patient's life is in great danger and the patient should be nursed in a private room in hospital. In porphyria variegata there is often a fall in the blood electrolytes and calcium. Perhaps the loss of salt is due to excess antidiuretic hormone. Additional salt should be given, intravenously if necessary, because it may be necessary to keep the stomach empty for a while by Wangensteen's method. Calcium gluconate (1 g. 12-hourly intramuscularly) lessens the risk of convulsions. Chlorpromazine (Largactil) (50 mg. 8-hourly) is a good tranquilizer. Digitalis is not very effective in slowing the pulse.

There should be no delay in doing a tracheostomy and using a mechanical respirator, such as a Birds or Engström machine, if respiration is embarrassed, because these patients develop peripheral paralysis with respiratory involvement. We have had one such patient on a respirator for three months and she eventually made a very good recovery. A respiratory unit can be the greatest help in this situation. Every effort is worth while because, if the patient can be kept alive until the acute attack subsides, eventual recovery is excellent, although it may take some months (Dean, 1967).

The skin sensitivity that occurs in porphyria cutanea tarda (symptomatic porphyria) and porphyria variegata is treated by avoiding the use of alcohol and toxic drugs and by protecting the exposed skin from sunlight as much as possible, for instance by using gloves and a hat out of doors.

Those who have inherited porphyria variegata, intermittent acute porphyria, or coproporphyria should carry a letter stating the evidence for the diagnosis and mentioning the extreme danger of barbiturates and sulphonamides. They should be strongly advised to show this letter to any doctor they consult.

• • •

The study of porphyria variegata has carried me back in a great adventure to the first settlement in South Africa by the Dutch East India Company, who started a revictualling station for their spice ships trading with the East. It has involved the study of the formation of a people, and has been an exciting voyage of discovery which still continues—the "everlasting Whisper. . .something hidden. Go and find it".

REFERENCES

Barnes, H. D. (1955) S. Afr. med. J. **29**:781.

Berger, H. & Goldberg, A. (1955) Br. med. J. **2**:85.

Brunsting, L. A. (1954) Arch. Derm. Syph. **70**:551.

Cam, C. (1959) Dirim (Istanbul) **34**:11.

Dean, G. (1956) S. Afr. med. J. **30**:377.

Dean, G. (1960) S. Afr. med. J. **34**:745.

Dean, G. (1963) The porphyrias: a story of inheritance and environment. Pitman Medical, London.

Dean, G. (1965) S. Afr. med. J. **39**:278.

Dean, G. (1967) Curr. Ther., p. 346.

Dean, G. (1969) S. Afr. med. J. (In press).

Dean, G. & Barnes, H. D. (1955) Br. med. J. **2**:89.

Dean, G. & Barnes, H. D. (1959) S. Afr. med. J. **33**:246.

Fischer, H., Hilmer, H., Lindner, F. & Putzer, B. (1925) Hoppe-Seyler's Z. physiol. Chem. **150**:44.

Fourie, P. J. J. (1936) Onderstepoort J. vet. Sci. Anim. Ind. **7**:535.

Garrod, A. E. (1923) Inborn errors of metabolism, 2nd ed. Frowde and Hodder & Stoughton, London.

Goldberg, A., Rimington, C. & Lochhead, A. C. (1967) Lancet **1**:632.

Haeger-Aronsen, B. (1962) Lancet **1**:1073.

Holti, G., Rimington, C., Tate, B. C. & Thomas, G. (1958) Q. J. Med. **27**:1.

Mauzerall, D. & Granick, S. (1958) J. biol. Chem. **232**:1141.

Rimington, C. (1936) Onderstepoort J. vet. Sci. Anim. Ind. **7**:567.

Waldenström, J. (1937) Acta med. scand., Suppl. 82.

Watson, C. J. & Schwartz, S. (1941) Proc. Soc. exp. Biol. Med. **47**:393.

chapter 11

Genes and development

Through their control of metabolic pathways, genes determine the pattern of morphological and physiological changes that constitute the *development* of an organism. For unicellular forms, all structural and functional modifications that may occur are obviously limited to the cell itself. However, for multicellular organisms, the changes that may occur can involve cell, tissue, and organ-system differentiation. It is clear that the orderly modifications necessary for the successful completion of a life cycle require a rigid regulation of gene action. This includes regulation of DNA replication, DNA transcription, mRNA translation, and all subsequent cell metabolism leading to tissue and organ-system differentiation.

REGULATION OF DNA REPLICATION

The normal pattern of a cell cycle includes the growth of the cell, the replication of its DNA, and its subsequent division into two equivalent cells. Inhibition of DNA synthesis blocks cell division, whereas inhibition of cell division does not necessarily block DNA synthesis. However, it was found by C. Lark in 1966 (Ref. 11-1) that DNA synthesis in *Escherichia coli* is controlled by the level of overall cell metabolism. The faster the growth of the cell, the greater is the rate of chromosome replication. Other experiments reported by K. G. Lark and his co-workers in 1963 (Ref. 11-2) and by Pritchard and Lark in 1964 (Ref. 11-3) demonstrated that the initiation of DNA replication requires the earlier synthesis of a protein that is hypothesized to act as an "attachment site" of the newly formed chromosome to the cell membrane. A review of the experiments dealing with the regulation of DNA replication in *E. coli* was published by K. G. Lark in 1966 (Ref. 11-4). The possibility that the mechanism by which a cell achieves its immunity to an invading virus is through regulation of the viral-DNA replication was raised in the experiments reported by Green and his co-workers in 1967 (Ref. 11-5). That paper is reprinted in this chapter.

REGULATION OF DNA TRANSCRIPTION

Control of gene action can depend on the regulation of mRNA formation. An example of regulation of DNA transcription in *E. coli* can be found in the discussion of regulation of gene action in Chapter 9. A demonstration that there is a change in the types of mRNA produced in succeeding developmental stages of a multicellular organism was reported in 1966 by Whiteley and co-workers (Ref. 11-6), whose paper is included in this chapter.

In some multicellular organisms the chromosomes of certain tissues exhibit enlarged areas called "puffs" where the genes are actively engaged in transcription. An examination of the chromosomes in different tissues shows that the pattern of some puffs varies with the tissues. The results of such a study in the dipteran fly *Chironomus tentans* was reported by Beermann in 1956 (Ref. 11-7). Later investigations, reported by Clever in 1964 (Ref. 11-8) and in 1966 (Ref. 11-9), showed that the puffing pattern of the chromosomes of each tissue changed with succeeding developmental stages. It was further discovered that the hormone ecdysone, which controls molting in insects, stimulates sequentially the formation of a different puff pattern in each chromosome. However, it was also found that the effects of ecdysone could be inhibited by actinomycin D, which is a DNA-transcription inhibitor, and by puromycin, an inhibitor of protein synthesis at the translational level.

A most interesting example of regulation of DNA transcription involves the genes located in the sex chromosomes. In most organisms the males and females have either different numbers or different types of sex chromosomes. Yet, it has been found that the amounts of gene end products produced by sex-linked genes are the same in both sexes. The equality of amounts of these gene products in both males and females has been termed *dosage compensation.* The mechanism by which dosage compensation is achieved in mammals was discussed in a paper by Davidson and his co-workers in 1963

(Ref. 11-10); the mechanism involved in *Drosophila* was considered in a paper by Mukherjee and Beermann in 1965 (Ref. 11-11).

REGULATION OF mRNA TRANSLATION

It is well known that protein synthesis varies during different stages of the cell cycle. A study of the changing rate of protein synthesis in dividing sea urchin eggs was reported in 1966 by Sofer and his co-workers (Ref. 11-12), whose paper is reprinted in this chapter. However, changes in protein synthesis could be caused by either modifications in DNA transcription or mRNA translation. In an attempt to avoid DNA transcription as an experimental factor, investigations involving enucleated cells have been performed. An example of this type of experiment, involving the production of a specific enzyme in the unicellular green alga *Acetabularia crenulata,* was reported by Spencer and Harris in 1964 (Ref. 11-13).

A study on hemoglobin synthesis in rabbit reticulocytes was reported by Dintzis in 1961 (Ref. 11-14). He found that there were three to five times more beta than alpha chains in each cell. This implied that the production of hemoglobin molecules was limited by the number of beta chains available. The rate of synthesis of beta chains could be a function of regulation of mRNA translation for the beta chains. However, here as in the above experiments, the possibility always exists that the results reflect the presence of different amounts of mRNA molecules that have previously been transcribed.

NUCLEUS-CYTOPLASM INTERACTION

In higher organisms cellular specialization, although obviously a reflection of chromosomal differentiation, is most apparent in the structural and functional characteristics of the cytoplasm. When the role of genes in development is examined, the question arises as to the degree of control that the nucleus has over a differentiated cytoplasm and whether a specialized cytoplasm can, in turn, control the nucleus. A technique that has been widely used to examine nucleus-cytoplasm interaction is that of nuclear transplantation.

An experiment in which nuclei from frog blastula cells were transplanted into enucleated frogs' eggs was reported by Briggs and King in 1952 (Ref. 11-15). It was found that completely normal embryos developed from the eggs, demonstrating that blastula nuclei are not irreversibly differentiated. However, later experiments reported by the same authors in 1953 (Ref. 11-16) showed that the transplantation of nuclei from later stages (i.e., gastrula and tail bud) produced a progressively increasing number of deaths and abnormalities. These results have been interpreted to indicate that the nuclei of older embryos become irreversibly differentiated and can no longer transcribe all the genetic information needed for normal development. A review of the experiments involving nucleus-cytoplasm interactions in the frog was published by Briggs and King in 1959 (Ref. 11-17).

In Chapter 2, the two types of cell division, mitosis and meiosis, were considered, as was the time of DNA synthesis during the cell cycle. Cell division, which in most organisms involves a breakdown of the nuclear membrane, could provide an opportunity for nucleocytoplasmic exchanges that perhaps could not take place during interphase. An examination of this possibility by Feldherr in 1966 (Ref. 11-18) demonstrated, however, that this was not the case. With regard to the effect of nucleus-cytoplasm interactions on DNA synthesis, a report by Prescott and Goldstein in 1967 (Ref. 11-19) is printed in this chapter.

TISSUE DIFFERENTIATION

Tissue differentiation involves the formation of distinctly different types of cells. One of the questions raised about differentiated cells is whether they can revert to a generalized state and become specialized in another direction. The ultimate in such a process would be the ability of a differentiated cell to produce an entire organism. Cells that could do this would be said to be *totipotent,* namely, they would have retained their ability to utilize their entire genome. No example of totipotency in animals has been demonstrated. However, totipotency in plants has been clearly shown to occur. The experimental techniques involved and results obtained were reviewed in a paper by Steward in 1964 (Ref. 11-20).

As stated at the beginning of this chapter, tissue differentiation includes chemical as well as morphological differences. An excellent example of chemical differentiation of tissues can be found in the study of the enzyme lactic acid dehydrogenase (LDH), which is involved in carbohydrate metabolism. LDH consists of four polypeptide chains, each of which can be one of two types, A or B. From this arrangement, five different proteins can be formed: A_4, A_3B_1, A_2B_2, A_1B_3, and B_4. These proteins are referred to as LDH-1, 2, 3, 4, and 5, respectively. Different molecular forms of an enzyme are called *isozymes.* A report was made by Markert and Ursprung in 1962 (Ref. 11-21) on the distribution of LDH isozymes in 20 different tissues or organs of the adult mouse. They found that each tissue or organ can be characterized by the specific proportions of the several isozymes present in the tissue. They also found that the proportions of the various isozymes in each tissue changed during development. A general review of the control of protein synthesis in differentiating cells was published by Markert in 1963 (Ref. 11-22).

The morphological and physiological changes that can occur in cells may involve the production of tumors and cancers. A number of these have been shown to be produced by DNA viruses, as reviewed by Eckhart in 1968 (Ref. 11-23). More recently, Aaronson and Todaro in 1969 (Ref. 11-24) reported the transformation of human cells solely by the DNA of such a virus. Their paper is reprinted in this chapter. A general review on the origin of cancer cells was published by Braun in 1970 (Ref. 11-25).

ORGAN-SYSTEM DIFFERENTIATION

In a study of the development of organs and systems, one is immediately faced with complex interactions of gene products, cells and tissues. The biochemical pathways involved are so complicated that in most cases it has not been possible to trace the effects of a gene to the original gene product. One very useful line of research of organ-system differentiation has been the study of *lethal genes.* A discussion of this type of gene and its usefulness in studying differentiation was published by Gluecksohn-Waelsch in 1963 (Ref. 11-26) and is the last article reprinted in this chapter. One of the lethal genes that has given us a good deal of information on the intricacies of the genetic control of organ-system development is the T-locus in mice. A detailed discussion of this gene was published in 1964 by Dunn (Ref. 11-27) and by Bennett (Ref. 11-28). A book reviewing organ-system development and lethal genes was published by Hadorn in 1961 (Ref. 11-29).

REFERENCES

Regulation of DNA replication

11-1. Lark, C. 1966. Regulation of chromosome replication and segregation in bacteria. Biochim. Biophys. Acta **119**:517-525.

11-2. Lark, K. G., Repko, T., and Hoffman, E. J. 1963. The effect of amino acid deprivation on subsequent DNA replication. Biochim. Biophys. Acta **76**:9-24.

11-3. Pritchard, R. H., and Lark, K. G. 1964. Induction of replication by thymine starvation at the chromosome origin in *Escherichia coli.* J. Mol. Biol. 9:288-307.

11-4. Lark, K. G. 1966. Regulation of chromosome replication and segregation in bacteria. Bacteriol. Rev. **30**:3-32.

11-5. Green, M. H., Gotchel, B., Hendershott, J., and Kennel, S. 1967. Regulation of bacteriophage lamda DNA replication. Proc. Natl. Acad. Sci. U. S. A. **58**:2343-2350.

Regulation of DNA transcription

11-6. Whiteley, A. H., McCarthy, B. J., and Whiteley, H. R. 1966. Changing populations of messenger RNA during sea urchin development. Proc. Natl. Acad. Sci. U. S. A. **55**:519-525.

11-7. Beermann, W. 1956. Nuclear differentiation and functional morphology of chromosomes. Cold Spring Harbor Symp. Quant. Biol. **21**: 217-232.

11-8. Clever, U. 1964. Actinomycin and puromycin: Effects on sequential gene activation by ecdysone. Science **146**:794-795.

11-9. Clever, U. 1966. Gene activity patterns and cellular differentiation. Am. Zool. **6**:33-41.

11-10. Davidson, R. G., Nitowsky, A. M., and Childs, B. 1963. Demonstration of two populations of cells in the human female heterozygous for glucose-6-phosphate dehydrogenase variants. Proc. Natl. Acad. Sci. U. S. A. **50**:481-485.

11-11. Mukherjee, A. S., and Beermann, W. 1965. Synthesis of ribonucleic acid by the X-chromosomes of *Drosophila melanogaster* and the problem of dosage compensation. Nature **207**:785-786.

Regulation of mRNA translation

11-12. Sofer, W. H., George, J. F., and Iverson, R. M. 1966. Rate of protein synthesis: Regulation during first division cycle of sea urchin eggs. Science **153**:1644-1645.

11-13. Spencer, T., and Harris, H. 1964. Regulation of enzyme synthesis in an enucleate cell. Biochem. J. **91**:282-286.

11-14. Dintzis, H. M. 1961. Assembly of the peptide chains of hemoglobin. Proc. Natl. Acad. Sci. U. S. A. **47**:247-261.

Nucleus-cytoplasm interaction

11-15. Briggs, R., and King, T. J. 1952. Transplantation of living nuclei from blastula cells into enucleated frogs' eggs. Proc. Natl. Acad. Sci. U. S. A. **38**:455-463.

11-16. Briggs, R., and King, T. J. 1953. Factors affecting the transplantability of nuclei of frog embryonic cells. J. Exptl. Zool. **122**:485-506.

11-17. Briggs, R., and King. T. J. 1959. Nucleocytoplasmic interactions in eggs and embryos. pp. 537-617. *In* J. Brachet and A. E. Mirsky [eds.] The cell, vol. I. Academic Press, Inc., New York.

11-18. Feldherr, C. M. 1966. Nucleocytoplasmic exchanges during cell division. J. Cell Biol. **31**:199-203.

11-19. Prescott, D. M., and Goldstein, L. 1967. Nuclear-cytoplasmic interaction in DNA synthesis. Science **155:**469-470.

Tissue differentiation

11-20. Steward, F. C. 1964. Growth and development of cultured plant cells. Science **143:**20-27.
11-21. Markert, C. L., and Ursprung, H. 1962. The ontogeny of isozyme patterns of lactate dehydrogenase in the mouse. Devel. Biol. **5:**363-381.
11-22. Markert, C. L. 1963. Epigenetic control of specific protein synthesis in differentiating cells. pp. 65-84. *In* M. Locke [ed.] Cytodifferentiation and macromolecular synthesis. Academic Press, Inc., New York.
11-23. Eckhart, W. 1968. Transformation of animal cells by oncogenic DNA viruses. Physiol. Rev. **48:**513-533.
11-24. Aaronson, S. A., and Todaro, G. J. 1969. Human diploid cell transformation by DNA extracted from the tumor virus SV40. Science **166:**390-391.
11-25. Braun, A. C. 1970. On the origin of the cancer cells. Am. Scientist **58:**307-320.

Organ-system differentiation

11-26. Gluecksohn-Waelsch, S. 1963. Lethal genes and analysis of differentiation. Science **142:**1269-1276.
11-27. Dunn, L. C. 1964. Abnormalities associated with a chromosome region in the mouse. I. Transmission and population genetics of the t-region. Science **144:**260-263.
11-28. Bennett, D. 1964. Abnormalities associated with a chromosome region in the mouse. II. Embryological effects of lethal alleles in the t-region. Science **144:**263-267.
11-29. Hadorn, E. 1961. Developmental genetics and lethal factors. John Wiley & Sons, Inc., New York.

53 Regulation of bacteriophage lambda DNA replication

Melvin H. Green
Barbara Gotchel
Judy Hendershott
Stephen Kennel*
Department of Biology
University of California at San Diego
La Jolla, California

When λ phage infects *Escherichia coli*(λ), the phage DNA is injected into the lysogenic cell, but it is quickly prevented from being transcribed into λ messenger RNA (mRNA).[1] The cI region of the prophage λ genome directs the synthesis of a repressor that maintains cell immunity via its ability to block λ mRNA synthesis.[1-3] Since at least three phage cistrons must function in order for λ DNA to replicate autonomously,[4,5] it is understandable that there is no λ DNA synthesis upon infection of an immune lysogen.[6]

Thomas and Bertani were the first to suggest that the λ repressor might also interact with the phage DNA to block replication even in the presence of the required enzymes.[7] *E. coli*(λ) was mixedly infected with λ and a closely related, but immunity-insensitive, phage, such as λvir or 434hy. The repressor-insensitive phage should provide all of the enzymes needed for λ DNA replication. However, the yield of λ phage was extremely low, never exceeding the number of phage used for infection.

It is possible that whereas few infectious phage of the immunity-sensitive type were produced, considerable DNA replication nevertheless occurred. The following alternatives must be considered before one can conclude that the λ repressor is able to block λ DNA replication in the presence of the replicating enzymes: In *E. coli*(λ) superinfected with λ and λvir, (1) infectious λ production might begin later than that of λvir; (2) λ DNA could replicate, but be packaged with low efficiency; (3) λ DNA could replicate and be packaged normally, but the resulting particles might not be infectious.

We shall now present experiments that examine each of these alternatives and allow us to conclude tentatively that the λ repressor does in fact prevent λ DNA synthesis in a cell containing all the enzymes needed for phage replication.

MATERIALS AND METHODS

Bacteriophages. λ, λb_2, $\lambda b_2 b_5 c$, λvir, and λcI, t1 were initially obtained from Dr. Jean Weigle (Cal Tech). λb_2 is a deletion mutant in which about 20% of the wild-type genome is missing.[8,9] The b_2 phage thus have a lower buoyant density than wild-type λ.[8] The b_5 region confers an altered immunity sensitivity on λ phage.[8] $\lambda b_2 b_5 c$ can develop upon infection of *E. coli*(λ), but not of *E. coli*(λb_5). The *c* mutation, which occurred spontaneously in λb_5, causes the phage to form clear plaques since it is unable to lysogenize *E. coli.*[10] The virulent phage, λvir, cannot lysogenize and is insensitive to prophage-conferred immunity.[2] The temperature mutant, λcI, t1 (or λt1), directs the synthesis of a thermolabile repressor,[1] thereby causing *E. coli*(λt1) to be heat-inducible.[11] λb_2vir was isolated in our laboratory by crossing λb_2 with λvir.

Bacterial strains. *E. coli* strains W3110, W3110(λ), and C600(λb_5) were obtained from Dr. Jean Weigle in 1963. *E. coli* W3110(λcI, t1) was derived from W3110 by infection.[1] The thymine-requiring derivative, W3110(λt1) Thy^-, was isolated by selection of spontaneous mutants that could grow in minimal medium containing thymidine (TdR, 50 μg/ml) and trimethopterin (10 μg/ml), an antibiotic that prevents thymine formation and causes thymineless death in *E. coli* Thy^+. This method was suggested to us by Dr. Fred Hickson. W3110 Thy^- was isolated from W3110(λt1) Thy^- by selection of heat-resistant colonies. This strain was then made lysogenic for λcI^+ by infection under conditions that favor lysogenization.[12]

Media. Media for the growth of cells and for phage assays have been described previously.[1] We now use K medium containing 0.5% casamino acids instead of 1.5%. For labeling phage with P^{32}, K-low PO_4 medium was employed. The PO_4^{-3} concentration is reduced to 10^{-3} *M* and the medium is buffered with 0.05 *M* tris, pH 7.4. Thymine-requiring strains were

From Proceedings of the National Academy of Sciences **58**:2343-2350, 1967. Used with permission.

Abbreviations: mRNA, messenger RNA; TdR, thymidine; BUdR, 5-bromodeoxyuridine.

This research was supported by U. S. Public Health Service research grant GM-14318-01.

*Supported by a predoctoral fellowship from the National Institutes of Health.

grown in K medium + TdR (50 $\mu g/ml$). To label λ with H^3-TdR the cells were grown to the desired concentration in K + 50 $\mu g/ml$ TdR, then changed to K-low TdR (2-5 $\mu g/ml$) before infection or induction. Heavy λ were formed in K medium containing 5-bromodeoxyuridine (BUdR) (50 $\mu g/ml$), TdR (1 $\mu g/ml$), and uracil (50 $\mu g/ml$). The cultures were kept dark to prevent inactivation of BU-containing λ by visible light.[13]

Radioactive chemicals were obtained from New England Nuclear Corporation, trimethopterin from Burroughs-Wellcome Co., Inc.

Phage purification and density analysis. Log phase cultures (2-4 × 10^8 cells/ml) were infected after being pelleted and resuspended in 0.01 *M* tris, pH 7.4, 0.01 *M* $MgSO_4$ at 2-4 × 10^9 cells/ml. At least 90% of the phage adsorbed in 15 min at 37°. The cells were diluted into K medium and aerated. Lysates were cleared by low-speed centrifugation (10 min at 7000 *g* in the Sorvall) and the phage were pelleted by ultracentrifugation (2 hr at 23,000 rpm in the Spinco no. 30 rotor). Pellets were gently resuspended in 0.01 *M* tris, pH 7.4, 0.005 *M* $MgSO_4$ (dilution medium), and cleared at low speed (10 min at 7000 *g*). For density analysis, the phage suspension was diluted to 3.5 ml with CsCl so that the final density was 1.49 gm/cc. The phage were then centrifuged for 16-18 hr at 33,000 rpm in the Spinco SW39L rotor. Fractions were collected from the bottom of the tube, either into sterile tubes containing dilution medium or directly onto Whatman 3MM paper or glass fiber disks. Samples on the 3MM strips were acid-precipitated by immersion in 10% trichloracetic acid (TCA).

A Nuclear-Chicago liquid scintillation spectrometer was used for all radioactivity measurements. Samples were counted in 10 ml of diluted Liquifluor (New England Nuclear Corp.).

RESULTS

(1) *Production of λ and λb_2b_5c upon mixed infection of E. coli(λ):* The ability of λb_2b_5c to stimulate a small amount of λ production in *E. coli*(λ) is apparent from the data shown in Table 1. The λ burst size was increased by b_2b_5c eightfold over that due to spontaneous induction. Infecting *E. coli*(λ) with λ as well as λb_2b_5c resulted in a further increase in the λ burst size, but the yield never exceeded the input number of λ genomes, including *ca.* two prophages per cell. Mixed infection of a nonlysogen by λ and b_2b_5c, or induction of a b_2b_5c-infected λ lysogen, resulted in phage yields approximately equal to the input multiplicity ratio. Therefore the restriction of λ in *E. coli*(λ) is most likely due to the prophage repressor.[2]

The rate of λ and b_2b_5c production upon mixed infection of *E. coli*(λ) is nearly identical (Fig. 1). It is thus apparent that the restricted maturation of λ phage cannot be due to a delay in their formation while the immunity-insensitive b_2b_5c are replicating.

Table 1. Production of λ by b_2b_5c in *E. coli*(λ)

Phage input (moi)		Yield (phage/cell)	
b_2b_5c	λcI	Total λ	b_2b_5c
0	0	0.2	0
5	0	1.6	73
5	5	5.1	55
5	15	5.5	45

Infected cultures of W3110(λ) were diluted 100-fold into K medium and aerated 90 min at 37°C. The phage yield is based on the total number of cells at the time of infection.

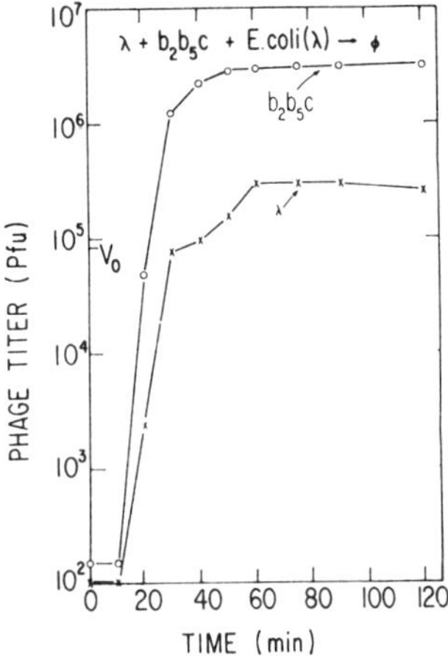

Fig. 1. Kinetics of λ and λb_2b_5c formation in *E. coli* (λ). A concentrated culture of W3110(λ) (see *Methods*) was mixedly infected with λ (moi = 2.4) and λb_2b_5c (moi = 6.5), then diluted into K medium aerating at 37° (t = 0). V_0 indicates the total number of cells after dilution (8 × 10^4/ml). The burst size for λ was 3.5; for b_2b_5c it was 37.

(2) *Efficiency of packaging λ DNA in sensitive and in lysogenic cells:* The ability of b_2b_5c to package λ DNA in immune lysogens was compared with the packaging efficiency in nonlysogenic cells. Log phase cultures of *E. coli* and *E. coli*(λ) were each mixedly infected with H^3-λcI, t1 (moi = 2) and b_2b_5c (moi = 5). The infected cultures were aerated for 90 minutes at 37°, during which time lysis occurred. Chloroform was then added to ensure complete lysis, and the debris was cleared by low-speed centrifugation (see *Methods*). An aliquot was removed for determination of total acid-precipitable radioactivity and for phage assays. Previous experiments had confirmed the report that no λ DNA is degraded following infection.[14] The cleared lysates thus contained all of the λ H^3-DNA that had been injected into the

cells. The total burst size was 102 from *E. coli* and 66 from *E. coli*(λ).

The phage in each lysate were concentrated by centrifugation, then banded in a CsCl equilibrium density gradient (see *Methods*). Two-drop fractions were collected directly onto Whatman glass filters, dried, then counted in a nuclear scintillation spectrometer. All of the radioactivity banded in the phage region of the gradient ($\rho = 1.5$).

Correcting for the small losses in total recovery (~15% loss), 42 per cent of the parental λ H^3-DNA was packaged into progeny phage in nonlysogenic *E. coli* and 37 per cent was packaged in *E. coli*(λ). It may therefore be concluded that the λ repressor does not influence the extent to which parental λ DNA can be packaged into progeny phage.

(3) *Are noninfectious phage produced by b_2b_5c or b_2vir in E. coli(λ)?* Since late λ protein is known to be required in order to convert noninfectious intracellular λ DNA to infectious DNA,[15] it was conceivable that λ DNA was actually replicated in *E. coli*(λ) by enzymes synthesized by the immunity-insensitive phage, but that the λ DNA became packaged as noninfectious phage particles. To test this possibility, we compared the incorporation of $P^{32}O_4^{-3}$ and H^3-TdR into λ phage in sensitive and lysogenic cultures that had been mixedly infected. The first experiment of this type was performed according to the following scheme:

		Yield (per cell) λ	b_2b_5c
(a) $5\ b_2b_5c + 4\ H^3\lambda t1 + E.\ coli(\lambda)$	$\xrightarrow{P^{32}}$	2.6	66
(b) $5\ b_2b_5c + 4\ H^3\lambda t1 + E.\ coli$	$\xrightarrow{P^{32}}$	13	34
(c) $5\ b_2b_5c + E.\ coli(\lambda)$	$\xrightarrow{P^{32}}$	0.9	65

The three infected cultures (see above) were aerated 105 minutes at 37° in K-low PO_4 medium containing 1 μc/ml $P^{32}O_4^{-3}$. The phage in the lysates were concentrated by centrifugation, treated with DNase (20 μg/ml) and RNase (5 μg/ml) for one hour at 37°, then banded in a CsCl equilibrium density gradient. $H^3\lambda t1$ was added as a density marker for part *c*. Three-drop fractions were collected, acid-precipitated, and filtered through nitrocellulose membranes.

As may be seen in Figure 2*A* and *C* (from cultures *a* and *c*, respectively), little, if any, P^{32} was incorporated into λ phage upon infection of *E. coli*(λ) by b_2b_5c. In the nonlysogenic

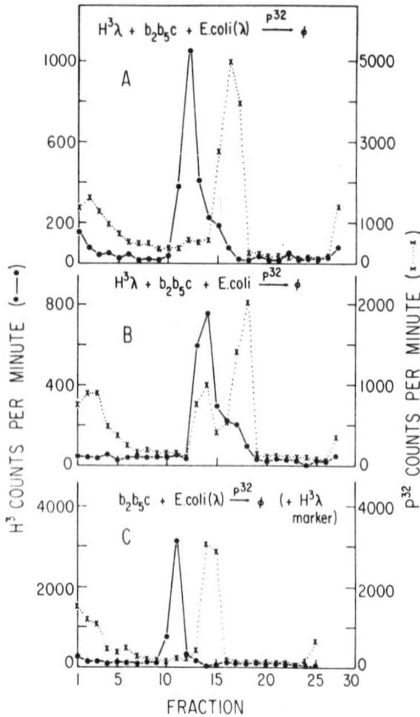

Fig. 2. Incorporation of $P^{32}O_4$ into λ and λb_2b_5c after mixed infection of *E. coli* (λ). In parts **A** and **B**, the H^3 profile represents parental $\lambda t1\ H^3$-DNA that was packaged into progeny phage. In part **C**, $H^3\lambda t1$ phage were added as a density reference. H^3 cpm (•——•); P^{32} cpm (X . . . X).

control culture (Fig. 2*B*), P^{32} was incorporated into λ and b_2b_5c in amounts corresponding to their respective burst sizes, listed above. The progeny λ in Figure 2*A* and *B* were labeled with parental H^3-DNA. The skewing of the H^3 profile toward the low-density region is most likely due to recombination between $H^3\lambda t1$ and b_2b_5c. From the extent of P^{32} incorporation into λ phage, it is concluded that few, if any, noninfectious λ particles were formed *de novo* in *E. coli*(λ).

A similar experiment was performed using λb_2vir in place of b_2b_5c. H^3-TdR was used to label the progeny phage. A schematic view of the procedure is shown below.

		Yield (per cell) λ	b_2 vir
(a) $5\ \lambda b_2vir + 2\ P^{32}\lambda +$ *E. coli*(λ) Thy⁻	$\xrightarrow{H^3\text{-TdR}}$	0.5	11
(b) $5\ \lambda b_2vir + 2\ P^{32}\lambda +$ *E. coli*(λ) Thy⁻, UV-induced	$\xrightarrow{H^3\text{-TdR}}$	45	30

After aerating the cultures for 90 minutes at 37°, the phage in the lysate were assayed, concentrated, then banded in CsCl. Two-drop fractions were collected on Whatman 3MM

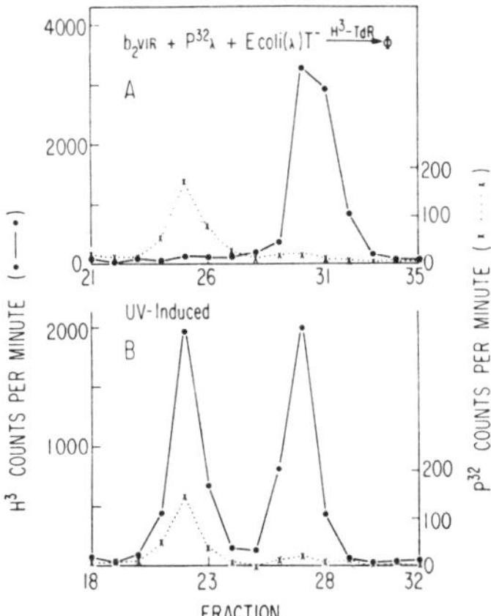

Fig. 3. Incorporation of H^3-thymidine into λ and λb_2 vir after mixed infection of *E. coli*(λ) Thy^-. In part **B**, the infected cells were irradiated for 25 sec with UV light in order to induce λ development. After infection and UV treatment, each culture was aerated for 90 min in K medium containing TdR at 5 μc/5 μg/ml. The P^{32} profile represents parental λ P^{32}-DNA that was packaged into progeny phage. Only the region of the CsCl gradient that contains the phage is shown. In **A** and **B**, ~45 and 42 fractions were collected, respectively. H^3 cpm (●——●); P^{32} cpm (X . . . X).

paper, acid-washed, dried, and assayed for radioactivity. In each culture the incorporation of H^3-TdR into λ and $b_2$vir reflects the relative number of infectious phage in the lysate (see Fig. 3). Noninfectious λ particles are apparently not formed in *E. coli*(λ) as a result of λ$b_2$vir infection. As in section (2), the packaging efficiency of parental λ DNA by λ and by the immunity-insensitive phage was nearly identical, this time with and without induction of $b_2$vir-infected lysogens. We conclude that the $b_2$vir and b_2b_5c enzymes are unable to promote λ DNA replication in *E. coli*(λ), subject to the quantitative limitations of this method (see below).

(4) *Can λ DNA be replicated once in E. coli(λ) by b_2b_5c or $b_2$vir?* From the previous experiments we can say that a very limited amount of λ DNA replication occurred in an immune cell containing all of the enzymes needed for λ DNA synthesis. The data of section (3) permit an upper limit to be set at 5 per cent of the amount of $b_2$vir DNA synthesis occurring in the same cells. In order to set a lower limit, namely, one round of replication, we carried out the mixed infection in a medium containing 5-bromodeoxyuridine (BUdR). After one round of replication, both strands of the parental λ DNA would become associated with heavy BUdR-containing strands since λ replication occurs semiconservatively.[6] Analysis of the progeny λ phage would thus reveal whether any replication had occurred before the λ DNA was packaged into phage particles. Since the packaging efficiency in *E. coli*(λ) is normal (see sections (2) and (3)), the progeny should reflect a random sampling of λ DNA molecules from the cells. Two kinds of experiments were performed in attempting to answer this question.

The first approach can be represented in the following manner:

		Phage yield (per ml) λ	b_2 vir
(a) b_2 vir + λ + *E. coli*(λ) Thy^-	$\xrightarrow{\text{BUdR}}$	6×10^6	4.7×10^8
(b) b_2 vir + λ + *E. coli*(λ) Thy^-, UV-induced	$\xrightarrow{\text{BUdR}}$	1.9×10^8	4.9×10^8
(c) λ + *E. coli*(λ) Thy^-	$\xrightarrow{\text{BUdR}}$	9×10^5	0

The three infected cultures were washed, diluted, and aerated for 90 minutes at 37°. Part *c* served as an indication of the extent of spontaneous induction under these conditions. Infection by $b_2$vir increased the λ yield *ca.* sixfold over that due to spontaneous induction. Phage from the lysate were spun in a CsCl gradient along with P^{32}λ as a density reference. The P^{32}λ had been killed by UV light so as not to interfere with subsequent phage assays. The density of the infectious λ progeny could thus be compared with that of the P^{32}λ marker, which previously had been shown to have a normal density despite the UV treatment.

As seen in Figure 4*A*, approximately two thirds of the total λ phage in the progeny of the noninduced culture were of the original density. The control (Fig. 4*B*) shows that UV induction caused all of the progeny λ to have a higher than normal density. It is difficult to determine the exact fraction of progeny λ produced as a result of spontaneous induction and that fraction due to $b_2$vir infection. It is certain that at least 70 per cent of the parental λ genomes never replicated before being packaged into phage particles.

Whereas the preceding experiment (Fig. 4) permitted us to analyze the density of all the λ phage produced by the entire cell population, the next one allowed us to disregard those phage produced by the small fraction of spontaneously induced cells. By using labeled phage for infection, only the replication of the parental λ DNA molecules was analyzed. The following diagram illustrates the procedure:

			Yield (per cell)	
			λ	b_2b_5c
(a) $P^{32}\lambda_{t1} + b_2b_5c + E.\ coli(\lambda_{t1})\ Thy^-$	$\xrightarrow[37°,\ 105\ min]{BUdR}$		3.8	35
(b) $P^{32}\lambda_{t1} + E.\ coli(\lambda_{t1}\ Thy^-$	$\xrightarrow[45°,\ 15\ min]{BUdR}$	$\xrightarrow[37°,\ 90\ min]{BUdR}$	27	0

In part *b*, the 15 minutes at 45° induced λ development in 90 per cent of the cells. The cI mutant, λ_{t1}, is known to synthesize a heat-labile repressor.[1] The phage in the lysates were concentrated by centrifugation, then banded in CsCl along with $H^3\lambda_{t1}$ as a density marker. Three-drop fractions were collected onto Whatman glass filters, which were then dried and assayed for radioactivity.

The results shown in Figure 5*A* clearly demonstrate that none of the parental λ P^{32}-DNA which was packaged into phage particles had replicated even once. Heat induction caused all of the parental λ P^{32}-DNA that was packaged to have a density greater than normal (Fig. 5*B*).

Since the packaging efficiency of λ by b_2b_5c is the same as by λ itself (see section (2)), it seems certain that this experiment affords an analysis of a random sampling of the intracellular λ DNA pool. We thus conclude tentatively (see *Discussion*) that the λ repressor present in lysogenic cells can prevent λ DNA replication even in the presence of all the enzymes necessary for such replication.

DISCUSSION

The experiments described in this report strongly support Thomas and Bertani's hypothesis that the λ repressor "blocks directly (mechanically or sterically) the replication of the phage chromosome."[7] We have seen that an immunity-insensitive λ phage can replicate itself without replicating an immunity-sensitive λ genome present in the same lysogenic cell. Before being certain that the repressor can

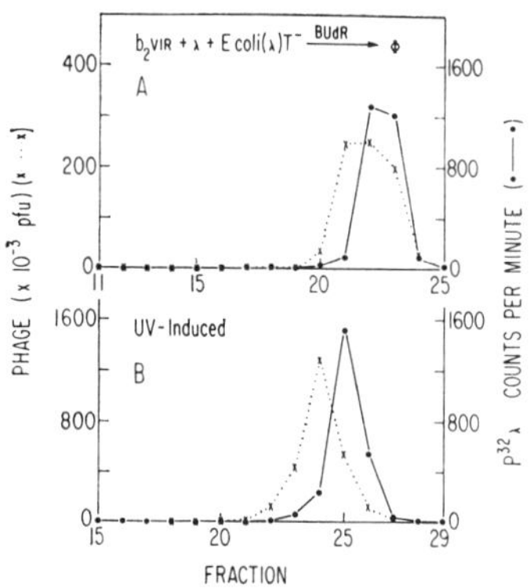

Fig. 4. Density labeling of λ and λb_2vir after mixed infection of *E. coli*(λ) Thy^-. UV-killed P^{32}-λ were used as a density reference. Two-drop fractions were collected into tubes containing dilution medium. Part **A** had 43 fractions; part **B** had 45. Aliquots (0.1 ml) were spotted on Whatman paper, acid-washed, and assayed for P^{32} activity. Phage were assayed on W3110, whereon λ forms turbid plaques and λb_2vir forms clear plaques. Only the λ assays are shown. Infectious λ (X . . . X); P^{32}λ (•——•).

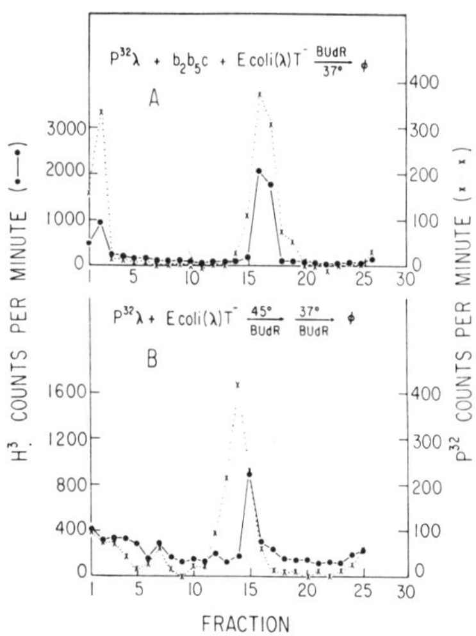

Fig. 5. Density labeling of λ phage containing parental DNA after infection of *E. coli*(λ_{t1})Thy^-. The P^{32} profile represents parental λ P^{32}-DNA packaged into progeny phage. H^3-λ_{t1} were added as a density reference in the CsCl gradient. H^3 cpm (•——•); P^{32} cpm (X . . . X).

directly interfere with λ DNA replication, the following questions must still be answered:

(1) Does the cI region of the prophage code for the product that interferes with λ DNA replication?

(2) Are all of the replicating enzymes coded by λvir diffusible in the cytoplasm, and can they act on a λ DNA molecule that did not code for their formation?

(3) Is the concomitant transcription of λ DNA into mRNA necessary for DNA replication?

In regard to these questions, the following evidence may be cited in support of our conclusion. No λ functions other than that of the cI region have been detected in lysogenic cells.[2,16] The great majority of the prophage genome is not transcribed into mRNA, nor is that of an immunity-sensitive superinfecting λ phage.[1,17] Complementation experiments with b_2b_5c and three λ *sus* mutants known to affect λ DNA replication (*sus* N, O, and P) indicate that the products of these cistrons can act on a heteroimmune phage present in the same cell (Green and Gotchel, unpublished data). As to the third question, we know that the λcI region is transcribed upon infection of an immune lysogen,[2,18-20] but that the λ genome is not replicated.[6] Under certain circumstances, regions other than cI may also be transcribed while replication appears to be inhibited by the λ repressor.[21]

The ability of the λ repressor to block λ DNA replication as well as mRNA synthesis could help to explain the existence of two classes of λcI temperature mutants. The class A lysogens are heat-inducible in the absence of protein synthesis, whereas the class B lysogens require protein synthesis at the elevated temperature in order to be heat-induced.[20,22] Heat causes the derepression of λ mRNA synthesis for both types of mutants, even in the absence of protein synthesis.[1]

One currently popular explanation for the two classes of cI temperature mutants is that the class B repressor becomes active very soon after the heated cells are returned to low temperature, whereas the class A repressor does not.[23,24] Thus the class B lysogens need to synthesize λ proteins at the elevated temperature in order to be irreversibly committed to phage production.

However, when the repressor assay is based on its ability to block λ mRNA synthesis (rather than its ability to confer immunity on cells), the data do not support this explanation. A class B lysogen, *E. coli*(λcI, t1), was found to remain derepressed for λ mRNA synthesis for at least two cell generations.[1] The cells had been infected with λcI, heated in the absence of protein synthesis, then restored to normal growth conditions. The failure of these cells to produce λ phage or to be killed by either the prophage or the superinfecting phage, even though λ mRNA synthesis was not repressed, testifies to the inadequacy of the above model.

Alternatively, if some step in the process of λ DNA replication were required in order for λ to kill *E. coli*,[25] the class B repressor might retain or regain the ability to block this event even after it had lost its ability to inhibit transcription. This model gains support from the fact that the class A and B mutants can complement each other, causing the appearance of physiologically heat-stable lysogens.[20,26] Thus the λ repressor might assume distinct activities depending on its state of aggregation or conformation. The mechanism whereby protein synthesis could permit heat induction of class B lysogens has been discussed previously.[1,20,26]

SUMMARY

Upon mixed infection of *E. coli*(λ) by immunity-sensitive phage (e.g., λ) and immunity-insensitive, but closely related, phage (e.g., λb_2b_5c or λb_2vir), the number of infectious λ produced never exceeded the input number of λ genomes. The restriction of λ was not caused by (1) an abnormal delay in their formation; (2) inefficient packaging of λDNA into phage particles; or (3) some step leading to the formation of noninfectious phage.

All of the progeny phage contained DNA which had not undergone replication. Since the immunity-insensitive phage provided all of the enzymes needed for the replication of λ DNA, it is tentatively concluded that the repressor coded by the prophage directly interferes with λ DNA replication. A basic assumption is that all of the requisite enzymes are diffusible and can act on a DNA molecule that did not code for their formation.

REFERENCES

1. Green, M. H., J. Mol. Biol. **16**:134 (1966).
2. Jacob, F., and J. Monod, J. Mol. Biol. **3**:318 (1961).
3. Attardi, G., S. Naono, J. Rouvière, F. Jacob, and F. Gros, in Cold Spring Harbor Symposia on Quantitative Biology, vol. 28 (1963), p. 363.
4. Brooks, K., Virology **26**:489 (1965).

5. Joyner, A., L. N. Isaacs, H. Echols, and W. S. Sly, J. Mol. Biol. **19**:174 (1966).
6. Wolf, B., and M. Meselson, J. Mol. Biol. **7**:636 (1963).
7. Thomas, R., and L. E. Bertani, Virology **24**:241 (1964).
8. Kellenberger, G., M. L. Zichichi, and J. Weigle, these Proceedings **47**:869 (1961); Nature **187**:161 (1960).
9. Burgi, E., these Proceedings **49**:151 (1963).
10. Kaiser, A. D., Virology **3**:42 (1957); Kaiser, A. D., and F. Jacob, Virology **4**:509 (1957).
11. Lieb, M., Science **145**:175 (1964).
12. Zichichi, M. L., and G. Kellenberger, Virology **19**:450 (1963).
13. Stahl, F. W., J. M. Crasemann, L. Okun, E. Fox, and C. Laird, Virology **13**:98 (1961).
14. Jacob, F., and E. L. Wollman, Sexuality and the Genetics of Bacteria (New York: Academic Press, 1961).
15. Dove, W. F., and J. J. Weigle, J. Mol. Biol. **12**:620 (1965).
16. Bode, V. C., and A. D. Kaiser, Virology **25**:111 (1965).
17. Taylor, K., Z. Hradecna, and W. Szybalski, these Proceedings **56**:1618 (1967).
18. Lieb, M., Virology **29**:367 (1966).
19. Horiuchi, T., and H. Inokuchi, J. Mol. Biol. **15**:674 (1966).
20. Green, M. H., in Edmonton Symposium on Molecular Biology of Viruses (New York: Academic Press, 1967).
21. Thomas, R., J. Mol. Biol. **22**:79 (1966).
22. Lieb, M., J. Mol. Biol. **16**:149 (1966).
23. Naono, S., and F. Gros, in Cold Spring Harbor Symposia on Quantitative Biology, vol. 31 (1966), p. 363; J. Mol. Biol. **25**:517 (1967).
24. Ogawa, T., and J. Tomizawa, J. Mol. Biol. **23**:225 (1967).
25. Weisberg, R. A., and J. A. Gallant, J. Mol. Biol. **25**:537 (1967).
26. Green, M. H., in IX International Congress for Microbiology (Moscow, 1966), Symposia, p. 509.

54 Changing populations of messenger RNA during sea urchin development

A. H. Whiteley
B. J. McCarthy
H. R. Whiteley*
Departments of Zoology, Microbiology, and Genetics
Friday Harbor Laboratories
University of Washington
Seattle, Washington

Protein synthesis is very rapidly initiated after fertilization of the sea urchin egg.[1,2] On the other hand, RNA synthesis is negligible before fertilization and increases very little, if at all, at fertilization.[3-5] The suggestion has been made that the onset of protein synthesis represents the activation of pre-existing messenger RNA molecules rather than the *de novo* synthesis of new messages. Support for this view has been provided by experiments in which enucleate egg fragments are found to carry on protein synthesis when artificially activated,[3,6,7] and by experiments with inhibitors of RNA synthesis.[8] Further, Maggio *et al.*[9] report that RNA fractions from unfertilized eggs will stimulate amino acid incorporation in rat liver test systems, and Monroy *et al.*[10] provide evidence that dormant protein-synthesizing machinery of the unfertilized egg may be activated *in vitro* without evident concomitant RNA synthesis.

The present communication demonstrates directly the presence of messenger RNA in the unfertilized sea urchin egg and compares the populations of messages present before fertilization with those isolated from early stages of development and from adult tissues. The attribute of messenger RNA used for its assay is the ability to form specific complexes with single-stranded DNA at a reasonably high RNA:DNA ratio. This property is due to complementarity in sequences of nucleotides between the nucleic acids and, therefore, reflects the primary expression of the genetic material. By injecting labeled phosphate into female sea urchins, Gross *et al.*[11] demonstrated the synthesis, during oögenesis, of labeled RNA capable of binding to DNA. In the present study, similarities in populations of RNA's have been measured by ascertaining the capacity of unlabeled samples from one stage or tissue to compete in the binding of labeled RNA. The labeled RNA was prepared from prism embryos so that the assay is limited to those messenger RNA molecules actively synthesized at that time.

MATERIALS AND METHODS

Cultivation of embryos. Eggs, obtained from the sea urchin *Strongylocentrotus purpuratus* by injection of isotonic KCl into the coelom or by electrical shock, were inseminated, washed several times with Millipore-filtered sea water, and cultured at 11-13°C in Millipore-filtered sea water containing 0.25 mg/ml streptomycin. The embryos were cultured at less than 0.5% concentration in jars with constant gentle stirring. Only cultures in which fertilization was 98% or better and in which development was morphologically normal were used. The embryos were harvested in a continuous-flow centrifuge at 255 × *g* and washed with filtered sea water containing streptomycin.

Unlabeled RNA was extracted from unfertilized eggs from which the jelly had been removed by brief exposure to sea water at pH 4.5 and from three developing stages. These, as defined by Whiteley and Baltzer,[12] were: 31-hr-old blastulae which were just hatching (hBl), 44-hr-old gastrulae (GaJ 1/4), and 74-hr-old prism embryos (Pr). In the late prism embryos the transverse spicules had not quite met medially and the stomodaeum had not yet broken through to form the mouth. An aliquot of these prisms was also pulse-labeled with P^{32} to label newly synthesized RNA. One female provided the eggs for the unfertilized sample, a second female provided the eggs for the blastulae and gastrulae, and a third female provided the eggs for the prism embryos. Sperm from one male was used for each insemination.

Adult tissue. The gut, including both stomach and intestine, of six adult starved sea urchins and the testes

---

From Proceedings of the National Academy of Sciences **55**:519-525, 1966. Used with permission.

This investigation was supported in part by a grant from the National Science Foundation, and U.S. Public Health Service research grants GM 12449 from the National Institute of General Medical Sciences and CA 03931 from the National Cancer Institute. Presented before the AAAS; abstract in Am. Zool. 5:709 (1965).

*Recipient of a Research Career Award, GM-K6-442.

of one unripe mature male were excised, washed in iced sea water, frozen in dry ice, and RNA was extracted.

Pulse labeling of RNA. Three ml of gently packed prism embryos were maintained suspended in a conical centrifuge tube in 20 ml of Millipore-filtered sea water, containing streptomycin, by gently bubbling air through the culture. One mc of P^{32} as orthophosphate was added and after 60 min at 13°C, the embryos were collected by low-speed centrifugation, the insignificant residual unadsorbed P^{32} was removed from the eggs by means of two sea water washes, and the RNA was extracted.

Extraction of RNA. One vol of embryos or tissue was homogenized at 0°C in 2 vol of 0.1 *M* sodium acetate–0.1 *M* NaCl–0.001 *M* $MgCl_2$ at pH 5.2 and containing 3-15 mg bentonite/ml. Two vol of the same solution containing 4% sodium lauryl sulfate were added and the homogenate was shaken for 15 min at 25°C with 5 vol of water-saturated phenol without preservative. The aqueous phase resulting from centrifugation of the emulsion at 25,000 × *g* for 10 min was reextracted with aqueous phenol and the nucleic acids were precipitated from the aqueous phase at –20°C with 3 vol of ethanol in the presence of 0.1 *M* NaCl. The sample was dissolved in 0.01 *M* sodium acetate–0.002 *M* $MgCl_2$, pH 5.2, and digested for 1 hr with deoxyribonuclease (50 μg/ml, Worthington Biochemical Corp., Freehold, N.J., electrophoretically purified) at 25°C. Residual protein was hydrolyzed for 1 hr at 25°C by adding pronase (50 μg/ml, Calbiochem, Inc., Los Angeles, Calif.) in the presence of 0.05 *M* Tris, pH 7.5.[13] The pronase was first self-digested at 37°C for 2 hr. After two additional phenol extractions, the RNA was reprecipitated with ethanol and dissolved in a minimal volume of 1/10 × SSC (1 × SSC is 0.15 *M* NaCl–0.015 *M* sodium citrate, pH 7.0).

Aliquots were put on a column (1.3 × 30-cm) of Sephadex G-50 overlaid with a 1-cm layer of Dowex-50; the column was equilibrated and eluted with 2 × SSC. Trials with unlabeled *S. purpuratus* RNA showed that with this column approximately 1/3 to 1/2 of the total sRNA was eluted in 3.0 ml, but that nucleotides were excluded. Bentonite (3-15 mg/ml) was immediately added to the sample which was then stored at –20°C.

RNA prepared as described was susceptible to ribonuclease and alkali digestion. In an experiment with P^{32}-labeled material, exposure for 30 min to 200 μg/ml of ribonuclease (Worthington) at 37°C or to 2 *N* NaOH at 60°C converted 98% of the label to a form not precipitable by 5% TCA.

Extraction of DNA. One gm of washed *S. purpuratus* sperm was suspended in 250 ml of 0.1 *M* EDTA–0.04 *M* Tris, pH 8.2, pronase (50 μg/ml) was added, and the preparation was incubated at 37°C for 8 hr. This preparation was shaken with 1 vol of water-saturated phenol and centrifuged. There was essentially no interphase pad. The DNA was precipitated from the aqueous phase with 3 vol of ethanol, washed, dissolved in 1/10 × SSC at a concentration of 500 μg/ml, heated at 100°C for 10 min, and cooled quickly to convert the DNA to the single-stranded form. The preparation was adjusted to a concentration of 1 × SSC, and the DNA was precipitated with alcohol and redissolved at the desired concentration in 1/10 × SSC.

The concentrations of DNA and RNA were estimated from the absorbancy at 260 mμ.

RNA-DNA binding. Aliquots (0.025 ml) of the P^{32}-labeled RNA, 0.05 ml of a bentonite suspension (20 mg/ml), competing RNA, and sufficient 2 × SSC to make a total volume of 0.475 ml, were mixed in a small test tube. The contents of the tube were transferred with a Pasteur pipette to a screw-cap siliconed vial containing 0.30 gm of DNA-agar prepared as described by McCarthy and Hoyer.[14] A single preparation of DNA-agar was used throughout. The contents of the vial were mixed with a Vortex mixer, and the vial was incubated for 48-60 hr in a 60°C water bath.

After incubation, the DNA-agar was washed at 60°C in 2 × SSC to remove unbound RNA, and then at 73°C in 1/100 × SSC to remove RNA bound to the DNA. Details are described by McCarthy and Hoyer.[14] The labeled RNA's were precipitated with TCA, collected on membrane filters, and assayed with a liquid scintillation counter. Per cent of binding of RNA to DNA is defined as the ratio of radioactivity eluted at 73°C to the total radioactivity precipitated from the two washes.

RESULTS

Properties of RNA and binding to DNA. The sedimentation profile of RNA obtained from prisms after a 1-hr exposure to radioactive phosphate is shown in Figure 1. Much of the incorporated label appears as a 4*S* component—presumably sRNA. Radioactivity appears not only in the regions of the bulk ribosomal and sRNA components, but also at intermediate positions and in a region indicating a size greater than 30*S*.

This labeled RNA preparation was incubated with DNA-agar to determine the amount required for optimal binding. Various amounts of RNA were incubated for 48 hr at 60°C with aliquots of the same DNA-agar preparation (Fig. 2). Four μg of labeled RNA, which gave an optimal binding of about 22 per cent, were used in most of the competition experiments to be described.

Figure 3 illustrates the kinetics of the binding reaction. The optimal period of incubation at 60°C is about 48 hr. Periods of incubation longer than 90 hr result in lower values, probably reflecting loss of DNA from the agar gel.[15]

Comparison of RNA's from developmental stages. Populations of RNA molecules isolated from four stages in the early development of the sea urchin were compared in competition experiments summarized in Figure 4. Each assay was performed with the same amount of P^{32}-labeled prism-stage RNA, a constant amount of DNA-agar, and various amounts of

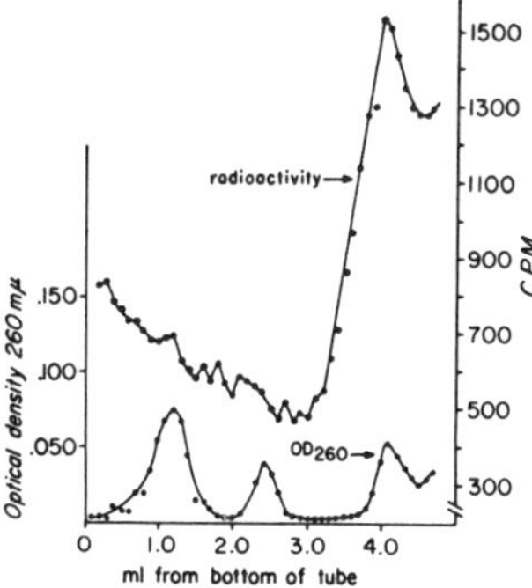

Fig. 1. Sedimentation properties of P^{32}-labeled *S. purpuratus* prism RNA. Eight μg of RNA were layered on 4.6 ml of a 5-20% sucrose gradient and centrifuged in the cold at 38,000 rpm in a SW39 rotor of a model L Spinco centrifuge for 5 hr. Optical density at 260 mμ and radioactivity were determined.

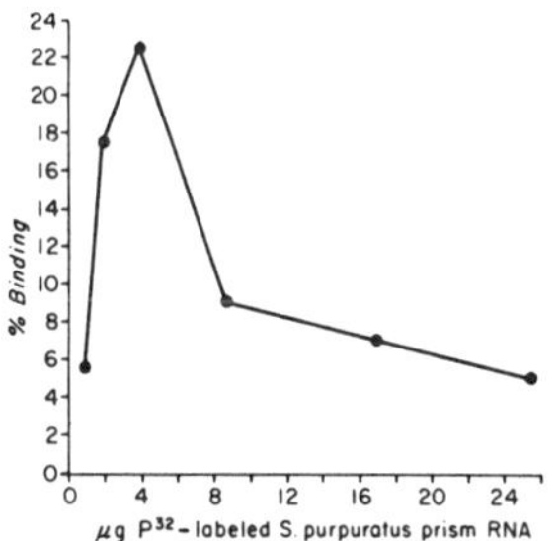

Fig. 2. Binding of different amounts of P^{32}-labeled *S. purpuratus* prism RNA to a constant amount of *S. purpuratus* DNA-agar. DNA-agar (0.30 gm) containing 110 μg DNA, varying amounts of labeled RNA, total volume = 0.475 ml, incubated at 60°C for 48 hr.

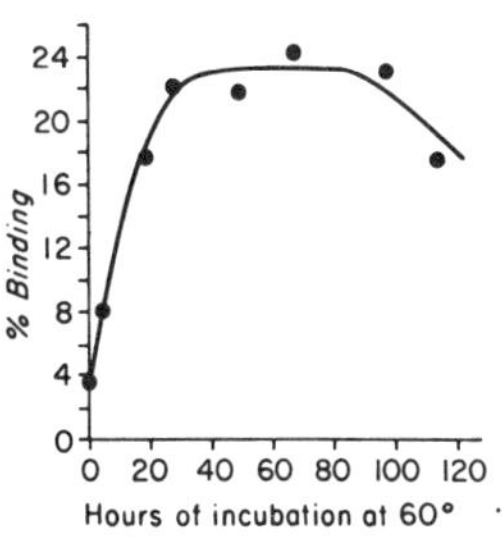

Fig. 3. Binding of P^{32}-labeled *S. purpuratus* prism RNA to *S. purpuratus* DNA-agar as a function of time. Four μg P^{32}-labeled RNA, 0.30 gm DNA-agar containing 110 μg DNA, total volume = 0.475 ml, incubated at 60°C for time shown.

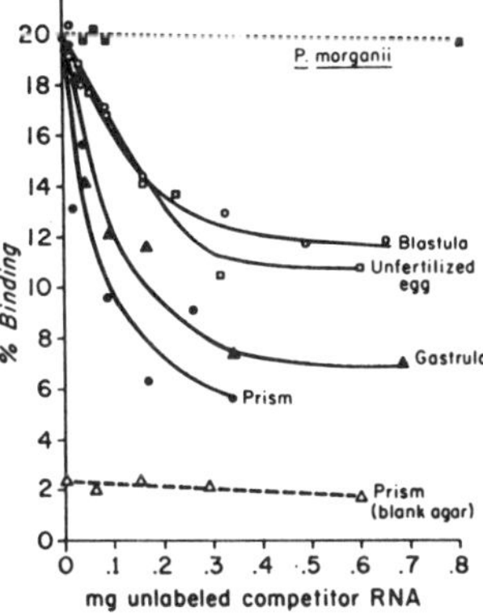

Fig. 4. Competition by unlabeled RNA from developmental stages in the binding of P^{32}-labeled prism RNA to DNA-agar. Four μg P^{32}-labeled *S. purpuratus* prism RNA, unlabeled competitor RNA as shown, 0.30 gm of *S. purpuratus* DNA-agar containing 110 μg DNA, or 0.30 gm agar lacking DNA (*bottom curve*), total volume = 0.475 ml, incubated at 60°C for 48 hr.

unlabeled RNA from one of the developmental stages. Each of the sea urchin RNA preparations compete in the binding, but to different extents. As would be expected, unlabeled RNA prepared from the stage identical to that of the pulse-labeled RNA is most effective in this competition. Competition by RNA extracted from unfertilized eggs is not inconsiderable. Similarly, RNA from blastulae at the time of hatching competes rather strongly and, as seen in Figure 4, this RNA is not clearly distinguishable from that of unfertilized eggs. RNA extracted from early gastrulae is a very effective competitor, nearly as effective as RNA of the homologous stage. In quantitative terms, depression of the binding by 50 per cent requires about 600 μg of unfertilized egg or blastula RNA, compared with about 150 μg of gastrula RNA and about 100 μg of homologous prism RNA.

The kinds of messenger RNA molecules present in different populations may be compared by a variant of the binding approach other than direct competition. Since the formation of DNA-RNA complexes is essentially irreversible at 60°C, it is possible to incubate the DNA sequentially with different RNA preparations. Molecules present in one sample may preempt sites on the DNA with which labeled molecules, subsequently added, may otherwise combine. This type of experiment was used to obtain the data of Table 1 which compares the four unlabeled RNA samples from the embryonic stages. It is clear that each group of RNA molecules is able to occupy positions on the DNA complementary to some

Table 1. Reaction of DNA-agar with competitor RNA before addition of labeled DNA

RNA added initially	RNA added after 36 hr, 60° C	Per cent binding	Per cent control
P^{32}-prism	None	9.6	100
P^{32}-prism + unlabeled prism	None	5.6	58
None	P^{32}-prism	7.1	100
Unlabeled prism	P^{32}-prism	4.2	59
Unlabeled gastrula	P^{32}-prism	4.2	59
Unlabeled blastula	P^{32}-prism	5.4	76
Unlabeled unfertilized	P^{32}-prism	5.8	82

P^{32}-labeled *S. purpuratus* prism RNA (8.5 μg), 0.6 mg each unlabeled competitor RNA, 0.30 gm of *S. purpuratus* DNA agar containing 110 μg DNA, total volume = 0.475 ml, incubated at 60°C for 60 hr as shown above.

P^{32}-labeled molecules. However, the blastula and unfertilized egg RNA's are considerably less effective than the RNA from prism and from gastrula. In addition, there is little difference between RNA's from blastulae and unfertilized eggs, and RNA from gastrulae and prisms are not distinguishable. Therefore, this experiment confirms the results shown in Figure 4. As seen in Table 1, preincubation of the DNA-agar with unlabeled prism RNA gives the same degree of competition (59% of the control) with subsequent binding of P^{32}-prism RNA as does simultaneous incubation with both labeled and unlabeled RNA (58% of the control), indicating that there is no destruction of nucleic acids during preincubation, and that complete saturation by unlabeled RNA was not achieved. It should be noted that this experiment was performed with 8.5 μg of P^{32}-RNA rather than 4.0 μg, so the per cent of binding is not the same as in the experiments of Figure 4.

RNA from adult tissues. The results of the experiments described suggest that there is a class of messenger RNA molecules which exist in the unfertilized egg and which are continually synthesized at least to the prism stage. These may be concerned with the synthesis of proteins of value to all types of cells as, for example, structural proteins and the enzymes of intermediary metabolism. In this connection, it would be of value to compare the RNA of adult tissues with the labeled prism RNA preparation. This comparison was undertaken by means of competition experiments with unlabeled RNA from adult gut and testis (Fig. 5). Again, these preparations act as competitors for at least some of the P^{32}-labeled messenger RNA molecules, but the level of competition is much less than with comparable amounts of prism RNA and, in fact, is less than that with RNA from unfertilized eggs and blastulae.

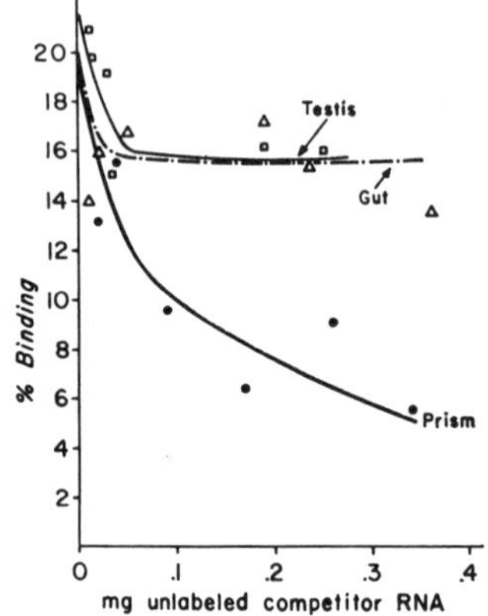

Fig. 5. Competition by unlabeled RNA from adult tissues in the binding of P^{32}-labeled prism RNA to DNA-agar. Conditions and amounts as for Fig. 4.

The competition curves presented in Figures 4 and 5 show a rapid decrease in binding at low levels of competitor RNA followed by a more gradual decrease with larger amounts of competitor. The form of this curve suggests the existence of two or more populations of RNA molecules in each preparation.[16] A similar effect may be seen in Figure 2 where the per cent of binding decreases rapidly with an increasing RNA:DNA ratio. Thus, one of these classes of molecules is either very abundant or represented by only a small fraction of sites on the DNA. This degree of reaction may, therefore, be attributed either to frequent messenger RNA molecules or ribosomal RNA. As shown in Figures 2, 4, and 5, other classes include RNA molecules at low concentrations so that large amounts of competitor must be added in order to decrease their binding. It may be that the frequent class is relatively less abundant in the gut and testis RNA than in the embryo RNA.

Specificity of RNA-DNA binding. RNA from a completely unrelated source, the bacterium *Proteus morganii,* does not bind to *S.*

purpuratus DNA, as shown by its failure to compete in the binding of *S. purpuratus* prism RNA, even when present in large quantities (Fig. 4). Likewise, little reaction occurs with the agar gel alone and, as shown in the bottom curve of Figure 4, when larger amounts of RNA are present, this nonspecific retention of radioactivity is not affected.

DISCUSSION

The present report demonstrates the existence of mRNA in the unfertilized sea urchin egg. It has been shown that some of this RNA is able to recognize sites in the DNA which are active in the synthesis of RNA at later stages of development. The former molecules compete in the binding of such newly synthesized RNA to DNA. Therefore, it is possible to conclude not only that this is messenger RNA but also that it belongs to a particular category of messages which are made both in oögenesis and during embryogenesis. By the nature of the experimental design it is, of course, impossible to say how much of the total message population of unfertilized eggs has this characteristic. It is not surprising, however, that such a core of messages exists, for all cells share a number of functions and structures requiring the synthesis of large groups of proteins held in common. Thus, it is to be expected that the populations of messenger RNA from groups of cells within an animal would overlap to some extent. This is apparently the case for messenger RNA isolated from various adult mouse tissues.[14]

The results also provide good evidence for the existence of messenger RNA molecules specific for a given developmental stage in accordance with the view that genes are sequentially activated during embryogenesis. For example, it is clear that many of the RNA molecules labeled at the prism stage are essentially unrepresented very early in development (unfertilized eggs and blastulae) or in two adult tissues. A similar conclusion has been reached for *Xenopus laevis.*[17]

RNA's from unfertilized eggs and hatching blastulae cannot be clearly distinguished. It is possible that different populations of messages exist at these two times and that the equal competitions are due to fortuitous proportions of these, but our experiments do not require this conclusion. The more direct interpretation is that, for the most part, relatively few new genes are transcribed in this developmental interval which is instead supported by masked messages already present before fertilization. This interpretation agrees with the host of observations,[18] dating from Boveri's early studies, that pregastrular development is maternal and that new nuclear influences begin to appear in the period between blastulation and gastrulation. Many biosyntheses are accelerated then. Between fertilization and hatching, few new differentiations appear: cilia, desmosomes, a few enzymes including the hatching enzyme, and phosphate transport carrier.[19] The formation of the gastrula, however, involves the differentiation of many specialized features in all three germ layers and the synthesis of numerous new enzymes. It also entails form changes. The early features may be supported by the utilization of pre-existing messages which would continue to be produced subsequent to hatching for maintenance purposes.

The latter differentiation may be expected to require new messages. Quantitatively we cannot distinguish clearly between the messenger populations at the onset of gastrulation and at the onset of the pluteus stage. Forty hr have elapsed between these two stages, whereas 10-13 hr elapse between hatching blastulae and early gastrulae. If it is established that the RNA at the onset of gastrulation has the same qualitative composition as that at the onset of the pluteus stage, it would follow that a burst of production of new messages occurs during the 10-hr interval just prior to gastrulation and that these are either activated at one time or stored and sequentially activated throughout this long interval.

SUMMARY

Experiments on the binding of RNA to DNA and competition in this binding by RNA's from various developmental stages have shown: (1) messenger RNA is present in unfertilized eggs of *Strongylocentrotus purpuratus;* (2) some of these kinds of molecules of mRNA continue to be synthesized as late as the prism stage; (3) mRNA from unfertilized eggs and blastulae were not distinguished; (4) adult tissues share some of the same molecules with prism embryos; and (5) other molecules, assembled at the prism stage, are much less abundant or absent at both earlier and later stages of development.

REFERENCES

1. Hultin, T., Exptl. Cell Res. **3**:494 (1952).
2. Nakano, E., and A. Monroy, Exptl. Cell Res. **14**:236 (1958).

3. Brachet, J., M. Decroly, A. Ficq, and J. Quertier, Biochim. Biophys. Acta **72**:660 (1963).
4. Nemer, M., these Proceedings **50**:230 (1963).
5. Wilt, F. M., Biochem. Biophys. Res. Commun. **11**:447 (1963).
6. Denny, P. C., and A. Tyler, Biochem. Biophys. Res. Commun. **14**:245 (1964).
7. Tyler, A., Am. Zool. **3**:109 (1963).
8. Gross, P. R., L. I. Malkin, and W. A. Moyer, these Proceedings **51**:407 (1964).
9. Maggio, R., M. L. Vittorelli, A. M. Rinaldi, and A. Monroy, Biochem. Biophys. Res. Commun. **15**: 436 (1964).
10. Monroy, A., R. Maggio, and A. M. Rinaldi, these Proceedings **54**:107 (1965).
11. Gross, P. R., L. I. Malkin, and M. Hubbard, J. Mol. Biol. **13**:463 (1965).
12. Whiteley, A. H., and F. Baltzer, Pubbl. Staz. Zool. Napoli **30**:402 (1958).
13. Gillespie, D., and S. Spiegelman, J. Mol. Biol. **12**:829 (1965).
14. McCarthy, B. J., and B. H. Hoyer, these Proceedings **52**:915 (1964).
15. Hoyer, B. H., B. J. McCarthy, and E. T. Bolton, Science **140**:1408 (1963).
16. McCarthy, B. J., and E. T. Bolton, J. Mol. Biol. **8**:184 (1964).
17. Denis, H., Ann. Rep. Carnegie Inst. Wash. **63**:505 (1964).
18. For a review, see J. Brachet, in Biochemical Cytology (New York: Academic Press, 1957), chap. 8.
19. Whiteley, A. H., and E. L. Chambers, in Symposium on Germ Cells and Development (Institut. Intern. d'Embryologie and Fondazione A. Baselli, 1960), p. 387.

55 Rate of protein synthesis: regulation during first division cycle of sea urchin eggs

W. H. Sofer*
J. F. George
R. M. Iverson
Department of Biology
Laboratory for Quantitative Biology
University of Miami
Coral Gables, Florida

Abstract. *Protein synthesis in fertilized sea urchin eggs, or in 12,000g supernatants derived from them, increased linearly during the period preceding prophase of the first mitotic cycle, dropped during metaphase and anaphase, and increased again after telophase. Similar results were observed for whole cells incubated in the presence of colchicine. These changes in the rate of protein synthesis during the mitotic cycle may be regulated at the translational level.*

The overall rate of protein synthesis in the fertilized sea urchin egg increases throughout the early cleavage stages until blastulation *(1)*. In addition to this general increase, fluctuations exist in the rate of protein synthesis during the first few division cycles *(2)*. Similar variations have been observed in dividing mammalian cells in tissue culture *(3)*. We report here that these variations in the rate of incorporation of labeled amino acids into protein occur in vivo, in vitro, and in the presence of colchicine, and they are detectable at the polysome level during the first division cycle of the sea urchin egg.

Gametes of the sea urchin *(Lytechinus variegatus)* were obtained by the KCl method described by Harvey *(4)*. Eggs were washed three times in filtered sea water, washed three more times in Millipore-filtered sea water, and then incubated in Millipore-filtered sea water containing 75 μg of streptomycin sulfate and 300 units of penicillin per milliliter.

The eggs in suspension were fertilized, and portions of the suspension were transferred to separate incubation vessels. After fertilization, eggs were exposed to C^{14}-amino acids for the times indicated in each experiment. Unfertilized control eggs were treated similarly. After incubation an equal volume of 10 percent trichloroacetic acid (TCA) was added and the resultant precipitates were homogenized. These homogenates were then centrifuged and washed three times with 5 percent TCA, heated to 90°C in 5 percent TCA for 30 minutes, again washed three times with 5 percent TCA, and extracted twice with a mixture of ethanol and ether (3 : 1) and once with acetone. Dried precipitates were dissolved in 88 percent formic acid, plated on either preweighed aluminum planchettes or Whatman GF/C glass filter pads, and dried. Radioactivity was measured with a Nuclear-Chicago gas-flow counter or a Packard scintillation counter (efficiencies, 33 percent and 63 percent, respectively). No corrections for self-absorption were required. The amount of protein was determined by weight.

In the experiments on in vitro synthesis of protein, eggs were first incubated at 30°C to the appropriate stage, rapidly cooled to 5°C, washed three times with a mixture of cold isotonic NaCl and KCl (19 : 1) and once with cold homogenizing medium (0.01*M* $MgCl_2$, 0.24*M* KCl, and 0.01*M* tris-HCl, *p*H 7.6), resuspended in four volumes of homogenizing medium, and homogenized gently in a Duall tissue homogenizer fitted with a Teflon pestle. Homogenates were centrifuged at 12,000*g* for 30 minutes, and the supernatants were used for incubation. One milliliter of the cell-free reaction mixture contained 20 μ*M* phosphoenolpyruvate, 0.4 μ*M* guanosine triphosphate, 0.1

Supported in part by NIH fellowship GM-24115 (W.H.S.), a Ford Foundation fellowship (J.F.G.), NIH grant GM-11156, NSF grant GB-2891, and NIH career development award GM-6809.

*Present address: Mergenthaler Laboratory for Biology, The Johns Hopkins University, Baltimore, Maryland.

mg of pyruvate kinase, 0.3 μc of C^{14}-leucine (specific activity, 67.4 mc/m*M*), 2 μ*M* adenosine triphosphate, and approximately 2 mg of supernatant protein. After incubation for 45 minutes at 30°C, an equal volume of 10 percent TCA was added. The precipitates were washed and analyzed as above.

The experiment designed to show differences in polysomes at various times after fertilization was performed as previously reported *(5)*. The $O.D._{260}$ was measured in a Gilford multiple sample absorbance unit with the use of a 1-cm flow-through cell.

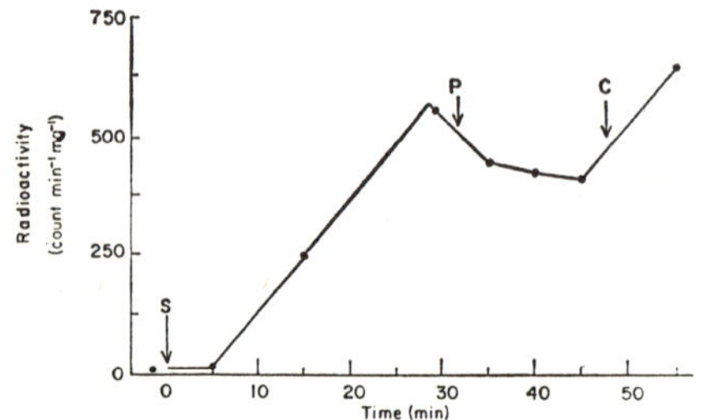

Fig. 1. Incorporation of C^{14}-leucine (231 mc/m*M*) into 5 percent trichloracetic acid–precipitable protein during the first division cycle. At the indicated time after fertilization the eggs (at 26°C) were "pulse" labeled for 2.5 minutes with C^{14}-leucine (0.03 μc/ml). Eggs were fertilized at zero time. Radioactivity is expressed as counts per minute per milligram of protein; time, as minutes after fertilization. *S* indicates when sperm was added; *P* is prophase; and *C*, cleavage.

Upon fertilization the rate of incorporation of C^{14}-leucine into protein increased linearly (Fig. 1) until about the time of nuclear membrane rupture. At that time, the rate of synthesis declined and remained constant to approximately the metaphase or anaphase stage, after which it again increased. Similar patterns occurred in the incorporation of C^{14}-valine and C^{14}-phenylalanine into protein *(6)*. This pattern of synthesis was also discernible by determining the amount of labeled protein associated with the polysomes (Fig. 2). At metaphase (Fig. 2b) the amount of radioactivity associated with the polysome region was lower than in comparable regions isolated from eggs before (Fig. 2a) or after (Fig. 2c) division.

Experiments were conducted to eliminate the possibility that observed changes in rates of synthesis were due to differences in permeability of the eggs to added amino acids at various times during the division cycle. Therefore, we measured the capacity of 12,000*g* supernatant fractions, derived from eggs that had been homogenized at different stages of division, to incorporate amino acids in vitro. The patterns observed in vitro (Fig. 3) were similar to those seen in vivo (Fig. 1).

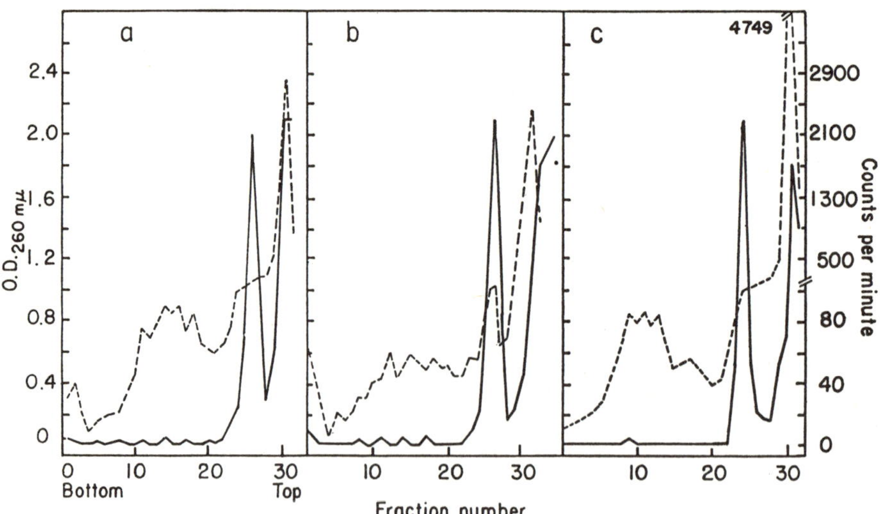

Fig. 2. Sucrose-gradient analysis of the 12,000*g* supernatant from eggs homogenized after exposure to C^{14}-leucine (231 mc/m*M*) at different stages during the first division cycle. Eggs were incubated in the presence of C^{14}-leucine from, **a**, 17 to 22 minutes; **b**, 28 to 33 minutes; and **c**, 43 to 48 minutes after fertilization. After fertilization eggs (at 30°C) were "pulse" labeled at the indicated times with C^{14}-leucine (0.025 μc/ml), then homogenized and centrifuged at 12,000*g* for 30 minutes; 0.25 ml of the supernatants of equal $O.D._{260}$ were layered on 29 ml of 15 to 50 percent sucrose gradients. The gradients were centrifuged for 3 hours at 24,000 rev/min in a SW-25 rotor. The polysome region lies between fractions 0 and 20. Solid line is $O.D._{260}$; dotted line is counts per minute.

Other experiments demonstrated that these fluctuations in protein synthesis occurred even in the absence of cytokinesis. Even though the cells do not divide in the presence of colchicine *(7)*, they exhibited a pattern of incorporation of amino acids (Fig. 4) similar to that of the controls (Fig. 1). Experiments represented in Figs. 1 and 4 cannot be quantitatively compared because eggs, incubation temperature, and specific activity of C^{14}-leucine were different.

Fluctuations in the rate of incorporation of labeled amino acids into protein during the first division cycle of sea urchin eggs are apparently due to some regulatory mechanism affecting the rate of protein synthesis. Observed findings do not appear to have been due to variations in permeability of C^{14}-leucine during the division cycle (Fig. 3). Nor can the results be attributed to a lack of either adenosine or guanosine triphosphate—required for protein synthesis—since an excess of nucleotide triphosphate and a regenerating system were added to each in vitro incubation mixture. Both the size of the amino acid pool *(8)* and the amounts of total protein *(9)* have been observed to be relatively constant during division.

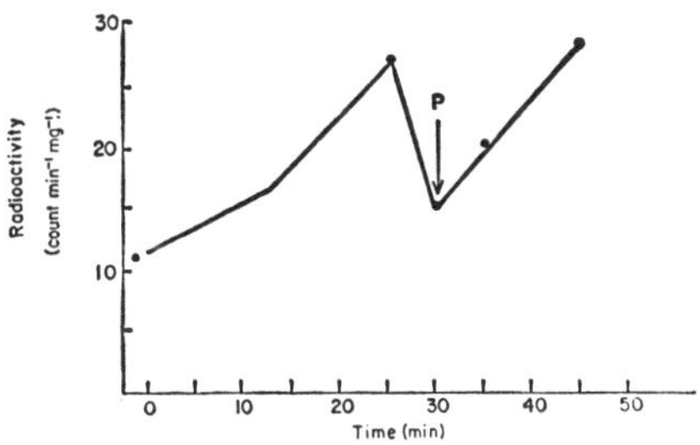

Fig. 3. Incorporating capacity in vitro of the 12,000*g* supernatant derived from eggs homogenized at different times after fertilization. Radioactivity is expressed as counts per minute per milligram of protein; time, as minutes after fertilization. *P* is prophase.

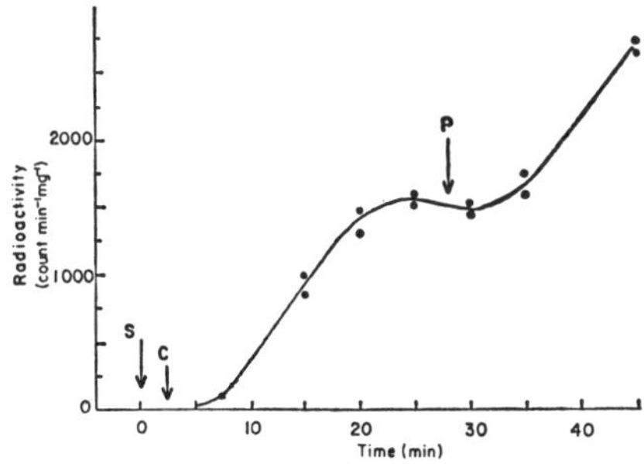

Fig. 4. Effects of colchicine on the incorporation of C^{14}-leucine (67.4 mc/m*M*) into precipitable protein during the first division cycle. Colchicine (1×10^{-4} *M*) was added 2 minutes after fertilization. At the indicated times after fertilization the eggs (at 30°C) were "pulse" labeled for 2.5 minutes with C^{14}-leucine (0.025 μc/ml). Radioactivity is expressed as counts per minute per milligram of protein; time, as minutes after fertilization. *S* indicates when sperm was added and *C*, when colchicine (1×10^{-4} *M*) was added; *P* is prophase.

Furthermore, the colchicine experiment (Fig. 4) suggests that neither the normal assembly and disassembly of the mitotic apparatus nor the cytokinetic process itself is solely responsible for regulating protein synthesis. In addition, the in vitro experiment (Fig. 3) demonstrates that structural integrity of the whole egg is not required for regulation.

Inhibition of RNA synthesis after fertilization in the sea urchin egg either by enucleation *(10)* or by treatment with actinomycin D *(11)* does not appear to limit the rate of protein synthesis during the early cleavage stages. Therefore, differences in rate of protein synthesis that we observed during the first division cycle may be due to variations in the rate of translation of the mRNA message or to some undescribed mechanism. Thus, changes in rate of protein synthesis during division may be additional examples of translational control of gene function *(12)*.

REFERENCES AND NOTES

1. T. Hultin, Exp. Cell Res. **3**:494 (1952); E. Nakano and A. Monroy, *ibid.* **14**:236 (1958).
2. P. R. Gross, J. Exp. Zool. **157**:21 (1964); D. W. Stafford, thesis, University of Miami, 1964.
3. D. M. Prescott and M. A. Bender, Exp. Cell Res. **26**:260 (1962); M. D. Scharff and E. Robbins, Science **151**:992 (1966).
4. E. B. Harvey, The American Arbacia and Other Sea Urchins (Princeton Univ. Press, Princeton, N.J., 1956), p. 57.
5. D. W. Stafford, W. H. Sofer, R. M. Iverson, Proc. Nat. Acad. Sci. U.S. **52**:313 (1964).
6. J. F. George, thesis, University of Miami, 1966.
7. H. W. Beams and T. C. Evans, Biol. Bull. **79**:188 (1940).
8. A. Monroy and R. Maggio, in Advances in Morphogenesis, M. Abercrombie and J. Brachet, Eds. (Academic Press, New York, 1964), vol. 3, p. 136.
9. W. H. Sofer, unpublished observation.
10. J. Brachet, A. Ficq, R. Tencer, Exp. Cell Res. **32**:168 (1963); P. Denny and A. Tyler, Biochem. Biophys. Res. Commun. **14**:245 (1964).
11. P. R. Gross and G. Cousineau, Biochem. Biophys. Res. Commun. **10**:321 (1963).
12. B. Colombo and C. Baglioni, J. Mol. Biol. **16**:51 (1966); S. A. Terman and P. R. Gross, Biochem. Biophys. Res. Commun. **21**:595 (1965); E. Bell, T. Humphreys, H. S. Slaton, C. E. Hall, Science **148**:1739 (1965).

56 Nuclear-cytoplasmic interaction in DNA synthesis

David M. Prescott
Lester Goldstein
Institute for Developmental Biology
University of Colorado
Boulder, Colorado
and
Department of Biology
University of Pennsylvania
Philadelphia, Pennsylvania

Abstract. *In* Amoeba proteus *the transplantation of a nucleus engaged in DNA synthesis into a G_2-phase (after DNA synthesis) cell results in inhibition of such synthesis. When the nucleus of a G_2 cell is transplanted into an S-phase (period of DNA synthesis) cell, such a nucleus may begin to synthesize DNA.*

Little is known about the cellular mechanisms that regulate the initiation of DNA synthesis. The problem is a major one in cell biology since control over DNA synthesis appears to be central in the regulation of cell proliferation. Protein synthesis is probably required for the initiation of replication *(1)*, and there is one piece of evidence that a replicating unit of DNA may control its own replication through a DNA product *(2)*. However, there is no reason a priori to suppose that the regulation of DNA replication is any less precise or strict in biochemical-genetic terms than is the regulation of DNA transcription; in fact, there are reasons to expect that regulation of replication may be more precise.

As a physiological approach to defining necessary conditions in vivo for DNA synthesis, we have begun the study of recombinations of nuclei and cytoplasms derived from cells in different stages of the life cycle in *Amoeba proteus* using the technique of nuclear transplantation. This report deals briefly with DNA synthesis following transplantation of an S-phase (period of DNA synthesis) nucleus into a late G_2 cell (when DNA synthesis has stopped) and a late G_2 nucleus into an S-phase cell.

We determined the period of DNA synthesis in *A. proteus* at 23°C by measuring incorporation of ^{3}H-thymidine during various intervals of the interphase period. So that cells would be in synchrony dividing cells were selected with a micropipette. The S period lasts 3 to 6 hours, beginning in late telophase. No G_1 period is detectable in the normal cell cycle. The G_2 period is 30 or more hours long, and the total generation time is roughly 36 hours. Autoradiographic assessment of ^{3}H-thymidine incorporation was made on nuclei that had been isolated individually in a nonionic detergent, treated with 1*N* HCl at 23°C for 5 minutes, and washed with water. Procedures for cultivation of amoebae *(3)* and transplantation of nuclei *(4)* have been described.

Four types of transplants were used: (i) an S nucleus into an S cell (a control); (ii) a G_2 nucleus into a G_2 cell (a control); (iii) an S nucleus into a late G_2 cell; and (iv) a late G_2 nucleus into an S-phase cell. From one experiment to the next, the S and G_2 cells were roughly the same ages within their respective subsections of the cell cycle. Immediately after nuclear transplantation, each cell was placed in a medium containing ^{3}H-thymidine (25 μc/ml, 15 c/mole) for 5 hours, after which the nuclei were isolated and assayed by autoradiography.

Incorporation of ^{3}H-thymidine by an S nucleus transplanted into an S-phase cell is not detectably different from that by a normal, nontransplanted S-phase nucleus (grain counts above 75 per nucleus) indicating that disturbances engendered by the operation do not affect DNA synthesis to a measurable extent. When an S nucleus is transplanted into a G_2 cell, the amount of ^{3}H-thymidine incorporated is clearly reduced (Table 1). When nuclei have 0 to 45 grains, it is doubtful that such counts can be accepted as a demonstration of DNA synthesis. Grain counts of this magnitude, as

From Science **155:**469-470, January 27, 1967.

Table 1. Incorporation of ^{3}H-thymidine after nuclear transplantation in *Amoeba proteus*

Nuclei (No.) having			
5-25 grains	26-45 grains	46-75 grains	>75 grains
G-2 nuclei transplanted to G-2 cytoplasm			
11	4	0	0
G-2 nuclei transplanted to S cytoplasm			
1	2	3	2
S nuclei transplanted to S cytoplasm			
0	0	1	12
S nuclei transplanted to G-2 cytoplasm			
4	4	1	1

some of our other experiments have shown, are found frequently for nonsynthesizing nuclei, and such activity is largely or probably entirely insensitive to deoxyribonuclease treatment. A nucleus synthesizing DNA is almost always easily recognized because its incorporated activity is so much greater than the nonspecific labeling.

In the case of the ten nuclei transplanted into G_2 cells, eight do not show significant labeling, and two are not different from controls (an S nucleus into an S cell). Obviously the new environment (a G_2 cell) for the S nucleus has a marked effect on the incorporation of ^{3}H-thymidine.

When G_2 nuclei are transplanted into G_2 cells, all have counts of 0 to 45 grains. This amount of labeling cannot be considered significant. When G_2 nuclei were transplanted into S-phase cells the results showed wide variation, but clearly two and probably five out of eight G_2 nuclei had incorporated ^{3}H-thymidine in their new environment.

Thus, an S nucleus placed in a G_2 cell slows or stops its DNA synthesis, and a G_2 nucleus placed in an S-phase cell may recommence DNA synthesis. These results could be explained if only S cells are capable of conversion of ^{3}H-thymidine to ^{3}H-thymidine triphosphate, but there is much evidence *(1)* that nucleoside triphosphates, while necessary for DNA synthesis, do not act in the control or regulation of synthesis. Alternatively, the results are interpretable either in terms of cytoplasmic inhibitors of DNA synthesis (present in G_2 cells but absent in S cells) or cytoplasmic initiators (present in S cells and absent in G_2 cells). Although the experiments summarized in Table 1 do not allow a choice between an inhibitor or an initiator situation, we favor the initiator hypothesis of Jacob, Brenner, and Cuzin *(2)*, not only on the basis of their work, but also because the sequential pattern for DNA synthesis in mammalian chromosomes seems more reasonably explained by a hypothesis that assumes sequential appearance of initiators.

Finally, the conclusion that the cytoplasm has an important influence on nuclear DNA synthesis may surprise no one, but there is perhaps merit in proving the existence of a phenomenon, the "evidence" for which has been primarily assumption.

REFERENCES AND NOTES

1. K. G. Lark, in Molecular Genetics, J. H. Taylor, Ed. (Academic Press, New York, 1963), part 1, pp. 153-206.
2. F. Jacob, S. Brenner, F. Cuzin, Cold Spring Harbor Symp. Quant. Biol. **28**:329 (1963).
3. D. M. Prescott and R. F. Carrier, in Methods in Cell Physiology, D. M. Prescott, Ed. (Academic Press, New York, 1964), vol. 1, pp. 85-95.
4. L. Goldstein; *ibid.*, pp. 97-108.

57 Human diploid cell transformation by DNA extracted from the tumor virus SV40

Stuart A. Aaronson
George J. Todaro
Viral Carcinogenesis Branch
National Cancer Institute
Bethesda, Maryland

Abstract. *The DNA isolated from simian virus 40 (SV40) can transform human fibroblast cells in tissue culture. Clonal lines of DNA-transformed human cells have been obtained that have properties characteristic of cells transformed by whole virus. They all contain SV40 T-antigen, and infectious virus can be recovered by cocultivation. This is the first demonstration of a permanent genetic alteration produced in human cells by purified DNA.*

Simian virus 40 (SV40), a small, DNA-containing tumor virus, has been shown to induce a morphological transformation of various cell types, including human cells, in tissue culture *(1)*. The virus-transformed cells are characterized by their epithelioid morphology and lack of sensitivity to contact inhibition of cell division *(2)*. They contain SV40-specific messenger RNA *(3)*, and infectious virus can generally be recovered *(4)*. The in vitro transformed cells are tumorigenic in the appropriate animal host *(1)*. SV40-infected cells produce a new antigen, the T-antigen, that can be detected in the cell nucleus *(5)*. The production of this antigen persists in transformed cells and in tumor cells.

The nucleic acid extracted from SV40 was shown by Gerber to be infectious for green monkey kidney cells *(6)*. The observation by McCutchan and Pagano *(7)* that the infectivity of SV40 DNA could be markedly enhanced with diethylaminoethyl-dextran led us to study the effect of SV40 DNA on human cells. The studies described below demonstrate that infectious nucleic acid can enter and permanently alter the genetic makeup of a normal human cell.

A human fibroblast strain (F.R.), obtained from a skin biopsy specimen of a clinically normal adult female, was tested. It had previously been shown to have high susceptibility to transformation by SV40 *(8)*. Other cells studied included green monkey kidney cells, the mouse cell line Balb/3T3 *(9)*, and a line of rat kangaroo fibroblasts. Cultures were maintained in 50-mm plastic petri dishes in Dulbecco's modification of Eagle's medium supplemented with 10 percent calf serum. For experiments with whole virus, a pool of plaque-purified SV40 small-plaque virus (SV-S) *(10)* was used.

Small-plaque SV40 DNA *(11)* was extracted with a mixture of chloroform and isoamyl alcohol in the presence of sodium dodecyl sulfate and sodium perchlorate *(12)*, and the double-stranded, twisted circular DNA (DNA I) was isolated by equilibrium density centrifugation in cesium chloride containing ethidium bromide *(13)*. One μg of DNA I was found to correspond to 3.0×10^4 plaque-forming units.

Cells in the logarithmic phase of growth were washed twice with serum-free medium and incubated with 0.2 ml of fluid containing SV40 DNA and 300 μg/ml of diethylaminoethyl-dextran in Earle's balanced salt solution without sodium bicarbonate. After 30 minutes at room temperature with frequent gentle agitation, the cultures were washed twice and fresh medium was added. Within 24 hours the cells were transferred both to dishes containing cover slips for SV40 T-antigen studies and to dishes at appropriate dilutions for the transformation assay *(8)*. Following exposure to SV40 DNA the cells were maintained in medium supplemented with 0.5 percent rabbit antiserum to SV40.

Cells containing SV40 T-antigen were stained by the indirect fluorescent antibody method

From Science **166:**390-391, October 17, 1969.

Supported by National Cancer Institute contract No. PH 43-66-641. We thank Dr. David Axelrod and Dr. Malcolm Martin for their advice and Elaine Rands and Lillian Killos for their excellent technical assistance.

(5). The fraction of T-antigen positive cells was scored at 72 hours, a time at which, under our conditions, the maximum number of positive cells was seen either with intact virus or with DNA. In a typical test 20,000 to 40,000 cells were counted to determine the fraction of cells that were T-antigen positive. Infectious virus was recovered from DNA-transformed clones by cocultivation with green monkey kidney cells in the presence of ultraviolet-inactivated Sendai virus *(4)*.

Human cells and green monkey kidney cells were infected with serial dilutions of DNA I. Figure 1 shows that the percentage of SV40 T-antigen positive cells rose with the concentration of DNA for both monkey and human cells. The efficiency of infection was comparable in spite of the fact that whole virus is over 100 times more efficient at inducing T-antigen in monkey cells than in human cells. The same efficiency of T-antigen induction by DNA was also found with Balb/3T3 cells, and with rat kangaroo cells. In the latter case the cells were fully resistant to T-antigen induction by whole virus.

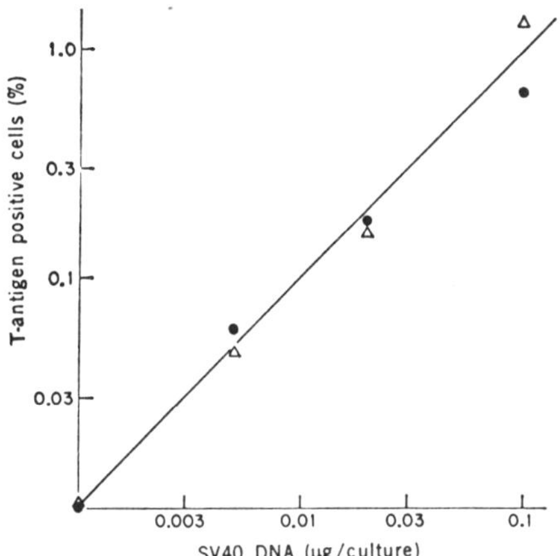

Fig. 1. Induction of SV40 T-antigen in monkey and human cells by SV40 DNA; relationship between micrograms of SV40 DNA I per infection and the fraction of T-antigen positive cells. Each point represents the data from two separate experiments; at least 20,000 cells were scored for each point. Triangles, green monkey kidney cells; dots, human diploid fibroblasts.

Transformed colonies were first observed in DNA-infected human cultures 2 to 3 weeks after infection. The transformation frequency was assayed at 3 to 4 weeks. In their tissue culture properties the transformed colonies that developed were typical of SV40-transformed cells. The cells were epithelioid, showed pronounced loss of contact inhibition of cell division, and readily formed multiple cell layers. They also showed persistent nuclear abnormalities and many giant cells.

In Table 1 virus and viral DNA were compared for their transforming activities. With the DNA I preparation transformed colonies were seen by using as little as 6.0×10^2 plaque-forming units or 0.02 μg per culture. When 5.0×10^4 infectious units were used, transformed colonies were readily seen with crude DNA or DNA I while no transformed colonies were found by using whole virus. At the lower DNA concentrations approximately 200 infectious units of DNA I were sufficient to produce one transformed human colony. At higher DNA concentrations there appeared to be a plateau, both for infectivity, as judged by the percentage of T-antigen positive cells, and for transformation. With intact SV40 small-plaque virus approximately 2.5×10^5 infectious units were required to produce one transformed colony. Viral DNA was, therefore, 1000-fold more efficient per infectious unit than whole virus. The inefficiency of whole virus was due to inefficiency at an early step in

Table 1. Efficiency of T-antigen induction and transformation in human cells by SV40 DNA and SV40

SV40	Infectious units added	SV40 T-antigen (% positive)	Transformed colonies/ 5×10^5 cells*	T-antigen/transformed colonies
DNA†	1.5×10^2	0.06	0	
DNA	6.0×10^2	.18	3	3.0×10^2
DNA	3.0×10^3	.68	21	1.6×10^2
DNA	1.5×10^4	1.2	26	2.3×10^2
DNA	5.0×10^4	1.5	21	3.6×10^2
Virus	5.0×10^4	$< .01$	0	
Virus	5.0×10^7	5.8	164	1.8×10^2

*Average of two separate experiments. At each concentration in each experiment 5×10^5 cells were exposed.
†One microgram of DNA I contained 3.0×10^4 infectious units.

human cell infection, for example, adsorption, penetration, or uncoating, or a combination of these. Once either the virus or the viral DNA reached the stage of inducing SV40 T-antigen there was a comparable but low probability that the cell would develop into a transformed colony. The T-antigen/transformation value ranged from 1.6 to 3.6×10^2 for DNA and was 1.8×10^2 for whole virus [see also *(8)*].

Several clones of DNA-transformed human cells have been isolated and grown up to mass culture in the presence of SV40 antiserum for at least 20 to 30 cell generations. Ten transformed clones, each the product of an independent DNA-cell interaction, were stained for T-antigen. With each clone practically every cell was found to be T-antigen positive. By cocultivation with green monkey kidney cells in the presence of ultraviolet-inactivated Sendai virus, infectious virus was recovered from each clone. As with transformation by intact virus, therefore, the entire viral genome is present and recoverable from the DNA-transformed cells.

Diderholm *et al.* described transformation by SV40 DNA in mass cultures of bovine embryo cells *(14)*, which were very resistant to transformation by high-titered intact virus. Black and Rowe have reported T-antigen induction by SV40 DNA in hamster BHK21 cells. These cells, too, were found to be insensitive to infection with virus *(15)*. In our present report it is also shown that SV40 DNA can induce T-antigen in marsupial rat kangaroo cells in which there is a complete block to infection by whole virus. The species differences in T-antigen induction with whole virus are, therefore, probably due to differences in attachment, penetration, or uncoating, rather than to differences in the ability of the exposed viral genome to begin functioning once inside the cell. The ultimate outcome of infection, however, remains under cellular control. SV40 DNA is highly cytopathic for monkey cells, weakly cytopathic for human cells, and noncytopathic for mouse cells, behaving in the same manner as whole virus.

While transformation of human cells by DNA is more efficient than by whole virus per infectious unit, it is still highly inefficient per molecule. When 5×10^5 cells were infected with 0.1 μg of SV40 DNA I (3.0×10^3 plaque-forming units), 21 cells were transformed. If the molecular weight of SV40 DNA is 2.5×10^6 *(1)*, it can be calculated that 10^9 DNA molecules are needed per transformation event. Transformation and T-antigen induction appear, at least over a limited range, to be linearly related to DNA concentration. This suggests that a single molecule of DNA is sufficient to induce SV40 T-antigen and also to permanently transform human cells.

There is considerable evidence that shows that SV40 DNA can become a permanent part of the host cell genome *(4, 16)*. Most of the SV40 DNA found in transformed cells is associated with the chromosomes *(17)*. Whether SV40 has an integration function similar to that of bacteriophages λ and P 22 *(18)* is not yet known. However, if the viral DNA's ability to integrate into the human cell genome can be separated from its oncogenicity, it then may be possible to use "integrating" viral DNA as a carrier to insert specific genetic information into human cells.

REFERENCES AND NOTES

1. P. H. Black, Annu. Rev. Microbiol. **22**:391 (1968).
2. H. M. Shein and J. F. Enders, Proc. Nat. Acad. Sci. U.S. **48**:1164 (1962); H. Koprowski, J. A. Ponten, F. Jensen, R. G. Ravdin, P. Moorehead, E. J. Saksela, J. Cell. Comp. Physiol. **59**:281 (1962).
3. Y. Aloni, E. Winocour, L. Sachs, J. Mol. Biol. **31**:415 (1968); K. Oda and R. Dulbecco, Proc. Nat. Acad. Sci. U.S. **60**:525 (1968); G. Sauer and J. R. Kidwai, *ibid.* **61**:1256 (1968).
4. P. Gerber, Virology **28**:501 (1966); H. Koprowski, F. Jensen, Z. Steplewski, Proc. Nat. Acad. Sci. U.S. **58**:127 (1967).
5. J. H. Pope and W. P. Rowe, J. Exp. Med. **120**:124 (1964).
6. P. Gerber, Virology **16**:96 (1962).
7. J. A. McCutchan and J. S. Pagano, J. Nat. Canc. Inst. **41**:351 (1968).
8. S. A. Aaronson and G. J. Todaro, Virology **36**:254 (1968).
9. ———, Science **162**:1024 (1968).
10. K. K. Takemoto, G. J. Todaro, K. Habel, Virology **35**:1 (1968).
11. Supplied by Dr. M. A. Martin, NIH.
12. J. Marmur, J. Mol. Biol. **3**:208 (1961).
13. M. A. Martin and D. Axelrod, Science **164**:68 (1969).
14. H. Diderholm, B. Stenkvist, J. Ponten, T. Wesslen, Exp. Cell. Res. **37**:452 (1965).
15. P. H. Black and W. P. Rowe, Virology **27**:436 (1965).
16. T. L. Benjamin, J. Mol. Biol. **16**:359 (1966); H. Westphal and R. Dulbecco, Proc. Nat. Acad. Sci. U.S. **59**:1158 (1968).
17. J. Sambrook, H. Westphal, P. R. Srinivasan, R. Dulbecco, Proc. Nat. Acad. Sci. U.S. **60**:1288 (1968).
18. E. R. Signer, Annu. Rev. Microbiol. **22**:451 (1968).

58 Lethal genes and analysis of differentiation

In higher organisms lethal genes serve as tools for studies of cell differentiation and cell genetics

Salome Gluecksohn-Waelsch
Albert Einstein College of Medicine
Yeshiva University
New York, New York

Among the many problems of genetics untouched by recent spectacular advances in molecular genetics is that of lethal genes. The reasons for this lie primarily in the fact that lethal-gene problems touch upon various areas of genetics all of which are concerned with gene effects on levels more or less removed from that of primary gene action. Among these areas are population genetics, radiation genetics, and developmental and physiological genetics.

The analysis of gene action at the molecular level stops necessarily at a point where the interaction of genes, of gene products, of cells, and of tissues begins—in short, at the phenotype level. It is, on the other hand, precisely at this level that problems of cell, tissue, and organ differentiation make their appearance, and that those phenomena occur which serve to characterize a gene as a lethal. Even further removed from the molecular level of gene action are problems of lethal genes in populations, and of their origin, whether spontaneous or induced—for example, by radiation.

Because of the great significance of lethal genes in populations, because of their importance in studies of radiation genetics, and because of their role in the analysis of development and differentiation of higher organisms, it appears worth while to attempt at this time a re-evaluation of this problem and to relate it to new concepts of genetics.

In this attempt emphasis will be put on the physiological genetics of lethal genes. The significance of lethal genes in populations has been the subject of numerous discussions *(1)*; an extensive literature deals with studies of radiation-induced lethals *(2)*. These topics are not taken up here. Problems considered in this article include the mode of action of lethal genes, the properties that distinguish them from other mutant genes, the physiology of the effects of lethal genes, and their utilization in analyzing cell, tissue, and organ differentiation in higher organisms.

DEFINITION OF LETHAL GENE

To begin with, it is necessary to define and to discuss briefly the concept of lethal genes.

The term *lethal* is applied to those changes in the genetic makeup of an organism which produce effects severe enough to cause death. The nature of such changes may vary widely, from alterations in the number of chromosomes to changes in chromosome structure and to gene mutations. Lethal gene mutations, with which this article is concerned, are distinguished from other gene mutations only by the severity of their effects. When speaking, therefore, of lethal genes we must realize that these do not represent a separate category but are mutations not different in principle from others compatible with life. Consequently, no sharp borderline can be drawn between "viable" and "lethal" genes. Any attempt to distinguish between them must remain arbitrary, and many variables, including those of the environment, determine whether a particular gene is "viable," having effects compatible with life, or "lethal," resulting in the death of the organism that carries it. This relative character of lethal genes clearly illuminates the problem of the roles of

---

From Science **142**:1269-1276, December 6, 1963.

This article is dedicated to Professor L. C. Dunn in friendship and admiration.

I thank the Weizmann Institute of Science, and particularly Professor E. Katchalski, for the hospitality extended to me during the preparation of this article. I also thank Dr. Gertrude Moser and Mrs. Anna C. Pai for their generous help in the preparation of the manuscript. Work in my laboratory has been supported by grants from the National Institutes of Health, the National Science Foundation, and the American Cancer Society.

nature and nurture in the expression of genetic traits.

In bacteria, a mutation is considered lethal if growth and survival of the mutant organism cannot be achieved by manipulation of the environment. If, on the other hand, it is possible to devise environmental conditions which permit the organism to survive, the mutation becomes "viable" and therefore may be called a "conditional lethal" mutation. This is the case, for example, when a mutation results in inability of the organism to synthesize an essential amino acid. If the organism is grown on a basic minimal medium, such a mutation will have a lethal effect; if, however, the missing amino acid is supplied from the outside—that is, if a nongenetic modification of the environment is produced—the same mutant becomes "viable." Similarly, it is possible to introduce into the genotype a suppressor gene specific for the mutation in question and thus to achieve a genetic modification of the lethal effect.

Similarly, a mutation in a higher organism which makes the organism unable to synthesize a certain indispensable metabolite must be termed "lethal" unless the vital substance is supplied from the outside. If, for example, a mammalian organism were to become unable, as the result of mutation, to synthesize glycine, this amino acid would have to be supplied from an external source, as is necessary, for example, in the case of the growing chick. If, for some reason, this outside supply of glycine failed, certain essential biosynthetic reactions requiring glycine could not be carried out by the mutant organism and death would ensue; the mutation would then have become "lethal." It has been suggested that a mutation which occurred in the course of evolution is responsible for the human requirement for ascorbic acid from the outside; primates and the guinea pig also have this requirement, whereas other animals appear to be able to synthesize this vitamin themselves. This mutation in man might be termed a conditional lethal mutation, since its effects lead to death unless ascorbic acid is provided in the diet.

Examples such as these emphasize the relative nature of most lethal mutations. The lethal or nonlethal character of a mutation is not an absolute and inherent property of the genetic material in question but is determined by the mechanism of the mutational effect, which may or may not permit compensation by other genes or the environment for deficiencies caused by the genetic alteration.

Lethal genes should, therefore, not be considered a separate class of genes with special characteristics inherent in their material make-up. They are distinguished, rather, by effects which are not compatible with life unless compensated for by other genetic or nongenetic factors. These lethal effects may differ only quantitatively from the effects of other detrimental genes and thus may represent the most extreme expressions of a wide range of mutational effects. Or, the lethal effect may differ qualitatively from the effect of other mutant genes by interfering with a function absolutely vital to the organism.

For purposes of illustration, several series of multiple alleles in the house mouse may be cited; some of the alleles are compatible with the survival of the organism carrying them in single or double dose, while others, in the homozygous condition, cause death. In the *W*-series of multiple alleles in the mouse, which affect the erythropoietic system, viable and lethal alleles seem to be distinguished by quantitative differences of their effects. In another multiple allelic series, that at the so-called agouti locus, the effects of all the alleles but one are compatible with life and relate to pigment formation only. The one lethal allele, yellow, appears to have, in the homozygous condition, an effect qualitatively different from that of the other alleles: it interferes with a function vital to implantation in the uterus and thus causes the death of the embryo. In this case, as well as in other cases, the distinction between qualitative and quantitative differences of allelic effects on the phenotype level is of course not always obvious or actually proved. This is particularly true, for example, in the *T*-series in the mouse, discussed later in some detail.

It may be argued that there exist genetic loci which are "immune" to the appearance of lethal mutations—that is, where no such mutations can be expected to occur. If, for example, the complete lack of a given locus as a result of a chromosomal deficiency is compatible with life, one would perhaps not expect a mutational change at that locus to have lethal effects, since, it might be argued, no change could be more severe than total loss of the genetic region. However, in terms of physiological function, total lack of a metabolite, as the result of the genetic deficiency, could conceivably be more easily compensated for than could the production of an abnormal metabolite by a mutated gene which might result, for example, in a chain of inhibitory reactions.

In the field of population genetics, those genes are considered "lethal" which do not contribute to the gene pool of the population; the main concern is not necessarily with the lethal gene's genetic or physiological mechanisms. Consequently, the term *lethal* is applied to those factors which either cause death before the organism is able to reproduce or which interfere in some way with reproductive activity, thus obstructing the transmission of the gene to the next generation.

The term *lethal gene,* then, as used in various areas of genetics, is so loose as to make it appear futile to attempt a definition which would fit all types of "lethal" genes, even approximately.

In view of the many and varied viewpoints from which the problem of lethal genes may be considered, it seems permissible to limit the discussion deliberately, unless a full treatise on lethal genes is planned, such as the exemplary one by Hadorn *(3)*. In this article I use the term *lethal genes* to mean genes that have lethal action at the phenotype level—action which does not necessarily indicate the nature of the change at the gene level and which is subject to the effects of other genetic factors and of nongenetic factors.

The physiological geneticist may have manifold reasons for being interested in lethal genes; the ultimate goal, however, is to learn their mode of action. In the course of such studies, lethal genes become tools for investigating special problems within the framework of interest of the researcher.

In many subdivisions of genetics (such as population, radiation, developmental, and cell genetics, to name just a few), and in other areas of biology, lethal genes have been found to be a rich source of experimental material. In the field of experimental embryology, for example, the causal analysis of developmental mechanisms, or knowledge of cell and tissue differentiation, particularly of higher organisms, has benefited and will continue to benefit from studies of the physiological genetics of lethal genes.

In the discussion that follows, particular emphasis is placed on lethal genes as tools for analyzing the development and differentiation of mammalian embryos and for studying the genetic control of these processes. The utilization of developmental abnormalities for analyzing genetic phenomena goes back to Boveri, who in his classical experimental studies of sea urchin development, around the turn of the century, established the individuality of the chromosomes and their essential role in development. Lethal genes have been shown to produce a wide variety of morphological, physiological, biochemical, and metabolic abnormalities which may lead to the death of the organism before or after birth—or even as late as maturity, thus not necessarily interfering completely with reproductive activity. Some of the mechanisms by which these lethal genes exert their effects are examined here, and possible experimental approaches to their analysis are discussed.

ORIGIN AND FREQUENCY OF LETHAL GENES

Lethal genes may originate in spontaneous mutations or they may be produced by mutagenic agents—for example, radiation or chemicals. The frequency of spontaneous lethal or sublethal recessive mutations has been measured in different species of *Drosophila:* 41.2 percent of chromosomes No. 2 and 32.1 percent of chromosomes No. 3 of South American populations of *Drosophila willistoni* were found to carry them *(4)*. The frequencies of such mutations cannot be ascertained easily in man or even in experimental mammals. They have been estimated, however, and according to one such estimate *(5)*, a normal person is heterozygous for an average of eight recessive lethal or sublethal genes.

For many reasons, some of them obvious, the chance of discovering recessive lethal genes is much smaller than that of discovering nonlethal mutations *(6)*. Nevertheless, studies, by Muller, of sex-linked lethal mutations in *Drosophila* revealed a large number of such mutations—a finding which led him to conclude that, in view of the great methodological difficulties of demonstrating lethals, "probably the majority, if not the vast majority, of mutants are lethals." These studies provided the experimental evidence for Muller's earlier prediction on theoretical grounds that most newly arising mutations would turn out to be lethal. Since, in the course of evolution, Muller argues, the organism had made use of all newly arising mutant factors to build up the best-adapted biological machinery, any new mutant change would be likely to lead to a severe disturbance of this complex mechanism and would thus be "lethal."

The approximate ratio of lethal to visible and viable mutations has been calculated by various workers. In spite of variations in different species, the frequency of lethal factors appears always to be several times that of visible mutations, among spontaneous and induced mutations alike.

Numerically, therefore, lethal genes make up a large portion of the genetic load of a population, which is defined by Crow *(7)* as "the proportion by which the population fitness (or whatever other trait is being considered) is decreased in comparison with an optimum genotype."

Dominant lethal genes come to the attention of the investigator only rarely under normal conditions, and they are eliminated from the population almost as soon as they appear; no studies have been made of their mode of action in higher organisms. The lethal genes considered here are all recessives.

In the course of analyzing the effects of lethal genes, inquiry must be made into the levels at which lethal gene action occurs. Although lethal gene action resides, of course, ultimately in an error on the molecular level, its expression may occur at any of various levels—the level of the cell, the tissue, the organ, or the organism.

The first approach to the study of lethal genes in higher organisms is necessarily descriptive, involving use of the methods of morphology, embryology, physiology, and biochemistry. In this way the foundation may be laid for subsequent experimental probing into causal mechanisms.

Frequently a lethal effect is the end result of a long chain of processes started off by one primary gene-controlled abnormality. In such a case, the search for this primary abnormality may reveal hitherto unknown links in the chain of normal processes, as well as interrelationships between different processes involved in the development and differentiation of a tissue or organ, and may thus contribute to the analysis of pleiotropy—that is, multiple effects of a gene. The phenomenon of pleiotropy in higher organisms has been the cause of many arguments, particularly in respect to its relevance to the analysis of primary gene action, and here reference is made only to several recent discussions *(8, 9)*.

LETHAL GENES AND ORGAN DIFFERENTIATION

The lethal effects of genes trace back to widely divergent causal factors and operate through equally divergent mechanisms. A relatively simple situation, from the point of view of analysis, exists when a mutation results in the suppression of normal development of a vital organ and in this way causes death of the organism. The causal analysis of such gene-controlled interference has thrown light on causal mechanisms of normal organ development.

An example of a lethal gene that produces its effect by suppressing differentiation of a normal organ is a semidominant mutation in the mouse: animals homozygous for the gene Sd are completely tailless and die soon after birth because of an almost total lack of kidneys (Fig. 1). Investigation of the steps by which the presence of this lethal gene in homozygous condition leads to the absence of kidneys has contributed to the causal analysis of kidney development in mammals (see *10*).

The descriptive study of abnormal kidney development in this mouse mutant could do no more than lead to the proposal of certain alternatives as causal mechanisms possibly re-

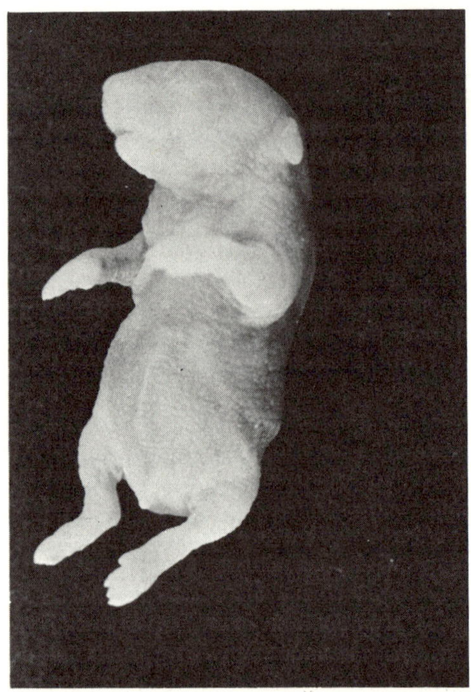

Fig. 1. A tailless and imperforate newborn mouse homozygous for gene *Sd* (×3).

sponsible for the abnormal kidney differentiation. In one of these proposals it was postulated that, as a result of the mutation, the development of the ureter was suppressed and, consequently, the ureter could not exert the necessary inductive effect on the kidney mesenchyme and further differentiation failed to take place. However, a decision between this and other possible interpretations of the observational results could only be made through experimental approaches. Among these, tissue- and organ-culture methods seemed to be best suited for this particular study and were applied in the analysis of kidney differentiation in the mutant embryos *(11)*.

To begin with, the developmental potencies of entire kidney rudiments from embryos homozygous for this lethal gene were tested in organ tissue culture. Contrary to the findings for the whole embryo in which kidney differentiation is suppressed, kidney tubule differentiation occurred in the intact "lethal" rudiments in vitro, although to a lesser degree than normal. The question tested next was: To what extent is the suppression of kidney differentiation in the mutant due to interference of the lethal gene with the inductive interaction between the two components of the kidney rudiment? Reciprocal combinations of normal and "lethal" ureter with normal and "lethal" kidney mesenchyme were made in vitro; the existence of inductive and reacting potencies in both "lethal" components was demonstrated by the differentiation of kidney tubules in all combinations, although both the degree and the rate of differentiation were reduced in comparison with findings for normal tissue. With this experimental approach, therefore, it could be shown that the lethal gene had not suppressed the kidney inducing factor in the kidney rudiments nor their ability to react. The failure of kidney differentiation in the embryos homozygous for the mutation might be explained in a different way. It seems that in normal development the rates of differentiation of the two kidney rudiment components, ureter and kidney mesenchyme, are geared to each other in such a way that inductive interaction is made possible at the required time and stage. A change in these rates may interfere with this essential synchronization of inductive and reacting processes and result in failure of normal differentiation. Such a change of rates may have been caused by the mutation that suppressed kidney development.

LETHAL GENES AND EARLY EMBRYONIC DIFFERENTIATION

The study of lethal genes has made possible the further analysis of various causal mechanisms of early embryonic differentiation in higher organisms. Central among the problems that confront any student of developmental genetics is that of the mechanisms by which genes exert their effects on normal embryonic differentiation and inductive interaction between various parts of the developing embryo. An extensive series of mutations in the mouse, with lethal effects in early stages of embryogenesis, has furnished a wealth of material for such studies by virtue of the interference of the mutations with normal embryonic differentiation and normal inductive phenomena.

Much of our knowledge of the causal morphology of early mammalian development is based on analysis of the effects of this large series of lethal genes located in an abnormal region of chromosome 9 of the mouse. Genetically, these mutations are of considerable interest for various reasons. Their relatively high rate of mutation appears to be similar to the rate of crossing over in this abnormal *T* region *(12)*. Consequently, it is assumed that rare crossing over changes the length or position of the abnormal region and results in alteration of the effects of a particular "mutated" *t* allele. Different lethal alleles result from different changes in the length of the abnormal chromosome region. Since lethal *t* alleles were first reported, a large number of new alleles have been discovered, all of them possibly the result of crossing over in the abnormal region of chromosome 9.

Individuals heterozygous for any two of these different lethal alleles are viable, and two such heterozygotes when mated with each other appear to breed true because of the operation of a balanced lethal system, in which both types of homozygotes are eliminated before birth by lethal gene effects and only the heterozygotes survive.

Balanced lethal systems have been demonstrated and described in *Drosophila (6)*. Such systems lead to "enforced heterozygosis," since both homozygotes are inviable. In contrast to an ordinary population, where inbreeding results in a gradual separation of homozygous strains, inbreeding of individuals with balanced lethal genes continues to produce heterozygotes only.

Aside from the peculiar genetic features of

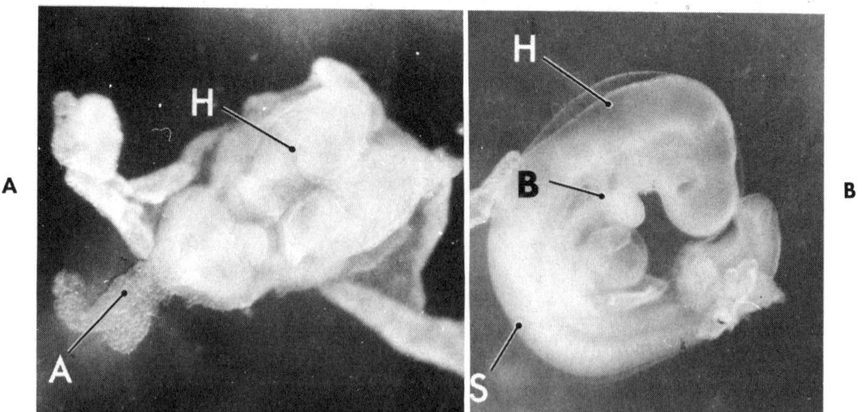

Fig. 2. A, Abnormal retarded 9½-day-old mouse embryo homozygous for gene t^9, with irregular head folds (*H*) and partially duplicated allantois (*A*) (about ×40). **B,** Normal littermate. *S,* somites; *B,* branchial arches; *H,* head folds (about ×12).

this series of *t* alleles in the mouse, the main interest in these lethal genes with effects on the embryo derives from the fact that, because of their interference with specific processes of differentiation, they are tools for studying the relationship between chromosomal or gene structure and embryonic differentiation *(13).*

Figure 2 illustrates an embryo homozygous for one of the *t* mutations. A comparison with its normal littermate shows general retardation, overgrowth and abnormal differentiation of head folds, reduction of mesodermal trunk areas—for example, complete absence of somites—and overgrowth and duplication of the allantois. This homozygote does not survive beyond the stage pictured here. The differentiation of neural tissue and structures, of somites, and of other mesodermal tissues is affected by this mutation; its analysis may illuminate many features of normal differentiation of the same elements.

The earliest lethal effects of any *t* allele were demonstrated in preimplantation stages where differentiation of morula to blastocyst was suppressed and where, on the cellular level, the concentration of RNA was decreased and the shape of the nucleolus was abnormal in cells of embryos homozygous for the allele.

Suppression of mesoderm formation and abnormal embryonic organization are characteristics of embryos homozygous for another lethal allele of this series. The correlation of abnormal notochord differentiation with abnormal leural tube development in embryos homozygous for the allele of this series that was studied first suggested the existence in mammals of an inductive relationship between notochord and neural ectoderm resembling that in lower vertebrates.

The problem of lethal gene action in this mutant was attacked experimentally *(14)* by explantation and in vitro culture of tissues from mouse embryos homozygous for the lethal gene, which ordinarily die around the age of 10 days of prenatal life. Such explanted tissues were capable of living and differentiating far beyond their potentialities for survival and differentiation in vivo.

Thus it was concluded that this mutation does not act as a "cell lethal," at least not in every cell of the homozygous embryo. The ability of tissues from this mutant to survive in vitro is in agreement with the demonstration *(15)* that the immediate cause of death of the homozygous embryo is the complete absence of umbilical and vascular connections with the mother and subsequent failure of processes of nutrition and waste-product removal. Consequently, it does not seem necessary to postulate an inherent defect of cells and tissues to account for the death of the embryo. It is impressive that in spite of the complete absence of umbilical connections with the mother, the homozygous embryo is able to develop up to the limb-bud stage—a finding which indicates the embryo's amazing independence of direct maternal blood supply during early stages of organ differentiation.

The appearance of duplications of various embryonic structures and rudiments in another homozygote of this series has been interpreted as a strong indication of the existence of organizer phenomena and inductive interrelationships during early processes of normal differentiation in mammals, similar to the phenomena and interrelationships demonstrated in lower vertebrates in the experimental analyses of Harrison and Spemann and members of their schools *(16)*.

Both in vitro explantation and transplantation to various sites of a host may prolong the survival time of parts of embryos homozygous for a lethal gene. Thus, the differentiation potentials of cells and tissues beyond the stage at which the whole embryo ordinarily dies may be examined. In this way, effects of a gene other than those expressed in vivo are revealed. For example, the lethal gene "Splotch" in the mouse, which interferes with the differentiation of derivatives of the neural crest, causes death of the homozygous embryo at a stage before the differentiation of all derivatives of the neural crest is completed in the normal embryo. Transplantation experiments made it possible to study the differentiation potentials of neural crest derivatives, from mutants, beyond the lethal stage and to demonstrate, for example, abnormalities of melanophore differentiation; thus, melanophores were added to the list of neural crest derivatives affected by this mutation *(17)*.

LETHAL GENES AND SPECIFIC TISSUE DIFFERENTIATION

In a large number of lethal genes with effects on the embryo, these effects are expressed in a multiplicity of embryonic tissues, although general cell lethality can be excluded as the cause of death of the embryo. There exist various other lethal genes whose effects are concentrated on a particular tissue; among them are genes, for example, which interfere with erythropoiesis to an extent incompatible with survival. An interesting series of such lethal alleles is that which produces anemia of varying degrees in the mouse *(8)*.

In man, the effect of a sex-linked conditional lethal gene seems to be restricted to the red blood cell, which in the mutant shows decreased activity of the enzyme glucose-6-phosphate dehydrogenase and is subject to premature breakdown and death. The conditional nature of this particular lethal gene appears from the fact that premature breakdown of the red cell occurs only in the presence of certain noxious agents, such as exist, for example, in the bean *Vicia fava* and in certain drugs *(18)*.

The restriction of the effect of this "lethal" to the red blood cell may, of course, be simply an expression of the state of the reacting system. A deficiency of the enzyme glucose-6-phosphate dehydrogenase can have an effect only in those tissues where the enzyme is essential, such as the red blood cell.

A striking example of a lethal gene that affects one particular tissue is the mutation called "fetal muscular degeneration," (fmd) which is being studied in the genetics laboratory of the Albert Einstein College of Medicine *(19)*. This recessive autosomal gene interferes, in the homozygous condition, with the early differentiation of the muscle cell to such an extent that normal skeletal muscle never develops. At birth, the lethal syndrome in the homozygote consists of various gross abnormalities of shape and structure (Fig. 3). Degeneration of the skeletal musculature, including the intercostal muscles, contributes to the inability to breathe and the subsequent asphyxia which is probably the immediate cause of perinatal death. The fetus is totally unable to respond to tactile or electrical stimuli, probably because of the muscular degeneration. In early developmental stages severe edema characterizes the abnormal fetus; the precise nature of the edema and its possible cause are as yet not known, but it may be correlated with the degeneration of differentiating muscle cells. Abnormalities of the myotubes in the earliest stages of muscle differentiation are actually the first observable deviations from normal. The correlation of abnormal differentiation of muscle with muscle degeneration and fetal edema makes this lethal syndrome of special interest for studies of differentiation of muscle, its biochemical basis, fetal muscle physiology and neurophysiology, and, possibly, the etiology of muscular dystrophies. It would be of great interest to apply methods and concepts developed recently in studies of the genesis of multiple forms of enzymes in the developing muscles of the chick embryo *(20)* in investigating the biochemical nature of the defect of muscle differentiation in mouse embryos homozygous for this lethal gene.

CELL LETHALITY AND AUTONOMY OF LETHAL EFFECTS

The concept of "cell lethal" mentioned in the discussion of explantation and transplantation experiments, usually refers to the expression of the lethal effect of a mutation on the cellular level in every cell. In the cases discussed, true cell lethality was thought to be excluded by the observation of survival and differentiation of explanted tissues beyond the survival and differentiation of the same tissues in vivo.

If transplantation of "lethal" tissues into a host of normal genotype, or explantation in vitro, fails to increase the survival time and the degree of differentiation, the lethal effect is referred to as "cell lethal" and considered to be autonomous. There seems to be a fallacy in an argument which correlates autonomous behavior of mutant cells—that is, the failure of correction of the defect by external factors—with lethal gene effect at the cellular level.

The autonomous or nonautonomous character of a lethal gene depends on many factors, such as the material nature of the genetic defect, or conditions extraneous to the gene effect, such as permeability, which permit or exclude correction of the deficiency through the supply of enzymes or substrates from outside the cell.

It is conceivable, for example, that a "cell lethal" with the phenotype of an essential-enzyme deficiency based on a structural gene mutation might show autonomous behavior and be incapable of being corrected under conditions in which substrate concentrations vary. The same variations of substrate concentration might, on the other hand, reveal nonautonomous behavior and correct for a defect in cells with the same lethal phenotype, caused, in this case, by mutation in a regulatory gene.

Both examples illustrate lethal gene effects on the cellular level—that is, effects of "cell lethals," one of them autonomous and unaffected by certain external conditions, the other nonautonomous and capable of being corrected by extraneous factors.

In unicellular organisms, differences of reaction of mutants of the same cistron in respect to suppressibility of phenotypic expression by external suppressors have been demonstrated in *Escherichia coli (21).* Certain alkaline phosphatase negative mutants in *E. coli* are suppressible—that is, they show nonautonomous behavior as the result of experimental manipulation; other mutants of the same cistron are not suppressible and are therefore autonomous. As Garen points out, different primary gene action in the case of these different mutations may account for the difference in their reactions.

When the problem is one of lethal gene action in higher organisms, the situation is obviously even more complex because of the interaction of various parts of the organism both during embryonic differentiation and in the mature state.

It is important to realize that results of

Fig. 3

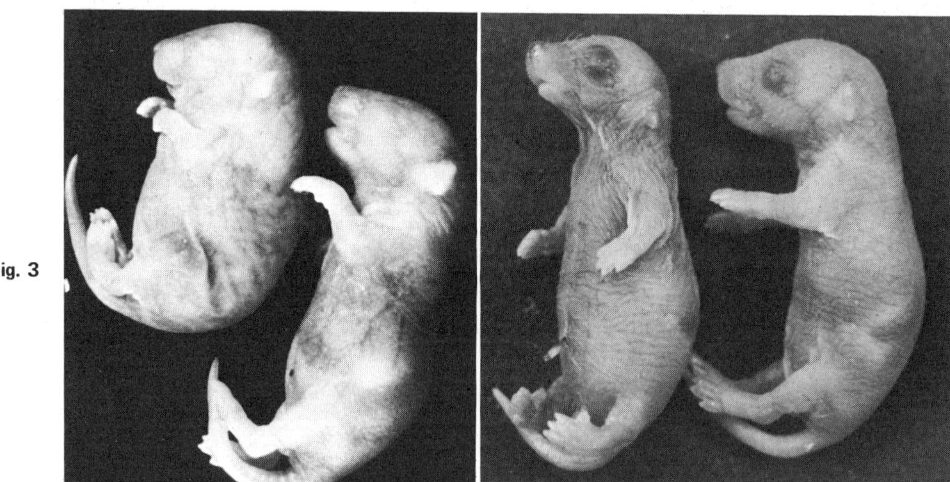

Fig. 4

Fig. 3. Newborn mouse (left), homozygous for gene *fmd,* showing effects of fetal muscular degeneration: abnormal shape, looseness of skin, typical limb position, microagnathia. Normal littermate (right) (×3).

Fig. 4. Newborn mouse (left), homozygous for gene *pc* (the "phocomelia" gene), with phocomelic limbs and an abnormally shaped head. Normal littermate (right) (about ×3).

transplantation or explantation experiments which change the mutant cell's environment serve to determine autonomy or nonautonomy of the lethal gene on the phenotype level only and in no way reflect the nature of the gene itself. On the other hand, studies, such as those just mentioned, of the suppressibility of mutations within one cistron are concerned with problems of autonomy on the gene level and have helped in the analysis of the nature of the mutated gene itself, not just of the nature of its phenotypic effect.

LETHAL GENES AND DIFFERENTIATION OF EMBRYONIC PATTERNS

In the thinking of embryologists interested in the causal analysis of development and differentiation, the phenomenon of aggregation of cells into definite patterns has played an important role. The possible role of genes in such pattern formation has been the object of many studies in the past. Recent methods of cell disaggregation and reaggregation have served to revive interest in the significance of developmental patterns; studies of the genetic basis of cell aggregation behavior appear to be promising. Much material that is valuable in this connection is provided by lethal mutations in which the main effect of the lethal genes appears to be a disturbance of the normal formation of aggregates of cells in early differentiation.

One such lethal gene is "phocomelia," which, in the homozygous condition, causes death of the newborn within a few hours after birth as the result of a cleft palate *(22)*. Phocomelic newborn mice show severe shortening of the extremities and other skeletal abnormalities (Fig. 4), reminiscent, to some extent, of those described in human babies of mothers who had taken thalidomide. Studies of the development of the limb rudiments in embryos homozygous for the lethal gene revealed, as the first observable deviation from normal, an abnormal pattern of mesodermal cell aggregates in the early limb rudiment. Subsequent differentiation showed no specific abnormalities, such as cell degeneration or deficiencies demonstrable by histochemical methods; abnormal differentiation expresses itself primarily in disturbances of the pattern of development in time and space. This lethal gene seems to produce its effect by interfering with the normal pattern of mesodermal cell aggregation in the early embryo. It would be interesting to examine this hypothesis further with the help of methods of cell disaggregation and reaggregation.

Recently a new mutation in the mouse called "lumbarless" was discovered in this laboratory—a conditional lethal mutation which causes the death of certain homozygotes in early adulthood as the result of paralysis of the posterior part of the trunk and the posterior extremities. Morphological studies revealed abnormalities of the lumbar spine of various degrees of severity: lumbar vertebrae may be missing altogether, or they may be so abnormal that they appear jumbled. The spinal cord is severely abnormal in the lumbar region, and much thinner than in the normal animal (Fig. 5). The paralysis seems to be a consequence of the abnormality of the spinal cord. Our interest in this particular mutation lies in the implications it may have for studies of the inductive relationship between neural tube and cartilage-forming somite material. It has been demonstrated *(23)*, with methods of organ tissue culture, that the spinal cord is able to induce, specifically, cartilage in somites; a nucleotide-containing component extracted from the

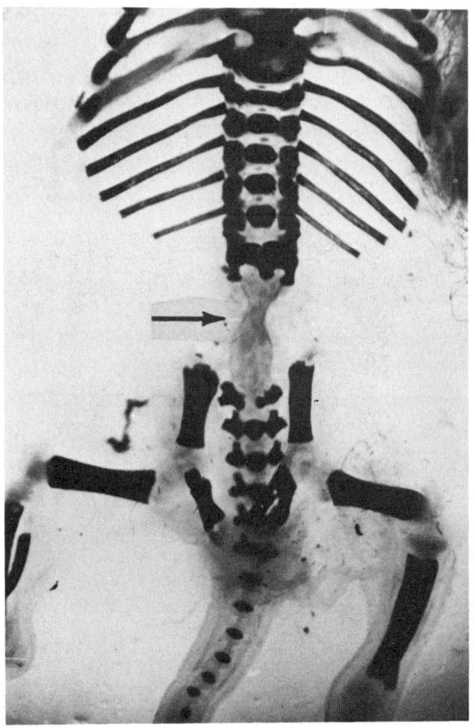

Fig. 5. Skeleton of newborn "lumbarless" mouse, stained with alizarin red. Several vertebrae of the lumbar region are missing, and the spinal cord shows a constriction, as indicated by the arrow (about ×7).

spinal cord seemed to be responsible for promoting such cartilage formation.

The lethal mutation affecting lumbar differentiation of spine and spinal cord may serve as a tool in further studies of inductive interaction between spinal cord and somites in the embryo. Questions to be asked might include the following: Does an abnormality of the neural tube in the lumbar region precede abnormalities of differentiation of the somites in the same region? Or do the somites in the lumbar region of the mutant embryo react abnormally to a normal inductive stimulus of the ventral spinal cord? If the latter is the case, is the eventual abnormality of the spinal cord, as found in the newborn, caused in turn by the abnormal somites? In this way various problems in the analysis of the normal system of inductive interaction between spinal cord and somites may be approached with the help of this lethal mutation.

PHENOCOPIES

Analysis of the mechanisms of effects of lethal genes has gained much from the use of so-called phenocopies. A phenocopy is an abnormality which, in a normal genotype, closely copies the appearance of a disturbance caused by an abnormal gene. It may arise sporadically, for unknown reasons, or it may be produced experimentally. Similarities between phenocopies and gene-controlled abnormalities open a way of approach to the question of pathways by which particular congenital malformations arise, whether they be gene-controlled or caused by external agents. With the use of phenocopies, an attempt was made to analyze the effects of a lethal mutation in the chick, the so-called Creeper mutation that results in severe skeletal abnormalities which can be "copied" experimentally by injecting insulin into the developing chick embryo *(24)*. Although it is not possible as yet to ascribe the effects of the lethal Creeper gene to a specific disturbance of carbohydrate metabolism, the experimental production of phenocopies by the injection of insulin has served, first of all, to focus attention on carbohydrate metabolism as the possible site of the lethal effect; it has furthermore contributed to knowledge of the chemical embryology of carbohydrate metabolism and has called attention to the chick embryo as an excellent experimental object for biochemical studies of differentiation.

Congenital abnormalities such as those produced in human babies by thalidomide are phenocopies of similar malformations resulting from abnormal gene effects. Studies of the responsible mechanisms in both groups may possibly lead toward identification of the causality of the defect.

CONCLUSIONS

In the foregoing discussion emphasis has been placed on the analysis of lethal genes on the phenotype level. It has been argued that lethal genes are fundamentally not different from other mutant genes but that in their effects they are the most extreme representatives of a wide spectrum of mutations. Lethality, therefore, is not inherent in the lethal gene; it depends to varying degrees on other genes and the environment.

Because of their effects on development and differentiation, lethal genes offer particular promise as tools in the investigation of problems of cell differentiation and cell genetics. Study of lethal genes in higher organisms has aspects different from such study in microorganisms, primarily because of the phenomenon of interaction between cells and tissues which in embryonic differentiation provides one of the most significant causal mechanisms. Similarly, the existence of regulation, another causal mechanism of differentiation, gives special characteristics to the problem of lethal gene effects in multicellular organisms, both during embryogeny and in adult life.

The study of lethal genes in higher organisms is concerned with the analysis of gene expression beyond the molecular level of gene action. By providing aberrant processes—frequently a means whereby normal processes are illuminated—lethal genes have already proved themselves valuable tools in the analysis of causal mechanisms of development. In the future, studies of cell differentiation, in the focus of interest at present, will profit greatly from the use of genetically controlled aberrations of normal differentiation, and investigation of somatic cells that have been steered into abnormal channels of differentiation by lethal genes will no doubt contribute to progress of the analysis of the molecular basis of somatic cell differentiation, as well as to somatic cell genetics.

In recent years it has become obvious to all those actively engaged in studies of molecular biology that the problem of cell differentiation is still far from being solved, and that any general theory or model of gene action must be able to account for phenomena of cell differentiation. In this connection, lethal genes in multicellular organisms may well turn out to be valuable tools of analysis, providing experimental material in which abnormal cell differentiation occurs in combination with phenomena of interaction and regulation during embryogenesis.

REFERENCE AND NOTES

1. H. J. Muller, Am. J. Human Genet. **2**:111 (1950); N. E. Morton, J. F. Crow, H. J. Muller, Proc. Natl. Acad. Sci. U.S. **42**:855 (1956); J. F. Crow, in Progress in Medical Genetics, A. G. Steinberg, Ed. (Grune and Stratton, New York, 1961), vol. 1, p. 1; T. Dobzhansky, Mankind Evolving (Yale Univ. Press, New Haven, Conn., 1962).
2. H. J. Muller, in Radiation Biology, A. Hollaender, Ed. (McGraw-Hill, New York, 1954), vol. 1, p. 351; "The Biological Effects of Atomic Radiation," Natl. Acad. Sci. Summary Rept. (1956; 1960).
3. E. Hadorn, Developmental Genetics and Lethal Factors (Methuen, London, 1961).
4. C. Pavan, A. R. Cordeiro, N. Dobzhansky, T. Dobzhansky, C. Malogolowkin, B. Spassky, M. Wedel, Genetics **36**:13 (1951).
5. H. J. Muller, Am. J. Human Genet. **2**:111 (1950).
6. ———, Selected Papers (Indiana Univ. Press, Bloomington, 1962).
7. J. F. Crow, Human Biol. **30**:1 (1958).
8. S. Gluecksohn-Waelsch, Am. J. Human Genet. **13**:113 (1961).
9. ———, Progr. Med. Genet. **2**:295 (1962).
10. S. Gluecksohn-Schoenheimer, Genetics **28**:341 (1943); **30**:29 (1945).
11. S. Gluecksohn-Waelsch and T. R. Rota, Develop. Biol. **7**:432 (1963).
12. L. C. Dunn, Cold Spring Harbor Symp. Quant. Biol. **21**:187 (1956); M. F. Lyon and A. J. S. Phillips, Heredity **13**:23, (1959).
13. S. Gluecksohn-Waelsch, Cold Spring Harbor Symp. Quant. Biol. **19**:41 (1954).
14. B. Ephrussi, J. Exptl. Zool. **70**:197 (1935).
15. S. Gluecksohn-Schoenheimer, Proc. Natl. Acad. Sci. U.S. **30**:134 (1944).
16. Detailed references to the literature of the *T*-series of lethal genes in the mouse are given in S. Gluecksohn-Waelsch, Cold Spring Harbor Symp. Quant. Biol. **19**:41 (1954).
17. R. Auerbach, J. Exptl. Zool. **127**:305 (1954).
18. A. Szeinberg, C. Sheba, A. Adam, Nature **181**: 1256 (1958).
19. Unpublished observations by Anna C. Pai, who has kindly permitted me to quote them.
20. R. D. Cahn, N. O. Kaplan, L. Levine, E. Zwilling, Science **136**:962 (1962).
21. A. Garen and O. Siddiqui, Proc. Natl. Acad. Sci. U.S. **48**:1121 (1962).
22. B. F. Sisken and S. Gluecksohn-Waelsch, J. Exptl. Zool. **142**:623 (1959).
23. H. Holtzer, in Synthesis of Molecular and Cellular Structure, D. Rudnick, Ed. (Ronald Press, New York, 1961), p. 35; J. W. Lash, F. A. Hommes, F. Zilliken, Biochim. Biophys. Acta **56**:313 (1962); G. Strudel, Develop. Biol. **4**:67 (1962).
24. W. Landauer, J. Exptl. Zool. **105**:1415 (1947); E. Zwilling, Arch. Biochem. Biophys. **33**:228 (1951).

chapter 12

Behavior genetics

Although the sciences of genetics and animal behavior have coexisted for some time, it is only recently that a widespread interest has developed in the study of the hereditary control over an organism's actions. The fusion of these two disciplines, now generally called *behavior genetics,* is in reality an extension of developmental genetics, since the ability to act must await the development of the structures involved in the action. The dependence of an action on a previously developed structure tends to complicate the genetic analysis of a behavioral trait because, in most cases, it removes the phenotype far from the primary gene product. This distant separation of phenotype from primary gene product permits a significant involvement of environmental factors, as well as gene-product interactions, in the determination of an organism's behavior. Still further problems arise because genes involved in behavior can affect any one or more of the following: (1) the sensory organs, thus changing information input; (2) an intermediate system (nervous, endocrine), thus altering coordination and perceptual capacities; and (3) the effector organs, thus influencing the response.

Three major lines of research have been pursued in studies on behavior genetics: (1) effects of single genes, (2) behavioral differences between inbred lines or strains of organisms, and (3) selection for extremes of a behavioral trait in an initially variable population.

EFFECTS OF SINGLE GENES

A mutation whose effects on behavioral traits is well known is the "albino" gene in mice. It has generally been assumed that the behavioral differences exhibited by albino mice, when compared to their pigmented counterparts, is due to the blindness or near blindness of albino mice and, hence, to the lack of proper information input. A paper by Fuller in 1967 (Ref. 12-1), which is reprinted in this chapter, examines some of the effects of the albino gene on mouse behavior and discusses the difficulty of ascertaining the mechanism by which this gene produces its effects.

Behavior patterns can also be altered by the action of a gene on an intermediate system of the body. An example of a mutation that affects an intermediate system is the "waltzer" gene in mice. Animals homozygous for this recessive gene run in circles for long periods of time. When at rest, they exhibit a head-shaking behavior; in addition, they are deaf. These abnormalities of behavior are usually referred to as the *shaker-waltzer* syndrome. The mice are found to have structural defects in the inner ear, including the cochlea and semicircular canals. Other, nonallelic, mutations have also been discovered. Some of the mice possessing these latter mutations show the circling and head-shaking characteristics but can hear; others, though deaf, do not circle or shake their heads. Analyses of the structural anomalies involved in different mutants were made by Deol in 1954 (Ref. 12-2) and in 1956 (Ref. 12-3). A paper by Deol in 1966 (Ref. 12-4) examines the probable mode of gene action in this type of mutant and concludes that the abnormal behavior is most likely due to lesions in the animal's nervous system. The inner ear defects are hypothesized to be independent effects of the gene and are not thought to be related to the mouse's behavior in a cause-and-effect manner.

Genes that affect behavior solely through their action on effector organs are difficult to demonstrate, since there is always the possibility that an intermediate system is actually the primary target organ of the gene. One possible example of a gene acting on an effector organ is "handedness" in man. The results of studies on the inheritance of left-handedness were reported by Trankell in 1955 (Ref. 12-5), Merrell in 1957 (Ref. 12-6), and Falek in 1959 (Ref. 12-7). These investigations were beset with the problem that a substantial proportion of left-handers are modified by cultural pressures to become right-handed. The human infant exhibits no hand dominance. Handedness develops as the infant acquires more and more skilled movements. Initially, responses are made with both hands, but those of one side are more efficient and hence differentially reinforced. As a result, hand dominance develops. However, parental pressure can force a child to perform his tasks with a given hand regardless of the child's tendency to use the other.

STUDIES ON INBRED LINES

Inbred lines are strains that have been produced by many generations of brother-sister matings. As a result of this inbreeding, each strain will have become homozygous for most of its genes. However, the various strains will, by chance, have become homozygous for different alleles at many loci. Studies on behavioral traits of inbred lines will not indicate how many or which genes are involved in any observed differences between the lines. Such information can sometimes come from data on the F_1, F_2, and backcrosses between the strains.

An experiment examining the sex drive in male mice from two different inbred strains and in the hybrid between these strains was reported by McGill and Tucker in 1964 (Ref. 12-8) and is reprinted in this chapter. A number of investigations have been performed on alcohol preference in mice. One of these studies, reported by Fuller in 1964 (Ref. 12-9), measured the alcohol preference of mice from four different inbred strains. The four strains used showed great variation in their alcohol preferences. This would indicate that the strains are genetically quite different from one another. The various hybrids between the different strains showed alcohol preferences that varied with the particular cross. This indicated that high preference for alcohol may be dominant, recessive, or neither, depending on the type of the mating. An experiment examining the relationship between alcohol tolerance and genotype in mice was reported by Kakihana and his co-workers in 1966 (Ref. 12-10).

Investigations of fighting behavior between male mice from different inbred strains were reported by Ginsburg and Allee in 1942 (Ref. 12-11) and by Levine and his co-workers in 1965 (Ref. 12-12). Experiments on the mating behavior of female guinea pigs from different inbred strains were reported by Goy and Young in 1957 (Ref. 12-13); strain differences in maternal behavior in rats were reported by McIver and Jeffrey in 1967 (Ref. 12-14).

STUDIES INVOLVING SELECTION

Selection experiments usually attempt to modify a given trait in opposite directions (e.g., "bright" versus "dull" rats) by mating individuals in each generation who exhibited extreme phenotypes. In general these experiments are successful, indicating some degree of genetic control over the trait being studied. However, in many cases, selection becomes effective only after a number of generations, implying a polygenic control of the trait. An early study of this type was reported in 1940 by Tryon (Ref. 12-15), who was able to select for maze-running ability in rats. After 8 generations of selection, there was no overlap in maze performance between the bright rats (smallest number of errors) and the dull rats (largest number of errors). The animals of both selected lines were tested for other behavioral traits, and the results were reported by Searle in 1949 (Ref. 12-16). It was found that bright rats learned hunger-motivation problems better than dull rats. However, the dulls were superior in escape-from-water tests. A study was undertaken to determine the biochemical basis for brightness in rats, and the results were reported by Rosenzweig and his co-workers in 1960 (Ref. 12-17). It was found that maze performance was linked to the amount of acetylcholine in the brain cells of the rats. A more recent example of selection for a behavioral trait in rats was reported in 1968 by Eriksson (Ref. 12-18), whose paper is reprinted in this chapter.

A number of selection experiments involving behavioral patterns have been conducted on *Drosophila.* It was reported by Manning in 1961 (Ref. 12-19) that lines of *D. melanogaster* could be selected for fast and slow mating speeds. The pattern of divergence of the selected lines was a gradual one, implying that a polygenic system controlled the behavior. Another behavior pattern of *D. melanogaster* that has been studied in selection experiments is "phototaxis." The results of such an experiment was reported in 1958 by Hirsch and Boudreau (Ref. 12-20), whose paper is included in this chapter. A later paper involving selection for phototaxis in this species was published by Hadler in 1964 (Ref. 12-21).

Yet another behavior of *Drosophila* that has been studied is the response of the flies to gravity, "geotaxis." It was reported by Erlenmeyer-Kimling and her co-workers in 1962 (Ref. 12-22) that lines of *D. melanogaster* could be selected for positive and negative geotaxis.

The extent of response to selection, together with the prolonged and gradual nature of the strain divergence, strongly suggested that a polygenic system was involved. An analysis was made of the influence of the three major chromosomes (in this species) on geotaxis. It was reported by Hirsch and Erlenmeyer-Kimling in 1962 (Ref. 12-23) that all three chromosomes contain genes that affect this behavior. Selection experiments involving geotaxis and phototaxis have also been conducted with another *Drosophila* species, *D. pseudoobscura.* The results of these experiments were summarized in a paper by Dobzhansky and Spassky in 1969 (Ref. 12-24).

ROLE OF NUCLEIC ACIDS AND PROTEINS IN LEARNING AND MEMORY

In all genetic studies, the ultimate goal is to trace a phenotype to a gene and study its action. This aim applies to behavioral characteristics no less than to structural or physiological traits and includes the type of behavior that is usually referred to as *intellectual.* The ability of organisms to learn tasks or solve problems and retain this "learning" for future use (i.e., memory) is well known. Experiments were reported by Hydén and Egyházi in 1964 (Ref. 12-25) which demonstrated that forcing a rat to permanently change from using his right paw in a given task to his left paw resulted in a change in the RNA base composition of the cortical neurons controlling the new behavior. The implication of these findings is that the forced change in behavior resulted in the formation of an mRNA, rather than rRNA or tRNA, with the presumed subsequent synthesis of some appropriate protein. It is not known what role such a protein might have in memory formation. A report by Shashoua in 1970 (Ref. 12-26) discussing the changes that occur in RNA metabolism in the goldfish brain during acquisition of new behavioral patterns is reprinted in this chapter.

As soon as it became apparent that learning and memory might be mRNA-protein dependent, experiments were performed to test whether learning and memory can occur in animals whose brain mRNA or protein synthesis is inhibited. Experiments on mice, involving inhibition of brain mRNA synthesis by actinomycin D, were reported by Cohen and Barondes in 1966 (Ref. 12-27). The behavioral task involved was a Y maze, which the animals had to traverse in order to avoid shock. The experiments tested the mouse's memory one hour and four hours after the training period. They found that despite a 94% to 96% reduction of brain RNA synthesis, the actinomycin-treated mice learned and remembered the maze as well as the controls. Since the experiments tested what is essentially "short-term" memory, the authors could only conclude that memory retention of recent events was not mRNA-protein dependent. They were unable to draw any conclusions about memory consolidation covering long periods of time. Experiments on both "short-term" and "long-term" memory, involving the use of puromycin (an inhibitor of protein synthesis) were reported in 1966 by Barondes and Cohen (Ref. 12-28) whose paper is included in this chapter. A general review of the molecular basis of memory was published by Gurowitz in 1969 (Ref. 12-29).

CHROMOSOME COMPLEMENT AND HUMAN AGGRESSION

In the field of human behavior, a controversy has developed over the possible effects of different karyotypes on aggression. It will be recalled that in Chapter 7 various types of aneuploidy were discussed. These included, for humans, males with an XXY karyotype and those with an XYY karyotype. One of the first reports of an association between XXY males and aggressive behavior, made by Casey and his co-workers in 1966 (Ref. 12-30), is included in this chapter. A similar association between aggressive behavior and the XYY male was made by Jacobs and her co-workers in 1965 (Ref. 12-31). A later paper on the XYY male, published by Price and Whatmore in 1967 (Ref. 12-32), is the last paper reprinted in this chapter.

Attorneys for XXY and XYY individuals accused of crimes, have been quick to point out the association of abnormal chromosome complement and aggressive behavior. In such cases, the defense has rested heavily on arguments of legal nonresponsibility because the defendant possessed a particular karyotype. A review of the information available on the relationship of the XYY karyotype to criminality was published by Kessler and Moos in 1970 (Ref. 12-33).

REFERENCES

Effects of single genes

12-1. Fuller, J. L. 1967. Effects of the albino gene upon behaviour of mice. Anim. Behav. **15**:467-470.

12-2. Deol, M. S. 1954. The anomalies of the

labyrinth of the mutants varitint-waddler, shaker-2 and jerker in the mouse. J. Genet. **52**:562-588.

12-3. Deol, M. S. 1956. The anatomy and development of the mutants pirouette, shaker-1 and waltzer in the mouse. Proc. Roy. Soc., Lond. B. **145**:206-213.

12-4. Deol, M. S. 1966. The probable mode of gene action in the circling mutants of the mouse. Genet. Res., Camb. **7**:363-371.

12-5. Trankell, A. 1955. Aspects of genetics in psychology. Am. J. Hum. Genet. **7**:264-276.

12-6. Merrell, D. J. 1957. Dominance of eye and hand. Hum. Biol. **29**:314-328.

12-7. Falek, A. 1959. Handedness, a family study. Am. J. Hum. Genet. **11**:52-62.

Studies on inbred lines

12-8. McGill, T. E., and Tucker, G. R. 1964. Genotype and sex drive in intact and in castrated male mice. Science **145**:514-515.

12-9. Fuller, J. L. 1964. Physiological and population aspects of behavior genetics. Am. Zool. **4**:101-109.

12-10. Kakihana, R., Brown, D. R., McClearn, G. E., and Tabershaw, I. R. 1966. Brain sensitivity to alcohol in inbred mouse strains. Science **154**: 1574-1575.

12-11. Ginsburg, B., and Allee, W. C. 1942. Some effects of conditioning on social dominance and subordination in inbred strains of mice. Physiol. Zool. **15**:485-506.

12-12. Levine, L., Diakow, C. A., and Barsel, G. E. 1965. Interstrain fighting in male mice. Anim. Behav. **13**:52-58.

12-13. Goy, R. W., and Young, W. C. 1957. Strain differences in the behavioural responses of female guinea pigs to alpha-oestradiol benzoate and progesterone. Behaviour **10**:340-354.

12-14. McIver, A. H., and Jeffrey, W. E. 1967. Strain differences in maternal behavior in rats. Behaviour **28**:210-216.

Studies involving selection

12-15. Tryon, R. C. 1940. Genetic differences in maze-learning ability in rats. Yearb. Nat. Soc. Stud. Educ. **39**:111-119.

12-16. Searle, L. V. 1949. The organization of hereditary maze-brightness and maze-dullness. Genet. Psychol. Monogr. **39**:279-325.

12-17. Rosenzweig, M. R., Krech, D., and Bennett, E. L. 1960. A search for relations between brain chemistry and behavior. Psychol. Bull. **57**:476-492.

12-18. Eriksson, K. 1968. Genetic selection for voluntary alcohol consumption in the albino rat. Science **159**:739-741.

12-19. Manning, A. 1961. The effects of artificial selection for mating speed in *Drosophila melanogaster.* Anim. Behav. **9**:82-92.

12-20. Hirsch, J., and Boudreau, J. C. 1958. Studies in experimental behavior genetics: I. The heritability of phototaxis in a population of *Drosophila melanogaster.* J. Comp. Physiol. Psychol. **51**:647-651.

12-21. Hadler, N. M. 1964. Genetic influence on phototaxis in *Drosophila melanogaster.* Biol. Bull. **126**:264-273.

12-22. Erlenmeyer-Kimling, L., Hirsch, J., and Weiss, J. M. 1962. Studies in experimental-behavior genetics: III. Selection and hybridization analyses of individual differences in the sign of geotaxis. J. Comp. Physiol. Psychol. **55**:722-731.

12-23. Hirsch, J., and Erlenmeyer-Kimling, L. 1962. Studies in experimental behavior genetics: IV. Chromosome analyses for geotaxis. J. Comp. Physiol. Psychol. **55**:732-739.

12-24. Dobzhansky, T., and Spassky, B. 1969. Artificial and natural selection for two behavioral traits in *Drosophila pseudoobscura.* Proc. Natl. Acad. Sci. U. S. A. **62**:75-80.

Role of nucleic acids and proteins in learning and memory

12-25. Hydén, H., and Egyházi, E. 1964. Changes in RNA content and base composition in cortical neurons of rats in a learning experiment involving transfer of handedness. Proc. Natl. Acad. Sci. U. S. A. **52**:1030-1035.

12-26. Shashoua, V. E. 1970. RNA metabolism in goldfish brain during acquisition of new behavioral patterns. Proc. Natl. Acad. Sci. U. S. A. **65**:160-167.

12-27. Cohen, H. D., and Barondes, S. H. 1966. Further studies of learning and memory after intracerebral actinomycin-D. J. Neurochem. **13**:207-211.

12-28. Barondes, S. H., and Cohen, H. D. 1966. Puromycin effect on successive phases of memory storage. Science **151**:594-595.

12-29. Gurowitz, E. M. 1969. The Molecular Basis of Memory. Prentice-Hall, Inc., Englewood Cliffs, N. J.

Chromosome complement and human aggression

12-30. Casey, M. D., Segall, L. J., Street, D. R. K., and Blank, C. E. 1966. Sex chromosome abnormalities in two state hospitals for patients requiring special security. Nature **209**:641-642.

12-31. Jacobs, P. A., Brunton, M., Melville, M. M., Brittain, R. P., and McClemont, W. F. 1965. Aggressive behaviour, mental sub-normality and the XYY male. Nature **208**:1351-1352.

12-32. Price, W. H., and Whatmore, P. B. 1967. Criminal behaviour and the XYY male. Nature **213**:815.

12-33. Kessler, S., and Moos, R. H. 1970. The XYY karyotype and criminality: A review. J. Psychiat. Res. **7**:153-170.

59 Effects of the albino gene upon behaviour of mice

John L. Fuller
The Jackson Laboratory
Bar Harbor, Maine

Winston & Lindzey (1964) showed an association in mice between the homozygous condition for albinism *(cc)* and delayed escape from a water tank. Meier & Foshee (1964) questioned the attribution of the performance deficit to the albino locus itself, because a random-bred albino stock was competent in the water escape task. They suggested that genes closely linked with the C-locus might be responsible for the performance decrement reported by Winston & Lindzey.

An opportunity to test the effects of the homozygous *cc* condition occurred as the result of a back mutation, *C* to *c*, in a pigmented inbred line C57BL/6J. The gene was maintained by a cross-intercross system in which albinos were crossed with ordinary inbred C57BL/6J *(CC)* to produce heterozygotes *(Cc)*. Sib matings of heterozygotes yield approximately ¼ albino and ¾ pigmented offspring. Considering the origin of the mutation and its maintenance in a brother-sister mating system with forced heterozygosis at the C-locus, one can eliminate linked genes as an explanation of differences in performance found between pigmented and albino animals of this stock.

METHOD

Subjects

The subjects, provided by Dr E. S. Russell, were thirty-four pigmented *(Cc* or *CC)* and thirty-four albino *(cc)* C57BL/6J from matings between heterozygotes. Sex was recorded, but has been disregarded in the analysis since no sex differences were found. Testing of all but five subjects commenced between 75 and 85 days of age. The five older subjects, maximum 121 days, did not deviate from the remainder of the sample and have been included in the data.

From Animal Behaviour **15**:467-470, 1967. Used with permission.

This research was supported in part by Public Health Service grant MH-11327. Jane Harris and Frank Clark provided technical assistance. Mice were provided by Dr Elizabeth S. Russell.

Procedures and apparatus

On the first day of testing the water escape test was given following the procedure of Winston & Lindzey. Subjects were dropped into a 9 cm depth of water in a galvanised tub 43 cm in diameter. The time required to emerge from the water on a ladder directly opposite the point of immersion was recorded. The sum of four escape times in seconds was scored. On the following day subjects were tested in an open field (18-in. diameter) in which activity was measured by counters whenever the subject interrupted one of four beams of light, two at right angles to the other two (Lehigh Valley Electronics, model B1497). The mice were then placed in modified Bruell (1962) rotors supplied with food and water and left for a minimum of 3 days. The total number of revolutions for days 2 and 3 was recorded, unless occasional mechanical difficulties made it necessary to substitute a later day. After a recuperative period of 2 or 3 days subjects were tested for black-white discrimination in a water-maze (Waller *et al.*, 1960; Wimer & Weller, 1964). Ten trials were given daily for 5 days using a predetermined irregular sequence of right and left placements. Subjects alternated in running to black or white. The score was the number of correct responses during the final 4 days of testing. Escape time on the final ten correct trials was computed as well.

RESULTS

The results of the tests are summarized in Table I. The pigmented mice performed better on the water escape test, thus confirming the findings of Winston & Lindzey. They were more active in the open field and made more correct responses in the water-maze. Swimming time in the maze was slightly longer in the albinos, although there was considerable overlap between the two groups. Albinos and pigmented subjects did not differ significantly in wheel-running activity measured over 48 hr.

One way of looking at this experiment is to

Table I. Median scores of *C* and *cc* mice

	C		*cc*		
	N	Median	N	Median	P*
Water escape time–sec five trials	34	51	34	112	<0.00006
Open field activity–15 min	34	317	31	266	<0.0014
Activity wheel–thousands of revolutions in 24 hr	33	13.1	31	14.5	0.2150
Water maze–correct trials out of forty. Black-white discrimination	32	36	24	24	<0.00006
Mean swimming time–last ten correct trials	32	2.6	24	2.8	<0.06

*Probability estimated by Mann-Whitney U test.

Table II. Intraclass correlations of four behavioural measures

Measure (X)	Transformation	Intraclass correlation	*P*
Water escape time in seconds	$1000/x$	0.217	<0.01
Open field crossings–15 min	$\sqrt{x}$	0.249	<0.01
Revolution of wheel–48 hr	$\sqrt{x}$	0.007	ns
Percentage correct discrimination	arc sin $\sqrt{x}$	0.747	<0.001

Table III. Reproductive performance of C57BL/6J albino stock

Genotype		Number of matings	Percentage fertile matings	Litters per fertile female	Mean size litter	Percentage survival to weaning	Percentage of albinos
Female	Male						
cc	*CC*	44	75.0	3.2	5.2	87.2	–
CC	*cc*	49	63.5	3.6	6.5	87.5	–
Cc	*Cc*	61	77.0	4.3	6.1	85.3	25.4

consider *cc* and *C* as two treatments applied to a mouse. The effect of the *c* for *C* substitutions can be quantitatively measured by the intraclass correlation coefficient, r_i, provided reasonable homogeneity of variance exists within the two classes. The intraclass correlation measures the ratio of between-group variance to the sum of the between- and within-group variances. In this particular experiment the correlations may be regarded as coefficients of genetic determination (Falconer, 1960), since all within-group variance is environmental and all between-group variance is genetic, if we assume no systematic environmental differences between the pigmented and albino subjects.

The results of this analysis are shown in Table II. The fact that the contributions of the gene substitution differ so widely among the four measures is not surprising. One thing is certain; these values should not be regarded as heritability estimates even though they were derived in a way which resembles that used for computation of heritability. The population studied is the outcome of a peculiar mating system arranged to maximize phenotypic variation. The contribution of the C-locus to variation in the behaviour of natural populations is close to zero because mutants are so rare. The justification for computing the coefficients must rest upon their value for the analysis of problems of behaviour rather than of population genetics.

Some difficulty was found in applying Winston & Lindzey's criterion for water escape time to the albinos as they often hung onto the base of the escape ladder without actually emerging. They appeared to be more fatigued by swimming than pigmented mice, and stress of testing may be responsible for the deaths (ten in thirty-four) which occurred during the experiment. In contrast, only two pigmented animals were lost during the experiment. These observations substantiate the opinion of the persons caring for this stock that the albinos are generally less vigorous.

Because of this indication that the albinos were less energetic than the pigmented mice, particularly on the swimming tasks, an analysis was made of Dr Russell's breeding records to see whether reproductive fitness under laboratory conditions was affected by the *c* gene as well. The results (Table III) show a small reduction in average litter size of *cc* as compared with *CC* females. A smaller proportion of *CC* females produced litters, but this may have been the fault of their *cc* consorts. Once the young are produced, pigmented and albino mothers appear to care for their offspring equally well, as shown by the proportion of young surviving to weaning. The proportion of albino offspring is precisely that predicted by the Mendelian laws. Because the breeding records were incomplete no statistical tests were performed on these data. However, it seems safe to conclude that the *c* for *C* substitution has little effect upon fitness as measured by reproductive and care-taking behaviour in the laboratory. Nevertheless, the failure of *c* to become established in wild populations despite its frequent occurrence through mutation points to a deleterious effect upon fitness.

DISCUSSION

This experiment clarifies one point which was at issue. Substitutions at the C-locus have behavioural effects in their own right and are more than markers for unspecified cryptic genes. The observation of Meier & Foshee that albino CF/1 mice escaped from water as well as some pigmented strains must indicate that the physiological deficiency produced by the *cc* genotype can be compensated by nonallelic genes under some circumstances. This report in conjunction with those cited above seems to confirm Merrell's (1965) suggestion that behaviour geneticists should look to mutants for experimental material. However, more is required than the demonstration of an association between a gene substitution and altered behaviour for mutants to be of any substantial value to behavioural research. What are the potentialities and limitations of mutants as behavioural subjects?

Consider, for example, whether observations of the behaviour of mutants in the laboratory can throw light upon the effects of the mutant gene on fitness in natural populations. The present experiment suggests that both lowered resistance to stress and poorer discriminative learning could account for the lower fitness of albinos in the wild. However, there seems to be no way of allocating the observed decrement in fitness among several pleiotropic effects of a gene, unless one can suppress one effect at a time while leaving all others unchanged. Furthermore, any set of observations on behaviour will be incomplete, and effects upon fitness of behavioural characteristics which were not measured can never be excluded. It follows that behavioural observations on mutants may suggest real mechanisms underlying the rather abstract concept of fitness, but they are not likely to lead to a rigorous quantitative apportionment of decrements in fitness between physiological and psychological levels of integration.

Another possible value of the study of mutants might be the definition of clusters of behavioural characteristics with some common biological basis. For example, in this experiment behaviours with high intraclass correlations may be assumed to be closely dependent upon the physiological processes affected by the *c* for *C* substitution. Such a cluster may not coincide with commonsense judgments of behavioural similarity. Open-field activity seems *a priori* to be more closely related to wheel-running than to discriminative learning in a water-maze, but the results of the genetic experiment contradict the *a priori* judgment (Table II).

One can conceive of a more elaborate experiment in which a battery of genic substitutions is applied to a battery of behavioural tests with the view of determining which parts of the test battery are dependent upon common biological substrates. An analysis of this kind will be essentially a 'black-box' procedure if the genes are specified only as the names of mutations and the results are scores on behavioural tests. It cannot provide information on the nature of the relationship between gene and behaviour. In the conceptual sense, black boxes come in many hues and shapes, for changing the colour or form of a mouse does not by itself provide a view of internal mechanisms.

If both physiological and psychological measurements are included in the test battery there is at least a possibility of inferring what goes on in the box (or the mouse). Physiological and behavioural measures with similar patterns of variation following genic substitutions are likely to have something in common. The difficulty with this strategy is that the researcher is faced with the choice of a potentially infinite variety of tests, but he can perform only a few. A useful guiding principle is that gene substitution should be used as a treatment in behavioural experiments when it produces a physiological change which could be accomplished by other means only with difficulty or not at all. It follows as an ideal that mutant genes in behavioural research should actually be methods of producing biochemical and anatomical effects which are related to important behavioural systems.

Applying this criterion one may ask whether the albino gene is promising material for additional psychophysiological investigation. A primary defect of *cc* animals is a diminution of tyrosinase and dopa oxidase activity in the skin (Silvers, 1961). Since dopa is the precursor of norepinephrine (Sourkes, 1962) it is conceivable that the observed behavioural effects are explicable by an imbalance of catecholamines in the central nervous system rather than by the peripheral effects of the gene. The evidence on this point is ambiguous. Maas (1962, 1963) found no differences in the norepinephrine content of the brains of C57BL/10 (pigmented) and BALB/c (albino) mice, although BALB/c was higher in serotonin. Since these strains differ at many loci the role of the C-locus is undetermined. The same indeterminateness attaches to the hypothesis that differences between Wistar albino and Long-Evans hooded rats are attributable to pleiotropic effects of the albino gene upon the brain (Meyer *et al.*, 1966). Human albinos do not seem to be deficient in pressor amines as determined by urine analysis (Fitzpatrick & Quevedo, 1966).

It appears that only an intensive investigation into the neural effects of the gene for albinism will determine whether its behavioural effects are a curious fact or a clue to the way in which genetic information programmes the behaviour of a complex organism.

SUMMARY

Albino mice otherwise congenic with strain C57BL/6J escaped more slowly from water, were less active in an open field and made more errors on a black-white discrimination task than their pigmented congeners. No difference was found in activity in a revolving cage nor in several measures related to reproduction and survival under laboratory conditions. Albinos appeared to be more greatly stressed by the test procedures.

The interpretation and significance of experiments which uncover pleiotropic behavioural effects of single genes are discussed. It is concluded that the greatest value of such findings is the possibility of a physiological analysis of a behavioural process by observing the effects of genes whose physiological effect is known. Albino mice are deficient in tyrosinase and may therefore differ from pigmented mice in the concentration of brain catecholamines.

REFERENCES

Bruell, J. H. (1964). Heterotic inheritance of wheel running in mice. J. comp. physiol. Psychol. **58**:159-163.

Falconer, D. S. (1963). Quantitative genetics. In Methodology in Mammalian Genetics (ed. by W. J. Burdette), pp. 193-216. San Francisco: Holden-Day.

Fitzpatrick, T. B. & Quevedo, W. C., Jr. (1966). Albinism. In The Metabolic Basis of Inherited Disease (ed. by J. B. Stanbury, J. B. Wyngaarden & D. S. Frederickson), 2nd ed., pp. 324-340. New York: McGraw-Hill.

Maas, J. W. (1962). Neurochemical differences between two strains of mice. Science **137**:621-622.

Maas, J. W. (1963). Neurochemical differences between two strains of mice. Nature **197**:255-257.

Meier, G. W. & Foshee, D. P. (1964). Albinism and water escape performance in mice. Science **147**: 307-308.

Merrell, D. J. (1965). Methology in behavior genetics. J. Hered. **56**:263-266.

Meyer, D. R., Yutzey, D. A. & Meyer, P. M. (1966). Effects of neocortical ablations on relearning of a black-white discrimination habit by two strains of rats. J. comp. physiol. Psychol. **61**:83-86.

Silvers, W. K. (1961). Genes and the pigment cells of mammals. Science **134**:368-373.

Sourkes, T. L. (1962). Biochemistry of the catecholamines. In Neurochemistry (ed. by K. A. C. Elliott, I. H. Page & J. H. Quastel), 2nd ed., pp. 590-619. Springfield, Ill.: Thomas.

Waller, M. B., Waller, P. F. & Brewster, L. A. (1960). A water maze for use in studies of drive and learning. Psychol. Rep. **7**:99-102.

Wimer, R. & Weller, S. (1965). Evaluation of a visual discrimination task for the analysis of the genetics of a mouse behavior. Percept. mot. Skills **20**:203-208.

Winston, H. D. & Lindzey, G. (1964). Albinism and water escape performance in the mouse. Science **144**:189-191.

60 Genotype and sex drive in intact and in castrated male mice

Thomas E. McGill
G. Richard Tucker
Department of Psychology
Williams College
Williamstown, Massachusetts

Abstract. *Male mice of two inbred strains and one hybrid strain were observed for sexual behavior for 42 consecutive days. Half the males of each strain were then castrated, and daily testing was continued until the ejaculatory reflex was lost. Strain differences were found in ejaculatory frequency both before and after castration.*

Male mice of different inbred strains differ significantly in several aspects of sexual behavior *(1, 2)*. The time required to recover sex drive after an ejaculation is one of the variables which has been shown to be affected by genotype *(2)*. More specifically, males of the inbred strain DBA/2J recovered sex drive (achieved a second ejaculation) in 1 hour while C57BL/6J males required a median recovery time of 4 days. Hybrid males resulting from a cross between the two inbred strains resembled DBA/2J males in that the time required to recover sex drive after an ejaculation was comparable. The previous studies, however, did not show that "fast-recovery" males were in fact capable of more ejaculations over an extended period of testing than were "slow-recovery" males.

One purpose of the present study was to test the foregoing hypothesis; the second purpose was exploratory in nature. One of the accepted generalizations from studies on sexual behavior is that the behavior of animals high on the phylogenetic scale is less dependent on gonadal hormones than is the sexual behavior of animals with a lower phylogenetic status *(3)*. For example, the sexual behavior of experienced, male cats and dogs *(4, 5)* persists much longer after castration than does the behavior of experienced, castrated rats and guinea pigs *(6)*. The second part of our experiment was designed to determine whether genetic differences within a species affect the persistence of sexual behavior after castration.

A total of 72 male mice was used, including 24 C57BL/6J males, 24 DBA/2J males, and 24 B6D2F$_1$ males. The last named strain results from crossing C57BL/6J females with DBA/2J males. Each male was housed with five other males of the same genotype in the intervals between the daily testing sessions.

Two hundred and fifty-two BALB/cJ females were used in the mating tests. Thirty-six of these females were brought into behavioral estrus each day by injections of estrogen and progesterone *(7)*.

All animals were 9 weeks old at the beginning of the experiment. The animals were maintained on a reversed light-dark cycle with the light phase lasting 13 hours. The dark phase began 2 hours before the onset of testing which occurred under normal room illumination between 8:30 a.m. and 2:00 p.m.

Males were placed individually in plastic cylinders 25 cm in diameter and 50 cm in height. In the early stages of the experiment, males were allowed 30 minutes to adapt to the cylinder prior to the introduction of an estrous female. This 30-minute adaptation period became unnecessary as the males gained experience in the test situation. A given male was allowed from 5 to 10 minutes to initiate mating with the estrous female. If the male did not gain intromission during this interval, the female was removed and a second female was presented to the male. When a male "refused" all three females, he was scored as "negative" for that day. In order to achieve a "positive" score, the male was required to mate with one of the three females until ejaculation occurred. Occasionally, a male would cease copulating before ejaculation. Such a test was scored as

From Science **145**:514-515, July 31, 1964.

Supported in part by research grant 07495 from the Institute of General Medical Sciences, PHS. One of us (G.R.T.) was a NSF research trainee in the Undergraduate Science Education Program; NSF grant No. 22864.

Table 1. Sexual performance of 24 intact males of each strain during 42 consecutive days of testing

No. of ejaculators*	No. of ejaculations per ejaculator		Day of first ejaculation		Days between ejaculations	
	Median	Range	Median	Range	Median	Range
			C57BL/6J			
10	2	1-9	17	4-31	6	1-38
			DBA/2J			
22	15	4-28	3	1-36	2	1-19
			B6D2F₁			
24	15	5-27	2	1-32	2	1-9

*Males that achieved ejaculation.

Table 2. Sexual performance after castration

Castrates (No.)	Preoperative ejaculators (No.)	Postoperative ejaculators (No.)	Total ejaculations after castration (No.)	Day after castration on which last ejaculation occurred	
				Median	Range
		C57BL/6J			
11	4	0			
		DBA/2J			
11	10	3	3	3	3-8
		B6D2F₁			
12	12	9	42	28	3-60

"negative" if 40 minutes elapsed without an intromission.

Daily testing continued for 42 days at which point half the males of each strain were castrated. Castrate groups and noncastrate groups were matched within strains on the basis of number of ejaculations for individual animals. One C57BL/6J male and one DBA/2J male, both of which had previously copulated, died as a result of the operation. Daily testing was resumed 72 hours after castration. Daily tests then continued for each group until at least 14 days had elapsed without the occurrence of the ejaculatory reflex in a castrated male.

The results (Tables 1 and 2) may be briefly summarized as follows: (i) DBA/2J males exhibited higher sex drive than C57BL/6J males. This is illustrated in the four measures presented in Table 1. All four of the measures revealed statistically significant differences between the two inbred strains *(8)*. (ii) The hybrid males, $B6D2F_1$, resembled the DBA/2J males in sex drive as defined by the measures presented in Table 1. They differed significantly from the C57BL/6J males, but not from the DBA/2J males, on all four measures. This finding agrees with a previous report concerning the "dominance" of the DBA/2J genotype over the C57BL/6J genotype in the determination of sex drive *(2)*. (iii) The data of Table 2 illustrate "hybrid vigor" in the persistence of sex drive after castration. The hybrid males retained the ejaculatory reflex in greater numbers and for a longer time than either inbred parent strain.

This study has shown that genotype has a definite effect on the sex drive of the intact male mouse and that high sex drive seems to be a "dominant" character. The hypothesis, based on a previous study *(2)*, that "fast-recovery" males are capable of more ejaculations over an extended period of testing than are "slow-recovery" males was supported.

Further, the study has demonstrated that the persistence of sex drive after castration varies with genotype *within* a species. Genetic homozygosity was associated with a rapid loss of the ejaculatory reflex in the castrated males; heterozygosity, on the other hand, resulted in a retention of this reflex for a maximum of 60 days after castration. This finding raises a question concerning the accepted generalization of an inverse correlation between dependence on gonadal hormones and phylogenetic status. For example, it may be hypothesized that the differences in decline of sexual behavior between carnivores and rodents is due not to their phylogenetic status, but rather to the amount of heterozygosity in the samples. The carnivores studied have been mongrel cats and dogs, while the rodent groups have been selected from relatively inbred laboratory colonies. Heterozygosity is doubtless greater in the carnivore groups and this may have accounted for the slower decline in sexual behavior. Support for this hypothesis is found in a recent study by Bruell *(9)* who tested several inbred strains of mice and their hybrids in a running wheel. Hybrid vigor occured for all crosses, and the degree of hybrid vigor was found to be proportional to the amount of suspected heterozygosity.

Genetic differences may also account for the observation *(4, 11)* that mongrel cats fall into three different types on the basis of decline of sexual behavior after castration.

There is one previous study regarding the decline of sexual behavior in mice after castration which does not support the foregoing hypothesis. Champlin, Blight, and McGill *(10)* castrated hybrid males of the $CD2F_1$ strain (BALB/c females X DBA/2J males) and found that the ejaculatory reflex was lost within 1 week. The discrepancy between that study and ours may be due to (i) genetic differences (the $CD2F_1$ strain may remain homozygous at critical loci), (ii) maternal effects, or (iii) any of several procedural differences between the two studies.

The hypothesis that heterozygosity determines persistence of sexual behavior is testable. Should experiments support this hypothesis, the search for the underlying physiological differences will be greatly simplified as only one species need be used.

REFERENCES AND NOTES

1. T. E. McGill, Anat. Rec. **138**:367 (abstr.) (1960); Behaviour **19**:341 (1962).
2. ——— and W. C. Blight, J. Comp. Physiol. Psychol. **56**:887 (1963).
3. F. A. Beach, Hormones and Behavior (Hoeber, New York, 1948).
4. J. S. Rosenblatt and L. R. Aronson, Behaviour **12**:285 (1958).
5. F. A. Beach, Ciba Found. Colloq. Endocrinol. **3**:3 (1952).
6. ———, J. Exptl. Zool. **97**:249 (1944); F. A. Beach and R. S. Pauker, Endocrinology **45**:211 (1949); J. A. Grunt and W. C. Young, *ibid.* **51**:237 (1952).
7. The hormone preparations, Progynon B and Proluton, were generously supplied by Dr. R. Richard McCormick, Schering Corporation, Bloomfield, New Jersey.
8. "Number of ejaculators" was tested by "the significance of the difference between two independent proportions" test from G. A. Ferguson, Statistical Analysis in Psychology and Education (McGraw Hill, New York, 1959). The comparisons of C57BL with DBA and of C57BL with $B6D2F_1$ were significant ($p<0.0002$). The DBA males did not differ from $B6D2F_1$ males. The other three measures were analyzed using the Mann-Whitney U test from S. Siegel, Nonparametric Statistics for the Behavioral Sciences (McGraw-Hill, New York, 1956). The comparisons of C57BL with DBA and those of C57BL with $B6D2F_1$ showed significant differences, with p values ranging from 0.02 to 0.002. The comparisons of DBA with $B6D2F_1$ did not reveal significant differences. Two-tailed tests were used in all cases.
9. J. H. Bruell, J. Comp. Physiol. Psychol., in press.
10. A. K. Champlin, W. C. Blight, T. E. McGill, Animal Behaviour **11**:244 (1963).
11. L. R. Aronson, in Comparative Endocrinology, A. Gorbman, Ed. (Wiley, New York, 1959), chap. 5.

61 Genetic selection for voluntary alcohol consumption in the albino rat

Kalervo Eriksson
Research Laboratories
State Alcohol Monopoly
Helsinki, Finland

Abstract. *By outbreeding Wistar rats and selecting for breeding animals that differ in their alcohol consumption, we have raised two genetically different lines. Marked differences between the sexes and the strains were evident by the eighth generation. Selection is reflected in the regression coefficient .754, which accounts for 65.9 percent of the variance. The heritabilities differ significantly in the two sexes,* h^2 *for the males being .263, and for the females .371; this difference seems mainly ascribable to sex-linkage of some of the genetic factors controlling voluntary consumption of alcohol.*

The underlying assumption in experiments on voluntary selection of alcohol is that heritable biochemical reactions which control physiological mechanisms also to some extent regulate alcohol consumption. This hypothesis has been the basis of numerous animal experiments used to clarify the heritability of voluntary consumption of alcohol. Reed *(1)* investigated the selection of alcohol by six different inbred lines of rats, and, although the investigation was not designed for genetical research, the results reveal differences between the strains. Mardones *(2)* set out to raise two inbred rat strains that would differ in their alcohol habits; he found the differences to be genetically determined, although the heritability was fairly low. These strains differed relatively little, however, and both must be considered comparatively moderate consumers of alcohol; moreover, individual differences in alcohol consumption were also small, ranging from 0.02 to 0.72 ml/100 g of absolute alcohol. That the original stock was apparently very homogeneous probably accounts for the lack of differences correlated with alcohol consumption between the sexes, and between the strains *(3)*. Rodgers and McClearn *(4, 5)* have studied the voluntary consumption of alcohol by mouse strains inbred for long periods, and have associated the differences found with metabolic data. But the mouse strains used had been selected for traits other than alcohol consumption, and they differed from each other considerably in many other respects; thus interpretation of the results is difficult. Apparently for that reason, the authors avoided quantitative estimation of heritability and contented themselves with the observation that different mouse strains differ widely in the voluntary consumption of alcohol. Because several years of cross-breeding our Wistar rat strain resulted in a very wide range of individual variation and in a rather high average preference for alcohol *(6)*, we started to raise two rat strains which would differ genetically in their alcohol preference.

To avoid restriction of individual variation at a given level, we at first intentionally avoided mating sibs. The first four outbred generations constituted a single line, and from the fifth generation on, there was an alternation of sib-mating and outbreeding. The purpose of this slow selection was to secure the maximum possible difference between strains. Inasmuch as previous work *(1-5)* warranted the assumption that heritability is rather low, the breeding animals were chosen in the manner proposed by Falconer *(7)*, by selection within the family so that about 25 percent of the animals were used for continued breeding.

Every animal in each generation was tested as follows. After puberty (at 4 months) the animals were isolated in individual cages, where for 10 days the only available fluid was a 10 percent (by volume) alcohol solution. After this period, the animals were given access to water as well as to the alcohol solution for 4 weeks during which the mutual location of the tubes was changed once a week; the rats were provided free access to standard laboratory food. The quantities of water and alcohol solution consumed were measured daily, food consumption was reckoned once a week, and the rats were weighed weekly. As previously shown *(8)*, fluid intake may vary greatly in animals consuming the same amount of alcohol per unit of body weight. For this reason, alcohol intake per body weight was chosen as the phenotypic measure; this is reduced to milliliters of absolute alcohol per 100 g of body weight per day. Fuller *(9)* criticizes the use of only one alcohol concentration for determination of the phenotype because inbred mouse strains differ in their concentration preferences. According to our experience, rats consume the same amount of absolute alcohol when offered the choice between water and solutions of alcohol (5 to 15 percent by volume). Rick and Wilson *(10)* have reported the same observation. On this basis, one alcohol concentration can be used to establish the phenotype.

As breeding has been selective, estimation of the classic heritability value h^2 from the regression coefficient between the mean of parents and the values of the offspring (mid-parent/offspring regression) is misleading. In Fisher's opinion *(11)*, however, a value that measures the effectiveness of selection can be calculated in this way. The value derived from the regression of the offspring to their F_7 parents was .754±.15 and is highly significant ($t = 4.32$, $P < .001$). The model accounts for 65.9 percent of the variance in the data. A phenomenon to be noted in all generations tested so far is that females consume markedly more alcohol per unit of body weight than males do as has been reported *(6)*. Although the means of the sexes differ, the coefficients of variation are similar. To obtain a rough estimate of heritability, we made 14 F_2 crosses in a manner permitting calculation of the h^2 value. Comparison of the means of the F_1 generation with the means of the two parental strains provides evidence that the genetic factors are, for the most part, additive in their effect. The mid-parent/offspring regression for the males, .263±.06 ($t = 4.12$, $P < .005$), was lower than the corresponding value for the females, .371±.09 ($t = 3.40$, $P < .01$). The higher heritability value of the females seems mainly ascribable to sex-linked inheritance. Rodgers and McClearn *(4)* likewise reported that the influence of the females is greater. They were not able to demonstrate maternal effect when they removed the young; neither was I. That the sexes differ with respect to the rate of alcohol elimination seems at least partly to explain the difference in preferences although this difference only appears in strains where alcohol consumption is high.

In eight generations selection has resulted in a curve of variation, with four peaks, constituted by the females and the males of the two strains (Table 1). Furthermore, even with due allowance for sex, the two strains differ in weight, drinkers being lighter than nondrinkers. Mardones *(3)* observed a similar difference in his strains, although it was not statistically significant. The calorie intake of the drinkers is significantly higher than that of the nondrinkers. The same applies to fluid intake. The greater fluid intake is probably due to the diuretic effect of alcohol. And with the same standard food, greater consumption of food usually leads to higher fluid intake; this may reflect a metabolic difference between strains. With respect to weight, such a marked difference between the strains may also signify an inherited metabolic difference, which is associated with alcohol consumption *(6)*. The sexes differ in the rate of alcohol elimination, which is higher in females *(6)*. Subsequent experiments have indicated that these strains do not differ in the rate of alcohol elimination or in alcohol dehydrogenase activity. The difference between the sexes, which is conspicuous in the drinker strain, is not so apparent in the nondrinker strain. In both strains the difference between sexes is greater for alcohol consumption than for alcohol preference. The alcohol consumption as a percentage of total fluid intake has generally been adopted as a measure of preference. In my experiments, the regression of preference values of alcohol intake per unit weight was determined for the F_7 generation, of which there were 210 animals. The regression model accounted for as much as 94 percent of the variance, the regression coefficient being .85. The two measures show a high correlation, but the preference value is at a disadvantage, because it does not vary freely but depends on total fluid intake. From the same data, the correlations of other variables with alcohol consumption were also calculated. Because the four-peaked curve of variation does not permit analysis of the total data, I have chosen drinker males as an example (Table 2), especially because their range of intake was wide but the range of other variables, within individuals, was slight during the experimental period. Consequently, the possible association

Table 1. The average values and standard deviations of the functions measured during a 4-week period in drinker and nondrinker strains in the F_8 generation. Statistical values of t and P also shown.

Subjects	Number	Mean weight of rats		Absolute alcohol consumption (ml/100 g body weight per day)		Alcohol preference (% total fluids)		Total calories/ 100 g body weight per day		Total fluids (ml/100 g body weight per day)	
Nondrinkers, males	49	363.7±35.4		0.18±.13		22.4±16.5		17.6±1.3		8.4±1.1	
Nondrinkers, females	42	221.6±16.0		.29±.26		29.3±22.7		22.3±1.3		10.6±2.1	
Drinkers, males	59	329.6±29.7		.48±.25		51.2±24.6		19.0±1.3		9.0±1.1	
Drinkers, females	60	207.1±15.3		.97±.34		75.3±22.3		24.1±1.4		12.5±2.4	
		Statistical validity between groups									
	df	*t*	*P*	*t*	*P*	*t*	*P*	*t*	*P*	*t*	*P*
Nondrinkers, males–females	89	24.9	<.001	2.4	<.025	1.6	<.200	17.4	<.001	6.1	<.001
Drinkers, males–females	117	28.5	<.001	8.9	<.001	5.5	<.001	20.4	<.001	10.3	<.001
Males, drinkers–nondrinkers	106	5.9	<.001	7.9	<.001	7.2	<.001	5.6	<.001	2.9	<.005
Females, drinkers–nondrinkers	100	4.5	<.001	11.3	<.001	9.9	<.001	6.7	<.001	4.2	<.001

Table 2. The correlation coefficients *(r)* between functions measured in the males of the drinker strain in the F_7 generation. The limit value of the statistical significance at $P = .05$ is .25; at $P = .01$ it is .32. Functions measured are absolute alcohol and total fluids in milliliters per 100 g of body weight per day, total calories per 100 g of body weight per day, weight at beginning of period, and growth in grams.

Functions	1	2	3	4	5
1 Absolute alcohol	1.00				
2 Total fluids	.65	1.00			
3 Total calories	.43	.58	1.00		
4 Weight	− .19	− .26	− .43	1.00	
5 Growth	.34	− .17	.16	.24	1.00

of some variables with alcohol consumption is easier to discern. The results confirm the validity of the sex and interstrain differences (Table 1), in that calorie intake, fluid intake, and alcohol consumption constitute a positively correlated group of variables. In contrast, the interstrain difference in initial weight shows only a weak negative correlation with alcohol consumption, and the difference is possibly fortuitous.

Rodgers and McClearn *(4)* have adopted the working hypothesis that alcohol consumption is an additively inherited polygenic trait. However, they have not constructed a heritability model, although they have shown that females exert a greater effect on the offspring. Construction of a heritability model on the basis of the hypothesis of self-selection requires that the inheritance be additive, that there be no dominance, that the quantitative effects of the genetic units involved be approximately equal, and that there be no linkage. The experiments of Reed *(1)* reveal large deviation within inbred strains with regard to drinking behavior. These results suggest that the role of inheritance is slight, either because of an inadequate phenotypic measure or of a predominating role of the environment. The model constructed here shows the importance of genetic factors in determining selection of alcohol, although the determining power of the environment still seems stronger. The model will undoubtedly be improved by an elaboration of measuring techniques.

REFERENCES AND NOTES

1. J. G. Reed, Univ. Texas, Publ. No. 5109 (1951), p. 144.
2. J. Mardones R., N. Segovia M., A. Hederra D., Quart. J. Studies Alc. **14**:1 (1953).
3. J. Mardones R., Int. Rev. Neurobiol. **2**:41 (1960).
4. D. A. Rodgers and G. E. McClearn, in Roots of Behavior, E. L. Bliss, Ed. (Harper, New York, 1962), p. 68.
5. ———, Quart. J. Studies Alc. **23**:26 (1962); *ibid.* **25**:26 (1964).

6. K. Eriksson and K. K. Malmström, Ann. Med. Exp. Biol. Fenniae (Helsinki), in press.
7. D. S. Falconer, Introduction to Quantitative Genetics (Oliver and Boyd, London, 1964).
8. K. Eriksson, Nature **213**:316 (1967).
9. J. L. Fuller, J. Comp. Physiol. Psychol. **57**:85 (1964).
10. J. T. Rick and C. W. M. Wilson, Quart. J. Studies Alc. **27**:447 (1966).
11. R. A. Fischer, Statistical Methods for Research Workers (Oliver and Boyd, London, ed. 12, 1954).

62 Studies in experimental behavior genetics: I. The heritability of phototaxis in a population of Drosophila melanogaster

Jerry Hirsch*
James C. Boudreau
Columbia University
New York, New York
and
University of California
Berkeley, California

It has frequently been observed that individual differences (IDs) in behavior can be inherited; e.g., Tryon (11) has reported on the inheritance of maze-learning ability, and Kallmann and Baroff (5) on the inheritance of behavior pathologies. The present paper extends the study of the inheritance of IDs in behavior to a part of the phylogenetic series at which experimental behavior genetic (BG) analysis is feasible, viz., the genus *Drosophila*. The behavior chosen for BG analysis is the reaction to light, phototaxis an apparently innate or unconditioned response. Taxes have the advantage of representing relatively constant S-R relationships: the repeated presentation of a single stimulus value appears to elicit, depending on the method of measurement, either a characteristic response or a characteristic probability of response. Both the characteristics of the response and the probability of the response have been shown to vary as a function of two parameters, the value of the stimulus presented and the strain of organisms stimulated (1, 8).

Brown and Hall have measured strain differences in phototaxis. The immediate purpose of the present study is to measure IDs in, and to estimate the heritability of, phototaxis within a single strain. (Roughly, "heritability," h^2, refers to that portion of the total variance due to additive genetic causes [6, p. 111].) This is one of several studies of *Drosophila* behavior in which an experimental attack is being made on the long unresolved question of whether abilities are under the control of one general factor (9) or many specific factors (10).

At present three *Drosophila* behaviors are under study: phototaxis, geotaxis (3), and eating rate (2).

From Journal of Comparative and Physiological Psychology **51**:647-651, 1958.

This work was initiated during the senior author's tenure as a National Science Foundation postdoctoral fellow at the University of California, Berkeley, and completed under Grant No. G-3846 from the National Science Foundation.

The authors gratefully acknowledge their indebtedness to Curt Stern and Th. Dobzhansky for generously supplying many useful facilities.

*Present address: Department of Psychology, University of Illinois, Urbana, Illinois.

METHOD

Experimental design

Individual differences in phototaxis were measured in a Y maze by the method of mass screening (4). The measurements consisted of ten mass-screening trials, which in the foundation population had the reliability, $r_{tt} = 0.673$ (4, Formula 6).

Selection pressure was applied, and a system of assortative mating was used, i.e., the highest-scoring animals within the high strain and the lowest-scoring animals within the low strain were bred together, respectively.

Behavioral analysis

Apparatus. Individual differences in the approach to light were studied in the Y maze shown in Figure 1. The maze consisted of three 5-in. lengths of acrylic tubing, *f, e, d,* having 1/2-in. inside diameter. These tubes were attached to a Y joint, *a,* having 1/2-in. outside diameter and 3/8-in. inside diameter. Tubes *d* and *e,* which served as the starting path and the lighted arm of the Y, respectively, were attached to the center unit, *a,* by plastic sleeves, *b* and *c.* Both sleeves were fitted with sliding plastic "doors" 0.02-in. thick to prevent premature approach to the choice point and retracing after a choice.

All parts of the maze were painted black on the outside with the exception of the starting tube, *d,* and the lighted tube, *e.* To eliminate reflections, the cotton plunger in the starting tube was dyed black. The attrahent was light reflected from cotton at the end of Arm *e.* Although the same tube (Arm *e*) was always illuminated, the illuminated side was varied after each block of two trials (tests have shown that the tubes themselves do not act as stimuli). This was accomplished by rotating the entire front section of

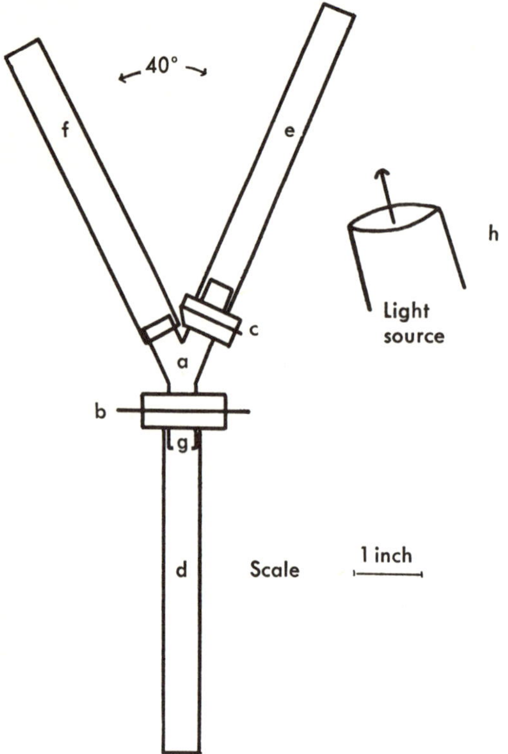

Fig. 1. Apparatus for measuring the reaction of *Drosphila* to light.

the maze 180° on the Y joint and shifting the light to the other side. A microscope light was placed at *h* and focused on the cotton at the end of *e*. The distance from the light source to the cotton was 6 in. The illumination at this distance is 100 ft-c., as measured by a Weston illumination meter, Model 756. Thus, the stimulating source of light was indirect. This was necessary because in preliminary studies using a direct source, i.e., a light shining through *e* to *d*, it was evident that secondary reflections were being set up in the starting arm, *d*, and that these constituted competing attrahents with the result that many flies never left the starting tube.

Procedure. The sexes were run separately on successive days. Each generation the males were run one day after hatching and the females two days after hatching. The role of heredity in determining IDs in behavior was assessed by the response of the population to selection. Virgin females are necessary for selective breeding. Since *melanogaster* females remain virgin for only the first 7 to 8 hr. after hatching, the cultures were cleared of all flies the day before an experiment, and only those flies that hatched in the following 7 hr. were tested. Occasionally, to obtain larger samples, flies were collected over two consecutive 7-hr. periods. Hence, the maximum age difference among the animals never exceeded 14 hr.

The behavior experiment consisted of introducing a group of flies into the starting tube of the Y maze and inserting a plug of black cotton behind them to seal the tube. The cotton was immediately pushed forward to within 0.5-in. of the door to the choice point so that all flies would be in the vicinity of the choice point at the start of a trial.

A trial lasted 30 sec. Both doors were opened at the beginning of a trial, and 30 sec. later the door to the illuminated arm of the Y was closed. In this way, on each trial the flies that approached the light were separated from the others. That is, on Trial 1 the initial group of flies was separated into two pass-fail subgroups. On Trial 2 both the pass group and the fail group from Trial 1 were tested and were in turn subdivided into pass-fail subgroups. In the method of mass screening, the subgroups obtained on one trial are retested separately on the next trial and further subdivided. A complete account of the method is given elsewhere (4). In this experiment ten mass-screening trials were used; therefore, the distribution of final scores ranges from 0 through 10.

Dyes were added to the medium on which the flies were raised. Bismark brown was used for the low strain and Nile blue for the high strain (tests have shown that reversing the colors does not affect the behavior). The colors were ingested along with the food. Since the females of the high and low strains could be distinguished by the colors they had absorbed, they were run in the apparatus together. It was necessary, however, to run the males in separate groups, since they did not show the colors clearly.

Genetic analysis

Subjects. The *S*s were 3,424 fruit flies, *Drosophila melanogaster*, Formosa wild type. The initial sample of animals was obtained from regular stocks in the genetics laboratory of the zoology department of the University of California, Berkeley. The flies were raised on standard *Drosophila* medium (to which color had been added) in 1/2-pt. culture bottles at 25°C.

Mating system. Selection pressure of variable intensity was applied under a system of restricted assortative mating. Animals were mated on the basis of phenotypic merit without regard to family relationship. In the foundation population animals with similar extreme phototactic scores were mated. In all filial generations the same selection criteria were applied with the further restriction that matings were always within and never between the two strains established by selection from the foundation population. Thus, the high and low strains were reproductively isolated, and inbreeding undoubtedly increased down through the generations.

If a sufficient number of animals received extreme scores, i.e., 0 or 10, only members of these classes were chosen for breeding. If not, individuals in adjacent classes were also used for breeding. The intensity of selection pressure increased because the percentage of animals receiving extreme scores increased as selection progressed. Selection was carried on over 29 generations with the exception of Generations 10, 11, 12, and 13, when mass mating was permitted (within the separate strains) and no behavior tests were made.[1]

---

[1] As the medium on which to establish the basic principles of BG, *Drosophila* is thus an animal quite superior to the laboratory mammals such as mice, rats, or guinea pigs. Compare the present results, obtained in about a year, with the Tryon study (11), which required over 15 years.

RESULTS

Male and female data have been combined except for Generations 7, 8, and 9, for which female data alone are presented. In these generations the males were given only enough test trials to identify the extreme scorers for breeding. Data are not available for Generations 10 through 13 or for Generation 16, when the apparatus broke.

Figure 2 presents the percentage of trials on which the light was approached by the high and low strains over 29 generations of selection. Clearly, there is an early response to selection, and despite fluctuations the expected values of the selected strains show progressive divergence from the foundation population value of 51.6% to asymptotic values of approximately 80.0% for the high strain and 15.0% for the low strain.

Figure 3 presents the distribution of phototactic scores for the foundation population and for filial Generations 1, 2, 7, and 29. Inspection of the figure reveals that selection effects marked changes in dispersion as well as in central tendency. The changes in dispersion are shown in Figure 4, where the ratio of the variance of each selected generation to that of the foundation population is plotted.

It has been predicted that the limits of selective breeding would depend upon the reliability of the ID measurements in the foundation population (4, p. 410), i.e., when the variance in the selected lines decreases to the size of the variance error of measurement in the foundation population, further selection should be ineffective, since at that point the method of observation is no longer discriminating among individuals. In the present study the foundation population measurements have a reliability $r_{tt} = 0.673$, a variance $\sigma_0{}^2 = 6.38$, and a variance error of measurement $\sigma_e{}^2 = \sigma_0{}^2 (1 - r_{tt}) = 6.38 (1 - 0.673) = 2.09$ (4, Formula 8). Hence the ratio of the variance of the selected strains to that of the foundation population should approach the asymptote:

$$\frac{\sigma_e{}^2}{\sigma_0{}^2} = 2.09/6.38 = 0.327$$

In Figure 4 it can be seen that the variance ratio for the high photo strain appears to be settling down near the line 0.327. The variance ratio for the low strain, however, has not stabilized enough yet to determine whether it is approaching the predicted asymptote.

Next, let us examine the extent to which IDs

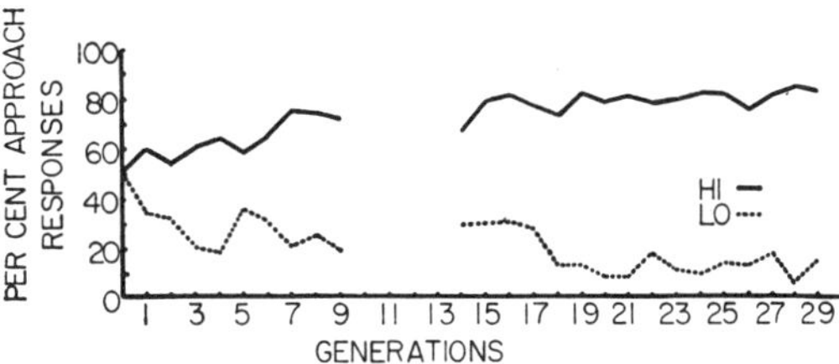

Fig. 2. Percentage of trials on which light was approached per generation.

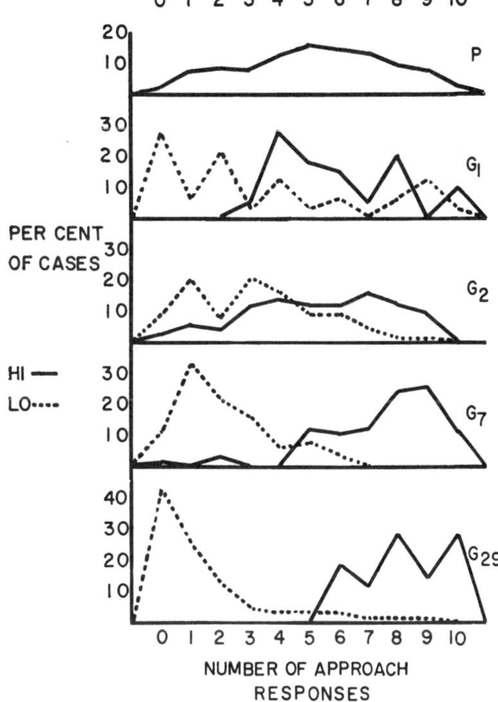

Fig. 3. Distribution of light-approach scores for Generations 0, 1, 2, 7, and 29.

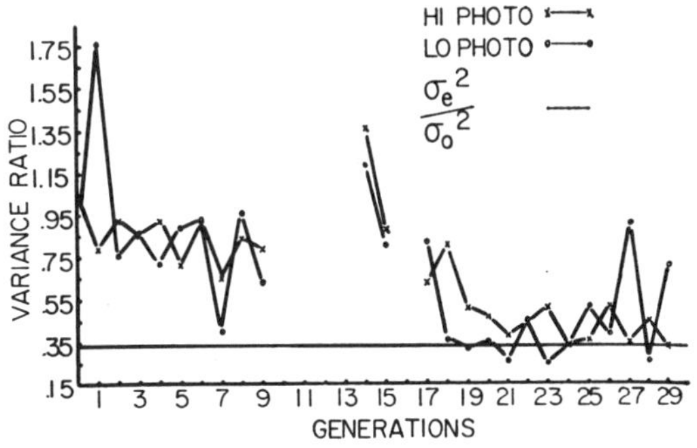

Fig. 4. The ratio of the variance of each selected generation to the variance of the foundation population, $\sigma_i^2/\sigma_0{}^2$, for the "high" and "low" phototactic strains. $\sigma_e{}^2/\sigma_0{}^2$ is the asymptote predicted from the reliability of the foundation population scores (see text).

in phototaxis are genetically determined. If it is assumed that the average of the variances of the two selected strains over Generations 28 and 29 represents an upper limit to the variability to be expected in an isogenic line, then we have available a conservative estimate of the heritability, h^2, of phototaxis in the foundation population under the present experimental conditions.

> All methods of estimating heritability rest on measuring how much more closely animals with similar genotypes resemble each other than less closely related animals do. . . .Variation within isogenic lines is wholly environmental. Comparing this with the variation in an otherwise similar random breeding population may give an estimate of heritability (7, p. 92).

For the high strain $\sigma_{28}^2 = 2.88$ and $\sigma_{29}^2 = 2.08$, for the low strain $\sigma_{28}^2 = 1.63$ and $\sigma_{29}^2 = 4.51$. The average of these four variances is $\overline{\sigma^2} = 2.77$. Hence: $h^2 = ({\sigma_0}^2 - \overline{\sigma^2})/{\sigma_0}^2 = (6.38 - 2.77)/6.38 = 0.566$, i.e., at least 57% of the phenotypic variance is genetic variance. This is a conservative estimate because the value of the reliability coefficient sets an upper limit to the values an estimate may take (as calculated, h^2 is 84% of the reliable phenotypic variance). Furthermore, h^2 contains only the additive portions of the genetic variance; it does not include variance due to dominance or to epistasis (i.e., dominance of nonallelic genes).

No estimate can be made of the heredity-environment interaction because only a single stimulus condition has been employed.

DISCUSSION

The aim of the present experiment has been both exploratory and descriptive. Its purpose has been to examine the possibility of studying IDs in behavior and their genetic bases in a species on which detailed genetic analysis can be performed.

The results which have been reported indicate that the study of *Drosophila* behavior is quite feasible, that IDs in performance can be measured in groups both reliably and efficiently by the method of mass screening, and that the ID's variance contains a large genetic component to which the techniques of experimental genetics may now be applied.

Since the present data have been obtained with a laboratory stock considered to be rather inbred[2] and therefore not very heterogeneous genetically, it is to be expected that a larger genetic variance would be found in a less inbred natural population. These findings have implications for psychological theory.

Theory testing in psychology is usually done on human *S*s or on laboratory strains of animals which, it is reasonable to assume, are genetically much less alike than our *Drosophila.* (The Formosa stock has been maintained in the laboratory in small cultures for more than 20 years. Inbreeding has, therefore, had over 700 generations in which to exercise its homogenizing influence. Within the same period of time laboratory strains of rats would have completed about 70 generations. Furthermore, *Drosophila melanogaster* have only 4 independently assorting pairs of chromosomes whereas rats have 21 and human beings at least 23.) If large genetic differences do exist in the populations now being studied by psychologists, it should be of interest to determine in what ways stimulus control of behavior depends upon the genotype under stimulation.

SUMMARY

Behavior genetic analysis of the unconditioned response, phototaxis, has been carried through several steps: Individual differences in phototaxis have been measured reliably and efficiently in a *Drosophila* population by the method of mass screening. The genetic determination of individual differences in behavior has been demonstrated by the response to selection, and the heritability has been estimated to be more than one-half the phenotypic variance.

REFERENCES

1. Brown, F. A., & Hall, V. A. The directive influence of light upon *Drosophila melanogaster* Meig and some of its eye mutants. J. exp. Zool., 1936, **74**:205-220.
2. Durkin, R. D. Eating behavior of *Drosophila melanogaster.* Unpublished honors study, Psychol. Dept., Columbia Univer., 1957.
3. Hirsch, J. Behavior genetic studies of individual differences in *Drosophila melanogaster.* Amer. Psychologist, 1956, **11**:450-451. (Abstract)
4. Hirsch, J., & Tryon, R. C. Mass screening and reliable individual measurement in the experimental behavior genetics of lower organisms. Psychol. Bull., 1956, pp. 402-410.
5. Kallmann, F. J., and Baroff, G. S. Abnormal psychology. Ann. Rev. Psychol., 1956, **6**:297-326.
6. Lerner, I. M. Population genetics and animal improvement. Cambridge: Cambridge Univer. Press, 1950.
7. Lush, J. L. Animal breeding plans. Ames, Iowa: Collegiate Press, 1945.

[2] Personal communication from Th. Dobzhansky.

8. Scott, J. P. Effects of single genes on the behavior of *Drosophila.* Amer. Naturalist, 1943, **77**:184-190.
9. Spearman, C. The abilities of man. New York: Macmillan, 1927.
10. Tryon, R. C. A theory of psychological components—an alternative to 'mathematical factors.' Psychol. Rev., 1935, **42**:425-454.
11. Tryon, R. C. Genetic differences in maze-learning ability in rats. In Yearb. nat. Soc. Stud. Educ., 1940, **39**(I):111-119.

63 RNA metabolism in goldfish brain during acquisition of new behavioral patterns

Victor E. Shashoua*
Department of Biology
Massachusetts Institute of Technology
Cambridge, Massachusetts

Abstract. *Base composition measurements were used as a criterion of RNA changes in goldfish brain. RNA synthesized during the acquisition of new swimming skills was found to have a uridine/cytidine ratio 20-80 per cent higher than that of RNA formed under non-learning conditions. A variety of behavioral situations were used to demonstrate that these RNA changes were not due to such factors as intense physical exertion, passive responses to stress, or the intense electrical activity of the brain and convulsive behavior produced by KCl injections. The RNA changes were produced by two behavioral situations; (1) during the process of acquiring new swimming skills, and (2) as a result of attempts to master an impossible task. The results suggest that the modified RNA synthesis taking place during the acquisition of new behavioral patterns is probably not specific as to the particular information content being stored, but may be required for the consolidation step of new information storage.*

INTRODUCTION

The attempts to correlate brain RNA synthesis with the acquisition of new behavior patterns have led to two types of observations: (1) a general increase in RNA synthesis,[1-3] and (2) the formation of new RNA species with a changed base composition.[4-7] In a previous paper,[6] it was suggested that the synthesis of an RNA with an altered base composition was characteristic of the information-gathering state of goldfish brain. This type of RNA appeared to be required for the consolidation of learned behavior but not for the repair of effects due to stress or to a generalized activity of the animals. When protein synthesis was inhibited by puromycin, the goldfish could acquire new swimming skills just as rapidly as in the absence of drug but could not remember the task 22 hours later. Under these conditions no decrease in brain RNA synthesis was observed and no alteration in base composition occurred. From these observations, a changed base composition was thought to be an index of the state of brain function during the acquisition phase of learning. It was, however, noted that the use of drugs to dissociate stress and work effects from those due to learning is subject to a number of difficulties, including observations that hippocampal seizures are produced by puromycin[8] and the possible occurrence of anoxia arising from the reported decrease in respiration of nervous tissue in the presence of puromycin and cycloheximide.[9] In the present study, the above features of the behavioral experiment were examined in detail and nonpharmacological methods were devised to dissociate the information-gathering state of goldfish brain from other states such as stress, intense work, or electrical activity. In all the experiments to be described, labeled orotic acid was used as the precursor of UTP and CTP for the synthesis of RNA in goldfish brain during the acquisition of new behavioral patterns.

EXPERIMENTAL METHODS

Behavioral experiments. Four types of behavioral experiments were used in this work. The main study was an experiment in which goldfish acquired a new swimming skill in a 4-hr training period as previously described.[6] Groups of seven experimental and seven control animals (7-8 cm long) received intraventricular injections of labeled orotic acid (10 μl, 100 μCi orotic acid 5-H^3 spec. act. 25 Ci/mM). Foamed polystyrene floats were sutured at the ventral surface of the

From Proceedings of the National Academy of Sciences **65**:160-167, 1970. Used with permission.

This work was initiated during tenure of a resident scientist appointment at the MIT Neurosciences Research Program.

This research was supported by grants from the National Institutes of Health.

The author wishes to thank Professor Alexander Rich for providing the opportunity to complete these experiments.

*Present address: McLean Hospital, Belmont, Massachusetts.

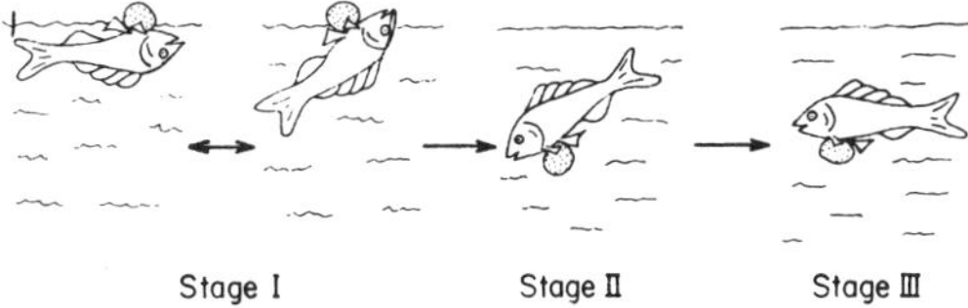

Fig. 1. Behavioral experiment—stages in the adaptation of a goldfish to the float.

experimental animals 1 mm caudal to the first pair of lateral fins; the control group had a suture at the same position but no float. Each group was placed in a separate 5-gal tank filled with 3 gal of water and equipped with an air bubbler and filter.

Figure 1 illustrates the sequence of adaptation of the experimental animals to the task. In stage I, the animals are upside down. This is followed by a period of struggling during which the animals train to swim at a 45° angle in an upright position to achieve stage II of the task. This usually requires 1 hr and is followed by a varying time interval of up to a total of 4 hr for the animals to assume stage III (horizontal swimming). After the 4-hr training period, the animals were decapitated and each group of brains was washed with distilled water and pooled for use in the RNA extraction procedure. Not all the groups of animals can adapt as rapidly to the standard size float of 0.7-cm cube. The time to reach stage III was noted for each experimental group, the training time was determined by assuming that stages II and III represent the 50 and 100% trained levels, respectively. At the end of a 4-hr period, the swimming behavior of each animal was observed and classified as types I, II, or III. The total score of the group was expressed as a per cent trained level. The "per cent trained" level was converted to a training time by assuming that 100% is equivalent to 240 min.

In previous studies[6] it was demonstrated that there is some retention of the acquired swimming skills for at least 22 hr. Thus, a group of ten animals which required 156 min to learn the task in trial 1 were found to relearn the task in 21 min in trial 2 administered 22 hr later.

Two variations (behavioral experiments 2 and 3) of the standard float-training experiment were used in this paper. The first was to use a very small float (0.3-cm cube). This provided a minimal challenge to the animals. They generally were able to attain stage III of the experiment within 5 min but had to continue swimming throughout the 4-hr experimental period in order to maintain a horizontal posture. This experiment was used as one of the controls for the effect of physical exertion on the brain RNA metabolism.

The second variation of the float-training experiment was to use a very large float (a 1.7-cm cube) in order to provide an impossible task. In this behavioral experiment the animals struggled against the float but were unable to avoid the upside-down position. Some animals gave up trying immediately, whereas others struggled for varying lengths of time. The time of struggling was recorded for each group studied. The total duration of this experiment was also 4 hr. This experiment was used as one of the controls for a generalized stress.

In the fourth behavioral experiment an intense exercise situation was created by placing the animals in a whirlpool. Groups of seven animals, after the intraventricular injection of the labeled orotic acid, were placed in one compartment of a 5-gallon tank containing 3 gallons of water. The tank was partitioned by a stainless steel screen into two compartments: one for a high-speed stirrer and one for the animals. The speed of the stirrer was adjusted so that the whirlpool created would pull the animals against the screen. The water flow was such that the animal could escape the situation by swimming vigorously. In this setting the animals immediately swam against the current to the position of minimum turbulence in the tank and continued swimming for the 4-hr duration of the experiment. This experiment was designed to require a large amount of physical work with some stress, and with essentially no training component.

Biochemical experiments. The pooled brains from seven animals in each group of the behavioral experiments were placed in ice-cold BSB buffer (4 ml) in a homogenizer. The BSB buffer contained 0.01 *M* KCl, 0.01 *M* $MgCl_2$, 0.01 *M* Tris, pH 7.9, and 50 mg hydrocortisone per liter. For isolation of total brain RNA content, 0.2 ml of 10% sodium dodecyl sulfate (SDS) was added to give a 0.5% SDS content prior to homogenization; 7 strokers in a Teflon-glass homogenizer with a 175 μ gap. The salt content of the homogenate was adjusted to 0.5 *M* with 2 *M* NaCl and the mixture was extracted with hot phenol (60°C) and purified according to published methods.[10]

For extraction of cytoplasmic RNA the homogenization was carried out in BSB buffer with 0.5 per cent tween 40 as the detergent and only four strokes were used with the Teflon-glass homogenizer at 4°C. The phenol extraction was performed at room temperature after adjusting the SDS content of the supernatant to 1 per cent and after removal of nuclear and membrane fragments by centrifugation at 5000 rpm in a Sorval for two 15-minute periods at 4°C. In both total RNA and cytoplasmic RNA preparations, the final supernatants after the phenol extractions were treated with 0.5 ml of 0.4 *M* EDTA in BSB buffer followed by addition of 2 vol of ethanol and cooling at −18°C for 24 hours to precipitate the RNA. The precipitate was isolated by centrifugation, washed twice with ethanol, and dried *in vacuo.*

For base composition analysis it was essential to further purify the RNA from adsorbed nucleotides. The RNA samples were dissolved in 2 ml of distilled water and a 0.2-ml aliquot was removed for total RNA content and total radioactivity measurements. The remaining sample was cooled to 4°C and mixed with 3 vol of 20 per cent trichloroacetic acid to precipitate the RNA. A small amount of yeast RNA was added as carrier. After standing for 30 min at 0°C, the RNA was isolated by centrifugation at 4°C. This was next dissolved in one drop of ammonia and immediately reprecipitated with cold trichloroacetic acid and centrifuged. The precipitate was washed three times with 20 per cent trichloroacetic acid and then with ethanol and finally with ether to give a dry RNA product free of adsorbed nucleotides. The purified RNA was then hydrolyzed by heating in 0.15 ml of 0.3 *M* KOH at 37°C for 22 hours. The product was purified by ion-exchange chromatography and analyzed by paper electrophoresis.[11]

Table 1. Correlation of training times with base composition changes

Expt. no.	Training* time (min)	*Up* (cpm)	*Cp* (cpm)	*U/C*	$(U/C)_E$ – $(U/C)_C$	Change (%)
1*E*	35†	416	90	4.6	1.2	+35
1*C*	–	420	124	3.4		
2*E*	40†	1474	258	5.7	1.1	+23
2*C*	–	384	83	4.6		
3*E*	125	5109	935	5.5	1.7	+45
3*C*	–	917	242	3.8		
4*E*	120	1825	242	7.6	2.0	+36
4*C*	–	3917	702	5.6		
5*E*	220	560	72	7.8	3.2	+70
5*C*	–	618	135	4.6		
6*E*	200	490	65	7.6	3.3	+78
6*C*	–	286	67	4.3		

Note: Each experiment had seven pooled brains for experimental *(E)* and control *(C)* groups. The total labeling time was 4 hr.

*Time to achieve stage III, 100% trained level.

†These experiments represent trial 2 training times (trial 1 was 24 hr earlier without label and required 180 and 220 min, respectively).

For the analysis of the metabolic pools the brains were homogenized in 0.4 *M* perchloric acid at 0°C. After removal of the precipitate by centrifugation, the supernatant containing the nucleotide pools was neutralized with 0.4 *M* KOH and cooled for 30 minutes at 0°C to precipitate potassium perchlorate. This step was repeated several times till no further salt was obtained on cooling. The supernatant was then heated at 100°C in 1 *N* HCl for 30 minutes to convert the nucleic acids to the monophosphates. The product was then dried *in vacuo,* dissolved in distilled water, and analyzed by paper electrophoresis.[11]

The RNA samples from goldfish kidneys were prepared by the same methods used for brain RNA.

RESULTS

Table 1 summarizes the base composition measurements as a function of training time for all animals in a given group to reach stage III, i.e., swimming in a horizontal position with the float on the ventral side (see Fig. 1). Some groups of animals learned faster than others. The results show that the changes in $\Delta(U/C)$ i.e., $((U/C)_E - (U/C)_C)$, are substantial ranging from an increase of 23 to 78 per cent. One interesting feature of these results is that the animals which seem to learn the slowest show the largest changes in $\Delta(U/C)$ values, i.e., a $\Delta(U/C)$ of 3 for experiment 6 compared with 1.1 for experiment 2. The results suggest that the magnitudes of the RNA changes are a function of the time required by the animals to acquire the new swimming skills necessary to negotiate the task, rather than the 4 hours spent in the behavioral situation. In addition, the RNA changes seem to be a transitory phenomenon (trial 2 data show smaller $\Delta(U/C)$ changes, see Table 1, expts. 1 and 2). In one experiment where the float training was performed for 24 hours, $\Delta(U/C)$ was zero. This was substantiated by an analysis of the sucrose density gradient pattern of RNA synthesis in goldfish brain. As shown in Figure 2, a four-hour pulse time results in the appearance of a newly-synthesized RNA with maximum labeling at regions between 18*S* and 4*S*, i.e., species not destined for use in the assembly of new ribosomal particles or for tRNA. After 24 hours labeling, however, all these RNA species are metabolized and the label appears at three peaks, 28*S*, 18*S*, and 4*S*, characteristic of ribosomal and transfer RNA components.

Table 2. Comparison of brain and kidney

	Brain		Kidney	
No.	$\Delta(U/C)$	Change (%)	$\Delta(U/C)$	Change (%)
7	1.3	80	0.1	2.8
3	1.7	45	0.6	9

Note: $\Delta(U/C) = (U/C)_E - (U/C)_C$; each group had 14 animals.

Table 2 demonstrates that the $\Delta(U/C)$ changes are specific for brain tissue and do not appear in the kidneys of the same group of animals. The changes in $\Delta(U/C)$ values in kidney were 2.8 and 9 per cent as compared to 80 and 45 per cent in the brains of the same animals.

Two types of behavioral situations were used to study the effect of physical exercise on brain RNA metabolism. Experiments 8*W* and 8*C* in

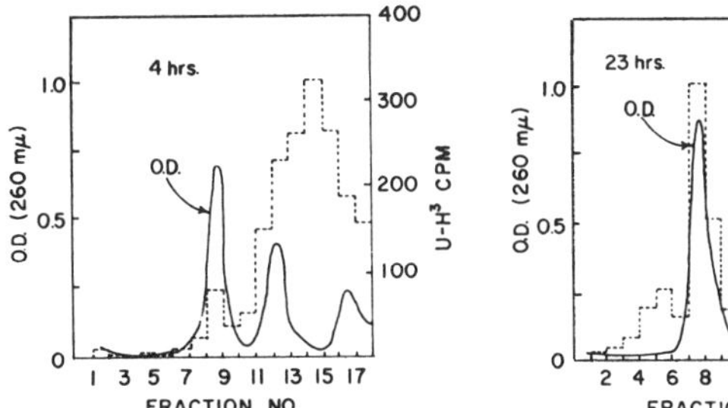

Fig. 2. Sucrose gradient density analysis of cytoplasmic RNA from goldfish brain. Labeling times—4 and 23 hr with uridine-H^3 (centrifugation was performed at 19°C for 17 hr in a 15 to 30% sucrose gradient in BSB buffer + 0.5% SDS, pH 7.4).

Table 3. Control experiments

Expt. no.	Type*	Behavior	U/C	$\Delta (U/C)$	Change (%)
8*W*	Small float	Constant swimming	3.7	0.1	2.9
8*C*	–		3.6		
3*W*	Whirlpool	Constant swimming	3.9	0.1	2.6
3*C*	–		3.8		
9*W*	Whirlpool	Constant swimming	3.5	0.1	2.9
9*C*	–		3.4		
10*K*	KCl (10 μl/0.25 *M*)	Convulsive, disoriented wild swimming	4.3	0.1	2.5
10*C*	–		4.2		
11*S*	Large float	Passive–upside down	4.6	0.1	2
11*C*	–		4.5		
11*E*	Normal float	Learned task (100 min)	6.3	1.7	38
12*S*	Large float	Struggled for 1 hr	5.8	0.8	16
12*C*	–		5.0		
3*S*	Large float	Struggled for 4 hr	7.5	3.7	98
3*C*	–		3.8		
3*E*	Normal float	Learned task (125 min)	5.5	1.7	45

Each group had seven pooled animal brains.
*Small float = $(0.3\text{ cm})^3$; normal float = $(0.7\text{ cm})^3$; large float = $(1.7\text{ cm})^3$.
W = work, *S* = stress, *K* = KCl, *E* = exp. learning group, *C* = control group.

Table 4. Metabolic pool measurements

	Counts Per Min					$\frac{U_P + U}{C_P + C}$
No.*	Orotic acid	U_P	C_P	U	C	
13*E*	45	1447	159	35	18	8.4
13*C*	45	1600	149	32	24	9.4
14*E*	1420	1188	367	4242	229	9.1
14*C*	639	1566	335	4671	192	11.8

*Each group had seven pooled animal brains.

Table 3 show the results for the use of a small float (a 0.3-cm cube) on the U/C measurements. With this float the animals mastered the task within five minutes but continued to swim at stage III for four hours. The results show a 2.9 per cent increase in the $\Delta(U/C)$ values indicating that, in this situation, work has little influence on the results. The second type of behavioral controls for work is the whirlpool experiments (3*W*, 3*C*, and 9*W* and 9*C*) in Table 3. Here the animals swam continuously against the current in a whirlpool throughout a 4-hour time period. This experiment appears more stressful than the small float situation, but in spite of this, the per cent changes in the $\Delta(U/C)$ are not significant (2.6 and 2.9 per cent, respectively), indicating that physical work is not a major factor in the observed RNA changes.

The effect of massive electrical activity on

brain RNA metabolism was examined in experiment 10*K* and 10*C* in Table 3. Here a dose of KCl was injected into the ventricles of seven experimental animals to produce convulsive seizures. The goldfish swam vigorously in a disoriented and wild manner through most of the four hour time period. A comparison of the *U/C* values of the RNA for these animals with controls showed no significant changes (i.e., 2.5 per cent) indicating that the type of massive electrical activity elicited by KCl does not result in base composition changes.

Table 4 shows the measurements of the metabolic pools for two groups of experimental animals 13*E* and 14*E* in comparison with control groups 13*C* and 14*C*. In both cases, radioactive orotic acid was found in the pools indicating that the labeled precursor and products may be at equilibrium. The $U_p + U/C_p + C$ values are quite different from the values obtained in comparable experiments where the labeled RNA was isolated and measured. Since it is not possible to measure the specific activity of the pools due to the presence of many interfering compounds which adsorb at the same wavelengths as the nucleotides, the pool measurements can not rule out pool changes. Moreover, the *U/C* values for the pools are in the reverse order to the findings for the RNA. In addition, the fact that deliberate alteration of the metabolic pools by feeding the animals at different schedules[7] did not eliminate the increase in $(U/C)_E - (U/C)_C$ results suggests that the RNA changes are probably not due to pool changes. Finally, the observation that the $\Delta(U/C)$ values reach a maximum at 4 to 8 hours and then become zero at 24 hours when the RNA labeling profiles (Fig. 2) become identical to the optical density profiles, likewise argues against the possibility that pool changes could be a major factor in the observed RNA changes.

The metabolic pool measurements give an average ratio of about 10/1 for the *U/C* values. This high bias in the labeling of uridine suggests that the pool amplifies the *U/C* ratio of the newly-synthesized RNA. The amplification effect may account for the large measured change produced by the behavioral challenge. On the basis of this finding it seems likely that the actual RNA base compositions must have a lower uridine than cytidine in the control experiments and that this ratio changes toward a more equal uridine to cytidine distribution under conditions of learning to master a new task.

The studies of the effects of stress on the brain RNA metabolism are summarized in Table 3 experiments 11*S*, 12*S*, and 3*S*. In each case a large float (a 1.7-cm cube) was sutured to the ventral surface of the experimental animals. This was an impossible task and the goldfish could either remain passive in an upside-down position or struggle unsuccessfully to attain a horizontal normal swimming posture. In experiment 11*S*, the animals struggled for a few minutes and quickly gave up to remain passive in an upside-down position throughout the 4-hr labeling period. In this case, the observed $\Delta(U/C)$ values showed no significant change (2 per cent) when compared with the 38 per cent change found in a parallel group (11*E*) of animals with normal size floats. In experiment 12*S*, however, the animals struggled for about one hour before giving up the task for the remainder of the 4-hour time. In this case there was a 16 per cent change in the $\Delta(U/C)$ values. In experiment 3*S*, the animals continued to struggle for a horizontal position throughout the 4-hour duration of the experiment. Here the $\Delta(U/C)$ values were quite large (98 per cent). A parallel group of animals with standard floats (0.7-cm cube) mastered the task in 100 minutes and showed a smaller change of 45 per cent. The results demonstrate that the struggling behavior or the attempt to learn the task produces changes in RNA composition similar to those observed in the float-training experiments. In both cases a change in $\Delta(U/C)$ is observed when there is a positive response to the behavioral situation, i.e., a struggling toward the preferred swimming posture. Since other forms of stress which accompany the convulsive behavior and intense work situation in the whirlpool experiment do not produce comparable RNA changes, the observed increases in the $\Delta(U/C)$ values can not be considered to be exclusively due to stress.

DISCUSSION

There are a number of explanations which can be suggested to account for the above results. One possibility is that both successful and unsuccessful attempts to master a given problem can result in a common biochemical demand on brain metabolism. The $\Delta(U/C)$ measurements could therefore be an index of the change produced by such a biochemical demand signal. The results described demonstrate that the metabolic signal resulting in the $\Delta(U/C)$ changes in the RNA is related to a limited number of behavioral situations, which do not include intense physical exertion, massive electrical activity of the nervous system, passive behavior, or the stress associated with convulsive seizures. Moreover, the RNA changes seem to be specific to brain; at any rate they do not occur in the kidneys of the goldfish. The observations suggest that the response of the animals, in situations where learning occurs–including the learning involved in attempting to perform an impossible task–can act as a trigger signal for the synthesis of a new RNA with the different base composition.

The biochemical demand signal for this RNA, for example, might arise from an extremely rapid rate of consumption of transmitter substances to a level beyond the set point of the system. This could act as a trigger signal to derepress a DNA gene or genes to elicit the synthesis of an RNA molecule coding for a critical enzyme or set of enzymes for the biosynthesis of transmitter substances. Alternatively the biochemical demand signal might be for an increase in the supply of a critical protein or proteins required for alteration or repair of existing membrane components or for growth of new membrane structures. The findings in this paper suggest that the changes in RNA synthesis in goldfish brain, during the acquisition of new behavioral patterns, are not specific with respect to the information content being stored but may be required for the consolidation step of information storage.

REFERENCES

1. For a general review see: Glassman, E., Ann. Rev. Biochemistry **38**:605 (1969).
2. Dingman, W., and M. B. Sporn, Science **144**:26 (1964).
3. Zemp, J. W., J. E. Wilson, K. Schlesinger, W. O. Boggan, and E. Glassman, these Proceedings **55**:1423 (1966).
4. Hyden, H. V., and E. Egyhazi, these Proceedings **48**:1366 (1962).
5. Hyden, H. V., and A. W. Lange, these Proceedings **53**:948 (1965).
6. Shashoua, V. E., Nature (London) **217**:238 (1968).
7. Shashoua, V. E., in The Central Nervous System and Fish Behavior, ed. David Ingle (Chicago: Chicago University Press, 1968), pp. 203-213.
8. Cohen, H. D., and S. H. Barondes, Science **157**:333 (1967).
9. Jones, C. T., and P. Banks, J. Neurochem. **16**:825 (1969).
10. Scherrer, K., and J. Darnell, Biochem. Biophys. Res. Commun. **7**:486 (1962).
11. Wagner, E. K., and V. M. Ingram, Biochemistry **5**:3019 (1966).

64 Puromycin effect on successive phases of memory storage

Samuel H. Barondes*
Harry D. Cohen
Research Laboratory
McLean Hospital
Belmont, Massachusetts
and
Department of Psychiatry
Harvard Medical School
Boston, Massachusetts

Abstract. *Mice injected bitemporally with puromycin 5 hours before training learned to escape or to avoid shock by choosing the correct limb of a Y-maze. When retested 15 minutes after training they had normal retention. In the ensuing 2¾ hours the animals injected with puromycin, unlike the controls, showed a progressive decrease of savings to less than 7 percent.*

Memory storage is generally believed to progress through two phases—a "short-term" or "labile phase" and a "long-term" or "stable phase." That the "short-term" phase may be mediated by a reversible molecular change (for example, a configurational change in a protein at the synapse) whereas the "long-term" phase would probably be mediated by a self-replicating biosynthetic process (for example, synthesis of a protein at the synapse) has been suggested *(1)*. We now report experiments to test this hypothesis and we believe the results are consistent with it.

Injections of puromycin into both temporal regions of the brain inhibit more than 80 percent of protein synthesis in this zone for from several hours to more than half a day after injection *(2)*. Furthermore if mice are trained to solve a Y-maze and are then injected intracerebrally with puromycin from 1 to 3 days after training they appear to have forgotten the solution to the maze when tested 3 days thereafter *(3)*. The foregoing experiments suggest that protein synthesis in the temporal region is required for maintenance of a memory within the period studied. They provide no information, however, on the time, during or after training, when the puromycin-sensitive process first becomes necessary for memory storage. We now report our attempts to answer this question.

Male Swiss albino mice (30 to 40 g, Charles River Breeding Co.) were lightly anesthetized with pentobarbital (40 mg/kg) and, when necessary, with small amounts of ether and mounted in a stereotaxic instrument. Their scalps were incised and reflected and a hole was made in each "temporal" site *(3)*. Ten microliters of a freshly prepared solution containing 90 μg of puromycin dihydrochloride (titrated to *p*H 6 with NaOH) was injected at each temporal site at a depth of 2.5 mm from the outer surface of the skull, perpendicular to its horizontal axis. The animals were awake within several hours of this procedure. Five hours after injection, a time at which protein synthesis in the temporal zone has already been inhibited more than 80 percent for several hours *(2)*, the mice were trained to choose the left limb of a Y-maze to escape or avoid shock (Fig. 1; *4*). The animals were trained to a criterion of nine out of ten correct responses.

Puromycin-injected animals learned the maze in an average of 17.7 trials. They did not differ significantly from controls that were injected with 0.06*M* NaCl, which approximates the NaCl concentration in the puromycin solution. Memory was evaluated by retraining the animals at one of a number of time intervals after initial learning and by comparing the number of trials to reach criterion on retraining with the number of trials to reach criterion in initial

From Science **151**:594-595, February 4, 1966.

Supported by USPHS grant No. 1-501-FR-05484-03. H. D. Cohen was a summer predoctoral fellow and is now at Tufts University, Medford, Massachusetts.

*Present address: Department of Psychiatry, School of Medicine, University of California at San Diego, La Jolla, California.

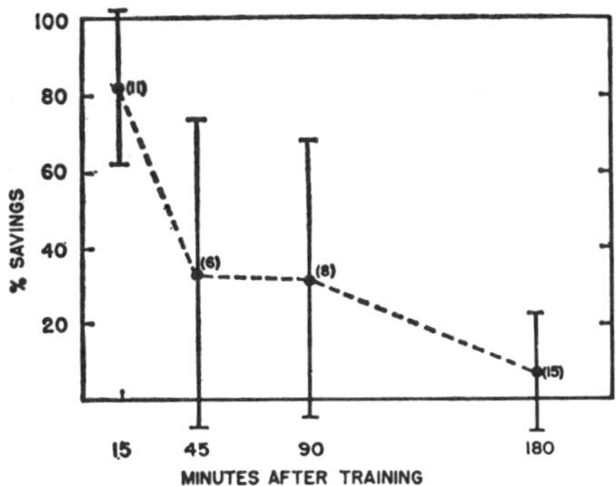

Fig. 1. Percentage savings after training. All animals were injected with puromycin at both temporal sites. Five hours later they were trained to choose the left limb of a Y-maze. The floor of the maze was a stainless steel grid and the left limb was insulated with colorless cellulose nitrate. Shock (approximately 0.5 ma) was administered 5 seconds after the animal was placed in the starting limb. A trial was considered correct if the animal did not enter the incorrect limb before reaching the correct limb. A trial was completed when the mouse entered the correct limb. The animal was allowed to remain on the safe area for 10 seconds before being returned to his home cage where he remained for about 1 minute before the next trial was begun. The animals learned in 17.7 ± 6.5 trials (mean ± S.D.). Savings was determined at the indicated times after completion of initial training. The results are mean ± S.D., and the number of animals in each group is shown in parentheses. The group tested at 15 minutes differs significantly from that at 3 hours ($P < .001$, t-test) and also from that at 45 minutes ($P < .05$).

Table 1. Learning and savings in control groups. Animals received bilateral injections of 10 μl of puromycin solution or 0.06*M* NaCl at frontal (F) or temporal (T) sites *(3)*. The results are mean ± S.D., and the number in each group is shown in parentheses.

Site	Period (hr) Training*	Period (hr) Tested†	Initial trials to criterion	Savings (%)
		Sodium chloride		
T	5	3	17.0±3.1(8)	91.2± 9.1
		Puromycin solution		
F	5	3	17.1±5.7(9)	86.7±20.4
T	8	¼	17.7±4.0(8)	89.4±14.5

*Hours after injection.
†Hours after training.

learning. The percentage of savings is $[(I - 9) - (R - 9)]/(I - 9) \times 100$, where I is the number of trials required to reach criterion in initial learning and R is the number of trials required to reach criterion on retraining. When puromycin-treated animals were evaluated 15 minutes after they had finished learning the task, they had savings percentages indistinguishable from those of controls injected with 0.06*M* NaCl. In the ensuing few hours the puromycin-injected animals showed a progressive loss of savings and, 3 hours after learning, their savings was less than 7 percent (Fig. 1).

In contrast to the loss of memory in the treated mice, injection of 0.06*M* NaCl at the temporal sites did not interfere with savings 3 hours after training (Table 1). This suggests that the impairment of memory is not due to some nonspecific effect of the temporal injections. Furthermore, injection of identical amounts of puromycin at "frontal" sites *(3)* does not interfere with memory storage (Table 1), an argument against the phenomenon being due to a nonspecific toxic effect of intracerebral puromycin. It was also established that animals trained 8 hours after bitemporal puromycin injections learned normally and showed normal savings when tested 15 minutes after training (Table 1). Thus, at a time when animals have forgotten what they learned 3 hours before, their capacity for learning and "short-term" memory is retained.

These experiments are consistent with the hypothesis that there is an initial phase of memory storage which is independent of protein synthesis in the temporal lobe and that this overlaps a second phase which is dependent on protein synthesis in this region of the brain. The effects of puromycin on memory in the goldfish *(5)* may also be interpreted in this way. Nevertheless puromycin may be exerting its effect on memory storage by some mechanism other than inhibition of protein synthesis. We have found *(4)* that injections of actinomycin D which inhibit cerebral RNA synthesis 94 to 96 percent do not interfere with retention of the solution to a Y-maze within 4 hours after training. This suggests that, if protein synthesis is indeed required for the second phase of memory storage, such protein synthesis is directed by a stable messenger RNA which was synthesized before and independent of acquisition.

It is customary to consider that memory storage has two phases—"short-term" and "long-term." Our experiments and those of Flexner *et al. (3)*, when considered together, suggest that there are at least three phases of memory storage in the shock-motivated maze learning which they and we studied. There appears to be an initial phase, uninfluenced by puromycin, which extends for a number of minutes after learning; a second phase, inhibited by temporal injections of puromycin, which may extend for several days; and a third phase, beyond these, which can be interfered with only by more diffuse intracerebral injections of puromycin. It is therefore more appropriate to consider memory storage a triphasic or multiphasic process rather than a biphasic one.

REFERENCES AND NOTES

1. S. H. Barondes, Nature **205:**18 (1965).
2. L. B. Flexner, J. B. Flexner, R. B. Roberts, G. de la Haba, Proc. Nat. Acad. Sci. U.S. **52:**1165 (1964); L. B. Flexner, J. B. Flexner, G. de la Haba, R. B. Roberts, J. Neurochem. **12:**535 (1965). We have confirmed the reported effect of temporal injections of puromycin on C^{14}-valine incorporation into the protein of temporal lobe cortex.
3. J. B. Flexner, L. B. Flexner, E. Stellar, Science **141:**57 (1963).
4. H. D. Cohen and S. H. Barondes, J. Neurochem., in press.
5. B. W. Agranoff, R. E. Davis, J. J. Brink, Proc. Nat. Acad. Sci. U.S. **54:**788 (1965).

65 Sex chromosome abnormalities in two state hospitals for patients requiring special security

M. D. Casey*
L. J. Segall
D. R. K. Street
C. E. Blank
Genetics Department
University of Sheffield
Centre for Human Genetics
Sheffield, England
and
Rampton Hospital
Retford, Nottinghamshire, England

The association between sex chromatin abnormality and mental sub-normality is well established. The incidence of sex chromatin abnormality in the new-born male and female population is about 0.2 per cent and 0.08 per cent respectively; in institutionalized mentally sub-normal male and female populations about 1 per cent and 0.4 per cent respectively. Court Brown[1] reported that a high proportion of sex chromatin positive males recognized in his unit had been committed to a mental defective institution because of anti-social behaviour. He suggested that an abnormal sex chromosome complement might predispose to delinquency. However, a survey by Wegman and Smith[2] on a group of socially disturbed males of relatively normal intelligence failed to show an increase in sex chromatin abnormality.

These observations prompted us to undertake a survey of the sex chromatin pattern in a group of the mentally sub-normal who had been 'institutionalized' because of anti-social behaviour. Since this study was begun a report of a similar survey has appeared[3].

Rampton and Moss Side Hospitals are State hospitals for patients who require special security on account of persistent violent or aggressive behaviour. The patients are mentally sub-normal and have been referred by the Courts or transferred from other mental institutions. Twenty-one sex-chromatin-positive males were detected among 942 male patients, an incidence of 2.2 per cent. Two of the 420 female patients examined had two sex chromatin bodies, an incidence of 0.49 per cent. On examining the chromosomes, twelve of the twenty-one sex-chromatin-positive males had 47 chromosomes and an *XXY* sex chromosome constitution. Seven had 48 chromosomes and an *XXYY* complement. Two male patients were presumed to have *XXY/XY* mosaicism. The majority of the cells analysed for chromosome complement in both female patients with two sex chromatin bodies had 47 chromosomes. Both female patients had a sufficiently large percentage of cells with 46 chromosomes to indicate the presence of *XXX/XX* mosaicism.

The frequency of sex-chromatin-positive males at Rampton and Moss Side Hospitals is compared with the frequency of similar males in other institutions for the mentally sub-normal in Table 1. There is a statistically significant excess of sex-chromatin-positive males at Rampton and Moss Side when compared with institutionalized mentally sub-normal populations where persistent violent or aggressive behaviour is not a feature ($\chi^2 = 21.7$, $n = 1$, $P < 0.001$). The incidence at Rampton and Moss Side does not differ significantly from that in a similar population in Sweden (Table 1). The frequency of females with two sex chromatin bodies is not significantly different from that observed in other institutions for the mentally sub-normal (Table 2).

It is thus tempting to refer the increased incidence of sex-chromatin-positive males to

From Nature **209:**641-642, 1966. Used with permission.

We thank Dr. J. N. McDougall, medical superintendent of Moss Side Hospital, for permission to study the patients in his care and to the medical and nursing staffs of Rampton and Moss Side Hospitals for their full co-operation. This investigation was supported in part by a grant from the National Spastic Society.

*Beit Memorial Research Fellow.

Table 1. Frequency of sex-chromatin-positive males in institutionalized mentally sub-normal populations

Population	Total No. of patients	No. of sex-chromatin-positive patients	Incidence (%)
Institutions for the mentally sub-normal without persistent violent or aggressive behaviour (refs. 4-6, 8-13)	10,685	90	0.84
Rampton and Moss Side. Hospitals for the mentally sub-normal requiring special security on account of persistent violent or aggressive behaviour	942	21	2.2
Criminal or hard to manage males of sub-normal intelligence. Forssman and Hambert (ref. 3)	760	15	2.0

Table 2. Frequency of females with two or more sex chromatin bodies in institutionalized mentally sub-normal populations

Population	Total No. of patients	No. of patients with 2 or more sex chromatin bodies	Incidence (%)
Institutions for the mentally sub-normal without persistent violent or aggressive behaviour (refs. 10-12, 14-16)	6,047	24	0.40
Rampton and Moss Side. Hospitals for the mentally sub-normal requiring special security on account of persistent violent or aggressive behaviour	420	2	0.49

Table 3. Types of abnormal sex chromosome complement found in sex-chromatin-positive males

Population	Chromosome complement			
	XXY	*XXYY*	Other	Total
Babies				
Maclean *et al.* (ref. 17)	12	1	5	18
Institutionalized mentally sub-normal				
Maclean *et al.* (ref. 10)	16	2	10	28
De la Chapelle (ref. 11)	5	0	1	6
Rampton and Moss Side	12	7	2	21

violent or aggressive behaviour. It was, however, noted that the patients at Rampton and Moss Side have a higher intelligence quotient (mean 77.3±14.07) than patients in other institutions for the mentally sub-normal. Since it is recognized that the incidence of sex-chromatin-positive males among the institutionalized mentally sub-normal is proportional to the intelligence quotient[4-7], being highest in the range of more than 50, it might be expected that a higher frequency of sex-chromatin-positive males would be found in Rampton and Moss Side than in other institutions for the mentally sub-normal. None the less the unusual distribution of the type of sex chromosome abnormality observed suggests that the observed high incidence is not simply a reflexion of higher intelligence quotient. There is an extraordinary excess of individuals with an *XXYY* chromosome complement (Table 3). Indeed, it may reasonably be suggested that this high incidence at Rampton and Moss Side is attributable entirely to an excess of patients with two *Y* chromosomes.

The nature of offences committed prior to admission by the 21 sex-chromatin-positive male patients at Rampton and Moss Side was compared with the behaviour patterns of a randomly selected group of controls from the remainder of the Rampton population. No statistically significant difference in behaviour pattern between these two groups of patients was observed. However, certain trends are suggested. The sex-chromatin-positive patients tended to abscond more frequently than the controls and they committed less homosexual or heterosexual offences. There were no significant differences in type of behaviour disorder between the *XXY* and *XXYY* patients.

Features of Klinefelter's syndrome were present in each of the patients with a simple *XXY* or *XXYY* chromosome complement. The patients with an *XXYY* sex chromosome complement were taller than those with an *XXY* complement, but the difference is not statistically significant. The difference in height is entirely due to the Pubic-Sole measurement and becomes significant if previously reported cases[18,20] are taken into account. The mean measurement for patients of karyotype *XXY* is 89.4 cm ($\sigma = 5.67$, $n = 76$) and for patients of karyotype *XXYY* 94.8 cm ($\sigma = 7.7$, $n = 16$). This confirms the suggestion by Barr[18] and Bray and Josephine[19] that the extra *Y* chromosome has some effect on growth. The two patients with *XXY/XY* mosaicism had normal-sized testes and normal secondary sex development.

REFERENCES

1. Court Brown, W. M., Lancet **ii**:508 (1962).
2. Wegman, T. G., and Smith, D. W., Lancet **i**:274 (1963).
3. Forssman, H., and Hambert, G., Lancet **i**:1327 (1963).
4. Barr, M. L., Evelyn, L. Shaver, Carr, D. H., and Plunkett, E. R., J. Ment. Defic. Res. **4**:89 (1960).
5. Ferguson-Smith, M. A., Acta Cytol. Symp. Sex Chromatin **73** (1962).
6. Mosier, D., Scott, L. W., and Cotter, L. H., Pediatrics **25**:291 (1960).
7. Miller, O. J., Amer. J. Obstet. Gynaecol. **90** (No. 7), Pt. 2:1121 (1964).
8. Ferguson-Smith, M. A., Lancet **i**:928 (1958).
9. Shapiro, A., and Ridler, M. A. C., J. Ment. Defic. Res. **4**:48 (1960).
10. Maclean, N., Mitchell, J. M., Harnden, D. G., Williams, J., Jacobs, P. A., Buckton, K. A., Baikie, A. G., Court Brown, W. M., McBride, J. A., Strong, J. A., Close, H. G., and Jones, P. D., Lancet **i**:293 (1962).
11. De la Chapelle, A., J. Ment. Defic. Res. **7**, Pt. 2:129 (1963).
12. Breg, W. R., Castilla, E. E., Miller, O. J., and Cornwell, J. G., Soc. Pediat. Res. (Abst.) **63** (No. 4), Pt. 2:738 (1963).
13. Anderson, I. F., Goeller, E. A., and Wallace, C., South Afr. Med. J. **38** (No. 17):346 (1964).
14. Fraser, Jean H., Campbell, J., MacGillivray, R. C., Boyd, E., and Lennox, B., Lancet **ii**:626 (1960).
15. Johnston, A. W., Ferguson-Smith, M. A., Handmaker, S. D., Jones, H. W., and Jones, G. S., Brit. Med. J. **ii**:1046 (1961).
16. Ridler, M. A. C., Shapiro, A., and McKibben, W. R., Brit. J. Psychiat. **109**:390 (1963).
17. Maclean, N., Harnden, D. G., Court Brown, W. M., Bond, J., and Mantle, D. J., Lancet **i**:286 (1964).
18. Barr, M. L., Carr, D. H., Soltan, H. C., Wiens, R. G., and Plunkett, E. R., Canad. Med. Assoc. J. **90**:575 (1964).
19. Bray, P. F., and Josephine, A., J. Amer. Med. Assoc. **63**, No. 6:1207 (1963).
20. Court Brown, W. M., Harnden, D. G., Jacobs, Patricia A., Maclean, N., and Mantle, D. G., M.R.C. Special Report, No. 305.

66 Criminal behaviour and the XYY male

W. H. Price*
P. B. Whatmore
M.R.C. Clinical Effects of Radiation Research Unit
Western General Hospital
Edinburgh, Scotland
and
State Hospital
Carstairs, Lanarkshire, Scotland

In 1965, Jacobs *et al.* published their preliminary findings of a chromosome survey conducted at a maximum security hospital, The State Hospital, Lanarkshire, Scotland.[1] The most remarkable finding in the completed survey was the discovery among 315 men of nine patients with an *XYY* sex chromosome constitution. Their behaviour, together with their pattern of crime, has now been closely studied. The full clinical details of this investigation will be published elsewhere by us, and this communication directs attention to the ways in which the *XYY* males differ from males with an *XY* sex chromosome complement at the same hospital.

All the patients admitted to this hospital have severely disordered personalities and they have been classified according to whether the cause is known or not. For example, some have brain damage which followed infections, others are epileptics, and others suffer from a psychosis. The largest group of patients have no known cause for their personality disorders. All the men with an *XYY* complement were classified in this category and eighteen other men have been randomly selected from this group for comparison with the nine *XYY* males. Seventeen of the eighteen control males were known to have an *XY* sex chromosome complement, the remaining being one of twenty-seven who had not been willing to be investigated when the chromosome survey was carried out.

There are three ways in which the *XYY* males differed importantly from the controls. First, although the patients in the two groups have penal records of comparable length, those of the *XYY* males include considerably fewer crimes of violence against persons. Thus, the nine *XYY* males had been convicted on a total of ninety-two occasions, but only eight of these convictions (8.7 per cent) had been for crimes against the person, while eighty-one (88.0 per cent) had been for crimes against property. In contrast, the eighteen control males had been convicted on 210 occasions, and forty-six of these (21.9 per cent) had been for crimes against the person while 132 (62.9 per cent) had been for crimes against property. Second, the disturbed behaviour of the *XYY* patients showed itself at an earlier age. This is reflected in a mean age at first conviction of 13.1 yr, compared with a mean age of 18 yr for the control patients, a difference which is significant at the 5 per cent level. Third, in the families of these patients the incidence of crime among the siblings of the *XYY* patients is significantly less than among those of the control patients. Thus, only one conviction is recorded among thirty-one sibs of the *XYY* patients while no less than 139 convictions are recorded for twelve of sixty-three sibs of the control patients.

The distribution of intelligence quotient among the *XYY* males probably reflected the distribution among the patients of the hospital as a whole. Seven were considered to be mentally sub-normal, but it is worth noting that the pattern of behaviour among the two whose intelligence quotients were not unusually low conformed with those of the other seven.

The picture of the *XYY* males that emerges from examination of those detained at the State Hospital is of highly irresponsible and immature individuals whose waywardness causes concern at a very early age. It is generally evident that the family background is

---

From Nature **213**:815, 1967. Used with permission.

We thank Dr. R. P. Brittain and Dr. W. M. Court Brown for help, and Mr. Peter Smith for statistical advice.

*Presently with M.R.C. Clinical and Cytogenetics Research Unit at the same hospital.

not responsible for their behaviour. They soon come into conflict with the law, their criminal activities being aimed mainly against property, although they are capable of violence against persons if frustrated or antagonized. Their failure to respond to corrective measures leads to a sentence of prolonged detention in safe custody at an earlier age than is usual for offences of this kind. All nine men with an *XYY* chromosome complement conform fairly closely to this broad description and it seems reasonable to suggest that their antisocial behaviour is due to the extra *Y* chromosome.

REFERENCE

1. Jacobs, P. A., Brunton, M., Melville, M., Brittain, R. P., and McClemont, W. F., Nature **208**:1351 (1965).

chapter 13

Genes in populations

Until now we have considered genes solely in terms of the individual. Now we want to examine the distribution of genes in populations. We shall no longer have as our point of reference a single genotype per individual but rather many genotypes per population. In addition, we shall have to consider the fates of these genotypes from generation to generation. A discussion of genes in populations, especially over a number of generations, will bring us to the subject of evolution. The study of the genetic changes involved in evolution constitutes the field of *population genetics.* Since the study of evolution is essentially a study of change, we can consider population genetics as having three facets of research: (1) investigations on the *origin of genetic diversity* (mutation in the broadest sense, including any heritable change); (2) investigations on the *patterning of genetic diversity* (selection, migration, genetic drift); and (3) investigations on the *segregation of genetic diversity* (race and species formation). In the present chapter, we shall consider the origin and patterning of genetic diversity. In Chapter 14, we shall study race and species formation.

GENOTYPE FREQUENCIES AND EQUILIBRIUM

The basic algebraic formula that describes the expected frequencies of various genotypes in a population was demonstrated in 1908 independently by Hardy (Ref. 13-1) and by Weinberg (Ref. 13-2). The similarity of their contributions went largely unnoticed until attention was drawn to both authors in 1943 by Stern (Ref. 13-3), whose paper is included in this chapter. From the Hardy-Weinberg theorem, one can demonstrate that the relative frequencies of various alleles in a population tend to remain constant, generation after generation. If the alleles of a population are constant in frequency, no evolutionary changes will occur. However, forces exist that modify the gene frequencies in populations. These forces are (1) mutation, (2) selection, (3) migration, and (4) genetic drift.

MUTATION

We reviewed the various methods of producing chromosomal rearrangements in Chapter 5 and point mutations in Chapter 9. For our present purposes, these two types of genetic modifications can be considered together, since they follow the same rules of population dynamics. Our interest is to examine the fates of mutations once they have arisen. Normally, all genes mutate. When an allele at a given locus is rare, its production of other alleles is hardly detectable because mutation rates tend to be small. However, as an allele becomes fairly frequent in a population, the mutations to and from it have to be taken into consideration. Eventually, a point is reached in which a reduction in frequency of a specific allele in a population due to mutation is offset by an increase in frequency of the same allele in the population due to reverse mutations from other alleles. When this occurs, a *mutation-equilibrium* results, and no further change in frequency of the particular allele will occur in subsequent generations. It follows from the above that the evolutionary changes caused by recurring mutations cease, once mutation-equilibrium has been achieved. The foregoing applies only if the other evolutionary forces such as selection, migration, and genetic drift are not operating to affect gene frequency in the population.

In Chapter 9, papers reviewing the production of mutations by chemicals (Ref. 9-5) and by radiations (Ref. 9-7) were reprinted. It is clear from those reports that mutation rates can be affected by environmental factors. In addition, it has been possible to demonstrate, in some instances, that mutation rates are under genetic control. It was discovered by Demerec in 1937 (Ref. 13-4) that an unusually high frequency of spontaneous sex-linked recessive lethals occurring in his Florida 10 stock of *Drosophila melanogaster* was a result of the stock's being homozygous for a recessive mutator gene on the second chromosome. Another example of a mutator gene was discovered in maize (corn) and reported in 1941 by Rhoades (Ref. 13-5), whose paper is reprinted in this chapter. Mutator genes have also been discovered in microorganisms. Goldstein and Smoot in 1955 (Ref. 13-6) described a wild strain of *Escherichia coli* in which there was a very high frequency of mutants (up to 15% of the total number of viable cells) in every generation. Experiments conducted by Zamenhof and reported in 1966 (Ref. 13-7) demonstrated that a single gene is responsible for the high mutability of these bacteria.

In addition to mutator genes, other genetic factors may be involved in increasing mutation rates. It was reported by Thompson in 1960 (Ref. 13-8) that an inversion in a chromosome of *D. melanogaster* causes an increase in the spontaneous lethal-mutation rate of the homologous normal chromosome. This was found to be the case for both second and third chromosomes of the fly.

SELECTION

Once a new allele has arisen in a population, the fate of that allele is determined by many factors. We have already noted one possible result based on forward and reverse mutation rates. Now we shall examine the survival or elimination of an allele due to *selection.* Selection is the process that determines the relative contribution that carriers of different genetic constitutions will make as parents of the next generation. Selection does not act on individual genes but rather on the organism possessing them. The relative measure of the survival and reproductive efficiency of the carriers of a particular genome, when compared to its competitors in a given environment, is called the organism's *adaptive value* or *Darwinian fitness.*

Whether a genotype is beneficial, neutral, or detrimental to the survival and reproduction of its carrier is determined by the environmental factors that prevail during the organism's life cycle. It was reported by Kalmus in 1941 (Ref. 13-9) that in three species of *Drosophila,* the mutant yellow body color is less resistant to dessication than the wild type. On the other hand, mutants that increase body pigmentation (black and ebony) are more resistant to dessication than wild type.

It is clear that selection against a dominant allele will be most effective, since all its carriers will exhibit the phenotype associated with the allele. Selection against a recessive allele will be most rapid when the allele is high in frequency in the population. This results from the fact that many recessive homozygotes will be formed and they will be subjected to whatever are the adverse environmental forces. Once the recessive allele becomes rare, it will most often be found in a heterozygous condition. Since its carriers will be unaffected by selection, the recessive allele will be "protected" from the action of the environment. Although selection against a recessive allele is not nearly as effective as against a dominant gene, one should find that all deleterious alleles are relatively low in frequency in the population. In some populations, however, one finds that a number of deleterious alleles are in much higher frequency than anticipated. An example of this type of situation in a human population was reported in 1954 by Allison (Ref. 13-10), whose paper is included in this chapter.

MIGRATION

Another force by which the frequencies of alleles in a population can be changed is *migration.* This can involve either a departure of individuals with certain genotypes from a population or an influx of individuals with different allele frequencies into a population. A situation of great consequence occurs when two widely different groups continuously exchange migrants. The most obvious effect of continued intergroup migration is to make the allele frequencies of the two groups more nearly alike and thus, in the absence of any countermeasures, render the total population more homogeneous. An example of continued intergroup migration can be found in the small but persistent number of marriages that occur in each generation between Caucasians and Negroes in the United States. The effect of this migration on allele frequencies was examined in 1953 by Glass and Li (Ref. 13-11), who calculated that 30% of the genes of the American Negroes have been acquired from the American white population. More recently, in 1969, this same subject was examined by Reed (Ref. 13-12), whose paper is reprinted in this chapter.

GENETIC DRIFT

The last force to be considered by which allele frequencies in a population can be changed is *genetic drift.* This factor represents the chance events that may result in the elimination (frequency of 0%) of an allele from a population or in the fixation (frequency of

100%) of that allele in a population. The basis for genetic drift lies in the fact that the frequencies of alleles among the gametes that give rise to the individuals of the next generation may not represent exactly the gene frequencies of the population from which these gametes were formed. This type of "sampling error," which can occur in a random-breeding population, has more chance of taking place in a small population than in a large one.

Laboratory experiments designed to test the occurrence of genetic drift were reported in 1954 by Kerr and Wright (Ref. 13-13), whose is the last paper reprinted in this chapter. A report on the possible occurrence of genetic drift in a natural population was made by Grüneberg in 1961 (Ref. 13-14). Five populations of Delhi (India) rats were studied. They were taken from grain shops in one part of the city and found to vary, in statistically significant fashion, in a number of skeletal characteristics. It was assumed that the rat population of each grain shop had its origin in a small number of foundation animals from some ancestral population and that the unique skeletal characteristics of each group of rats reflected the occurrence of genetic drift. Human populations have also been studied for evidence of genetic drift. An example of such an investigation was reported by Glass and his co-workers in 1952 (Ref. 13-15). A clear and excellent coverage of the mathematics of population genetics, including mutation, selection, migration, and genetic drift, was published by Li in 1955 (Ref. 13-16).

REFERENCES

Genotype frequencies and equilibrium

13-1. Hardy, G. H. 1908. Mendelian proportions in a mixed population. Science **28**:49-50.

13-2. Weinberg, W. 1908. Uber den Nachweis der Vererbung beim Menschen. Jahreshefte Ver. vatrl. Naturk. Württemberg **64**:368-382. (Translated *in* S. H. Boyer, IV [ed.] Papers on human genetics. Prentice-Hall, Inc., Englewood Cliffs, N. J.).

13-3. Stern, C. 1943. The Hardy-Weinberg Law. Science **97**:137-138.

Mutation

13-4. Demerec, M. 1937. Frequency of spontaneous mutations in certain stocks of *Drosophila melanogaster*. Genetics **22**:469-478.

13-5. Rhoades, M. M. 1941. The genetic control of mutability in maize. Cold Spring Harbor Symp. Quant. Biol. **9**:138-144.

13-6. Goldstein, A., and Smoot, J. S. 1955. A strain of *Escherichia coli* with an unusually high rate of auxotrophic mutation. J. Bacteriol. **70**:588-595.

13-7. Zamenhof, P. J. 1966. A genetic locus responsible for generalized high-mutability in *Escherichia coli*. Proc. Natl. Acad. Sci. U. S. A. **56**:845-852.

13-8. Thompson, P. E. 1960. Effects of some autosomal inversions on lethal mutations in *Drosophila melanogaster*. Genetics **45**:1567-1580.

Selection

13-9. Kalmus, H. 1941. The resistance to dessication of *Drosophila* mutants affecting body color. Proc. Roy. Soc. London B **130**:185-201.

13-10. Allison, A. C. 1954. Protection afforded by sickle-cell trait against subtertian malarial infection. Brit. Med. J. **1**:290-294.

Migration

13-11. Glass, B., and Li, C. C. 1953. The dynamics of racial intermixture–an analysis based on the American Negro. Am. J. Hum. Genet. **5**:1-20.

13-12. Reed, T. E. 1969. Caucasian genes in American Negroes. Science **165**:762-768.

Genetic drift

13-13. Kerr, W. E., and Wright, S. 1954. Experimental studies of the distribution of gene frequencies in very small populations of *Drosophila melanogaster:* I. Forked. Evolution **8**:172-177.

13-14. Grüneberg, H. 1961. Evidence for genetic drift in Indian rats (*Rattus rattus* L.) Evolution **15**:259-262.

13-15. Glass, B., Sacks, M. S., Jahn, E. F., and Hess, C. 1952. Genetic drift in a religious isolate: An analysis of the causes of variation in blood group and other gene frequencies in a small population. Am. Nat. **86**:145-159.

13-16. Li, C. C. 1955. Population genetics. University of Chicago Press, Chicago.

67 The Hardy-Weinberg law

Curt Stern*
Department of Zoology
University of Rochester
Rochester, New York

One of the basic relations in the genetics of populations is expressed by the statement that in a very large random-mating population in which two alleles A and A′ occur in the frequencies p and q (= 1 − p) the three types AA, AA′ and A′A′ are expected to remain in equilibrium from generation to generation at frequencies of p^2, 2pq and q^2, in the absence of mutation or selection. This theorem, of which a special case was discovered by Pearson (1904), is known in its general formulation as Hardy's law, or Hardy's formula (*e.g.*, Sinnott and Dunn, 1939, Sturtevant and Beadle, 1939, and Dobzhansky, 1941). It is the purpose of this note to point out that the important population formula was independently and simultaneously recognized by the Stuttgart physician, Wilhelm Weinberg (1862-1937). On January 13, 1908, Weinberg gave a lecture before the "Verein für vaterländische Naturkunde in Württemberg" under the title "Über den Nachweis der Vererbung beim Menschen." In the course of a keen exposition of both the difficulties to be met by students of human heredity and of statistical approaches which should help to overcome these difficulties, he derived the equilibrium law. The full lecture was printed in the Jahreshefte of the Verein, Volume 64: 368-382 (1908), and appeared sometime before the fall of 1908 as judged by the stamped entry on the title page of the volume which I have consulted: "Academy of Natural Sciences of Philadelphia, Sept. 28, 1908." Hardy's note in Science is signed April 5, 1908, and is published in the July 10, 1908, number.

The following is a translation of the relevant section of Weinberg's communication making corrections for three minor typographical errors:

Quite different is the situation when one considers Mendelian inheritance under the influence of panmixis. I start from the general premise that there are originally present m each of pure male and female representatives of type A and correspondingly n of each pure representatives of type B. If these cross at random one obtains, by applying the symbolism of the binomial theorem, the following composition of the filial generation:

$$\frac{(m\,AA+n\,BB)^2}{(m+n)^2}=\frac{m^2}{(m+n)^2}AA+\frac{2\,m\,n}{(m+n)^2}AB+\frac{n^2}{(m+n)^2}BB$$

or if $m + n = 1$

$$m^2\,AA+2\,m\,n\,AB+n^2\,BB.$$

If now the male and female members of the first generation are crossed at random among themselves one obtains the following frequencies of the various cross combinations:

$$\begin{aligned}&m^2\cdot m^2\cdot(AA\times AA)=m^4\,AA\\&4m^2\,m\,n\,(AA\times AB)=2\,m^3n\,AA+2\,m^3n\,AB\\&2m^2n^2\,(AA\times BB)=2\,m^2n^2\,AB\\&4\,(mn)^2\,(AB\times AB)=m^2n^2\,AA+2\,m^2n^2\,AB+m^2n^2\,BB\\&4m\,n\,n^2\,(AB\times BB)=2\,m\,n^3\,AB+2\,m\,n^3\,BB\\&n^2n^2\,(BB\times BB)=n^4\,BB\end{aligned}$$

or the relative frequencies

$$\begin{aligned}&AA:m^2\,(m+n)^2\\&AB:2m\,(m+n)^2\,n\\&BB:(m+n)^2\,n^2\end{aligned}$$

and the composition of the second filial generation is again

$$m^2\,AA+2\,m\,n\,AB+n^2\,BB.$$

Thus we obtain under the influence of panmixis in each generation the same proportion of pure and hybrid types. . . .

While Weinberg's paper, like Mendel's, appeared in an obscure journal, its failure to be recognized can not be ascribed to this fact alone. His later contributions dealing with extensions of the statistical treatment of the genetics of populations are found in the "regular" journals. These papers have received some attention (*e.g.*, Sewall Wright, 1930) and in them Weinberg refers to his 1908 pioneer work. However, both Weinberg and Hardy were ahead of contemporary thought and similar problems were not generally considered for at least eight years. At that time perhaps Hardy's

From Science **97**:137-138, February 5, 1943. Used with permission.

*Present address: Department of Zoology, University of California, Berkeley, California.

name and the prominent place of his publication both helped to leave Weinberg's contribution neglected.

Hardy as a mathematician did not follow up his discovery by any further consideration of its genetic implications. Weinberg in 1909 reformulated his theorem in terms valid for multiple alleles—at a time when no case of multiple alleles had been discovered in man nor in plants and even Cuénot's demonstration of multiple alleles in the mouse had remained unnoticed. He also for the first time investigated polyhybrid populations and recognized their essentially different method of attaining equilibrium. Considering these facts it seems a matter of justice to attach the names of both the discoverers to the population formula.

68 The genetic control of mutability in maize

M. M. Rhoades
Department of Botany
Indiana University
Bloomington, Indiana

Genes with a high spontaneous mutation rate are called "unstable" or mutable genes. Some investigators, notably Correns and Goldschmidt, hold that these highly mutable genes are sick or diseased—in other words their high mutability arises from some intrinsic and presumably fundamental difference not shared with more stable loci. Goldschmidt believes that any conclusions reached concerning the phenomenon of mutation from a study of unstable loci are not applicable to an understanding of the mutation process in general but apply only to the unstable loci. Inasmuch as mutable loci offer one of the most favorable objects for the study of mutation it is essential that this supposed difference between stable and unstable genes be subjected to careful scrutiny. The purpose of this paper is to present previously published data bearing on this question as well as some unpublished data recently obtained.

The spontaneous mutation rate of various genes in maize differs widely. On the one extreme there are those loci with certain alleles having a very low mutation rate, measured in terms of the rate of change of the dominant allele to a recessive condition. Stadler measured the mutation rate of seven genes in maize, which were unselected save that they affected the color or composition of the endosperm and aleurone. Four of the seven tested genes, namely *Wx*, *Sh*, *Y* and *Su*, proved to be extremely stable and may be considered examples of genes with low spontaneous rates. The *R* allele had a mutation rate several hundred times greater than that of the four genes listed above. It mutated to the *r* allele at the rate of 492 per million gametes while no mutations of *Wx* to *wx* occurred in 1,503,744 gametes and only 3 of *Sh* to *sh* were found in 2,469,285 gametes. The mutation rates of *Pr* and *I* were intermediate between these two extremes. Stadler's data indicate that there is no sharp demarcation between genes in terms of mutation rate although too few were studied to establish this definitely.

In addition to the foregoing data on the frequency of germinal mutations there exist in maize a number of cases of variegation in which it has been demonstrated that the mosaic tissue is due to the instability of certain genes. The classic example of an allele with a high spontaneous mutation rate is the unstable P^{vv} allele which mutates with a high frequency to a dominant allele producing color in the pericarp. Emerson (1914, 1917, 1929) presented conclusive evidence that the color change from colorless to colored pericarp found in the variegated strains of maize arose by mutation. Since variegation is a widespread phenomenon throughout the plant kingdom Emerson's demonstration that variegation in maize pericarp color is due to the mutation of the P^{vv} allele to *P* suggests that mutation may be responsible for many cases of variegation. Sturtevant and Beadle (1939) are of the opinion that Emerson's work with variegated pericarp constitutes one of the mile posts in the progress of genetics—an accolade to which the writer wholly subscribes, since Emerson was the first to prove that variegation is the result of a genetic change which is heritable when it occurs in germinal tissue. Demerec (unpublished) has analyzed a situation in maize in which a waxy strain has many sectors showing the dominant starchy condition and has shown that the waxy-starchy mosaicism is due to the mutation of an unstable *wx* to a stable *Wx* allele. Hadjinov (1939) has described an allele at the *R* locus which apparently mutates from a dominant to a recessive condition with high frequency. The published data on this case are not completely convincing, however.

Although the P^{vv} and mutable *wx* genes are considered as unstable recessives which frequently mutate to dominant alleles there exist at both these loci other recessive alleles which have never mutated. One should speak of stable or unstable alleles rather than of loci since at

Reproduced from Vol. IX; Cold Spring Harbor Symposia, 1941. Used with permission.

certain loci both stable and unstable alleles occur.

There exists then in maize a wide, and possibly gradual, difference in the frequency of mutation of various genes. The writer agrees with Demerec's statement that there is no discontinuous range in mutation rate frequency and that the mutation for different genes may be considered as falling at various places along a continuous spectrum. If this be true it follows that mutation of an unstable gene is not a phenomenon *sui generis.* The writer has recently (1936, 1938, 1939) described a situation in maize where an extremely stable gene is made highly unstable when subjected to a specific genetic environment. The results previously published are briefly presented below.

The *A* locus in chromosome 3 is concerned with anthocyanin formation in the aleurone, plant and pericarp colors. Of the four alleles at this locus described by Emerson and Anderson (1932) the lowest member of the allelic series is the recessive *a* allele which when homozygous produces colorless aleurone, brown plant color, and a recessive brown pericarp. Stocks homozygous for *a* and the complementary dominant factors necessary for aleurone color, *A2, C* and *R,* are known as *a*-testers and have been extensively used by maize geneticists for many years. In so far as the writer is aware the *a* allele has never mutated to a higher member of the allelic series; there is every reason to believe it has an extremely low mutation rate. These *a* strains so widely used in genetical experiments, and indeed all strains of maize so far tested save one, are homozygous for the recessive allele *dt* which is in chromosome 9. However in a stock of Black Mexican sweet corn the dominant allele *Dt* was discovered. The same *a* alleles so stable when present in the same nucleus with *dt* become highly unstable when the *Dt* allele is substituted, and mutate to various members of the allelic series. Mutations of *a* in the presence of *Dt* occur in every tissue in which this locus is concerned with anthocyanin formation, but the size of the mutant areas is small, indicating that the mutations occur late in development in all tissues. Especially in the aleurone is this true, and the slight variation in size of the colored areas further suggests that the time of mutation is restricted to a fairly definite stage in the maturation of the tissue. It is as if the cellular environment conducive to mutability becomes effective at a relatively late stage for each tissue. Since the tissues in which mututations of *a* can be detected mature at different times during the development of the plant, and since in each tissue mutations occur late in ontogeny, it is probably that mutation takes place when each tissue has reached a certain physiological condition as it approaches maturity. Occasionally a mutation occurs early enough so that a large portion of the aleurone is colored, but the great majority of the colored spots are of the same approximate size, indicating that mutations are most likely to occur at a definitely restricted stage in the development of the aleurone. No extremely small mutant areas are found in the aleurone and large ones are rare.

The phenotypic effect produced by the mutations of *a* in the various tissues is given below. The outline of the colored areas indicates the planes of cell division followed by the descendants of the mutated cell.

The below color changes in the aleurone, plant and pericarp colors are indicative of the mutations of the *a* allele to *A* but do not in themselves constitute proof that mutations have occurred since other hypotheses could be invoked to account for the colored areas. The direct proof of the mutation hypothesis was possible because in the germinal tissue an occasional mutation occurred which could be subjected to genetic tests in subsequent generations. The proof of the mutation hypothesis is discussed in some detail in the writer's 1938 paper.

It has been demonstrated that when a mutation of an *a* allele occurs in a cell of *a a Dt Dt* constitution there is no accompanying

	Aleurone color (with *A2, C, R*)	Plant color (with *B, Pl*)	Pericarp color (with *P*)
A dt or *A Dt*	colored	dominant purple	dominant red
a dt	colorless	recessive brown	recessive brown
a Dt	colorless background with numerous small, round colored areas	brown background with narrow, longitudinal purple stripes	brown background with red stripes converging at the point of silk attachment

change or mutation of either of the *Dt* alleles. The *Dt* allele acts as a catalyst since it stimulates the mutability of *a* but remains unchanged when the mutations take place.

Since the mutability of the *a* allele is under the control of *Dt* the classification of *a* as a mutable or stable allele depends wholly upon the presence or absence of *Dt*. If all maize stocks were homozygous for *Dt* then the *a* allele would be classed as a mutable gene. It is possible that some cases of unstable genes might prove to be stable in the absence of certain factors affecting or even controlling their mutability.

DOSAGE EFFECTS

As the aleurone is triploid tissue, two of the haploid sets of chromosomes being contributed by the embryo sac and one coming from the pollen grain, it was possible to determine the frequency of mutation when one, two and three *a* alleles were present. The test of the dosage effect of *a* was technically feasible by the use of the a^p allele which produces a pale colored aleurone, recessive to the deep color of the *A* allele but dominant to the colorless condition of *a*. The mutability of the a^p allele, or of any allele at the *A* locus save *a*, is not affected by *Dt*. Seeds carrying *Dt* and heterozygous for *a* and a^p have the pale color due to the a^p allele and in addition show deep colored dots which are produced by mutations of the *a* alleles to *A*. By substituting the a^p for the *a* allele it was possible to obtain seeds with one, two and three *a* alleles. Extensive data presented in the 1938 paper show that the effect of increasing the dosage of the *a* allele is a linear or arithmetic increase in the number of mutations. Seeds with one *a* allele have one-half as many mutations as do seeds with two *a* alleles, and one-third as many as seeds with three *a* alleles.

Previous data published in 1936 show that a non-linear effect is obtained with different doses of *Dt* while the dosage of *a* is held constant. However the later demonstration that several modifying factors exist which affect the *a-Dt* reaction made it necessary to repeat the experiment on the dosage effect of *Dt*, using lines which were relatively isogenic. Such a stock was obtained through repeated self-fertilization of a *Dt dt* strain; heterozygous *Dt dt* seed being used in every generation to further the inbreeding. After five years of selfing the F_6 seed was classified into *Dt Dt, Dt dt* and *dt dt* classes. For the dosage relation between one and two *Dt* alleles exact reciprocal crosses were made between *Dt Dt* and *dt dt* individuals. It was necessary to self *Dt Dt* plants to obtain data on the effect of three *Dt* genes. The data obtained are presented in Table 1. At least 50 seeds were used for each determination. The figures represent the average number of mutations (dots of color) in the aleurone layer. The mutation frequency in the three *Dt* class is too low. There is considerable overlapping and

Table 1. Effect of different dosages of *Dt* on the mutability of the *a* allele

		Mean number of mutations per seed		
Pedigree		*Dt dt dt* class (1 *Dt*)	*Dt Dt dt* class (2 *Dt*)	*Dt Dt Dt* class (3 *Dt*)
6134-13×6131-7	reciprocally	6.8	19.5	–
6134-6 ×6131-14	"	5.9	19.6	–
6134-1 ×6131-2	"	7.8	19.9	–
6134-2 ×6131-9	"	9.1	23.9	–
6385-24×6386-13	"	6.7	24.9	–
6385-9 ×6386-19	"	8.3	26.6	–
6389-11×6390-17	"	8.4	24.1	–
7261-11×7260-6	"	6.1	24.0	–
7261-2 ×7260-1	"	6.0	17.2	–
		Mean ratio for 1 *Dt*:2 *Dt*=7.2:22.2		
6131-18 selfed		–	–	110.1
6131-8 selfed		–	–	126.7
6386-2 selfed		–	–	128.7

1 *Dt* allele gives 7.2 mutations per seed.
2 *Dt* alleles give 22.2 mutations per seed.
3 *Dt* alleles give 121.9 mutations per seed.

fusion of colored areas with such high numbers of dots. Error also enters from the fact that an earlier mutation of one *a* allele in a cell obscures a later mutation of a second or third allele. In the case of one and two doses of *Dt* this is not a significant matter, but it should be taken into account in considering the data from three doses of *Dt*. No corrections, however, have been made to the raw data given in Table 1. Due to the extreme difficulty in counting accurately the number of dots on seeds with three *Dt* alleles only three ears were classified. They were in no way different in appearance from numerous uncounted ears of the same constitution. The data show an exponential increase in mutation rate as the dosage of the *Dt* allele is increased. These data confirm the earlier conclusion reached in 1936 that the effect is non-linear.

MUTATIONS OF *a* TO DIFFERENT ALLELES

Emerson and Anderson (1932) described four alleles at the *A* locus. Two of them, *A* and A^b produce deep colored aleurone, a^p gives a pale aleurone color and *a* produces colorless aleurone. The effects of these four alleles on aleurone, plant and pericarp colors are shown in the upper part of the table shown below.

The *a* allele is recessive to the other alleles in all respects. The a^p allele is dominant to *A* in pericarp color but is recessive to it in aleurone and plant colors. The *A* and A^b alleles are alike in their effect on aleurone and plant color but the brown pericarp color produced by A^b is dominant to the red of *A*.

Germinal mutations of *a* to higher alleles can be detected in purple anthers found on *a B Pl* plants or in seeds with self-colored aleurone that are occasionally found on *a Dt* ears. All of the mutations isolated from purple anthers gave the deep self-colored aleurone similar to that produced by the *A* and A^b alleles, and all mutations from seeds with deep colored aleurone gave purple plants with the complementary *B* and *Pl* factors. Since the *A* and A^b alleles differ only in their effect on pericarp color a number of mutations identical to both *A* and A^b as far as aleurone and plant color are concerned were subjected to further tests. Twenty-nine such mutations have now been tested. Twenty-seven of them produced red pericarp color; they may be considered as identical to the *A* allele. One of the remaining two gave a recessive brown pericarp color with *P* in contrast to the dominant red of the *A* allele and the dominant brown of the A^b allele. This represents a mutation of *a* to a new, previously undescribed allele. It has been designated A^{br}. The other mutation produced a reddish-brown pericarp color with *P* which is dominant to the brown of *a* but recessive to the red of *A*. This is also a new allele. It has the symbol A^{rb}.

The pale aleurone color produced by the a^p allele is clearly distinguishable from the deep color of the *A* and A^b alleles. Therefore it was possible to determine the frequency with which *a* mutates to an allele or alleles giving pale aleurone color as compared with the frequency to alleles giving deep aleurone by inspection of the color of the aleurone dots. In the 1938 paper data were reported showing that the frequency of mutation to alleles giving deep aleurone color was a thousand times as great as to alleles producing pale colored dots. Since these mutations occur in somatic tissue it is impossible to ascertain the nature of these mutations as far as their effect on plant and pericarp colors are concerned. One germinal

Allele	Aleurone color	Plant color	Pericarp color
		Old alleles	
a	colorless	recessive brown	recessive brown
a^p	pale	red-brown	dominant brown
A	deep	purple	red
A^b	deep	purple	dominant brown
		Alleles derived by mutation	
A	deep	purple	red
A^{br}	deep	purple	recessive brown
A^{rb}	deep	purple	recessive red-brown
a^{br}	pale	red-brown	recessive brown
a^s	colorless	recessive brown	recessive brown

The a^s allele differs from *a* in that it is more stable with *Dt*.

mutation giving pale aleurone color was found however and tested. It was similar to a^p in aleurone and plant color but gave a recessive brown pericarp color instead of the dominant brown produced by a^p. This is also a new allele. It is designated a^{br}.

A number of self pollinations were made in lines homozygous for *a* and *Dt* to obtain a number of germinal mutations. Families 6641 and 6642 had a combined total of 389 ears of which 19 were segregating dotted and colorless seeds in approximately 3 : 1 ratios. Six of the 19 segregating ears were tested further and in five of them the occurrence of the colorless seeds was due to mutations of the *a* allele to a new allele or alleles indistinguishable from it save that they have much reduced mutation rates with *Dt.* In all five tested mutations the seeds homozygous for the new stable alleles appear macroscopically to be devoid of mutant colored areas. Two of these mutations have been subjected to further study. One of them appears to be completely stable with *Dt* since no mutations have been found on several hundred seeds examined under a binocular microscope. The other mutation so tested is to an allele with a markedly reduced rate of mutation with *Dt* but is not as stable as the first mutation since an average of 0.4 mutations per seed was found. The size of these infrequent dots is smaller than those on seeds with the original *a* allele indicating that the time of mutation occurs later in development. The wholly stable allele may be considered to be different from the second allele with a low but still detectable mutation rate but these two as well as the other three cases have been, pending further study, collectively designated by the symbol a^s.

The sixth tested ear proved to be heterozygous for *Dt dt.* It is possible that the *dt* allele came from contamination rather than from mutation of *Dt* to *dt* but no such explanation can account for the presence of the a^s alleles since all *a* stocks available to the writer were mutable with *Dt.*

It is highly probable that the *Dt* allele caused the mutations of *a* to a^s, since the observed mutation rate is so great that the a^s allele would have replaced the *a* allele in the genetic stocks if this mutation occurred with *dt,* and all *a* stocks are, as previously stated, mutable with *Dt.* It is possible that the a^s mutations observed in families 6641 and 6642 stem from only two mutations occurring in sporogenous cells which gave rise to a considerable number of the sex cells but this is not likely since the wholly stable allele and the allele with a low mutation rate came from the same family.

The relatively high frequency of the a^s alleles indicates that mutations to these alleles occur more frequently than to other allelic forms. While a relatively small number of germinal mutations producing color of one grade or another have been tested it appears that mutations to the *A* allele occur with the highest frequency since 27 of the 29 tested mutations giving deep aleurone color proved to be to this allele and the frequency of mutation to alleles giving pale colored aleurone is much less than to alleles giving deep color. Since the a^{br}, A^{br}, and A^{rb} alleles have been found only once nothing can be said about the relative frequency with which they arise.

A total of five different alleles have been found in these experiments. Four of these alleles are new, previously undescribed ones and it may be expected that additional new alleles will be found as these studies are continued.

LINKAGE RELATIONS OF *Dt*

In the 1938 paper it was suggested that *Dt* was linked with the *C* locus in chromosome 9. This statement was based on the ratio of dotted to colorless kernels in the F_2 of plants segregating for *A a Dt dt C c.* Since the *C* allele must be present for the expression of aleurone color, linkage between *C* and *Dt* would manifest itself as distorted ratios of dotted and colorless seeds. Linkage data of this type are unsatisfactory so that linkage tests of *Dt* with other factors in chromosome 9 were made holding the *C* allele homozygous. According to the data summarized in Emerson, Beadle and Fraser (1935) the linear order and map positions of certain genes in chromosome 9 is as follows:

knob	yg^2	*C*	*sh*	*bp*	*wx*	*v*
0	2	21	24	39	54	66

The knob is located at the end of the short arm of chromosome 9 and all of the above loci except *v* fall in this arm. The linkage data of *Dt* with *Yg2, Sh* and *Wx* are given in table 2. These data definitely prove that *Dt* is in chromosome 9 and further indicate that *Dt* lies close to the *Yg* locus near the end of the short arm. The order is unquestionably *Dt Sh Wx* and may be *Dt Yg2 Sh Wx* with *Dt* about ten units beyond *Yg2.* Creighton found approximately two per-

Table 2. Linkage data of *Dt* with various loci in chromosome 9

Genes	Linkage Phase	XY	Xy	xY	xy	Total	% Recom-bination
Dt Sh	C S	679	100	156	138	1073	27
Dt Sh	C B	617	266	305	588	1776	32
Dt Wx	R S	1663	525	690	118	2996	41
Dt Wx	C B	682	465	472	677	2296	41

$\frac{Dt\ Yg\ Wx}{dt\ yg\ wx}$ selfed

Dt Yg Wx	1450	*dt yg Wx*	385	Total 2793	
Dt yg wx	38	*dt Yg Wx*	223	Recombination values:	
Dt Yg wx	360	*dt Yg wx*	63	*Dt Yg*	11 percent
Dt yg Wx	36	*dt yg wx*	238	*Dt Wx*	42 percent
				Yg Wx	37 percent

$\frac{Dt\ Yg\ Sh\ Wx}{dt\ yg\ sh\ wx}$ selfed

Dt Yg Sh Wx	387	*dt Yg Sh Wx*	52	Total: 779	
Dt yg Sh Wx	7	*dt yg Sh Wx*	84	Recombination values:	
Dt Yg Sh wx	59	*dt Yg sh Wx*	2	*Dt Yg*	10 percent
Dt yg Sh wx	2	*dt yg sh Wx*	49	*Dt Sh*	27 percent
Dt Yg sh wx	35	*dt Yg Sh wx*	1	*Dt Wx*	44 percent
Dt yg sh wx	10	*dt yg Sh wx*	3	*Yg Sh*	22 percent
Dt Yg sh Wx	15	*dt Yg sh wx*	9	*Yg Wx*	38 percent
Dt yg sh Wx	3	*dt yg sh wx*	61	*Sh Wx*	20 percent

cent recombination between *Yg* and the terminal knob on the short arm of 9 so there is some doubt if *Dt* is beyond *Yg* and if it is that the crossover distance is as great as ten units. It should be noted that in selfing a *Dt dt* plant three classes of dotted seed are obtained, that is, *Dt Dt Dt*, *Dt Dt dt* and *Dt dt dt* classes. In the latter class possessing a single *Dt* allele the mutation rate is so low that a considerable number of *Dt dt dt* seeds are classified as *dt* because no dots (mutations) are evident. This fact introduces some error into the recombination values and would tend to increase the *Dt Yg* value. Nevertheless the data indicate that *Dt* is close to the terminal knob which is presumably composed of heterochromatin. While the placing of the *Dt* locus close to the knob is probably without significance it is of interest in view of Schultz's demonstration that loci normally stable when removed from heterochromatin show a type of variegation suggestive of mutation when brought into proximity with heterochromatin. The situation here is different since the *a* allele made mutable by *Dt* is in another chromosome and both *Dt* and *dt* containing chromosomes have terminal knobs on the short arm of 9 which do not appear to differ in size.

LINKAGE RELATIONS OF ALLELES DERIVED BY MUTATION

Goldschmidt holds that gene mutations are position effects, that is that every so-called point mutation results from some chromosomal rearrangement such as translocation, inversion, etc. The writer (1938) has presented several reasons which seem to negate the possibility that mutations induced by *Dt* at the *a* locus are due to breaks in chromosome 3 at or near this locus. Unless all mutations involve inversions so minute as to be beyond the resolving power of the microscope we should expect to find, in some mutations at least, evidence of gross structural changes. No such aberrations have been found. Chromosomes 3 carrying *A* alleles derived by mutation are not visibly different from homologues bearing the *a* allele. Minute inversions would be more difficult than translocations to detect cytologically or genetically. Inversions, when heterozygous, should reduce the amount of crossing over between the *A* locus and neighboring genes in chromosome 3 if they are of considerable length. Unpublished data of the writer give 30 percent recombination between *A* and *lg2*. Four different mutations to the *A* allele were tested for linkage with *lg2*. All of them showed approximately 30 percent recombination, a fact which argues against the presence of any but small inversions. It is difficult to visualize any mechanism by which the *Dt* allele located in chromosome 9 could cause breaks to arise with an extremely high frequency at the *a* locus of chromosome 3 while all other chromosomes remain unaffected. And the fact that the phenomenon of position effect is unknown in maize although many reciprocal translocations and some inversions have been studied makes it difficult to accept the view that the mutations at the *a* locus induced by *Dt* are due to minute rearrangements.

EFFECT OF *Dt* ON THE P^{vv} ALLELE

The *Dt* gene affects only the mutability of the *a* allele. A number of endosperm, aleurone and seedling recessives have been tested for increased mutability without any indication that their mutation rates were in any way changed. The tested genes have a low mutation rate, however, and it seemed desirable to test *Dt* against a gene with a high spontaneous mutation rate such as P^{vv} or the unstable waxy gene. Plants heterozygous for *Dt* and possessing the variegation allele were backcrossed to *dt p* individuals. The F_1 seed was classified into *Dt* and *dt* classes and the resulting ears graded for degree of variegation in a similar manner to that employed by Emerson in his studies on variegation. The data obtained are as follows:

Family	*Dt* seed Number ears	Mean variegation grade	*dt* seed Number ears	Mean variegation grade
1	23	4.09	34	4.12
2	22	4.09	19	4.05
3	22	3.82	31	3.87
4	17	4.18	11	4.36
5	32	4.06	35	4.00
6	30	3.67	38	3.68
7	21	4.67	28	4.68
	167		196	
Mean grade		4.08		4.11

The above data show that there is no effect of the *Dt* gene on the mutability of the unstable P^{vv} allele. Similar studies were made of the effect of *Dt* on the unstable waxy allele and no change in mutability was found.

The *A2* locus in chromosome 5 closely resembles the *A* locus in its phenotypic effect. The *A2* allele gives deep colored aleurone and purple plant color as does the *A* allele and the recessive *a2* allele is like *a* in producing colorless aleurone and brown plant color. They differ in that the *a2* allele does not change the pericarp color from red to brown as does the *a* allele. Despite the similarity between these two loci the *Dt* gene does not affect the mutability of the *a2* allele.

TEMPERATURE EFFECT

Although Demerec has shown that the mutability of the unstable miniature-3 gamma gene of *Drosophila virilis* is little affected by temperature a similar experiment was conducted with the *a-Dt* material. Two different *a Dt* strains were grown at a temperature of approximately 21°C until flowering. Immediately after pollination the plants of each strain were divided at random into two lots; one lot was placed in a greenhouse maintained around 15.5°C and the second lot put in an adjoining house at or near a temperature of 27°C. The two lots of plants were left in the two houses until seed had matured. The mutation rates at the two temperature levels were determined by counting the number of dots in the aleurone of 50 seeds from each plant except for those ears marked by asterisks in the table below where less than 50 seeds were available. The data obtained from one strain are given below.

	Average number mutations per seed	
	15.5° C	27° C
	50.2	2.9
	47.2	9.0
	37.5	11.5
	41.2	9.9
	44.9*	3.7
	29.5*	14.5*
		13.5
Total	250.6	65.0
Mean	41.8	9.3

The above data clearly show a depressive effect of a rise in temperature upon the mutability of *a*. However the other strain failed to show a striking effect with a rise of temperature although this statement is based on inspection of the ears as no detailed counts were made. Since the effect of an increase in temperature is beyond statistical doubt for the one strain it is possible that the second strain carried a modifying gene or genes which opposed the depressing action of temperature. This problem must be prosecuted further. Beale and Fabergé have recently (1941) described a similar depressive effect of temperature upon the mutability of an unstable gene in Portulaca.

DISCUSSION

It is doubtful if the mechanism by which *Dt* induces the *a* allele to become highly mutable is the same as that responsible for increased mutation rates found following treatment with short wave radiation and temperature.

These two agents cause a general rise in mutation rate while *Dt* is specific in its effect. Further, chromosomal aberrations are produced in great abundance by irradiation but none has been found in the *Dt* strains. Some investigators hold that the effect of irradiation is to cause a single alteration in the equilibrium position of an atom or electron in the gene molecule. This change in equilibrium may occur as a result of heat vibrations or of ionization following treatment with X-rays, etc. The writer is not willing to judge the validity of this hypothesis. It seems, however, that the effect of the *Dt* allele on the mutability of the *a* allele is best accounted for by assuming that *Dt* alters the cellular environment in some way so that the *a* allele becomes mutable. Simply put, the effect of *Dt* seems to be a chemical rather than a physical phenomenon. While the situation reported here probably is unique, it is possible that a similar case exists—also in maize. Emerson has shown that the *Bh* allele located in chromosome 6 causes kernels which are homozygous for the recessive *c* allele, and therefore normally colorless, to have patches of color. The *Bh* allele affects only the *c* gene and it may be that the color found is produced by mutation of *c* to *C*. This has not yet been demonstrated.

REFERENCES

Beale, G. H., and Fabergé, A. C., 1941, Nature (in press).
Demerec, M., 1935, Bot. Rev. **1**:233-248.
Emerson, R. A., 1914, Amer. Nat. **48**:87-115.
1917, Genetics **2**:1-35.
1929, Genetics **14**:488-511.
Emerson, R. A., and Anderson, E. G., 1932, Genetics **17**:503-509.
Emerson, R. A., Beadle, G. W., and Fraser, A. C., 1935, Cornell Univ. Agric. Exp. Sta. Mem. 180. 83 pp.
Hadjinov, M. I., 1939, Comptes Rendus (Doklady) **23**:366-369.
Rhoades, M. M., 1936, J. Genet. **33**:347-354.
1938, Genetics **23**:377-395.
1939, Proc. Seventh Int. Genetical Congress 247-248.
Sturtevant, A. H., and Beadle, G. W., 1939, An introduction to genetics, 391 pp. Philadelphia: W. B. Saunders Co.
Schultz, Jack, 1936, Proc. Nat. Acad. Sci. **22**:27-33.

DISCUSSION

Demerec: How does the *Dt* gene affect the new alleles of *a?*

Rhoades: They are stable.

Delbrück: The spots in one strain are uniform in size, in another strain they are of different sizes in the *A* material.

Rhoades: The range in all cases is small, however, though some range of size occurs in all stocks.

Demerec: Then modifying factors determine the time of mutation.

Rhoades: Yes, a recessive gene delays the time of mutation and the size of the spots in this strain are extremely small.

Plough: In the temperature effect, are you dealing with an effect on the time of development? In other words, do spots occur at a certain time of development and has high temperature shortened that time?

Rhoades: This is possible but it probably does not account for all the difference. Fabergé and Beale have found a similar depressive effect of a rise in temperature in the case of an unstable gene in *Portulaca grandiflora* and their account will appear shortly in Nature.

Child: You would expect this ıt the *Dt* gene is effective late and only during a certain period.

Muller: The rate of development is changed by temperature but the mutation rate must be affected much more.

Schultz: The temperature effect is the sort of thing obtained in the variegated types in Drosophila.

Rhoades: I do not think the locus of the *Dt* gene near the knob of chromosome 9 has any special significance, although this remains to be determined.

69 Protection afforded by sickle-cell trait against subtertian malarial infection

A. C. Allison*
Clinical Pathology Laboratory
Radcliffe Infirmary
Oxford, England

The aetiology of sickle-cell anaemia presents an outstanding problem common to both genetics and medicine. It is now universally accepted that the sickle-cell anomaly is caused by a single mutant gene which is responsible for the production of a type of haemoglobin differing in several important respects from normal adult haemoglobin (Pauling *et al.*, 1949; Perutz and Mitchison, 1950). Carriers of the sickle-cell trait who are heterozygous for the sickle-cell gene have a mixture of this relatively insoluble haemoglobin and normal haemoglobin; hence their erythrocytes do not sickle *in vivo*, whereas some at least of the homozygotes, who have a much greater proportion of sickle-cell haemoglobin, have sickle cells in the circulating blood, with inevitable haemolysis and a severe, often fatal, haemolytic anaemia. There is also a much smaller group of sickle-cell anaemia patients who are heterozygous for the sickle-cell gene as well as for some other hereditary abnormality of haemoglobin synthesis (Neel, 1952).

It is thus possible to approach the problem from the clinical or the genetical side. From the clinical point of view it is important to distinguish between carriers of the sickle-cell trait who show no other haematological abnormalities and patients with sickle-cell anaemia, who have a haemolytic disease which can reasonably be attributed to sickling of the erythrocytes. From the genetical point of view the main distinction is to be drawn between those who are homozygous and those who are heterozygous for the sickle-cell gene. In the great majority of instances two classifications coincide—that is, most individuals with the sickle-cell trait are heterozygous and most cases of sickle-cell anaemia, in Africa at least, are homozygous for the sickle-cell gene.

The sickle-cell trait is remarkably common in some parts of the world. Among many African negro tribes 20% or more of the total population have the trait, and frequencies of 40% have been found in several African tribes (Lehmann and Raper, 1949; Allison, 1954). In parts of Greece frequencies of 17% have been described (Choremis *et al.*, 1953), and as many as 30% of the population in Indian aboriginal groups are affected (Lehmann and Cutbush, 1952).

Wherever the sickle-cell trait is known to occur sickle-cell anaemia will also be found. For a time it was thought by some workers that sickle-cell anaemia was rare among African negroes, but so many cases have been described during the past few years that this view is no longer tenable (Lambotte-Legrand and Lambotte-Legrand, 1951; Foy *et al.*, 1951; Edington, 1953; Vandepitte and Louis, 1953).

The main problem can be stated briefly: how can the sickle-cell gene be maintained at such a high frequency among so many peoples in spite of the constant elimination of these genes through deaths from the anaemia? Since most sickle-cell anaemia subjects are homozygotes, the failure of each one to reproduce usually means the loss of two sickle-cell genes in every generation. It can be estimated that for the lost genes to be replaced by recurrent mutation so as to leave a balanced state, assuming that the sickle-cell trait—that is, the heterozygous condition—is neutral from the point of view of natural selection, it would be necessary to have

---

From British Medical Journal **1**:290-294, 1954. Used with permission.

This investigation was made possible by a grant received from the Colonial Office at the recommendation of the Colonial Medical Research Committee. Acknowledgment is made to the Directors of Medical Services, Kenya and Uganda, to Dr. E. J. Foley, Nairobi, to the staff of the Mulago Hospital, Kampala, and to the volunteers for their kind co-operation. To Dr. G. I. Robertson, Nairobi, a special debt of gratitude is due for constant help and advice. Drs. A. E. Mourant, R. G. Macfarlane, and A. H. T. Robb-Smith read the manuscript and made valuable comments and suggestions.

*Staines Medical Research Fellow of Exeter College, in receipt of a grant from the Medical Research Council.

a mutation rate of the order of 10^{-1}. This is about 3,000 times greater than naturally occurring mutation rates calculated for man and, with rare exceptions, in many other animals—3.2×10^{-5} in the case of haemophilia (Haldane, 1947). A mutation rate of this order of magnitude can reasonably be excluded as an explanation of the remarkably high frequencies of the sickle-cell trait observed in Africa and elsewhere.

POSSIBILITY OF SELECTIVE ADVANTAGE

Of the other explanations which can be advanced to meet the situation, one has received little attention: the possibility that individuals with the sickle-cell trait might under certain conditions have a selective advantage over those without the trait. It was stated for many years that the sickle-cell trait was in itself a cause of morbidity, but this belief seems to have been based upon unsatisfactory criteria for distinguishing the trait from sickle-cell anaemia. The current view is that the sickle-cell trait is devoid of selective value. Henderson and Thornell (1946) found that in American negro air cadets who had passed a searching physical examination the incidence of the sickle-cell trait was the same as in the general negro population of the United States. Lehmann and Milne (1949) were unable to discover any correlation between haemoglobin levels and the presence or absence of the sickle-cell trait in Uganda Africans. And Humphreys (1952) could find no evidence that the sickle-cell trait was responsible for any morbidity in Nigerian soldiers.

However, during the course of field work undertaken in Africa in 1949 I was led to question the view that the sickle-cell trait is neutral from the point of view of natural selection and to reconsider the possibility that it is associated with a selective advantage. I noted then that the incidence of the sickle-cell trait was higher in regions where malaria was prevalent than elsewhere. The figures presented by Lehmann and Raper (1949) for the frequency of the sickle-cell trait in different parts of Uganda lent some support to this view, as did the published reports from elsewhere. Thus the trait is fairly common in parts of Italy and Greece, but rare in other European countries; in Greece the trait attains its highest frequencies in areas which are conspicuously malarious (Choremis *et al.*, 1951).

RELATION BETWEEN MALARIA AND SICKLE-CELL TRAIT

Other reports appeared suggesting more directly that there might be a relationship between malaria and the sickle-cell trait. Beet (1946) had observed that in a group of 102 sicklers from the Balovale district of Northern Rhodesia only 10 (9.8%) had blood slides showing malaria parasites, whereas in a comparable group of 491 non-sicklers 75 (15.3%) had malaria parasites. The difference in incidence of malaria in the two groups is statistically highly significant (χ^2 = 19.349 for 1 d.f.)*; hence Beet's figures imply strongly that malaria is less frequent among individuals with the sickle-cell trait than among those without the trait. The difference in malarial susceptibility between sicklers and others seemed to be most pronounced at the time of the year when malaria transmission was lowest.

Later, in the Fort Jameson district of Northern Rhodesia, Beet (1947) found that the same difference was present, although it was much less pronounced. Of 1,019 non-sicklers, 312 (30.6%) had blood slides with malaria parasites, whereas of 149 sicklers 42 (28.2%) showed malaria parasites. This difference is not statistically significant. However, among the sicklers from Fort Jameson enlarged spleens were less common than among non-sicklers. In a series of 569 individuals there were 87 with the sickle-cell trait; 24 (27.9%) of these had palpable spleens, as compared with 188 (39.0%) with splenomegaly out of 482 non-sicklers. This difference is again statistically significant (χ^2 = 4.11 for 1 d.f.). Beet concluded that Africans with the sickle-cell trait were probably liable to recurrent attacks of thrombosis, with resultant shrinkage of the spleen.

Brain (1952a), also working in Rhodesia, confirmed Beet's observation that the spleen is palpable in a much lower proportion of sicklers than of non-sicklers; he went on to suggest that the finding might be explained by diminished susceptibility to malaria on the part of the sicklers. Moreover, Brain (1952b) compared the proportion of hospitalized cases in groups of African mine-workers with and without the sickle-cell trait. He found that the sicklers actually spent less time in hospital, on an average, than did the control group of non-sicklers. The incidence of malaria and pyrexias

*These and other statistics in this paper are my own, using available figures.

of unknown origin was much lower in the group with sickle cells.

It became imperative, then, to ascertain by more direct methods of investigation whether sickle cells can afford some degree of protection against malarial infection, thereby conferring a selective advantage on possessors of the sickle-cell trait in regions where malaria is hyperendemic. An opportunity to do this came during the course of a visit to East Africa in 1953.

INCIDENCE OF MALARIAL PARASITAEMIA IN AFRICAN CHILDREN WITH AND WITHOUT THE SICKLE-CELL TRAIT

The observations of Beet and of Brain on differences in parasite rates and spleen rates are open to criticism because the populations were heterogeneous, and were drawn from relatively wide areas. It was decided, therefore, to carry out similar tests on a relatively small circumscribed community, where all those under observation belong to a single tribe. Children were chosen rather than adults as subjects for the observations so as to minimize the effect of acquired immunity to malaria. The recorded incidence of parasitaemia in a group of 290 Ganda children, aged 5 months to 5 years, from the area surrounding Kampala (excluding the non-malarious township) is presented in Table I. The presence of sickling was demonstrated by chemical reduction of blood with isotonic sodium metabisulphite (Daland and Castle, 1948). Fresh reducing solutions were made up daily.

It is apparent that the incidence of parasitaemia is lower in the sickle-cell group than in the group without sickle cells. The difference is statistically significant (χ^2 = 5.1 for 1 d.f.). In order to test as many families as possible only one child was taken from each family. There is no reason to suppose that these groups are not comparable, apart from the presence or absence of the sickle-cell trait.

The parasite density in the two groups also differed: of 12 sicklers with malaria, 8 (66.7%) had only slight parasitaemia (group 1 on an arbitrary rating), while 4 (33.3%) had a moderate parasitaemia (group 2). Of the 113 non-sicklers with malaria, 34% had slight parasitaemia (group 1), the parasite density in the remainder being moderate or severe (group 2 or 3).

It may be noted, incidentally, that of the four cases in the sickle-cell group with moderate parasitaemia three had *P. malariae,* even though this species is much less common than *P. falciparum* around Kampala. It seems possible from these and other observations that the protection afforded by the sickle-cell trait is more effective against *P. falciparum* than against other species of plasmodia, but much further work is necessary to decide the point.

These results suggest that African children with the sickle-cell trait have malaria less frequently or for shorter periods, and perhaps also less severely, than children without the trait. Further evidence regarding the protective action of the sickle-cell trait could be obtained only by direct observation on the course of artificially induced malarial infection in volunteers.

PROGRESS OF MALARIAL INFECTION IN ADULT AFRICANS WITH AND WITHOUT THE SICKLE-CELL TRAIT

Fifteen Luo with the trait and 15 Luo without the trait were accepted for this investigation. All the volunteers were adult males who had been away from a malarious environment for at least 18 months. The two groups were of a similar age and appeared to be strictly comparable apart from the presence or absence of the sickle-cell trait. Two strains of *P. falciparum* were used—one originally isolated in Malaya and one from near Mombasa, Kenya; in Table II these are marked with the subscripts 1 and 2 respectively. The infection was introduced by subinoculation with 15 ml. of blood containing a large number of trophozoites (B in the table) or by biting with heavily infected *Anopheles gambiae* (M in the table). At least 3 out of the 10 mosquitoes applied had bitten each individual, and the presence of sporozoites was confirmed by dissection of the mosquitoes.

The cases were followed for 40 days. Parasite counts for each case were made by comparison with the number of leucocytes in 200 oil-immersion fields of thick films, the absolute leucocyte counts being checked at intervals. The abbreviated results of these counts are shown in Table II. In the few cases in which parasitaemia was pronounced and the symptoms were relatively severe the progress of the disease was arrested. At the end of the period of observation in every case a prolonged course of antimalarial chemotherapy was given.

Table I

	With parasitaemia	Without parasitaemia	Total
Sicklers	12 (27.9%)	32 (72.1%)	43
Non-sicklers	113 (45.7%)	134 (53.3%)	247

DISCUSSION

It is apparent that the infection has become established in 14 cases without the sickle-cell trait. The parasitaemia is relatively light, however, when compared with that observed in non-immune populations—for example, the Africans described by Thomas *et al.* (1953). This is to be expected: the Luo come from a part of the country where malaria is hyperendemic, and they have acquired a considerable immunity to malarial infection in childhood. This factor makes the interpretation of the observations rather more difficult; however, it could not be avoided, since all the East African tribes who have high sickling rates come from malarious areas, and the acquired immunity should operate with equal force in the groups with and without sickle-cells. The acquired immunity was actually an advantage, since the symptoms were mild and the chances of complication very slight.

In the group with sickle cells, on the other hand, the malaria parasites have obviously had great difficulty in establishing themselves, in spite of repeated artificial infection. Only two of the cases show parasites, and the parasite counts in these cases are comparatively low. The striking difference in the progress of malarial infection in the two groups is taken as evidence that the abnormal erythrocytes in individuals with the sickle-cell trait are less easily parasitized than are normal erythrocytes.

It can therefore be concluded that individuals with the sickle-cell trait will, in all probability, suffer from malaria less often and less severely than those without the trait. Hence in areas where malaria is hyperendemic children having the trait will tend to survive, while some children without the trait are eliminated before they acquire a solid immunity to malarial infection. The protection against malaria might also increase the fertility of possessors of the trait. The proportion of individuals with sickle cells in any population, then, will be the result of a balance between two factors; the severity of malaria, which will tend to increase the frequency of the gene, and the rate of elimination of the sickle-cell genes in individuals dying of sickle-cell anaemia. Or, genetically speaking, this is a balanced polymorphism where the heterozygote has an advantage over either homozygote.

The incidence of the trait in East Africa has recently been investigated in detail (Allison, 1954), and found to vary in accordance with the above hypothesis. High frequencies are observed among the tribes living in regions where malaria is hyperendemic (for example, around Lake Victoria and in the Eastern Coastal Belt), whereas low frequencies occur consistently in the malaria-free or epidemic zones (for example, the Kigezi district of Uganda; the Kenya Highlands; and the Kilimanjaro, Mount Meru, and Usumbara regions of Tanganyika). This difference is often independent of ethnic and linguistic grouping: thus, the incidence of the sickle-cell trait among Bantu-speaking tribes ranges from 0 (among the Kamba, Chagga, etc.) to 40% (among the Amba, Simbiti, etc.). The world distribution of the sickle-cell trait is also in accordance with the view presented here that malarial endemicity is a very important factor in determining the frequency of the sickle-cell trait. The genetical and anthropological implications of this view are evident.

The fact that sickle cells should be less easily parasitized by plasmodia than are normal erythrocytes is presumably attributable to their haemoglobin component, although there may be other differences, not yet observed, between the two cell-types. Sickle-cell haemoglobin is unlike normal adult haemoglobin in important physico-chemical properties, notably in the relative insolubility of the sickle-cell haemoglobin when reduced (Perutz and Mitchison, 1950). The malaria parasite is able to metabolize haemoglobin very completely in the intact red cell, the haematin pigment remaining as a by-product of haemoglobin breakdown (Fairley and Bromfield, 1934; Moulder and Evans, 1946). That plasmodia are greatly affected by relatively small differences in their environment is suggested by their remarkable species specificity. Thus the difficulty of establishing an infection in monkeys with human malaria parasites, and vice versa, is generally recognized.

How far species differences in the haemoglobins themselves, known from immunological and other studies, are responsible for the species specificity of plasmodia it is impossible

Table II

No.	Mode of infection and strain	Day after infection																
		8	10	12	14	16	18	20	22	24	26	28	30	32	34	36	38	40
								Luo with no sickle-cells										
1	M_2 B_1	0.03	–	0.07	2.5	5.0	2.5	5.0	1.2	0.4	0.02	0.01	–	–	0.1	0.01	0.01	ST
2	M_2 B_1	–	–	–	–	–	–	–	0.03	0.13	0.41	–	–	–	–	0.03	–	ST
3	M_2 B_1	–	–	–	–	–	–	–	0.1	0.02	0.20	5.0	2.5	1.25	1.67	0.2	5.0	2.0 S
4	M_2 B_1	–	–	–	–	0.02	0.02	0.5	0.83	0.12	0.2	1.0	1.0	0.83	0.25	0.17	–	ST
5	M_2 B_2	–	–	–	–	0.05	1.0	1.67	0.25	0.05	0.07	0.25	1.2	1.0	0.03	–	–	ST
6	B_1	0.02	5.0	10.0	10.0	1.0	0.1	0.01	ST	–	–	–	–	–	–	–	–	–
7	B_2	–	–	–	–	15.0	50.0	ST	–	–	–	–	–	–	–	–	–	–
8	B_2	–	–	0.13	5.0	1.67	0.33	–	–	ST	–	–	–	–	–	–	–	–
9	B_1	–	–	–	–	5.0	–	0.1	0.5	2.5	–	1.0	0.1	2.5	10.0	5.0	0.5	ST
10	B_2	–	–	–	–	–	–	0.05	0.05	–	–	0.67	–	0.1	0.05	5.0	5.0	ST
11	B_2	–	0.05	0.3	–	–	–	0.2	ST									
12	B_2	–	–	0.3	0.3	0.3	0.1	0.3	ST									
13	B_2	–	–	–	–	–	–	–	–	–	–	–	–	–	–	–	ST	
14	B_2	2.0	1.7	2.0	60.0	5.0	0.6	ST										
15	B_2	0.05	0.3	–	0.4	0.1	0.3	ST										
								Luo with sickle-cell trait										
1	M_1 B_2	–	–	–	–	–	–	–	–	–	–	–	–	–	–	–	–	ST
2	M_1 B_2	–	–	–	–	–	–	–	–	–	–	–	–	–	–	–	–	"
3	M_1 B_1	–	–	–	–	–	–	–	–	–	–	–	–	–	–	–	–	"
4	M_1 B_1	–	–	–	–	–	–	–	–	–	–	–	–	–	–	–	–	"
5	M_1 B_1	–	–	–	–	–	–	–	–	–	–	–	–	–	–	–	–	"
6	M_1 B_1	–	–	–	–	–	–	–	–	–	–	–	–	–	–	–	–	"
7	M_1 B_1	–	–	–	–	–	–	–	–	–	–	–	–	–	–	5.0	0.5	"
8	M_1 B_1	–	–	–	–	–	–	–	–	–	–	–	–	–	–	–	–	"
9	M_1 B_1	0.7	–	–	–	–	–	–	–	–	–	0.03	0.1	0.03	0.03	–	–	"
10	M_1 B_1	–	–	–	–	–	–	–	–	–	–	–	–	–	–	–	–	"
11	B_2 M_2	–	–	–	–	–	–	–	–	–	–	–	–	–	–	–	–	"
12	B_2 M_2	–	–	–	–	–	–	–	–	–	–	–	–	–	–	–	–	"
13	B_2 M_2	–	–	–	–	–	–	–	–	–	–	–	–	–	–	–	–	"
14	B_2 M_2	–	–	–	–	–	–	–	–	–	–	–	–	–	–	–	–	"
15	B_2 M_2	–	–	–	–	–	–	–	–	–	–	–	–	–	–	–	–	"

Figures represent parasite counts in hundreds per $mm.^3$ of blood.
ST = Stopped by chemotherapy.

to say. However, the physico-chemical differences between human adult haemoglobin and monkey haemoglobin appear to be less pronounced than the differences between either type and sickle-cell haemoglobin. It is clear that the natural resistance to malaria among individuals with the sickle-cell trait is relative, not absolute. This is perhaps attributable to differences in the expressivity of the sickle-cell gene, which may be responsible for the production of from nearly 50% to only a very small amount of sickle-cell haemoglobin (Wells and Itano, 1951; Singer and Fisher, 1953). Moreover, the sickle-cell haemoglobin may not be evenly distributed in the cell population: most observers recognize that there are cases in which only some of the red cells are sickled even after prolonged reduction. However, even a relative resistance to malaria may be enough to help those with the sickle-cell trait through the dangerous years of early childhood, during which an active immunity to the disease is developed.

The above observations focus attention upon the importance of haemoglobin to plasmodia in the erythrocytic phase. Hence it is worth considering whether erythrocytes containing other specialized or abnormal types of haemoglobin might be resistant to malaria also. Thus, human foetal haemoglobin differs from human adult haemoglobin in many properties. Red cells containing foetal haemoglobin continue to circulate in the newborn for the first three months of life, after which they are quite rapidly replaced by cells containing normal adult haemoglobin. It has long been known that the newborn has a considerable degree of resistance to malarial infection: Garnham (1949), for instance, found that in the Kavirondo district of Kenya at the end of the second month of life only 10% of babies were infected; after this age the percentage affected rises rapidly, until by the ninth month practically all children have the disease. The correspondence between the appearance of cells containing normal adult-type haemoglobin and malarial susceptibility is illustrated in the Chart. The correspondence may of course be fortiuitous, but it is striking enough to merit further investigation, even though other factors, such as a milk diet deficient in *p*-aminobenzoic acid (Maegraith *et al.,* 1952; Hawking, 1953) and immunity acquired from the mother (Hackett, 1941) may play a part in the natural resistance of the newborn to malaria.

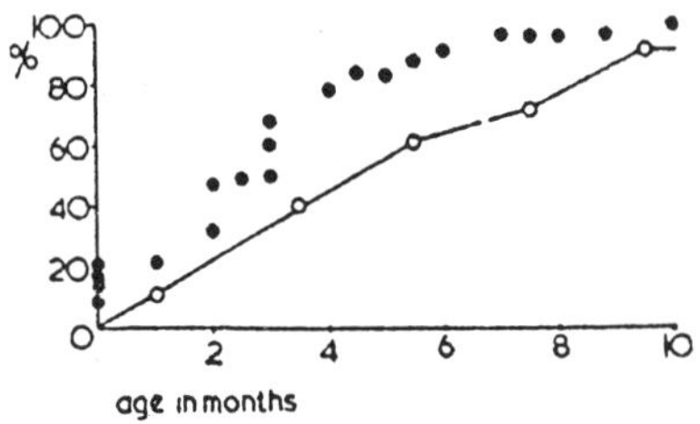

The apparent relationship between the appearance of adult-type haemoglobin (dots) and malarial infection (circles) in the newborn. Each dot represents a test on a single individual, using an alkali denaturation technique (Allison, unpublished observations); the circles represent the percentage of Luo children showing malaria parasites (Garnham, 1949).

Finally, it is possible that the explanation offered above for the maintenance of the sickle-cell trait may also apply to thalassaemia. The problems presented by the two diseases are very similar; many homozygotes, and possibly some heterozygotes, are known to die of thalassaemia, and yet the condition remains remarkably common in Italy and Greece, where as many as 10% of the individuals in certain areas are affected (Bianco *et al.,* 1952). Greek and Italian authors have commented that cases of thalassaemia usually come from districts severely afflicted with malaria (Choremis *et al.,* 1951). Perhaps those who are heterozygous for the thalassaemia gene suffer less from malaria than their compatriots: the fertility of the heterozygotes appears to be greater (Bianco *et al.,* 1952). Selective advantage of the heterozygote with the sickle-cell gene, and possibly the heterozygote with the thalassaemia gene also, would explain why such high gene frequencies can be attained in the case of these conditions while other genetically transmitted abnormalities of the blood cells remain uncommon, not very much above the estimated mutation rate—for example, hereditary spherocytosis (Race, 1942).

SUMMARY

A study has been made of the relationship between the sickle-cell trait and subtertian malarial infection. It has been found that in indigenous East Africans the sickle-cell trait affords a considerable degree of protection against subtertian malaria. The incidence of parasitaemia in 43 Ghanda children with the sickle-cell trait was significantly lower than in a comparable group of 247 children without the trait. An infection with *P. falciparum* was established in 14 out of 15 Africans without the sickle-cell trait, whereas in a comparable group of 15 Africans with the trait only 2 developed parasites.

It is concluded that the abnormal erythrocytes of individuals with the sickle-cell trait are less easily parasitized by *P. falciparum* than are normal erythrocytes. Hence those who are heterozygous for the sickle-cell gene will have a selective advantage in regions where malaria is hyperendemic. This fact may explain why the sickle-cell gene remains common in these areas in spite of the elimination of genes in patients dying of sickle-cell anaemia. The implications of these observations in other branches of haematology are discussed.

REFERENCES

Allison, A. C. (1954). Trans. roy. Soc. trop. Med. Hyg. In press.

Beet, E. A. (1946). E. Afr. med. J. **23**:75.

____ (1947). Ibid. **24**:212.

Bianco, I., Montalenti, G., Silvestroni, E., and Siniscalo, M. (1952). Ann. Eugen. Lond. **16**:299.

Brain, P. (1952a). British Medical Journal **2**:880.

_____(1952b). S. Afr. med. J. **26**:925.

Choremis, C., Zervos, N., Constantinides, V., and Zannos, Leda (1951). Lancet **1**:1147.

________ Ikin, Elizabeth W., Lehmann, H., Mourant, A. E., and Zannos, Leda (1953). Ibid. **2**:909.

Daland, G. A., and Castle, W. B. (1948). J. Lab. clin. Med. **33**:1082.

Edington, G. M. (1953). British Medical Journal **2**:957.

Fairley, Sir N. H., and Bromfield, R. J. (1934). Trans. roy. Soc. trop. Med. Hyg. **28**:141.

Foy, H., Kondi, A., and Brass, W. (1951). E. Afr. med. J. **28**:1.

Garnham, P. C. C. (1949). Ann. trop. Med. Parasit. **43**:47.

Hackett, L. W. (1941). Publ. Amer. Ass. Advanc. Sci. **15**:148.

Haldane, J. B. S. (1947) Ann. Eugen., Lond. **13**:262.

Hawking, F. (1953). British Medical Journal **1**:1201.

Henderson, A. B., and Thornell, H. E. (1946). J. Lab. clin. Med. **31**:769.

Humphreys, J. (1952). J. trop. Med. Hyg. **55**:166.

Lambotte-Legrand, J., and Lambotte-Legrand, C. (1951). Institut Royal Colonial Belge, Sect. d. Sci. Nat. et Med., Memoires, Tome XIX. fasc. 7, p. 98.

Lehmann, H., and Cutbush, Marie (1952). British Medical Journal **1**:404.

________ and Milne, A. H. (1949). E. Afr. med. J. **26**:247.

________ and Raper, A. B. (1949). Nature, Lond. **164**:494.

Maegraith, B. G., Deegan, T., and Jones, E. S. (1952). British Medical Journal **2**:1382.

Moulder, J. W., and Evans, E. A. (1946). J. biol. Chem. **164**:145.

Neel, J. V. (1950). Cold Spr. Harb. Symp. quant. Biol., **15**:141.

____(1952). Blood **7**:467.

Pauling, L., Itano, H. A., Singer, S. J., and Wells, I. C. (1949). Science **110**:543.

Perutz, M. F., and Mitchison, J. M. (1950). Nature, Lond. **166**:677.

Race, R. R. (1942). Ann. Eugen., Lond. **11**:365.

Singer, K., and Fisher, B. (1953). Blood **8**:270.

Thomas, A. T. G., Robertson, G. I., and Davey, D. G. (1953). Trans. roy. Soc. trop. Med. Hyg. **47**:338.

Vandepitte, J. M., and Louis, L. A. (1953). Lancet **2**:806.

Wells, I. C., and Itano, H. A. (1951). J. biol. Chem. **188**:65.

70 Caucasian genes in American Negroes

Measurement of non-African ancestry is difficult, but it is worthwhile for several genetic reasons

T. Edward Reed
Departments of Zoology, Anthropology, and Paediatrics
University of Toronto
Toronto, Canada

It is very difficult to describe the genetic history of a large, defined human population in a meaningful way. As a result there have been few opportunities, at the population level, to study the consequences of known genetic events in the recent past of modern populations. The Negro population of the United States, however, is one of the exceptions to these generalizations. The American individual to whom the term *Negro* is applied is almost always a biracial hybrid. Usually between 2 and 50 percent of his genes are derived from Caucasian ancestors, and these genes were very probably received after 1700. While it is obviously of social and cultural importance to understand Negro hybridity, it is less obvious that there are several pertinent genetic reasons for wishing to know about the magnitude and nature of Caucasian ancestry in Negroes. Recent data, both genetic and historical, now make possible a better understanding of American Negro genetic history than has been possible heretofore. Here I review and criticize the published data on this subject, present new data, and interpret the genetic significance of the evidence.

From Science **165**:762-768, August 22, 1969.

Preparation of this article was begun while I was engaged in work for the Child Health and Development Studies (Division of Biostatistics, School of Public Health, University of California, Berkeley, and the Kaiser Foundation Research Institute, Oakland, California), on leave from the University of Toronto, and was supported there by U.S. Public Health Service research grants HD 00718 and HD 00720 from the National Institutes of Health. The analysis was supported in part by a grant from the Medical Research Council of Canada. I thank Professor Philip D. Curtin for making unpublished data available and for commenting on the manuscript, Dr. Arthur E. Mourant and Mrs. K. Domaniewska-Sobczak for recent references to African blood group distributions, and Professors Curt Stern, Donald Rucknagel, and Peter Carstens for their comments. Dr. H. Gershowitz and Dr. M. Shapiro made available unpublished data on Duffy blood groups in Negroes.

In order to put the genetic data in proper context, I must first give a little of the history of American slavery. The first slaves were brought to what is now the United States in 1619. Importation of slaves before 1700 was negligible, however, but after that date it proceeded at a high rate for most of the 18th century. Importation became illegal after 1808 but in fact continued at a low rate for several more decades *(1, 2)*. The total number of slaves brought into the United States was probably somewhat less than 400,000 *(3)*. Charleston, South Carolina, was the most important port of entry, receiving 30 to 40 percent of the total number *(4)*. More than 98 percent of the slaves came from a very extensive area of West Africa and west-central Africa—from Senegal to Angola—and, in these areas, from both coastal and inland regions. Shipping lists of ships that brought slaves to the United States—and to the West Indies, often to be sent later to the United States—provide a fairly detailed picture of the geographic origins of the slaves and a less complete picture of their ethnic origins. Table 1 gives the approximate proportions of American slaves brought from the eight major slaving areas of Africa. The contribution from East Africa is seen to be negligible, whereas the area from Senegal to western Nigeria contributed about half the total and the region from eastern Nigeria to Angola contributed the other half. An earlier tabulation for entry at Charleston alone *(5)* is quite similar, except that the contribution from the Bight of Biafra is much less (0.021 as compared to 0.233) and that from "Angola" is appreciably greater (0.396 as compared to 0.245).

At some early point in American slavery,

Table 1. African origins of slaves imported into the North American mainland [data of Curtin *(37)*]. Distribution by areas is approximate and is an average of data for Virginia (1710-1769), for South Carolina (1773-1807), and for the British slave trade (1690-1807).

Coastal region of origin	Approximate present area	Peoples	Approximate proportion from region
Senegambia	Senegal and Gambia	Mainly Bambara and Malinke (from interior)	0.133
Sierra Leone	Sierra Leone	Sierra Leone, Guinea, Portugese Guinea peoples, plus Bambara and Malinke	.055
Windward Coast	Ivory Coast, Liberia	Various peoples of area	.114
Gold Coast	Ghana	About ¾ Akan people from southern part, the rest from northern part	.159
Bight of Benin	Togo, Dahomey, Nigeria west of Benin river	Peoples of Togo, southern Dahomey, and western Nigeria	.043
Bight of Biafra	Nigeria (east of Benin river) to 1°S (Gabon)	About ¾ Ibo, the rest Ibibio and people from Cameroon	.233
"Angola"	1°S to southwest Africa (Gabon, Congo, Angola)	Many peoples of the area, from the coast to far inland	.245
Mozambique and Madagascar			.016
Region unknown			.002

matings between slaves and Caucasians began to occur. Quantitative data are lacking, and we can say only that most of these matings occurred after 1700. Our concern here is the genetic consequences of the matings—the introduction of Caucasian genes into the genome (or total complement of genetic material) of the American Negro. We could, in theory, estimate the Caucasian contribution to American Negro ancestry in a very simple way *if* certain strict criteria were met. In practice it is not possible to show that all these criteria are met, but this fact has not stopped geneticists, including myself, from making estimates.

The usual estimation procedure is simple and direct. Consider some gene—say the allele *A* of the ABO blood group locus, whose frequency was q_a in the African ancestors of American Negroes and q_c in the Caucasian ancestors, while in modern American Negroes the frequency is q_n. If *M* is the present proportion of genes at this genetic locus (and, ideally, at every other locus too) which are derived from Caucasians, and if race mixture is the only process affecting q_n, then, by definition,

$$q_n = Mq_c + (1 - M)q_a \quad (1)$$

and therefore

$$M = (q_n - q_a)/(q_c - q_a) \quad (2)$$

This formula for *M*, or an algebraic equivalent, was used for all estimates of *M* given in Table 2 except one. [This one differed only in that three alleles were used simultaneously at one locus to obtain a maximum likelihood estimate for *M;* for each allele an equation of the type of Eq. 1 could be written, and used to estimate *M (6)*]. We see that if we know q_a, q_c, and q_n (for a defined area) without error and if there were no factors affecting q_n other than race crossing, estimation of *M* would be simple. Unfortunately, such is not the case.

CRITERIA FOR CRITICAL ESTIMATION OF M

Critical evaluation of estimates of *M* requires complete specification of the needed criteria and judgment on the degree to which these criteria are met. These criteria are simple and obvious, but the demands they make have not always been appreciated. They are as follows.

1) The exact ethnic compositions of the two ancestral populations, African Negro and Caucasian, are known.

2) No change in gene frequency (for the gene in question) between ancestral and modern populations either of African Negroes or of American Caucasians has occurred.

3) Interbreeding of the two ancestral popu-

lations is the only factor affecting gene frequency in U.S. Negroes—that is, there has been no selection, mutation, or genetic drift.

4) Adequate samples (that is, samples that are unbiased, from correct populations, with small standard error) of the modern descendants of the ancestral African Negroes and U.S. Caucasians, and of modern U.S. Negroes, are available.

It should be said immediately that none of these criteria has been shown to be fully met in any study. In particular, point 1 is not met, because the detailed ethnic origins of slaves from the various slaving areas are unknown *(4)*. Point 2 can never be met because ancestral gene frequencies are unknown and point 3, at best, can only be inferred from indirect evidence. Point 4 cannot be fully met for African

Table 2. Published estimates of the proportion *(M)*, in American Negroes, of genes that are of Caucasian origin. All estimates except those based on genes Fy^a, Gm^1, $Gm^{1,2}$, or Gm^5 (and perhaps AK^2) require an estimate of African gene frequency appreciably different from zero. Within regions, localities are listed in chronological order of the estimates. Standard errors for *M* were not given (except for reference *6*).

Region* and locality	Gene(s)†	Sample size: Negro	Sample size: Caucasian	*M*	Reference
	Estimates for M *presumed by their authors to be valid*				
Non-southern					
Baltimore	R^0	907	7,317	0.306	*(7)*
Baltimore	R^0	907	7,317	.216	*(15)*
Five areas	R^0, R^1, Jk^b, T, S	96 to 3,156	189 to 7,317	~.20	*(16)*
Cleveland and Baltimore	Gm^1, Gm^5	623	249	.310	*(11)*
Various	R^0, R^1, R^2, r, M, S, Jk^b, k, Fy^b			.232-.261	*(17)*
Chicago	AK^2‡	1.063	1,315	.13	*(14)*
Washington, D.C., Baltimore, New York City	R^0, R^1, Fy^a			.20-.24	*(8)*
Oakland, Calif.	Gm^1, $Gm^{1,2}$, Gm^5	260	478	.273±.037	*(6)*
Southern					
Evans and Bullock counties, Ga.	R^0, R^1	340	331	.104	*(9)*
Evans and Bullock counties, Ga.	Gm^1, $Gm^{1,2}$	189	295	.073	*(12)*
	Gc^1	231	292	~.10	
Charleston, S.C.	R^0, R^1, Fy^a	515		.04-.08	*(8)*
James Island, S.C. and Evans and Bullock counties, Ga.	R^0, R^1, Fy^a	394		.09-.12	*(8)*
	Estimates of M *presumed by their authors to indicate selection*				
Non-southern §					
Four areas, mainly Seattle	Hp^1	936	865(?)	~.40	*(21)*
Seattle	Hp^1	1,657	?	.478	*(8)*
Seattle	Gd^{A-}	658♂♂		.490	*(8)*
Southern					
Evans and Bullock counties, Ga.	T	285	314	.466	*(9)*
	Hp^1‖	167	145	.42-.70	
	Gd^{A-}	76♂♂		.34-.44	
	Hb^S	247		.46-.69	
	Tf^{D1}‖	133	107	.495	
Memphis	Gd^{A-}	97♂♂		.175	*(8)*

*An estimate of 0.34, from Hb^S data on 10,858 Negroes, is based on 11 sources in both the North and the South *(38)*. It is therefore not placed in a regional category. †Locus and alleles used are as follows. Blood groups: Rh *(R^0, R^1, R^2, r)*, Kidd *(Jk^b)*, M-N-S-s *(M, S)*, Kell *(k)*, Duffy *(Fy^a, Fy^b)*; serum protein genes: Gm *(Gm^1, $Gm^{1,2}$, Gm^5)*, haptoglobin *(Hp^1)*, Gc *(Gc^1)*, transferrin *(Tf^{D1})*; hemoglobin: HbS *(Hb^S)*; red cell enzymes: adenylate kinase *(AK^2)*, glucose-6-phosphate dehydrogenase deficiency *(Gd^{A-})*; phenylthiocarbamide tasting *(T)*. ‡Newly investigated gene. The African frequency of AK^2 is poorly known, but it is assumed to be zero. The 95-percent confidence interval for *M* is 0.03-0.23, according to my calculation. §Seven non-southern estimates ranging from 0.270 to 0.685, obtained by Workman *(8)* (using Hp^1 or Gd^{A-}) on small samples (79 to 238 Negroes) are omitted here. ‖"Possibly" reflecting selection.

Negroes, since the proportions of various ethnic contributions are only roughly known. The problem is simpler for U.S. Negroes and Caucasians, although marked heterogeneity in values of *M* between different Negro populations is now known to complicate the matter.

Somewhat more affirmative views on these criteria can also be given, however. *If* it can be shown that gene frequencies in neighboring modern tribes and in populations of adjacent former slaving areas do not differ appreciably, point 1 becomes less important. For example, this appears to be the situation for the ABO blood groups, the best-known genetic system throughout the slaving area. With regard to point 2, since the populations concerned usually were, and are, large, it is probable that this criterion is quite well satisfied. If point 1 is satisfied in the way suggested, point 4 may be met by using large, carefully collected samples. Unfortunately, it is less easy to overcome the problem posed by point 3. This is discussed below.

REVIEW OF PUBLISHED ESTIMATES OF M

Table 2 is a tabulation of published estimates of *M* for American Negroes, beginning with the well-known estimate of 0.31 for Baltimore Negroes given by Glass and Li in 1953 *(7)*. The estimates are grouped according to the authors' statements as to their validity or lack of validity (due to selection) as estimates of *M*. They are further classified as "southern" (estimates for Georgia, South Carolina, and Tennessee) or "non-southern." As has been noted elsewhere *(6, 8, 9)*, among the presumed valid estimates, all "non-southern" estimates are greater than "southern" estimates. Also, the estimates presumed to indicate selection are usually appreciably higher than the estimates presumed to be valid. Among the "valid" estimates of *M*, that of Glass and Li *(7)* is by far the best known, and is often quoted as "the" estimate for the amount of Caucasian ancestry in "the" American Negro (see, for example, *10-14*). A revision of this estimate from 0.31 to 0.216 *(15)* appears to have escaped general notice.

The estimates of Table 2 must be considered in the light of the four criteria given above. As already noted, criterion 1 cannot be strictly satisfied for any estimate because the detailed ethnic origins of the slaves are unknown. The estimates for *M* in Table 2, however, do not even roughly meet criterion 1, since none of them is based on quantitative information on distribution of origins, such as is given in Table 1. Typically, data from only one or two regions of West Africa are taken to represent the whole slaving area. Ironically, for the best-known estimate, that of Glass and Li *(7)*, Rh blood group data from East and South Africa, as well as from Ghana, were used to represent ancestral Rh blood group frequencies because better data were not then available. Glass, for his revised estimate *(15)*, used only Rh data from Nigeria and Ghana. Of the 540 individuals from Nigeria studied *(15)*, 105 were Ibos, who may be representative of ancestral inhabitants of the Bight of Biafra region, the area of origin of about 23 percent of American slaves (Table 1); the remaining 435 individuals from Nigeria may be representative of the slaves (4 percent) who came from the Bight of Benin. The 274 individuals from Ghana studied *(15)* may be representative of the slaves (16 percent) from that region. The slaves (57 percent) from areas other than Nigeria and Ghana are unrepresented in Glass's revised estimate. These same Rh blood group data were used by later investigators in arriving at their own estimates *(8, 9, 16, 17)*. These critical comments on the best-known estimate are made to illustrate the nature of the problem: similar comments could be made about each of the other estimates of Table 2.

With regard to criterion 4 (adequacy of samples), one can distinguish between (i) adequate representation (by the mean gene frequency used) of the entire slaving area and (ii) adequate sample size (as shown by a small standard error for *M*). If the gene used has a uniform frequency over the entire slaving area, any large sample from one part of the area could adequately represent the whole. The problem, of course, is to demonstrate uniformity. If, as one would expect, gene frequencies vary from region to region of the slaving area, appropriate samples over the whole area are needed if one is to obtain a properly weighted mean frequency. Neither of these approaches has been used in making any of the estimates. [I made an attempt to confirm the belief that the frequency of certain Gm alleles is near zero in African populations *(6)* but found that not enough surveys had been made.]

To make the problem more concrete, let us consider Glass's estimate of *M (15)* in the light of more recent Rh data. For the R^0 allele of the Rh locus, he used 0.5512 for the frequency in West Africa (on the basis of the data from Nigeria and Ghana). The frequencies in present-day U.S. Negroes and Caucasians were found to be 0.4381 and 0.0279, respectively, so that, from Eq. 2, we estimate M to be (0.5512–0.4381)/(0.5512–0.0279), or 0.216. However, the frequency of R^0 in Liberia is 0.60 *(18)*, and in Bantu of the Congo (Leopoldville) it is also about 0.60 *(19)*. If the true overall value for the slaving area were 0.60, the estimate for M would be 0.283.

With regard to the purely statistical accuracy of the estimates of M, as shown by standard errors, calculation of the standard errors for several pertinent estimates indicates that they may be much larger than the authors may have suspected *(20)*. The standard error for Glass's estimate *(15)*, for example, is 0.042, giving a 95-percent confidence interval of 0.133 to 0.299. The estimate in Table 2, of 0.13 for M for gene AK^2 (the lowest estimate for the non-southern region) has a standard error of 0.053, producing a 95-percent confidence interval of 0.025-0.234, overlapping Glass's estimate. This large error seems particularly surprising at first, in view of the large sample sizes, but it is explained by the very low AK^2 gene frequencies (<5 percent). The standard errors of the other estimates appear to be of comparable size or larger (due to smaller sample sizes).

I have said enough to show the deficiencies of most of the estimates of Table 2 with regard to both African gene frequency and statistical accuracy. I should also comment on the classification of M estimates as "valid" (not affected by selection) or as indicating the effects of selection. Classification of an estimate in this way requires a "standard" M that is thought to be free from the effects of selection. Such a "standard" can then be used to determine whether an M estimated for some other gene demonstrates selection. The M estimates from Rh genes R^0 and R^1 have been assigned this role of "standard" by various investigators [Parker *et al.* *(21)* chose R^0 alone; Workman and his associates *(8, 9)* chose R^0 and R^1 in combination]. In addition, M estimates from frequencies of the Fy^a allele of the Duffy blood group locus *(8)* and the Gm^1 and Gm^5 alleles of the Gm serum group locus *(21)* have been considered as possible standards. Yet, as discussed above, it is not possible to prove directly that selection has not affected a particular gene frequency in American Negroes, and no evidence in support of the belief that it has not has been offered. We can only draw inferences of varying degrees of rigor as suitable data become available. I attempt in the remainder of this article to draw and apply such inferences.

AN APPROACH TO A MORE CRITICAL ESTIMATE OF M

To constitute a critical estimate in the light of the four criteria listed above, an estimate of M should substantially meet three of them—1, 3, and 4 (2 is, of course, untestable). This means that we should (i) have good survey data on gene frequency from most or all of the seven West African and west-central African slaving areas of Table 1; (ii) be able to calculate a mean African gene frequency properly weighted according to the origins shown in Table 1; (iii) have adequate data on Caucasians and U.S. Negroes; (iv) have samples large enough to give an acceptably small standard error for M; and, very importantly, (v) have some evidence that in U.S. Negroes the gene in question is not subject to strong selection. With regard to points (i) and (ii), an ideal situation is to have a gene which can be shown to be absent or rare in all parts of the slaving area but common in Caucasians. The problem of finding "the" African-ancestor gene frequency is then eliminated, and M is simply q_n/q_c. The Caucasian gene contribution is then directly determinable. It has been claimed that Gm alleles Gm^1, $Gm^{1,2}$, and Gm^5 are of this type *(22)*; it is quite likely that they are, but not enough of the slaving area has been surveyed for Gm alleles for us to be sure *(6)*.

The Fy^a gene may be almost an ideal "Caucasian gene" for estimating M. Available survey data for regions from Liberia to the Congo (Leopoldville), presented in Table 3, show that in this region (from which about 56 percent of the ancestral slaves came) the mean frequency of Fy^a is probably not over about 0.02. The mean frequency for all Africans of the slave area is probably less than 0.03. The frequency for U.S. Caucasians is about 0.43 (Table 4). Moreover, recent extensive studies in a population of California Negroes revealed no evidence for natural selection (evidence pertaining to fetal and infant growth and viability and to adult growth and fertility) associated with Duffy blood group phenotypes *(23)*. Strong selection due to this locus seems excluded, so there is some protection against bias in the estimation of M. Table 4 presents available Fy^a frequency data for U.S. Negroes and for some U.S. Caucasians, and the resulting M estimates. The M estimates for the three non-southern regions studied do not differ significantly, so the estimate 0.2195±0.0093 for California Negroes—the largest of the three samples—may tentatively be used as the best estimate of M for a non-southern area. The very small standard error of this estimate reflects both the discrimination power of this "Caucasian gene" and the large sample sizes for the Negro and Caucasian populations. The two estimates from the "Deep South" do differ significantly and should be kept separate. The smaller one, 0.0366±0.0091 from Charleston, appears to justify the statement that these Gullah Negroes have an unusually small amount of Caucasian ancestry *(5)*. It is clear that the data of Table 4 are especially useful in comparing M for different U.S. Negro populations, because the same gene, Fy^a, is used as the basis for all estimates. Any bias due to selection should operate quite similarly in the different Negro populations. The difference between "southern" and "non-southern" M values evident in Table 2 is also marked in Table 4 and must be regarded as real.

Thus Fy^a, for the reasons given, may be the best gene presently available for estimating M. When more African survey data are available, the "Caucasian" alleles Gm^1, $Gm^{1,2}$, and Gm^5 of the Gm locus, used jointly, may be as good. The AK^2 gene (Table 2) may be of some use if further African data establish a general zero frequency, but the low frequency, 0.047, of the AK^2 gene in Caucasians considerably reduces its discrimination power. The K gene of the Kell blood group system is sometimes thought of as a "Caucasian gene," but this is not strictly the case. This gene was present in 8 of 1202 Africans from the Liberia-Dahomey *(18)* and western Nigeria *(24)* region, at a mean frequency of 0.0033. The California Negroes of Table 4 ($N = 3146$) have a K gene frequency of about 0.0083, and the California Caucasians, a K gene frequency of about 0.046 *(25)*. If we consider q_a to be zero, we obtain an estimate of 0.181±0.026 for M for this population—clearly a maximum estimate and not reliable. This

Table 3. Frequencies of Duffy blood group FY (a+) in West African and Congo (Leopoldville) populations.

Region	Sample size *(N)*	Proportion of FY (a+)*	Reference
Liberia (many tribes)	661	0.00	*(18)*
Ivory Coast	163	.043†	*(18)*
Upper Volta	75	.00	*(18)*
Dahomey	20	.00	*(18)*
Ghana (Accra) and Nigeria (Lagos)	37	.00	*(39)*
Congo (Bantu)	501	.078‡	*(40)*

*Reacting positively with anti-Fy^a, indicating a genotype of Fy^aFy (most likely), or Fy^aFy^b, or Fy^aFy^a (rare) *(39)*. †The true proportion is probably zero because the Ivory Coast positive reactions with anti-Fy^a are believed to be incorrect. ‡The gene frequency for Fy^a is 0.040.

Table 4. Estimates of M derived from Fy^a gene frequencies for American Negroes from various areas. The frequency of this gene in the African ancestors of American Negroes is assumed here to be zero; if it is not zero, these are *maximum* estimates. N = number in sample, q = Fy^a gene frequency, S.E. = standard error of q (all estimates by T. E. Reed).

Region and locality	Negroes		Caucasians			
	N	q±S.E.	N	q±S.E.	M±S.E.*	Reference
Non-southern						
New York City	179	0.0809±0.0147			0.189 ±0.034	*(39)*†
Detroit	404	.1114± .0114			.260 ± .027	*(41)*
Oakland, Calif.	3146	.0941± .0038	5046	0.4286±0.0058	.2195± .0093‡	*(25)*
Southern						
Charleston, S.C.	515	.0157± .0039			.0366± .0091	*(5)*
Evans and Bullock counties, Ga.	304	.0454± .0086	322	.422 ± .0224	.106 ± .020	*(9)*

*The q for Oakland Caucasians (who are of West European ancestry) was used in all estimates. $M = q_n/q_e$. †Two other New York City studies *(42)* are omitted because they involved selection for dark skin color. The data used here were grouped with both anti-Fy^a and anti-Fy^b. The observed distribution of four Duffy phenotypes differs from the Hardy-Weinberg expectation at the 0.025 level of significance. ‡If the frequency of Fy^a in the African ancestors were 0.02, this estimate would be 0.181.

Table 5. Frequencies of genes *A* and *B* of the ABO blood-group system in surveys in the major slaving areas of Africa (see Table 1); *p* = frequency of *A* gene, *q* = frequency of *B* gene.

Region	Peoples or population	Sample size *(N)*	*p* ± S.E.*	*q* ± S.E.*	Reference
Senegambia	Bambara, Malinke	2,120	0.159 ± .006	0.174 ± .006	*(43)*
Sierra Leone	Gbah-Mende	1,015	.159 ± .009	.151 ± .008	*(44)*
Liberia	> 18 tribes†	2,337	.143 ± .005	.148 ± .006	*(18)*
Gold Coast	Unspecified, from Accra	1,540	.130 ± .006	.122 ± .006	*(45)*
Bight of Benin	Yoruba of Lagos, Ibadan	1,003	.130 ± .008	.141 ± .008	*(46)*
Bight of Biafra	Ibo ("Eastern")	572	.161 ± .011	.089 ± .009	*(47)*
"Angola"	"Bantu"–8000 (mainly Bakongo) near Leopoldville and 8000 from Angola	16,000	.152 ± .002	.138 ± .002	*(48)*
Mean frequencies‡ over the entire slaving area			.150	.131	

*Maximum-likelihood estimate *(49)*. †Exclusive of Americo-Liberians. ‡Calculated from values for *p* and *q* given in the body of the table, weighted by the proportions of Table 1 (after the removal of values for Mozambique, Madagascar, and "region unknown").

maximum does not differ significantly from the Fy^a estimate for this same population. The relatively large standard error here again reflects the low Caucasian gene frequency.

Although a zero q_a is generally preferable, there is one situation where a q_a value appreciably different from zero might yield a useful estimate of *M*. This could occur when there are sufficiently extensive and detailed data on African gene frequency to make it possible to calculate a mean African gene frequency, with weighting of regional gene frequencies according to the proportions of Table 1. At present, the ABO blood groups provide the only such usable genetic marker [the gene for hemoglobin S is known to be affected by selection, and much less information is available for other loci *(26)*; for selection data on hemoglobin S, see *(27)*]. Table 5 gives the gene frequencies for genes *A* and *B* of the ABO system from relevant surveys in the seven major slaving areas of Table 1.

These extensive surveys reveal an overall uniformity in gene frequency, with the one exception of a somewhat low *B* frequency for the Bight of Biafra (Ibos). From these mean values for African frequencies of genes *A* and *B* and from extensive data on ABO-system distribution in California Negroes and Caucasians *(25)*, a maximum likelihood estimate for *M* of 0.200±0.044 was obtained *(28)*. This estimate is not greatly affected by the accuracy of the proportions given in Table 1 or by the exactness of the values for individual regional gene frequencies *(29)*. A good fit of the observed number of individuals in each of the eight race and blood-group classes with the corresponding number expected from the estimated parameters (gene frequencies and *M* values) tested by the chi-square method, indicates both that the estimation is reasonable and that there are no large selective differences between genes *A* and *B* in U.S. Negroes *(28)*. This procedure therefore seems justified for the case of ABO blood groups. Practically, however, the large standard error for *M* indicates that, in spite of large samples, the estimate for this locus is too imprecise to be very useful.

Since there are now three different estimates of *M*, and since extensive data on other aspects of the problem, including selection, are available for this one large California population of Negroes, these estimates are presented in a single table, Table 6. We note that they do not differ significantly from each other; this is due at least in part to the relatively large standard errors for the Gm and ABO estimates. The marked superiority, for estimating *M*, of Fy^a over *A* and *B* for samples of equal size is evident *(30)*, whereas, if the sample sizes were the same for Fy^a and the three Gm alleles, it would be found that these are equally efficient for estimating *M*. An extensive search for evidence of natural selection due to the presence of ABO blood groups in these Negroes, similar to the search reported above for the Duffy blood group, also failed to reveal any consistent selective effect *(23)*. This finding, plus the good chi-square fit in the estimation of *M*, which implies that the *A* and *B* genes are not very different with respect to their selective values in U.S. Negroes, gives some assurance

Table 6. Estimates of *M* calculated from data on Gm serum groups, Duffy blood group, and ABO blood group from Negroes and Caucasians of the Oakland, California, area. [Data of the Child Health and Development Studies *(6, 25)*.]

Locus	Alleles used	Sample size (N)		M
		Negroes	Caucasians	
Gm	Gm^1, $Gm^{1,2}$, Gm^5	260	478	0.273 ± 0.037*
Duffy	Fy^a	3146	5046	.220 ± .009†
ABO	*A,B*	3146	5046	.200 ± .044†

*See *(6)*.
†See text.

that the ABO estimate is not greatly disturbed by selection *(28)*. No selection studies for Gm were made on these California Negroes, but extensive studies on a Brazilian population which was about 30 percent Negro, 11 percent Indian, and 59 percent Caucasian *(13)* revealed no evidence of selective effect *(31)*. Further evidence is provided by the good chi-square fit in the multi-allelic estimation obtained with the three Gm alleles *(6)*. It seems reasonable to conclude that strong selective effects on these three estimates of *M* may be excluded. The existence of weaker effects, however, still sufficient to bias these estimates appreciably, cannot be ruled out. As more independent estimates on these and other genes become available, each having regard to the criteria listed above and including some safeguard against a strong bias due to selection and having a relatively small standard error (say, less than 0.02), it should become possible to obtain a "consensus" on the true value of *M* (for specified Negroes). Estimates biased upward or downward by selection will be separated from those little affected by selection, and so, in time, the former can be identified and rejected.

USE OF M TO DETECT SELECTION

Several investigators *(8, 9, 21, 32)* have argued that selection for or against a gene may be clearly inferred from the *M* value that the gene produces. From the foregoing section it is clear that if (i) the true (unbiased) value of *M* (say, M_0) is known, (ii) the estimate in question (M_e) is calculated with regard to the criteria given above, and (iii) M_e differs significantly from M_0, then we may reasonably suspect that selection has caused the observed deviation. These conditions have not been met. In particular, we have no M_0. The *M* estimates obtained with R^0 *(8, 9, 21)*, R^1 *(8, 9)*, and Fy^a *(8)* were considered to be valid estimates unbiased by selection, but no objective evidence was offered to support these views. With one or more of these *M* estimates used as reference standards, it has been claimed that the deviant *M* estimates of the following genes demonstrate selection on these genes in U.S. Negroes: Hp^1, *T*, Gd^{A-}, Hb^S, and Tf^{D1} (see Table 2). These results can, at present, be considered only suggestive, but it must be admitted that the usually high *M* estimates obtained with Hp^1 and Gd^{A-} argue for an effect of selection *(27)*.

A different approach was used to show that *M* estimates obtained with *r*, R^0, and R^1 alleles of the Rh locus ranked in this (decreasing) order of size for a Georgia population and also for two Brazilian populations *(32)*. Accepted at face value, this is evidence of differences between *M* values from different Rh alleles. The investigators attribute these differences to selection. This same approach in these populations also indicates that *M* for the *B* allele is greater than 1.5*M* for the *A* allele *(32)*. African Rh and ABO gene frequencies, weighted by slaving-area origins, were not used, however, although the African areas of origin of Brazilian Negroes are known *(2)*. Again, these findings are interesting and suggestive but far from conclusive.

Workman *(8)*, from inspection of A_1, A_2, and *B* allele frequencies in various West African, U.S. Negro, and U.S. Caucasian populations, concludes that there has been strong selection in U.S. Negroes against A_1 and for A_2. He identifies the various African data only as "West Africa," and does not use significance tests. Since Workman and also Hertzog and Johnson claim to find selection in the ABO system, it is pertinent here to recall that the *M* estimate obtained from ABO-system distributions that is discussed earlier in this article (an estimate based on *large* populations and good estimates for African gene frequency) did not suggest selective differences between the *A* and *B* alleles.

This critical review of claims for selection would be incomplete if I did not mention that there *is* an important theoretical reason to look for selection in hybrid populations such as the American Negro. As has been previously recognized *(6, 8, 32)*, selection in U.S. Negroes over several generations can produce a cumulative effect in present-day individuals appreciably greater than the effect of a single generation of selection—the type of data usually available. There is thus a possibility of detecting, in hybrids, selection due to common polymorphisms which is too small [usually less than 5 to 10 percent of the mean *(23)*] to be detectable by ordinary one-generation studies. This possibility, together with the probability that some of the genes are selective [because it is most unlikely that a new genotype (the hybrid) in a new environment would be exactly neutral in selective value], makes the search for selection here especially worthwhile. Some of these selective genes may already have been identified.

OTHER USES OF M ESTIMATES

In addition to the definite likelihood of their yielding valuable information on the action of natural selection in human populations, good estimates of the amount of Caucasian ancestry in U.S. Negro populations have at least two other "uses."

1) They provide objective information about the genetic heterogeneity among various populations of U.S. Negroes. Evidence of marked differences between southern and non-southern Negroes with respect to the amount of Caucasian ancestry, as shown in Tables 2 and 4, is the first clear result from this use of *M* estimates. As more good estimates from defined U.S. Negro populations become available, we may expect further heterogeneity to be revealed.

2) They provide an understanding of the distribution in American Negroes of those genetic traits, including diseases, that are due primarily to genes of Caucasian origin. There are few examples of such genes at present, but, aside from common genetic polymorphisms, like blood groups, few genes have been sufficiently studied to permit possible identification of racial differences in gene frequency. One probable example of such a genetic trait is phenylketonuria—a condition resulting from homozygosity for a rare autosomal recessive gene, producing a deficiency of phenylalanine hydroxylase and resulting (if untreated) in severe mental defect. This occurs in about 1 in 10,000 births of persons of North European ancestry *(33)* but appears to be much rarer in U.S. Negroes *(34)*. This rarity is understandable if the gene frequency in African Negroes is much lower than that in Caucasians (about 0.01). For example, if U.S. Negroes have, on the average, 20-percent Caucasian ancestry, the frequency of occurrence of phenylketonuria at birth in U.S. Negroes would be only 1/25th that in Caucasians, or roughly 1 in 250,000—rare indeed.

An example of a disease which is not simply inherited but which may show a similar racial distribution is cirrhosis of the liver. A study in Baltimore Negro cirrhotics revealed, relative to Negro controls, a significant increase in Fy (a+b+) Duffy blood group phenotype and a decrease in Fy (a−b−) phenotype, whereas Caucasian cirrhotics showed no such difference from Caucasian controls *(35)*. The simplest interpretation is that the disease is more frequent in Caucasians, and that Negroes with some degree of Caucasian ancestry, as shown by their Duffy blood group, are more likely to develop the disease than those lacking such ancestry *(35)*. Other examples of traits whose frequency of occurrence in U.S. Negroes is affected by the amount of their Caucasian ancestry will surely be reported *(36)*. Accurate information on *M* will be clinically useful here.

SUMMARY

Published estimates of the proportion, in American Negroes, of genes which are of Caucasian origin are critically reviewed. The criteria for estimating this proportion *(M)* are discussed, and it is argued that all estimates published to date have either deficiencies pertaining to the African-gene-frequency data used or statistical inaccuracies, or both. Other sources of error may also exist.

Evidence is presented that the Fy^a gene of the Duffy blood group system may be the best gene now available for estimating *M*. Estimates based on Fy^a frequencies have been obtained for Negroes in three non-southern and two southern areas. The value of *M* is found to be appreciably greater in non-southern areas, the best estimate being 0.2195±0.0093 (Oakland, California). This estimate is still subject to some uncertainty. The value of *M* in the South is appreciably less.

Natural selection can introduce a bias in the estimate of *M*. Claims that selection acting on certain genes in American Negroes have been demonstrated are reviewed, and it is concluded that they are not yet proved. The approach discussed here may be valuable in the future as a sensitive method for detecting the action of natural selection. In addition, knowledge of the amount of Caucasian ancestry may be of medical value in explaining the frequencies of occurrence of certain hereditary diseases in Negroes.

REFERENCES AND NOTES

1. J. H. Franklin and T. Marshall, in World Book Encyclopedia (Field Enterprises Educational Corporation, Chicago, 1966), vol. 14, p. 106.
2. P. D. Curtin, personal communication (1969).
3. J. Potter, in Population in History: Essays in Historical Demography, D. V. Glass and D. E. Eversley, Eds. (Univ. of Chicago Press, Chicago, 1965), p. 641.
4. P. D. Curtin, personal communication (1968).
5. W. S. Pollitzer, Amer. J. Phys. Anthropol. **16**:241 (1958).
6. T. E. Reed, Amer. J. Hum. Genet. **21**:71 (1969).
7. B. Glass and C. C. Li, *ibid.* **5**:1 (1953).
8. P. L. Workman, Hum. Biol. **40**:260 (1968).
9. ———, B. S. Blumberg, A. J. Cooper, Amer. J. Hum. Genet. **15**:429 (1963).
10. C. Stern, Principles of Human Genetics (Freeman, San Francisco, ed. 2, 1960), p. 356.
11. A. G. Steinberg, R. Stauffer, S. H. Boyer, Nature **188**:169 (1960).
12. B. S. Blumberg, P. L. Workman, J. Hirschfeld, *ibid.* **202**:561 (1964).
13. H. Krieger, N. E. Morton, M. P. Mi, E. Azevêdo, A. Freire-Maia, N. Yasuda, Ann. Hum. Genet. **29**:113 (1965).
14. J. E. Bowman, H. Frischer, F. Ajmar, P. E. Carson, M. K. Gower, Nature **214**:1156 (1967).
15. B. Glass, Amer. J. Hum. Genet. **7**:368 (1955) (non-D^u-tested data for Africans, Negroes, and Caucasians; use of D^u-tested data for Africans and non-D^u-tested data for others gives $M = 0.281$).
16. D. F. Roberts, *ibid.*, p. 361. The estimate is "provisional"; 66 separate estimates were made, ranging from 0.0404 to 0.3341.
17. ——— and R. W. Hiorns, *ibid.* **14**:261 (1962). No sample sizes are specified.
18. F. B. Livingstone, H. Gershowitz, J. V. Neel, W. W. Zeulzer, M. D. Solomon, Amer. J. Phys. Anthropol. **18**:161 (1960).
19. P. V. Tobias, in The Biology of Human Adaptability, P. T. Baker and J. S. Weiner, Eds. (Clarendon Press, Oxford, 1966), p. 161.
20. The following formula for the standard error (S.E.) of a ratio $R = y/x$ was used:

$$\text{S.E.}\,R = R\left[\frac{V_y}{y^2}+\frac{V_x}{x^2}-\frac{2C_{xy}}{xy}\right]^{1/2}$$

where the variance of y is V_y, that of x is V_x and the covariance between x and y is C_{xy} [see, for example, L. Kish, Survey Sampling (Wiley, New York, 1965), p. 207]. This formula is adequate for large or moderate-sized samples when it is unlikely that x is near zero. In terms of Eq. 2 for M,

$$\text{S.E.}\,M = M\left[\frac{V(q_a-q_n)}{(q_a-q_n)^2}+\frac{V(q_a-q_c)}{(q_a-q_c)^2}-\frac{2V(q_a)}{(q_a-q_n)\,(q_a-q_c)}\right]^{1/2}$$

where V represents the variance of the adjoining quantity in parentheses. The covariance between numerator and denominator of Eq. 2, due to the presence of q_a in both, is allowed for in this standard error.
21. W. C. Parker and A. G. Bearn, Ann. Hum. Genet. London **25**:227 (1961). The total number of American, Canadian, and British individuals tested in the study reported is 865. A weighted estimate for the frequency of gene Hp^1 in U.S. Negroes, based on the data of Parker and Bearn, is 0.55; this value gives an M of 0.53.
22. A. G. Steinberg, in Symposium on Immunogenetics, T. J. Greenwalt, Ed. (Lippincott, Philadelphia, 1967), pp. 75-98.
23. T. E. Reed, Amer. J. Hum. Genet. **19**:732 (1967); *ibid.* **20**:119 (1968); *ibid.*, p. 129.
24. B. S. Blumberg, E. W. Ikin, A. E. Mourant, Amer. J. Phys. Anthropol. **19**:195 (1961).
25. T. E. Reed, Amer. J. Hum. Genet. **20**:142 (1968).
26. For gene distributions, see J. Hiernaux, La Diversité Humaine en Afrique Subsaharienne (Institut de Sociologie, Université Libre de Bruxelles, Brussels, 1968), figs. 2, 3, 7, 8, 12, 14.
27. There are good a priori reasons, entirely separate from M values, for expecting, in U.S. Negroes, a decrease in the frequency of the genes for sickle-cell hemoglobin, Hb^S, and for glucose-6-phosphate dehydrogenase deficiency, Gd^{A-}. (i) There is good evidence that in Africa the high frequency of the Hb^S gene is due to a selective advantage of heterozygotes for Hb^S in regions where malaria is endemic [see, for example, F. B. Livingstone, Abnormal Hemoglobins in Human Populations (Aldine, Chicago, 1967), pp. 105-107; A. C. Allison, in Abnormal Haemoglobins in Africa, J. H. P. Jonxis, Ed. (Davis, Philadelphia, 1965), pp. 369-371; D. L. Rucknagel and J. V. Neel, in Progress in Medical Genetics, A. G. Steinberg, Ed. (Grune & Stratton, New York, 1961), vol. 1, pp. 158-260]. There is strongly suggestive evidence that the Gd^{A-} gene in Africa is similarly kept at high frequencies due to selective advantage in malarious areas [see F. B. Livingstone, Abnormal Hemoglobins in Human Populations (Aldine, Chicago, 1967); A. G. Motulsky, in Abnormal Haemoglobins in Africa, J. H. P. Jonxis, Ed. (Davis, Philadelphia, 1965), pp. 181-185)]. (ii) Both genes are known to have selective disadvantages which can explain their rarity in nonmalarious areas. It is therefore to be expected that Negroes moved from their highly malarious homelands to the less malarious, and now nonmalarious, regions of North America would have lower frequencies of these two genes. This selective decrease would raise M estimates above the true value.
28. The computer program [see T. E. Reed and W. J. Schull, Amer. J. Hum. Genet. **20**:579 (1968)] estimated M and Caucasian A and B gene frequencies, given the two African mean frequen-

cies as constants and the two California populations determined by the gene frequencies to be estimated, subject to the constraints that, for *both* A and B, $q_n = Mq_c + (1-M)q_a$. This equation is Eq. 1 applied to both alleles and is true when there is simple gene mixture without selection (see *6*). Comparison of the observed numbers of the eight race and blood-group classes (2 races × 4 groups) with the corresponding numbers expected on the basis of parameter estimates gives a chi-square value of 5.910 for 3 d.f., $P>.10$.

29. When the negligible contribution from Mozambique, Madagascar, and "Unknown" is excluded, the proportions of Table 1, column 4, become (in order): 0.135, 0.056, 0.116, 0.162, 0.044, 0.237, and 0.249. The corresponding proportions for South Carolina (1773-1807) are 0.197, 0.068, 0.164, 0.134, 0.016, 0.021, and 0.399 [data of Curtin *(4)*], yielding overall African mean values of 0.149 and 0.144 for p and q. These two series differ appreciably with respect to the final two values, yet when the South Carolina series is used, the estimate of M is 0.256±0.042, a difference of just over one standard error. Also, q for the Bight of Biafra is the only markedly variant gene frequency among the frequencies for the seven regions, but replacing the p and q for this region by the p and q for the Bight of Benin or for "Angola" does not significantly change M (0.281±0.040 or 0.251±0.042, respectively, when corrected proportions of Table 1 are used).
30. The Fy^a estimate is based on $(0.044)^2/(0.0093)^2$, or 22 times as much statistical information as the ABO estimate.
31. N. E. Morton, H. Krieger, M. P. Mi, Amer. J. Hum. Genet. **18**:153 (1966).
32. K. P. Hertzog and F. E. Johnston, Hum. Biol. **40**:86 (1968).
33. V. A. McKusick, Mendelian Inheritance in Man (Johns Hopkins Press, Baltimore, ed. 2, 1968), p. 346.
34. For example, H. P. Katz and J. H. Menkes [J. Pediat. **65**:71 (1964)] report the first definite case of phenylketonuria in a U.S. Negro. R. G. Graw and R. Koch [Amer. J. Dis. Child. **114**:412 (1967)] report two "pale skinned" Negro brothers with phenylketonuria, bringing the total for U.S. Negroes at that time to five.
35. N. C. R. W. Reid, P. W. Brunt, W. B. Bias, W. C. Maddrey, B. A. Alonso, F. L. Iber, Brit. Med. J. **2**:463 (1968).
36. Differences between Caucasians and Japanese with respect to gene frequency are instructive here. Phenylketonuria appears to be much rarer (perhaps a tenth as frequent) among Japanese than among Caucasians [K. Tanaka, E. Matsunaga, Y. Hanada, T. Murata, K. Takehara, Jap. J. Hum. Genet. **6**:65 (1961)]. Another single-gene trait, Huntington's chorea, also appears to be about ten times as frequent in Caucasians as in Japanese, according to T. E. Reed and J. H. Chandler, Amer. J. Hum. Genet. **10**:201 (1958). Examples of congenital abnormalities which are commoner in Caucasians than in Japanese, and vice versa, are given by J. V. Neel, Amer. J. Hum. Genet. **10**:398 (1958).
37. P. D. Curtin, The Atlantic Slave Trade: A Census (Univ. of Wisconsin Press, Madison, in press).
38. J. V. Neel and W. J. Schull, Human Heredity (Univ. of Chicago Press, Chicago, 1954), p. 255; J. V. Neel, personal communication (1969). The estimate refers to "non-Negro" ancestry.
39. R. R. Race and R. Sanger, Blood Groups in Man (Blackwell, Oxford, ed. 5, 1968).
40. M. Shapiro and J. M. Vandepitte, Int. Congr. Blood Transfusion, 5th, Paris (1955), p. 243; M. Shapiro, personal communication (1969).
41. H. Gershowitz, unpublished data.
42. E. B. Miller, R. E. Rosenfield, P. Vogel, Amer. J. Phys. Anthropol. **9**:115 (1951); R. Sanger, R. R. Race, J. Jack, Brit. J. Haematol. **1**:370 (1955).
43. R. Koerber, C. R. Seances Soc. Biol. **141**:1013 (1947); R. Koerber and J. Linhard, Bull. Soc. Anthropol. Paris **2**:158 (1951). The frequencies for Bambara and Malinke do not differ significantly.
44. P. Julien, Z. Rassenphysiol. **9**:146 (1937).
45. G. M. Edington, West. Afr. Med. J. **5**:71 (1956).
46. H. R. Muller, Proc. Soc. Exp. Biol. **24**:437 (1927); J. P. Garlick, quoted in A. E. Mourant, A. C. Kopeć, K. Domaniewska-Sobczak, The ABO Blood Groups (Blackwell, Oxford, 1958), p. 173; J. P. Garlick and N. A. Barnicot, Ann. Hum. Genet. **21**:420 (1957). The frequencies in these three surveys do not differ significantly from each other.
47. J. Hardy, Roy. Anthropol. Inst. **92**:223 (1962). I have not used Hardy's data on "Onitsha Ibo" (sample size, 228) because I consider the subjects to be probably not of Ibo origin. I have not used data of J. N. M. Chalmers, E. W. Ikin, and A. E. Mourant [Ann. Eugenics **17**:168 (1953)] on "southeastern" Nigeria (105 Ibo and 1 Tiv) because information on the type of Ibo was not given. Table 1 describes this region as "about ¾ Ibo," and I could find no suitable data for the remaining quarter.
48. G. van Ros and R. Jourdain, Ann. Soc. Belge Med. Trop. **36**:307 (1956); L. Mayor, Bull. Clin. Statistics **7**, No. 3, suppl. 126 (1954). The first study (for the Congo) gave values of 0.1556±0.0030 for p and 0.1244±0.0027 for q. The second (for Angola) gave values of 0.1486±0.0029 and 0.1514±0.0030, respectively. An unweighted average of these values was used. The values for Angola do not differ significantly from the mean p and q values for seven central-coastal named tribes in Angola (total population, 1285) tabulated by Hiernaux (see *26*).
49. T. E. Reed and W. J. Schull, Amer. J. Hum. Genet. **20**:579 (1968).

71 Experimental studies of the distribution of gene frequencies in very small populations of Drosophila melanogaster: I. Forked

Warwick E. Kerr
Sewall Wright
Departments of Genetics and Mathematics
Faculdade de Medicine de Ribeirão Prêto
Universidade de São Paulo
São Paulo, Brazil
and
University of Chicago
Chicago, Illinois

INTRODUCTION

For quantitative study of the sort of random drift due to inbreeding it is desirable to experiment with segregating genotypes that can be classified without risk of error and with populations that are all of a very small definitely known size.

With respect to the first condition, it is much the most satisfactory if *all* segregating genotypes can be distinguished accurately. This, however, limits severely the number of loci that can be studied. Something can be done where only dominants and recessives can be accurately classified.

Most pairs of alleles with such conspicuous difference in effect as to satisfy the first condition turn out to be subject to such enormous differences in selection that the accumulation of random deviations, implied by the term random drift, is largely prevented. Selection pressure and the effects of random processes are roughly comparable in magnitude at a given gene frequency if the change in frequency in a generation which the former tends to bring about (Δq) and the variance increment due in a generation to the latter ($\sigma^2_{\delta q}$) are of the same order (Wright, 1931, 1948). Where the ratio ($\Delta q/\sigma^2_{\delta q}$) is as high as 10 in absolute value, there is not much accumulation of random deviations and hence little random drift. The size of experimental populations in studies of the latter should be small enough to meet this condition. They should be constant for convenient comparison with theory.

The present experiments (all with *Drosophila melanogaster*) were designed to meet these conditions. Three series were performed. About 120 lines were started in each. In the first series, the sex-linked mutation forked (f) competed with its type allele. Four females (1 f/f, 2 f/+, 1 +/+) and 4 males (2 f/0, 2 +/0) were put in each vial. The second series involved the sex-linked semidominant mutation Bar (B) and its type allele. Each initial vial contained 4 B/+ females and 4 males (2 B/0, 2 +/0). The third series was with the autosomal alleles aristapedia, ss^a, and spineless ss, which produce a heterozygote that is close to type. Each initial vial contained 4 ss^a/ss females and 4 ss^a/ss males. The alleles were thus equally frequent at the beginning of each experiment.

The cultures were allowed to develop until about 2 to 4 days after the offspring began to emerge. The flies hatched up to this time were discarded (in an evening). The flies which appeared next morning (if enough had emerged) were etherized and from among them 4 females and 4 males were taken at random and served as progenitors of the following generation. The etherized flies were put on a porcelain plate and the first 4 males and the first 4 females that happened to be closest to the right end of the plate were the flies taken. It was often, however, necessary to wait to the second and sometimes to the third day to obtain 4 of each sex. This procedure was

From Evolution 8:172-177, 1954. Used with permission.

Experimental data by W. E. Kerr, under a fellowship of the Rockefeller Foundation, mathematical analysis by S. Wright. Acknowledgment is made to Dr. Th. Dobzhansky for material and for his hospitality during the conduct of these experiments. Acknowledgment is also made to Dr. J. Crow for material and suggestions. Analysis was aided by a grant from the Wallace C. and Clara A. Abbott Memorial Fund of The University of Chicago.

Editor's Note: This paper is the first of a series of three. The others will be published in successive issues of Evolution.

repeated in every following generation in every line. The first series was carried 16 generations, the second 10 and the third 9 generations. In general, all 8 parents in each culture lived until their progeny started to emerge. In some instances one or more died a few days after the culture was started. No new flies were substituted. Lines were discontinued if fixation was attained. A few were discarded on account of mite infection and other accidents. These latter strains have not been included in the statistics presented here. The present paper will be confined to the experiments with the forked.

FORKED

Table 1 shows the most important results. Among the 96 lines with initial gene frequency (.50 f + .50 f^+), 26 were still unfixed at the end of 16 generations. Wild type had become fixed in 41 and forked in 29. There is here only an insignificant suggestion that forked was at a selective disadvantage. There seems, however, to have been a shift toward increasing disadvantage of forked as the experiment proceeded. In the first 8 generations, wild type became fixed in 17 lines, forked in 23, while in the last 8 generations, wild type became fixed in 24 lines, forked in only 6, a difference with a probability from accidents of sampling of less than .01. Nevertheless it is clear that the selection against forked must have been slight to have permitted as much fixation as occurred against its pressure.

This result is in agreement with those of previous experiments. In Ludwin's (1951) experiments, the initial gene frequency of .50 in cultures which contained on the average about 44 males, 51 females but with enormous variations, fell to .30 in about 2 or 3 months and was still about the same at 6 months. In experiments by Merrill (1953) in populations that rarely exceeded 100 adults and in many cases were down to less than 10 flies, the frequency of forked similarly fell from .50 at the beginning to .30 by 99 days and averaged .33 in counts from 125 days to 270 days. There was no indication of increasing selection against forked in those experiments. Both authors found evidence of important differences in gene frequency among individual cultures which they attributed to random drift.

From inspection of table 1, it appears that the percentage of fixation (including both that of type and forked) rose to generation 4 but

Table 1. The amount of fixation of forked and its type allele in 96 lines, each consisting of 4♂'s and 4♀'s in each generation and carried 16 generations unless fixed earlier. The amount of fixation from generation 4 to 16 is compared with the expected amount at a constant rate of 8.9% per generation.

	Type newly fixed	Not fixed	Forked newly fixed	Total	Observed (fixed) No.(o)	Observed (fixed) %	Calc. No.(c)	(o-c)	$\frac{(o-c)^2}{c}$
1	1	94	1	96	2	2.1			
2	0	92	2	94	2	2.1			
3	1	87	4	92	5	5.4			
4	5	79	3	87	8	9.2	7.7	+0.3	.01
5	3	70	6	79	9	11.4	7.0	+2.0	.57
6	1	66	3	70	4	5.7	6.2	−2.2	.78
7	5	59	2	66	7	10.6	5.9	+1.1	.21
8	1	56	2	59	3	5.1	5.3	−2.3	1.00
9	3	52	1	56	4	7.1	5.0	−1.0	.20
10	4	47	1	52	5	9.6	4.6	+0.4	.03
11	5	39	3	47	8	17.0	4.2	+3.8	3.44
12	2	37	0	39	2	5.1	3.5	−1.5	.64
13	3	34	0	37	3	8.1	3.3	−0.3	.03
14	3	30	1	34	4	11.8	3.0	+1.0	.33
15	1	29	0	30	1	3.3	2.7	−1.7	1.07
16	3	26	0	29	3	10.3	2.6	+0.4	.06
1-3	2	273	7	282	9	3.2			
4-8	15	330	16	361	31	8.6	32.1	−1.1	
9-16	24	294	6	324	30	9.3	28.9	+1.1	
4-16	39	624	22	685	61	8.9	61.0		8.37

did not change consistently thereafter. The average rate for generations 4 to 16 was 8.91% (standard error 1.09%). Assuming theoretical constancy at this figure, the differences between observed and calculated numbers for unfixed and for newly fixed lines yields $\chi^2 = 8.4$, 12 degrees of freedom, probability .70-.80 of being exceeded by accidents of sampling.

The theoretical rate of fixation for a neutral sex-linked gene after a steady rate has been attained has been given as approximately $(2 N_M + N_F)/(9 N_M N_F)$ in which N_M is the effective number of males and N_F that of females. (Wright 1933). This approximation only applies, however, if N_M and N_F are moderately large. The panmictic index P (= 1-F where F is the inbreeding coefficient) measures the amount of heterozygosis relative to that in a random bred stock. The exact recurrence formula derived by the method of path coefficients was given as follows in the paper cited. Primes refer to preceding generations.

$$P = P' - C_1 (2 P' - P'') + C_2 (2 P'' - P''')$$

where

$$C_1 = (N_F + 1)/(8 N_F),$$
$$C_2 = (N_M - 1) (N_F - 1)/(8 N_M N_F)$$

Table 2. The theoretical values of P (=1-F), P/P′ and the proportional rate of change of P in a population consisting of 4 males and 4 females under sex linked inheritance and no selection

Generation	P(=1-F)	P/P′	ΔP/P′
0	1		
1	1	1	0
2	1	1	0
3	.914062500	.914062500	.085937
4	.854980469	.935363248	.064637
5	.788848877	.922651341	.077349
6	.731885910	.927789759	.072210
7	.677245259	.925342665	.074657
8	.627418808	.925427759	.073572
9	.580946889	.925931581	.074068
10	.538047138	.926155468	.073845
11	.498260630	.926053861	.073946
12	.461439099	.926099858	.073900
13	.427329066	.926079014	.073921
14	.395744515	.926088456	.073912
15	.366492734	.926084178	.073916
16	.339403833	.926086117	.073914
17	.314316879	.926085236	.073915
18	.291084347	.926085636	.073914
19	.269568980	.926085455	.073915
20	.249643934	.926085538	.0739145

In the present case

$$C_1 = 5/32, C_2 = 9/128.$$

$$P = .6875 P' + .296875 P'' - .0703125 P'''$$

The correlation $(1 - P_0)$ between the gametes that united to produce the foundation females was zero and the correlation $(1 - P_1)$ between the gametes that unite to produce their daughters is also zero. If there were no differential fecundity among the foundation flies there would be no correlation between mating males and females and hence $(1 - P_3) = 0$. From this point, however, the inbreeding coefficient rises. For calculation of P_3, P' (= P_2), P'' (= P_1) and P''' (= P_0) are all assigned the value 1. Table 2 shows the values of P, the ratio of successive values P/P′ and the percentage change in P per generation (100 (P - P′)/(P′) for 20 generations. It may be seen that P/P′ oscillates about the value .9260855. This is approximated to 4 places by the 11th generation and to 7 places by the 19th. Values of P beyond this point can thus be calculated to 6 places by the formula P = .9260855 P′.

The rate of decrease of heterozygosis after it reaches stability can also be derived at once by equating P/P′, P′/P″ and P″/P‴ and expressing in terms of $\chi = \Delta P/P'$ (Wright 1933).

$$\chi^3 + \chi^2 (2 + 2 C_1) + \chi (1 + 3 C_1 - 2 C_2) + (C_1 - C_2) = 0$$

The solution is .0739145, the limit about which ΔP/P′ oscillates in the successive generations.

The rate of fixation of lines approaches this same value although somewhat more slowly. Actually it has been noted that a practically constant rate 8.91% is attained by F_4. As this is 20.5% larger than the theoretical rate, it is implied that the actual variance due to random processes in each generation was 20.5% greater than expected. As the difference is only 1.5 times its standard error, it is not certainly significant. Taking it at face value, the excess might conceivably be due to fluctuating selection but as there is very little average selection it more probably means that the effective size of population is about 83% of the theoretical value. In an autosomal diploid population the rate of decrease of heterozygosis is ΔP/P′ = 1/(2N) (Wright, 1931). If effective size is defined as P′/2ΔP, the theoretical effective size with sex linkage and 4 females and 4 males per generation is 6.7646 while the effective size of

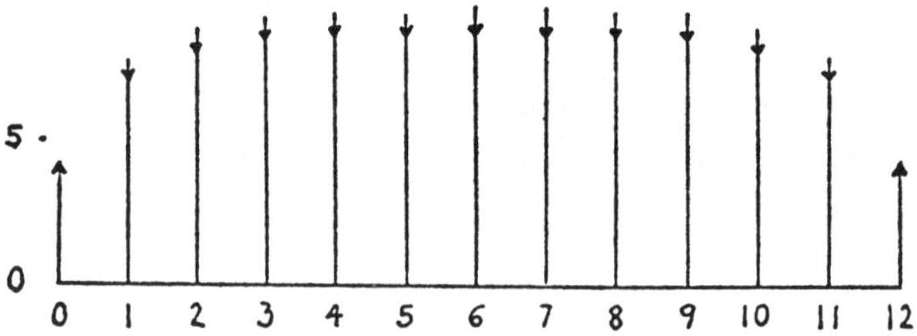

Fig. 1. The theoretical distribution of gene frequencies (including newly fixed classes) after attainment of stability of form in a population in which 2N = 12 and selection is absent. Rough estimates for the unfixed classes are given by the ordinates of the curve, y = 1/12, unit area (base 12) derived from indefinitely large 2N. The frequencies (1/24) of the newly fixed classes are given exactly by half the terminal ordinates of this curve.

Table 3. The distribution of unfixed classes after stability of form has been reached in a population in which 2N = 12, no selection; and the distribution in the following generation, including newly fixed classes (each 1/24 of the total unfixed classes of the preceding generation). The frequency in each of the unfixed classes is 11/12 of its value in the preceding generation, thus maintaining stability of form.

	Unfixed classes	Following generation
0		.04167
1	.07881	.07224
2	.08985	.08236
3	.09317	.08541
4	.09475	.08685
5	.09554	.08758
6	.09576	.08778
7	.09554	.08758
8	.09475	.08685
9	.09317	.08541
10	.08985	.08236
11	.07881	.07224
12		.04167
	1.00000	1.00000

the experimental population was 5.61. The difference if real can easily be accounted for if 1 or 2 of the 8 flies fail completely to reproduce in each generation.

The distribution of gene frequencies, during the period of constant rate of fixation, must have practically reached equilibrium of form. The actual distribution was not determined because of the lack of visible distinction between +/+ and +/f. It is of some interest, however, to consider what it must have been. It may suffice to give the distribution for an effective population of 6 and ignore the indicated slight selective differential (Wright, 1931). The standard is considered a population of 6 monoecious diploid individuals with completely random union of gametes, as it has been shown that the distribution is nearly the same for given N_e, irrespective of the system of mating (Wright, 1931).

Let f(q) be the frequency of gene frequency q. The class with this gene frequency contributes to the classes of the next generation according to the expansion of $[(1-q)(f^+) + q(f)]^{12} f(q)$ and thus to the class with gene frequency q_1 by

$$\frac{12!}{(2Nq)!\,[2N(1-q)]!}\, q^{2Nq_1}(1-q)^{2N(1-q_1)} f(q).$$

The total frequency of q is the sum of all contributions from the 11 values of q from 1/12 to 11/12. This must be 11/12 of its value in the preceding generation if there is equilibrium of the form of the distribution. The equation can be solved algebraically as in a number of simple examples given in the reference cited[1] but it is probably simplest in this case to start from a rough approximation and iterate, rating up each generation by 12/11, until stability is reached. The resulting distribution is shown in table 3 and figure 1.

While this solution applies to a situation with a slightly lower rate of fixation (1/24 for each allele) than that observed, the form of the

[1] We will note that one of these is incorrect as published, viz., that for irreversible mutation in populations of 3 monoecious individuals (p. 118). The correct percentages are 42.40% for q_a = 1/6, 21.07% for q_a = 2/6, 15.54% for q_a = 3/6, 11.99% for q_a = 4/6, 9.00% for q_a = 5/6, total 100% for unfixed classes. The rate of fixation of A is 17.05% and of a, 4.53% of the unfixed classes.

Table 4. The history of a line in which an eye color mutation appeared in generation 5 and drifted into fixation in generation 12

	F_4		F_5		F_6		F_7		F_8		F_9		F_{10}		F_{11}		F_{12}	
	f	w^a	f	w^a	f	w^a	f	w^a	f	w^a	f	w^a	f	w^a	f	w^a	f	w^a
♀ recessive	2	–	1	1	0	0	1	3	0	0	0	2	0	4	0	3	0	4
dominant	2	–	3	3	4	4	3	1	4	4	4	2	4	0	4	1	4	0
♂ recessive	1	–	1	3	0	4	1	0	0	4	0	2	0	3	0	4	0	4
dominant	3	–	3	1	4	0	3	4	4	0	4	2	4	1	4	0	4	0

distribution apart from the newly fixed classes is substantially correct for 4 ♀'s, 4 ♂'s (12 representative of the locus) for any effective size of parental population not too remote from 6. Indeed with an indefinitely large population in each generation, the form of the distribution is not very different, being approximately uniform for all gene frequencies (Wright, 1931).

A situation that arose in one line (No. 109) in this series is interesting. In the 5th generation the flies taken to be parents of the 6th generation were found to be segregating for an eye color found to behave as an allele of white. This gene drifted in frequency until it became fixed in the 12th generation. The frequencies of f and this gene (apparently w^a) were as shown in table 4.

Thus this line had probably become fixed for the type allele of forked by the 9th generation and after the 12 generation for the new eye color mutant.

DISCUSSION

Populations of 4 males and 4 females per generation are so exceedingly small that experiments such as the present may seem to have no implications for evolution in nature. It must be borne in mind, however, that changes in the underlying multifactorial genetic structure of species probably occur so slowly that an appreciable change in a thousand generations must be considered as an explosively rapid process.

Study in the laboratory of the factors that can contribute to such change is practicable only by stepping up the rates by at least one hundred fold. Thus the interaction between a weak selective advantage of one isoallele over another and a slight random drift, due to inbreeding, can be simulated by using alleles with selective differentials of ten percent or more instead of perhaps only one tenth of a percent or even one hundredth of a percent in populations of only one percent or even one tenth of one percent of the size of a typical natural deme.

In the case of forked, the selective differential is clearly much less than ten percent so that the results of the present paper illustrate random drift from inbreeding in an almost pure form. More complicated situations will be considered in the later papers of this series.

SUMMARY

Ninety-six lines of flies *(Drosophila melanogaster)* were started, each from 4 females (1 f/f, 2 f/+, 1 +/+) and 4 males (2 f/0, 2 +/0), and continued to fixation or to the sixteenth generation by random selection of 4 females and 4 males as parents of each new generation. Type (f^+) became fixed in 41 lines, forked (f) in 29 lines, and 26 lines were still unfixed at the end. The amount of selection against forked was thus slight, although there was evidence that it was greater in the later generations than at first.

The rate of fixation (of both alleles combined) reached approximate constancy by the fourth generation at 8.9% per generation. This would imply an effective size of population 83 per cent of that expected under sex linkage with 4 females and 4 males per generation but the reduction is of doubtful significance.

In one line, an eye color mutation, probably apricot, appeared in F_7 and became fixed in F_{12}, three generations after fixation of the type allele of forked.

LITERATURE CITED

Ludwin, I. 1951. Natural selection in *Drosophila melanogaster* under laboratory conditions. Evolution **5**:231-242.

Merrell, D. J. 1953. Gene frequency changes in small laboratory populations of *Drosophila melanogaster.* Evolution **7**:95-101.

Wright, S. 1931. Evolution in Mendelian populations. Genetics **16**:97-159.

______. 1933. Inbreeding and homozygosis. Proc. Nat. Acad. Sci. **19**:411-420.

______. 1948. On the roles of directed and random changes in gene frequency in the genetics of populations. Evolution **2**:279-294.

chapter 14

Race and species formation

In Chapter 13 we considered the origin of genetic variation (mutation) and the factors (selection, migration, genetic drift) that result in genetic differentiation of the various populations of a species. As a consequence of the varied distribution of alleles and chromosomal rearrangements among the geographical groups of a species, it is possible to characterize different populations by their genetic compositions. Genetically distinguishable populations are called *races.* Races can become so divergent genetically that reproduction between their members is difficult or impossible. The populations involved are then classified as *species.* In the present chapter, we shall examine the genetic composition of some natural populations and the experimental evidence for the operation of various factors in race formation. We shall also consider the methods that enable species to maintain their genetic integrity.

GENETICS OF POPULATIONS

A species whose populations have been well studied for their genetic diversity is *Drosophila psuedoobscura.* This species of fly has five pairs of chromosomes in its genome. These chromosomes can easily be studied in late larval salivary gland preparations. It is found the chromosomes contain patterns of bands of different characteristic widths, and the distribution of bands along any one chromosome is fixed. The third chromosome pair in this fly varies in its band sequence, depending on the population from which the fly was obtained. Some 15 separate more or less common band patterns have been found, distributed within the geographic range of the species. These 15 band sequences were analyzed and found to form an interlocking series of overlapping inversions whose phylogenetic relationships are shown in the figure below.

The results of a study of various populations of *D. pseudoobscura* was reported in 1947 by Dobzhansky (Ref. 14-1), whose paper is reprinted in this chapter. This paper also discusses a set of laboratory experiments which indicate the role of selection in maintaining the genetic diversity of these populations.

As a result of the discovery of a persistence of inversion types *(chromosomal polymorphism)* both in natural and experimental populations of *D. pseudoobscura,* a series of investigations on factors that could affect the establishment and maintenance of this polymorphism were initiated. One of the factors considered was temperature, as indicated in the

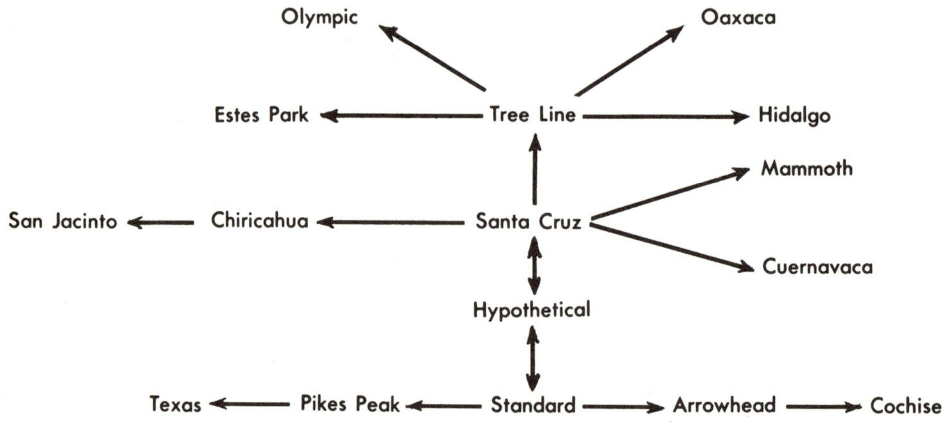

The phylogeny of the gene arrangements in the third chromosome of *Drosophila pseudoobscura.* (From Dobzhansky, T., and Epling, C. 1944. Carnegie Inst. Wash. Publ. **554**:1-183. Courtesy Carnegie Institution of Washington.)

above-mentioned article. A later series of experiments, reported by Van Valen and his co-workers in 1962 (Ref. 14-2), showed that when a chromosomally polymorphic population was maintained at a temperature of 22° C., the polymorphism sometimes disappeared, with the concomitant establishment of chromosomal monomorphism. An experiment that investigated the effect on chromosomal polymorphism of changing the sequence of temperatures at which a population is maintained was reported in 1964 by Levine and Van Valen (Ref. 14-3), whose paper is included in this chapter. Investigations on the effects of humidity on different karyotypes were reported by Heuts in 1948 (Ref. 14-4). He found that the humidity conditions during the pupal stage affect differentially the percentage of hatching of the various karyotypes and also the longevity of the adult flies hatching from the pupae. The role of "mating success" in the maintenance of chromosomal polymorphism in a population was reported by Ehrman in 1968 (Ref. 14-5).

It is clear from our knowledge of genes and their functions that differences in adaptive values of various karyotypes must reflect differences in the biochemistry of the organisms involved. A study of protein differences among flies carrying different gene arrangements was reported by Prakash and Lewontin in 1968 (Ref. 14-6).

In human populations, chromosomal rearrangements seem to be relatively rare, with no evidence that chromosomal polymorphisms play an important role in human evolution. However, a great number of examples have been found of gene polymorphisms in different human populations. Unfortunately, in most cases, the evolutionary forces that produced the polymorphisms are completely unknown. The blood groups provide one of the best illustrations we have of the genetic diversity that exists in human populations. The blood group gene that shows the most regular geographic pattern of distribution is *B*. Its highest frequency is observed in central Asia, with the incidence decreasing in all directions as one moves out from this center. In the indigenous populations of Australia and North and South America, the *B* gene is either completely absent or is found in rather low frequencies. Among native populations in Africa, the *B* allele has its highest frequency in the central Congo region and then decreases in both north and south directions. Blood group gene *A* has a more universal but less regular distribution than *B*. Allele *A* has its highest frequencies in western Europe, Australia, and certain areas of North America. There are isolated localities with high frequencies of *A* among the Tibetans, the Bushmen in South Africa, Congo Pygmies, Negritos in the Philippine Islands, and some "hill tribes" of southern India. Blood group gene *O* is the most frequent allele of the series and is found in high frequency over most of the world. Some American Indian tribes (e.g., Kwakiutl of British Columbia, Utes of Montana, and Toba of Argentina) are almost 100% homozygous for the *O* allele. Other tribes (e.g., Blackfeet of Montana and Navaho of New Mexico) have lesser amounts of *O*, with *A* as the other allele.

The great variation in the frequencies of different blood group genes has raised the question whether these alleles bestow any selective advantage or disadvantage on their carriers in different parts of the world. Thus far, no evidence exists that any of these alleles make their carriers more fit for a given environment. If we cannot explain the observed gene frequencies by selection, we must consider other possibilities. The most plausible alternative to selection appears to be genetic drift, based on the assumption that until rather recently man was a nomadic hunter. Under such conditions, it is very likely that many of the modern-day populations originated from small migrant groups that became isolated from the parental population.

A study of the alleles at the Rh-locus also reveals differences in frequencies in different human populations. However, no gradual gradients in frequency, such as that found for the *ABO* genes, have been discovered for the alleles at the Rh-locus. As in the ABO situation, there is no evidence that any of the Rh alleles make their carriers better adapted for a given environment. Once more, genetic drift appears to be the most plausible explanation for the observed frequencies of the various alleles. A detailed consideration of blood groups in man was published by Race and Sanger in 1968 (Ref. 14-7). An article reviewing enzyme and protein polymorphisms in human populations was published by Harris in 1969 (Ref. 14-8) and is reprinted in this chapter.

The concept of race applies to the geographical populations comprising the human species just as it does to *Drosophila*, etc. However, a situation has arisen for human populations that does not exist in *Drosophila*. In man, the development of different cultures has allowed human races, which in the past were restricted to different territories, to live side by side in the same locality without immediate fusion. Nevertheless, the inevitable meeting of individuals from different racial groups living in the same geographical area does lead to some gene exchange. As a result of this gene exchange, human populations are being fused into larger, more genetically variable groups. Should this process continue for a long period of time, the subdivision of the human species into distinct races will disappear.

REPRODUCTIVE ISOLATION

Genetic differences between populations, when sufficiently great, will result in structural, physiological, and behavioral differences in the members of one race as compared to another. When these differences become so great that they prevent the members of two races from breeding with one another, we say that we have two species. The populations involved are *reproductively isolated* from one another, each population forming a "closed genetic system." The large number of mechanisms by which reproductive isolation can be achieved falls into three categories: (1) mechanisms that prevent the meeting of potential mates, (2) mechanisms that prevent the formation of zygotes, and (3) mechanisms that handicap species hybrids.

An example of a mechanism that prevents the meeting of potential mates was reported by Valentine in 1948 (Ref. 14-9). He found two primrose species in England that were reproductively isolated from one another by preferences for different types of soils. This form of reproductive isolation is called *ecological isolation.* A different situation that prevents the meeting of potential mates was reported by Blair in 1941 (Ref. 14-10). He found two species of toads whose ranges overlap but who are reproductively isolated from one another because of differences in breeding seasons. This type of reproductive isolation is called *seasonal* or *temporal isolation.*

Should members from two different species meet during the reproductive stages of their lives, mating will occur and offspring will be produced unless hindered by some mechanism that prevents the formation of zygotes. Zygote formation can be avoided by differences in courtship patterns that permit only conspecific matings to occur. This type of reproductive isolating mechanism has been variously called *sexual, psychological,* or *ethological isolation.* A review of the differences in courtship patterns in various *Drosophila* species was published by Spieth in 1952 (Ref. 14-11). An example of a different mechanism that can prevent the formation of zygotes was reported by Federley in 1932 (Ref. 14-12). He found that in some matings between members of different moth species, the males are unable to withdraw the penis. This results in the death of the males upon eventual separation of the copulating pair, and, in addition, results in the retention of the male genitalia in the vagina of the female, making egg deposition impossible. This type of reproductive isolation is called *mechanical isolation.* If members of two species meet and successfully copulate, zygote formation may still not occur because of physiological differences that prevent the union of the gametes involved. An example of this was reported by Patterson in 1946 (Ref. 14-13). He reported that following some interspecific matings in *Drosophila,* the vagina of the females swelled, obstructing the passage of eggs and rendering the female sterile. The reproductive isolation, in this case, is said to be due to *gametic* or *physiological isolation.*

If an interspecific mating does occur and a zygote is formed, other mechanisms may still serve to keep the species reproductively isolated from one another. Experiments dealing with this problem were reported by Moore in 1949 (Ref. 14-14). He found that, in most cases, crosses between different species of frogs resulted in hybrids that died during development. This type of isolating mechanism is called *hybrid inviability.* One final possibility exists for keeping species genetically separated from one another. The interspecific hybrid may be sterile. Possibly the best known example of this type of reproductive isolation are the hybrids produced as a result of a cross between a female horse and a male donkey (a mule) or between a female donkey and male horse (a hinny). A report on the chromosome complements of horse, donkey, and their hybrids was published by Benirschke and his co-workers in 1962 (Ref. 14-15). A review of hybrid sterility as an isolating mechanism in *Drosophila* was published by Ehrman in 1962 (Ref. 14-16).

A number of experiments have been conducted with the aim of producing new species. The greatest successes have been achieved with plants where the induction of polyploidy can result in a new species. A paper which reported an example of this was reprinted in Chapter 7 (Ref. 7-10). Another example of the successful production of a new species through polyploidy was reported by Karpechenko in 1927 (Ref. 14-17). He crossed a radish and a cabbage and was able to obtain a hybrid which produced some gametes with complete haploid sets of chromosomes from both species. When fusion occurred between such "diploid" gametes, an allotetraploid was formed, which was reproductively isolated from both parental species.

Speciation without polyploidy must rely on genetic variability and selection to achieve reproductive isolation. This is a difficult process to demonstrate in the laboratory. However, certain results have been obtained which indicate that given the time, energy, and resources, the process of speciation without polyploidy could be effected in laboratory populations. Experiments were reported by Koopman in 1950 (Ref. 14-18) which showed that the strength of reproductive isolation between two species of *Drosophila* could be increased by eliminating, in each generation, the interspecific hybrids. An experiment involving selection for reproductive isolation within a species was reported in 1956 by Knight and his co-workers (Ref. 14-19), whose paper is the final article reprinted in this chapter. A detailed review of the genetics of race and species formation was published by Dobzhansky in 1970 (Ref. 14-20).

Our discussion of reproductive isolation has omitted the human species. This results from the fact that any earlier evolutionary trends toward speciation in humans has been reversed by modern man's migrations and concomitant interracial gene exchanges. Undoubtedly, future populations of man will be even larger and genetically more variable than those we find today, and the singleness of the human species will be continued.

REFERENCES

Genetics of populations

14-1. Dobzhansky, T. 1947. Adaptive changes induced by natural selection in wild populations of *Drosophila*. Evolution **1**:1-16.

14-2. Van Valen, L., Levine, L., and Beardmore, J. A. 1962. Temperature sensitivity of chromosomal polymorphism in *Drosophila pseudoobscura*. Genetica **33**:113-127.

14-3. Levine, L., and Van Valen, L. 1964. Genetic response to the sequence of two environments. Heredity **19**:734-736.

14-4. Heuts, M. J. 1948. Adaptive properties of carriers of certain gene arrangements in *Drosophila pseudoobscura*. Heredity **2**:63-75.

14-5. Ehrman, L. 1968. Frequency dependence of mating success in *Drosophila pseudoobscura*. Genet. Res., Camb. **11**:135-140.

14-6. Prakash, S., and Lewontin, R. C. 1968. A molecular approach to the study of genic heterozygosity in natural populations. III. Direct evidence of coadaptation in gene arrangements of *Drosophila*. Proc. Natl. Acad. Sci. U. S. A. **59**:398-405.

14-7. Race, R. R., and Sanger, R. 1968. Blood groups in man. 5th ed. Blackwell Scientific Publications, Oxford.

14-8. Harris, H. 1969. Enzyme and protein polymorphism in human populations. Brit. Med. Bull. **25**:5-13.

Reproductive isolation

14-9. Valentine, D. H. 1948. Studies in British primulas. II. Ecology and taxonomy of primrose and oxlip (*Primula vulgaris* Huds. and *P. elatior* Schreb.). New Phytologist **47**:111-130.

14-10. Blair, A. P. 1941. Variation, isolation mechanisms, and hybridization in certain toads. Genetics **26**:398-417.

14-11. Spieth, H. T. 1952. Mating behavior within the genus *Drosophila* (Diptera). Bull. Amer. Mus. Nat. Hist. **99**:399-474.

14-12. Federley, H. 1932. Die Bedeutung der Kreuzung für die Evolution. Jenaische Zeits. Naturwiss. **67**:364-386.

14-13. Patterson, J. T. 1946. A new type of isolating mechanism in *Drosophila*. Proc. Natl. Acad. Sci. U. S. A. **32**:202-208.

14-14. Moore, J. A., 1949. Patterns of evolution in the genus *Rana*. pp. 315-338. *In* Jepsen, G. L., Mayr, E., and Simpson, G. G. [eds.] Genetics, paleontology, and evolution. Princeton University Press, Princeton, N. J.

14-15. Benirschke, K., Brownhill, L. E., and Beath, M. M. 1962. Somatic chromosomes of the horse, the donkey and their hybrids, the mule and the hinny. J. Reprod. Fertil. **4**:319-326.

14-16. Ehrman, L. 1962. Hybrid sterility as an isolating mechanism in the Genus *Drosophila*. Quart. Rev. Biol. **37**:279-302.

14-17. Karpechenko, G. D. 1927. Polyploid hybrids of *Raphanus sativus* L. X *Brassica oleracea* L. Bull. Appl. Botany **17**:305-408.

14-18. Koopman, K. F. 1950. Natural selection for reproductive isolation between *Drosophila pseudoobscura* and *Drosophila persimilis*. Evolution **4**:135-148.

14-19. Knight, G. R., Robertson, A., and Waddington, C. H. 1956. Selection for sexual isolation within a species. Evolution **10**:14-22.

14-20. Dobzhansky, T. 1970. Genetics of the evolutionary process, Columbia University Press, New York.

72 Adaptive changes induced by natural selection in wild populations of Drosophila

Th. Dobzhansky*
Columbia University
New York, New York

INTRODUCTION

The theory of the origin of adaptations through natural selection is more than a century old, if one takes as its inception the date of Darwin's first essay written in 1842. Nevertheless, no agreement as to the rôle played by natural selection in evolution has as yet been reached. Weismann called natural selection "all powerful," but, during the first quarter of the present century, the idea fell into disrepute because of a failure to comprehend the meaning of the mutation theory and Johannsen's experiments on pure lines. So wide a divergence of opinion has been possible because the theory of natural selection has rested either on deductions from very general propositions or on inference from indirect evidence. That adaptive evolution in nature is too slow a process to be observed within a human lifetime has been taken for granted almost universally. Furthermore, selection pressures which act upon non-pathological traits of wild species have been assumed to be small.

Recent observations have shown, however, that natural populations, even of higher organisms, sometimes undergo rapid adaptive changes. Some wild species react to seasonal alterations in their environment by cyclic modifications of their genetic structure. Knowing these facts, direct observation and experimentation on natural selection has become possible. Controlled experiments can now take the place of speculation as to what natural selection is or is not able to accomplish. Furthermore, we need no longer be satisfied with mere verification of the existence of natural selection. The mechanics of natural selection in concrete cases can be studied. Hence, the genesis of adaptation, which is possibly the central problem of biology, now lies within the reach of the experimental method.

From Evolution **1**:1-16, 1947. Used with permission.

*Present address: Rockefeller University, New York, New York.

The first discovery of cyclic changes in the genetic composition of populations of wild species was made by Timofeeff-Ressovsky (1940). European as well as North American and Asiatic populations of the beetle *Adalia bipunctata* vary greatly in the elytral color pattern. Two color types, red and black, can easily be distinguished. They are known to differ by a single Mendelian gene. Near Berlin, where the species produces two and possibly three generations per year, the black type increases in relative frequency from about 37 per cent to 59 per cent from spring until autumn. During the winter the frequency of the black type is reduced and that of the red is increased. The beetle hibernates as an adult, the mortality among hibernating individuals being high. Only 4 per cent of black individuals survive hibernation, while about 11 per cent of the reds survive. The inference can be drawn that the black type is selected during the warm season.

The coloration of the Adalia beetle is a visible character easy to work with. In the corresponding Drosophila work, a character is used which is discernible only by microscopic examination of the larval salivary glands, namely the gene arrangement in the chromosomes. Variation of this character in species of Drosophila is due almost entirely to inversion of chromosome segments. Two or more such gene arrangements frequently occur in the same population. Since the carriers of different arrangements interbreed freely, some individuals have paired chromosomes with the same gene arrangements (inversion homozygotes) and others with unlike gene arrangements (inversion heterozygotes). Because the inversion homozygotes and heterozygotes are indistinguishable in external appearance, there was no reason to suppose that inversions are other than adaptively neutral characters. Not until Dobzhansky (1943, in *D. pseudoobscura*) and Dubinin and Tiniakov (1945, 1946, in *D. funebris*) found that populations which live in different habitats often differ in the relative frequencies of their gene arrangements, and that the composition of a single population may vary appreciably from season to season, was it realized that carriers of different gene arrangements may be favored or discriminated against by different environments.

The seasonal changes in the composition of the Adalia and Drosophila populations are adaptive responses of the living species to the succession of seasonal environments. It is important to note that such seasonal changes occur not only in organisms which, like the ones named above, produce several generations per year, but also in the longer lived ones. Gershenson (1945) has found changes of this sort in the hamster, *Cricetus cricetus.* Black and agouti individuals occur in Russian populations of this mammal, and the relative frequencies of these coat colors differ significantly in different seasons and in different places. The difference between the two color forms is due to a single gene.

Observations and experiments on natural selection in *Drosophila pseudoobscura* will be reviewed in the following pages. Previously published as well as unpublished data are discussed briefly; the latter will be presented in more details elsewhere.

LOCAL RACES

Fifteen different gene arrangements are known in the third chromosome of *Drosophila pseudoobscura.* None of them occur in the entire distribution area of the species. Hence,

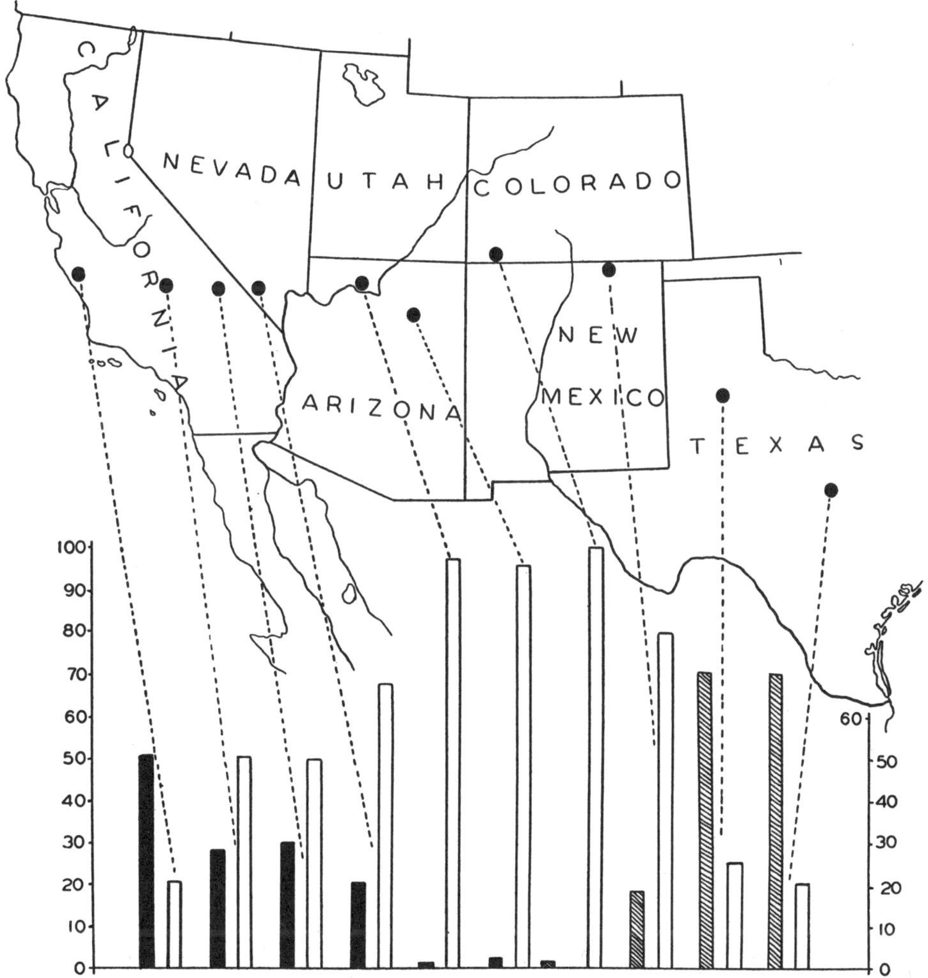

Fig. 1. Frequencies (in per cent) of Standard (black columns), Arrowhead (white columns), and Pikes Peak (hatched columns) chromosomes in populations of ***Drosophila pseudoobscura*** in certain localities in the western United States.

there is no "normal" or "wild-type" gene arrangement. On the other hand, the populations of most localities contain more than one, and up to seven, gene arrangements. Because of the free interbreeding of the carriers of different arrangements, many, frequently a majority, of wild individuals are inversion heterozygotes. The population of any locality can be described in terms of relative frequencies of different gene arrangements. The frequencies may differ in populations of different localities. Sometimes the differences are more or less proportional to the distances which separate the localities. Geographic gradients or "clines" are thus formed (Dobzhansky and Epling, 1944). An example of such clines is given in figure 1.

About 50 per cent of third chromosomes in populations of south Coast Ranges of California have the so-called Standard gene arrangement, represented by black columns in figure 1. But the frequency of Standard falls to between 20 and 30 per cent in the Sierra Nevada and in the Death Valley regions which lie to the east of the Coast Ranges. Further east, in Arizona, the frequency falls to less than 5 per cent, and still further east Standard chromosomes occur but rarely. The Arrowhead gene arrangement (white columns in figure 2) is very common in Arizona and New Mexico, so much so that populations of some localities seem to be homozygous for it. But its frequency decreases eastward as well as westward from Arizona, reaching about 20 per cent in central Texas and in coastal California. The Pikes Peak gene arrangement (hatched columns in figure 1) is common in Texas, but rapidly decreases in frequency westward.

The transect across the southwestern United States shown in figure 1 is roughly 1200 miles long. Differences in the frequencies of gene arrangements may be observed however between populations which live in localities only a dozen or so miles apart. For example, three localities on Mount San Jacinto, California, were sampled repeatedly between 1939 and 1946. The approximate distances between these localities are 10 to 15 miles. One, Keen Camp, lies at an elevation of about 4500 feet in the ponderosa pine belt, the second, Piñon Flats, lies at 4000 feet in the much drier piñon forest, and the third, Andreas Canyon, lies at 800 feet on the desert's edge. The frequencies of gene arrangements in the populations of these localities are shown in table 1. (Chromosomes with Standard gene arrangement are henceforth denoted as ST, Arrowhead as AR, and Chiricahua as CH.)

Table 1 shows that ST chromosomes are most frequent in the lowest locality, Andreas, and least frequent in the highest locality, Keen. CH chromosomes show the opposite relationship. No significant differences appear for AR chromosomes in the three localities (Dobzhansky, 1943). How common such altitudinal gradients are is an open question. Preliminary data suggest the existence of gradients among populations that occur at different elevations in the region of the Yosemite National Park, in the Sierra Nevada in California. Here, however, the ST and AR, and not CH, chromosomes vary in frequencies from locality to locality. Thus, at Jacksonville, elevation about 800 feet, 40 to 45 per cent of third chromosomes have ST and 20 to 25 per cent have AR gene arrangement. At Lost Claim Campground, elevation about 3000 feet, the frequencies of both ST and AR are between 30 and 35 per cent. At Mather, elevation 4600 feet, ST fluctuates between 20 and 40 per cent and AR between 30 and 45 per cent. Finally, at Aspen Valley, elevation about 6800 feet, ST falls to 20 and AR rises to almost 50 per cent. The frequencies of CH chromosomes are between 15 and 20 per cent in all the localities. The horizontal distance between the farthest localities, Jacksonville and Aspen Valley, is about 35 miles.

Altitudinal gradients in the frequencies of gene arrangements suggest that the differences between the inhabitants of different elevations on the same mountain range are adaptive and

Table 1. Frequencies (in per cent) of chromosomes with different gene arrangements in populations of localities on Mount San Jacinto (California)

Locality	Gene arrangements				Numbers of chromosomes studied
	ST	AR	CH	Others	
Keen Camp	33.7	23.8	38.0	4.5	6634
Piñon Flats	40.7	25.1	29.1	5.1	4853
Andreas Canyon	57.6	24.0	15.3	3.0	3818

are produced by natural selection. This hypothesis is strengthened by the observations on seasonal changes and on experimental populations discussed in the following paragraphs.

SEASONAL CYCLES

The repeated samplings of the populations in the three localities on Mount San Jacinto (see above) have disclosed a very interesting fact, namely that the composition of a population may change quite significantly from month to month (Dobzhansky, 1943). Furthermore, these changes are regular and follow the annual cycle of seasons. In two of the three localities, namely at Piñon Flats and at Andreas Canyon, the changes are qualitatively similar. Figure 2 gives a summary of the data for Piñon Flats. In this figure, the observations for all six years of collecting are grouped by months. It can be seen that in spring (March) the population contains about 50 per cent of ST chromosomes (shown in figure 2 by circles), and slightly more than 20 per cent CH chromosomes (shown by triangles). From March to June the frequency of ST declines to less than 30 per cent and that of CH increases to just below 40 per cent. During the summer, from June to September, the reverse change takes place, namely ST increases in frequency, and CH decreases, to about the same values which these gene arrangements had during the spring. The

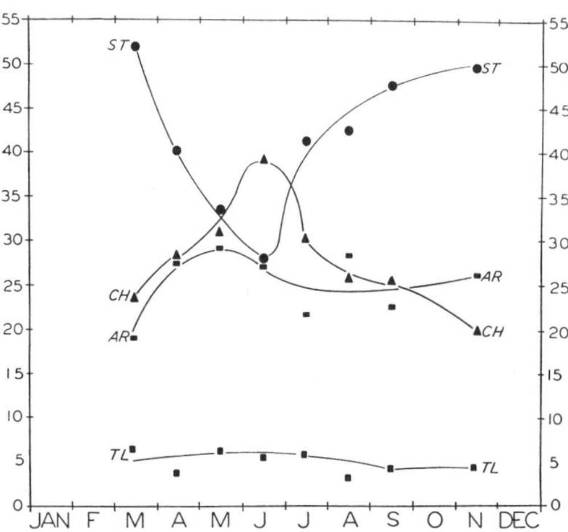

Fig. 2. Changes in the frequencies of chromosomes with Standard (circles), Chiricahua (triangles), Arrowhead (horizontal rectangles), and Tree Line (squares) gene arrangements in the population of Piñon Flats, California. Ordinate—frequencies in per cent; abscissa—months. Combined data for six years of observation.

changes of the frequencies of AR chromosomes (rectangles in figure 2) are less regular than those of ST and CH, but on the whole AR seems to follow the same path as CH. No regular changes occur in the frequency of Tree Line chromosomes (squares in figure 2).

The changes at Andreas Canyon run parallel to those at Piñon Flats. From autumn till early spring the frequency of ST chromosomes keeps on a high level, and that of CH chromosomes on a low one. From March to June ST wanes and CH waxes in frequency. During the hot part of the summer very few *D. pseudoobscura* flies can be collected at Andreas Canyon. But when the population begins to build up in numbers in September, the ST chromosomes are found to have recovered their high frequency, while CH have dwindled to about the winter level. Curiously enough, no significant changes from month to month are detectable in the Keen Camp population. Because of the climate, the breeding season of the flies is clearly shorter at Keen Camp than at the other localities, but nevertheless fly samples have been collected at Keen Camp from April to September and even October. Such time intervals are amply sufficient to detect changes in the Piñon and Andreas populations, but no cyclic changes have been found at Keen Camp (Dobzhansky, 1943).

Interestingly enough, a different kind of change has taken place in the Keen Camp population during the period of observation, from 1939 to 1946. Namely, there seems to exist a non-cyclic, or at any rate non-seasonal, trend toward decreasing frequencies of AR and CH and increasing ones of ST chromosomes. In 1939, only 28 per cent of the third chromosomes found in the Keen locality had the ST gene arrangement; in 1942 the frequency rose to 36 per cent, and in 1946 to 50 per cent. The frequencies of AR and CH chromosomes in 1939 were 30 and 38 per cent respectively. In 1946 only 15 per cent of the chromosomes were AR and only about 28 per cent CH. No such directional trends of change have appeared at Piñon Flats or at Andreas Canyon, although statistically significant differences in the composition of the populations from year to year have been recorded also in these localities (Dobzhansky, 1943, and a paper in press).

It will obviously be important to ascertain how general are the phenomena of cyclic seasonal and non-seasonal changes not only in different populations of *Drosophila pseudo-*

obscura but in other species as well. The fact that cyclic changes occur at Piñon Flats and at Andreas Canyon, but not at Keen Camp only some 15 miles away, makes generalizations at this time decidedly premature. Data are however available which show cyclic seasonal changes in the population of a locality in central Texas. Here the Arrowhead and Pikes Peak gene arrangements in the third chromosome of *D. pseudoobscura* are involved (Dobzhansky and Epling, 1944; cf. also figure 1). Unpublished data suggest that changes in the frequencies of ST, AR, and perhaps of CH chromosomes occur in populations of at least some of the localities in the Sierra Nevada.

NATURAL SELECTION AS A CAUSE OF THE SEASONAL CHANGES

The regular and cyclic nature of the changes observed in the populations of *D. pseudoobscura* on Mount San Jacinto can be most reasonably accounted for by natural selection as the prime causative factor. If during the spring the carriers of CH chromosomes leave more surviving progeny on the average than the carriers of ST chromosomes, then the frequency of CH will increase and that of ST will decrease. This is what happens from March to June (Figure 2). The reversal of the change during the summer months points toward the hypothesis that, in the summer environments, the carriers of ST chromosomes survive or reproduce on the average more often than do the carriers of CH chromosomes. The absence of changes during the autumn and winter at Andreas Canyon suggests that flies of different chromosomal types are equivalent in reaction to the environments prevailing during these seasons.

But the great rapidity of the observed changes constitutes an apparently serious argument against accounting for them on the ground of natural selection. Indeed, at Piñon Flats the frequency of ST chromosomes falls from about 50 per cent in March to 28 per cent in June, and increases again to about 48 per cent in September (figure 2). Even though Drosophila is a rapidly breeding insect, time intervals such as these can correspond to at most two to four generations. The selective forces that are necessary to bring about changes so swift as these must be very strong.

It should be remembered however that very little is known about the intensity of selective forces which operate in natural populations. The wide-spread opinion that these forces are generally weak, and their effects negligible except in terms of quasi-geological time is only an opinion and has no basis in factual data. To find in natural populations great selective pressures and the rapid changes produced by them may be unexpected but not inherently impossible. On the other hand, the occurrence of changes does not in itself prove that they are produced by natural selection. Such proof would be very difficult to adduce from observations of natural populations alone. The difficulty lies in the fact that, despite persistent effort, very little has been learned as yet about the food and shelter requirements of *D. pseudoobscura* in its natural habitats. Proof of selection by a method analogous to that employed by Timofeeff-Ressovsky in Adalia is still out of the question in Drosophila.

Nevertheless, the postulated high selective advantages and disadvantages of the carriers of different gene arrangements in different environments has made practicable a still more ambitious project: to demonstrate the occurrence of natural selection by means of laboratory experiments. For this purpose, a modification of the population cage devised for Drosophila by l'Héritier and Teissier is used. These cages are wooden boxes with glass or wire screen sides and a detachable glass top (figure 3). The bottom has 15 circular openings closed by corks which carry glass containers with culture medium. Wire loops hold the containers in place (figure 4). Several hundred flies are introduced into the cage at the beginning of the experiment. These flies are a mixture of individuals with different gene arrangements in desired proportions. Within a single generation, the population of the cage increases to the maximum compatible with the amount of food given. This is usually between two and four thousand flies. The numbers of eggs deposited in a population cage are tens to hundreds of times greater than the numbers of adult flies that hatch. The competition for survival is intense. (For a more detailed description of population cage, see Wright and Dobzhansky, 1946.)

Once a month, or at other suitable intervals, samples of the eggs which have been deposited in the cages are taken, and the larvae which emerge from these eggs are grown in regular culture bottles. Salivary gland chromosomes of fully grown larvae are then examined.

It is known that many chromosomes in

natural populations carry recessive lethals, semilethals, or viability modifiers. We are, however, interested not in the effects of individual chromosomes on the survival of the flies, but in the selective values of ST, AR, CH, and other chromosomes as classes. In other words, in our experiments it is desired to have flies which are genetically heterogeneous, regardless of whether they are inversion homozygotes or heterozygotes. Accordingly, the initial population of a cage is always made from a mixture of several strains of each of the inversion types to be studied. As a result flies homozygous for any given individual chromosome are relatively rare in such population cages.

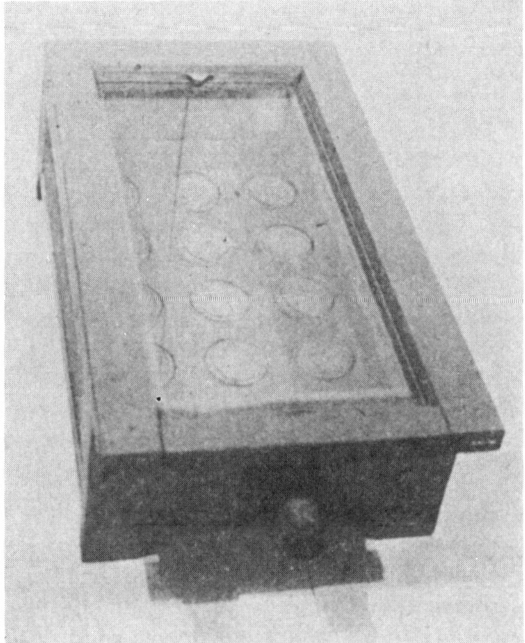

Fig. 3. A population cage.

SELECTIVE DIFFERENTIALS BETWEEN CHROMOSOMES FROM PIÑON FLATS

Twenty-nine experiments have been either completed or are now under way, employing populations the chromosomes of which were derived from ancestors collected at Piñon Flats. Some cages contained mixtures of ST and CH chromosomes, others of ST and AR, or of AR and CH, or of all three gene arrangements. Some were kept in incubators or constant temperature rooms at 25° C., others at 16½° C., and still others at variable room temperatures between 20° and 26° C.; some were exposed to alternation of day and night, others were kept in the dark. Two types of results have been obtained. First, at 16½° C. no significant changes in the frequencies of the gene arrangements have taken place. The relative frequencies present in the original population of the cage have been retained generation after generation. Second, at 25° C. or at room temperatures, the relative proportions of the gene arrangements have changed with time until certain definitive equilibrium proportions have been attained.

Figure 5 shows an example of changes in the frequency of ST chromosomes observed in the population cage No. 35 at 25° C. On March 1, 1946, a population was introduced into this cage containing 10.7 per cent ST and 89.3 per cent CH chromosomes of Piñon Flats origin. In about a month, in early April, the frequency of ST has approximately doubled (21.7 per cent), in early May nearly trebled (28.3 per cent), and in early June nearly quadrupled (37.7 per cent).

Fig. 4. A cork and a jar with culture medium used in population cages.

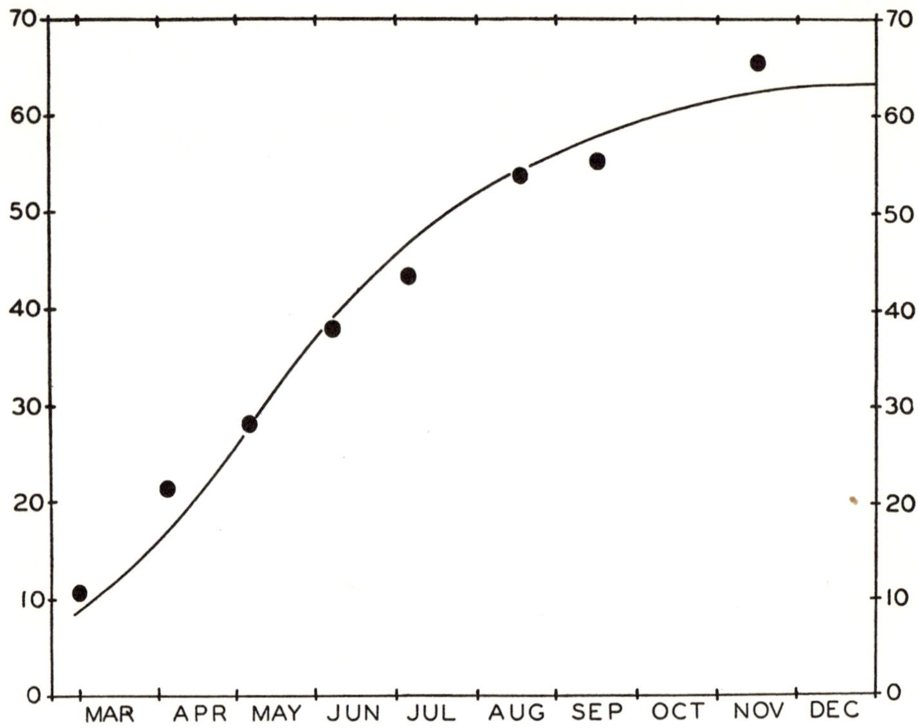

Fig. 5. Frequency of Standard chromosomes (in per cent) in different months in the population cage No. 35.

By mid-November ST reached the frequency of 66.7 per cent and by the end of December 71.0 per cent. It can be seen, first, that the changes are greater when ST chromosomes are rare but relatively slight when they become frequent, and, second, that ST chromosomes never supplant entirely CH chromosomes. The final result of selection is a stable mixture of both ST and CH chromosomes.

Wright (in Wright and Dobzhansky, 1946) has analyzed mathematically the results of experiments like that illustrated in figure 5. He has concluded that the simplest hypothesis to account for the data is that the highest adaptive value exists in ST/CH heterozygotes, and that both homozygous classes, ST/ST and CH/CH, are inferior to the heterozygotes. Furthermore, that the adaptive value of ST/ST homozygotes is higher than that of CH/CH homozygotes. The curve shown in figure 5 is calculated on the assumption that the adaptive values of the ST/CH, ST/ST, and CH/CH classes of individuals are as 1.0:0.7:0.4, and that a fly generation had an average span of one month in population cage No. 35. The observed values fit the theoretical curve remarkably well.

Since the heterozygotes, ST/CH, are superior in adaptive values to both homozygotes, the final result of selection is not elimination of either CH or ST but establishment of a certain equilibrium at which both ST and CH gene arrangements occur in the population. Since ST/ST are superior to CH/CH homozygotes, the equilibrium frequency of ST is higher than that of CH. With the adaptive values indicated above, the population would be expected to reach equilibrium at about 67 per cent ST and 33 per cent CH. The results obtained in the experiment shown in figure 5 agree with the expectation.

All the experiments in which mixtures of ST and CH chromosomes were kept at temperatures above 20° C. have led to ST becoming more frequent than CH; whenever the experiments were continued long enough, equilibria were reached at values close to 70 per cent ST and 30 per cent CH. Under similar conditions, populations containing ST and AR have reached equilibria in which ST chromosomes are also more frequent than AR. The adaptive values of the three possible genotypes must therefore be ST/AR>ST/ST>AR/AR. Experiments with mixtures of AR and CH show that the hierarchy of adaptive values is AR/CH>

AR/AR>CH/CH. Some experiments have been made in which population cages have contained mixtures of three gene arrangements, ST, AR, and CH. Equilibrium proportions are indicated at, roughly, 50-55 per cent ST, 30-35 per cent AR, and 10-15 per cent CH.

The lack of perceptible changes in the population cages kept at 16½° C. indicates that at that temperature the adaptive values of the inversion heterozygotes and homozygotes are more nearly similar than they are at higher temperatures. Such changes in the relative adaptive values of different genotypes at different temperatures have been observed experimentally (Dobzhansky and Spassky, 1944).

STAGE OF THE LIFE CYCLE AT WHICH SELECTION TAKES PLACE

The experiments summarized above demonstrate that, in some environments, the adaptive values of inversion heterozygotes and homozygotes are strikingly unlike. But these experiments tell us nothing of the stage of the life cycle at which the differential survival or reproduction take place. Natural selection may operate in a variety of ways. The chromosomal types may be characterized by differential mortality, or differential longevity, or fecundity, or differences in sexual activity, or combinations of two or more of these and other variables. The adaptive value of a chromosomal type is the net effect of interaction of all the variables.

Perhaps the simplest, though by no means the only possible, hypothesis would assume a differential mortality of the different chromosomal types among the crowded larvae in the population cages. Let the frequencies of the ST and CH gene arrangements in a population cage be q and $(1-q)$ respectively. Provided that the flies mate at random with respect to gene arrangement, the proportions of heterozygotes and homozygotes among the eggs deposited in a population cage will be:

$$q^2 \text{ ST/ST} : 2q(1-q) \text{ ST/CH} : (1-q)^2 \text{ CH/CH}.$$

If, however, the larvae which hatch from these eggs survive and reach the adult stage in a proportion 0.7 ST/ST:1 ST/CH:0.4 CH/CH, then the relative frequencies of the chromosomal types of the adult flies developed in a population cage will be:

$$0.7q^2 \text{ ST/ST} : 2q(1-q) \text{ ST/CH} : 0.4(1-q)^2 \text{ CH/CH}.$$

The frequencies of the ST/ST, ST/CH, and CH/CH types have been determined in samples of larvae hatching from the eggs deposited in population cages but grown in regular culture bottles under approximately optimal conditions. The deviations from the $q^2:2q(1-q):(1-q)^2$ proportions are found to be relatively small in such samples. A sample of adult flies hatched in a population cage was now taken, and the chromosomal constitution of these flies was determined with the aid of a suitable method (Dobzhansky, 1947b). The numbers of ST/ST, ST/CH, and CH/CH flies which would be expected to occur in this sample if there were no differential mortality between the egg and the adult stage were calculated with the aid of the Hardy-Weinberg formula $q^2:2q(1-q):(1-q)^2$. The observed and the expected values are as follows:

	ST/ST	ST/CH	CH/CH
Observed	57	169	29
Expected	78.5	126.0	50.5
Deviation	−21.5	+43.0	−21.5

Among the adult flies, the heterozygotes, ST/CH, are considerably more frequent, and the homozygotes less frequent, than expected on the basis of the Hardy-Weinberg formula. Since, as stated above, the Hardy-Weinberg proportions are approximately realized among the eggs deposited in the population cages, a differential elimination of ST/ST and CH/CH homozygotes at some time between the egg and the adult stage may be regarded established.

EXPERIMENTS ON CHROMOSOMES OF DIFFERENT GEOGRAPHIC ORIGIN

The above described experiments on the behavior of chromosomes with different gene arrangements in population cages have been made with flies the ancestors of which were collected at Piñon Flats, Mount San Jacinto. The ST, AR, and CH gene arrangements occur, however, in populations from most of the western United States, from British Columbia, and from Lower California. The question that naturally arises is whether or not the biological properties of these chromosomes are constant throughout the geographic area in which a given gene arrangement occurs (Mayr, 1945).

Population cage experiments are now in progress on mixtures of flies with different gene arrangements from Keen Camp and from Mather, California. These localities are respectively about 12 and 300 miles from Piñon Flats. The data so far obtained leave no doubt in that the adaptive properties of chromosomes with the same gene arrangement but of different geographic origin may be different. The relevant data will be published in more detail later; the following comparison will suffice as an illustration.

On March 2nd, 1946, population cage No. 36 was started with a mixture of about 12 per cent ST and 88 per cent AR chromosomes of Piñon Flats origin. In late November of the same year this population cage contained about 67 per cent ST and 33 per cent AR. It is obvious that when populations of ST and AR of Piñon Flats origin reach equilibrium, ST chromosomes are decidedly more common than AR. This is confirmed by the experiment No. 34, in which the initial population of a cage consisted on February 25th, 1946, of 85 per cent ST and 15 per cent AR. By late summer and autumn of the same year the frequency of AR in this cage fluctuated between 20 and 25 per cent AR. But the results of experiments on ST and AR chromosomes of Mather origin are quite different. On December 22nd, 1945, population cage No. 29 was started with about 70 per cent ST and 30 per cent AR chromosomes from Mather. In late June and late July of 1946, when the experiment was terminated, the frequency of ST had fallen to 54-55 per cent, and that of AR had risen to 45-46 per cent. In a parallel experiment started simultaneously, the initial population of cage No. 32 contained 19 per cent ST and 81 per cent AR chromosomes of Mather origin. By late summer and autumn of 1946, the frequency of ST had risen in this cage to 50-54 per cent, and the frequency of AR had fallen to 46-50 per cent (figure 6). It seems safe to conclude that mixtures of ST and AR chromosomes of Mather origin reach equilibria at about 50-55 per cent ST and 45-50 per cent AR. This is significantly different from the behavior of ST and AR chromosomes of Piñon Flats origin. All the experiments described in this paragraph were carried out in a temperature-controlled room at 25° C.

Chromosomes with the same gene arrangement but of different geographic origin may, then, behave differently. This fact has a bearing on the problem of the genetic nature of the adaptive differences between chromosomes

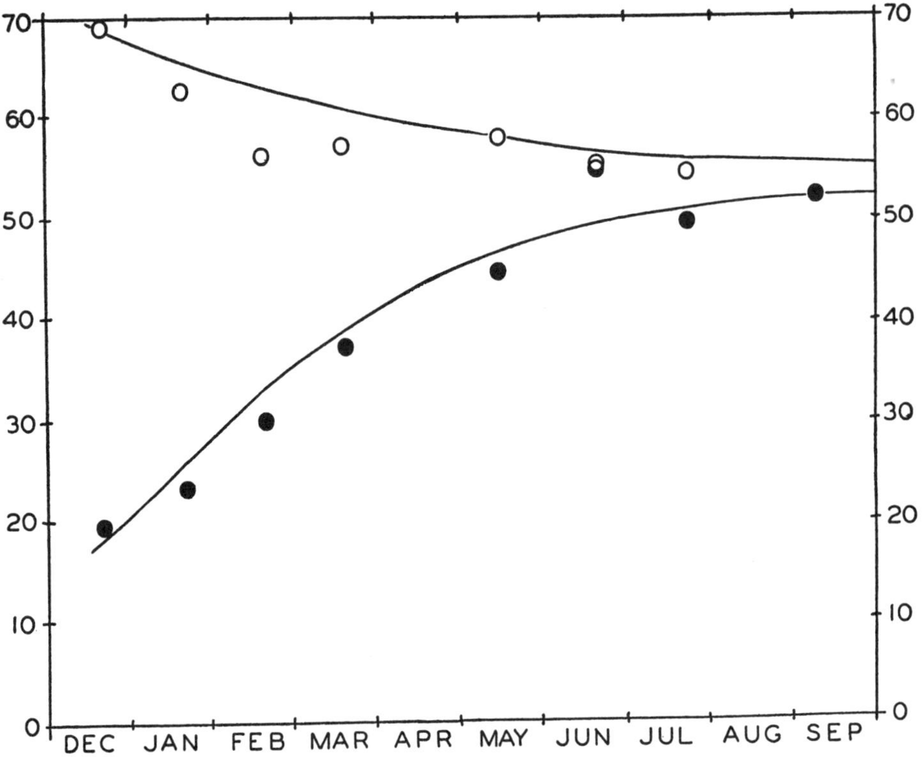

Fig. 6. Frequency of Standard chromosomes (in per cent) in different months in the population cages Nos. 29 (white circles) and 32 (black circles).

with different gene arrangements. The properties of a chromosome may, in general, depend on its gene contents, or on the gene arrangement itself (position effect), or on a combination of both factors. The different behavior of the ST and AR chromosomes of Piñon Flats and Mather origin proves that the gene contents of these chromosomes are different. We are evidently dealing with gene complexes evolved under the control of natural selection which confer upon their carriers ecological properties which adapt them to certain seasonal and local environments. Since inversions modify the gene arrangement and diminish the effective crossing over between the chromosomes involved, their biological function is binding together of the gene complexes of proven adaptive worth. In this sense, the effect of natural selection on the chromosomal inversions is indirect. For the purposes of this discussion, it must be borne in mind that the inversions may be simply structural characters marking chromosomes with different gene contents. Of course, the possibility still exists that the biological properties of chromosomes with different gene arrangement differ not only because of the dissimilarities of their gene contents but because of position effects as well.

HETEROSIS

In the experiments discussed above, the populations of the cages reached, or tended to reach, equilibria at which all the gene arrangements introduced into a cage at the beginning were preserved in certain proportions in the final population. Natural selection does not lead to the complete elimination of some and to fixation of other gene arrangements. The final outcome of the selective process is the establishment of those relative frequencies of the competing gene arrangements at which the average adaptive level of the population as a whole is the highest one attainable. This outcome of the selective process is explained by the fact that inversion heterozygotes, ST/CH, ST/AR, and AR/CH, possess adaptive values greater than the corresponding homozygotes ST/ST, CH/CH, and AR/AR. The populations at equilibrium contain the greatest possible proportions of the well adapted heterozygotes compatible with the lowest possible proportions of the relatively ill adapted homozygotes.

The superiority of the heterozygotes over the homozygotes is indicated in all but one experiment carried out at temperatures above 20° C. The single exceptional experiment, No. 31, was started in December 23, 1945. The initial population of the experimental cage contained 45 per cent of ST chromosomes and 55 per cent of chromosomes with the Tree Line gene arrangement, all the chromosomes descended from flies collected at Mather, California. Tree Line is a gene arrangement rather widespread geographically but relatively rare in most natural populations. In the population of Mather its frequency is approximately 10 per cent of the total. The changes in population cage No. 31 were rapid. Within a single month, in late January of 1946, the frequency of Standard chromosomes had risen to 62 per cent, while Tree Line fell to 38 per cent. By mid-May of 1946 the corresponding frequencies were 82 and 18 per cent, in late July 93 and 7 percent, and in mid-November 99 and 1 per cent. It seems clear that no equilibrium is being reached in cage No. 31; instead, the Tree Line chromosomes were being eliminated and entirely supplanted by ST chromosomes. This is possible only because ST/ST homozygotes seem to have an adaptive value higher not only than Tree Line homozygotes but also than Standard/Tree Line heterozygotes.

The Tree Line gene arrangement seems to be deleterious to its carriers in homozygous as well as in heterozygous condition. The question naturally arises: why are Tree Line chromosomes retained in natural populations rather than eliminated by natural selection? A possible answer is provided by experiment No. 33, as yet uncompleted. The initial population of this cage, started on January 16th, 1946, consisted of 34 per cent Arrowhead and 66 per cent Tree Line chromosomes of Mather origin. Within a month, in mid-February, the proportions had changed to 58 and 42 per cent respectively. The frequencies of AR chromosomes rose to 69 per cent in mid-April, 75 per cent in mid-July, and 80 per cent in early November. It appears that, in this experiment, the Tree Line chromosomes are not eliminated entirely; instead, an equilibrium at about 80 per cent AR and 20 per cent Tree Line chromosomes is indicated.

Arrowhead/Tree Line heterozygotes are, under the conditions of this experiment, superior to both Arrowhead and Tree Line homozygotes. This explains not only the fact that the Tree Line gene arrangement is relatively rare but that it is nevertheless retained in natural populations. Tree Line chromosomes form adaptively valuable heterozygotes with AR but not with ST chromosomes present in the same populations. This assures the retention, but prevents excessive increase of the incidence, of the Tree Line gene arrangement in the species.

The classical theory of heterosis assumes that the restoration of vigor following intercrossing of inbred lines is due to the covering up of deleterious recessives by favorable dominants. The greater vigor observed in inversion heterozygotes as compared with homozygotes requires a different explanation. Inversion homozygotes found in natural populations of Drosophila are evidently no more "inbred" than are the heterozygotes in the same population. The lower adaptive value of homozygotes is, therefore, not due to manifestation of deleterious recessives normally covered up by their more favorable dominant alleles. The higher adaptive value of heterozygotes is, in this case, a product of the action of natural selection on the heterozygous genotypes, which are more widespread in natural populations than inversion homozygotes. In view of their greater frequency, the adaptive qualities of heterozygotes may be maintained by natural selection on a level higher than that in homozygotes.

ADAPTATION AND PLASTICITY

The conflict between adaptive fitness and genetic variability was pointed out by Haldane (1937), Dobzhansky (1937, 1938), and Mather (1943). The mutation process furnishes the raw materials without which adaptive changes cannot be constructed, but the same process also unavoidably produces a multitude of poorly adapted variants. Restriction of the supply of heritable variability might permit a species to reach a higher level of immediate fitness, but it jeopardizes its adaptability to changing environments. When an adaptive change does occur it generally uses up a part of the available supply of variability and thus limits the possibilities of further change.

Observations and experiments on the gene arrangements in third chromosome of *Drosophila pseudoobscura* demonstrate that a remarkable adaptive mechanism exists in this species. The plasticity of the species is so great that it reacts by adaptive reconstructions of the genotype to environmental changes of even so ephemeral a nature as the succession of the year's seasons. There is no doubt that environmental changes in time and in space elicit adaptive responses of a more durable kind as well. And yet all the responses occur without expenditure of genetic variability stored in the populations. This great efficiency of the adaptive mechanism is made possible by the fact that the inversion heterozygotes are endowed with survival values greater than the inversion homozygotes.

We have seen that in the Piñon Flats population the frequencies of ST chromosomes increase and those of CH chromosomes decrease during the hot part of the summer (figure 2). In summer environments the adaptive values of the carriers of ST chromosomes are evidently higher than those of CH. Yet, during the spring months the carriers of CH seem to be superior to the carriers of ST chromosomes. An uncommonly hot or prolonged summer can not eliminate all the CH chromosomes and thus endanger the welfare of the species during the ensuing spring. This is because the completion of the selective process in the summer environment would lead not to elimination of CH chromosomes but to establishment of a certain equilibrium of ST and CH gene arrangements. The genetic variability is, therefore, preserved intact and the species remains capable of immediate and rapid response to new environmental changes. This resembles the balanced polymorphism studied by Fisher (1930) and Ford (1940).

Adaptive mechanisms which permit very rapid reconstructions of the population genotype in accordance with the demands of the seasonal environment will probably be most important and widespread in species which inhabit the temperate and frigid climatic zones, as well as tropical territories in which great seasonal changes regularly occur. Inversion systems in which the heterozygotes possess adaptive values higher than the homozygotes are capable of responding very rapidly to alterations in the environment. The biological significance of heterozygosis for inversions is, however, not necessarily confined to seasonal changes. The high adaptive value of inversion heterozygotes may be desirable because it permits rapid adjustments to various microenvironments which are found in the same region and which are not connected with seasonal phenomena. Thus, inversions may be found also in tropical countries characterized by relatively invariant climates. Indeed, fragmentary data now available show that inversion heterozygotes occur very frequently in natural populations of some tropical species of Drosophila (Pavan, 1946, and unpublished data of the writer). Inversion heterozygosis may be now connected with seasonal changes also in some temperate zone species. Carson and Stalker (1947) found no seasonal changes in frequencies of inversions in *D. robusta* near St. Louis, and the present writer has ground to suppose that such changes are absent in populations of *D. persimilis* in the Sierra Nevada of California. The inversion mechanism may facilitate adaptation to different ecological niches within the same geographic environment. Here is a promising field for further studies.

SUMMARY

The gene arrangement in the third chromosome of *Drosophila pseudoobscura* is variable. Each gene arrangement occurs in populations of a definite geographic area. However, two or more gene arrangements may occur together in many populations. Inversion homozygotes and heterozygotes occur frequently in natural populations.

The relative frequencies of various gene arrangements in some populations undergo seasonal cyclic changes. These changes are produced by natural selection, and represent adaptive reconstructions of the population genotype, thus facilitating survival in different seasonal environments.

Some of the changes taking place in nature can be reproduced experimentally in "population cages" (figure 3). Populations containing desired proportions of chromosomes with different gene arrangements are introduced into the cages, samples of the eggs deposited by the flies in the cages are taken from time to time, and the incidence of the chromosomes of different types determined in these samples. Most experiments have been made with chromosomes derived from parents collected at Piñon Flats, Mount San Jacinto, California, and some from Mather, California.

The relative proportions of chromosomes with different gene arrangements remain constant in population cages kept at 16½° C. Changes are frequently observed in cages kept at temperatures above 20° C.: the incidence of some gene arrangements increases and of others decreases. However, the final outcome of the selective process is rarely a complete replacement of one gene arrangement by another. Instead, an equilibrium is usually reached at which all gene arrangements present in the initial population of an experimental cage are retained, but often with frequencies very different from the initial ones. The establishment of equilibria in the populations indicates that individuals heterozygous for different gene arrangements (inversion heterozygotes) are characterized by the highest adaptive values, while homozygotes are relatively inferior in survival and reproduction. The correctness of this interpretation is demonstrated by means of observations on deviations from the Hardy-Weinberg proportions of heterozygotes and homozygotes among the adult flies developed in the population cages (see page 467).

The balanced polymorphism, resulting from the superiority of the heterozygotes over the corresponding homozygotes, permits the species to react adaptively to changes in its environment. The species is "buffered" against environmental change; at the same time, the adaptive responses do not consume or deplete the store of hereditary variability present in the species.

LITERATURE CITED

Carson, H. L., and H. D. Stalker. 1947. A seasonal study of gene arrangement frequencies and morphology in *Drosophila robusta.* Genetics **32**:44.

Dobzhansky, Th. 1937. Genetics and the origin of species. Columbia Univ. Press, New York.

__________. 1938. The raw materials of evolution. Sci. Monthly **46**:445-449.

__________. 1943. Genetics of natural populations. IX. Temporal changes in the composition of populations of *Drosophila pseudoobscura.* Genetics **28**:162-186.

__________. 1947a. Genetics of natural populations. XIV. A response of certain gene arrangements in the third chromosome of *Drosophila pseudoobscura* to natural selection. Genetics **32**:142-160.

__________. 1947b. A non-seasonal change in the genetic constitution of a natural population of *Drosophila pseudoobscura.* Heredity **1**:51-62.

Dobzhansky, Th., and C. C. Epling. 1944. Contributions to the genetics, taxonomy, and ecology of *Drosophila pseudoobscura* and its relatives. Carnegie Inst. Washington Publ., **554**:1-183.

Dobzhansky, Th., and B. Spassky. 1944. Genetics of natural populations. XI. Manifestation of genetic variants in *Drosophila pseudoobscura* in different environments. Genetics **29**:270-290.

Dubinin, N. P., and G. G. Tiniakov. 1945. Seasonal cycles and the concentration of inversions in populations of *Drosophila funebris.* Amer. Natur. **79**:570-572.

_______. 1946a. Structural chromosome variability in urban and rural populations of *Drosophila funebris.* Amer. Natur., **80**:393-396.

_______. 1946b. Natural selection and chromosomal variability in populations of *Drosophila funebris.* Jour. Heredity **37**:39-44.

_______. 1946c. Inversion gradients and natural

selection in ecological races of *Drosophila funebris.* Genetics **31**:537-545.

Fisher, R. A. 1930. The genetical theory of natural selection. Clarendon Press, Oxford.

Ford, E. B. 1940. Polymorphism and taxonomy. Huxley's New Systematics, Clarendon Press, Oxford.

Gershenson, S. Evolutionary studies on the distribution and dynamics of melanism in the hamster (*Cricetus cricetus* L.). II. Seasonal and annual changes in the frequency of black hamsters. Genetics **30**:233-251.

Haldane, J. B. S. 1937. The effect of variation on fitness. Amer. Natur. **71**:337-349.

Mather, K. 1943. Polygenic inheritance and natural selection. Biol. Reviews **18**:32-64.

Mayr, E., 1945. Symposium on age of the distribution pattern of the gene arrangements in *Drosophila pseudoobscura.* Some evidence in favor of a recent date. Lloydia **8**:70-83.

Pavan, C. 1946. Chromosomal variation in *Drosophila nebulosa.* Genetics **31**:546-557.

Timofeeff-Ressovsky, N. W. 1940. Zur Analyse des Polymorphismus bei *Adalia bipunctata.* Biol. Zblt. **60**:130-137.

Wright, S., and Th. Dobzhansky. 1946. Genetics of natural populations. XII. Experimental reproduction of some of the changes caused by natural selection in certain populations of *Drosophila pseudoobscura.* Genetics **31**:125-156.

73 Genetic response to the sequence of two environments

Louis Levine
Leigh Van Valen
Biology Department
The City College of the City University of New York
New York, New York
and
Department of Vertebrate Paleontology
American Museum of Natural History
New York, New York

Studies on chromosomal polymorphism in *Drosophila pseudoobscura* have shown that the relative fitness of the different gene sequences varies under different temperature conditions. At 25° C., populations composed of the gene sequences Arrowhead (AR) and Chiricahua (CH) establish a stable polymorphic system with characteristic equilibrium frequencies of 25-30 per cent. CH and correspondingly 75-70 per cent. AR (Dobzhansky, 1948; Levine, 1955; Beardmore, Dobzhansky and Pavlovsky, 1960; Mourad, 1962; Strickberger, 1963). At slightly lower temperatures, there appears a partly non-determinate situation. In one set of experiments at 22° C., CH appeared to have established itself at the characteristic 25-30 per cent. level. In a second set of experiments at 22° C., and in experiments at 17-27° C. fluctuating temperatures, CH tended toward elimination (Van Valen, Levine and Beardmore, 1962). In the populations in which CH tended toward elimination, the direction of the selection process became evident when the populations were one year old. At that time, subsamples of four populations were taken and placed in population cages at 25° C. In three of the four derived populations, CH continued to tend toward elimination (Van Valen, Levine and Beardmore, 1962). The inability of these three derived populations to establish and maintain the chromosomal polymorphism so characteristic at 25° C. has led us to study the effects of reversing the sequence of temperatures.

An experimental population, #173, containing AR and CH, has been maintained at 25° C. for some seven years (Beardmore, Dobzhansky and Pavlovsky, 1960; M. Druger, personal communication). This population was started in January, 1957, in a wooden cage, with initial frequencies of 80 per cent. CH and 20 per cent. AR. In a little more than a year and a half, the population reached the expected frequency of

Table I. Percentage CH in population #173 and its derivative

#173			#173-22° C.	
Month	Percentage CH	Dates	Month	Percentage CH
0	80.0	Jan. 1957		
Data for this period reported by Beardmore *et al.*, 1960				
26	25.3	Mar. 1959		
34	28.0	Nov. 1959		
46	25.3	Nov. 1960		
53	25.9	June 1961		
55	–	Aug. 1961	started from #173	
–	–	Jan. 1962	5	22.7
62	19.0	Mar. 1962	–	–
–	–	May 1962	9	10.7
–	–	Aug. 1962	12	7.7
–	–	Jan. 1963	17	9.3
74	19.0	Mar. 1963	–	–
–	–	June 1963	22	5.0
–	–	Aug. 1963	24	6.3
81	26.3	Oct. 1963	26	7.0
–	–	Jan. 1964	29	4.3

From Heredity **19**:734-736, 1964. Used with permission.

These experiments were conducted in the laboratory of Professor Th. Dobzhansky at Columbia University; we gratefully acknowledge his interest, encouragement and hospitality. During this time, one of us (L.V.) was the recipient of a Boese Postdoctoral Fellowship from Columbia University. During its long history, population #173 was maintained at various times by Mrs. O. Pavlovsky, Dr A. M. Mourad and Dr M. Druger. The authors also thank Dr Druger for his recent sampling of population #173 and for his permission to use the datum in this paper.

25-30 per cent. CH. In August, 1961, a subsample was taken from this population and placed in a lucite population cage. The methods used have been described elsewhere (Van Valen, Levine and Beardmore, 1962). The population cage containing the subsample was placed in a 22° C. incubator. Eggs from both populations were sampled at varying intervals, 300 chromosomes being examined at each sample. The data are presented in table I. It can be seen that the frequency of CH dropped initially in both populations. In population #173, CH decreased to 19 per cent., a frequency which was maintained for one year. The most recent sample shows that CH has returned to its characteristic level. In contrast to the parental population, population #173-22° C. appears to be in the process of slowly eliminating CH.

Based on earlier findings, these results were unexpected. In previous studies it was found that populations maintained at 22° C. or 17-27° C. fluctuating temperatures for only one year (about 11 generations) tended to lose their ability to respond to a 25° C. environment. Reversing the sequence of temperatures gives different results. A population maintained at 25° C. for 4½ years (over 50 generations) can still respond to a 22° C. environment as seen in the elimination or near elimination of CH. This finding indicates that the effects on a population of a series of environments may depend on the order in which the environments are experienced. These results also support our earlier findings (Van Valen, Levine and Beardmore, 1962) that 22° C. represents an environment in which chromosomal polymorphism may be eliminated.

REFERENCES

Beardmore, J. A., Dobzhansky, Th., and Pavlovsky, O. A. 1960. An attempt to compare the fitness of polymorphic and monomorphic experimental populations of *Drosophila pseudoobscura.* Heredity **14**:19-33.

Dobzhansky, Th. 1948. Genetics of natural populations. XVIII. Experiments on chromosomes of *Drosophila pseudoobscura* from different geographic regions. Genetics **33**:588-602.

Levine, L. 1955. Genotypic background and heterosis in *Drosophila pseudoobscura.* Genetics **40**:832-849.

Mourad, A. M. 1962. Effects of irradiation in genetically coadapted systems. Genetics **47**:1647-1662.

Strickberger, M. W. 1963. Evolution of fitness in experimental populations of *Drosophila pseudoobscura.* Evolution **17**:40-55.

Van Valen, L., Levine, L., and Beardmore, J. A. 1962. Temperature sensitivity of chromosomal polymorphism in *Drosophila pseudoobscura.* Genetica **33**:113-127.

74 Enzyme and protein polymorphism in human populations

Harry Harris*
The Galton Laboratory
and
Medical Research Council
Human Biochemical Unit
University College
London, England

A very large number of different enzymes and proteins are made in the human organism and there are now good reasons to believe that the amino-acid sequence of each of their polypeptide chains is coded in the DNA of a separate gene locus. So there is a vast array of so-called "structural" gene loci in the genetic constitution of each individual. Furthermore it has been shown that, at certain loci, many different alleles determining structurally distinct versions of the corresponding polypeptide may exist in human populations. Most of these are quite rare. But in some cases certain alleles at a particular locus may be sufficiently frequent as to give rise to what is often referred to as genetically determined polymorphism. That is a situation in which individual members of the population can be categorized into two or more separate types, each relatively common and each characterized by the distinctive manner in which they synthesize the particular enzyme or protein.

In the present paper we will be mainly concerned with considering the incidence and distribution of such alleles at different loci, and with the general question of how they may have come to assume the frequencies that we observe. An idea of the general nature and dimensions of the problem can perhaps be best obtained by considering briefly some well-studied examples.

From British Medical Bulletin **25**:5-13, 1969. Used with permission.

*Present address: Department of Human Genetics and Biometry, The Galton Laboratory, University College, London, England.

1. SOME EXAMPLES OF MULTIPLE ALLELISM AND POLYMORPHISM

a. The haemoglobin variants

During the past twenty years a considerable number of genetically determined variants of haemoglobin have been discovered by the investigation of patients with various haematological abnormalities and also by random surveys of individuals in different populations (Lehmann & Huntsman, 1966). The great majority of these appear to differ from so-called normal haemoglobin by a single amino-acid substitution in one or other of the constituent polypeptide chains, and they can be attributed to mutational events which have resulted in only a single base change in the corresponding gene (Crick, 1967). In fact more than 40 different β-chain variants in which the specific amino-acid substitution has been defined are now known, and also at least 20 different α-chain variants. No doubt many more occur.

Many of these mutant alleles can be regarded as abnormal or deleterious because they produce some degree of haematological disease (sometimes very severe) in either the heterozygous or homozygous state. But some are relatively harmless even in homozygotes, and it is likely that these may be somewhat under-represented in the sample because of a bias against their selection in favour of those leading to obvious haematological disease.

One can readily classify the various alleles into two main groups in terms of their incidence. The first and much the smallest group comprises those alleles which are found relatively frequently in one or another human population. In the case of the β locus, it includes the allele that determines Hb S (or sickle-cell haemoglobin), which is very common in tropical Africa, where in different populations it may occur in 5-40% of all individuals; the allele determining Hb C, which is more localized in West Africa; the allele determining Hb E, which is common in many populations in South-East Asia; and the allele determining Hb D Punjab, which occurs in appreciable frequency in certain populations in India. (For detailed references see Livingstone (1967).)

The second group of alleles are all rare. They have turned up in an irregular manner in a wide variety of populations and the majority have been seen in members of only a single family. It is obviously difficult to get estimates of their incidence, but it appears from population surveys carried out in Europe that perhaps 1 in 1,000 people may be heterozygous for one or other of the many alleles which give rise to electrophoretically detectable variant of adult haemoglobin (see Lehmann & Carrell, p. 14 of this Bulletin). They may represent mutants of either the α or β locus. In the case of the α-chain locus, all the structural variants so far defined appear to belong to this group of rare alleles.

The incidence of these various haemoglobin alleles has generally been accounted for in terms of classical theory. It is supposed that the primary amino-acid sequences of the polypeptide chains of so-called normal haemoglobin have been evolved by natural selection and are now more or less optimal for the species. But fresh mutations resulting in structural alterations of proteins are always occurring at a finite rate and essentially at random. In the vast majority of cases, the altered protein structure is expected to be in some degree functionally inferior to the normal type and so is at a selective disadvantage. Consequently the incidence of the rare variants is assumed to be mainly determined by a balance between the rate of recurrence of fresh mutations and their elimination by natural selection. Very occasionally, however, a variant occurs which, at least in the particular environment obtaining at the time, confers some advantage on people carrying it, and so it tends to spread. Usually the peculiar advantage is confined to heterozygotes, and the homozygotes are less fit. Under these circumstances the frequency of the allele tends to approach an equilibrium situation, a so-called balanced polymorphism, in which the relatively increased contribution of the allele to the next generation by heterozygotes is offset by the relatively reduced contribution from homozygotes. The classic example of this is of course the sickle-cell allele in Central Africa. Homozygotes with sickle-cell anaemia generally die in early life and contribute virtually nothing to the next generation. Yet the allele is very common. This is evidently owing to the fact that the heterozygotes have a better chance of surviving to adult life and so of leaving more children than normal homozygotes, because they are less likely to die from malaria, which is a major cause of mortality in this part of the world (Allison, 1964). Presumably, as malaria is eradicated the incidence of the sickle-cell allele will progressively decline and eventually become very rare, like most other β-chain alleles.

The special features of the sickle-cell polymorphism are a very severe pressure of selection against the homozygote, and yet a very high frequency of the allele. This means that the selective advantage favouring the heterozygotes has to be quite considerable to maintain the polymorphism. But even in these circumstances it proved to be quite difficult to demonstrate the effect and, although the evidence is now very convincing, it is still in certain aspects incomplete. In other common polymorphisms, selection against the homozygote appears to be much less marked and so the increased fitness of the heterozygote required to maintain the allele frequency need be only small, and may be virtually undetectable by available methods. Indeed although the polymorphisms of the other haemoglobin variants, such as Hb C in West Africa, and Hb E in South-East Asia, are often also assumed to have been due to selective survival of the heterozygotes with respect to malaria, which is endemic also in these areas, substantial direct evidence in support of this has not yet been obtained.

b. Glucose-6-phosphate dehydrogenase (G6PD)

The structure of G6PD is coded at a gene locus on the X chromosome, and more than 20 different variant forms, each apparently determined by a different allele at this locus, have been identified (for references, see World Health Organization (1967)). They have been shown to differ from the normal or standard form of the enzyme and from one another in such properties as electrophoretic mobility,

Michaelis constants, thermostability and pH optima and, although the nature of the structural differences has so far been defined in only one case (Yoshida, 1967), it seems very probable that most or all of them are due to single amino-acid substitutions in the protein, similar to those found in the haemoglobins. An important point is that these structural differences often result in quite marked differences in the level of activity of the enzyme in red cells and other tissues of individuals carrying the various alleles.

Mostly these alleles are very rare, but certain of them have an unusually high incidence in particular populations and give rise to characteristic polymorphisms. For instance, besides the allele Gd^{B} which determines the so-called normal or standard form of the enzyme, two other alleles $Gd^{\mathrm{A}-}$ and $Gd^{\mathrm{A}+}$ both occur in many African populations with gene frequencies of between about 0.1 and 0.2, though they are rare or absent elsewhere. The variant protein determined by $Gd^{\mathrm{A}-}$ causes the well-known Negro form of G6PD deficiency which is the basis of primaquine sensitivity and certain other adverse drug reactions. However, apart from this drug idiosyncrasy, individuals carrying this allele appear to be in other respects quite healthy. The other common variant in Negroes that is determined by the allele $Gd^{\mathrm{A}+}$ is associated with only a very slight reduction in enzyme level, and this is apparently harmless.

In many populations living in Southern Europe and the Middle East, a different sort of G6PD polymorphism occurs because of the high incidence of the allele $Gd^{Mediterranean}$. This determines another striking form of G6PD deficiency, and it predisposes to the haemolytic disease known as favism, which may occur when affected individuals eat fava beans, a common feature of the diet in this part of the world. There are probably also other G6PD alleles which occur commonly in particular areas, for example Gd^{Canton} in South-East Asia and Gd^{Athens} in Greece, though their distributions have not yet been worked out in detail.

Because populations which have a high incidence of one or another form of G6PD deficiency come from areas in which malaria is or has been in the past a major cause of mortality, it has been suggested (Motulsky, 1964) that here, as in the case of the sickle-cell gene, malaria may have been an important selective agent in determining the prevalence of particular G6PD alleles (e.g., $Gd^{\mathrm{A}-}$ in Negro populations, and $Gd^{Mediterranean}$ in Southern European and Middle Eastern populations). The malaria parasite might proliferate less well in individuals whose red cells were G6PD-deficient and therefore metabolically abnormal. There is some, though as yet not very extensive, evidence (Gilles, Fletcher, Hendrickse, Lindner, Reddy & Allan, 1967) to suggest that these alleles may indeed confer some selective advantage in terms of malarial morbidity or mortality. But one must also note that the $Gd^{\mathrm{A}+}$ allele, though as prevalent as the $Gd^{\mathrm{A}-}$ allele in Africa and similarly rare or absent elsewhere, does not result, like the $Gd^{\mathrm{A}-}$, in a marked enzyme deficiency.

c. The haptoglobin variants

Another extensively studied example is the serum protein, haptoglobin (for recent review and references, see Giblett (1968)). There are two sorts of polypeptide chains, α and β, and most of the variations which have been observed can be attributed to multiple alleles at the α-gene locus. The findings here, however, are in striking contrast to those obtained with haemoglobin and G6PD, because there are three alleles ($Hp^{1\mathrm{S}}$, $Hp^{1\mathrm{F}}$ and $Hp^{2\mathrm{FS}}$) which are common and wide spread. In European and African populations all three are found, though with differing frequencies, and in Asiatic populations $Hp^{1\mathrm{S}}$ and $Hp^{2\mathrm{FS}}$ both have a significant incidence but $Hp^{1\mathrm{F}}$ is rare or absent. As with haemoglobin and G6PD, a number of very rare alleles at the haptoglobin loci (both α and β) occur.

A special point of interest about the haptoglobin polymorphism is that it is possible to infer from the structural differences in the protein something about the origin of the alleles (Smithies, Connell & Dixon, 1962). The α polypeptides determined by $Hp^{1\mathrm{F}}$ and $Hp^{1\mathrm{S}}$ each contain 83 aminoacid residues and differ in only a single one (Black & Dixon, 1968). The α polypeptide determined by $Hp^{2\mathrm{FS}}$ is nearly twice as long (142 residues) and appears to represent an end-to-end fusion of the hp1Fα and the hp1Sα polypeptides, with a sequence of 24 residues missing at the site of fusion. It presumably originated as the result of a mutational event involving a chromosomal re-arrangement in an individual who happened to be heterozygous for $Hp^{1\mathrm{F}}$ and $Hp^{1\mathrm{S}}$. In other words, the new allele probably arose in a population already polymorphic for the $Hp^{1\mathrm{F}}$

and Hp^{1S} alleles. Furthermore the peculiar structure of the polypeptide results in a rather characteristic polymerization of the haptoglobin molecule which is readily detected by starch-gel electrophoresis and, since this effect has not been seen in haptoglobins in other species, including higher apes (Parker & Bearn, 1961), it seems quite likely that the mutational event giving rise to the Hp^{2FS} allele occurred only after the separation of the human line. Nevertheless it has spread throughout the species and today is the commonest of the three alleles in most human populations.

So one might imagine that the Hp^{2FS} allele conferred some distinctive selective advantage. Yet it is difficult to see from what is known of the differences in the properties and function of the proteins (Giblett, 1968) what exactly this might be, or for that matter how selective forces affecting the Hp^{1F} Hp^{1S} polymorphism might be operating. Certainly individuals of the different common haptoglobin types do not appear different in any obvious way in fitness. Hence one must assume either that the selective differences, if they occur, are very slight, or that they were for some unknown reason much more significant in the past but have been minimized and rendered trivial by subsequent changes in the environment.

d. Phosphoglucomutase

Many forms of phosphoglucomutase enzyme protein with apparently similar catalytic specificities have been demonstrated in different individuals (Hopkinson & Harris, 1966, 1968). There appear to be at least three different gene loci (designated PGM_1, PGM_2, and PGM_3) which are separate and not closely linked, and each locus determines a distinct set of two or three enzyme proteins (so-called isozymes). Since the electrophoretic properties of the several members of a particular set, but not of the other sets, are all similarly affected by allelic substitutions at the corresponding locus, one presumes that they contain a common polypeptide chain, which is not present in the isozymes of the other sets.

Multiple alleles determining electrophoretically distinct variants of the corresponding isozymes have been shown to occur at each of the three loci, but their incidence and distribution differ considerably from locus to locus. They have been mainly studied in populations of European or of African origin, though a number of other groups have also been examined. At locus PGM_1, two common alleles occur. In Europeans their frequencies are about 0.76 and 0.24, and they are both present with very similar frequencies among Africans and indeed in certain other population groups. So the polymorphism appears to be wide spread and much the same in very different areas. Besides these common alleles, a number of others have also been found. They are, however, individually extremely rare. At locus PGM_3, two common alleles also occur both in Europeans and in Africans, but their relative frequencies differ strikingly. In Europeans the frequencies of these two alleles are about 0.74 and 0.26, whereas in Africans they are about 0.34 and 0.66 respectively. At locus PGM_2 there is one allele which predominates in both Europeans and Africans, and the other alleles that have been detected are all much less common. In Africans, however, there is some degree of polymorphism because, besides this common or standard allele, another occurs with a frequency of about 0.01 and, since this has not been detected in the many thousands of Europeans that have been tested, there must be a considerable difference in its incidence between the two groups.

These phosphoglucomutase variants were discovered in the course of an electrophoretic screening programme deliberately aimed at searching for common polymorphic differences. The individuals studied were, in the main, normal and healthy, and there was no indication that the common variant types of the enzyme were associated with any marked functional differences which might be of selective significance. It seems therefore that, if such differences do occur, they are probably very subtle and relatively small in magnitude.

2. THE EXTENT OF POLYMORPHISM

These examples illustrate something of the degree of allelic variation that may occur at different gene loci. At some loci, although multiple alleles may be demonstrable, there is one allele that can be regarded as the standard or normal form and is almost universally present, while all the others are extremely rare. At other loci (e.g., the β-haemoglobin locus and the G6PD locus), although a standard allele occurs and is recognizable as such, there are in some populations, but not in others, alleles which are present in sufficient frequency as to give rise to a common polymorphism. At still other loci (e.g., the α-haptoglobin locus and the

PGM_1 and PGM_3 loci), polymorphism is the rule. Two or more alleles occur relatively frequently and are widely distributed in many different populations. Indeed in some cases there appears to be no valid reason for regarding one allele rather than another as the so-called normal or standard form.

A question which obviously arises is: what is the relative incidence of these various situations among gene loci in general, and in particular how often do polymorphisms occur? Do the haemoglobin, G6PD, haptoglobin, phosphoglucomutase and other polymorphisms represent special and rather unusual forms of variation not typical of enzymes or proteins in general, or are they examples of a relatively common phenomenon? It would clearly be of some interest to know what fraction of the very large number of proteins and enzymes which are formed in the human organism exhibits these sorts of variation. Since the structure of each protein is presumed to be determined by at least one gene locus, we are in effect asking: at what proportion of this vast array of gene loci do two or more relatively common alleles occur in different human populations?

In principle, it should be possible to get an approximate answer to this question by examining in detail a series of arbitrarily chosen enzymes and proteins in randomly selected individuals and preferably in several different populations. In each case one would aim to see whether or not common polymorphism is demonstrable, and in what proportion of the enzymes or proteins it occurs. One might also hope to obtain some information about the occurrence of rare variants.

During the past few years the Medical Research Council's Human Biochemical Genetics Research Unit, working in London, has been engaged in this kind of project (Harris, 1966, 1967). We have mainly studied enzymes which occur in red cells, because these provide a convenient source of material if one wishes to examine large numbers of different individuals. But in most cases the enzymes are also found in other tissues and there is no reason to believe that they are unrepresentative of enzymes in general, at least as far as the incidence of inherited variation is concerned. More recently the placenta has been found to be a useful source of material for such investigations.

The enzymes studied were chosen in an essentially arbitrary fashion, the only criteria adopted being whether they happened to occur in blood or placenta, and whether a suitably sensitive method could be devised by which a search for structural variants could be conveniently conducted in a reasonably sized sample of the general population. The work has mainly been done on English people, but samples from other population groups, mainly Africans, have been studied, though less extensively. Where variant forms of an enzyme were found, detailed family studies were carried out to determine their genetic basis.

The general procedure used to search for enzyme differences was starch-gel electrophoresis.[1] This method is particularly powerful in the detection of differences in molecular charge, such as those that might be produced by the substitution of a basic or acidic amino acid for a neutral amino acid or vice versa. It is not, however, very sensitive to other types of molecular difference (e.g., the substitution of one neutral amino acid by another) and so, even if the electrophoretic conditions employed were optimal, this method could be expected to detect only a proportion of all the possible forms of variation in enzyme structure that might occur.

So far 18 different enzymes have been examined in varying degrees of detail in the course of this project, and a considerable number of genetically determined electrophoretic variants have been identified. Many of the alleles involved are relatively infrequent. But, in six of the enzymes, evidence was obtained for two or more alleles, each with a frequency of at least 0.01 in the particular population. The relatively common individual differences so produced can be regarded as polymorphisms in the sense that this term is conventionally used in genetics. In the case of one of these enzymes, phosphoglucomutase, as has been previously mentioned, polymorphism attributable to independent allelic variation at three separate and unlinked loci was identified. Thus altogether eight polymorphic loci were discovered during this investigation.

The frequencies of different alleles found at these loci are summarized in Table I. In five cases (red-cell acid phosphatase, phosphoglucomutase PGM_1 and PGM_3, adenosine deaminase, and peptidase D), the polymorphism occurs in both Europeans and Africans, although the allele frequencies vary somewhat between these

[1] See McDougall & Synge, Br. med. Bull. 1966, **22**:115.–ED.

Table I. Results of an electrophoretic survey of eighteen arbitrarily chosen enzymes in European and Negro populations. Only common alleles (frequency>0.01) are listed. Many "rare" alleles were also detected.

Enzymes	Europeans			Negroes			References
	Allele 1	Allele 2	Allele 3	Allele 1	Allele 2	Allele 3	
Red-cell acid phosphatase	0.36	0.60	0.04	0.17	0.83	–	Hopkinson, Spencer & Harris (1964)
Phosphoglucomutase							
Locus PGM_1	0.76	0.24	–	0.79	0.21	–	Spencer, Hopkinson & Harris (1964)
Locus PGM_2	1.00	–	–	0.99	0.01	–	Hopkinson & Harris (1966)
Locus PGM_3	0.74	0.26	–	0.34	0.66	–	Hopkinson & Harris (1968)
Adenylate kinase	0.95	0.05	–	1.00	–	–	Fildes & Harris (1966)
Peptidase A	1.00	–	–	0.90	0.10	–	Lewis & Harris (1967)
Peptidase D (prolidase)	0.99	0.01	–	0.95	0.03	0.02	Lewis & Harris, unpublished work
Adenosine deaminase	0.94	0.06	–	0.97	0.03	–	Spencer, Hopkinson & Harris (1968)

Other enzymes studied were: phosphohexoseisomerase, malate dehydrogenase, isocitrate dehydrogenase, red-cell hexokinase, lactate dehydrogenase, methaemoglobin reductase, red-cell pyrophosphatase, pyruvate kinase, placental acid phosphatase, peptidases B and C, and a red-cell "oxidase". None of these showed common electrophoretic polymorphism, though a number of rare variants were identified.

contrasting population groups. At one locus (adenylate kinase), polymorphism was found only among the Europeans, and in two (peptidase A and PGM_2) only among the Africans. Thus about one-quarter of the enzymes in this arbitrarily selected series exhibited polymorphism in each population group. Clearly the phenomenon is a relatively common one.

If anything, the results must under-estimate the true incidence of polymorphism simply because the enzymes were scrutinized only for electrophoretic differences. Furthermore the discriminative power of even this technique has been found to vary considerably from enzyme to enzyme because of technical problems, and it is quite possible that in some cases polymorphic variation has been missed.

It is difficult to arrive at a satisfactory estimate of the total number of different enzymes and proteins which occur in the human organism, and which are presumably coded at separate gene loci. But they must certainly number many thousands. If the results of the enzyme survey can be taken as at all representative, then in any given population of individuals there must be a significant fraction of loci at which two or more relatively common alleles occur. So probably thousands of polymorphisms, each involving a different enzyme or protein, exist. It is not without significance that essentially the same conclusion has been reached by Lewontin & Hubby (1966) on the basis of a similar electrophoretic survey of enzymes and proteins in a quite different species, *Drosophila pseudo-obscura.*

An interesting point which emerges concerns the degree of individual diversity which must actually occur in human populations. Some idea of this can be obtained by considering together the several enzyme polymorphisms that have already been demonstrated in a single population. Relevant data on ten loci involving eight different enzymes in the English population are given in Table II. Since each of these polymorphisms appears to occur independently of the others, it follows that a very large number of combinations of enzyme phenotypes may occur among individuals in the general population. By combining the frequencies given in column three of the table, one finds that the most frequent combination of phenotypes will occur in less than 2% of the population. Furthermore, from column four one can show that the chance that two randomly selected people in the population would have exactly the same combination of enzyme types is about 1 in 200. Column five provides another way of looking at the data. It gives the proportion of individuals in the population who are heterozygous for alleles at each of the loci. It appears that approximately 97% of people in this population must be heterozygous at at least one of the ten loci listed in the table. Thus quite a high degree of individual differentiation in

Table II. Enzyme polymorphism in the English population (data on ten loci)

Enzymes	Number of alleles with frequency greater than 0.01	Frequency of commonest phenotype	Probability of two randomly selected individuals' being of same type	Proportion of population who are heterozygous	References
Red-cell acid phosphatase	3	0.43	0.34	0.51	Hopkinson *et al.* (1964)
Phosphoglucomutase					
Locus PGM_1	2	0.58	0.47	0.36	Spencer *et al.* (1964)
Locus PGM_3	2	0.55	0.45	0.38	Hopkinson & Harris (1968)
Placental alkaline phosphatase	3	0.41	0.30	0.50	Robson & Harris (1967)
Liver acetyltransferase	2	0.50	0.50	0.50	Price Evans & White (1964)
Adenylate kinase	2	0.90	0.82	0.10	Fildes & Harris (1966)
Serum cholinesterase					
Locus E_1	2	0.96	0.92	0.04	Kalow & Staron (1957)
Locus E_2	2	0.90	0.82	0.10	Robson & Harris (1966)
Phosphogluconate dehydrogenase	2	0.96	0.92	0.04	Parr (1966)
Adenosine deaminase	2	0.88	0.79	0.11	Spencer *et al.* (1968)
All enzymes combined	–	0.018	0.005	–	

enzyme make-up can be demonstrated from this quite limited series of examples. This must surely represent only the tip of the iceberg, and one may plausibly conclude that, in the last analysis, every individual will be found to have a unique enzyme constitution.

An obviously important problem is how far these common enzyme variations are reflected functionally. Not much information is as yet available about this. But, in a few cases, significant differences in activity levels between the several common phenotypes of a particular enzyme have been detected. In the case of red-cell acid phosphatase, for example, there are six electrophoretically distinct phenotypes that can be distinguished, and they represent the homozygous and heterozygous combinations of three common alleles. These phenotypes, presumably because of the structural differences in the enzyme present, also differ in their average total level of red-cell phosphatase activity (Hopkinson, Spencer & Harris, 1964). For instance, homozygous individuals with the so-called type B form of the enzyme show on average about 50% greater activity than homozygous individuals of type A, and the heterozygote type BA is intermediate in this respect. Similar differences are seen with the other phenotypes. Differences in activity have also been noted in the case of the common serum cholinesterase phenotypes (Kalow & Staron, 1957; Harris, Hopkinson, Robson & Whittaker, 1963) and in the phosphogluconate dehydrogenase phenotypes (Parr, 1966).

Another interesting example is provided by the polymorphism of a liver enzyme which behaves as an acetyltransferase (Price Evans & White, 1964). This polymorphism was discovered when it was found that people differed very markedly in the rate at which they acetylate and therefore inactivate the drug isoniazid, which is used in the chemotherapy of tuberculosis. About 50% of Europeans are so-called rapid inactivators because they have a relatively high level of the transferase whereas, in the other 50%, the level of activity of the enzyme is very low and the drug is inactivated very slowly. This polymorphic difference is due to two common alleles, the slow inactivators being homozygous for one, and the rapid inactivators heterozygous or homozygous for the other. It is of interest that in Europeans the allele resulting in very low enzyme activity is more than twice as common as the allele determining the active enzyme, the gene frequencies being roughly 0.7 and 0.3.

3. RARE VARIANTS

Besides the relatively common alleles which give rise to the so-called polymorphisms, a considerable number of rare alleles determining different variant forms of particular enzymes or proteins have also been found during the course of population surveys. Thus evidence for 5-10 different rare alleles at loci determining the structures of the enzymes phosphohexoseisomerase (Detter, Ways, Giblett, Baughan, Hopkinson, Povey & Harris, 1968), placental alkaline phosphatase (Robson & Harris, 1966), peptidase A (Lewis & Harris, 1967, and unpublished work), lactate dehydrogenase (Kraus & Neely, 1964; Davidson, Fildes, Glen-Bott, Harris, Robson & Cleghorn, 1965), and phosphoglucomutase (Hopkinson & Harris, 1966) has been obtained in the course of electrophoretic studies in which samples from several thousand individuals have been examined. And numerous other examples of rare variants of particular enzymes or proteins picked up in the course of routine investigations could be cited. In fact it is beginning to appear that, if virtually any enzyme or protein is examined by sufficiently sensitive methods in a large enough number of individuals, one or more rare variants are likely to be detected. The multiplicity of rare haemoglobins mentioned earlier illustrates the same phenomenon, and shows how very many different rare alleles at single loci may be demonstrable if a protein is subjected to particularly intensive investigation.

The precise incidence of any particular one of these rare alleles is obviously difficult to determine. Some appear to occur with gene frequencies of between 10^{-2} and 10^{-3} in certain populations, but in the majority of cases the individual gene frequencies are probably much lower. Nevertheless at any single locus so many different rare alleles may exist that an appreciable fraction of the population can be heterozygous for one or another of them. For instance, from the presently available data it seems that the fraction of the population which is heterozygous for one or other of the rare alleles determining electrophoretically detectable variants of such enzymes as phosphohexoseisomerase, phosphoglucomutase (PGM_1 and PGM_2), the red-cell peptidases A and B, and lactate dehydrogenase, is in each case in the order of 1 in 300 to 1 in 700. Since presumably only a proportion of the structural variants of any given enzyme will be detectable electrophoretically, the true fraction is in fact likely to be somewhat higher. If, as is not improbable, there are many other loci at which there is a similar incidence of "rare" alleles, the multiplicity of rare enzyme and protein variants so produced could in toto contribute quite significantly to the diversity among individuals in the population.

Because these "rare" variants have mainly been discovered in the course of random population surveys, they have usually been observed only in heterozygotes, who synthesize the common form of the enzyme or protein (or, where there is polymorphism, one or other of the common forms), as well as the variant. Such individuals are usually normal and healthy. However, in some cases the variant enzyme or protein may be functionally defective, so that in the homozygous state it could result in significant clinical abnormality. In fact most rare "recessive" diseases can probably be attributed to rare alleles of this sort (for further discussion, see Harris (1968)).

It is not yet clear what proportion of the rare alleles at any particular locus may lead to obvious clinical abnormality in the homozygous state. Probably it varies considerably from locus to locus according to the structure and also the function of the enzyme or protein involved. However, judging from the properties of many of the rare enzyme and protein variants as they have been observed in heterozygotes, it seems that the proportion that is likely to be markedly deleterious in the homozygous state is, at many loci, quite small.

4. SOME GENERAL CONSIDERATIONS

If we extrapolate from these various observations, the picture that is beginning to emerge can perhaps be formulated in the following way. As a result of mutations in the remote or more recent past, there is likely to be present in any sizeable human population more than one allele at virtually every gene locus which codes for a specific enzyme or protein, and at many loci a considerable number of such alleles probably occur. In general they produce structurally distinct forms of the enzyme or protein and these in most cases probably differ from one another by single amino-acid substitutions. The majority of these alleles are very rare. However, at perhaps 25% or more of all such loci, at least two alleles—each present in an appreciable fraction of the population (at least 2%)—may occur.

The incidence and distribution of these alleles in populations must depend essentially on three main factors: (i) the rate at which fresh mutations occur; (ii) differential selection due to the effects, exerted by the enzyme and protein variants that the alleles determine, on the survival and fitness of individuals who carry them; and (iii) chance factors which may lead fortuitously to the elimination of particular alleles from the population or to their spread.

a. Mutation rates

Gene mutations are thought to occur more or less at random. In most cases, though certainly not in all, they apparently involve simply the change of a single base in the sequence of several hundred or more that are present in the DNA of the particular gene. A typical polypeptide chain may contain 100-500 amino acids, any one of which can be replaced by one of several others as a result of such a mutational event. So a vast number of distinct structural variants of any one protein may be generated by recurrent mutations at a single gene locus. Some of these variants, because of the alteration in their properties and functional activity induced by the specific amino-acid substitution, can be expected to give rise in either the homozygous or heterozygous state to some frank clinical abnormality. But many of the possible alterations in structure which could occur are likely to be of only moderate or minor functional significance and others may have no obvious consequences at all.

Considered in these terms, the significance of the mutation rates often quoted for human genes is difficult to assess. They have mainly been derived from studies on the population and familial incidence of rare inherited abnormalities such as haemophilia, retinoblastoma and neurofibromatosis (for review and discussion, see Penrose (1961)), and lead to estimates of mutation rates per gene locus per generation of around 10^{-5}. That is to say, they suggest that a fresh mutation for a gene determining one of these three conditions may be present in one in every 100,000 sperm or ova. If, however, one supposes that the mutations involved in causing these abnormalities usually represent single base changes resulting in specific defects in a particular enzyme or protein, one would expect that at any such locus there might be a number of different sites where the substitution of one base by another could cause a defect in the corresponding protein capable of inducing the particular clinical abnormality observed. There would probably also be a great many other sites in which a single base change would not have the same consequences. Furthermore, at any one site the effect that follows from a mutation will depend on the particular base that happens to be substituted. Thus only a fraction of all mutations involving single base changes at a particular gene locus might lead to the disease in question. And we have very little idea of what the magnitude of this fraction might be in specific cases. In general, however, it would appear likely that the mutation rates often cited for rare abnormalities in man are probably higher than the rate at which a single base at a particular site in a gene is altered by mutation. On the other hand, they probably under-estimate the total mutation rate per locus per generation, since this would include base changes at all possible sites within the gene.

Watson (1965) suggests that the average probability of an error's giving rise to the insertion of an incorrect base during DNA replication may under optimal conditions be around 10^{-8} or 10^{-9}; and Kimura (1968) points out that this could imply a mutation rate for base substitutions of perhaps 5×10^{-7} or 5×10^{-8}/base pair/generation, assuming the number of cell divisions along the germ line from the fertilized ovum to a gamete in man to be roughly 50. Using such estimates, a number of interesting if speculative calculations are possible. For example, if we suppose that the mutation rate/base pair/generation is 5×10^{-8}; that 80% of base changes in a gene coding for a polypeptide chain result in single amino-acid substitutions; that the average polypeptide chain contains 300 amino acids and so is coded by a DNA sequence 900 bases long; and that there are perhaps 20,000 different polypeptide chains synthesized in the organism; then, on average, every newborn infant may be expected, as the result of a fresh mutation's having occurred in either of its parents, to synthesize at least one new structurally variant enzyme or protein (i.e., $2 \times (5 \times 10^{-8})(0.80)(9 \times 10^2)(2 \times 10^4)=1.44$). This kind of calculation (see also Kimura, 1968) is of course largely guesswork, but the assumptions appear not unrealistic, and the result perhaps illustrates the extent to which fresh mutations may be continuously generating enzyme and protein diversity among individuals in a population.

b. Chance effects

Quite apart from the question as to whether a particular mutant is relatively deleterious and so tends to be eliminated by natural selection, or confers some kind of selective advantage and so tends to spread, the odds against any new mutant allele's persisting in a population for many generations are very considerable. The new allele will on average be transmitted to only half the children of the individual who first receives it. There is, therefore, a distinct chance that it will not be transmitted to the next generation, and the odds in favour of its being lost by chance are compounded in successive generations. In a reasonably large stable population where each pair of parents is on average replaced by two children who become parents in the next generation, the probability that a new mutant will still be present after, say, 15 generations is only about 1 in 9 (Fisher, 1930). The odds in favour of the persistence of a mutant are somewhat greater if the population happens to be increasing in numbers when it appears, and are less if the population is declining. But, in general, the majority of new mutant alleles that appear are likely to be eliminated in the course of the next 10 or 20 generations in a more or less random manner. However, as we have seen, even if the mutation rates are low (e.g. 5×10^{-8}/base pair/generation), there is always going to be an appreciable though changing reservoir or pool of rare variant forms of different enzymes or proteins in any sizeable human population at any given time. Their nature and incidence will be largely a matter of chance.

Furthermore the individual frequencies of those mutant alleles that do happen to persist in a population may, because of such random phenomena, vary considerably, quite apart from any selective effects that may be superimposed. So, very occasionally, one or other variant might become relatively common in a particular population purely by chance. This is particularly likely to occur if the population is small and relatively isolated. For instance, a variant form of serum albumin has recently been observed in as many as 25% of members of a group of North American Indians known as the Naskapi (Blumberg, Martin & Melartin, 1968) and in several closely related tribes. This variant has not been seen anywhere else in the world, although it would quite easily have been picked up by very widely used procedures. A number of similar examples of other peculiar enzyme and protein variants occurring with an unexpectedly high incidence in odd communities could be readily cited. They are probably most simply accounted for in terms of such chance fluctuations or what is known as "drift".

Chance effects may also be of great importance in situations where the numbers of an established population are severely reduced by some epidemic or other disaster and then subsequently increase again. The sample of alleles which happen to be carried by the survivors and so form the gene pool from which the population is reconstituted is unlikely to be in all respects exactly representative of the alleles in the original population. Similarly if a small group of individuals from one population migrate elsewhere and found a new community which expands, particular alleles may by chance be over- or under-represented.[2] Under such circumstances marked changes in allele frequency may occur quite quickly. Relatively uncommon alleles in the original population may by chance become common in the population derived from it. Other alleles present in an appreciable frequency in the original population may be lost fortuitously and not appear in the derived one.

The role of selective forces in determining the incidence and distribution of the multiplicity of enzyme and protein variants that we observe in populations has therefore to be evaluated against this background of fortuitous and essentially haphazard effects inherent in the nature of the genetical structure of human populations and of the mutation process itself. Even if, as has often been extensively argued in the past, no allele is selectively neutral, the part that selective forces have played in determining its frequency may in many cases be effectively obscured by such chance effects.

c. Selection

Where two or more alleles determining structurally distinct forms of a particular enzyme or protein each occur with a relatively high incidence in some or all of the major ethnic population groups, it seems reasonable to suppose that selective forces at least to some degree have been important in determining their incidence and distribution. It seems biologically implausible to suppose that more than a small proportion of the many different enzyme and protein polymorphisms that evidently occur have come about purely fortuitously, although one may suppose that chance effects or drift could have been responsible for many of the detailed peculiarities in their distributions.

However, the elucidation of the selective forces which may have given rise to any particular polymorphism is turning out to be among the most difficult and intractable problems in human genetics. Various approaches seem possible but none has as yet been particularly rewarding, mainly perhaps because selective differences that may have been important in determining the polymorphism are possibly not of an order of magnitude capable of being demonstrated by currently available

---

[2] See Dean, p. 48 of this Bulletin.—ED.

methods. The principal exception to this is the sickle-cell polymorphism and this is now beginning to emerge as a perhaps rather special and very unusual situation.

One way of approaching the matter is by the direct investigation of the functional properties of the structurally distinct forms of a polymorphic enzyme or protein, in the hope that this might define differences which are likely to be significant metabolically or in some other way which could be selectively important in certain sorts of environment. In a number of cases, it has been shown that common structural variants of a particular enzyme are indeed associated with marked differences in activity. But there is as yet little or no indication of the possible significance of such differences in relation to selection, except perhaps in the case of G6PD mentioned earlier. An obvious difficulty is that selection, as it affects most individuals, is presumably directed at complex physiological variables dependent on many different enzymes acting together. The important thing may therefore be the constellation of enzyme phenotypes of the individual rather than the characteristics of any single one of them.

Another approach is to try and find out whether particular alleles render individuals more or less susceptible to the development of particular disorders or disabilities, especially common ones. The general method is to compare the incidence of the allele in individuals affected by the particular condition with the total incidence in the population of which they are a part. A now well-established example of this kind of effect is the association of the ABO blood groups with certain gastrointestinal disorders (Fraser Roberts, 1959). Blood-group A individuals are somewhat more susceptible to gastric cancer than the group O individuals. Group O individuals are more susceptible to peptic ulceration than group A. But the effects are relatively small and their significance in relation to selection for the different alleles in this polymorphism is difficult to assess. Nor have we any clear idea of exactly how these different blood-group antigens influence susceptibility to these diseases. A large number of different diseases, as well as differences in response to particular disease states such as acute and chronic infections, could well be studied in this way. And there are an increasing number of polymorphisms which might be tested. But generally there is no particular reason to expect one association rather than another, so that progress in this direction at the present time is likely to be somewhat fortuitous.

Where very wide variations in the frequencies of particular alleles are observed between different populations, it is of interest to ask whether their distribution is correlated in any obvious way with specific differences in the environments in which the populations live, or with any characteristic patterns of morbidity, in the hope that this might indicate the nature of the critical selective factors. This approach was valuable in directing attention to malaria in the case of the sickle-cell polymorphism, and provides suggestive evidence for the same kind of selection in the other haemoglobin polymorphisms as well as in the G6PD polymorphisms occurring in Africa and the Middle East. But it has not as yet appeared to be particularly helpful in other cases. And it might well in certain circumstances be rather misleading, if attention were arbitrarily focused on just one out of the multiplicity of environmental differences that frequently exist.

Another major line of attack is primarily demographic. The aim is to categorize individuals in one or more populations in terms of the various common allelic differences that are known, and then search for differences between them in the main parameters involved in selection, such as mortality and morbidity rates at various ages, and fertility; also by family analysis to investigate possible disturbances in segregation ratios and so on. Such data, although they may give only indirect information about specific selective factors in relation to particular polymorphisms, should in principle provide an assessment of the magnitude of any selective effects that are actually occurring in the given environmental situations. And this of course is fundamental to the whole problem. Such surveys are, however, extremely hard to mount on a scale which is both sufficiently large as to be likely to yield significant results, and yet sufficiently detailed and exact in the determination of the various demographic parameters as to yield precise answers. So far, although much suggestive information has been obtained (e.g., Morton, 1964; Reed, 1967, 1968a, 1968b), the results have perhaps mainly served to emphasize the difficulties involved in arriving at any certain conclusions about the biological implications of any particular polymorphism.

A general and inherent source of uncertainty in all these studies arises from the fact that the environment in which human populations live today or even during the last two or three generations is very different in many important aspects from that in which they lived in the past. In particular the incidence and age distribution of mortality and morbidity and its main causes have changed and are changing profoundly. Thus what may have been important selective agents in the past, and may well have shaped many of the polymorphisms that we see today, may now be of only minor or no significance. We are only looking at what is inevitably a changing situation over a very narrow period of time. Furthermore, as a general rule we have no means of knowing whether in any particular polymorphism we are, as is often assumed, dealing with a situation close to stable equilibrium due to heterozygous advantage, or with the steady increase of one particular allele at the expense of another, or with its progressive disappearance.

Thus, although there is little doubt about the occurrence of enzyme and protein polymorphisms as a wide-spread and general phenomenon in human populations, we have very little idea so far about the detailed nature of the selective effects that may have brought them about.

REFERENCES

Allison, A. C. (1964) Cold Spring Harb. Symp. quant. Biol. **29**:137.

Black, J. A. & Dixon, G. H. (1968) Nature, Lond. **218**:736.

Blumberg, B. S., Martin, J. R. & Melartin, L. (1968) J. Am. med. Ass. **203**:180.

Crick, F. H. C. (1967) Proc. R. Soc. B **167**:331.

Davidson, R. G., Fildes, R. A., Glen-Bott, A. M., Harris, H., Robson, E. B. & Cleghorn, T. E. (1965) Ann. hum. Genet. **29**:5.

Detter, J. C., Ways, P. O., Giblett, E. R., Baughan, M. A., Hopkinson, D. A., Povey, S. & Harris, H. (1968) Ann. hum. Genet. **31**:329.

Evans, D. A. Price & White, T. A. (1964) J. Lab. clin. Med. **63**:394.

Fildes, R. A. & Harris, H. (1966) Nature, Lond. **209**:261.

Fisher, R. A. (1930) The genetical theory of natural selection. Clarendon Press, Oxford.

Giblett, E. R. (1968) Ser. haemat., Kbh. n.s. **1**:3.

Gilles, H. M., Fletcher, K. A., Hendrickse, R. G., Lindner, R., Reddy, S. & Allan, N. (1967) Lancet **1**:138.

Harris, H. (1966) Proc. R. Soc. B **164**:298.

Harris, H. (1967) In: Crow, J. F. & Neel, J. V., ed. Proc. III int. Congr. hum. Genet., Chicago, Illinois, September 5-10, 1966, p. 207. Johns Hopkins Press, Baltimore, Md.

Harris, H. (1968) Br. med. J. **2**:135.

Harris, H., Hopkinson, D. A., Robson, E. B. & Whittaker, M. (1963) Ann. hum. Genet. **26**:359.

Hopkinson, D. A. & Harris, H. (1966) Ann. hum. Genet. **30**:167.

Hopkinson, D. A. & Harris, H. (1968) Ann. hum. Genet. **31**:359.

Hopkinson, D. A., Spencer, N. & Harris, H. (1964) Am. J. hum. Genet. **16**:141.

Kalow, W. & Staron, N. (1957) Can. J. Biochem. Physiol. **35**:1305.

Kimura, M. (1968) Nature, Lond. **217**:624.

Kraus, A. P. & Neely, C. L., jr (1964) Science, N.Y. **145**:595.

Lehmann, H. & Huntsman, R. G. (1966) Man's haemoglobins. North-Holland, Amsterdam.

Lewis, W. H. P. & Harris, H. (1967) Nature, Lond. **215**:351.

Lewontin, R. C. & Hubby, J. L. (1966) Genetics, Princeton **54**:595.

Livingstone, F. B. (1967) Abnormal hemoglobins in human populations. Aldine, Chicago.

Morton, N. E. (1964) Cold Spring Harb. Symp. quant. Biol. **29**:69.

Motulsky, A. G. (1964) Am. J. trop. Med. Hyg. **13**:147.

Parker, W. C. & Bearn, A. G. (1961) Ann. hum. Genet. **25**:227.

Parr, C. W. (1966) Nature, Lond. **210**:487.

Penrose, L. S., ed. (1961) Recent advances in human genetics. Churchill, London.

Reed, T. E. (1967) Am. J. hum. Genet. **19**:732.

Reed, T. E. (1968a) Am. J. hum. Genet. **20**:119.

Reed, T. E. (1968b) Am. J. hum. Genet. **20**:129.

Roberts, J. A. Fraser (1959) Br. med. Bull. **15**:129.

Robson, E. B. & Harris, H. (1966) Ann. hum. Genet. **29**:403.

Robson, E. B. & Harris, H. (1967) Ann. hum. Genet. **30**:219.

Smithies, O., Connell, G. E. & Dixon, G. H. (1962) Nature, Lond. **196**:232.

Spencer, N., Hopkinson, D. A. & Harris, H. (1964) Nature, Lond. **204**:742.

Spencer, N., Hopkinson, D. A. & Harris, H. (1968) Ann. hum. Genet. **32**:9.

Watson, J. D. (1965) Molecular biology of the gene. Benjamin, New York.

World Health Organization (1967) Tech. Rep. Ser. Wld Hlth Org. No. 366.

Yoshida, A. (1967) Proc. natn. Acad. Sci. U.S.A. **57**:835.

75 Selection for sexual isolation within a species

G. R. Knight
Alan Robertson
C. H. Waddington
Institute of Animal Genetics
Edinburgh, Scotland

Two mechanisms have been advanced for the origin of reproductive isolation between species. Muller (1939), dealing in the main with barriers to crossing in the later stages of species divergence, such as hybrid inviability and infertility, suggests that these arise almost by chance as a product of change in the genetic background either by genetic drift or as adaptation to different biological situations. This would lead to accelerating divergences as the process continues, or, as Muller puts it, "ever more pronounced immiscibility as an inevitable consequence of non-mixing." Dobzhansky's suggestion (1937), which is perhaps complementary rather than antagonistic to Muller's, is that when sufficient divergence between two species has arisen so that the hybrids are less well adapted for any available habitat than either parental type, there will be selection for sexual isolation. That is to say—if mating can take place and if the resulting hybrids are inviable or infertile, then natural selection will operate to reduce the chance that mating will occur, either by reducing the chance of encounter or the chance of mating with members of the other species when they are encountered.

Some writers, in discussing the mechanism proposed by Dobzhansky have suggested that "natural selection will favour any mechanism which prevents the wastage of gametes involved in unsuccessful hybridisation." This seems to be unduly teleological. Natural selection will only tend to suppress crossbreeding if those individuals which hybridise will in consequence pass on fewer gametes in the form of pure-bred offspring. It would seem probable that this would be more often the case in females than in males. In *Drosophila melanogaster,* for instance, females seem reluctant to mate again for a period of two or three days after an effective mating. If the first mating has been heterogamic, this will reduce the number of purebred offspring that she will produce in her lifetime. Gestation in mammals will have a similar effect. But the male, who must on the average have the same number of effective matings in his life as the female, is usually capable of many more if willing females are available. It follows then that willingness to cross-breed, which may merely be a sign of greater general sexual activity, will not necessarily reduce the number of purebred progeny that a male will leave. If Dobzhansky's mechanism for the establishment of sexual isolation is correct, it follows that it should be in the main a matter of female preference. Merrell (1954) has recently presented evidence that it is the female which exercises discrimination in matings between *D. pseudoobscura* and *D. persimilis.*

Koopman (1950) has shown that selection leads to an intensification of the sexual isolation between these two species. Using marked stocks of the two species, he selected continually for pure-bred flies—the progeny of parents that had mated homogamically. He showed that the proportion of hybrids emerging declined dramatically after a few generations of selection. More recently, Wallace (1950) and King (private communication) attempted to demonstrate the production of sexual isolation by selection within a species. They used two stocks of *D. melanogaster,* from widely separated localities, which had each been marked by a different recessive gene. After 12 generations, when the experiment was first reported, little change in the proportion of wild-type flies emerging had been observed, but in subsequent generations the proportion declined significantly, showing that sexual isolation had been to some extent established. This was confirmed by observation of individual matings.

Our own experiment on very similar lines was started before we were aware of Wallace's work, and as our work was slightly different in conception, we decided to proceed with it. In

From Evolution **10:**14-22, 1956. Used with permission.

Wallace's experiment, the mutants were used solely as markers, the stocks because of their origin presumably differing in many genes. As it happens, we had used in our work stocks marked with the autosomal recessive genes, ebony and vestigial, which has been extracted from a population in which the two had been segregating for many generations. The original stocks making up this population were actually those used by Rendel (1951) in his work on the effect of light on the mating of these mutants. Our two foundation stocks, both of which contained a considerable amount of genetic variability, were thus probably genetically very similar except for the marker genes. These genes were chosen because of the ease of scoring but they do react differently to light and, as Rendel has shown, ebony males mate more frequently in the dark than in the light.

In the first experiment of this type that we carried out, there appeared in the seventh generation some flies that were both ebony and vestigial, indicating that in previous generations either a non-virgin female or else a wild-type heterozygote had been used as a parent with the result that each mutant stock was contaminated with the other gene. Theoretical consideration of the effect of this showed that the proportion of double recessive flies should increase by a factor of four each generation until they reached a level of 11% of all flies emerging. At that point, the proportion of flies in each mutant stock that were heterozygous for the other gene would be 2/3. There would then be a continual interchange of genes between the two stocks. In addition, one-third of the apparently pure mutants used as parents would be derived from heterogamic matings, thereby reducing the selection for sexual isolation. We therefore discarded the line and started afresh with stringent precautions against non-virginity, parents being collected over a 7 hour period. In the two experiments presented here in detail, no double recessive flies were ever observed.

DESCRIPTION OF EXPERIMENTS

Box experiment

Two mutant strains of *D. melanogaster* homozygous for the genes ebony and vestigial respectively were used. They had been extracted from the same population, after segregation for many generations. At the start, 54 males and 54 virgin females from each of the stocks were put together into a breeding box (size 18″ X 18″ X 7″) which contained 10 unstoppered ¼ pint bottles of maize meal–molasses–yeast–agar medium. Flies were etherised for counting, but were not put into the box until three hours after complete recovery from anaesthesia. The box was then placed in a constant temperature room at 25° C. All phases of the experiment were done at this temperature. The box was always put in the same part of the room, where, due to the direction of the light, two sides of the box near the edge were in slight shadow. The ebony flies, immediately the box was positioned, migrated towards the light source, that is, towards the shaded edge. After some time, the majority of them moved more freely about the cage.

After six days of mating, the ten food bottles were removed, cleared of any flies which remained inside, and stoppered. The parents were discarded. The count of the next generation was started five days afterwards, i.e., on the eleventh day after the parents were put into the box. Three types of flies emerged; hybrids from heterogamic matings, and the two mutants ebony and vestigial from homogamic fertilizations. For 3½ days every fly which emerged was counted. The culture bottles were completely cleared at 10 A.M. Flies which emerged by 5 P.M. on the same day were segregated and mutants were kept in separate vials to be used as parents for the next generation when 1 to 4 days old. When insufficient virgins were obtained, those collected were bred with their own kind, and the experiment carried on from their progeny. In the box experiment, this was done three times in 38 generations.

In order to ensure that any changes in external conditions had not affected the course of the experiment, controls were done on the box experiment in the later generations. Parent virgin flies were obtained from the original stocks and put into a box of identical proportions to the experimental one. The control box and the experimental one received exactly the same treatment throughout. This was done seven times between the 25th and 35th generations.

Jar experiment

An experiment on similar lines was run in conjunction with the cage one. A 2 lb. glass jar containing approximately 1″ of food was used as the breeding chamber. The number of parent flies employed in this case was between 20 and

30 of each sex of the mutants. Again, it was sometimes necessary to mate the virgins with their own kind to produce sufficient numbers for the next preferential mating. This was done three times in 33 generations. From generations 1 to 12 the parent flies were still under ether when put into the jar, as it was thought that they might otherwise escape. This was found, however, to be unsatisfactory. So from the 13th generation onwards the parents were introduced into the jar three hours after recovery from the ether. The jar was put into the same constant temperature room and at the same time as the box. Thereafter, all operations, such as clearing parents from the jar, counting and segregating flies of each generation, etc., were carried out at the same time and in an exactly similar manner to the box experiment.

Because of the small capacity of the jar compared with that of the box and the fact that there was little or no variation in the light within the jar, it was assumed that any tendency towards an ecological isolation between the ebony and vestigial flies would be eliminated.

RESULTS

One of the first impressions at the start of the experiments was of the great fluctuation in results from generation to generation. The jar experiment was in fact started to try to remove this by having all flies developing in one food mass. Our criterion of isolation has been the ratio of wild-type flies, produced by heterogamic matings, to the total number of mutants produced by homogamic matings. The standard deviation of this ratio due to chance fluctuation, determined from the mean square difference between successive generations, was 0.16 for the box experiment and 0.24 for the jar. This fluctuation is equal to that produced by random sampling of 160 units and 70 units respectively from a population made up of two types of objects with equal frequency. The total count was actually of the order of 2,000 flies in both cases. But the number of female parents was 108 and 50 in the box and jar respectively. The observed fluctuations suggest that the effective units are the initial inseminations of the individual females. In this respect, it is of interest that of the individual platings of females taken from the box after six days of mating, 660 gave offspring all of the same type and only 75 had mixed offspring. However, whatever the reason, it is still true that too many flies were counted each generation and a sample of a quarter of the size that we took would have been quite adequate.

Box experiment

The results of the box experiment are set out graphically in figures 1 and 2. The graphs are moving averages over 5 generations to smooth out fluctuations. In figure 1, the number of hybrids is expressed as a percentage of the sum of the ebony and vestigial emergences. Figure 2 shows separately the numbers of the three types of flies emerging. From the first to the eighteenth generation a more or less steady decline in the percentage of hybrids is noted. The lowest percentage of hybrids in any individual generation was 10.3% at the eighteenth, with emergences of ++246, e736, vg1640. Only once afterwards, at the 23rd generation, does the vestigial line graph fall as low as the control mean for this mutant. Thereafter the values remain high for vestigial emergences. The hybrid figure drops, and is lowest between the 16th and 18th generations, only rising a little and slowly towards the end of the experiment. During the whole 38 generations the emergence values for ebony alternate slightly above and below the figure for the control mean. This suggests that the sexual isolation, after the 18th generation, is due mainly to the increase in the number of homogamic matings of the vestigial flies.

The average values for the seven control generations are also shown in figures 1 and 2. The proportion of wild-type flies to mutants averages 0.66, compared to the proportion in the selected population at the same period of 0.38. The figures for the individual mutants show that the change is due to a decrease of wild-type flies and an increase of vestigial.

It has been shown by Rendel (1951) that ebony reacts to light intensity in its mating behaviour. It seemed possible that the sexual isolation was due to an accentuation of this response. Towards the end of the experiment, therefore, duplicates of the selection box were made up from parents from the selected stock but were kept instead in complete darkness. The ratio of wildtype to mutant offspring was 0.48 compared to 0.46 for the three contemporary generations in the light. It seems therefore that the demonstrated sexual isolation is not concerned with phototropic response. However, there were many more ebony

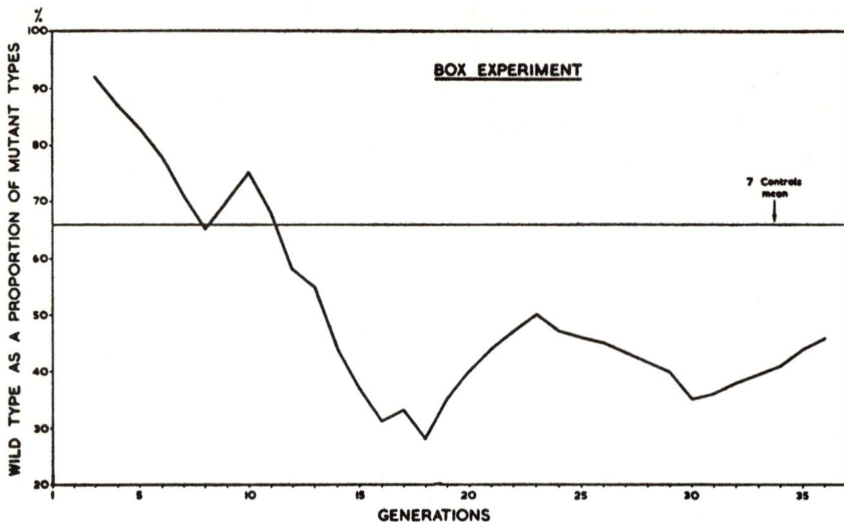

Fig. 1. Results of box experiment. Number of hybrids expressed as a percentage of sum of ebony and vestigial emergences.

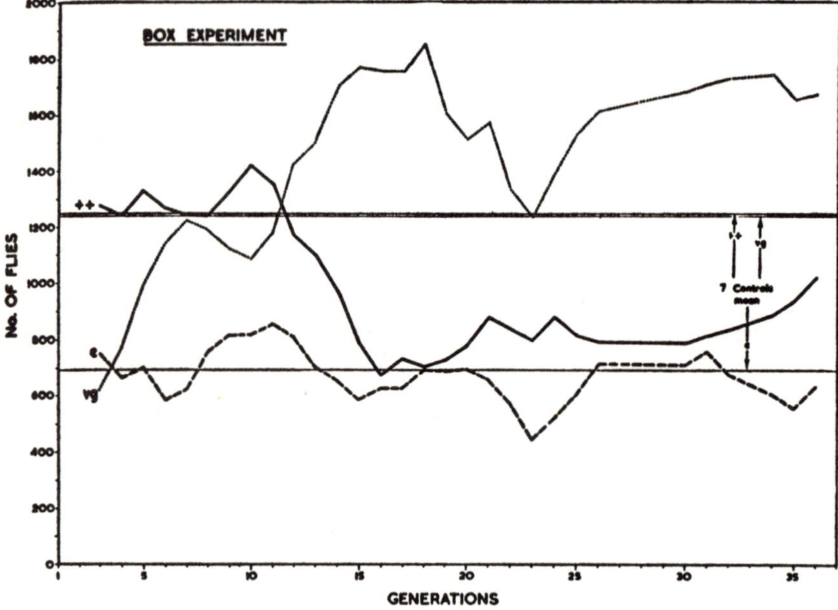

Fig. 2. Results of box experiment. Numbers of 3 types of flies emerging shown separately.

flies in the dark boxes—in fact the average of the three tests (1102 flies) had only once been exceeded by a single generation in the light, and the average in the last few generations of the latter was about 650. There was correspondingly a shortage of vg flies, but the proportional effect was not so great. This agrees with Rendel's observation that ebony males show greater sexual activity in the dark.

Between the 20th and 30th generations, the

Table 1

Controls ♀	Inseminated by e	Inseminated by vg
e	71	69
vg	41	63
Selected stocks		
e	151	108
vg	77	142

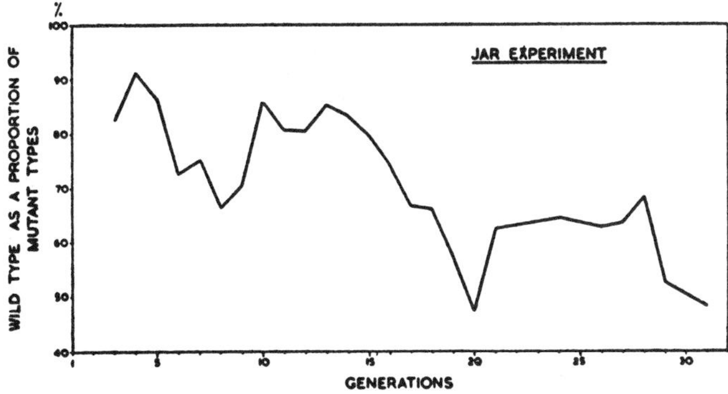

Fig. 3. Results of jar experiment. Compare figure 1.

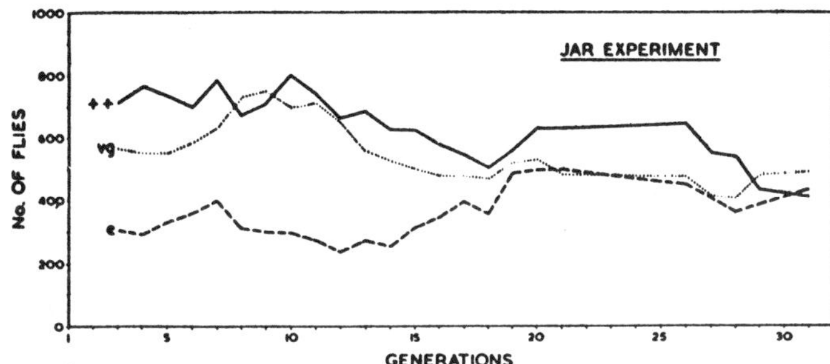

Fig. 4. Results of jar experiment. Compare figure 2.

females were placed in individual vials after they had been removed from the box, and their progeny were examined on emergence. This was done with 6 generations of the selected stocks and with three of the controls. The results in terms of effective matings are given in table 1.

There is a slight tendency to homogamic mating in the controls but the heterogeneity χ^2 is only 2.15. In the selected stocks the tendency is much more marked and the χ^2 value is 27.14. This is confirmatory evidence that some degree of sexual isolation has been obtained in the selected population.

It might have happened that this type of selection, picking out always the mutant flies, would have affected the segregation ratio by selecting those genes favouring the survival from egg to adult of the mutant types. However, a check based on several thousand flies at the end of the experiment showed no differences between control and selected stock in the segregation ratio for either mutant.

Jar experiment

The results for the jar experiment are given in figures 3 and 4. In the jar, the light intensity was much more uniform than in the box and in addition the flies were more confined—the volume of the jar being of the order of 1/50 of that of the box. Here again there is a decline in the proportion of wild-type flies as the experiment proceeds, although the proportion at the end is a trifle higher than in the box experiment. However, the ratio of wild-type to mutant flies has declined from 80% to between 50% and 60%. As was noted above, there was a change in method in the middle of this experiment. Up to generation 12, flies were put into the jar etherised but afterwards they were put in an active state. This change does not appear to have affected the sexual isolation. In the five generations before the change, the average ratio of wild-type to mutants was 0.86 and in the five after the change it averaged 0.80. It may however have affected the separate types. The numbers of vg and wild-type flies decline by about one-third as a result of the change, whereas the ebony count is unchanged. The subsequent change in the wild-type/mutant ratio appears in this case to be due to an increase in the number of ebony flies. Control experiments were not carried out on the jar population, as the latter was subsidiary to the main experiment in the box.

Sexual preferences in inbred lines and closed populations

It is convenient to present here a small amount of data on sexual preferences between lines and populations chosen at random, with a bearing on the "chance" occurrence of sexual isolation. These experiments were carried out by the usual "male choice" method in which males are given equal numbers of two types of female, one of which is recognisable—in our case by a spot of silver paint. The females are then examined for the presence of sperm in the seminal tract. In the first case, two wild-type inbred lines of completely different origin were used, given in our stock list the symbols W20 and K7. The results are given in table 2, which shows the proportion of females inseminated. In all cases, the ♂♂ were equal in number to the ♀♀ of each of the separate lines.

A similar experiment was then done using lines which, although of common origin, had been selected in different directions without inbreeding for 20 generations for number of chaetae on the 4th and 5th abdominal sternites. There was no overlap in chaetae count between the high and low lines used, so that this character could be used for identifications. The results are given in table 3, for matings between one high line, H1, and two low lines, L4 and L5.

In these rather meagre results, there is little suggestion of sexual isolation having developed by chance in either the inbred experiment or in that with the selected lines. In the latter, it is of interest that the selection for the quantitative character has caused a differentiation in mating ability. The H1 ♂♂ are poorer than those from the two low lines but on the other hand the high ♀♀ seem to be better. But this seems to be a general change in sexual drive, not specifically adapted to the other sex of the same line. Wallace (1955 in press, and personal communication) has also tested whether the mere isolation of two populations is sufficient to cause sexual isolation to arise between them. His populations had been separated for 80-100 generations, and differed in certain morphological characters (primarily abdominal pigmentation). In an extensive series of tests, no tendency towards preferentially mating could be detected nor does there seem to have been any evidence of differences in the intensity of general sexual drive.

Table 2

♀s marked	♂	W20 ♀	K7 ♀	Duration of mating
W20	W20	22/23	11/25	40 mins.
	K7	17/23	9/25	40 mins.
W20	W20	17/19	16/20	70 mins.
	K7	4/20	1/19	70 mins.
K7	W20	9/10	5/10	30 mins.
	K7	8/14	9/14	60 mins.
K7	W20	10/20	2/19	30 mins.
	K7	10/20	5/20	45 mins.

Table 3

♂	H1 ♀	L4 ♀	Duration of mating
H1	3/10	1/9	30 mins.
L4	9/14	3/14	30 mins.
	H1 ♀	L5 ♀	
H1	11/20	5/19	60 mins.
L5	20/24	17/25	60 mins.

DISCUSSION

Our results may be summarised in the statement that some sexual isolation developed when we selected for a tendency towards homogamic matings, but that none was found to have arisen by chance in a few lines which had been selected for abdominal chaeta number or inbred. Laboratory experiments on evolutionary mechanisms can, of course, only be indicative and not demonstrative—they can show what might happen in wild populations, rather than what has happened. As far as they go, our experiments lend support to the mechanism suggested by Dobzhansky rather than that discussed by Muller. But when attempting to apply these results to occurrences in nature, one must bear in mind the ways in which artificial populations may fail to imitate conditions in the wild.

It is perhaps misleading to put Muller's hypothesis of the chance origin of sexual isolation in antithesis to that of Dobzhansky, which attributes it to the action of selection. In all probability, both mechanisms have operated in the wild in different cases. Some evidence supporting Dobzhansky's hypothesis comes from Dobzhansky and Koller (1938) who found, in an analysis of crosses between *Drosophila pseudoobscura* and *D. miranda*, that the isolation was greatest between races close to each other in their range. King (1947) has similar evidence from the *guarani* group. However, even between races of the two species widely separated in origin, the isolation was considerable. One has the suspicion that sexual isolation is common between species which have never had the opportunity to crossbreed, though the evidence is rarely conclusive since seldom if ever do we know the full evolutionary history of the populations.

It is perhaps not surprising that differences in sexual behaviour arose in the experiments involving selection, but were not found in the comparison of populations which has originated independently. If they had occurred in the latter, they could only have appeared by chance, or as a correlated response. It might be expected that random changes in sexual behavior would be slow, even though they arose as a secondary response to an adaptive change in the population. Mating involves the cooperation of the two sexes and it seems unlikely that a genetic shift in the population causing a change of sexual behaviour in the female, perhaps by a modification of the pattern by which a male recognises an animal of his own species, would also change male behaviour in a compensating manner (although an exception to this might be in habitat preference). An individual with aberrant sexual behaviour is not likely to leave many progeny. A population gradually changing its genetic situation could only change its pattern of mating by the selection of males capable of responding to the altered female behaviour. This must constitute a brake on the change of mating behaviour either by chance changes in the genetic situation or even as a correlated response to an adaptive change. This will be particularly true of inbred lines in which selection between potential mates is small or non-existent. Reproductive behaviour, excepting perhaps choice of habitat, would therefore be more stable than other physiological systems to genetic changes.

Both the hypotheses that we have discussed demand the development of a previous geographic isolation before sexual isolation can be established. In this sense, the selection hypothesis is perhaps clumsy, since in the formation of a new species showing sexual isolation with the parent species it demands first a geographic isolation and then an overlapping of the species range so that members of the two species can be selected for refusal to crossbreed. It seems to us that sexual isolation instead of being a consequence of geographic isolation, may be a contributory factor in its establishment. The spread of a population into new territory will often involve the occurrence of genotypes with new hereditary habitat preferences. The existence of such preferences amongst Drosophila stocks has recently been shown by Waddington, Woolf and Perry (1954). In organisms such as birds, in which rather sudden changes in geographical or ecological range are well-known, learning may play a part, but this may also have an important genetic component. In the genetic constitution of a sub-population which has broken out of the original species boundaries and is spreading into new territory, one must expect to find that a number of adjustments are occurring simultaneously. There is most likely to be, in the first place, an evolution of a new system of habitat-preferences and/or of general activity; in the second, the adaptive characters and general fitness of the migrating group will be attuned to the new circumstances which it has to meet. Both these necessary modifications of the gene pool will be made more easily if the genetic constitution of the sub-population is prevented from continual intermingling with that of the original stay-at-home group. Thus any tendency for preferential mating within the migrating group, and sexual isolation between it and the main population, will acquire selective value. It seems rather probable that a species may be able to spread into new territory even if no sexual isolation develops between the main population and the migrating one; but if the increase in species-range demands considerable adjustment of the genotype to fit the new environment, the evolution of some degree of mating-barrier will undoubtedly be of considerable advantage. Our experiments show that the necessary genetic variability is likely to be present in a population; and the fact that the change of environment involves alterations to the behaviour pattern of the migrating animals makes it more likely that their preferences for sexual partners as well as for habitats, will exhibit evolutionary flexibility. Thus species-spread and sexual isolation will tend to act synergistically.

SUMMARY

1. Partial sexual isolation (between two stocks of *D. melanogaster* differing only in marker genes) has been established by selection of the offspring of flies mating with their own type. This has been demonstrated by a reduction in the number of cross-bred offspring

found and also by examination of the progeny of individual females.

2. In a small series of tests, no tendencies towards preferential mating were found to have arisen by chance in a sample of lines which had been inbred or selected for number of abdominal chaetae, although there were differences in intensity of sexual drive.

3. Changes in reproductive behaviour brought about by selection are more likely to affect female than male behaviour. Willingness to cross breed, in a male, will not materially reduce the number of his pure-bred offspring but in a female it usually will.

4. It is argued that selection pressure against cross-breeding of two partially separated populations, although probably effective when it occurs, is not likely to be the only mechanism by which sexual isolation between taxonomic groups develops in nature. It is suggested that an important part in the origin of such isolation may be played by the factors (e.g. changes in hereditarily controlled behaviour patterns) which bring about the spread of an initial panmictic population into new geographical or ecological situations.

LITERATURE CITED

Dobzhansky, T. 1937. Genetics and the Origin of Species. Columbia Univ. Press, New York.

Dobzhansky, T., and P. C. Koller. 1938. An experimental study of sexual isolation in Drosophila. Biol. Zentral. **58**:589-607.

King, J. C. 1947. Interspecific relationships within the guarani group of Drosophila. Evolution **1**:143-153.

Koopman, K. F. 1950. Natural selection for reproductive isolation between Drosophila pseudoobscura and Drosophila persimilis. Evolution **4**: 135-148.

Merrell, D. J. 1954. Sexual isolation between Drosophila persimilis and Drosophila pseudoobscura. Am. Nat. **88**:93-100.

Muller, H. J. 1939. Reversibility in evolution considered from the standpoint of genetics. Biol. Rev. **14**:261-280.

Rendel, J. M. 1951. Mating of ebony, vestigial and wild-type Drosophila melanogaster in light and dark. Evolution **5**:226-230.

Wallace, B. 1950. An experiment on sexual isolation. D.I.S. **24**:94-96.

Waddington, C. H., B. Woolf, and M. Perry. 1954. Environment selection by Drosophila mutants. Evolution **8**:89-96.

Index of first authors